Free Student Aid.

Log on.

Tune in.

Succeed.

To help you succeed in introductory botany, your professor has arranged for you to enjoy access to a great media resource, the Botany Place. You'll find that these resources that accompany your textbook will enhance your course materials.

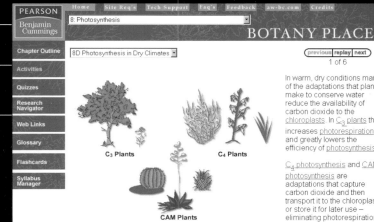

What your system needs to use these media resources:

WINDOWS™
266 MHz CPU
Windows 98, 2000, NT, XP
64 MB RAM
1024 x 768 screen resolution
Thousands of colors
Browsers: Internet Explorer 5.0 and higher; Netscape 4.7, 7.0
Plug-ins: Shockwave Player 8, Flash Player 5, QuickTime 4
Internet Connection: 56k modem minimum for Website
NOTE: Use of Netscape 6.0 and 6.1 are not recommended due to a known compatibility issue between Netscape 6.0 and 6.1 and the Flash and Shockwave plug-ins.

MACINTOSH™
266 MHz CPU
OS 9.X, OS X
64 MB RAM minimum
1024 x 768 screen resolution
Thousands of colors
Browsers:
   OS 9.X: Netscape 4.7, 6.2, or higher; (Web only: Internet Explorer 5 or higher)
   OS X: Internet Explorer 5.2 or higher; Netscape 6.2 or higher
Plug-ins: Shockwave Player 8, Flash Player 5, QuickTime 4
NOTE: Use of Netscape 6.0 and 6.1 are not recommended due to a known compatibility issue between Netscape 6.0 and 6.1 and the Flash and Shockwave plug-ins.

Got technical questions?
For technical support, please visit www.aw.com/techsupport and complete the appropriate online form. Technical support is available Monday-Friday, 9 a.m. to 6 p.m. Eastern Time (US and Canada).

Here's your personal ticket to success:

How to log on to www.botanyplace.com:

1. Go to www.botanyplace.com.
2. Click "Register."
3. Scratch off the silver foil coating below to reveal your pre-assigned access code.
4. Enter your pre-assigned access code exactly as it appears below.
5. Complete the online registration form

to create your own personal user Login Name and Password.
6. Once your personal Login Name and Password are confirmed by email, go back to www.botanyplace.com, type in your new Login Name and Password, and click "Log In."

Your Access Code is:

If there is no silver foil covering the access code above, the code may no longer be valid. In that case, you need to either:
• Purchase a new student access kit at your campus bookstore.
• Purchase access online using a major credit card.
Go to www.botanyplace.com, and click Buy Now.

Important: Please read the License Agreement, located on the launch screen before using the Botany Place. By using the website, you indicate that you have read, understood, and accepted the terms of this agreement.

INTRODUCTION TO
BOTANY

Murray W. Nabors

University of Mississippi

PEARSON

Benjamin
Cummings

San Francisco   Boston   New York
Cape Town   Hong Kong   London   Madrid   Mexico City
Montreal   Munich   Paris   Singapore   Sydney   Tokyo   Toronto

Development Editors: John Burner, Matt Lee
Art Development Editor: Carla Simmons
Project Editor: Susan Minarcin
Senior Art Editor: Donna Kalal
Photo Program Manager: Travis Amos
Associate Project Editor: Alexandra Fellowes
Editorial Assistant: Alissa Anderson
Senior Production Editor: Corinne Benson
Senior Marketing Manager: Josh Frost
Market Development Manager: Susan Winslow
Managing Editor, Production: Erin Gregg
Manufacturing Buyer: Pam Augspurger
Compositin: Thompson Type
Text Design: Mark Ong
Cover Designer: Mark Ong
Illustrations: Imagineering

On the cover: Photograph of False agave (*Hechtia texensis*), by Kevin Schafer. Courtesy of Imagebank.

Library of Congress Cataloging-in-Publishing Data
Nabors, Murray W.
       Introduction to Botany / Murray W. Nabors.
          p. cm.
       Includes bibliographical references (p.   ).
       ISBN 0-8053-4416-0
       1. Botany—Textbooks.   I. Title.

QK47.N226 2004
580—dc22   2003065632

www.aw-bc.com   4 5 6 7 8 9 10 -DOW-07 06

About the Author

This book is dedicated to my wife,

Adriana,

and to my children,

Cyrus, Darius, Mariaelena,

Alexanaka, and Amandamikkel.

Murray W. Nabors has been teaching Botany for more than 30 years. He began his teaching career at the University of Oregon before moving to the University of Santa Clara. In 1972, he transferred to Colorado State University, where he taught for 27 years, serving as the director of the Honors Program and the assistant chair of the Department of Biology. Next, Dr. Nabors went on to James Madison University, where he served as the head of the Department of Biology for 4 years. Currently, you will find him at the University of Mississippi in Oxford, where he serves as the chair of the Department of Biology.

Dr. Nabors's research interests, which focus on the use of biotechnology to improve crop tolerance, led his lab to be the first in the nation to obtain plants, grown from tissue-cultured cells, with an increased tolerance to salt. His current research centers on isolating the genes that enhance the ability of cultured cells to regenerate into whole plants.

Dr. Nabors's dedication to his students and the classroom experience have shaped the development of this text. He is grateful to have guided nearly two dozen students through their graduate programs in botany, and he has enjoyed watching the progress of numerous undergraduate research students who have contributed to ongoing research projects and publications. Dr. Nabors has published more than 50 refereed scientific papers, as well as numerous reports and abstracts. His broad interest in writing generally has culminated in the publication of short stories and a book of cowboy–Western Christmas stories.

Bringing Botany to Life

Introduction to Botany, A NEW TEXT BY MURRAY NABORS, CAPTURES STUDENTS' ATTENTION BY DEMONSTRATING TO STUDENTS THAT PLANTS ARE A FASCINATING AND ESSENTIAL PART OF THEIR EVERYDAY LIVES, AND BY FOCUSING ON FOUR MODERN THEMES— PLANTS AND PEOPLE, CONSERVATION BIOLOGY, EVOLUTION, AND BIOTECHNOLOGY.

 4

Roots, Stems, and Leaves: The Primary Plant Body

Each chapter begins with a lively photograph and a chapter outline.

Found in Central America, *Gunnera insignis* is known as sombrilla del pobre or poor man's umbrella.

Roots

Taproot systems usually penetrate more deeply than fibrous root systems

Root development occurs near the root tip

The root cap protects the root apical meristem and helps the root penetrate the soil

Absorption of water and minerals occurs mainly through the root hairs

The primary structure of roots relates to obtaining water and dissolved minerals

Some roots have specialized functions in addition to anchoring the plant and absorbing water and minerals

Roots have cooperative relationships with other organisms

Stems

Botanists have developed zone and cell-layer models to describe stem growth

In primary growth of most stems, the vascular tissue forms separate bundles

A transition region maintains vascular continuity between the root and stem

Leaf primordia form in specific patterns on the sides of shoot apical meristems

Stem variations reflect different evolutionary pathways

Some stems have specialized functions in addition to support and conduction

Leaves

A leaf primordium develops into a leaf through cell division, growth, and differentiation

The leaf epidermis provides protection and regulates exchange of gas

Mesophyll, the ground tissue in leaves, carries out photosynthesis

The vascular tissue in leaves is arranged in veins

Leaf shapes and arrangements have environmental significance

Abscission zones form in the petioles of deciduous leaves

Some leaves have specialized functions in addition to photosynthesis and transpiration

As you read in Chapter 3, growth produced by the apical meristems at the tips of shoots and roots results in what is called the *primary plant body.* The roots, stems, leaves, and reproductive structures of plants are all derived originally from apical meristems. Even the lateral meristems that produce secondary growth, making woody trunks and roots thicker, form from cells produced by apical meristems. Here we will focus on the primary growth of roots, stems, and leaves, exploring how these organs in vascular plants develop and function together.

Vascular plants that live only one year or two years, called *annuals* and *biennials,* often have only primary growth. Those that live longer, called *perennials,* add new primary growth every year, lengthening shoots and roots and also replacing damaged or dead tissue. Although many perennial plants such as trees and shrubs have secondary growth, a few trees, such as palms, consist solely of primary growth. That is, they have no lateral, or secondary, meristems.

In a sense, primary growth is about getting from one place to another. Plants cannot move through their environment as animals do, but they can grow through their surroundings to obtain what they need. Roots absorb water and mineral nutrients by growing through the soil to

new regions of resources from regions that have been depleted. Meanwhile, stems and leaves acquire the solar energy needed for photosynthesis by growing toward regions of greater illumination.

The growth of roots, stems, and leaves is interrelated. For example, seedlings usually have more roots than shoots because a germinating seed contains a supply of food but needs water for elongation to allow the photosynthetic shoot to develop. As photosynthesis becomes the main source of the plant's food, the root-to-shoot ratio is reduced. Throughout a plant's life, the ratio of shoots and roots changes as necessary so that the light and CO_2 collected by leaves enter the plant in the correct proportions with the water and minerals collected by roots.

Evolutionary changes have resulted in modified roots, stems, and leaves that have contributed to survival in various environments. In some plants, for instance, enlarged roots and stems have evolved that store water, helping the plants to survive droughts, dry seasons, or dry climates. Roots and stems may also store food, producing reserves that can be used when decreases in photosynthesis result from shading or from leaf damage by wind, cold, disease, and predation. Sometimes modified leaves fulfill unusual roles, as in the case of plants like the Venus's flytrap, which "eats" insects to compensate for lack of nitrogen in the soil.

In short, roots, stems, and leaves do not function in isolation but instead work together, not only in producing, transporting, and storing nutrients but also in providing structural support and protection for the plant. As we examine what makes each of these organs unique, we will also explore how they relate to and depend on each other.

The Venus flytrap is a dramatic example of leaf adaptation.

An introductory overview with art that relates to the central topic of the chapter draws students into the material.

Capturing student interest

THE INTRIGUING WORLD OF PLANTS

Black Pepper: Savior of Rotting Meat

It might surprise you to learn that European settlement of the Americas came about largely as a result of the search for black pepper and other spices. Black pepper comes from the dried, ground fruits of the shrub *Piper nigrum,* native to the Malabar Coast of southwestern India. The fruits are green and have a white interior, with the black coating resulting from fungal action.

Why was pepper so important? In those pre-refrigeration days, salt preserved meat by keeping bacteria and fungi generally at bay, but this also made the meat largely inedible. Adding spices such as black pepper made salted meat palatable, which is why sailors often carried small bags of peppercorns.

During the Middle Ages, traders brought spices from Asia to Europe along trade routes through the Middle East. Caravans of camels laden with black pepper, cloves, cinnamon, nutmeg, ginger, and other spices had been making the trip for a thousand years. Beginning in 1470, however, the Turks blocked these overland routes, and Europeans looked to the ocean for an alternate passage to Asia. Christopher Columbus won financial support from the Spanish court to seek a new route to China and India. Landing in the Caribbean, he believed he had reached islands off the coast of India, referring to the inhabitants as "Indians" and the islands as the "Indies." Although he found no black pepper, it is not surprising that the spicy hot fruits he did find were later known as "peppers," even though they were members of the genus *Capsicum,* an entirely different group of plants. Today we distinguish between the two plants by calling one black pepper and the other hot peppers, such as jalapeños and habaneros.

Black pepper. The fruits of *Piper nigrum* become black pepper.

---

selves are also plant products, having been formed hundreds of millions of years ago primarily from fossilized remains of plants.

Wood is still the main source of construction material, and it is used as the framework for most houses and many other buildings. Even steel construction relies indirectly on plants because the super-hot furnaces used to make steel out of iron are powered mainly by fossil fuels.

Another important plant product is paper, which can be made from a variety of plants. Most of the paper we use is made of pulp derived from woody plants such as fir and pine.

These are just a few examples of how plants provide a variety of useful products—in addition, of course, to being our basic source of food and oxygen. Throughout this book, the main narrative and "Plants and People" box features will explore how plants affect human life.

Conservation biology is a critical area of research

Since plants supply so many of our needs, we must make sure that we have enough of them and that valuable plant species do not become extinct. We are the conservators, or caretakers, of Earth's resources. **Conservation biology** is an important multidisciplinary field of science that studies ways to counter the widespread extinction of species and loss of critical habitats. It studies the impact of human activities on all facets of the environment and searches for less ecologically destructive ways to harvest trees, build cities, and generally interact with the biosphere's biological resources, such as forests.

The large and rapidly increasing human population has consumed wood much more rapidly than it has been replaced. Therefore, it is not surprising that wood will command higher prices as supplies dwindle. About half of the yearly harvest of trees in the United States is used to make paper, most of which is not recycled. Only a few remnants of old-growth forests exist, mostly in national forests, national parks, and private preserves (see the *Conservation Biology* box on page 9).

Worldwide, about half of the original forests are gone, replaced by cities, farms, or other human activities. Much rain forest is destroyed in what is called "slash-and-burn" agriculture, in which the forest is cleared (Figure 1.5), the land is used for crops for a few years until soil nutrients are depleted, and then the land is abandoned. In view of the importance of plants, it is chilling that human

Integrating key themes

BIOTECHNOLOGY

Using Plants to Battle Bacteria

Imagine this nightmare: Bacteria have evolved to become totally resistant to antibiotics that once readily eliminated them. Diseases such as tuberculosis, once treated by antibiotics, reach nearly epidemic levels. Does this sound unbelievable? Well, this "science fiction" scene is coming closer to harsh reality.

When antibiotics first became widely

alkaloids such as berberine. Long used by Native Americans as a medicine, berberine is already sold as a relief for arthritis, diarrhea, fever, hepatitis, and rheumatism. Stermitz and Lewis found berberine in the leaves and sap of several varieties of a plant known as Oregon grape, in the genus *Berberis*. Initial test-tube studies on the berberine were disappointing, indicating weak antibiotic activity when used as a pure extract. However, combining berberine with other compounds from the plant greatly increased its antibiotic activity, rivaling the strongest antibiotics available against *Staphylococcus*.

What exactly made the berberine more effective? Stermitz and Lewis discovered that Oregon grape produces a second compound, 5'-methoxyhydnocarpin, which prevents bacteria from pumping out the berberine, thereby allowing the antibiotic to kill the bacteria. Now researchers are testing whether this combination works in living animals. Eventually, vaccines and new antibiotics, perhaps derived from plants, may bring this danger...

PLANTS & PEOPLE

A Taste of Tea History

Rich in caffeine and mildly addictive, tea is the world's most popular beverage. The drink originated in China more than 5,000 years ago. Legend says that Emperor Shen Nong observed tea leaves accidentally fall into boiling water. Being curious, he asked for a taste, and the beverage was born.

After being introduced to Europe in the 1580s, tea became immensely popular. By the mid-1600s Great Britain, with a superior navy, had cornered the European tea trade. By the early 1800s, tea had become so popular that the British began importing opium from India and selling it to China to pay for tea, as well as other Chinese products, such as silk and porcelain.

Tea was also immensely popular in the American colonies. The Americans avoided the British... importing tea from China in exchange for op... Ottoman Empire in Constantinople. By the r... ...thriving in the colonies...

CONSERVATION BIOLOGY

The Challenge of Forest Conservation

Before the arrival of large numbers of Europeans in the 1700s, North America was extensively forested. By the late 1800s, more than half of those forests were gone, and people were beginning to realize that agricultural and forest resources were not unlimited.

President Theodore Roosevelt and reformer Gifford Pinchot were among the earliest advocates of conservation,

the movement to preserve natural resources through wise use. Pinchot saw a nation "obsessed by a fury of development . . . exploiting the riches of the richest of continents," aided by the sale of government-owned forest reserves to the highest bidders. Appointed chief of the U.S. Division of Forestry in 1898, Pinchot worked to preserve the nation's forests. After assuming the presidency in 1901, Roosevelt warned that without conservation the nation's forests would be gone in 60 years. In 1905, Congress transferred forest reserves to the new U.S. Forest Service, with Pin[chot as] first chief. Later renamed national forests, these [lands] were to be managed for the greatest good [and greates]t number of people in perpetuity.

[Th]e idea of national forests was very unpopular, [especially in] the West. Even today, despite having national [forests, we a]re far from achieving a sustainable use of our [resour]ces. The very notion of conservation must be [renewed] in the minds and actions of each new genera[tion of Ameri]cans.

[Since the] time of Roosevelt and Pinchot, the world has [become a dif]ferent place. In many cases it is hard to bal[ance current] needs with the possibility of future shortages. [As we] have increased pressures to use natural re[sources, our] standards of living...

EVOLUTION

Leaves That "Eat" Insects

Why do some plants have leaves that "eat" insects? The answer has to do with nitrogen and the environment. Plants typically get nitrogen from the soil, often relying on nitrogen-fixing bacteria. However, swamps and bogs have little nitrogen available in the soil because their acidic environment is unsuitable for the growth of nitrogen-fixing bacteria. Epiphytes, which often grow on other plants high in the rain forest, sometimes lack sufficient nitrogen as well. In more than 200 species of plants, modified leaves have evolved that trap insects as an alternative source of nitrogen. Among these plants are the bladderwort, sundew plant, Venus's flytrap, and pitcher plant.

Bladderwort (*Utricularia vulgaris* and related species) is an aquatic plant that produces tiny bladders with a trap door. The bladders, typically less than a half-centimeter in diameter, open their doors when an aquatic insect brushes against one of four stiff trigger hairs. Water [rushes into] the bladder, along with the insect. T...

Sundew plant.

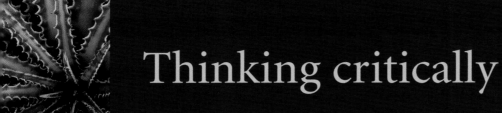

Thinking critically

Section Review Questions throughout the chapter reinforce key concepts and test students' understanding of what they have read.

Section Review

1. Describe the general characteristics of a plant.
2. Describe the major types of plants.

Questions for Thought and Discussion

1. Why do you think many people take plants for granted? How would you respond to them?
2. Explain what is meant by the statement "Plants do not need us, but we need plants."
3. In what sense would you say that plants also need us?
4. Choose one of the GM plants mentioned in the chapter and discuss possible benefits and problems that should be considered in determining whether and how to use that plant.
5. Some people argue that all foods should be labeled to indicate whether they contain genetically modified ingredients. Do you agree? Explain.
6. Why must a hypothesis be potentially falsifiable?
7. Is the statement "A bachelor is an unmarried man" a hypothesis? Explain.
8. Review the "steps" of the scientific method. Why do you think scientists do not always follow these steps in a strict order? Give some examples.
9. Formulate a question about plants and describe a scientific approach to answering that question. For example, a question might be "How does pruning a branch affect tree growth?"
10. How does the statement "Evolution is just a theory" reflect a misunderstanding of the scientific meaning of the term *theory*?

Questions for Thought and Discussion strengthen critical thinking skills by encouraging students to apply the concepts they have just learned.

Evolution Connection

Consider the following types of plants: mosses, ferns, conifers, flowering plants. Beginning with the ferns and then moving on o the conifers and flowering plants, list diagnostic features present in each group that are not present in the preceding group. Which of these features represent evolutionary trends in the plant kingdom? Explain.

 Sketch a food chain for a water environment that begins with a primary producer and has at least four levels of consumers. See Figure 1.1 for an example.

Drawing exercises actively engage students in learning plant structure.

Learn More sections feature recommended annotated reading selections for students who want to learn more about topics covered in the chapter.

To Learn More

Visit The Botany Place at *www.thebotanyplace.com* for quizzes, exercises, and Web links to new and interesting information related to this chapter.

Bacon, Francis, and Peter Ubach. *Novum Organum: With Other Parts of the Great Instauration.* Chicago: Open Court, 1994. A new classroom edition of Bacon's great work of 1620, in which he argues that scientists need to make detailed observations rather than relying on tradition.

Desowitz, Robert S. *The Malaria Capers: More Tales of Parasites and People, Research and Reality.* New York: W. W. Norton, 1993. Despite quinine, malaria still kills at least one million

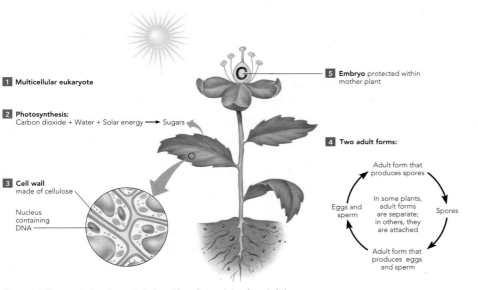

1 Multicellular eukaryote

2 Photosynthesis:
Carbon dioxide + Water + Solar energy → Sugars

3 Cell wall
made of cellulose

Nucleus
containing
DNA

5 Embryo protected within
mother plant

4 Two adult forms:

Adult form that
produces spores

In some plants,
adult forms
are separate;
in others, they
are attached

Spores

Eggs and
sperm

Adult form that
produces eggs
and sperm

Figure 1.8 Characteristics of a typical plant. These characteristics of a typical plant are numbered to match the description in the text.

	Bacteria	Algae	Plants	Fungi	Animals
Cell type	Prokaryotic; single-celled but may form colonies	Eukaryotic; single-celled and multicellular	Eukaryotic; multicellular		
Cell wall	Cell walls do not contain cellulose	Cell walls of some species contain cellulose	Cell walls composed mainly of cellulose		
Mode of nutrition	Various; some photosynthetic autotrophs	Photosynthetic autotrophs	Photosynthetic autotrophs		
Reproduction	Mostly asexual	Sexual and asexual; some species have two adult forms: one that produces spores and one that produces eggs and sperm	Sexual and asexual; two adult forms: one that produces spores and one that produces eggs and sperm; embryo protected within female parent		
Growth	Indeterminate	Indeterminate or determinate	Indeterminate or determinate		

Figure 1.9 Comparing plants with some other organisms. As you can see, plants share one or more characteristics with each of these other types of organisms: bacteria, algae, fungi, and animals. However, plants differ from each of these groups in one or more aspects of cell structure, nutrition, reproduction, and growth.

The text and art work together to help guide students through complex processes, often with examples that relate to their everyday lives.

1 Plants are multicellular ("many celled") eukaryotes. In modern classification, the most basic distinction between organisms is at the level of the cell—between eukaryotes and prokaryotes. Plants are among the **eukaryotes,** organisms whose cells have a **nucleus,** an enclosed structure that contains the cell's DNA. Eukaryotes also include animals, fungi, and protists such as algae. **Prokaryotes** are organisms whose cells do not have an enclosed nucleus, such as bacteria. You will read more about prokaryotes, protists, and fungi later in the book.

2 Almost all plants are capable of photosynthesis. Since plants can make their own food through photosynthesis, they are known as **autotrophs** ("self-feeders"). By contrast, animals and fungi are **heterotrophs** ("other feeders"), meaning that they obtain food from other organisms. Animals ingest their food, while fungi absorb food.

3 Plants have cell walls composed mainly of cellulose. **Cellulose** consists of a chain of glucose molecules. Cellulose-rich cell walls help to distinguish plants from other eukaryotes because cell walls of algae and fungi are usually composed mainly of other substances, while animals do not have cell walls.

4 Plants have two adult forms that alternate in producing each other. One of these adult forms makes **spores,** reproductive cells that can develop into adults without fusing with another reproductive cell. The other adult form makes either **sperm**

reproduce through **sex** volves fertilization of an offspring that are differe trast, plants can produc production or by **asexu** single parent produces itself. In addition, plant animal growth. Plants c out their lives. Since t known as **indetermina** mal basically stops grow pattern known as **dete** compares key characteri algae, and bacteria.

Having looked at the g we will take a brief trip life. If the history of Eart day, plants would have ex a half hours. However, th mans, who would have e and a half. Most people i flowers, but flowers evolv in the following overvi mosses, ferns, conifers (p plants (Figure 1.10).

Mosses are a
types

Mosses were among th

INSTRUCTOR SUPPLEMENTS

Instructor's Manual/Test Bank 0-8053-4417-9

This comprehensive resource includes chapter overviews, detailed lecture outlines, teaching tips, and media tips. The Test Bank portion of the manual features more than 50 multiple-choice, fill-in-the-blank, and short essay/critical thinking questions per chapter.

Computerized Test Bank 0-8053-4418-7

This dual-platform CD-ROM contains all questions from the printed Test Bank. Instructors can export questions into their own tests and print questions in a variety of formats.

Instructor's Art CD-ROM 0-8053-4455-1

This cross-platform CD-ROM includes 450 pieces of art from the book in PowerPoint® and jpeg files.

Transparency Acetates 0-8053-4421-7

Includes 300 selected illustrations from the book.

STUDENT SUPPLEMENT

The Botany Place

www.thebotanyplace.com

This interactive Website includes chapter outlines and objectives, lecture outlines, interactive tutorials, quizzes, Web links, flashcards, and glossary terms. Tutorials are topic specific and incorporate art from the text, helping students master difficult topics. Website quiz questions ask students to think conceptually about what they have learned. The Botany Place includes access to Research Navigator™, a new online academic research destination that combines three databases—The New York Times Search by Subject Archive, EBSCO's ContentSelect™ Academic Journal and Abstract Database, and Pearson Education's Link Library's "Best of the Web"—providing access to topical content from a variety of sources.

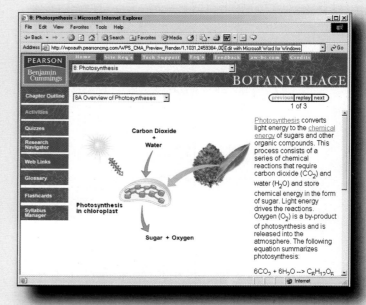

Preface

Plants are worth getting to know for many reasons. They are amazing and fascinating life-forms that transform and enhance our lives in both aesthetic and practical ways. The beauty of bouquets and the warmth of campfires are just two examples of the interest and the utility that plants bring to our lives. In fact, plants are involved in every aspect of human civilization: The remains of ancient photosynthetic organisms have produced all fossil fuels. Plants provide foods and medicines, fiber for clothing, and wood for fuel and for shelter. In the most basic sense, plants make life on Earth possible. Along with other photosynthetic organisms, plants act as biological sun catchers to trap solar energy. In the process of photosynthesis, they produce virtually all of the oxygen and food for planet Earth. They synthesize all of the structural organic molecules, which are modified and recycled by nonphotosynthetic organisms throughout all food webs.

Approach

Introduction to Botany has grown out of my experience teaching a variety of botany and biology classes over the past 31 years. During this time, the amount of basic information in the plant sciences has increased tremendously. Some textbooks in botany and biology attempt to keep pace by becoming encyclopedic. Others retreat to short, simple expositions of selected areas of botany. This book follows another approach. I have attempted to write a text of moderate length with an even coverage of all the central topics in modern botany, while giving emphasis to important, cutting-edge areas. I stress science as a way of knowing and the scientific method as an ongoing process for learning about plants.

Goals of the Book

The overall goal of this book is to provide important, up-to-date, and essential information about plants and about modern plant biology in an interesting, readable fashion. While showing what plants have in common with other living creatures, I hope to demonstrate conclusively that the unique features of plants and other photosynthetic organisms make them necessary for the survival of all life-forms, including humans.

Throughout the book, I focus on four important themes: **Evolution, Biotechnology, Conservation Biology,** and **Plants & People.** Evolution by natural selection is the most important paradigm of modern botany. The anatomical and biochemical features of plants, as well as their behaviors in populations and ecosystems, are rendered understandable by Darwin's theory. Central dogma, the flow of information from genes, to enzymes, to structure and function, is a second important botanical paradigm. Plant biotechnology and genetic engineering provide important demonstrations of both the possibilities and problems that arise from an understanding of plant molecular biology. Plant biotechnology and the related fields of plant genomics and proteomics are and will increasingly be in the news. Understanding these areas of plant biology is important for students who are educated for functional success in the current world. As the human population continues to increase in the face of declining natural resources and increasing environmental damage, knowledge of the role of plants in the biosphere is essential. Conservation biology is a rapidly developing field of study and of practical action. It depends on an understanding of biology at all levels and on the role played by plants in both natural ecosystems and ecosystems altered by human activity. Both biotechnology and conservation biology relate to an understanding of plants by both scientists and citizens. Humans depend on plants in many different ways, which are explored for both basic knowledge and human interest in the plants and people theme.

Organization of the Book

The chapter outline covers all of the traditional topics found in a botany text. Each chapter contains at least one box relating to one of the four themes of the book (**Evolution, Biotechnology, Conservation Biology,** and **Plants & People**), as well as one box entitled **The Intriguing World of Plants,** which is meant to engage students and motivate them to learn more about botany. In addition, the four central themes are integrated into the text throughout all chapters.

The introductory chapter is meant to be read for its content and does not serve as a summary of the book. It contains discussions of both the basic distinguishing characteristics of plants as well as their importance. The scientific method is covered using a case study relating to the discovery of phototropism by Charles Darwin and his son. Cell and plant anatomy are covered in Chapters 2–6 in the first unit of the book entitled **The Structure of Plants.** A basic knowledge of plant structure is important for all other areas of botany, although these chapters could be postponed for later depending on the wishes of the instructor. In the anatomy chapters, primary growth of stems and roots is covered in Chapter 4 and secondary growth of stems and roots is covered in Chapter 5. This arrangement eliminates the duplication involved in discussing both primary and secondary growth in separate chapters about stems and roots. However, the chapters are written in such a fashion that roots and stems are covered separately within each chapter, should the instructor wish to focus on roots and shoots separately. The basic aspects of plant reproduction, including life cycles, and the structures of cones, flowers, fruits, and seeds, are discussed in Chapter 6, so that students will be on familiar ground when they encounter life cycles in the chapters on plant diversity.

The second unit of the book, entitled **The Functions of Plants,** includes chapters on biochemistry (Appendix A covers basic chemistry), photosynthesis, and respiration, as well as coverage of water relations and mineral nutrition and plant growth and development. Biochemistry is covered first in Chapter 7, since knowledge of the molecules of plants facilitates an understanding of the chapters on energetics and physiology.

Genetics and Gene Expression, the third unit of the book, includes chapters on genetics and gene expression, as well as a chapter on plant biotechnology (Chapter 14). Since plant biotechnology is a theme of the book, material on the topic is found throughout, with the more complex topics introduced after Chapter 14.

The fourth unit of the book, **Evolution and Diversity,** includes chapters on evolution and on classification, as well as seven chapters covering photosynthetic creatures and other organisms such as fungi, which are usually considered in botany texts. Since evolutionary principles are basic to a phylogenetic discussion of organisms, evolution has been introduced before the diversity chapters. While viruses and prokaryotes have been combined into one chapter, the algae, as well as each group of land plants, are covered in separate chapters, which are designed to high-

light the distinctive features of each group. Finally, in the book's fifth unit, **Ecology** is introduced, first in terms of the biosphere, then ecosystems, and finally disturbed ecosystems, or conservation biology.

Pedagogical Features

Each chapter begins with an outline of major chapter headings, followed by an engaging story that relates to material covered in the chapter. No new terms are introduced in the stories, which serve to stimulate interest in the specific topics and information to come.

The outstanding art and photo selections for each chapter illustrate key points. Explanatory text is included whenever possible in the art rather than in the legend. Background shading has been minimized in favor of a style that emphasizes details of the figure itself, enhanced by a bright, appealing color palette. Photos have been selected to provide information on subjects likely to be new to students or that require additional emphasis.

End-of-chapter materials include a **Summary** of each section, **Review Questions,** and **Questions for Thought and Discussion,** as well as a **To Learn More** section that focuses on materials available, whether for research or simply to learn more about an interesting topic. The evolution theme is continued in a special question under the heading **Evolution Connection,** which encourages students to consider how evolution has shaped the development of plant life. Also, the last Review Question for each chapter features a drawing question (indicated by an appropriate pencil icon) to help students more actively think about the structures they are seeing. You will also find information on Web-based materials, available at www.thebotanyplace.com, which details materials available for each chapter.

Supplements

For Students
The Botany Place
www.thebotanyplace.com

This interactive Website includes chapter outlines and objectives, lecture outlines, interactive tutorials, quizzes, Web links, flashcards, and glossary terms. Tutorials are topic specific and incorporate art from the text, helping students master difficult topics. Website quiz questions ask students to think conceptually about what they have learned. The Botany Place includes access to Research Navigator™, a new online academic research destination that combines three databases—New York Times Search by Subject Archive, EBSCO's ContentSelect™ Academic Journal and Abstract Database, and Pearson Education's Link Library's "Best of the Web"—providing access to topical content from a variety of sources.

For Instructors

Instructor's Manual/Test Bank. This comprehensive resource includes chapter overviews, detailed lecture outlines, teaching tips, and media tips. The Test Bank portion of the manual features more than 50 multiple-choice fill-in-the-blank, and short essay/critical thinking questions per chapter.

Computerized Test Bank. A dual-platform CD-ROM with all questions from the printed Test Bank are provided. Instructors can export questions into their own tests and print questions in a variety of formats.

Instructors Art CD-ROM. This dual-platform CD-ROM includes 450 pieces of art from the book in Power-Point and jpeg files.

Transparency Acetates. This set includes 300 selected illustrations from the book.

Reviewers and Class Testers

Without reviewers, no manuscript would come to fruition as a published text. I am most grateful for the intelligent and thoughtful comments of my colleagues teaching botany courses across the country. I am also grateful to both the students and the professors who took time out of their semesters to class-test chapters for us; their feedback has been crucial in shaping this material. Thank you to all for their invaluable contributions to this text.

Reviewers

David Aborne, *University of Tennessee-Chattanooga;* Richard Allison, *Michigan State University;* Bonnie Amos, *Angelo State University;* Martha Apple, *Northern Illinois University;* Kathleen Archer, *Trinity College;* Robert M. Arnold, *Colgate University;* Ellen Baker, *Santa Monica College;* Susan C. Barber, *Oklahoma City University;* T. Wayne Barger, *Tennessee Tech University;* Marilyn Barker, *University of Alaska, Anchorage;* Linda Barnes, *Marshall-town Community College;* Paul Barnes, *Texas State University-San Marcos;* Susan R. Barnum, *Miami University;* Terese Barta, *University of Wisconsin-Stevens Point;* Hans Beck, *Northern Illinois University;* Donna Becker, *Northern Michigan University;* Maria Begonia, *Jackson State University;* Jerry Beilby, *Northwestern College;* Tania Beliz, *College of San Mateo;* Andrea Bixler, *Clarke College;* Allan W. Bjorkman, *North Park University;* Catherine Black, *Idaho State University;* J.R. Blair, *San Francisco State University;* Allan J. Bornstein, *Southeast Missouri State University;* Paul J. Bottino, *University of Maryland College Park;* Jim Brenneman, *University of Evansville;* Michelle A. Briggs, *Lycoming College;* Beverly J. Brown, *University of Arizona;* Judith K. Brown, *Nazareth College of Rochester;* Patrick J. P. Brown, *University of South Carolina;* Beth Burch, *Huntington College;* Marilyn Cannon, *Sonoma State University;* Shanna Carney, *Colorado State University;* Gerald D. Carr, *University of Hawaii;* J. Richard Carter, *Valdosta State University;* Youngkoo Cho, *Eastern New Mexico;* Jung Choi, *Georgia Tech;* Thomas Chubb, *Villanova University;* Ross Clark, *Eastern Kentucky University;* W. Dennis Clark, *Arizona State University;* John Clausz, *Carroll College;* Keith Clay, *Indiana University;* Liane Cochran-Stafira, *Saint Xavier College;* Deborah Cook, *Clark Atlanta University;* Anne Fernald Cross, *Oklahoma Chapter Trustee;* Billy G. Cumbie, *University of Missouri-Columbia;* Roy Curtiss, *Washington University;* Paul Davison, *University of North Alabama;* James Dawson, *Pittsburg State University;* John V. Dean, *DePaul University;* Evan DeLucia, University of Illinois at *Urbana-Champaign;* Roger del Moral, *University of Washington;* Ben L. Dolbeare, *Lincoln Land Community College;* Valerie Dolja, *Oregon State University;* Tom Dudley, *Angelina College;* Arri Eisen, *Emory University;* Brad Elder, *University of Oklahoma;* Inge Eley, *Hudson Valley Community College;* Sherine Elsawa, *Mayo Clinic;* Gary N. Ervin, *Mississippi State University;* Elizabeth J. Esselman, *Southern Illinois University, Edwardsville;* Frederick Essig, *University of South Florida;* Richard Falk, *University of California, Davis;* Diane M. Ferguson, *Louisiana State University;* Jorge F.S. Ferreira, *Southern Illinois University-Carbondale;* Lloyd Fitzpatrick, *University of North Texas;* Richard Fralick, *Plymouth State College;* Jonathan Frye, *McPherson College;* Stephen W. Fuller, *Mary Washington College;* Stanley Gemborys, *Hampden-Sydney College;* Patricia Gensel, *University of North Carolina;* Daniel K. Gladish, *Miami University,* David Gorchov, *Miami University;* Govindjee, *University of Illinois at Urbana Champaign;* Mary Louise Greeley, *Salve Regina University;* Sue Habeck, *Tacoma Community College;* Kim Hakala, *St. John's River CC;* Michael Hansen, *Bellevue CC;* Laszlo

Hanzely, *Northern Illinois University;* Suzanne Harley, *Weber State University;* Neil A. Harriman, *University of Wisconsin, Oshkosh;* Jill Haukos, *South Plains College;* Jeffrey Hill, *Idaho State University;* Jason Hoeksema, *University of Wisconsin, Oshkosh;* Scott Holaday, *Texas Tech University;* J. Kenneth Hoober, *Arizona State University;* Patricia Hurley, *Salish Kootenai College;* Stephen Johnson, *William Penn University;* Elaine Joyal, *Arizona State University;* Grace Ju, *Gordon College;* Walter Judd, *University of Florida;* Sterling C. Keeley, *University of Hawaii;* Gregory Kerr, *Bluefield College;* John Z. Kiss, *Miami University;* Kaoru Kitajima, *University of Florida;* Kimberley Kolb, *Califonia State University, Bakersfield;* Ross Koning, *Eastern Connecticut State University;* Robert Korn, *Bellarmine College;* David W. Kramer, *Ohio State University;* Vic Landrum, *Washburn University;* Deborah Langsam, *University of North Carolina;* A. Joshua Leffler, *Utah State University;* David E. Lemke, *Texas State University-San Marcos;* Manuel Lerdau, *State University of New York;* Alicia Lesnikowska, *Georgia Southwestern State University;* Gary Lindquester; *Rhodes College;* John F. Logue, *University of South Carolina, Sumter;* Marshall Logvin, *South Mountain Community College;* A. Christina W. Longbrake, *Washington & Jefferson;* Steven Lynch, *Louisiana State University, Shrevport;* Linda Lyon, *Frostburg State University;* Carol C. Mapes, *Kutztown University of Pennsylvania;* Michael Marcovitz, *Midland Lutheran College;* Bernard A. Marcus, *Genesee Community College;* Diane Marshall, *University of New Mexico;* Tonya McKinley, *Concord College;* Laurence Meissner, *Concordia University at Austin;* Elliot Meyerowitz, *California Institute of Technology;* Mike Millay, *Ohio University;* David Mirman, *Mt. San Antonio College;* John Mitchell, *Ohio University;* L. Maynard Moe, *California State University, Bakersfield;* Clifford W. Morden, *University of Hawaii at Manoa;* Beth Morgan, *University of Illinois at Urbana Champaign;* Dawn Neuman, *University of Las Vegas;* Richard Niesenbaum, *Muhlenberg College;* Gisele Miller-Parker, *Western Washington University;* Carla Murray, *Carl Sandburg College;* Terry O'Brien, *Rowan University;* Jeanette Oliver, *Flathead Valley Community College;* Clark L. Ovrebo, *University of Cental Oklahoma;* Fatima Pale, *Thiel College;* Lou Pech, *Carroll College;* Charles L. Pederson, *Eastern Illinois University;* Carolyn Peters, *Spoon River College;* Ioana Popescu, *Drury University;* Calvin Porter, *Texas Tech University;* David Porter, *University of Georgia;* Daniel Potter, *University of California, Davis;* Mike Powell, *Sul Ross State University;* Elena Pravosudova, *Sierra College;* Barbara Rafaill, *Georgetown College;* V. Raghavan, *Ohio State University;* Brent Reeves, *Colorado State University;* Bruce Reid, *Kean University;* Eric Ribbens, *Western Illinois University;* Stanley Rice, *Southeastern Oklahoma State University;* Steve Rice, *Union College;* Todd Rimkus, *Marymount University;* Michael O. Rischbieter, *Presbyterian College;* Matt Ritter, *Cal Poly San Luis Obispo;* Laurie Robbins, *Emporia State University;* Wayne C. Rosing; *Middle Tennessee State University;* Rowan F. Sage, *University of Toronto;* Thomas Sarro, *Mount Saint Mary College;* Neil Schanker, *College of the Siskiyous;* Rodney J. Scott, *Wheaton College;* Bruce Serlin, *Depauw University;* Harry Shealy, *University of South Carolina;* J. Kenneth Shull, *Appalachain State;* Susan Singer, *Carleton College;* Don W. Smith, *University of North Texas;* James Smith, *Boise State University;* Steven Smith, *University of Arizona;* Nancy Smith-Huerta, *Miami University;* Teresa Snyder-Leiby, *SUNY New Paltz;* Frederick Spiegel, *University of Arkansas;* Amy Sprinkle, *Jefferson Community College Southwest;* William Stein, *Binghamton University;* Chuck Stinemetz, *Rhodes College;* Steve Stocking, *San Joaquin Delta College;* Fengjie Sun, *University of Illinois at Urbana-Champaign;* Marshall D. Sundberg, *Emporia State University;* Walter Sundberg, *Southern Illinois University-Carbondale;* Andrew Swanson, *University of Arkansas;* Daniel Taub, *Southwestern University;* David Winship Taylor, *Indiana University Southeast;* Josephine Taylor, *Stephen F. Austin State University;* David J. Thomas, *Lyon College;* Stephen Timme, *Pittsburg State University;* Leslie Towill, *Arizona State University;* Gary Upchurch, *Southwest Texas University;* Staria S. Vanderpool, *Arkansas State University;* C. Gerald Van Dyke, *North Carolina State University;* Susan Verhoek, *Lebanon Valley College;* Beth K. Vlad, *Elmhurst College;* Tracy Wacker, *University of Michigan-Flint;* Charles Wade, *C.S. Mott Community College;* D. Alexander Wait, *Southwest Missouri State University;* Tom Walk, *Pennsylvania State University;* Patrick Webber, *Michigan State University;* Richard Whitkus, *Sonoma State University;* Donald Williams, *Park University;* Paula Williamson, *Southwest Texas University;* Dwina Willis, *Freed-Hardeman University;* MaryJo A. Witz, *Monroe Community College;* Jenny Xiang, *North Carolina State University;* Rebecca Zamora, *South Plains College.*

Class Testers

David Aborne, *James Madison University;* Ellen Baker, *Santa Monica College;* Susan C. Barber, *Oklahoma City University;* Susan Barnum, *Miami University;* Dr. Tania Beliz, *San Mateo College;* Catherine Black, *Idaho State University;* Jorge Ferreira, *Southern Illinois University;* Patricia Gensel, *University of North Carolina;* Dr. Michael

Hanson, *Bellevue Community College,* Kim Hakala, *St. John's River Community College;* Davide Lemke, *Southwest Texas State University;* Brent Reeves, *Colorado State University;* Michael Rischbieter, *Presbyterian College;* Pat Webber, *Michigan State University;* Gerald Van Dayke, *North Carolina State University;* Rebecca Zammoraege, *South Plains College.*

Acknowledgments

I would like to thank Elizabeth Fogarty for her inspiration and help with the original conception of this book. The editors and production staff at Benjamin Cummings have my everlasting regard for their talent and professionalism. In particular, I'd like to thank the specific editors who worked with me to write this book. Developmental Editors John Burner and Matt Lee have superb editorial talents as word crafters and creative consultants. Both have the rare ability to convey complicated concepts in understandable language. Developmental Artist Carla Simmons provided consistently high-quality art as well as excellent ideas for using art to make botany understandable and interesting. I appreciate Travis Amos's unceasing efforts to find the best photos and to supply beautiful and unique photos from his endless supply of possibilities. Chalon Bridges helped the book in many ways, most recently as Acquisitions Editor. Project Editor Susan Minarcin deserves considerable credit and thanks for the often thankless job of keeping us focused on the vision of the book as well on the schedule and for keeping track of the many interacting facets of publishing. Director of Development Kay Ueno kept the entire project on just the right course between idealism and realism.

Associate Editor Alexandra Fellowes and Publishing Assistant Alissa Anderson provided much valuable help, often on short notice, to keep the project going. I would also like to thank Managing Editor Erin Gregg and Senior Production Editor Corinne Benson, Senior Art Editor Donna Kalal, and Marketing Manager Josh Frost for their hard work and many contributions. I am grateful to Mark Ong for his elegant design. I would also like to thank Professors Victoria E. McMillan and Robert M. Arnold for contributing the text drawing questions and the Evolution Connection questions; their attention to detail is greatly appreciated. I would like to thank the authors who created the excellent supplements that accompany this text; their hard work greatly contributes to this text as a teaching tool. I would like to thank Laszlo Hanzely of *Northern Illinois University* and Deborah Cook of *Clark Atlanta University* for their excellent work on the Instructor's Manual. Robert Arnold and Victoria McMillian from *Colgate University* deserve much credit for their work on the Test Bank.

Several faculty members have helped me along the way with advice, photos, and rapid answers to my incessant questions. These people include, particularly, Dr. Jennifer Clevinger, Mr. Curtis Clevinger at *James Madison University,* Drs. Paul Kugrens, Paul Lee and *Brent Reeves at Colorado State University,* and Dr. Conley McMullen.

Undergraduate student editors have played a huge and important role in the preparation and review of this book. I would especially like to thank Sarah Javaid, Rianna Barnes, and Pauline Adams, recent graduates of James Madison University, for their help over several years. At JMU I would also like to particularly thank Molly Brett Hunter, as well as Jacqueline Brunetti, Melissa Spitler, David Evans, and Sarah Jones. At Colorado State University, I wish to thank Heather Stevenson, Callae Frazier, Hillary Ball, Tiffany Sarrafian, and Nicola Bulled, all of whom provided incalculable and much appreciated assistance. At both universities, students in my classes made many suggestions and gave me considerable advice for drafts of several chapters, for which I am grateful.

Murray W. Nabors
University of Mississippi

BRIEF CONTENTS

Contents

Unit One
THE STRUCTURE OF PLANTS   23

3 An Introduction to Plant Structure 48

4 Roots, Stems, and Leaves: The Primary Plant Body 70

5 Secondary Growth in Plants   99

Unit Two
THE FUNCTIONS OF PLANTS 147

8 Photosynthesis   175

9 Respiration   195

Unit Three
GENETICS AND GENE EXPRESSION 259

12 Genetics 261

13 Gene Expression and Activation 277

14 Plant Biotechnology 299

Unit Four
EVOLUTION AND DIVERSITY 321

15 Evolution 323

23 Angiosperms: Flowering Plants 480

Unit Five ECOLOGY 503

24 Ecology and the Biosphere 505

25 Ecosystem Dynamics: How Ecosystems Work 524

26 Conservation Biology 548

1
The World of Plants

Woman picking coffee beans in Thailand.

The Importance of Plants

Photosynthesis sustains life on Earth

Plants are our fundamental source of food

Many medicines come from plants

Plants provide fuel, shelter, and paper products

Conservation biology is a critical area of research

Biotechnology seeks to develop new plant products

Plant Characteristics and Diversity

A set of characteristics distinguishes plants from other organisms

Mosses are among the simplest types of plants

Ferns and their relatives are examples of seedless vascular plants

Pine trees and other conifers are examples of nonflowering seed plants

Most plants are flowering plants with seeds protected in fruits

Botany and the Scientific Method

Botanists, like other scientists, test hypotheses

Botany includes many fields of study

Botanists also study algae, fungi, and disease-causing microorganisms

A Seattle stockbroker gulps a cup of coffee. A smoker in Tulsa contemplates quitting cigarettes. In a Denver cafe, students enjoy salsa and chips. In the highlands of Indonesia, people take quinine pills to avoid contracting the malaria that is common in the region. In a hospital in Washington, D.C., a patient receives a potent anticancer drug, vinblastine. In New York City, an allergy sufferer takes an antihistamine, while a few blocks away the same compound is used to make a dangerous street drug, methamphetamine.

All these scenes have something in common—plant chemicals called alkaloids. More than 12,000 alkaloids have been identified, including such familiar ones as caffeine, nicotine, cocaine, morphine, strychnine, quinine, and ephedrine. In plants, alkaloids deter predators because they taste bad or are poisonous. At low doses, many alkaloids stimulate the human nervous system. Caffeine, for example, enables certain nerve impulses to continue when they would otherwise be inactive. However, doses make the difference between a mild stimulant and a poison. For instance, insects that make a major meal of coffee leaves may consume a lethal dose of caffeine. Similarly, a little caffeine can have a pleasant effect on a person, but a lot can be dangerous and even deadly.

Of course, caffeine and other alkaloids are just a few examples of how plants affect our lives. While caffeine simply serves as a stimulant, we all actually depend on plants in a fundamental way— for our very survival. If all the plants on Earth were to suddenly die, they would quickly be followed by all the animals, including humans. However, if all the animals died, plants would still be able to survive. Why are humans and other animals so dependent on plants, rather than the other way around? This chapter will answer that question, while also providing an overview of the diversity of plant life and the importance of botany.

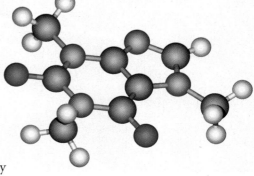

The molecular structure of caffeine with carbon, gray; nitrogen, purple; hydrogen, white; oxygen, red.

The Importance of Plants

The word *botany* comes from the Greek word for "plant." What does the word *plant* bring to mind—perhaps trees, wood, leaves, flowers, fruits, vegetables, and grains? If asked to define what a plant is, you might say it is an organism that is usually green, typically does not consume other organisms, and grows but does not move from place to place. As you might expect, and as you will see later in the chapter, a scientific definition is more formal and not quite so simple. However, these characteristics still serve as a very general, informal definition of a plant—whether it be a bush, tree, vine, fern, cactus, or any other type.

Since plants are such a familiar part of the scenery, we may not think about what makes plants unique or how they are vital to human life. Why do we need plants in order to survive, and why can they survive without us? The answer is photosynthesis. **Photosynthesis** is the process by which plants and certain other organisms use solar energy to make their own food by transforming carbon dioxide and water into sugars that store chemical energy. Animals and other organisms unable to make their own food can survive only by obtaining their food directly or indirectly from plants. In Chapter 8, you will see how photosynthesis works and why it is responsible for plants being generally green.

Photosynthesis sustains life on Earth

Almost all life on Earth relies on water and solar energy. However, only plants, algae, and photosynthetic bacteria can directly use these ingredients to survive. With sunlight, carbon dioxide, water, and a few soil minerals, a plant can make its own food, but no animal could live on those ingredients. Even with an unlimited supply of water, a person could survive for only a few weeks. In contrast, photosynthesis enables plants and other photosynthetic organisms to be solar-powered food factories. Almost one quarter of the approximately 1.5 million known species of living organisms are photosynthetic. Plants, bacteria, and algae carry out almost all of the photosynthesis on the planet, giving them a central role in the **biosphere,** the thin layer of air, land, and water occupied by living organisms. Plants are the main source of photosynthesis on land, while algae—ranging from microscopic organisms to seaweeds—contribute to photosynthesis in aquatic environments, along with photosynthetic bacteria.

Photosynthesis supports life in three ways:

1. Scientists now believe that photosynthesis produces almost all of the world's oxygen. During photosynthesis, plants break down molecules of water (H_2O) and produce oxygen (O_2). Most or-

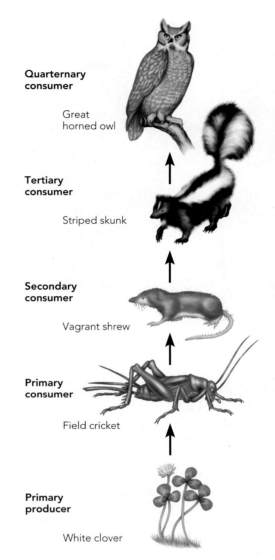

Figure 1.1 Food chain. This example of a terrestrial food chain includes four levels of consumers. Some food chains have fewer levels of consumers, but all terrestrial food chains have plants as their foundation.

Quarternary consumer
Great horned owl

Tertiary consumer
Striped skunk

Secondary consumer
Vagrant shrew

Primary consumer
Field cricket

Primary producer
White clover

ganisms, including plants and animals, need oxygen to release the energy stored in food.

2. Almost all organisms get their energy directly or indirectly from photosynthesis. Animals and most other nonphotosynthetic organisms obtain energy by eating plants or by eating animals that ate plants. In this way, a plant or other photosynthetic organism is the start of any **food chain,** a sequence of food transfer from one organism to the next, beginning with an organism that produces food (Figure 1.1). For example, a mountain lion eats a deer that ate grass. As organisms that make their own food, plants and other photosynthetic organisms are known as **primary producers.** By directly or indirectly supporting

Figure 1.2 Ancient agriculture. This Egyptian tomb wall painting, from around 1500 B.C., depicts workers using sickles to cut grain.

all levels of consumers, primary producers form the foundation of a food chain. Plants are the primary producers in terrestrial food chains, while algae and photosynthetic bacteria are the primary producers in aquatic food chains.

3. The sugars produced by photosynthesis are the building blocks of life. Plants produce sugars and related molecules through photosynthesis and subsequent processes, combining these products with soil minerals to make a wide variety of compounds. A plant uses these compounds to make its structural and physiological features. By eating plants, or by eating animals that eat plants, an animal receives compounds originally produced by photosynthesis and uses them to produce its own structures.

Since photosynthesis supplies the oxygen we breathe, the food we eat, and the very molecules of our being, we are all solar-powered organisms. Without photosynthesis, life on Earth would be extremely limited, if not impossible.

Plants are our fundamental source of food

Humans were originally hunter-gatherers, moving from place to place as the seasons dictated where food would be available. They ate almost anything they could find, track down, dig up, pick, or kill. Our teeth give evidence of this heritage, with large molars for grinding and chewing and sharper canines and incisors for biting and tearing. Around 12,000 to 14,000 years ago, some groups of humans began

to live in the same location year-round, cultivating crops and raising animals for food rather than depending solely on what nature provided. This domestication of plants and animals occurred in many different places and in many different ways. As cities that depended on food from farming arose throughout these regions, agriculture became a foundation for human civilization, making possible the development of culture, art, and government (Figure 1.2).

Early farmers saw that some types of food plants could be cultivated more successfully than others. Through trial and error, they learned how to harvest and store seed for the next year, when to plant, and how to nourish plants to a successful harvest. They noticed that some individual plants among a given type grew better than others. Over the years, they saved and planted seeds from these plants to increase food production, thereby becoming the first plant breeders.

Plant breeding is now a formal field of study, with state and national governments sponsoring research on how to increase crop yields (Figure 1.3). Plant breeders typically focus on improving a particular crop for their region. For example, wheat breeders in the northern Great Plains are looking for plants that grow as fast as possible because the

Figure 1.3 Modern plant breeding. Rice is a particularly important food source and is therefore the subject of intense experimentation to improve yields, nutritional quality, and disease resistance.

PLANTS & PEOPLE

A Taste of Tea History

Rich in caffeine and mildly addictive, tea is the world's most popular beverage. The drink originated in China more than 5,000 years ago. Legend says that Emperor Shen Nong observed tea leaves accidentally fall into boiling water. Being curious, he asked for a taste, and the beverage was born.

After being introduced to Europe in the 1580s, tea became immensely popular. By the mid-1600s Great Britain, with a superior navy, had cornered the European tea trade. By the early 1800s, tea had become so popular that the British began importing opium from India and selling it to China to pay for tea, as well as other popular Chinese products, such as silk and porcelain.

Tea was also immensely popular in the American colonies. The Americans avoided the British monopoly by importing tea from China in exchange for opium from the Ottoman Empire in Constantinople. By the mid-1700s, tea smuggling was thriving in the colonies because of the British tax on imports. To discourage smugglers, Britain lowered the tax. However, it left a small tax to stress its right to tax the colonists, a decision that sparked the Boston Tea Party protest in 1773, one of the events leading to the American Revolution. A boycott of British tea began in earnest and, partly as a result, coffee eventually became the favorite beverage of Americans.

For centuries tea had been grown only in China and was highly protected by the Chinese. Then, in the 1840s, Scottish botanist Robert Fortune smuggled tea seeds out of China. Now grown in 25 countries, the plant requires a warm, wet climate and an acidic soil high in organic matter.

Tea leaves. Tea is made from the evergreen leaves of *Camellia sinensis*, a small shrub.

Since leaves are handpicked, harvesting calls for abundant labor.

The early 1900s saw two advances in the use of tea. At the 1904 St. Louis World's Fair, a tea plantation owner dumped ice into his samples, increasing their popularity during that hot summer. Such was the origin of iced tea. Tea bags came along in 1908.

Today, 90% of the tea sold in the United States is black tea, in which leaves are fermented before being fired. Firing reduces the moisture content of leaves from 45% to 5%. Green tea, the remainder of regular tea sales to Americans, is not fermented. In addition, there are herbal teas made from the flowers, berries, roots, seeds, and leaves of other plants. Herbal teas are usually caffeine-free and have a surprising array of flavors.

---

growing season is short. They also seek plants that are resistant to strong winds, droughts, and common diseases.

Despite the domestication of many wild plants, most human food comes from just a few crops—primarily corn, rice, and wheat. Corn, which is probably native to Mexico, became the most important crop in North and South America, while rice originated in Asia and is the predominant crop there. Wheat was first cultivated in the Middle East and is the most important crop in Europe, the Middle East, and parts of North America and Africa. The grains of corn, rice, and wheat are rich in nutrients and are easy to store under dry conditions. Overall, six crops provide 80% of human caloric intake: wheat, rice, corn, potatoes, manioc or cassava, and sweet potatoes. An additional eight crops supply a considerable proportion of the remaining 20%: bananas, beans, soybeans, sorghum, barley, coconuts, sugarcane, and sugar beets.

In addition to being our major source of food, plants are used to produce many beverages. Alcoholic drinks such as wine and beer are made from sugar-rich plant material. Coffee and tea are drinks derived from beans (technically, berries) of the coffee plant (*Coffea arabica*) and leaves of the tea plant (see the *Plants & People* box above). Also, high-fructose syrup made from corn and other plants is the sweetener of choice for many soft drinks.

QUINQUINA GRIS.

Figure 1.4 The cinchona tree and quinine. Cinchona bark was used by Native Americans in the Andes Mountains of South America to cure a variety of fevers. Spanish explorers noticed the practice, which led to the discovery of quinine as a treatment for malaria.

Dried herbs and spices are used for a variety of purposes, including cooking ingredients, room fresheners, and medicines. The term *spice* typically refers to dried parts of tropical and subtropical plants, such as cinnamon, cloves, ginger, and black pepper (see *The Intriguing World*

of Plants box on page 7). Some commonly used herbs include marjoram, mint, parsley, rosemary, sage, and thyme.

Many medicines come from plants

Since ancient times, people have noticed that plants can alleviate symptoms of many medical conditions. For instance, tea made from the bark of willow trees can cure headaches. Now we know that this bark contains salicylic acid, very similar in structure to acetylsalicylic acid, better known as aspirin. For centuries, such knowledge was passed by word of mouth and accumulated by naturalists who collected plants. During the 1500s, books called *herbals* began to list practical uses of plants and attempted to classify and scientifically name plants, which was a useful endeavor because a particular plant might have many common names.

As modern chemistry developed in the 1700s and 1800s, plant extracts containing alkaloids and other useful compounds became more commonly available. The alkaloid quinine has played a particularly important role in human history. By the late 1600s, physicians realized that the bark of the cinchona tree and the white quinine powder extracted from it could be used to treat malaria (Figure 1.4). Despite the widespread use of quinine and related drugs, malaria remains one of the world's most devastating diseases. In tropical countries it kills between one million and three million people, mostly children, each year. Another commonly used alkaloid is ephedrine, which is produced by shrubs of the genus *Ephedra* and is a powerful antihistamine. Alkaloids also can affect animal physiology by disrupting cell division. Researchers have used two alkaloids produced by the rosy periwinkle plant, vinblastine and vincristine, to disrupt the division of cancer cells and kill the cells. Thousands of other plant products are useful in human medicine, with about 25% of all prescriptions written in the United States containing at least one product derived from plants.

Plants provide fuel, shelter, and paper products

In some long-lived plants, the stems become woody trunks, sometimes hundreds of feet tall. What we call wood consists almost entirely of dead cells that enable the trunks of some woody plants to become quite thick and strong.

Wood remains the most important source of fuel for cooking and heating because much of the world's population has little or no access to electricity and fossil fuels such as petroleum, coal, and natural gas. Fossil fuels them-

THE INTRIGUING WORLD OF PLANTS

Black Pepper: Savior of Rotting Meat

It might surprise you to learn that European settlement of the Americas came about largely as a result of the search for black pepper and other spices. Black pepper comes from the dried, ground fruits of the shrub *Piper nigrum,* native to the Malabar Coast of southwestern India. The fruits are green and have a white interior, with the black coating resulting from fungal action.

Why was pepper so important? In those pre-refrigeration days, salt preserved meat by keeping bacteria and fungi generally at bay, but this also made the meat largely inedible. Adding spices such as black pepper made salted meat palatable, which is why sailors often carried small bags of peppercorns.

During the Middle Ages, traders brought spices from Asia to Europe along trade routes through the Middle East. Caravans of camels laden with black pepper, cloves, cinnamon, nutmeg, ginger, and other spices had been making the trip for a thousand years. Beginning in 1470, however, the Turks blocked these overland routes, and Europeans looked to the ocean for an alternate passage to Asia. Christopher Columbus won financial support from the Spanish court to seek a new route to China and India. Landing in the Caribbean, he believed he had reached islands off the coast of India, referring to the inhabitants as "Indians" and the islands as the "Indies." Although he found no black pepper, it is not surprising that the spicy hot fruits he did find were later known as "peppers," even though they were members of the genus *Capsicum,* an entirely different group of plants. Today we distinguish between the two plants by calling one black pepper and the other hot peppers, such as jalapeños and habaneros.

Black pepper. The fruits of *Piper nigrum* become black pepper.

---

selves are also plant products, having been formed hundreds of millions of years ago primarily from fossilized remains of plants.

Wood is still the main source of construction material, and it is used as the framework for most houses and many other buildings. Even steel construction relies indirectly on plants because the super-hot furnaces used to make steel out of iron are powered mainly by fossil fuels.

Another important plant product is paper, which can be made from a variety of plants. Most of the paper we use is made of pulp derived from woody plants such as fir and pine.

These are just a few examples of how plants provide a variety of useful products—in addition, of course, to being our basic source of food and oxygen. Throughout this book, the main narrative and "Plants and People" box features will explore how plants affect human life.

Conservation biology is a critical area of research

Since plants supply so many of our needs, we must make sure that we have enough of them and that valuable plant species do not become extinct. We are the conservators, or caretakers, of Earth's resources. **Conservation biology** is an important multidisciplinary field of science that studies ways to counter the widespread extinction of species and loss of critical habitats. It studies the impact of human activities on all facets of the environment and searches for less ecologically destructive ways to harvest trees, build cities, and generally interact with the biosphere's biological resources, such as forests.

The large and rapidly increasing human population has consumed wood much more rapidly than it has been replaced. Therefore, it is not surprising that wood will command higher prices as supplies dwindle. About half of the yearly harvest of trees in the United States is used to make paper, most of which is not recycled. Only a few remnants of old-growth forests exist, mostly in national forests, national parks, and private preserves (see the *Conservation Biology* box on page 9).

Worldwide, about half of the original forests are gone, replaced by cities, farms, or other human activities. Much rain forest is destroyed in what is called "slash-and-burn" agriculture, in which the forest is cleared (Figure 1.5), the land is used for crops for a few years until soil nutrients are depleted, and then the land is abandoned. In view of the importance of plants, it is chilling that human

(a)

(b)

Figure 1.5 Rain forest destruction. The remnants of slash-and-burn agriculture in Mexico's Yucatan peninsula **(a)** stand in stark contrast to **(b)** an undisturbed rain forest in Carrillo National Park, Costa Rica**.**

activities, including most importantly the destruction of tropical rain forests, are causing the extinction of large numbers of species of plants and animals. Extinction estimates vary considerably. In his book *The Diversity of Life,* Harvard biologist Edward O. Wilson conservatively estimates that 2,700 species of organisms become extinct each year. Other scientists estimate that between 5 percent and 40 percent of all species may become extinct between 1990 and 2020, including many plants that may contain medically useful compounds of incalculable value. Twenty percent of all tropical and semitropical plant species may have already become extinct between 1952 and 1992. The theme of conservation biology will be stressed throughout the narrative of this text, in box features, and in the concluding chapter.

Biotechnology seeks to develop new plant products

Since prehistoric times, humans have looked for improved plants to better meet their needs. The science of botany developed because of basic curiosity about these organisms that are so necessary to our lives. Efforts to obtain improved plants and plant products using scientific techniques are known as **plant biotechnology.** In a broad sense, such endeavors have a long history, dating back to the first farmers who experimented with different seeds.

Modern biotechnology involves both chemistry and biology (see the *Biotechnology* box on page 10).

Like all organisms, plants contain hereditary information in the form of **DNA (deoxyribonucleic acid)** molecules. Specific DNA sequences containing hereditary information are called **genes,** which determine an organism's physical characteristics. In recent years, scientists have been able to move and modify genes to produce plants with desired traits—a relatively new method of biotechnology known as **genetic engineering.** The development of genetically modified (GM) plants has been both revolutionary and controversial, arousing so much debate that many people associate biotechnology exclusively with genetic engineering. Following are a few examples of uses of genetic engineering to develop GM plants.

CONSERVATION BIOLOGY

The Challenge of Forest Conservation

Before the arrival of large numbers of Europeans in the 1700s, North America was extensively forested. By the late 1800s, more than half of those forests were gone, and people were beginning to realize that agricultural and forest resources were not unlimited.

President Theodore Roosevelt and reformer Gifford Pinchot were among the earliest advocates of conservation,

The Gifford Pinchot National Forest. Located in Washington State, the forest includes the Mt. Saint Helens area, an important location for studying natural revegetation after volcanic eruptions.

the movement to preserve natural resources through wise use. Pinchot saw a nation "obsessed by a fury of development . . . exploiting the riches of the richest of continents," aided by the sale of government-owned forest reserves to the highest bidders. Appointed chief of the U.S. Division of Forestry in 1898, Pinchot worked to preserve the nation's forests. After assuming the presidency in 1901, Roosevelt warned that without conservation the nation's forests would be gone in 60 years. In 1905, Congress transferred forest reserves to the new U.S. Forest Service, with Pinchot as its first chief. Later renamed national forests, these resource areas were to be managed for the greatest good for the largest number of people in perpetuity.

Initially, the idea of national forests was very unpopular, particularly in the West. Even today, despite having national forests, we are far from achieving a sustainable use of our forest resources. The very notion of conservation must be reestablished in the minds and actions of each new generation of Americans.

Since the time of Roosevelt and Pinchot, the world has become a different place. In many cases it is hard to balance today's needs with the possibility of future shortages. Three factors have increased pressures to use natural resources: high standards of living in developed countries, accelerating population growth, and struggles of developing nations to attain acceptable standards of living. The efforts to meet the challenges of conservation biology will be a focus throughout this book.

"Golden Rice"

Like other important crops, rice does not contain adequate amounts of some necessary nutrients and is particularly deficient in vitamin A. Each year in regions where rice is the chief component of the diet, vitamin A deficiency results in blindness in nearly half a million children and the deaths of one million to two million pregnant women and people with a weakened immune system. Plant geneticists in Switzerland have begun to address this problem by adding two genes from the daffodil and one bacterial gene to rice. With these additional genes, this new "Golden Rice" plant can make golden-colored beta-carotene, a source of vitamin A (Figure 1.6).

Pest-resistant plants

Addition of a bacterial gene called the *Bt* gene makes crops resistant to particular insects, allowing farmers to avoid using pesticides.

Figure 1.6 "Golden Rice." Ingo Potrykus, of the Swiss Federal Institute of Technology, envisioned a genetically modified strain of rice that could help feed children all over the world. By injecting rice embryos with genes from daffodils and using a bacterium called *Erwinia uredovora*, Potrykus created genetically modified "Golden Rice" that has the ability to make beta-carotene, from which vitamin A can be derived.

BIOTECHNOLOGY

Using Plants to Battle Bacteria

Imagine this nightmare: Bacteria have evolved to become totally resistant to antibiotics that once readily eliminated them. Diseases such as tuberculosis, once treated by antibiotics, reach nearly epidemic levels. Does this sound unbelievable? Well, this "science fiction" scene is coming closer to harsh reality.

When antibiotics first became widely available in the 1940s, they were hailed as miracle drugs, eliminating bacteria with little or no harm to the infected person. However, bacteria have steadily become more resistant to antibiotics, often pumping out the antibiotics before they can harm the bacterial cell. Strains of *Staphylococcus* and other bacterial species capable of causing life-threatening illnesses are already resistant to more than 100 antibiotics.

Now that we are in this worrisome position, what can we do? Among the researchers looking for solutions are Frank Stermitz of Colorado State University and Kim Lewis of Tufts University, who are studying the chemistry of plants, particularly disease-fighting properties of alkaloids such as berberine. Long used by Native Americans as a medicine, berberine is already sold as a relief for arthritis, diarrhea, fever, hepatitis, and rheumatism. Stermitz and Lewis found berberine in the leaves and sap of several varieties of a plant known as Oregon grape, in the genus *Berberis*. Initial test-tube studies on the berberine were disappointing, indicating weak antibiotic activity when used as a pure extract. However, combining berberine with other compounds from the plant greatly increased its antibiotic activity, rivaling the strongest antibiotics available against *Staphylococcus*.

What exactly made the berberine more effective? Stermitz and Lewis discovered that Oregon grape produces a second compound, 5'-methoxyhydnocarpin, which prevents bacteria from pumping out the berberine, thereby allowing the antibiotic to kill the bacteria. Now researchers are testing whether this combination works in living animals. Eventually, vaccines and new antibiotics, perhaps derived from plants, may bring this dangerous bacterium under control.

Oregon grape plant. Berberine is extracted from this plant.

Edible vaccines

Humans and other mammals kill invading disease organisms by producing molecules called *antibodies* that mobilize the immune system. When a gene from a disease-causing organism is added to a plant, humans who eat the plant may produce antibodies against that disease. For example, a particular strain of the bacterium *Escherichia coli* causes a diarrhea that kills two million children a year. Potatoes containing a gene from this bacterium caused an immune response that made people resistant to the disease. The gene could be added to bananas or other crops commonly available in many developing nations.

Toxin-resistant plants

More than 40% of cultivated land contains excess toxic ions, which reduce and even prevent crop production. Adding certain bacterial genes makes plants resistant to toxic ions such as mercury compounds, aluminum salts, and sodium chloride. In some cases, plants that accumulate specific toxic ions can be grown and then removed, thereby purifying contaminated soils.

Herbicide-resistant plants

Weeds decrease crop production by competing with crops for nutrients, water, and growing space. Scientists have added genes to crops to make them resistant to herbicides (chemicals used to kill weeds), which can then be used to remove weeds without damaging the crops. Many of these herbicides have no known effect on animals or humans and can therefore be safely used on crops.

In some cases, plants can play an unusual role in genetic engineering experiments on animals. For example, scientists seeking ways to control pain have experimented with responses of mice to capsaicin, the alkaloid that makes peppers hot to the taste. The genes of these mice have been modified so that the mice no longer avoid hot peppers but eat them freely (Figure 1.7). Enabling mice to eat hot salsa may not seem like a scientific breakthrough, but cell receptors that sensitize people to hot peppers also respond to heat and other pain-inducing stimuli. Further study of these mice and their altered responses to capsaicin may yield clues on how to control certain types of pain.

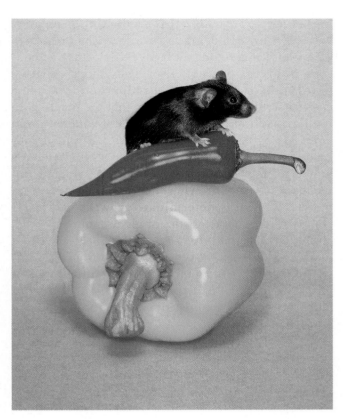

Figure 1.7 Using plants to modify animal responses. This mouse has been genetically modified to lack the receptor for capsaicin, the ingredient that makes chile peppers hot. The mouse eats spicy habanero peppers and also shows a reduced sensitivity to acidity and heat. A study of this animal's high tolerance for chiles may yield information that will help humans control pain.

Although the science of genetic engineering has advanced rapidly, the social implications are still unclear. GM crops must be tested and regulated to ensure that they are safe for human consumption and do not endanger animals or the environment. For example, some people may be allergic to gene products introduced into crops from other organisms. Herbicide-resistant genes could potentially escape into other plants, producing so-called super weeds. Pollen from insect-resistant crops could be eaten by and kill harmless or useful insect populations.

In the debate over the use of GM plants, it is important to understand how they are produced and what they can accomplish. Legitimate concerns need to be separated from fears that GM plants are the botanical equivalent of Frankenstein monsters. Plant breeders have been producing genetically modified crops for centuries using traditional methods. Genetic engineering techniques speed that process and also make it possible to introduce genes from a wide variety of organisms not related to plants.

Chapter 14 discusses plant biotechnology in general and genetic engineering in particular. In addition, biotechnology will be a recurring theme in the narrative and box features throughout the book.

Section Review

1. **What are three ways in which photosynthesis supports life on Earth?**
2. **How do plants meet everyday human needs?**
3. **What are major challenges facing conservation biology?**
4. **What are some potential benefits of GM crops? What are some possible problems?**

Plant Characteristics and Diversity

From the time of the ancient Greeks until the mid-1800s, organisms were simply classified as either plants or animals. They were considered to be plants if they did not move, were green, and did not eat other organisms. Those that moved and ate other organisms were called animals. Accordingly, the category of plants included not only such organisms as mosses, ferns, pine trees, and flowering plants, but also algae and fungi. Algae were included basically because they were green, and fungi were included largely because they did not move. Today, however, botanists consider plants to be fundamentally different from fungi, and plants are usually defined as being separate from algae.

In distinguishing plants from other organisms, scientists look at the history of **evolution,** all the changes that have transformed life from its beginnings to the present diversity of organisms. As organisms have evolved over time, differences in their genes have emerged, which are often reflected in physical characteristics. However, attempts to classify organisms based on physical characteristics and behavior always seem to run into exceptions. After examining how most botanists define plants, we will look at examples of plant diversity.

A set of characteristics distinguishes plants from other organisms

The following five characteristics, summarized in Figure 1.8, are most commonly used to identify plants and to distinguish them from other living organisms:

1 Plants are multicellular ("many celled") eukaryotes. In modern classification, the most basic distinction between organisms is at the level of the cell—between eukaryotes and prokaryotes. Plants are among the **eukaryotes,** organisms whose cells have a **nucleus,** an enclosed structure that contains the cell's DNA. Eukaryotes also include animals, fungi, and protists such as algae. **Prokaryotes** are organisms whose cells do not have an enclosed nucleus, such as bacteria. You will read more about prokaryotes, protists, and fungi later in the book.

2 Almost all plants are capable of photosynthesis. Since plants can make their own food through photosynthesis, they are known as **autotrophs** ("self-feeders"). By contrast, animals and fungi are **heterotrophs** ("other feeders"), meaning that they obtain food from other organisms. Animals ingest their food, while fungi absorb food.

3 Plants have cell walls composed mainly of cellulose. **Cellulose** consists of a chain of glucose molecules. Cellulose-rich cell walls help to distinguish plants from other eukaryotes because cell walls of algae and fungi are usually composed mainly of other substances, while animals do not have cell walls.

4 Plants have two adult forms that alternate in producing each other. One of these adult forms makes **spores,** reproductive cells that can develop into adults without fusing with another reproductive cell. The other adult form makes either **sperm** (male reproductive cells) or **eggs** (female reproductive cells). A sperm fertilizes an egg to produce an **embryo** that develops into an adult organism. In Chapter 6 you will see how the two adult forms in plants alternate in producing each other. For now, suffice it to say that plants have intriguing sexual lives.

5 Plants have a multicellular embryo protected within the female parent. Protected embryos evolved as an adaptation to life on land, preventing the embryo from drying out. This characteristic distinguishes plants from algae.

Each of these characteristics by itself is not unique to plants, but together they are useful for distinguishing plants from other organisms. Two additional characteristics help in particular to distinguish plants from most animals. Unlike most animals, plants can reproduce in two ways. Once mature, most animals can only reproduce through **sexual reproduction,** which involves fertilization of an egg by a sperm and results in offspring that are different from either parent. In contrast, plants can produce offspring by either sexual reproduction or by **asexual reproduction,** in which a single parent produces offspring that are identical to itself. In addition, plant growth is quite different from animal growth. Plants can continue to grow throughout their lives. Since this growth is unlimited, it is known as **indeterminate growth.** In contrast, an animal basically stops growing after becoming an adult, a pattern known as **determinate growth.** Figure 1.9 compares key characteristics of plants, animals, fungi, algae, and bacteria.

Having looked at the general characteristics of plants, we will take a brief trip through the evolution of plant life. If the history of Earth could be compressed into one day, plants would have existed for about the last two and a half hours. However, they are ancient compared to humans, who would have existed for only the last minute and a half. Most people immediately associate plants with flowers, but flowers evolved relatively late, as you will see in the following overview of major types of plants: mosses, ferns, conifers (plants with cones), and flowering plants (Figure 1.10).

Mosses are among the simplest types of plants

Mosses were among the first plants, evolving from algae-like ancestors around 450 to 700 million years ago. They belong to a group of small, nonflowering plants known as **bryophytes** (from the Greek *bryon,* moss, and *phyton,* plant), which are simpler in structure than other plants. You may be familiar with mosses that grow on rocks or form soft carpets on the forest floor. They never grow more than a few inches tall because they are inefficient at carrying water up the plant.

Sphagnum moss, which grows in bogs and swamps, is important economically in many regions of the world (Figure 1.10a). When decomposed, it forms a substance known as peat, which can be burned as fuel and also used as fertilizer. In addition, sphagnum moss absorbs a considerable amount of carbon dioxide that would otherwise reside in the atmosphere and could contribute to global warming. Chapter 20 discusses the evolution and characteristics of mosses and other bryophytes.

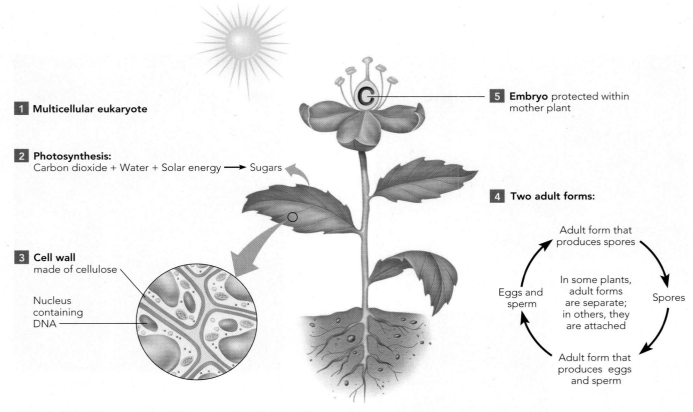

1 Multicellular eukaryote

2 **Photosynthesis:**
Carbon dioxide + Water + Solar energy ⟶ Sugars

3 **Cell wall**
made of cellulose

Nucleus containing DNA

5 **Embryo** protected within mother plant

4 **Two adult forms:**

Adult form that produces spores

In some plants, adult forms are separate; in others, they are attached

Eggs and sperm

Spores

Adult form that produces eggs and sperm

Figure 1.8 Characteristics of a typical plant. These characteristics of a typical plant are numbered to match the description in the text.

	Bacteria	Algae	Plants	Fungi	Animals
Cell type	Prokaryotic; single-celled but may form colonies	Eukaryotic; single-celled and multicellular	Eukaryotic; multicellular	Eukaryotic; multicellular	Eukaryotic; multicellular
Cell wall	Cell walls do not contain cellulose	Cell walls of some species contain cellulose	Cell walls composed mainly of cellulose	Cell walls composed mainly of chitin	No cell walls
Mode of nutrition	Various; some photosynthetic autotrophs	Photosynthetic autotrophs	Photosynthetic autotrophs	Heterotrophs that absorb food	Heterotrophs that ingest food
Reproduction	Mostly asexual	Sexual and asexual; some species have two adult forms: one that produces spores and one that produces eggs and sperm	Sexual and asexual; two adult forms: one that produces spores and one that produces eggs and sperm; embryo protected within female parent	Sexual and asexual	Mostly sexual, some asexual; embryo protected within female parent in some species, including most mammals
Growth	Indeterminate	Indeterminate or determinate	Indeterminate or determinate	Indeterminate or determinate	Determinate

Figure 1.9 Comparing plants with some other organisms. As you can see, plants share one or more characteristics with each of these other types of organisms: bacteria, algae, fungi, and animals. However, plants differ from each of these groups in one or more aspects of cell structure, nutrition, reproduction, and growth.

(a) Mosses are the most common bryophytes. This sphagnum moss is producing brown sporangia containing spores.

(b) Ferns are the largest group of seedless vascular plants.

(c) Most gymnosperms are conifers, which have seeds produced in cones. Conifer forests are common in cooler, mountainous climates.

(d) Flowering plants, known as angiosperms, are by far the largest group of modern plants.

Figure 1.10 The diversity of plants. These images represent the four main groups of plants: bryophytes, seedless vascular plants, gymnosperms, and angiosperms.

Ferns and their relatives are examples of seedless vascular plants

The vast majority of plants are **vascular plants,** which evolved from algae-like ancestors around the same time as bryophytes. Vascular plants have highly organized and efficient **vascular tissue,** which consists of cells joined into tubes that transport water and nutrients throughout the plant's body. Being generally larger than bryophytes, vascular plants are usually more noticeable, although they vary from tiny to towering. The simplest vascular plants are **seedless vascular plants,** which began to evolve about 450 to 700 million years ago.

If you could travel 350 million years back in time, you would find the continents lying together as one landmass along the equator, covered with steamy jungles and swamps. Amphibians, some quite large, would rule the world of animals, without a reptile, bird, or mammal in sight. Nor would you find any plants with cones, flowers, or fruits. Instead, a variety of seedless vascular plants would dominate the landscape, some looking much like modern ferns. However, others would be huge forest trees, some with fernlike branches, but quite unlike anything found today.

The most common surviving seedless vascular plants are ferns, which live in moist regions because their sperm have microscopic tail-like structures and must swim through a layer of water to reach the eggs (Figure 1.10b). Chapter 21 explores the evolution and characteristics of seedless vascular plants.

Pine trees and other conifers are examples of nonflowering seed plants

In most vascular plants, the embryo is contained in a **seed,** a structure that includes not only the embryo but also a store of food, packaged together within a protective coat. There are two general types of seed plants—nonflowering seed plants and flowering seed plants.

Nonflowering seed plants, known as **gymnosperms** (from the Greek *gymnos,* naked, and *sperma,* seed), first evolved around 365 million years ago. Their most familiar modern descendents are the woody plants called *conifers* (from the Latin *conifer,* cone-bearing), whose seeds develop on cones. The seeds are "naked" only in the sense that they are not completely sealed within a protective layer. Conifers are typically evergreen trees with needlelike leaves, such as fir, pine, and redwood trees (Figure 1.10c). Redwood trees are the world's tallest plants, with heights often exceeding 90 meters (300 feet).

During the Mesozoic Era (245 to 65 million years ago), when reptiles were the ruling animals, gymnosperms were the dominant plants and included thousands of species. However, today they consist mainly of only a few hundred species of conifers, commonly seen in colder regions nearer the poles and in mountains. Chapter 22 discusses the evolution and characteristics of gymnosperms.

Most plants are flowering plants with seeds protected in fruits

Flowering seed plants are called **angiosperms** (from the Greek *angion,* container, and *sperma,* seed) (Figure 1.10d). Unlike gymnosperm seeds, the seeds of flowering plants are enclosed in ovaries, which are called *fruits* when they mature. Although angiosperms evolved relatively recently, around 145 million years ago, they represent most of the plant varieties we see today. In fact, there are about 20 times as many types of flowering plants as there are ferns and conifers. One reason why angiosperms have adapted more successfully in many environments is that they have a more rapid system for moving water throughout their bodies, as you will see in Chapter 3. Another reason they are so widespread is that their seeds are enclosed in fruits, which provide added protection and may also aid in seed distribution. You will read more about flowering plants in Chapter 23.

Bryophytes, seedless vascular plants, gymnosperms, and angiosperms reflect four major evolutionary developments: the origin of land plants from algae, the origin of vascular tissue, the origin of seeds, and the origin of flowers and fruits. Throughout this book, the narrative and "Evolution" box features will explore how characteristics have evolved in various types of plants. To see how plants are scientifically classified into groups, look at the list in the Appendix at the end of the book.

Section Review

1. **Describe the general characteristics of a plant.**
2. **Describe the major types of plants.**

Botany and the Scientific Method

During the 1600s, botanists began to carry out experiments to determine how plants grew. They noted, for example, that a potted plant gained considerable weight over time, even though the amount of soil in the pot remained about the same. They correctly concluded that water must contribute greatly to plant growth. The study of plants, like other sciences, always involves making observations, forming questions about what is observed, developing possible answers, and then testing how well those answers are supported by facts. This approach has been commonly known as *the scientific method,* although there is no formal procedure that all scientists always follow.

Botanists, like other scientists, test hypotheses

The word *science* comes from the Latin *scientia,* meaning "having knowledge." Scientific knowledge is a particular kind of knowledge that is different from other types, such as knowledge based on faith. Science is the pursuit of knowledge based on direct observation of the natural world and experimentation to test conclusions. Any conclusion that can be investigated by observing natural phenomena is within the realm of science. For example, the proposition that acorns grow into petunias can be tested by planting acorns and observing that oak trees result. In contrast, any statement that cannot be tested by observation or experiment is outside the realm of science. For instance, the statement that the world is the result of divine creation is outside the realm of science because no experiment can be designed to test its truth. Note, however, that science itself rests on certain basic beliefs that cannot be tested, such as the belief that the laws of chemistry, physics, and biology will be the same tomorrow as they are today.

Here is an example of an experimental approach. If you wanted to know whether sunlight makes grass grow, you could conduct an experiment of covering a patch of lawn with a box that blocks out all light. You predict that if sunlight makes grass grow, then the grass underneath the box should grow less than the uncovered grass. After a week, you turn the box over and discover that the grass under it has grown more than the grass outside. In this instance, apparently the grass grew more in the absence of sunlight. (We will discuss these surprising results later in the chapter.)

An early advocate of using an experimental approach was Sir Francis Bacon (1561–1626), who believed that scientists of his day relied too heavily on the works of the Greek philosopher Aristotle (384–322 B.C.) (Figure 1.11). While Aristotle made many observations and wrote volumes about the sciences, particularly zoology, he was not always a careful observer and seldom did experiments to check his conclusions. For example, he believed plants did not reproduce sexually.

Many of Bacon's contemporaries accepted Aristotle's conclusions without testing them, simply applying those general conclusions to specific facts. This reasoning from the general to the specific is known as **deductive reasoning.** Here is an example of deductive reasoning: Aristotle developed a generalization (also called *a major premise*) that plants have no sexual reproduction. He would then identify a specific organism as a plant (a statement called *a minor premise*). Without in-depth observation of that plant, he then concluded that since this organism was a plant, therefore it had no sexual reproduction—a conclusion based on deduction, not on observation.

Bacon thought it was useless to simply gather data and then not form conclusions based on that data. He had even less sympathy for people who drew conclusions without bothering to support them with data. He described such people as spiders spinning webs, working out conclusions only in their minds like "cobwebs of learning, admirable for the fineness of thread and work, but of no substance or profit." Instead, Bacon admired people who gathered and analyzed data to reach conclusions that they then applied, comparing them to bees who visit many flowers and then use what they have collected to make something useful: honey.

Figure 1.11 Sir Francis Bacon. Bacon was a firm believer in the power of experimentation and the scientific method to uncover truth.

Bacon believed that scientists should start with specific observations and then form general conclusions based on those observations, a process known as **inductive reasoning.** Consider, for example, the observations that many plants produce pollen that bees transport from the flowers of one plant to the flowers of another. Later observations reveal that part of the pollinated flower develops into a fruit containing seeds that give rise to the next generation, while unpollinated flowers do not develop fruits or seeds. Considering this information, a tentative conclusion might be that plants have sexual reproduction.

Here is another example from botany to illustrate how inductive reasoning works. We might observe that flowering plants grow toward the light, and we might wonder if it is the leaves that detect light (Figure 1.12). Therefore, we cover the leaves so that the light does not reach them and then observe that the plant still grows toward the light. We then try covering the flowers, with the same result. However, when we cover just the top of the stem, or apex, we observe that the plant no longer grows toward the light. From these observations, we conclude that the apex detects light. Actually, this discovery was made by

(a)

(b)

Figure 1.12 Charles Darwin's experiment with stems detecting light. Charles Darwin published his results on a number of different scientific studies in addition to his works on evolution. Through experimentation, he discovered that the tip of a stem detects light. **(a)** If the tip is covered, the stem does not grow toward the light. **(b)** If uncovered, the tip bends in the direction of the light.

the famous English naturalist Charles Darwin, who published a book called *The Power of Movement in Plants* in 1880. Later, it was discovered that the apex responds specifically to blue light.

Although scientists do not all follow a strict set of steps, the following activities, summarized in Figure 1.13, generally describe how scientific knowledge is acquired:

1. Observation and data gathering. Scientists often collect information without at first trying to explain it. For example, many types of plants in various locations all seem to grow toward the light. In terms of our example, the light shines on many parts of the plant, at least one of which must detect light. We might shade various parts of the plant to see if it still turns toward light.

   Observations, however, may be inaccurate because of changes in the environment or problems with the observer or equipment. As you know, an observation of a distant lake in the desert can be an optical illusion known as a *mirage*. Also, different observers may have markedly different descriptions of the same event.

2. Critical question formation. Scientists ask critical questions, both before and after they make observations. In our example, a critical question is "Which part of the plant detects light?"

3. Hypothesis formation. A **hypothesis** is a tentative answer to a question, attempting to link together data in a cause-and-effect relationship. You might think of a hypothesis as an educated guess that can be tested. In our example, the first hypothesis is that the leaf detects light and causes the plant to grow toward the light.

   If one event follows another, that does not demonstrate a cause-and-effect relationship. If the bell in a clock tower sounds every morning just after you walk into class, this does not mean that you cause the bell to sound. Neither are roosters responsible for the Sun rising, although they may well think so. By contrast, a legitimate cause-and-effect relationship is a bee's visit to a flower and the flower's seed formation. Bees carry pollen from one plant to another, which may lead to fertilization and the subsequent development of the embryo. However, only a careful series of observations supported this cause-and-effect relationship.

4. Hypothesis testing. Each hypothesis is put to critical tests that may or may not support it. Moreover, the hypothesis must be at least potentially falsifiable or incorrect. The statement that "leaves enable plants to detect light" is a hypothesis because it can be tested. To see if the leaves detect light, we can remove them or cover them with foil and see if the plant still grows toward the light.

5. Hypothesis acceptance, modification, or rejection. To determine what part of a plant detects light, we might have to go through several rounds of proposing, testing, and rejecting hypotheses. For example, neither leaves nor stems turn out to be

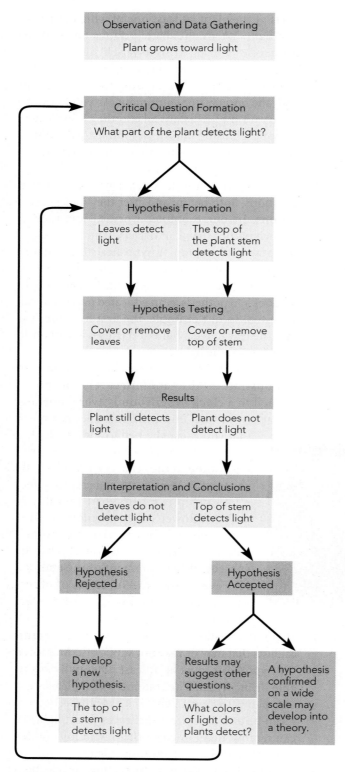

Figure 1.13 Forming and testing hypotheses. The scientific method consists of observations that lead to a question. One or more hypotheses tentatively answer the question. Each hypothesis is then tested by experiments, which either support it or do not support it. Based on experimentation, the hypothesis is either accepted or rejected.

the part of the plant that detects light. Only when Darwin shaded the very top of the stem did the plant cease to grow toward the light.

Often a hypothesis must be tested under different conditions or for a longer period of time. An example is the experiment that seemed to indicate that grass grows faster in darkness than in light. Actually, this is true only for the short term. If we place a box that blocks out all light over part of a lawn, the grass under it elongates more rapidly for a period of days. In response to the darkness, the stems grow longer toward the light, but if left in the darkness the grass would die. Grass needs sunlight and photosynthesis to produce food for long-term growth, even though growth in the light is slower than growth in the dark on a day-to-day basis.

A supported hypothesis can raise new questions. For instance, how does the detection of blue light by the apex cause a flowering plant to grow toward the light? What is the blue light receptor in the flowering plant, and why does the plant detect blue light but not green light? Do all plants detect light in the same way?

If a hypothesis survives all challenges, it may be the correct explanation, but it may have only limited application. After all, hypotheses are formed based on a number of specific experiments. For example, experiments on plant responses to light might be done on some flowering plants, but do the results apply to all plants? If a hypothesis supported on a small scale (in this case, for flowering plants) is later supported on a wide scale (in this case, for all types of plants), it is considered a **theory.** In other words, a theory has a much broader application than a supported hypothesis. For example, evolution is considered a theory because it has been supported by many repeated observations and experiments relating to many different types of organisms. In scientific terms, then, a theory is very well-supported by evidence, in contrast with the informal use of the term to mean a proposed explanation that is not necessarily well supported.

Most scientific hypotheses should have a short life because they are refined and perhaps discarded by testing. Although ideally scientists should be objective, they sometimes make subjective judgments, such as being reluctant to discard a favored hypothesis because it just seems to them that it should be correct. Good scientific work is often like solving a mystery or being a successful detective. You do not necessarily know how "it will all come out in the end," but if you are logical and careful, you will learn a lot and have fun in the process.

Figure 1.14 A botanist at work. This botanist is studying the effects of compounds that regulate the growth and development of plants.

Botany includes many fields of study

Today, botanists study many aspects of plants (Figure 1.14):

◆ Plant systematists study the evolutionary history of plants. They give scientific names to newly discovered plants and participate in studies to identify and preserve species that are endangered and threatened with extinction.

◆ Plant physiologists study how plants function, focusing on such aspects as photosynthesis, flowering, and the action of hormones. Physiologists study gene action and methods for isolating useful genes and moving them from plant to plant.

◆ Plant anatomists analyze how the structure of plants relates to function. Paleobotanists study the anatomy of fossil plants to determine their evolutionary relationship to living plants. Some anatomists look for genes that determine the development of particular structures and types of cells.

◆ Plant morphologists are interested in plant life cycles and particularly in how plants reproduce. They also study the evolution of plants and how life cycles and anatomy have changed over time.

◆ Plant ecologists determine how plants relate to other organisms in their environment and how plants meet their needs while coping with variables like temperature and rainfall. Some ecologists study human effects on the environment and how humans have contributed to increased rates of extinction of plant species.

◆ Plant geneticists investigate the transfer of genetic information from one generation to the next. As you have read, plant biotechnology and genetic engineering can improve agricultural crops.

In recent years, these traditional botanical disciplines have become interrelated. For example, a plant ecologist may include physiology and genetics in experiments by studying a particular gene that produces an alkaloid that deters insect predators. A morphologist may use physiological and biochemical techniques to investigate an enzyme that controls the development of sperm from the pollen of a flowering plant. An anatomist, physiologist, and geneticist might work together to learn how a mutation prevents efficient water uptake by a plant. Discoveries in one field of botany often have implications for other fields.

Botanists also study algae, fungi, and disease-causing microorganisms

In addition to studying the variety of plants, botanists investigate other photosynthetic organisms such as algae (Chapter 18) and photosynthetic bacteria. Studying algae provides insights into the evolution of the first plants from algae-like ancestors. Studying other photosynthetic organisms also leads to a greater knowledge of the mechanism and role of photosynthesis. In fact, algae and photosynthetic bacteria carry out about half of the world's photosynthesis.

Fungi are not photosynthetic and differ from plants in many other ways (Chapter 19). Indeed, many scientists believe that fungi are more closely related to animals than to plants, based on analysis of DNA. Nevertheless, botanists often study fungi because they have similarities to plants and also affect plants and humans. Some help roots absorb minerals, while others are important in food production for humans or as sources of antibiotics. Botanists also study how fungi cause diseases in plants and animals. In addition, they explore how microorganisms like bacteria cause plant diseases and also how viruses affect plants (Chapter 17).

Botanists are involved in a wide variety of occupations. In the academic world, they are teachers, professors, and researchers in schools, colleges, and universities. They also work as foresters, park rangers, ecological consultants, and wildlife biologists. Genetic engineering and pharmaceutical companies hire botanists to help develop useful types of plants and medicines. They become farmers, landscape architects, and horticulturalists—botanists who specialize in growing fruits, vegetables, flowers, and ornamental plants. They sometimes specialize in studying particular types of plants such as herbs, wine grapes, cut flowers, native grasses, or drought-tolerant plants for lawns and gardens.

As you have seen, botanists engage in many areas of research on an amazing variety of plants and associated organisms that affect our lives in myriad ways. Plants are fascinating life-forms in and of themselves, but you might also consider yourself to be studying plants as if your life depended on them—which it does.

Section Review

1. What is the difference between deductive reasoning and inductive reasoning?
2. Describe in general how scientific knowledge is acquired.
3. What is the difference between a supported hypothesis and a theory?
4. What are some fields of study for botanists?

SUMMARY

The Importance of Plants

Photosynthesis sustains life on Earth (pp. 3–4)
Photosynthesis supports all life on Earth in three ways: (1) producing oxygen, which most organisms need to release energy stored in food; (2) providing energy directly to plants and therefore indirectly to other organisms in food chains; and (3) producing sugars and other molecules, which are the building blocks of life.

Plants are our fundamental source of food (pp. 4–6)
Agriculture enabled early humans to establish year-round settlements supported by farming. Early farmers improved crops by breeding plants. Modern plant breeders improve crop yields by developing faster-growing, disease-resistant plants. About 80% of the world's food comes from wheat, rice, corn, potatoes, sweet potatoes, and manioc (cassava). Plants are sources of beverages such as coffee, tea, and soft drinks, as well as dried herbs and spices.

Many medicines come from plants (p. 6)
Plants have served as medicines for centuries. Modern chemistry has made available more plant extracts containing alkaloids and other useful compounds. About 25% of all prescriptions in the United States use plant products.

Plants provide fuel, shelter, and paper products (pp. 6–7)
Wood, consisting mostly of dead cells from tree trunks, is the world's main source of fuel for cooking and heating. Fossil fuels consist primarily of fossilized remains of plants. Wood provides construction material and paper products.

Conservation biology is a critical area of research (pp. 7–8)
Conservation biology studies ways to maintain biological resources. About half of the world's original forests are gone, and recent destruction of rain forests may cause extinction of thousands of useful species.

Biotechnology seeks to develop new plant products (pp. 8–11)
Efforts to improve plants and plant products are known as plant biotechnology, which includes genetic engineering—methods of improving plants by moving and modifying genes. Genetically modified (GM) plants provide sources of nutrition and vaccines and are resistant to pests, toxic soil minerals, and herbicides. Legitimate concerns and GM plants must be separated from unfounded fears.

Plant Characteristics and Diversity

A set of characteristics distinguishes plants from other organisms (pp. 11–12)
Plants are multicellular eukaryotes that have cell walls composed mainly of cellulose; are almost all capable of photosynthesis; have a spore-producing adult form and an adult form that produces eggs and sperm; and have a multicellular embryo protected within the female parent. Together these characteristics help distinguish plants from animals, fungi, algae, and bacteria.

Mosses are among the simplest types of plants (pp. 12–13)
Bryophytes, consisting of mosses and related small nonflowering plants, were among the first land plants.

Ferns and their relatives are examples of seedless vascular plants (pp. 14–15)
Most plants are vascular plants, which are plants with vascular tissue—tubes of cells that transport water and nutrients. Ferns are the most common seedless vascular plants.

Pine trees and other conifers are examples of nonflowering seed plants (p. 15)
Most vascular plants have seeds, structures that include an embryo and a store of food. Gymnosperms are nonflowering seed plants such as pine trees. The name means "naked seed" because the seeds are not completely sealed within a protective layer.

Most plants are flowering plants with seeds protected in fruits (p. 15)

Flowering seed plants, known as *angiosperms*, are the most common type of plant. Angiosperms are more adaptable to many environments because of their more rapid system of water transport and the protection of seeds within fruits. There are four main evolutionary developments in plants: the establishment of plants on land, origin of vascular tissue, origin of seeds, and origin of flowers and fruits.

Botany and the Scientific Method

Botanists, like other scientists, test hypotheses (pp. 15–18)

Scientific knowledge comes from direct observation and experimentation to test hypotheses. The process involves: (1) observation and data gathering, (2) critical question formation, (3) formation of a hypothesis, (4) testing the hypothesis, and (5) accepting, modifying, or rejecting the hypothesis. A well-supported hypothesis may be tested on a wider scale and become a theory, such as the theory of evolution.

Botany includes many fields of study (p. 19)

Botanists study many aspects of plants, such as evolution, functions, structure, life cycles and reproduction, ecology, and genes. Botanists are involved in areas such as teaching, forest and park management, ecology, genetic engineering, pharmaceutical work, farming, landscaping, and horticulture.

Botanists also study algae, fungi, and disease-causing microorganisms (pp. 19–20)

Studying algae and other photosynthetic organisms helps botanists learn more about photosynthesis. Botanists study how fungi help and harm plants and humans and how microorganisms and viruses cause diseases in plants and humans.

Review Questions

1. How does photosynthesis make life on Earth possible?
2. In what way is farming a foundation for human civilization?
3. What are some examples of how plants are used for medicine?
4. Explain how plants, directly or indirectly, are the source of most of the world's energy.
5. What is wood made of, and what are some common uses of wood and wood products?
6. How does conservation biology relate to plants?
7. Is plant biotechnology the same as genetic engineering? Explain.
8. Why are botanists developing GM crops? What are some potential problems with GM crops?

9. What are five characteristics commonly used to define plants?
10. What are major evolutionary changes in the development of plants?
11. Briefly describe each of these four main types of plants: bryophytes, seedless vascular plants, gymnosperms, and angiosperms.
12. What is the basis for scientific knowledge?
13. What is the role of inductive reasoning in the scientific method?
14. What are some subfields within botany?

Questions for Thought and Discussion

1. Why do you think many people take plants for granted? How would you respond to them?
2. Explain what is meant by the statement "Plants do not need us, but we need plants."
3. In what sense would you say that plants also need us?
4. Choose one of the GM plants mentioned in the chapter and discuss possible benefits and problems that should be considered in determining whether and how to use that plant.
5. Some people argue that all foods should be labeled to indicate whether they contain genetically modified ingredients. Do you agree? Explain.
6. Why must a hypothesis be potentially falsifiable?
7. Is the statement "A bachelor is an unmarried man" a hypothesis? Explain.
8. Review the "steps" of the scientific method. Why do you think scientists do not always follow these steps in a strict order? Give some examples.
9. Formulate a question about plants and describe a scientific approach to answering that question. For example, a question might be "How does pruning a branch affect tree growth?"
10. How does the statement "Evolution is just a theory" reflect a misunderstanding of the scientific meaning of the term *theory*?
11. Sketch a food chain for a water environment that begins with a primary producer and has at least four levels of consumers. See Figure 1.1 for an example.

Evolution Connection

Consider the following types of plants: mosses, ferns, conifers, flowering plants. Beginning with the ferns and then moving on to the conifers and flowering plants, list diagnostic features present in each group that are not present in the preceding group. Which of these features represent evolutionary trends in the plant kingdom? Explain.

To Learn More

Visit The Botany Place at www.thebotanyplace.com for quizzes, exercises, and Web links to new and interesting information related to this chapter.

Bacon, Francis, and Peter Ubach. *Novum Organum: With Other Parts of the Great Instauration.* Chicago: Open Court, 1994. A new classroom edition of Bacon's great work of 1620, in which he argues that scientists need to make detailed observations rather than relying on tradition.

Desowitz, Robert S. *The Malaria Capers: More Tales of Parasites and People, Research and Reality.* New York: W. W. Norton, 1993. Despite quinine, malaria still kills at least one million people a year, mostly children. This book includes many stories of people who have malaria and who fight malaria.

Desowitz, Robert S. *Who Gave Pinta to the Santa Maria?: Torrid Diseases in a Temperate World.* San Diego: Harcourt Brace, 1998. A captivating book on tropical diseases, especially malaria, with a focus on American history.

Hobhouse, Henry. *Seeds of Change. Six Plants That Transformed Mankind.* New York; Harper and Row, 1999. Contains fascinating details and stories about how plants have influenced human history and affairs.

Simpson, B. B., and M. Conner-Ogorzaly. *Economic Botany: Plants in Our World,* 2d ed. New York: McGraw-Hill, 1995. Provides many examples of plants that are important to humans in many different ways.

The Structure of Plants

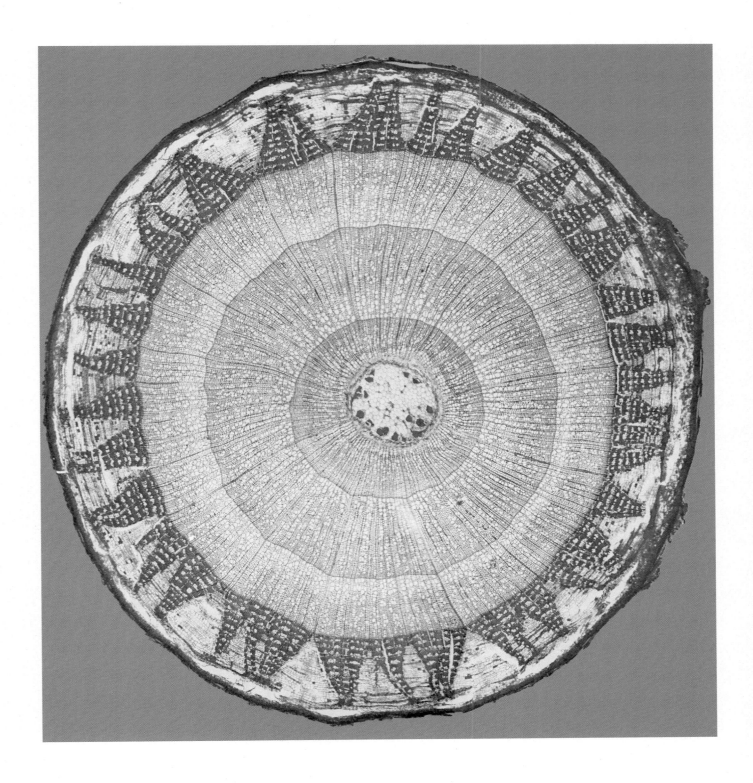

2

Cell Structure and the Cell Cycle

This common pond alga (*spyrogyra*) grows in long strands of cells.

An Overview of Cells

Microscopes reveal the world of the cell

The cell is the basis of an organism's structure and reproduction

All cells are either prokaryotic or eukaryotic

Cells produce nucleic acids, proteins, carbohydrates, and lipids

Major Plant Cell Organelles

The nucleus provides DNA "blueprints" for making proteins

Ribosomes build proteins

The endoplasmic reticulum is the site of most protein and lipid synthesis

The Golgi apparatus completes and ships cell products

Chloroplasts in green plant cells convert solar energy into stored chemical energy

Mitochondria convert stored energy into energy to power the cell

Microbodies aid in chemical reactions

Vacuoles play a variety of roles in cell metabolism and cell shape

The Cytoskeleton: Controlling Cell Shape and Movement

Microtubules play an important role in cell movements

Microfilaments help living cells change shape

Motor proteins, or "walking molecules," cause movement

Intermediate filaments help determine the permanent structure of cells

Membranes and Cell Walls

Membranes are gatekeeping barriers around and within cells

Cell walls protect plant cells and define cell shape

Plasmodesmata are channels that connect plant cells

The Cell Cycle and Cell Division

The cell cycle describes the phases of a cell's life

Mitosis and cell division are involved in growth and reproduction

Mitosis produces two daughter nuclei, each containing the same chromosome number as the original cell

New cells typically become specialized

Before the 1600s, no one knew that large organisms are made up of many small, living units that we call *cells*. Since most cells are microscopic, we cannot see them with the naked eye. Cells are typically between 1 and 300 micrometers (μm) in diameter. A micrometer is one-millionth of a meter. To put that size into perspective, a diameter of 300 μm is only about one third of a millimeter, or about one hundredth of an inch. The invention of the first microscopes, during the late 1500s, opened up a new frontier for science.

English scientist Robert Hooke (1635–1703) was the first person to view a cell (see the *Plants & People* box on page 28). In 1665 he observed plant cells in a slice of bark from an oak tree, using a multi-lens (compound) microscope he had designed and constructed. That year he published *Micrographia* ("Small Drawings"), a book that included his illustrations of his microscopic discoveries. Hooke used the term *cell* to refer to the tiny compartments within the oak bark because they reminded him of monks' cells. The word *cell* itself comes from the Latin *cella*, meaning "small room."

Over the years, new discoveries about cells have captured the interest of scientists and nonscientists alike. Cells are indeed wonderful and exciting to observe. In this chapter, we will examine the world of cells to see how they are the basis of the life of an organism, how they are organized, and how they divide to allow for growth and reproduction.

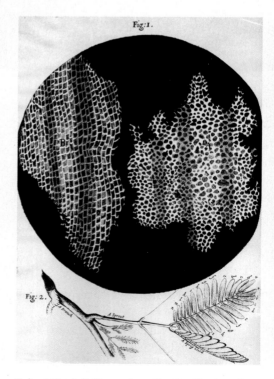

Robert Hooke's illustration of cork cells viewed under his microscope.

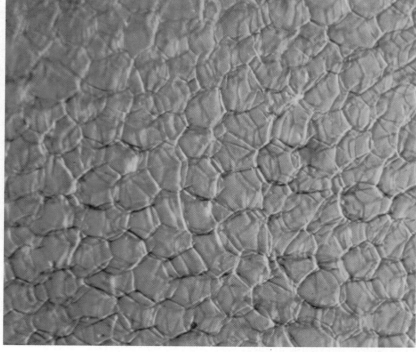

This is cork viewed through a modern light microscope.

An Overview of Cells

All living organisms are composed of one or more cells. Most cells are microscopic and are either barely visible or not visible with eyesight alone. Why are cells so small? The ratio of a cell's surface area to its volume is the main factor. As a cell becomes larger, its radius (r), volume, and surface area all increase. However, its volume (r^3) increases more than its surface area (r^2), a ratio that holds regardless of cell shape. A cell must remain relatively small so that its surface area, which controls the entry of oxygen, water, and nutrients, can supply the needs of its internal regions. If a cell grows too large, its genetic material cannot supply information fast enough to meet the needs of the cell.

Although they are small, cells are far from simple, carrying out a wide variety of functions that sustain the lives of organisms. Understanding the world of plants, or any other organisms, involves first exploring the structure and function of cells.

Microscopes reveal the world of the cell

Microscopes, of course, serve as our windows into the cell. The magnifying power of light microscopes has improved considerably over the years, but the basic principle remains the same. A **light microscope** uses glass lenses to bend the path of visible light, thereby producing a magnified image (Figure 2.1a). A modern light microscope can resolve objects 200 nanometers (nm) apart, or 1,000 times better than the human eye.

The development in 1939 of the **electron microscope,** which focuses electrons with magnetic lenses, made pos-

sible a wealth of discoveries about cells. The **transmission electron microscope (TEM)** reveals structures within cells by passing electrons completely through a thin section of tissue (Figure 2.1b). A TEM can magnify objects up to about 100,000 times, enabling scientists to view cell structures as small as 2 nm, a great advance over the light microscope. A second type of electron microscope, the **scanning electron microscope (SEM),** bounces electrons off a specimen to reveal the surface structure, often a detailed 3-D view (Figure 2.1c). An SEM can magnify up to about 20,000 times, giving detailed views of cells, groups of cells, and small organisms or parts of organisms.

Overall, light microscopes and scanning electron microscopes are most useful for viewing structures that are just beyond the resolution of the human eye. Transmission electron microscopes are best for obtaining very high magnifications of structures within a cell. Throughout this book, a scale will appear for size reference next to each microscopic image, or micrograph. The photo legend will identify the image as either a light micrograph (LM), transmission electron micrograph (TEM), or scanning electron micrograph (SEM) and also note if it is color enhanced. Figure 2.2 provides an overview of what types of objects can be observed with each type of microscope.

The cell is the basis of an organism's structure and reproduction

During the 1600s and 1700s, as scientists used light microscopes to observe details of organisms, ideas about

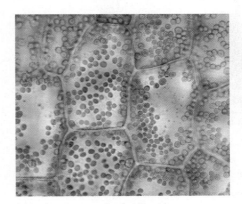
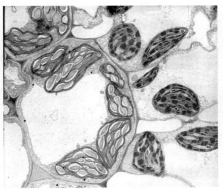
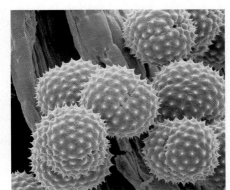

(a) Light micrograph of cells with chloroplasts

(b) Transmission electron micrograph of cells with chloroplasts

(c) Color enhanced scanning electron micrograph of pollen

Figure 2.1 Basic micrograph types and microscope ranges. (a) This micrograph was produced by a modern light microscope, in which one lens first magnifies the image, which is re-magnified and again inverted by the eyepiece lens. **(b)** This micrograph was produced by a transmission electron microscope, which uses the same basic design as the light microscope, with magnetic lenses instead of glass lenses. **(c)** This micrograph was produced by a scanning electron microscope, which bounces electrons off the surface of an object. A computer that analyzes the trajectories of the electrons produces an image.

Pioneers of Microscopy

Robert Hooke was a scientific virtuoso, making discoveries in such diverse fields as biology, physics, and astronomy. According to a biographer in 1705, Hooke was "an active, restless, indefatigable Genius . . . and always slept little to his death, seldom going to sleep till two, three, or four a Clock in the Morning."

Hooke maintained an association with a Dutch amateur scientist named Antonie van Leeuwenhoek (pronounced LAY-vuhn-hook) (1632–1723). Van Leeuwenhoek made hundreds of tiny, single-lens microscopes only a few inches high. To us they look more like miniature violins than microscopes, but they provided magnifications of up to 500 times with minimal distortion. These microscopes differed dramatically from instruments commonly used at the time, which were tubes up to three feet tall. Such tubes contained two or more lenses usually made by blowing glass into various shapes, which were then intentionally dropped on the floor in the hope of obtaining fragments with lens-like optical properties. These lenses did indeed magnify, but they also distorted, and the addition of a sec-

ond lens simply magnified the distortion. In contrast, van Leeuwenhoek became an important figure in microscopy, the use of microscopes, because he was the first person to make single-lens microscopes with high magnifications that did not distort what was being observed.

Van Leeuwenhoek found microbes virtually everywhere he looked. "There are," he said, "more animals living in the scum on the teeth in a man's mouth than there are men in a whole kingdom." In November 1677, Hooke confirmed van Leeuwenhoek's startling claim that many tiny animals lived in drops of pond water. Van Leeuwenhoek was not surprised that many people doubted his observations. "I can't wonder at it," he wrote, "since 'tis difficult to comprehend such things without getting a sight of 'em." In his confirmation, Hooke wrote that the animals were "perfectly shaped" with "such curious organs of motion as to be able to move nimbly, to turn, stay, accelerate, and retard their progress at pleasure." Van Leeuwenhoek's and Hooke's contributions to microscopy essentially opened up a new dimension of science.

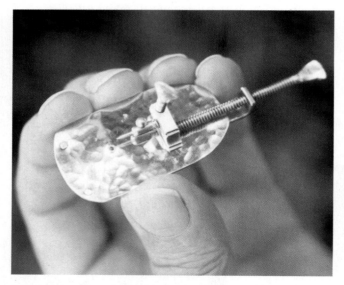

Microscopes by Antonie van Leeuwenhoek (L) and Robert Hooke (R). This is a replica of one of Leeuwenhoek's single-lens microscopes, which resembled tiny violins and

magnified up to 500 times. Robert Hooke's microscopes, made by London instrument maker Christopher Cock, were large tubular structures with illumination provided by an oil lamp.

cells continued to develop. Not until the mid-1800s, however, did a general theory emerge about the nature and significance of cells. In 1838, German botanist Matthias Schleiden concluded that microscopic studies of plant structure confirmed that all plant parts are composed of cells. Indeed, the cell walls of plant cells make the bound-

aries between cells readily visible, even with simple light microscopes. Animal cells, which do not have cell walls, could not be distinguished as clearly. However, in 1839, German biologist Theodor Schwann confirmed that animals are indeed also composed of cells. In 1855, another German biologist, Rudolf Virchow, made a further conclusion—that new cells could only arise from existing cells. Together, these observations have become known as the **cell theory,** which can be summarized as three main conclusions:

1. All organisms are made up of one or more cells.
2. The cell is the basic unit of structure of all organisms.
3. All cells arise only from existing cells.

In short, the cell theory declares that the cell is the basis of an organism's structure and reproduction. Take a moment to consider the amazing implications of this simple theory. Microscopic cells create organisms that range in size from single-celled algae to towering redwood trees. Through the process of cell division, a single cell can develop into a multicellular organism with trillions of cells. The cell theory revolutionized our knowledge of living things.

All cells are either prokaryotic or eukaryotic

As you saw in Chapter 1, all organisms are classified as either prokaryotes or eukaryotes, based on their cell type. Prokaryotes were the first forms of life on Earth, dating to at least 3.5 billion years ago. For about 1.4 billion years, no other organisms existed, until eukaryotes evolved from prokaryotes. Eukaryotic cells are more complex and generally larger than prokaryotic cells. Also, unlike a prokaryotic cell, the DNA in a eukaryotic cell is contained in an enclosed nucleus. The terms *prokaryote* (from Latin words meaning "before the nucleus") and *eukaryote* (from Latin words meaning "true nucleus") reflect this evolutionary change. Furthermore, a typical eukaryotic cell has at least 1,000 times as much DNA as the typical prokaryotic cell. DNA in a prokaryote is organized into one circular chromosome, while DNA in eukaryotes occurs in one or more linear chromosomes.

Eukaryotic cells are generally larger than prokaryotic cells, ranging in size between 5 and 300 μm, while a typical prokaryotic cell ranges from 1 to 10 μm in diameter (Figure 2.3). It would take 1,000 of the smallest prokaryotic cells lined up end-to-end to equal one millimeter. Most known prokaryotes are single-celled organisms called *bacteria.* Of course, in a bacterium or in a single-celled eukaryote, the cell *is* the organism, while many eukaryotes are multicellular.

An important basic difference between prokaryotes and eukaryotes concerns how they deal with the many functions

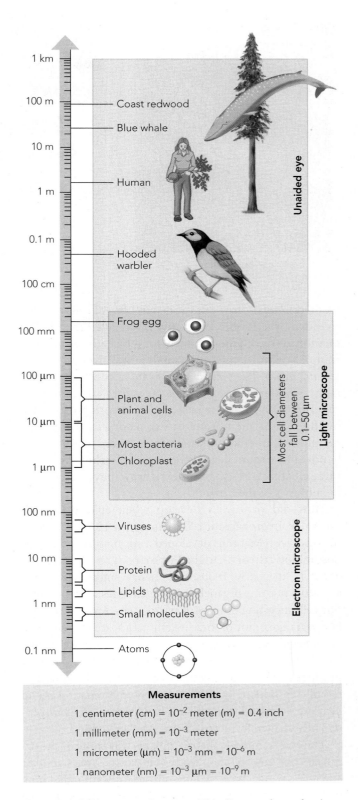

Measurements
1 centimeter (cm) = 10^{-2} meter (m) = 0.4 inch
1 millimeter (mm) = 10^{-3} meter
1 micrometer (μm) = 10^{-3} mm = 10^{-6} m
1 nanometer (nm) = 10^{-3} μm = 10^{-9} m

Figure 2.2 Microscope images. This diagram shows the size range of some typical objects as seen with the unaided eye and with the help of light and electron microscopes. The scale shown is exponential, meaning that each unit is ten times larger than the unit preceding it.

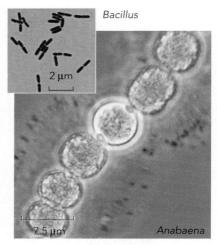

Bacillus

2 µm

7.5 µm   *Anabaena*

Prokaryotic

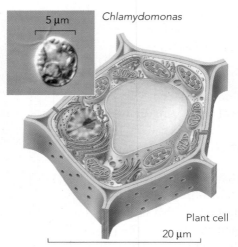

5 µm   *Chlamydomonas*

Plant cell

20 µm

Eukaryotic

Figure 2.3 A comparison of prokaryotic and eukaryotic cells. Although prokaryotic cells and eukaryotic cells can overlap in size, eukaryotic cells are generally larger.

that a cell must carry out to sustain life. All cells are surrounded by a flexible, protective layer called the **plasma membrane** or **cell membrane** (from the Latin *membrana*, skin), also known as the **plasmalemma** (from the Greek *lemma*, husk). The plasma membrane controls the movement of water, gases, and molecules into and out of the cell. In prokaryotes, the basic cell functions are handled by the plasma membrane or extensions of the plasma membrane. In eukaryotes, however, these processes are handled by **organelles** ("little organs"), which are separate cell structures that almost always have one or two membranes of their own.

How did organelles develop when eukaryotic cells evolved from prokaryotic cells? During the 1970s, American scientist Lynn Margulis developed a strong case for the **endosymbiotic theory.** Originally proposed by a Russian scientist in the early 1900s, the theory suggests that the ancestors of some organelles evolved as a result of prokaryotic cells ingesting other prokaryotic cells. Around two billion years ago, some of these ingestions apparently remained permanent because they helped both cells to survive (Figure 2.4). For example, if a bacterium that efficiently and rapidly produced energy was ingested by a cell that efficiently and rapidly reproduced by cell division, the combination would be more successful than either cell on its own. The term *endosymbiosis* literally means "living together within."

The endosymbiotic theory explains why some of the organelles that you will soon read about, such as mitochondria and chloroplasts, have two membranes, a feature that long puzzled scientists. The inner membrane may be the original membrane of the ingested bacterium, while the outer membrane formed from the host bacterium as it engulfed the other bacterium. Also, the theory explains why some organelles have chromosomes similar to the loops of DNA in bacteria, in contrast to the linear DNA in the nucleus.

Cells produce nucleic acids, proteins, carbohydrates, and lipids

Despite their differences in size and structure, prokaryotic and eukaryotic cells are similar in their basic **metabolism,** or chemical reactions. They all produce four general kinds of **macromolecules** (large molecules made up of smaller molecules) that organisms need in order to survive: nucleic acids, proteins, carbohydrates, and lipids. However, only the cells of autotrophs ("self-feeders") like plants and other photosynthetic organisms can synthesize these macromolecules without the help of nutrients from other organisms. The cells of heterotrophs such as animals and fungi must get the ingredients directly or indirectly from plants. Chapter 7 will discuss the chemical reactions in cells in more detail. For now, we will just look briefly at the structures and functions of nucleic acids, proteins, carbohydrates, and lipids.

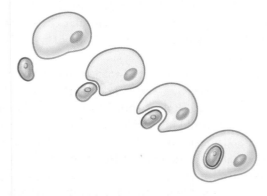

Figure 2.4 Endosymbiosis. Organelles such as the mitochondrion and the chloroplast probably have double membranes because they originated as prokaryotes that invaded another cell. Recent research indicates that the nucleus also arose by endosymbiosis.

Nucleic acids comprise the cell's genetic information. As you know from Chapter 1, DNA (deoxyribonucleic acid) stores sequences of genetic information known as *genes.* The other type of nucleic acid is **RNA (ribonucleic acid),** which is transcribed from the DNA for use in directing functions within the cell. Chapter 13 will explore how genetic information is stored and used.

Genetic information in the form of RNA directs the synthesis of **proteins,** which consist of chains of amino acids. There are 20 kinds of amino acids, and they can be used to make countless proteins. For instance, the human body has tens of thousands of different proteins, each consisting of a unique sequence of amino acids. Some proteins are structural building blocks, while others called **enzymes** help regulate chemical reactions. Proteins also store amino acids, respond to chemical stimuli, transport substances, protect against disease, and carry out a variety of other functions. An organism's proteins also define its physical characteristics. Reflecting the key roles of these macromolecules, the name *protein* comes from the Greek *proteios,* meaning "holding first place."

Carbohydrates, such as sugars and starches, are macromolecules composed of carbon, hydrogen, and oxygen. Carbohydrates supply and store energy and can serve as building blocks for larger molecules, such as cellulose in a plant cell wall.

Lipids are macromolecules, such as fats, that are insoluble in water. Indeed, the word *lipid* comes from the Greek *lipos,* meaning "fat." Fats basically store energy, while phospholipids are important as building blocks of membranes.

Section Review

1. **What are the main types of microscopes, and how do they differ?**
2. **How do eukaryotic cells differ from prokaryotic cells?**
3. **Describe the basic functions of nucleic acids, proteins, carbohydrates, and lipids.**

Major Plant Cell Organelles

Almost all of the organelles in a typical plant cell are also found in the cells of other eukaryotes. The exceptions are chloroplasts and a large central vacuole, which are not typically found in other photosynthetic organisms, fungi, or animals.

The interior of a plant cell—that is, everything except the cell wall—is called the **protoplast.** The protoplast consists of the nucleus and the **cytoplasm**—all the parts of the cell within the plasma membrane except the nucleus. (*Cyto* refers to cell, and *plasm* means "formed material.") To visualize a plant cell's organelles and their functions, imagine the cell's interior as a miniature factory, with the nucleus as the "supervisor" directing the work that goes on in the cytoplasm. Figure 2.5 provides an overview of a typical plant cell.

The nucleus provides DNA "blueprints" for making proteins

In a plant cell, as in other eukaryotic cells, the DNA in the nucleus is organized into complex, threadlike structures called **chromosomes.** Each chromosome consists of many genes, with each gene being the "blueprint" for making a particular protein. They are called *chromosomes* (from the Greek *chroma,* color, and *soma,* body) because they can be dyed for viewing with a light microscope when they shorten and thicken prior to cell division. A number of proteins associated with the DNA in chromosomes are involved in determining whether or not certain genes are active in a particular cell.

The nucleus also contains **nucleoli** (singular, *nucleolus*), which appear as round structures associated with chromosomes. The typical nucleus has one or two nucleoli, which synthesize subunits that then come together in the cytoplasm to form ribosomes. The role of ribosomes will be described shortly.

The nucleus has surrounding membranes, known together as the **nuclear envelope** (Figure 2.6). Pores in the nuclear envelope control movement of substances into and out of the nucleus. For many years, scientists have debated the origin of the nucleus. Recently, scientists studying the structure of DNA have proposed that the nucleus arose by endosymbiosis of two different types of prokaryotes. While the details remain to be determined, it appears that two types of bacterium invade another cell, and the genetic material became combined in a nucleus. As in all endosymbiotic events, the inner nuclear membrane is derived from the ingested cell, while the outer nuclear membrane probably derived from the plasma membrane of the surrounding host.

Ribosomes build proteins

Ribosomes are organelles that are formed in the cytoplasm and direct the synthesis of proteins, using genetic instructions in the form of messenger RNA (ribonucleic acid, hence the name *ribosome*). Some scientists prefer not to call ribosomes organelles because they are much smaller than other organelles, lack membranes, and appear also in prokaryotic cells. However, eukaryotic ribosomes are distinctly different from prokaryotic ribosomes, being generally larger and having different types of RNA. In Chapter 13 you will read about how ribosomes build proteins.

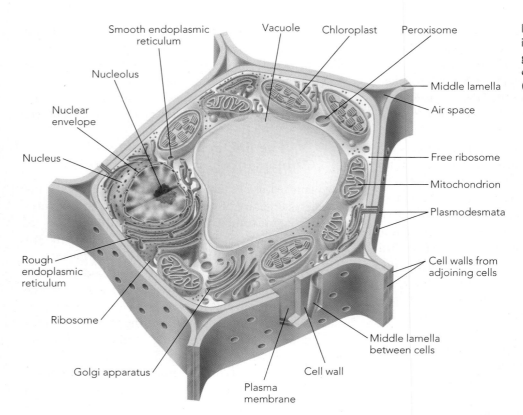

Figure 2.5 Structure of a typical plant cell. This view of a generalized plant cell shows the characteristic eukaryotic features. (Chloroplast not drawn to scale.)

The endoplasmic reticulum is the site of most protein and lipid synthesis

Throughout the cytoplasm is a network of connected membranes called the **endoplasmic reticulum (ER)** (see Figure 2.6). The name is a mouthful, but *endoplasmic* simply refers to the organelle's location ("within the plasm"), and *reticulum* is Latin for "little network." The ER, which is formed from and is continuous with the outer nuclear envelope, serves as a synthesis and assembly site for making proteins, lipids, and other molecules that are either exported from the cell or used to make the cell's membranes. Cells that produce substances for export to other cells have considerably more ER than cells that are not exporters.

The ER has two parts, the smooth ER and the rough ER, so named because of how they appear under the microscope. The **smooth ER,** which is usually tubular in shape, makes lipids and modifies the structure of some carbohydrates. The surface of the **rough ER** is dotted with protein-synthesizing ribosomes, resulting in the rough appearance of the exterior, which typically consists of flattened, interconnected sacs called **cisternae** (singular, *cisterna*). The space within the cisternae and tubules of the ER is known as the *lumen.*

The Golgi apparatus completes and ships cell products

The lipids, proteins, and other substances produced in the ER are packaged in membrane-surrounded structures called **transport vesicles.** The transport vesicles then separate from the ER and move to the **Golgi apparatus,** also known as the **Golgi complex** (named after its discoverer, Italian scientist Camillo Golgi). The Golgi apparatus consists of a number of separate stacks of cisternae, known as Golgi stacks, which originate from membranes produced in the ER (see Figure 2.6). The Golgi apparatus of a cell may include anywhere from several to hundreds of these Golgi stacks. Unlike the ER cisternae, however, the cisternae of a Golgi stack are not connected to each other. The side of a Golgi stack nearest to the ER receives transport vesicles from the ER. New transport vesicles are formed on the side of the Golgi stack that is away from the ER. These transport vesicles move to the cell's membrane, fuse with it, and then release their contents to the outside of the cell, to the structure of the cell membrane, or to various other cellular locations. In plant cells, Golgi stacks are also called **dictyosomes** (from the Greek *diktyon,* "to throw"), a name that reflects their role of moving products.

You can think of the stacks of the Golgi apparatus as receiving manufactured goods and then storing, modifying, packaging, and shipping them out of the cell or to various

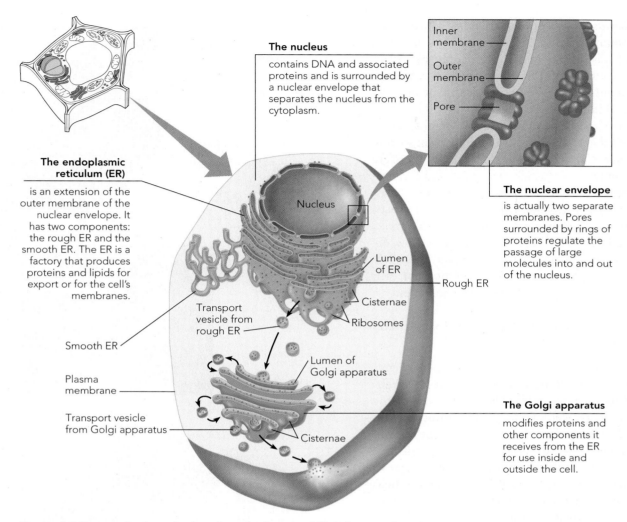

The nucleus

contains DNA and associated proteins and is surrounded by a nuclear envelope that separates the nucleus from the cytoplasm.

The endoplasmic reticulum (ER)

is an extension of the outer membrane of the nuclear envelope. It has two components: the rough ER and the smooth ER. The ER is a factory that produces proteins and lipids for export or for the cell's membranes.

Nucleus

Smooth ER

Plasma membrane

Transport vesicle from Golgi apparatus

Transport vesicle from rough ER

Lumen of ER

Rough ER

Cisternae

Ribosomes

Lumen of Golgi apparatus

Cisternae

Inner membrane

Outer membrane

Pore

The nuclear envelope

is actually two separate membranes. Pores surrounded by rings of proteins regulate the passage of large molecules into and out of the nucleus.

The Golgi apparatus

modifies proteins and other components it receives from the ER for use inside and outside the cell.

Figure 2.6 The nucleus, endoplasmic reticulum, and Golgi apparatus. As you can see in this cutaway view, the nucleus is connected to the endoplasmic reticulum (ER), and the smooth ER is continuous with the rough ER. The cell's genetic "blueprints" reside in the DNA in the nucleus. Ribosomes attached to the rough ER use transcribed genetic information in messenger RNA to direct protein synthesis. Transport vesicles then transfer proteins, lipids, and other products (green dots) from the ER to the Golgi apparatus, where they are tailored and shipped. Unlike the rough ER cisternae, Golgi cisternae are not connected. Transport vesicles move products from one Golgi cisterna to the next and, finally, to the plasma membrane if exported outside the cell. Changes in the color and size of the dots indicate that the Golgi apparatus modifies the products.

locations in membranes. The types of goods received and released by the Golgi apparatus depend on the particular cell. For example, in some cells the products may be cell wall components, while in others they may be proteins.

Chloroplasts in green plant cells convert solar energy into stored chemical energy

Organelles called **chloroplasts,** which contain green chlorophyll pigments, are the site of photosynthesis in plant cells (Figure 2.7). The word itself comes from the Greek *chloros,* meaning "greenish yellow." Not all plant cells carry out photosynthesis, though. Chloroplasts are found in the cells of

green parts of a plant, such as green stems and especially leaves. Chloroplasts are about 5 μm in diameter and can be spherical or quite elongated. Some photosynthetic cells have only one chloroplast, while others may have dozens.

The endosymbiotic theory may explain the origin of chloroplasts, which have two outer membranes and a small circular chromosome. Bacteria that trapped solar energy would have made a useful endosymbiotic addition to cells that lacked photosynthesis. Eventually, after millions of years, chloroplasts evolved from such an event. The structure of chloroplasts reflects their function in trapping solar energy. In addition to two outer membranes, chloroplasts have a series of internal, membrane-bound sacs called **thylakoids.** Stacks of thylakoids are called **grana** (singular, *granum*). The part of photosynthesis involving conversion of solar energy to chemical energy takes place within the membranes of thylakoids. The fluid surrounding the thylakoids, called the **stroma,** is the site of sugar production and storage.

Chloroplasts are a type of **plastid,** a general term for plant organelles involved either in making or in storing food or pigments. In addition to chloroplasts, there are two other main types of plastids: leucoplasts and chromoplasts. **Leucoplasts** (from the Greek *leukos,* white) consist of plastids that lack pigments. They include amyloplasts, which store starch. **Chromoplasts** (from the Greek *chroma,* "color") contain pigments responsible for the yellow, orange, or red colors of many leaves, flowers, and fruits. Depending on light exposure and the needs of the plant, each type of plastid can change into one of the other types.

Mitochondria convert stored energy into energy to power the cell

After chloroplasts convert solar energy into stored chemical energy, the plant cell needs to convert this stored energy to fuel its own activities. That is the role of the "powerhouse" of the cell's factory—the **mitochondria** (singular, *mitochondrion*) (Figure 2.8). A plant cell can have one mitochondrion or thousands, but typically one hundred or so. The mitochondria break down sugar to store its chemical energy in **ATP (adenosine triphosphate),** an organic molecule that is the main energy source for cells. You will read more about ATP in Chapter 9.

Almost all eukaryotes, with the exception of some protists, have mitochondria, which have two membranes and a small circular chromosome, evidence of their endosymbiotic origin. A typical mitochondrion is smaller than a chloroplast, measuring about 1 to 5 μm in length and about 0.5 to 1 μm in diameter. Mitochondria contain their own ribosomes, which produce proteins for use in the mitochondria and elsewhere in the cell. The infolding of a portion of the mitochondrion's inner membrane, called the **crista** (plural, *cristae*), markedly increases the surface area of the inner membrane, providing space for the critical proteins that reside there. The inner membrane and the liquid part inside it, called the **matrix,** contain the metabolic machinery that processes the energy in sugars into ATP.

Microbodies aid in chemical reactions

Microbodies are small, spherical organelles about 1 μm in diameter that contain enzymes. They were so named

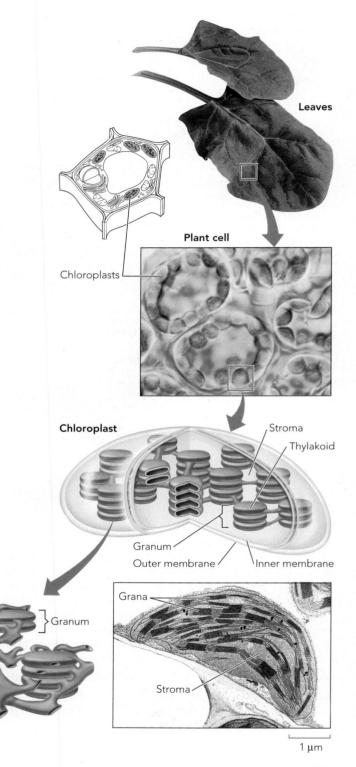

Figure 2.7 Chloroplasts. Chloroplasts are the organelles that carry out photosynthesis. In plant cells, the photosynthetic chlorophyll pigments are found in the membranes of sacs called *thylakoids,* which are stacked like coins. These stacks, called *grana,* are the only part of the plant cell that is green. The fluid outside the thylakoids, called the *stroma,* is where sugars are produced and stored.

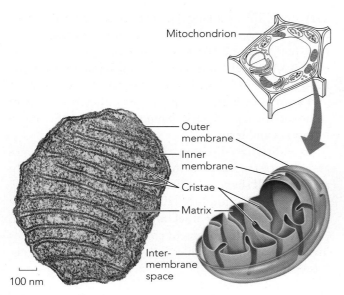

Figure 2.8 Mitochondria. Mitochondria are the organelles in which energy in sugar is used to make ATP. Eukaryotes produce energy storage molecules for all metabolic energy in their mitochondria. Mitochondria have two membranes, with the inner membrane folding into cristae that contain respiratory enzymes.

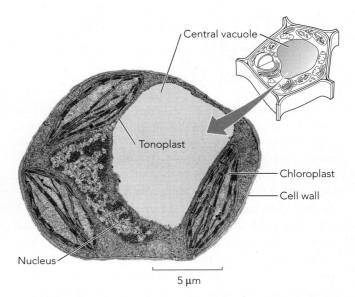

Figure 2.9 Vacuoles. Particularly large vacuoles, occupying up to 90% of a plant cell's volume, can be found in mature plant cells. Vacuoles serve as repositories for unwanted molecules and cell pieces, controlling water and salt balance in the cell and sometimes containing pigments that contribute red and blue colors to fruits and flowers. The membrane surrounding the vacuole of a plant cell is called the *tonoplast.*

because at first their function was unknown. Later, scientists discovered that certain microbodies generate and then break down hydrogen peroxide, formed by adding hydrogen to water. Since hydrogen peroxide is potentially toxic to other parts of the cell, confining these chemical reactions within particular organelles helps protect the rest of the cell. Accordingly, these microbodies were named **peroxisomes** ("peroxide bodies"), a term that some scientists use to refer to all microbodies.

Another type of microbody, called a **glyoxysome,** contains enzymes that assist in converting stored fats into sugars. These reactions, which do not occur in animals, are particularly important in seeds, which store fats as a food supply but need sugar during germination.

Microbodies have only one surrounding membrane and did not arise by endosymbiosis. They may originally have arisen from membranes of the endoplasmic reticulum. They replicate but do not have chromosomes or make their own proteins.

Vacuoles play a variety of roles in cell metabolism and cell shape

In many mature plant cells, up to 90% of the volume is occupied by a large central **vacuole.** The name comes from the Latin *vacuus,* meaning "empty," because vacuoles appear empty under the microscope (Figure 2.9). Actually, the central vacuole is filled with water and waste products and plays a variety of important roles, some relating to cell metabolism. For instance, it removes salts from the cytoplasm and controls the water balance of the cell. It also serves as a detoxification center for harmful substances, assists in breaking down large macromolecules, and helps regulate salt concentrations. The vacuole may also store toxic ions and ions needed only at certain times for particular chemical reactions. In addition, it helps maintain cell shape by pressing the rest of the cytoplasm contents against the cell wall.

Animal cells contain small vacuoles, but the large central vacuole is a distinctive characteristic of adult plant cells. It begins as small vacuoles produced by the ER with proteins contributed by the Golgi apparatus. These little vacuoles gradually fuse to produce a larger vacuole that has a surrounding membrane known as a **tonoplast.**

Section Review

1. **What is the role of the nucleus?**
2. **Which organelles are involved in protein synthesis, and how?**
3. **What is the main function of the Golgi apparatus?**
4. **Why are both chloroplasts and mitochondria essential in providing energy for a plant cell?**
5. **What functions does the large central vacuole serve?**

The Cytoskeleton: Controlling Cell Shape and Movement

As you have seen from our survey of plant cell organelles, cells are dynamic entities. Many of the cell contents are constantly on the move, such as when proteins and other molecules are transported from the rough ER to the Golgi apparatus. One major structure of the cell, the **cytoskeleton** ("cell skeleton"), helps maintain the dynamic activities and shape of the cell. For centuries, scientists could not identify the components of the cytoskeleton because methods of preparing cells for viewing generally destroyed them, and the microscopes lacked sufficient magnification. The advent of the electron microscope and the development of new preparation methods helped scientists understand the structure of the cytoskeleton.

Three types of threadlike proteins make up the cytoskeleton: microtubules, microfilaments, and intermediate filaments. These proteins extend throughout the **cytosol,** the fluid part of cytoplasm, serving as structural units to maintain the shape of the cell and its components. You might picture them as railroad tracks that guide the movement of various cell components to their destinations while controlling the shape of the cell. They help to anchor many organelles so that the organelles are not floating completely free throughout the cytoplasm.

Microtubules play an important role in cell movements

Microtubules are long, hollow tubes in the cytoskeleton that move cell components such as molecules, organelles, and chromosomes from one place to the other like a cellular postal service. They also move cells, and sometimes multicellular organisms, through water.

Microtubules are threads about 25 nm in diameter that vary in length from about 200 to 150,000 nm, depending on their function. Each thread is made of spherical proteins called alpha and beta **tubulin,** which are stacked in 13 rows in a helix around the hollow center. As Figure 2.10 shows, the structure of a microtubule looks less complicated than it sounds. Individual microtubules frequently serve as tracks to direct movement within the cell.

In some cells, microtubules not only form part of the cytoskeleton but also can be combined as part of external propelling appendages called **cilia** and **flagella.** Both cilia and flagella are composed of nine pairs of microtubules arranged in a circle with two microtubules in the center. This so-called 9 + 2 pattern occurs in all eukaryotes. Cilia and flagella have the same basic structural pattern, except that cilia are short and flagella are long (see Figure 2.10). Cilia move back and forth like oars, while flagella undulate in a snakelike motion. Although cilia and flagella are more common in protist and animal cells than in plant cells, they do occur in some mobile reproductive plant cells, such as the sperm of mosses and ferns.

In addition to directing the movement of organelles and other structures within cells, microtubules help maintain cell shape. In a typical plant cell, for example, the enzymes that produce cellulose in the cell wall move along a microtubular track much like a monorail train (Figure 2.11). All movement controlled by microtubules involves "walking molecules," which are discussed later in this chapter.

Microfilaments help living cells change shape

Microfilaments are another group of long filaments in the cytoskeleton that move cells or cell contents and help determine cell shape. Microfilaments are made of the globular protein **actin,** organized into two helical chains twisted around each other (see Figure 2.10). Much thinner than microtubules, microfilaments measure about 7 nm in diameter. Many biology students are familiar with microfilaments because they are part of the structure of vertebrate muscles. Actin causes movement or changes in cell shape by forming associations with other molecules that walk along it, as will be described later. In plant cells, the patterns of cell elongation and cell wall formation determine cell shape. In plants, microfilaments also help move cell contents around the central vacuole in a circular motion known as **cytoplasmic streaming,** also known as **cyclosis.**

Motor proteins, or "walking molecules," cause movement

Microtubules and microfilaments cause movement by an association with several types of **motor proteins,** also known as **"walking molecules"** (see Figure 2.11). These proteins require energy in the form of ATP to shift from one position to another and back again, in a manner reminiscent of legs as they walk. In general, microtubules and microfilaments provide tracks that guide walking molecules to particular destinations. The walking molecules attach to the structure to be moved, such as when moving a transport vesicle from the ER to the Golgi apparatus using a microtubular track. A walking molecule may also attach to two microtubules and move one relative to the other.

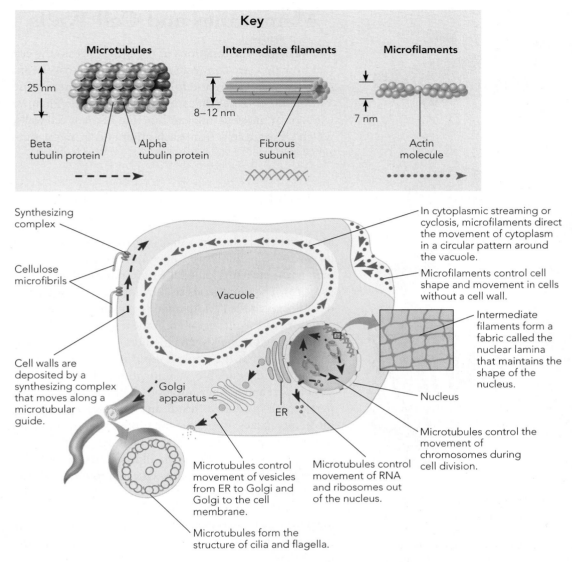

Figure 2.10 The cytoskeleton. This diagram shows several aspects of the cytoskeleton in a typical plant cell. The cytoskeleton, composed of microtubules, microfilaments, and intermediate filaments, regulates cell shape in cells without a cell wall, cell movement in motile cells, and the movement of cell contents in all cells. Microfilaments are made of actin, microtubules are made of tubulin, and intermediate filaments are made of various fibrous proteins.

Many types of movement in cells involve motor proteins. These movements include (1) vesicles moving from the ER to the Golgi apparatus along microtubular tracks; (2) chloroplasts moving from the bottom to the side of a cell along microtubules; (3) cyclosis, in which myosin walks along microfilaments to cause the cytoplasm to stream; and (4) cell wall synthesis in which an enzyme complex that makes cellulose microfibrils walks along a microtubule. All walking molecules use energy from ATP to move. The legs of these proteins move, release, reattach, and move again by using the energy in ATP.

Intermediate filaments help determine the permanent structure of cells

Intermediate filaments, the third component of the cytoskeleton, are so named because they are thicker than microfilaments but thinner than microtubules, being approximately 10 nm in diameter (see Figure 2.10). Several types of linear proteins combine to form intermediate filaments. In animals, intermediate filaments produce hair, nails, feathers, and scales and are also involved in muscles and neurons. Less is known about their role in plant cells,

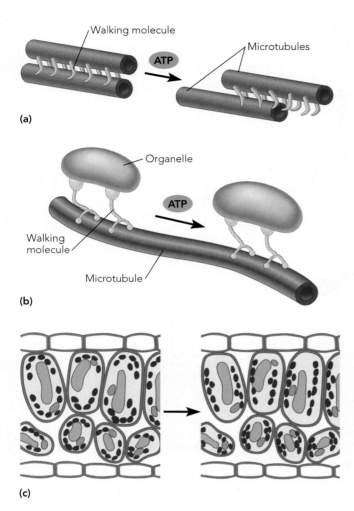

Figure 2.11 Motor proteins ("walking molecules").
(**a**) Walking molecules move one microtubule past another to initiate movement of flagella or cilia. (**b**) Walking molecules walk along a microtubule carrying an organelle. (**c**) In response to bright light, walking molecules move chloroplasts to the side of cells where they receive less light.

where they appear to be involved in maintenance of a more rigid, permanent type of internal cell structure. For example, they usually help in holding the nucleus in its permanent position in the cell and controlling the shape of the nucleus itself.

Section Review

1. What is the general purpose of the cytoskeleton?
2. What are the roles of microtubules and microfilaments?
3. What are intermediate filaments?

Membranes and Cell Walls

You have seen the functions of organelles within the cell and how they are connected through the cytoskeleton. Here we will explore the role of the plasma membrane and the membranes that surround organelles. Then we will look at the structure and function of plant cell walls, which separate cells from each other and from the outside world.

Membranes are gatekeeping barriers around and within cells

Membranes are barriers that control what enters and leaves the cell. Biologists have developed a model of membrane structure called the **fluid mosaic model** (Figure 2.12). The model consists of a double layer, or bilayer, of molecules called *phospholipids,* which form the basic structure of the plasma membrane and the membranes surrounding organelles. The membrane structure is fluid in the sense that the phospholipid molecules are quite movable and bendable. A phospholipid has a water-soluble part and a water-insoluble part. The water-soluble part, which is called a "head" and includes a phosphate molecule, makes it possible for some substances to pass through the plasma membrane. The water-insoluble part consists of two fatty acid "tails." Illustrations typically depict a phospholipid macromolecule as a round bead (the phosphate head) with two long tails attached. In the plasma membrane, the water-soluble phosphate heads form the inner and outer surfaces of the membrane, while the water-insoluble tails hang toward the center of the membrane.

A variety of proteins attach to the plasma membrane surface or are embedded in the plasma membrane to various degrees in a patchwork or mosaic pattern. Some proteins do not extend through the membrane, often binding molecules that arrive at the membrane. Other proteins or groups of proteins extend entirely through the plasma membrane, carrying out functions such as:

- providing channels across the membrane for various molecules to enter or leave the cell;
- controlling the transport of water and other substances through the membrane;
- serving as receptor sites for molecules being transported, or even other cells;
- identifying other cells and even disease organisms; and
- serving as attachment sites for molecules that control the cell's structure and function.

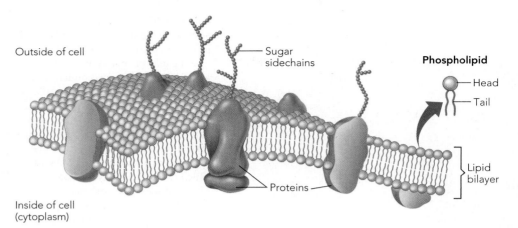

Outside of cell

Sugar sidechains

Phospholipid

Head

Tail

Lipid bilayer

Proteins

Inside of cell (cytoplasm)

Figure 2.12 Membrane structure. Cell membranes are composed primarily of phospholipids oriented with their water-insoluble fatty acid tails toward the inside and their water-soluble phosphate heads toward the two membrane surfaces. Proteins are embedded in various locations: attached to the membrane surface, penetrating completely through the membrane, and surrounding membrane pores.

The proteins and the double layer of phospholipids enable the plasma membrane to be **selectively permeable,** or partially permeable. That is, these structures transport some molecules but not others across the plasma membrane. Chapter 10 discusses the mechanisms by which molecules are transported across membranes.

Cell walls protect plant cells and define cell shape

Unlike animal cells, a plant cell has a cell wall that surrounds the plasma membrane, preventing excessive water from entering the cell. Water enters cells through the plasma membrane by a process called *osmosis,* which is discussed in Chapter 10. When a plant cell has sufficient water, its cell wall surrounds it like a close-fitting jacket. This jacket has one or two components. Plant cells produce a **primary cell wall** that is thicker than the plasma membrane and lies just outside it (Figure 2.13). Many mature plant cells, especially in woody plants, produce a thicker **secondary cell wall** between the primary wall and the plasma membrane. In some places, the secondary cell wall becomes thinner or disappears, leaving regions known as pits that allow the rapid transfer of water and minerals from cell to cell. The secondary wall provides most of the structural support for woody plants, constituting more than 90% of the weight of large trees.

Plant cell walls are mainly composed of cellulose, a carbohydrate consisting of many glucose sugars linked together. Many long cellulose molecules lie side by side to form cylindrical **microfibrils** (10 to 25 nm in diameter). The cellulose molecules in microfibrils are linked in some places to form crystalline subunits called **micelles.** This linking is done by proteins, jellylike substances called **pectins,** and gluelike or gumlike carbohydrates called

hemicelluloses. The microfibrils are then twisted together to form larger structures called *macrofibrils,* which can be 0.5 µm in diameter and 4 µm long. You might think of microfibrils as smaller threads twisted together to make larger threads, which are the macrofibrils. Other threads of pectins and hemicelluloses weave among the macrofibrils to bind them together into a cell wall fabric, which is stabilized by proteins at particular locations. You will read more about cellulose structure in Chapter 7.

In vascular plants, rigid molecules called **lignins** strengthen cell walls, replacing water and coating or encrusting the cellulose somewhat like tar. Nonvascular plants, such as mosses, do not have lignin.

Cell walls of algae and fungi are usually structurally different from those of plants. The cell walls of most algae contain cellulose but also include many different types of secondary compounds not found in the plant cell wall. Also, some algae have glasslike cell walls in which silica is the principal component. The cell walls of fungi are made primarily of **chitin,** a carbohydrate similar in structure to cellulose.

Between the primary cell walls of adjacent cells is a thin layer called a **middle lamella** (from the Latin *lamina,* "thin plate"). The middle lamella is composed mainly of pectins, which hold the cells together. Pectins are used in cooking to thicken jellies and jams.

Plasmodesmata are channels that connect plant cells

Plant cell walls provide barriers between cells, but connections between the cytoplasm of cells are necessary for a multicellular organism. Therefore, numerous channels called **plasmodesmata** (singular *plasmodesma;* from the Greek *desma,* "bond") maintain direct contact between cells (Figure 2.14). Thin strands of cell contents flow

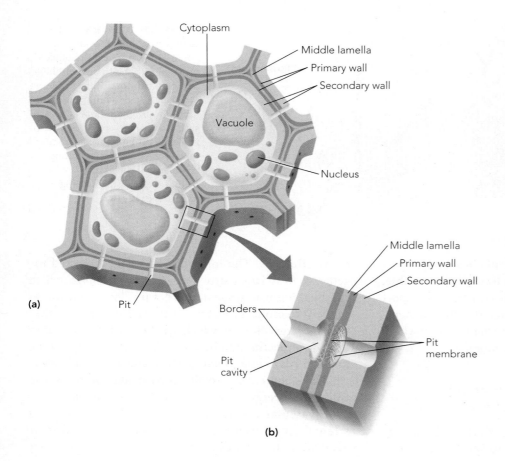

Figure 2.13 Primary and secondary cell walls. (a) Plant cells produce a primary cell wall just outside the plasma membrane. Later, many cells produce a secondary cell wall between the primary cell wall and the plasma membrane. (b) Often, secondary cell walls are quite thick compared with primary walls. Secondary walls have regions called pits where the secondary wall is thin or absent, which speed the transfer of water and dissolved minerals from one cell to the next. Cell walls are composed primarily of cellulose, with many cellulose molecules making up a microfibril and many microfibrils making up a macrofibril. Other components such as lignin, hemicellulose, and proteins link the macrofibrils together.

through these plasmodesmata. Each plasmodesma is surrounded by a plasma membrane and usually contains a connection, called a **desmotubule,** between cells. Plasmodesmata allow small molecules and even macromolecules to pass between cells.

Section Review

1. Explain how the plasma membrane is selectively permeable.
2. Why is it important for plant cells to have a cell wall?
3. Describe the basic structure of plant cell walls.
4. How are plant cells linked with one another?

The Cell Cycle and Cell Division

So far in this chapter, we have looked at the structure of a typical plant cell in the prime of its life. Like organisms, cells have a life cycle that includes youth, maturity, and old age. At the end of their life cycle, cells can either divide or die. For plant cells, death is not necessarily the end of their usefulness because the cell walls of dead cells become permanent parts of the plant body. In large trees, for example, more than 99% of the cells in the trunk are dead. These cells serve a variety of useful functions, including conducting water and providing the structural support necessary to lift the leaf canopy into the sunlight.

Although dead cells can play important roles, the growth of a multicellular organism depends on cell division. As you know, all multicellular organisms, regardless of size, start as a single cell. A relatively small number of consecutive cell divisions can eventually change a single cell into an organism with many billions of cells. For example, if a cell and its descendents continue through 50 consecutive divisions, more than a trillion (10^{15}) cells result. As you can see in the *Biotechnology* box on page 42, cell division can have important implications for plant biotechnology. In this section, we will look at how cell division is involved in the growth and reproduction of a eukaryotic organism.

somes, the DNA of the chromosomes must be replicated, or copied. In mature eukaryotic cells, the DNA of each chromosome is replicated during a short **S phase** (with *S* standing for DNA *synthesis*), producing two sister strands of DNA called **chromatids.** (Chapter 13 will describe the process of DNA replication.) The S phase is followed by the **G$_2$ phase,** when the cell continues normal functioning and begins to prepare for cell division.

The stage of cell division is known as the **mitotic phase,** or **M phase,** which refers to the type of division of the nucleus called *mitosis.* **Mitosis** is the division of a single nucleus into two genetically identical daughter nuclei. This process will be described shortly. The division of a single cell into new cells, or **daughter cells,** involves two processes. One process is the division of the nucleus. The other is the separation of the cytoplasm and the new nuclei into daughter cells, a process known as **cytokinesis** (which basically translates as "cell movement"). Figure 2.15 provides an overview of the cell cycle.

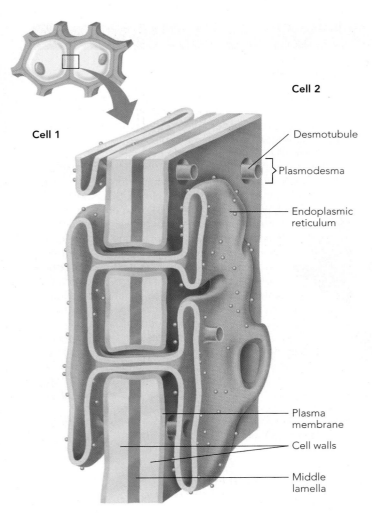

Figure 2.14 Plasmodesmata. In this diagram of plasmodesmata, note the continuity of the endoplasmic reticulum between the two cells by means of desmotubules.

The cell cycle describes the phases of a cell's life

The **cell cycle** is the sequence of events from the time a cell first arises as a result of cell division until the time when that cell itself divides. In eukaryotes, the cell follows a cell cycle consisting of periods of growth and development into a particular cell type, DNA replication, preparation for division, and finally cell division itself. The cells resulting from this division then begin the cell cycle anew.

Approximately 90% of a cell's life consists of a period known as *interphase,* in which the cell is not engaged in cell division. The name indicates that this is the period between cell divisions. The first part of interphase is a relatively long **G$_1$ phase** (with *G* standing for *gap*) when the cell grows, develops, and functions as a particular cell type. For the cell to eventually divide and still have a complete set of chromo-

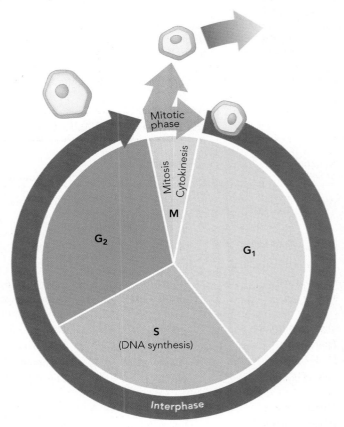

Figure 2.15 The eukaryotic cell cycle. Interphase, the period between cell divisions, consists of early cell growth and development (G$_1$ phase), DNA replication (S phase), and continued growth and preparation for cell division (G$_2$ phase). The actual nuclear division and cell division, known as the M phase or mitotic phase, make up only about 10% of a typical cell's life.

THE INTRIGUING WORLD OF PLANTS

Using Plant Cell Cultures

Each cell of a plant typically contains all of the genetic material and codes for that plant. This means that each cell in a plant has the potential to produce an entire plant. We say such cells are totipotent. A really interesting feature of plant cells is that if they are cultured on a medium containing nutrients in a test tube, we can easily encourage the expression of totipotency. This ability of plant cells to produce plants has many uses.

In addition to producing entire plants, individual plant cells can express the individual traits of a plant. For example, unique compounds that are typically produced by whole plants can be synthesized by cultures of plant cells that act as chemical factories. This is important because specific compounds such as alkaloids that protect plants from diseases and predators can also kill disease organisms that attack humans—some even destroy cancer cells.

Despite their usefulness, many plant compounds are expensive to obtain when whole plants are used. For example, the rosy periwinkle plant (*Catharanthus roseus*) produces the alkaloids vinblastine and vincristine, which are useful as anti-cancer drugs. These alkaloids constitute only 0.0005% of the plant by weight. Using whole plants as a source, vinblastine currently costs $1 million per kilogram (2.2 pounds), while vincristine costs $3.5 million. Enough of either drug to treat a patient would cost well over $10,000 using this method, and both molecules are too complex to economically synthesize outside the plant.

Scientists are experimenting with cell cultures to yield such alkaloids, but production costs have not yet improved significantly over using whole plants. Still, some techniques are promising. One method is to stimulate production through plant hormones. In another method scientists are engineering cells to make extra copies of genes that promote production. Also, mutant cells may be selected that produce high concentrations of the alkaloids. Finally, scientists can search for cells that secrete the desired alkaloid outside the cell, thus avoiding the need to kill the cell.

Some useful compounds obtained from plants are listed in the table below. Currently, only a few can be produced more economically in cell cultures, but the future of plant cell culture is bright. As scientists learn more about how cells produce, store, and release compounds, yields from cell cultures will improve, resulting in new drugs at reasonable costs, with no environmental damage involved in production.

Examples of Medically Important Compounds Produced by Plants

Scientific Name of Source	Common Name of Source	Compound	Use of Compound
Aloe spp.	Aloe	Crysophanic acid	Promotes healing of burns
Atropa	Deadly nightshade	Atropine, scopolamine, and hyoscamine	Anesthetic and antispasmodic drugs
Catharanthus roseus	Rosy periwinkle	Vinblastine and vincristine	Antitumor drugs
Cephaelis ipecacuanha	Ipecac	Compounds in rhizome	Induces vomiting
Curcuma longa	Turmeric	Compounds in rhizome extract	Lowers cholesterol
Datura spp.	Jimson weed	Atropine, hycosamine, and scopolamine	Anesthetic and antispasmodic drugs
Digitalis purpurea	Foxglove	Digitalis	Heart stimulant
Ephedra spp.	Ephedra	Ephedrine	Lowers blood pressure
Ginkgo biloba	Ginkgo	Compounds in leaf extracts	Improves memory
Lobelia inflata	Lobelia	Lobelline sulfate	Treatment of respiratory disorders
Mucana pruriens	Picapica	L-dopa	Treatment of Parkinson's disease
Papaver somniferum	Opium poppy	Morphine and codeine	Analgesic properties (pain relieving)
Podophyllum peltatum	May apple	Podophyllotoxin	Antitumor drug
Rauvolfia serpentina	Rauvolfia	Reserpine	Treatment of mental illness and hypertension
Taxus brevifolia	Pacific yew	Taxol	Anticancer drug

Mitosis and cell division are involved in growth and reproduction

Cell division involving mitosis plays a role in the growth of all eukaryotes. It can also be involved in reproduction. Mitosis occurs:

◆ during growth and development of an embryo and then an adult.

◆ to replace cells in an adult organism.

◆ during asexual reproduction of various types. Asexual reproduction is the formation of new organisms that are genetically identical to the parent.

◆ during some phases of sexual reproduction in plants. Sexual reproduction is the formation of new organisms that are genetically different from the two parents involved.

In Chapters 4 and 5, you will see how mitosis is involved in plant growth. Chapter 6 will explain the role of mitosis in plant reproduction and will compare mitosis with meiosis, a type of nuclear division that is only involved in reproduction.

Mitosis produces two daughter nuclei, each containing the same chromosome number as the original cell

Mitosis occurs after the G_2 phase of the cell cycle. Although a continuous process, mitosis can be described as having four main phases: prophase, metaphase, anaphase, and telophase. Figure 2.16 shows the most distinguishing characteristics of each phase.

1 The main sign of **prophase** is that the chromosomes have shortened and thickened enough to be visible under a light microscope. Each chromosome appears as two sister chromatids, the replicated DNA strands produced during the S phase. The chromatids are joined together by a narrow chromosonal region called a **centromere.** Meanwhile, the nucleoli disappear. A spindle of microtubules (not shown in the prophase figure) begins to form at two locations called **centrosomes,** or microtubule-organizing centers, which move to opposite ends of the cell. In animal cells, each centrosome contains two centrioles, composed of microtubules, but these do not occur in plant cells. In late prophase, also called *prometaphase,* the nuclear envelope breaks into pieces and disappears, and the spindle microtubules

begin to move the chromosomes toward the center of the cell. At its centromere, each chromatid forms a complex structure of proteins called a **kinetochore,** and some microtubules are attached to the kinetochore.

2 During **metaphase,** the chromosomes move to an imaginary plane called the **metaphase plate,** which extends across the diameter of the cell, equidistant from the two poles of the now fully formed spindle. The microtubules attached to the kinetochores cause this movement.

3 In **anaphase,** the sister chromatids of each chromosome move apart, so that each chromatid is now a separate chromosome. Motor proteins "walk" the chromosomes toward the opposite poles of the cell. The spindle microtubules attached to the kinetochores shorten, while the other microtubules lengthen, pushing the poles of the cell farther apart.

4 **Telophase,** the last phase of mitosis, begins when the two groups of chromosomes reach the opposite poles of the cell. Telophase is the reverse of prophase. That is, the nuclear envelope reforms in each cell, the chromosomes uncoil, and the spindle disappears.

5 **Cytokinesis,** the division of the cytoplasm, usually starts during late anaphase or early telophase. In animal cells, which have no cell wall, cytokinesis involves a sort of pinching apart of the cytoplasm. The "waist," called a *cleavage furrow,* becomes narrower and narrower, eventually separating one cell into two. In plants and in some algae, cytokinesis occurs by means of a phragmoplast, which forms a cell plate. The **phragmoplast** is a cylinder consisting of microtubules that are derived from the spindle and aligned between the daughter nuclei. The **cell plate,** consisting of two new plasma membranes and cell walls, forms between the nuclei and in the center of the phragmoplast. The cell plate gradually extends to divide the cell into two daughter cells.

New cells typically become specialized

You know that mitosis is necessary for the formation of new plants. By itself, however, mitosis would simply divide available cellular space into smaller and smaller subunits. In order to produce a functioning multicellular organism, mitosis must be followed by cell

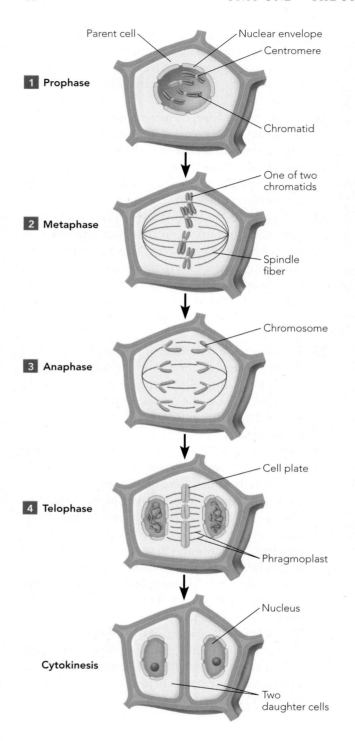

1 **Prophase**

Parent cell
Nuclear envelope
Centromere
Chromatid

2 **Metaphase**

One of two chromatids
Spindle fiber

3 **Anaphase**

Chromosome

4 **Telophase**

Cell plate
Phragmoplast

Cytokinesis

Nucleus
Two daughter cells

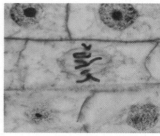

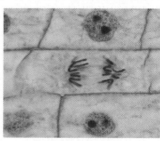

Figure 2.16 Mitosis. The process of mitosis is typically described as having four main stages: prophase, metaphase, anaphase, and telophase. The main changes are the shortening and thickening of the chromosomes during prophase, chromosome alignment during metaphase, chromosome migration to opposite poles during anaphase, and nuclear envelope formation and uncoiling of chromosomes during telophase. Cytokinesis, the division of the cytoplasm, usually starts during late anaphase or early telophase. In plant cells, a cylinder of microtubules called a phragmoplast forms the cell plate that separates the cell into two daughter cells.

growth and development of specialized cells for particular purposes. New cells typically mature into different types of adult cells appropriate to their location and function in the organism. In the next chapter, we will look at how plant cells become specialized for various purposes.

Section Review

1. List each phase of the cell cycle and describe what happens during each phase.
2. What are the purposes of mitosis and cell division?
3. Describe in general what happens during mitosis.

SUMMARY

An Overview of Cells

Cells are microscopic, enabling adequate absorption of oxygen, water, and nutrients through the cell's surface to meet the needs of internal regions.

Microscopes reveal the world of the cell (p. 27)

Light microscopes bend light to produce magnified images. Electron microscopes focus electrons with magnetic lenses to produce images. The transmission electron microscope (TEM) passes electrons through a specimen to produce television images. The scanning electron microscope (SEM) bounces electrons off a specimen to yield a 3-D image.

The cell is the basis of an organism's structure and reproduction (pp. 27–29)

The cell theory states that all organisms are made up of one or more cells, that the cell is the basic structural unit, and that all cells arise from existing cells.

All cells are either prokaryotic or eukaryotic (pp. 29–30)

Eukaryotic cells evolved from prokaryotic cells and are generally larger and more complex, with an enclosed nucleus and other membrane-bound organelles. Most known prokaryotes are bacteria, which are single-celled organisms. Eukaryotes can be single-celled, but most are multicellular. Cellular functions in eukaryotes are handled by organelles. The endosymbiotic theory states that some organelles evolved as a result of prokaryotic cells ingesting other prokaryotic cells.

Cells produce nucleic acids, proteins, carbohydrates, and lipids (pp. 30–31)

Nucleic acids (DNA and RNA) contain genes and transcribed genes. Genetic instructions in RNA direct synthesis of proteins, which are composed of amino acids and define an organism's physical characteristics. Some proteins are structural components, while others called *enzymes* regulate chemical reactions. Proteins are also involved in transport and responses to stimuli. Carbohydrates such as sugars and starch are sources of energy and structural building blocks. Lipids such as fats and phospholipids serve as building blocks and protective layers.

Major Plant Cell Organelles

The cells of plants, algae, fungi, and animals have essentially the same types of organelles. The interior of a plant cell (everything inside the cell wall) is called the *protoplast.* The interior consists of two parts: the nucleus and the cytoplasm.

The nucleus provides DNA "blueprints" for making proteins (p. 31)

The DNA in the nucleus is organized into complex threadlike chromosomes. A chromosome consists of many genes, each coding for a specific protein. Nucleoli are round structures in some chromosomes that synthesize subunits for ribosomes. The nuclear envelope has pores to allow movement of substances to and from the cytoplasm.

Ribosomes build proteins (p. 31)

Ribosomes are tiny particles that use RNA to synthesize proteins. Unlike other organelles, ribosomes lack a membrane and are also present in prokaryotic cells.

The endoplasmic reticulum is the site of most protein and lipid synthesis (p. 32)

The endoplasmic reticulum (ER), a network of membranes attached to the nuclear membrane, is an assembly site for making and exporting lipids (in the smooth ER) and proteins (in the rough ER).

The Golgi apparatus completes and ships cell products (pp. 32–33)

Also known as the Golgi complex, this organelle consists of stacks of membrane-bound sacs called *cisternae,* which store and modify products from the ER before exporting them from the cell through transport vesicles.

Chloroplasts in green plant cells convert solar energy into stored chemical energy (pp. 33–34)

Chloroplasts are photosynthetic organelles in cells of green parts of plants. Solar energy is converted to chemical energy in the thylakoids, while sugar production and storage occurs in the surrounding fluid called the *stroma.* Chloroplasts are a type of plastid, an organelle that makes and stores food or pigments for plant cells. Colorless plastids called *leucoplasts* often contain starch, and chromoplasts contain yellow, orange, or red pigments.

Mitochondria convert stored energy into energy to power the cell (p. 34)

The mitochondria are organelles that are the cell's "power plant," converting stored chemical energy into ATP, the organic molecule that fuels the cell's activities.

Microbodies aid in chemical reactions (pp. 34–35)

Microbodies are small, spherical organelles containing enzymes. Peroxisomes are microbodies that confine the processing of hydrogen peroxide for chemical reactions. Plants also have microbodies called *glyoxysomes,* which contain enzymes to convert fats to sugars.

Vacuoles play a variety of roles in cell metabolism and cell shape (p. 35)
Most of a mature plant cell's volume consists of a large central vacuole, which stores water and waste products, regulates salt concentrations, and helps maintain cell shape. The vacuole's membrane is called a *tonoplast*.

The Cytoskeleton: Controlling Cell Shape and Movement

The cytoskeleton ("cell skeleton") maintains cell shape and keeps organelles from floating freely. The cytoskeleton has three types of threadlike proteins: microtubules, microfilaments, and intermediate filaments, which extend like tracks throughout the cytosol, or cytoplasm fluid.

Microtubules play an important role in cell movements (p. 36)
Microtubules, the longest protein threads in the cytoskeleton, direct movement of molecules, organelles, and chromosomes within the cell. Microtubules can also be components of cilia and flagella, the propelling appendages found in many animal cells and in mobile reproductive cells in plants.

Microfilaments help living cells change shape (p. 36)
Microfilaments are the thinnest protein threads in the cytoskeleton. Composed of the protein actin, they cause movement or changes in shape. In plants, they move cell contents around the central vacuole in a circular motion called cytoplasmic streaming (cyclosis).

Motor proteins, or "walking molecules," cause movement (pp. 36–37)
Microtubules and microfilaments provide tracks that guide motor proteins ("walking molecules"), which attach to structures to move them to destinations within the cell.

Intermediate filaments help determine the permanent structure of cells (pp. 37–38)
Intermediate filaments are thicker than microfilaments but thinner than microtubules. In plant cells they may help maintain cell structure.

Membranes and Cell Walls

Membranes are gatekeeping barriers around and within cells (pp. 38–39)
Cells are surrounded by a plasma membrane (cell membrane) that controls what enters or leaves the cell. The fluid mosaic model describes the plasma membrane's structure as a double layer of phospholipids that make the membrane selectively permeable.

Cell walls protect plant cells and define cell shape (p. 39)
Plants, unlike animals, have a cell wall external to the plasma membrane. In addition to a primary cell wall that all plant cells produce, some cells in woody plants produce a thicker secondary cell wall that provides structural support between the primary cell wall and plasma membrane. Plant cell walls are composed mainly of long cellulose molecules. Rigid molecules called *lignins* strengthen cell walls in vascular plants. A thin layer called the *middle lamella* joins cell walls of adjacent cells.

Plasmodesmata are channels that connect plant cells (pp. 39–40)
Plasmodesmata allow molecules to pass between cells.

The Cell Cycle and Cell Division

At the end of their life cycle, cells either divide or die. Dead cells often provide structural support and conduct water. Growth and reproduction of multicellular organisms depend on cell division.

The cell cycle describes the phases of a cell's life (p. 41)
The cell cycle has two main periods: interphase and the mitotic phase. Interphase consists of 90 percent of the cell cycle, comprising the G_1 phase, DNA replication (S phase), and G_2 phase. During the S phase, the DNA of each chromosome is duplicated to produce two sister strands of DNA called chromatids. The mitotic phase, or M phase, consists of cell division, during which the chromatids become separate chromosomes and the nucleus and cytoplasm divide to form separate cells. Cell division consists of the division of the nucleus and cytokinesis, the separation of the cytoplasm and the new nuclei into daughter cells.

Mitosis and cell division are involved in growth and reproduction (p. 43)
Mitosis is involved in growth, replacement of cells in adult organisms, sexual reproduction in plants, and asexual reproduction.

Mitosis produces two daughter nuclei, each containing the same chromosome number as the original cell (p. 43)
Mitosis consists of prophase (appearance of each chromosome as a pair of sister chromatids and the beginning of the formation of a spindle of microtubules, breakup of the nuclear envelope, and movement of chromatids toward the center of the cell), metaphase (in which chromatids align across the center between the two poles of the spindle), anaphase (separation of chromatids to form individual chromosomes, which move to the opposite poles), and telophase (the reverse of prophase, with nuclear membranes reforming around each daughter cell). Cytokinesis usually starts during late anaphase or early telophase. In plant cell cytokinesis, a cell plate extends to divide the cell into two daughter cells.

New cells typically become specialized (pp. 43–44)
Cell division divides available cellular space but does not produce growth. For a multicellular organism to develop, cell division must be followed by cell growth and cell differentiation, the creation of specialized cells for particular purposes.

Review Questions

1. Describe in general how each of the following works: light microscope, transmission electron microscope, scanning electron microscope.

2. What are the three basic conclusions of the cell theory?
3. List the differences between prokaryotic and eukaryotic cells.
4. How are the functions of prokaryotic and eukaryotic cells similar?
5. How does endosymbiosis relate to the evolution of organelles?
6. In what sense is the nucleus the "supervisor" of a eukaryotic cell?
7. Where does protein synthesis occur within the cell?
8. What is the relationship between the endoplasmic reticulum and the Golgi apparatus?
9. What is the relationship between chloroplasts and mitochondria?
10. Describe several functions of the large central vacuole in a plant cell.
11. Compare and contrast plant and animal organelles.
12. What would happen if there were no cytoskeleton?
13. Compare and contrast microtubules and microfilaments.
14. Explain what makes a cell wall strong and rigid.
15. How are plant cells linked?
16. What occurs in each phase of the cell cycle?
17. Explain what happens during each phase of mitosis.

Questions for Thought and Discussion

1. Microscopes opened up a new world that most people had not even imagined. Recently, a few philosophers have suggested that science has discovered almost everything and that only details remain to be worked out. Do you agree? Explain.
2. In what ways is a cell like a factory?
3. Is a plant cell like a miniature organism? Explain.
4. If each plant cell contains all of a plant's genes, what keeps cells from turning into individual plants on their own? Why is it, for example, that each leaf does not form thousands of buds?
5. In what ways are the cytoskeleton and human skeleton similar? In what ways are they different?
6. What problems would large organisms like trees and humans encounter if cell division did not occur and they were simply one giant cell?
7. Prokaryotes are quite small cells compared with eukaryotes. Why do you think there are few large prokaryotic cells?
8. By means of a sequence of drawings, (a) show how the ingestion of a small prokaryotic cell by a larger prokaryotic cell would result in the cytoplasm of the ingested cell being surrounded by two membranes. Indicate which membrane is the original plasma membrane of the smaller prokaryote and which is the membrane of the host's digestive vacuole. (b) Now continue the diagram sequence to show how similar processes of ingestion could give rise to an ingested body surrounded by four membranes; this is the arrangement we find in the chloroplasts of brown algae. [Hint: What would be the result if the cell produced in part (a) were itself ingested?] Identify the origin of each of the four membranes.

Evolution Connection

Based on information about the similarities and differences between eukaryotic and prokaryotic cells described in this chapter, list the features you would expect from mitochondria and chloroplasts to possess if the endosymbiotic theory that these evolved from ingested prokaryotic cells is correct. Have these features been found? Does the fact that neither mitochondria nor chloroplasts can be grown in pure culture contradict the theory? Why or why not?

To Learn More

Visit The Botany Place at *www.thebotanyplace.com* for quizzes, exercises, and Web links to new and interesting information related to this chapter.

Recently a new kind of microscope called an atomic force or scanning-tunneling microscope has come into common use in science labs. This microscope provides a surface area view but has much greater magnification than a scanning electron microscope. Look up information on these microscopes on the Web and tell how they work.

Dekruif, Paul, and F. Gonzalez-Crussi. *Microbe Hunters.* Fort Washington, PA: Harvest Books, 1995. This account of important and interesting figures in the history of microbiology gives you the impression that you are there as exciting discoveries are made.

Drake, Ellen Tan. *Restless Genius: Robert Hooke and His Earthly Thoughts.* New York: Oxford University Press, 1996. What we do know of Robert Hooke indicates the many facets of his genius.

Medicine Man, a movie starring Sean Connery and Lorraine Bracco. This exciting movie describes rainforest destruction and the rain forest's potential to produce medically useful substances. Sean Connery plays a botanist!

Nichols, Richard. *Robert Hooke and the Royal Society.* Philadelphia, PA: Trans-Atlantic Publications, 1999. Robert Hooke had fascinating interactions with the Royal Society. Issac Newton, who headed the Society after Hooke's death, fought frequently with Hooke and tried to erase all memory of him from the Society's records.

Rasmussen, Nicolas. *Picture Control: The Electron Microscope and the Transformation of Biology in America, 1940–1960.* Stanford, CA: Stanford University Press, 1997. A fascinating account of an important period in the history of science.

Thomas, Lewis. *The Lives of a Cell: Notes of a Biology Watcher.* New York: Penguin, 1995.

Thomas, Lewis. *The Medusa and the Snail: More Notes of a Biology Watcher.* New York: Penguin, 1995. Books by Lewis Thomas are well written, easy to read, and thought-provoking.

3

An Introduction to Plant Structure

Coast redwood (*Sequoia sempervirens*).

Basic Types of Plant Cells

Parenchyma cells are the most common type of living differentiated cell

Collenchyma cells provide flexible support

Sclerenchyma cells provide rigid support

Tissues of Vascular Plants

The dermal tissue system forms the plant's outer protective covering

The vascular tissue system conducts water, minerals, and food

Ground tissue usually forms between dermal and vascular tissues

An Overview of Vascular Plant Organs

Stems position leaves for maximum photosynthesis

Leaves function in both photosynthesis and transpiration

Roots anchor the plant and absorb water and minerals

An Overview of Plant Growth and Development

Embryos give rise to stems, leaves, and roots of adult seed plants

Meristems enable plants to continue growing throughout their lives

Apical meristems initiate primary growth that makes roots and shoots longer

Botanists are discovering how genes control the formation of apical meristems

Apical meristems give rise to primary meristems, which produce primary tissues

Secondary growth from lateral meristems makes roots and stems thicker

Some plants live for one growing season, while others live for two seasons or longer

More than 200 million years ago, when dinosaurs ruled the animal world, gymnosperms dominated the world of plants—with conifers being the largest. Among their descendants are the tallest of living things—the coast redwoods (*Sequoia sempervirens*). The first part of the scientific name honors Sequoyah, creator of the Cherokee writing system, while *sempervirens* is Latin for "always green." Coast redwoods grow naturally only in a narrow coastal area of northern California and southern Oregon. They can grow elsewhere but will never reach the astounding heights found in their natural setting. The tallest living tree today is the Stratosphere Giant in Humboldt Redwoods State Park, California. At 112.34 meters (308.62 feet), it is five stories taller than the Statue of Liberty. Coast redwoods are also wide, with trunks of mature trees being typically 3.0 to 6.1 meters (10 to 20 feet) in diameter. They are fast-growing, with young trees 4 to 10 years old gaining as much as 2 meters (6.5 feet) in height in a growing season.

Sometimes *sempervirens* is mistranslated as "always living." Indeed, the Stratosphere Giant is estimated to be between 600 and 800 years old, and the oldest verified age of a coast redwood is at least 2,200 years. The secret of their longevity lies partly in their bark, which is very thick—as much as 30 centimeters (1 foot) in old trees—and resistant to insects, fungi, and fire.

How do coast redwoods grow so tall and remain standing? How do they transport water and minerals from the soil to the highest branches, hundreds of feet in the air? How are they able to live for so long? The answers to such questions can be found by exploring some basic characteristics of plant structure and growth. In this chapter we will look in general at how cells, tissues, and organs form the plant and at how cells can be specialized for such purposes as transport, support, and protection. Then we will look at basic patterns of plant growth.

Basic Types of Plant Cells

Like all multicellular organisms, a plant begins as a single cell. In Chapter 2 you saw how cell division through mitosis provides new cells for plant growth. Most of this mitosis takes place within meristematic cells—unspecialized cells that can divide indefinitely to produce new cells. Regions of these meristematic cells that produce new growth are called **meristems** (from the Greek *meristos,* "divided") and are present in all types of plants—from mosses to towering trees. Meristematic cells called **initials** remain within meristems as sources of new growth. You might think of them as "mitosis machines" that function as a "fountain of youth," enabling a plant to continue growing throughout its life. Later in the chapter, we will look at two types of meristems: apical meristems that add length at the tips of stems and roots and lateral meristems that add width to woody stems and roots.

When an initial divides by mitosis, one daughter cell remains an initial at the same location within the meristem. The other daughter cell, called a **derivative,** is pushed out of the meristem and either divides again or begins elongation and **differentiation,** the processes by which an unspecialized cell develops into a specialized cell. In this way, a plant always has a supply of initials, unspecialized meristematic cells to generate new growth, while at the same time developing new specialized cells to serve particular functions. Since meristematic cells are unspecialized, they are said to be undifferentiated cells. Specialized cells, which have specific structures and purposes, are called *differentiated cells.* Plant cells have varying degrees of differentiation—that is, some are more specialized than others. In plants, the process of differentiation can sometimes even be reversed, with a differentiated cell reverting to become an undifferentiated meristematic cell. Such "dedifferentiation" to a meristematic state occurs during the development of lateral meristems and in laboratory tissue cultures.

In this section, we will compare the three basic types of differentiated cells that occur commonly in plants: parenchyma, collenchyma, and sclerenchyma cells. Later we will look at more specialized cells that conduct water, minerals, and food in vascular plants.

Parenchyma cells are the most common type of living differentiated cell

Most living plant cells are **parenchyma** (pair–RENK–kuh–muh) cells, the general-purpose "workhorse" cells of a plant. They are the most common type of living differentiated cell and are also the most common type of cell in most plants (Figure 3.1). They are the least specialized type of plant cell, usually undergoing relatively little differentiation before assuming their role as mature plant cells. Some botanists view parenchyma cells as the immediate precursor of all other differentiated types of plant cells, while others see meristematic cells as the immediate precursor of all differentiated cell types, including parenchyma.

Parenchyma cells have thin primary cell walls and usually do not have secondary walls. Their thin walls enable them to grow into various shapes to fill the space available, but they are usually spherical, cubical, or elongated. Since they typically lack secondary cell walls and therefore contain less cellulose, parenchyma cells are relatively "inexpensive" for a plant to produce. Accordingly, they often fill space or provide structure to parts of the plant that must be frequently replaced, such as leaves. In fact, the term *parenchyma* comes from the Greek *parenchein,* meaning "to pour in beside."

Although often serving as space-fillers and structural components, parenchyma cells also have other purposes. Most photosynthetic cells are specialized parenchyma called **chlorenchyma.** Parenchyma cells also store food and water in roots, stems, leaves, seeds, and fruits. When you eat a fruit, most of what you consume is probably parenchyma. You will encounter parenchyma cells in many locations as we survey the structure of a plant. They are alive at maturity and can usually divide, allowing them to develop into different kinds of more specialized cells, as you will soon see.

Collenchyma cells provide flexible support

The main function of **collenchyma** (cole–LEN–kuh–muh) cells is to provide flexible support. Collenchyma cells are usually elongated and can grow into various shapes because they are alive at maturity and lack secondary walls. Unlike parenchyma cells, however, collenchyma cells have primary cell walls that are thickened in some places by additional cellulose, usually in the corners of the cell wall. The name *collenchyma* comes from the Greek *kolla,* meaning "glue," a reference to these thick cellulose layers, which enable collenchyma to provide more support than parenchyma while still remaining somewhat flexible (Figure 3.2). Researchers have discovered that plants living in windy environments or placed under artificial mechanical stress produce much more collenchyma, which allows them to bend without breaking.

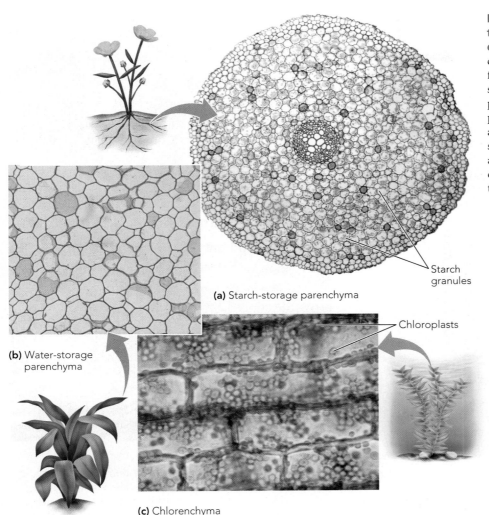

Figure 3.1 Parenchyma cells are the most common type of plant cell. Parenchyma are generalized plant cells that can be involved in a variety of functions, such as carrying out photo-synthesis, storing food and water, and providing structure. **(a)** Starch storage parenchyma cells occur in leaves, stems, and roots. **(b)** Parenchyma can also serve as water storage cells in roots and stems. **(c)** Chlorenchyma are chloroplast-containing parenchyma cells that are primarily found in leaves.

Starch granules

(a) Starch-storage parenchyma

Chloroplasts

(b) Water-storage parenchyma

(c) Chlorenchyma

In order to provide support, parenchyma and collenchyma cells must become **turgid**—that is, swollen or enlarged—as a result of being full of water. Consider the collenchyma cells that make up most of the structural "ribs" in a stalk of celery. As a result of water loss, the stalk will become limp after sitting in your refrigerator but regains turgidity if placed in water. If you have ever inflated an air mattress or filled a waterbed, you know how plant cells become turgid. Just as an inflated air mattress or full waterbed can support your weight, a plant cell filled with water helps support the plant.

Sclerenchyma cells provide rigid support

Unlike collenchyma cells and most parenchyma cells, **sclerenchyma** (sklair–RENK–kuh–muh) cells have secondary walls, often hardened with lignin. The word *sclerenchyma* comes from the Greek *skleros,* meaning "hard,"

reflecting the fact that the thick secondary cell wall provides rigid support. Indeed, sclerenchyma cell walls are much harder than those of collenchyma or parenchyma cells (Figure 3.3). Sclerenchyma cells are more "expensive" for a plant to produce than parenchyma or collenchyma cells because of the additional cellulose required to build the secondary cell walls. Accordingly, sclerenchyma cells are considerably less common in smaller plants than parenchyma or collenchyma. Unlike parenchyma and collenchyma cells, sclerenchyma cells are typically dead at maturity. They provide structural strength in regions that have stopped growing in length and no longer need to be flexible. If the plant wilts—loses turgidity from lack of water—sclerenchyma cells can still provide support because of their hardened cell walls.

There are two main types of sclerenchyma cells that provide support and protection: fibers and sclereids. **Fibers** are elongated cells with thick secondary walls reinforced

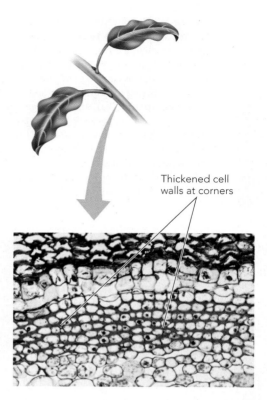

Thickened cell walls at corners

Figure 3.2 Collenchyma cells provide flexible support.
Collenchyma cells have cell walls that are thickened in some places with additional cellulose, enabling them to provide more support than parenchyma while still allowing living stems to be flexible.

with lignin, making them flexible as well as strong (see *The Intriguing World of Plants* box on page 53). Usually occurring in groups, they enable stems—including tree trunks—to move in the wind without snapping. **Sclereids** vary more in shape than fibers but are often cubical or spherical. They make structures such as nutshells and fruit pits typically rock-hard and inflexible. The gritty sandlike pieces in the fleshy part of a pear are also sclereids, commonly called *stone cells.*

Section Review

1. How do meristematic cells differ from parenchyma, collenchyma, and sclerenchyma cells?
2. Compare the three basic types of differentiated cells.

Tissues of Vascular Plants

All plants have cells that carry out photosynthesis, store and transport water and nutrients, and provide support. In vascular plants—which are the vast majority of

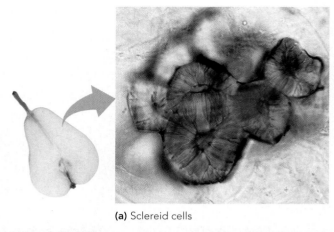

(a) Sclereid cells

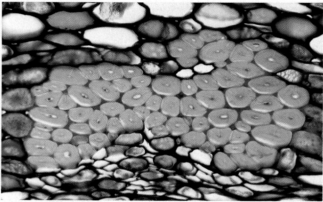

(b) Fibers

Figure 3.3 Sclerenchyma cells provide rigid support.
(a) Thick, hard cell walls of sclereid cells, also known as *stone cells,* provide the structural support in fruit pits and nutshells and also give a rough texture to the flesh of a pear. **(b)** This stem has long, tapering sclerenchyma cells called *fibers,* which provide sturdy support for a plant. Sclerenchyma cells are usually 20 to 50 μm in width and can be up to 70 mm in length in the case of fibers.

plants—many parenchyma, collenchyma, and sclerenchyma cells are modified into more specialized cells for transport, support, and protection. In contrast, mosses and other bryophytes do not have as many types of highly specialized cells for these functions. Here we will look at the structures and functions of the more specialized cells in vascular plants.

A group of cells with a common function is called a **tissue.** Unlike zoologists, botanists distinguish between two types of tissues: simple and complex. A **simple tissue** is composed of one type of cell. For instance, parenchyma, collenchyma, and sclerenchyma cells can each form a simple tissue. A **complex tissue** is composed of several cell types, such as a mixture of parenchyma, sclerenchyma, and water-conducting cells. In plants, various

THE INTRIGUING WORLD OF PLANTS

Flexible Fibers

The strong, elastic nature of fibers makes them economically useful. Bast fibers, also called soft fibers, contain only small amounts of lignin and come from the phloem tissue of stems of plants called dicots, such as flax (*Linum usitatissimum*), hemp (*Cannabis sativa*), and jute (*Corchorus capsularis*). (You will read about phloem later in the chapter.) Flax is typically used to make paper and linen, while jute is used for rope. Since hemp can be made into a variety of products, including paper, cloth, and rope, recent decades have seen calls for more widespread hemp cultivation to save forest resources. However, the varieties of the species *Cannabis sativa* cultivated for fiber look similar to those used to cultivate marijuana as a drug. Therefore, law enforcement officials worry that someone claiming to be growing hemp for rope, fabric, or paper might actually have other marketing plans.

Plant leaves yield another type of fiber known as *hard fibers,* which are typically stiffer and coarser than bast fibers because their cell walls contain more lignin. Hard fibers typically come from the xylem tissue of plants called *monocots.* (You will read about xylem later in the chapter.) Artisans use hard fibers to make strong, coarse ropes (cordage), such as those made in the Philippines from fibers of abaca leaves or Manila hemp (*Musa textilis*). Hard fibers also can be obtained from leaves of pineapple (*Ananas comosus*) and sisal (*Agave sisalana*). Many cultures have adapted leaves or stems of various plants to produce fiber and fabric.

Flax plant. Flax has been grown around the world since about 3,000 BC.

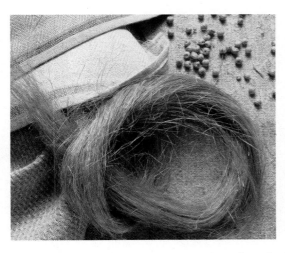

Flax. Seeds, fibers, and linen. Now most paper is made out of cellulose from ground up trees. However, flax fibers used to be the common source for both paper and linen cloth.

---

simple and complex tissues are organized into three functional units known as **tissue systems** that are continuous throughout the plant. Figure 3.4 shows the three tissue systems in vascular plants: the dermal tissue system, the vascular tissue system, and the ground tissue system, all of which originate from meristematic cells.

The dermal tissue system forms the plant's outer protective covering

The **dermal tissue system** (from the Greek *derma,* "skin") is the outer protective covering of a plant. Dermal tissue begins as parenchyma cells, which are then modified to form various types of cells that protect the plant from physical damage and desiccation, or drying out. In the first year of growth, plants typically have one layer of dermal tissue, called the **epidermis,** with the cells close together to produce a secure boundary. In plants that live more than one growing season, the epidermis of the stems and roots is replaced by a protective tissue known as **periderm** (from the Greek words meaning "the skin around"), which mainly consists of nonliving cork cells that protect the plant from predators and water loss. As you will see in Chapter 5, woody plants typically have multiple layers of periderm.

Dermal cells can become modified through the production of hair-like extensions called **trichomes** (TRI–cohms). The word *trichome* comes from the Greek word for hair. Examples are the hairs that extend from leaves and cotton seeds (see the *Plants & People* box on page 55).

Dermal tissue also helps to control the exchange of gases, including water vapor. For example, many plant stems and leaves produce a **cuticle,** a layer outside the cell wall composed of wax and a fatty substance called cutin, which helps limit water loss. Drought-tolerant plants sometimes produce large quantities of wax, which can play a crucial role in a plant's ability to survive and flourish in regions prone to drought and wind. The primary

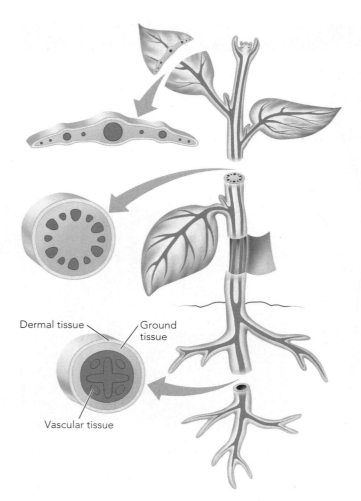

Figure 3.4 The three tissue systems. The dermal system consists of the protective outer coating of tissue. The vascular tissue system contains the tissues that conduct water, minerals, and food. The ground tissue system usually fills in between the dermal tissue and the vascular tissue. Notice that each system is continuous throughout all parts of the plant.

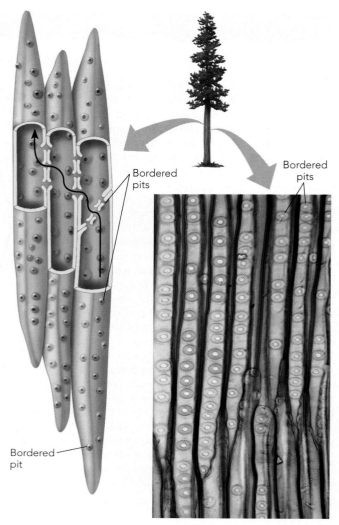

Figure 3.5 Tracheids. In almost all nonflowering plants, such as conifer trees, the conducting cells of xylem are tracheids. Some flowering plants produce conducting tissue consisting solely of tracheids, but most produce a mixture of tracheids and vessels. Tracheids are stacked with pits allowing for movement of water and minerals from one tracheid to the other.

dermal tissue in the aboveground parts of a plant also includes pores that can open to allow for gas exchange or close to prevent water loss (Chapter 4).

The vascular tissue system conducts water, minerals, and food

The **vascular tissue system**—a continuous system of tissues that conduct water, minerals, and food—consists of two complex tissues: xylem and phloem. **Xylem** (ZY–lem) brings water and mineral nutrients from the roots to the rest of the plant. **Phloem** (FLO–em) moves sugars and other organic nutrients from the leaves to the rest of the plant. In other words, phloem carries the food produced

by photosynthesis. The contents transported by xylem and phloem are known as **sap.** In Chapter 4, you will see how xylem and phloem are organized into bundles and other arrangements.

Xylem: water-conducting tissue

The xylem of all vascular plants contains **tracheids** (TRAY–key–ids), which are long cells with tapered ends (Figure 3.5). Tracheids were the first water-conducting cells to evolve in vascular plants and are typically the only type of water-conducting cell in ferns, conifers, and most other

Cotton through the Centuries

Trichomes can serve a variety of plant needs, such as shading leaves from excess light, dispersing wind, protecting from insects, aiding in nutrient uptake, and spreading seeds. In the case of cotton, trichomes have served both plant and human needs, playing a significant role not only in the evolution of a specific plant but also in human history. Cotton trichomes arise from the outer cell layer of cotton seeds and resemble long, hollow hairs. They stick to animal fur and other surfaces, helping transport seeds to new locations. For humans, hundreds of these trichomes can be twisted together to make thread, which is then woven into cloth.

Native to several continents, cotton has played a pivotal role in human history for more than 4,000 years. For thousands of years, cotton thread was made by hand, a process typically beginning by wrapping an unorganized mass of cotton fibers around a stick called a *distaff*. The thread was made by pulling fibers out of the distaff and attaching them to a wooden rod called a *spindle*, which was used to straighten, spin, and bind the threads together. Weavers then attached the thread to a wooden frame called a *loom*. The cloth was slow to make, not strong, and of uneven consistency.

A major problem with the production of cotton cloth was the slow and tedious removal of seeds from the cotton fibers, a process known as *ginning*. Eli Whitney, an American, solved this problem in the late 1700s by inventing the mechanical cotton gin, a machine that drags metal claws through cotton to remove debris.

The cotton gin had an impact on slavery in the United States. Slavery in the Americas had gotten its primary impetus from the need for workers on sugar plantations in the Caribbean and in the southern United States. By the early 1800s, slavery in the United States was in decline. The cultivation of sugar beets in Europe reduced the market for American sugar and led to a decline in demand for slave labor. With the cotton gin, however, southern planters turned increasingly to cotton, and the labor-intensive crop created a greater demand for slaves. By 1820, cotton had overtaken both sugar cane and tobacco as the South's main crop. After the abolition of slavery, cotton farming declined until industrial advances once again revitalized its production.

The cotton and textile industries were also linked to the widespread use of child labor in Great Britain and the United States. In 1769, an Englishman named Richard Arkwright patented a spinning frame that made 128 strands of uniform thread at once and required only a few unskilled people to operate. As his spinning frame became larger and more complex, however, Arkwright reorganized production to increase the number of unskilled workers and have them perform one task over and over. Unfortunately, these workers were usually children, beginning the massive use of child labor that continued well into the 1800s in Great Britain and was vividly illustrated and condemned in the novels of Charles Dickens. Child labor continued in the United States until the early 1900s.

Child labor. During the 1800s, Great Britain and the United States relied heavily on child labor to produce cotton products, a practice that continued in the United States until the early 1900s. This youth worked in a fabric mill in New England.

---

nonflowering vascular plants. Most botanists believe that tracheids are derived independently from meristematic cells, but some consider them to be a highly differentiated type of sclerenchyma. Like sclerenchyma cells, tracheids are dead at maturity, with only the cell walls remaining. The thick secondary cell walls of a tracheid surround the space that was previously occupied by the living cell contents.

Tracheids align with each other to form a continuous water-conducting system. The secondary cell wall of a tracheid has thinner regions called **pits,** in which only the primary wall is present. Pits in adjacent tracheids are

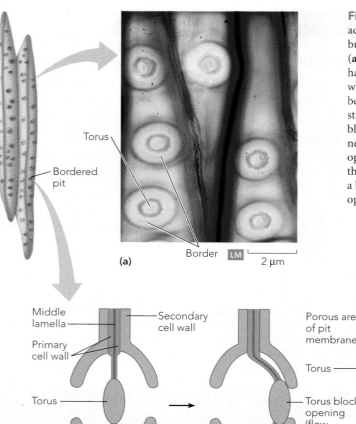

Torus

Bordered pit

Torus

Border LM |⎯ 2 µm

(a)

Figure 3.6 Pits are gaps in the secondary cell wall. In addition to regular pits, many plants have bordered pits, in which bulges in the secondary cell wall make the opening narrower. (**a**) In conifers and some primitive angiosperms, bordered pits have a thick area, called a *torus,* in the center of the pit membrane, which acts like a valve to control the flow of water and minerals between cells. When the torus is in the center, the flow is unobstructed, but when the pit membrane moves to one side, the torus blocks the opening. (**b**) This drawing shows the torus and the thinner surrounding area of the pit membrane, as viewed through the opening. As you can see, water and minerals can easily pass through the porous areas of the pit membrane. (**c**) This micrograph shows a bordered pit, viewed from the perspective of looking through the openings.

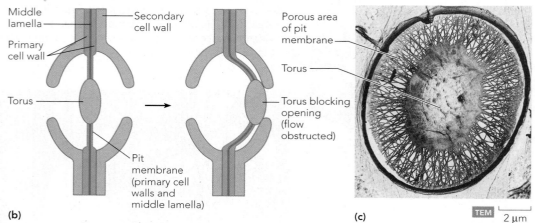

Middle lamella

Primary cell wall

Secondary cell wall

Torus

Pit membrane (primary cell walls and middle lamella)

(b)

Porous area of pit membrane

Torus

Torus blocking opening (flow obstructed)

(c) TEM |⎯ 2 µm

usually aligned, allowing water and minerals to flow from one tracheid to another one above, below, or next to it. In some plants, pits are often bordered by bulges in the secondary cell walls, which strengthen the opening and also make it narrower, slowing down the flow. Figure 3.6 shows how the pit membrane, consisting of the porous primary cell walls and the thin middle lamella, regulates the flow through bordered pits. In conifers and some primitive angiosperms, the middle of the pit membrane is a thicker area called a torus that acts like a valve. If the membrane moves to the side, the torus blocks the pit opening, thereby slowing the flow.

In addition to tracheids, the xylem of most flowering plants and a few gymnosperms contains other water-conducting cells called **vessel elements,** which transport water and minerals more rapidly than tracheids (Figure 3.7). Most botanists believe that vessel elements are de-

rived independently from meristematic cells, but some consider them to be a highly differentiated type of sclerenchyma. Like tracheids, vessel elements are dead at maturity, with the cell walls forming hollow tubes, but vessel elements are generally wider, shorter, and less tapered than tracheids. They have the largest diameters of all conducting cells—up to 100 micrometers (µm), compared with 10 µm for tracheids—and can carry about 100 times as much water and minerals as tracheids. Vessel elements lose some or most of their cell walls at each end, leaving perforation plates that allow water to flow through while still providing support. In this way, vessel elements are joined to form a continuous pipe, or **vessel.** Vessel elements also have pits, which allow lateral flow from vessel to vessel.

While vessels move water and minerals more rapidly, they can also pose a danger to the plant, in comparison

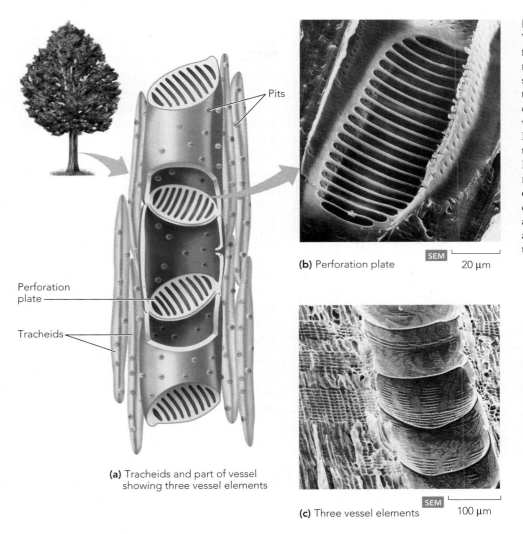

Pits

Perforation plate

Tracheids

(a) Tracheids and part of vessel showing three vessel elements

(b) Perforation plate   SEM   20 µm

(c) Three vessel elements   SEM   100 µm

Figure 3.7 Vessel elements. Vessel elements, found in most flowering plants and a few gymnosperms, are large cells with secondary walls that are linked together to make vessels. As you can see, vessel elements are much wider than the adjacent tracheids. In fact, they can be around ten times the diameter of tracheids. In vessels, the end walls of the elements are incomplete, consisting of some secondary wall material called a *perforation plate*, which allows water to flow through while also providing some support for the vessel.

with tracheids. If an air bubble forms in a single tracheid, the flow of water is broken in only that one cell, and overall water movement is scarcely slowed. In moving through tracheids, the water attaches to the cell wall of a relatively small cell, so there is less chance that the flow will be interrupted, and probably only one tracheid will be affected. In a vessel, however, the column of water is less well supported by the secondary cell walls in the vessel elements because they are wider, making the formation of air bubbles more likely. If even a single vessel element is blocked by an air bubble, the entire vessel may cease to conduct water. Also, vessels are more vulnerable to damage from freezing because ice crystals in one vessel element will block flow in the entire vessel, whereas ice has to form separately in each tracheid.

The cell structures of tracheids and vessel elements enhance both support and conduction. The rigid sec-ondary walls provide increased support, while transport is facilitated by the cells being hollow and having perforated walls.

Phloem: food-conducting tissue

In vascular plants, other types of specialized cells form phloem, which transports food. The phloem of flowering plants consists of cells called **sieve-tube members,** also called *sieve-tube elements* (Figure 3.8). Unlike tracheids and vessel elements, sieve-tube members remain alive and active at maturity. Stacked end-to-end to form **sieve tubes,** they conduct organic nutrients from the leaves to other parts of the plant. Most botanists think sieve-tube members developed independently from meristematic cells, but some believe them to be a highly differentiated type of parenchyma.

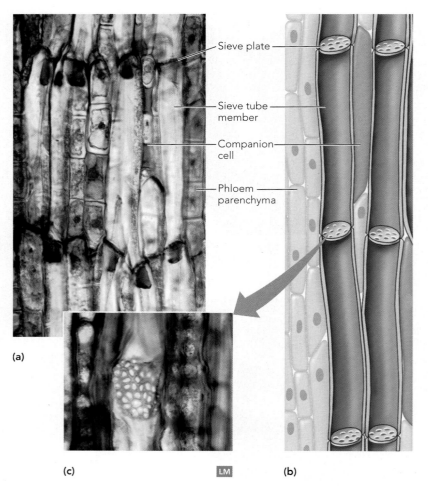

(a)

(c)   LM   (b)

Sieve plate

Sieve tube member

Companion cell

Phloem parenchyma

Figure 3.8 Sieve-tube members. In flowering plants, the cells that conduct food are called *sieve-tube members*. Each sieve-tube member has an associated nonconducting companion cell, which forms from the same parenchyma cell as the sieve-tube member. The companion cell provides proteins for the sieve-tube member. **(a)** This light micrograph shows sieve-tube members and companion cells. **(b)** This diagram shows stacked sieve-tube members, with adjoining companion cells and phloem parenchyma. **(c)** Light micrograph of a sieve plate.

A distinctive feature of sieve-tube members is the presence of **sieve plates,** cell walls with membrane-lined pores that allow materials to pass from cell to cell without crossing the plasma membranes and cell walls. Essentially, sieve-tube members form a continuous connection of cytoplasm from the top to the bottom of the plant. Another feature of a sieve-tube member is that at maturity it has no nucleus, depending instead on an adjacent **companion cell,** which has a nucleus and can therefore supply proteins for the sieve-tube member. The companion cell forms from the same parenchyma cell as the sieve-tube member.

When disturbed or damaged, sieve-tube members form a carbohydrate molecule called **callose** around the components of the cell wall in each sieve plate. Meanwhile, the pores in the sieve plates may become plugged with a substance called P-protein (*P* for *phloem*), which may also be a response to injury by preventing loss of cell contents.

In nonflowering vascular plants such as ferns and conifers, phloem consists of a more primitive type of water-conducting cell called a **sieve cell.** Rows of sieve cells function much like sieve-tube members, but the ends of sieve cells lack sieve plates and overlap rather than forming continuous tubes. This difference is somewhat like the difference between the overlapping tracheids and the continuous tube of vessel elements. Like sieve-tube members, sieve cells lack a nucleus when they are mature. Each sieve cell has an associated albuminous cell, which has a nucleus and appears to serve the same function as the companion cell does for a sieve-tube member.

The food-conducting cells and the cells that support them are not the only components of phloem tissue. In addition, phloem contains parenchyma and fibers.

Ground tissue usually forms between dermal and vascular tissues

The **ground tissue system,** also called the *fundamental tissue system,* consists of all the tissues other than the vascular tissue system and the dermal tissue system. It includes three simple tissues: parenchyma, collenchyma, and sclerenchyma, although parenchyma predominates. The cells in ground tissue often carry out photosynthesis and store nutrients. For example, the photosynthetic parts of a plant consist mostly of parenchyma tissue and are part of the ground tissue system.

The ground tissue system also fills up the space not occupied by either the dermal or vascular tissues. Usually ground tissue forms between the dermal and vascular tissues, where it is known as **cortex.** However, sometimes it also appears to the inside of the vascular tissues, where it is called **pith.**

Table 3.1 summarizes the main types of tissues and cells. As you will see in the next chapter, the distribution of ground and vascular tissues can vary, depending on the part of the plant or type of plant.

Table 3.1 Tissue systems, tissues, and cell types

Tissue Systems	Tissues	Cell Types	Locations of Cell Types	Main Functions of Cell Types
Dermal	Epidermis	Parenchyma and modified parenchyma	Outer layer of cells throughout the primary plant	Protection; prevention of water loss
	Periderm	Cork and parenchyma	Outer layers of cells throughout woody plants	Protection
Ground	Parenchyma	Parenchyma	Throughout the plant	Photosynthesis; food storage
	Collenchyma	Collenchyma	Under stem epidermis, near vascular tissue; along veins in some leaves	Flexible support in primary plant body
	Sclerenchyma	Fiber	Throughout the plant	Rigid support
		Sclereid	Throughout the plant	Rigid support and protection
Vascular	Xylem	Tracheid	Xylem of angiosperms and gymnosperms	Conduction of water and dissolved minerals; support
		Vessel element	Xylem of angiosperms and a few gymnosperms	Conduction of water and dissolved minerals; support
	Phloem	Sieve-tube member	Phloem of angiosperms	Conduction of food and other organic molecules
		Companion cell	Phloem of angiosperms	Metabolic support for sieve-tube members
		Sieve cell	Phloem of gymnosperms	Conduction of food and other organic molecules
		Albuminous cell	Phloem of gymnosperms	Metabolic support for sieve cells

Section Review

1. How does dermal tissue protect the plant?
2. What are the two complex tissues that form the vascular system, and how do they differ?
3. What are the purposes of the ground tissue system?

An Overview of Vascular Plant Organs

Simple and complex tissues form structures known as *organs*. An **organ** consists of several types of tissues adapted as a group to perform particular functions. Vascular plants have three types of organs: stems, leaves, and roots. Bryophytes and some seedless vascular plants have structures that can be called "stemlike," "leaflike," and "rootlike" but are not considered to be true stems, leaves, and roots.

The list of plant organs is much shorter than the list of animal organs. While some botanists refer to reproductive structures such as seeds, cones, and flowers as organs, most reserve the term solely for stems, leaves, and roots. Chapter 6 will discuss different reproductive structures. Here we will look briefly at the general functions of stems, leaves, and roots, with more detailed discussions to follow in Chapters 4 and 5.

Stems position leaves for maximum photosynthesis

A **stem** is any part of a plant that supports leaves or reproductive structures. Stems can vary greatly in size, such as a slender stalk supporting a small flower or a huge tree trunk dozens of feet in diameter and hundreds of feet high. In woody plants, what we commonly call *branches* are shorter stems attached to longer ones. Regardless of size, all stems display leaves in the best position for photosynthesis. Leaves are botanical sun catchers and food producers, and stems help ensure their success by providing not only support but also pathways for the transport of water, minerals, and food between leaves and roots. In addition, the very height of many stems can help protect leaves from the best efforts of many predators. The stems of woody plants develop bark that protects against predators and physical trauma.

Leaves function in both photosynthesis and transpiration

The earliest land plants were leafless photosynthetic stem systems. Eventually their flattened stems grew close together, evolving into continuous structures known as leaves. The **leaf** is the main photosynthetic organ of modern plants. In fact, to a casual observer, many plants seem to be principally composed of leaves.

As in the organs of all organisms, the structure of leaves closely relates to function. Since photosynthesis is usually the primary function of leaves, they are usually flat to maximize the surface area exposed to sunlight. In some cases, though, other functions are more important. In the desert, for instance, water retention drives the adaptation of cactus leaves, which evolved as thin spines that lose less water to evaporation. Most photosynthesis in cacti occurs in the thick stems, which also store water (Figure 3.9).

Through their veins, which contain both xylem and phloem, leaves are an extension of the plant's vascular tissue system. A beautiful and characteristic pattern of veins highlights the leaves of each plant species. The veins receive water and minerals from the stems and transport food to the stems.

Leaves not only conduct water but also provide most of the pressure that actually forces the water through the plant's body. Some pressure is a "push" from the roots, but a "pull" from the leaves is the main force that moves water in most plants. This pull comes from loss of water through the pores in leaves. This process of **transpiration,** the evaporation of water from a plant, pulls water and mineral nutrients up from the roots to the leaves. The chemistry of water helps explain how transpiration moves large amounts of water through a plant (Chapter 10).

Roots anchor the plant and absorb water and minerals

A **root** has two main functions: anchoring the plant in the soil and absorbing water and minerals. The absorp-

Figure 3.9 Cactus stem and leaf adaptations. The stem and leaves of cacti are well adapted for the hot, dry climate of the desert. The thick stems serve as water storage, while the small surface area of the thin spines (leaves) loses less water to evaporation.

tion takes place only near the very tips of roots, through trichomes called **root hairs,** which are tiny extensions of root epidermal cells that greatly increase the amount of root surface area (Figure 3.10). Even in large trees, the majority of the roots serve simply to get root hairs to regions in the soil where moisture and nutrients are available. In addition to anchorage and absorption, many roots store food for the plant. Some, such as carrots and sweet potatoes, are also important human foods. Although roots frequently store food, they do not actually produce it. They are usually nonphotosynthetic and in fact usually grow away from the light.

As noted previously, the vascular tissue system is continuous between roots, stems, and leaves, with all three organs depending on each other. Roots need sugars and other organic nutrients produced in leaves, while stems and leaves need water and minerals obtained by the roots. Xylem carries water and minerals from the roots, while phloem carries food from the leaves to the rest of the plant. In the next two chapters, we will look at the different arrangements of xylem and phloem in roots and stems.

Section Review

1. Describe the purpose of stems.
2. What are the two main functions of leaves?
3. Explain how the vascular tissues of roots, stems, and leaves are interrelated.

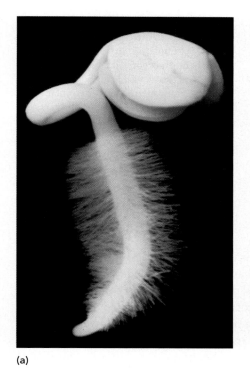

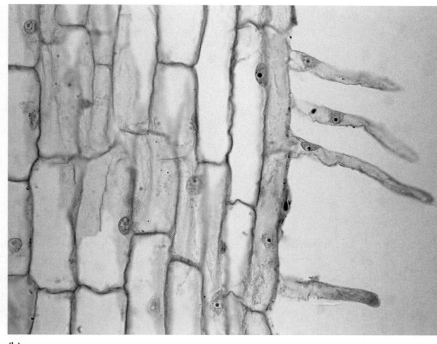

(a)   (b)

Figure 3.10 Root hairs. Root hairs are found within a centimeter of the tips of roots just above the region of cell elongation. They are specialized to take up both water and ions needed by the plant. (**a**) Here are root hairs on a radish seedling. (**b**) This micrograph provides a close-up view of root hairs.

An Overview of Plant Growth and Development

At every structural level—from cells to tissues to organs—a plant is a complex, dynamic organism. As you know, plants and other multicellular organisms develop from a single cell as the result of many cells dividing, elongating, and becoming specialized for different functions. To see in general how a typical plant grows, we will look at seed plants, which comprise about 90% of all living plants. As you may recall, seed plants are either gymnosperms or angiosperms. In gymnosperms (plants with "naked seeds," such as conifers), the seeds are exposed, typically on cones. In angiosperms (plants with "seeds in a container"), the seeds are enclosed within fruits. The vast majority of seed plants are angiosperms—the flowering plants.

The body of a typical plant can be described as having two connecting systems: a root system and a shoot system. The **root system** consists of all the roots, which are usually belowground. The **shoot system** consists of all the stems, leaves, and reproductive structures, which are usually aboveground. A **shoot** is any individual stem and its leaves, as well as any reproductive structures that extend

from the stem, such as flowers. A shoot that has leaves but no reproductive structures is called a *vegetative shoot*. Here we will focus on a brief overview of how the stems, leaves, and roots of a typical seed plant develop, starting from an embryo inside a seed. In Chapter 11 we will look in detail at the early stages of plant development.

Embryos give rise to stems, leaves, and roots of adult seed plants

A seed plant begins as a fertilized egg, or zygote, that grows into an embryo within a seed. The typical seed plant embryo includes the following embryonic "organs" that develop into the roots and shoots (Figure 3.11):

◆ one or more "seed leaves" called **cotyledons,** which are usually the largest, most visible parts of the embryo. Flowering plant seeds have either one or two cotyledons. Gymnosperm seeds have two or more cotyledons. In many plants the cotyledons store food for the germinating seed and may be thickened or "fleshy."

◆ an embryonic "root" called a **radicle** (from the Latin *radix,* "root"), which is always prominent.

◆ an embryonic "shoot" called a **plumule** (from the Latin *plumula,* "soft feather"), which is usually scarcely

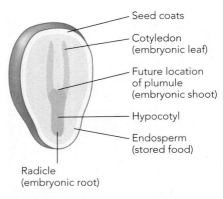

(a) A dicot seed (tobacco)

Seed coats
Cotyledon (embryonic leaf)
Future location of plumule (embryonic shoot)
Hypocotyl
Endosperm (stored food)
Radicle (embryonic root)

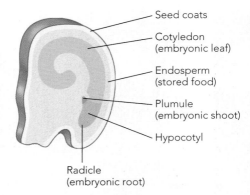

(b) A monocot seed (onion)

Seed coats
Cotyledon (embryonic leaf)
Endosperm (stored food)
Plumule (embryonic shoot)
Hypocotyl
Radicle (embryonic root)

Figure 3.11 Dicot and monocot seeds. The embryonic shoot (plumule), embryonic root (radicle), and embryonic leaves (cotyledons) are clearly visible in the embryo. Gymnosperm seeds are similar in structure to dicot seeds but may have more cotyledons.

developed. The plumule gives rise to the **epicotyl,** a portion of the embryonic "stem" located over the cotyledon—hence its name. In some seeds the plumule cannot yet be identified, while in others it is a distinct structure. The plumule develops into an epicotyl as the seed sprouts.

◆ a portion of the embryonic "stem" called a **hypocotyl,** which is located under the cotyledon and above the radicle and may be either short or long.

The embryo is surrounded by a storage tissue, called **endosperm** in flowering plants, that breaks down to provide nourishment to the developing embryo. Sometimes endosperm also supplies nutrition to the sprouting seed.

The seed may remain dormant for a time before **germination**—the process of sprouting. Environmental conditions such as temperature, light, or water may initiate germination by triggering production of **hormones** (from the Greek *hormon,* "to stir up"), organic compounds that cause developmental or growth responses in target cells. Germination occurs when the radicle breaks through the outer layers of the seed. Once in the soil, the radicle begins to take up water and mineral nutrients, and the process of forming a seedling has begun. Cotyledons, and sometimes endosperm, contain stored energy and organic building blocks in the form of starch, proteins, and lipids. These macromolecules break down to nourish the germinating seed until the plumule has begun photosynthesis. During germination, the seed often relies on food stored in the cotyledons. These embryonic seed leaves eventually shrivel and drop from the seedling, which then relies on photosynthesis and minerals from the soil for nutrition.

In flowering plants, germination results in different types of seedlings, depending on the numbers of cotyledons and the length of the hypocotyls. In fact, flowering plants have traditionally been divided into two main types—monocots and dicots—based in part on the num-

ber of cotyledons in their embryos. **Monocots** are flowering plants that have only one cotyledon. Some examples are orchids, lilies, palms, onions, and members of the grass family, such as corn, rice, and wheat. **Dicots** are flowering plants that have two cotyledons. Most flowering plants, especially most large flowering plants, are dicots. Some examples of dicots are beans, peas, sunflowers, roses, and oak trees. In Chapter 4, we will look at how adult monocots and dicots also differ in root, stem, and leaf structure.

Recent DNA comparisons have revealed that even though the plants traditionally called *dicots* have structural similarities, they are not all closely related and therefore should not be considered one group in terms of evolution. However, most of them, now called **eudicots** ("true" dicots), are indeed one group in both evolution and structure. Since the traditional distinction between dicots and monocots remains useful for describing structural differences, we will use those terns when comparing flowering plant structures.

Meristems enable plants to continue growing throughout their lives

As you know, plant growth differs fundamentally from that of animals. Most animals—including all mammals—have determinate (limited) growth. As animals change from infants to adults, every part of the body becomes larger. In humans, growth continues through the teenage "growth spurt" but then tapers off and usually ceases in an adult. Cell division, growth, and differentiation continue to occur for normal cell replacement, the production of white and red blood cells, wound healing, and the formation of eggs and sperm, but they do not occur to increase the size of adult animals. For instance, we may put on or take off weight and look older, but fundamentally our bodies do not increase in size or change radically in proportion or number of organs.

In contrast, meristems enable plants to have potentially indeterminate growth—the ability of an organism to continue growing as long as it lives. Through the action of apical meristems at the tips of roots and shoots, roots and stems grow longer, and new leaves continue to appear (Figure 3.12). In woody plants, lateral meristems cause roots and stems to become thicker. While some plants do grow throughout their lives, in many plants growth ceases when a genetically determined size is attained. Also, when apical meristems become flower-producing meristems, their indeterminate growth ceases.

Starting with the growth of a zygote into an embryo, meristems produce cells that develop into the three main tissue systems of a vascular plant: the dermal tissue system, the vascular tissue system, and the ground tissue system. Here we will look in general at the two main types of growth they produce: primary growth and secondary growth.

Apical meristems initiate primary growth that makes roots and shoots longer

Primary growth is growth in the length of roots and shoots, which is caused by apical meristems at the tip, or **apex** (plural, *apices*), of each root or shoot. These meristematic cells are organized into shoot and root **apical meristems.** In many seedless vascular plants, such as horsetails and some ferns, the apical meristem consists of one initial that is shaped like an inverted pyramid and divides repeatedly along its three sides to produce tissues. In seed plants, an apical meristem consists of around a hundred or few hundred initials that form a microscopic dome about 0.1 mm or less in diameter at the tip of a root or stem. As a result of cell divisions in the apical meristems and subsequent cell growth and development, a seedling becomes an adult plant. The body of the plant produced by shoot and root apical meristems is called the **primary plant body.**

Suppose you tied a rope around the stem of a young plant. As the plant grew taller, the rope would stay in the same vertical location because new stem growth only comes from the shoot apical meristems at the tip of each shoot. In the same way, new root growth comes only from the root apical meristems at the tip of each root.

While the root system is growing, shoot apical meristems are lengthening stems and producing leaves at periodic points. Leaves originate as small bulges called **leaf primordia** (singular, *primordium*) on the sides of shoot apical meristems (Figure 3.13). A primordium is a structure in its earliest stage of development. Different species of plants vary in leaf shape and arrangement. A thin, stemlike **petiole** attaches the leaf to the stem at a point called a **node.** Sections of the stem between leaves are

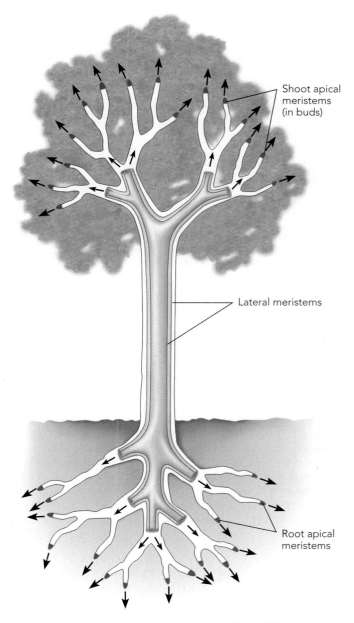

Figure 3.12 Locations of apical and lateral meristems. A plant produces new cells by division in meristems. Cell division in apical meristems at the tips of shoots and roots causes shoot growth and root growth. Lateral meristems, responsible for growth in thickness, occur as concentric cylinders on the inside of woody stems and roots.

known as **internodes.** At each node, a bud forms on the upper angle, or axil, where each petiole joins the stem. Each of these buds, known as an **axillary bud,** consists of an apical meristem and leaf primordia. When an axillary bud begins to grow, it will become a new shoot.

The growth of axillary buds is suppressed by the hormone **auxin,** which is produced in or near apical meristems. This phenomenon is called **apical dominance**

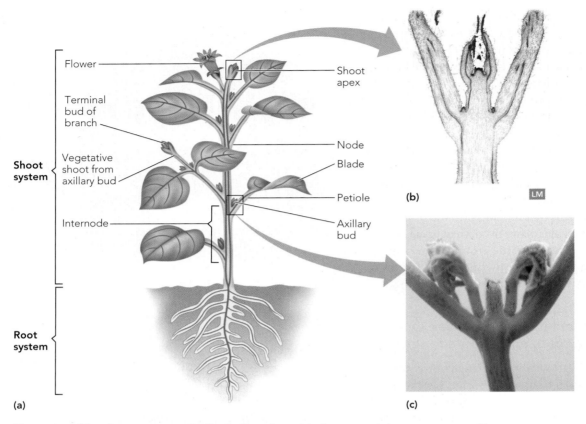

Figure 3.13 The shoot system of a flowering plant. (**a**) The roots and shoots are connected by a continuous vascular system. The shoot system consists of stems, leaves, flowers, and fruits. Leaves are attached to the stem at nodes, which are separated by internodes. Expanded leaf blades are attached to the stem by petioles. The root system of this plant has a primary tap root and lateral roots. (**b**) Leaves arise from bulges of rapidly dividing cells that form on the sides of a shoot apical meristem. The cells elongate into leaf primordia and then into leaves as the apical meristem grows up and away from them. (**c**) Branches form from axillary buds on the upper surface of the angle between the petiole and the stem.

(Figure 3.14). If the apical meristem is injured or re-moved, axillary buds begin to grow. They also frequently develop as the apical meristem grows farther away and the auxin concentration decreases. Since axillary buds develop into branches with their own apical meristems, a plant may have many shoot apical meristems, depending on the amount of branching. An apical meristem may cease leaf production and instead produce cones or flowers.

The numbers and growth of axillary buds determine the ultimate shape of the aboveground portion of a plant. Oak tree branches, for example, spread out fairly evenly in all directions as a result of many equally active shoot apical meristems. By contrast, the apical meristem at the top of fir and spruce trees, which are commonly used as Christmas trees, is much more active than those of the other branches.

Botanists are discovering how genes control the formation of apical meristems

Apical meristems develop early during the heart-shaped stage of plant embryo formation. The future location of the root meristem can be identified at the bottom of the heart, while the future location of the shoot meristem is a very small region of cells between the wings of the heart, which form the cotyledons. By observing mutations in germinating seeds and seedlings, botanists have identified genes that control the formation of apical meristems, and thereby root and shoot growth in general. Names assigned to these genes—such as *scarecrow, hobbit,* and *pinnochio*—often relate to their effect on growth, as well as the sense of humor of the discovering botanist.

By observing seedling growth, botanists have discovered that some mutations cause an increase, reduction, or

Axillary buds

"Stump" after removal of apical bud

Lateral branches

Figure 3.14 Axillary buds and apical dominance. Auxin produced by the principal shoot apex causes dormancy in axillary buds. When the apex is removed, auxin production ceases and the axillary buds begin to grow. Cytokinins produced in the roots also stimulate axillary bud growth. Even if the apical meristem remains, axillary buds distant from the shoot apex will begin to grow due to lowered auxin concentrations.

absence of either ground tissue or vascular tissue. Others appear to control the rate at which meristematic cells form, resulting either in apical meristems that are 2 to 1,000 times larger than normal or apical meristems that function for only a short time before disappearing. Mutations can also result in no shoot or root growth. By studying these mutations, botanists are learning more about precisely how genes control primary root and shoot growth.

Apical meristems give rise to primary meristems, which produce primary tissues

Earlier we looked at the basic types of plant cells and tissue systems. Now, let us examine briefly how they relate to regions of cell division known as **primary meristems,** which produce the tissues of the primary plant body (Figure 3.15). Root and shoot apical meristems give rise to primary meristems known as the *protoderm, procambium,* and *ground meristem.* The **protoderm** produces the primary dermal tissue system, or epidermis. The **procambium**

produces the primary vascular tissue system, consisting of xylem and phloem. The **ground meristem** produces the ground or fundamental tissue system, consisting of those parts of the plant that are neither conducting tissue nor dermal tissue. The cells produced by the primary meristems elongate and differentiate into the cells that make up the primary tissues. In the next chapter you will see how apical meristems and primary meristems produce new growth.

Secondary growth from lateral meristems makes roots and stems thicker

Many plants that live beyond one growing season are woody plants, such as trees and shrubs. What makes these plants woody is the development of **lateral meristems,** also known as *secondary meristems,* which cause a thickening of the stems and roots (see Figure 3.12). Lateral meristems are single-cell layers of meristematic cells that form cylinders running lengthwise along a stem or root. These meristematic cells were formerly differentiated

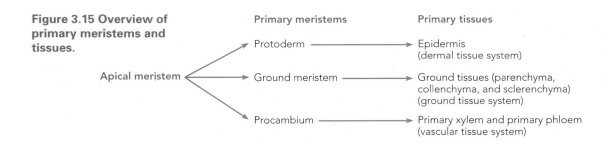

Figure 3.15 Overview of primary meristems and tissues.

Primary meristems

Apical meristem

Protoderm → Epidermis (dermal tissue system)

Ground meristem → Ground tissues (parenchyma, collenchyma, and sclerenchyma) (ground tissue system)

Procambium → Primary xylem and primary phloem (vascular tissue system)

Primary tissues

(a)

(b)

Figure 3.16 Classification of plants by length of life. Plants are classified as annuals, biennials, or perennials, depending on how long they live and when they reproduce. **(a)** Sunflowers are a good example of an annual. Plants do not survive the winter, and new plants are initiated the following season from seeds. **(b)** The foxglove is a typical biennial in which flowers appear during the second year of growth. The first year's growth is vegetative. **(c)** Trees are woody perennials. This photo shows a deciduous tree that loses its leaves each season.

(c)

parenchyma cells that have reverted to become undifferentiated meristematic cells.

Each lateral meristem produces new growth toward both the inside and outside of the cylinder, thereby widening a stem or root. This growth in thickness produced by lateral meristems is called **secondary growth.** Secondary growth is common in conifers and other gymnosperms and also in dicots but is rare in monocots. In

Chapter 5 we will look in more detail at how lateral meristems produce secondary growth.

Some plants live for one growing season, while others live for two seasons or longer

Plants that have significant secondary growth are known informally as *woody plants,* while those with little or no secondary growth are known as **herbaceous** plants. Both woody plants, such as trees and shrubs, and herbaceous plants can live for more than one growing season. However, even though meristems enable indeterminate growth, plants do not live forever. Depending on how long they live, they can be categorized into three distinct groups, typically determined by growing seasons: annuals, biennials, and perennials (Figure 3.16).

An **annual** is a plant that completes its life cycle during a single growing season, which in some annuals is less than a year. Annuals must be reinitiated from seeds or cuttings each growing season and are typically herbaceous. Marigolds, beans, and corn are examples of annuals. Many plants that are annuals in climates with freezing winters can grow for many years in tropical or subtropical climates.

A **biennial** is a plant that usually requires two growing seasons to complete its life cycle. The primary growth of stems, leaves, and roots occurs during the first season. During the second growing season, the plant produces flowers, sets seeds, and dies. Biennials are typically herbaceous and include such examples as carrots, beets, and cabbage. Most gardeners do not see the second season of growth because they harvest the plants during the first year.

A **perennial** is a plant that grows for many years. It may flower each year or only after many years. Most perennials are woody plants like trees, but perennials also include herbaceous plants like lilies and many grasses. In either case, enough of the plant survives the winter to reestablish itself the following spring.

Section Review

1. How does plant growth differ from animal growth?
2. How do shoot apical meristems affect the shape of a plant?
3. How does primary growth differ from secondary growth, and what are herbaceous plants?
4. What is the difference between annuals, biennials, and perennials?

SUMMARY

Basic Types of Plant Cells

Plant growth originates with meristems, groups of undifferentiated cells that can divide indefinitely, allowing a plant to continue growing its whole life. The three basic types of differentiated cells are parenchyma, collenchyma, and sclerenchyma cells.

Parenchyma cells are the most common type of living differentiated cell (p.50)
Usually spherical, cubical, or elongated, parenchyma cells are alive at maturity, with thin primary walls and typically no secondary wall. Parenchyma cells serve for photosynthesis and storage.

Collenchyma cells provide flexible support (pp. 50–51)
Alive at maturity and lacking secondary walls, collenchyma cells can attain a variety of shapes and have thicker primary walls than parenchyma cells. Parenchyma and collenchyma cells must be filled with water to become turgid enough for support.

Sclerenchyma cells provide rigid support (pp. 51–52)
Sclerenchyma cells are usually dead at maturity and have secondary walls, often hardened with lignin. They can provide support without being filled with water. There are two main types: fibers and sclereids.

Tissues of Vascular Plants

A tissue—a group of cells that performs a function—may be simple (one cell type) or complex (multiple cell types).

The dermal tissue system forms the plant's outer protective covering (pp. 53–54)
A single layer of epidermis is replaced during the second year of growth by periderm, mainly composed of nonliving cells. Dermal cells can be modified into hairlike trichomes. The aboveground dermal tissue often forms a waxy protective layer called a *cuticle*.

The vascular tissue system conducts water, minerals, and food (pp. 54–58)
The vascular system consists of two complex tissues: xylem (for transporting water and minerals) and phloem (for transporting food). All xylem contains tracheids, which align to conduct water through pits of cell walls. Most flowering plant xylem also has vessel elements, which conduct water more rapidly. In flowering plants, phloem consists of sieve-tube members, which form sieve tubes. Phloem of nonflowering vascular plants has rows of overlapping sieve cells.

Ground tissue usually forms between dermal and vascular tissues (p. 58)
Ground tissue fills up space not occupied by vascular or dermal tissue, carries out photosynthesis, and stores nutrients.

An Overview of Vascular Plant Organs

An organ is a group of several types of tissues that together perform certain functions. Plant organs are stems, leaves, and roots.

Stems position leaves for maximum photosynthesis (pp. 59–60)
Stems support leaves and reproductive structures, transport water and nutrients, and protect the plant.

Leaves function in both photosynthesis and transpiration (p. 60)
Leaves are the main photosynthetic organ. Through transpiration, they provide most of the pressure that moves water through the plant.

Roots anchor the plant and absorb water and minerals (p. 60)
Roots absorb water and minerals through root hairs. The vascular system is continuous between the roots, stems, and leaves.

An Overview of Plant Growth and Development

A typical plant has a root system (usually belowground) and a shoot system (usually aboveground) consisting of all stems, leaves, and reproductive structures.

Embryos give rise to stems, leaves, and roots of adult seed plants (pp. 61–62)

Seed plant embryos typically include one or two cotyledons (seed leaves), a radicle (embryonic root), plumule (embryonic shoot), and epicotyl and hypocotyl (parts of the embryonic "stem"). A seed may be dormant before germination. Monocots have one cotyledon, while dicots have two.

Meristems enable plants to continue growing throughout their lives (pp. 62–63)

Animals have determinate growth, basically stopping growth at adulthood. Meristems enable plants to grow throughout their lives, a pattern called *indeterminate growth.*

Apical meristems initiate primary growth that makes roots and shoots longer (pp. 63–64)

Primary growth (growth in length) comes from apical meristems at tips of roots and shoots. Leaves originate on shoot apical meristems as leaf primordia. Each mature leaf is attached to a stem by a petiole at a point called a *node.* Internodes are stem sections between leaves. Auxin suppresses axillary bud growth near the apical meristem. Apical meristems can generate reproductive structures.

Botanists are discovering how genes control the formation of apical meristems (pp. 64–65)

Gene mutations can increase, reduce, or prevent the growth of ground tissue, vascular tissue, apical meristems, and organs.

Apical meristems give rise to primary meristems, which produce primary tissues (p. 65)

The protoderm produces dermal tissue, the procambium produces vascular tissue, and the ground meristem produces ground tissue.

Secondary growth from lateral meristems makes roots and stems thicker (pp. 65–66)

Secondary growth, which occurs in woody plants, is common in gymnosperms and dicots, but rare in monocots.

Some plants live for one growing season, while others live for two seasons or longer (pp. 66–67)

Annuals live for one growing season, biennials for two seasons, and perennials for many years. Annuals and biennials are typically herbaceous (nonwoody) plants, while perennials are typically woody but also include many herbaceous plants.

Review Questions

1. What is the role of meristematic cells, and how do they differ from other cells?
2. Describe the characteristics of parenchyma, collenchyma, and sclerenchyma cells.
3. How do simple and complex tissues differ?
4. What are the purpose and basic structure of the dermal tissue system?
5. What are trichomes?
6. What tissues make up the vascular tissue system, and what is the purpose of each one?
7. Compare and contrast tracheids and vessel elements.
8. How do water-conducting cells differ from cells that transport food?
9. Compare and contrast sieve-tube members and sieve cells.
10. What are the roles of the ground tissue system?
11. Define the term *organ* and identify the main functions of each plant organ.
12. What does it mean to say that a plant's vascular system is continuous?
13. Describe the parts of a typical seed plant embryo.
14. What are monocots and dicots?
15. How do apical meristems affect the growth of a plant?
16. Describe the aboveground structure of a typical plant.
17. How does primary growth differ from secondary growth?
18. What is the difference between an annual, a biennial, and a perennial?

Questions for Thought and Discussion

1. Can meristems make a plant immortal? Explain.
2. If differentiated cells could not become meristematic cells again, what would be the effect on a plant?
3. Why do you think compounds that repel predators are located inside tiny hairs often found on leaves, rather than inside the leaves?
4. Why do you think cells that transport food are living cells, while cells that transport water are dead and hollow?
5. Why do you think that vessel elements did not evolve among the largest gymnosperms, which are the largest of plants?
6. Explain why it is wrong to say that xylem is similar to human arteries and phloem is similar to human veins.
7. Can a seed plant embryo be described as a miniature plant? Explain.
8. Why do you think annuals and perennials can both be found in the same environment?
9. When examining thin cross sections of plant tissues, it is difficult to distinguish small vessels from tracheids. These two cell types are, however, readily distinguish-

able in longitudinal section. Draw labeled diagrams of a tracheid and a vessel of the same diameter to illustrate how they appear in cross and longitudinal sections, making clear their similarity in the former view and their differences in the latter view.

Evolution Connection

Under what environmental conditions and habitat types do you think natural selection might favor plants with (a) an annual life cycle and (b) a perennial life cycle?

To Learn More

Visit The Botany Place at *www.thebotanyplace.com* for quizzes, exercises and Web links to new and interesting information related to this chapter.

Duke, James A., and Steven Foster. *A Field Guide to Medicinal Plants and Herbs of Eastern and Central North America.* Boston: Houghton Mifflin, 2000. A fascinating field guide exploring the wide variety of native plants with medical potential.

Esau, Katherine. *Anatomy of Seed Plants.* 2nd ed. New York: John Wiley and Sons, 1977. This classic book is the standard for modern plant anatomy.

Hogan, Linda, and Brenda Peterson, eds. *The Sweet Breathing of Plants: Women Writing on the Green World.* New York: North Point Press, 2002. The second volume in a trilogy about women and the natural world, containing poems and essays about how women view the place of plants in nature and society.

West, Keith. *How to Draw Plants: The Techniques of Botanical Illustration.* Portland, OR: Timber Press, 1996. This book is full of hints and creative ways of illustrating plants.

Wilson, Edward O., and Burkhard Bilger. *The Best American Science & Nature Writing 2001.* Boston: Houghton Mifflin, 2001. This annual anthology highlights many styles of nature writing.

4

Roots, Stems, and Leaves: The Primary Plant Body

Found in Central America, *Gunnera insignis* is known as sombrilla del pobre or poor man's umbrella.

Roots

Taproot systems usually penetrate more deeply than fibrous root systems

Root development occurs near the root tip

The root cap protects the root apical meristem and helps the root penetrate the soil

Absorption of water and minerals occurs mainly through the root hairs

The primary structure of roots relates to obtaining water and dissolved minerals

Some roots have specialized functions in addition to anchoring the plant and absorbing water and minerals

Roots have cooperative relationships with other organisms

Stems

Botanists have developed zone and cell-layer models to describe stem growth

In primary growth of most stems, the vascular tissue forms separate bundles

A transition region maintains vascular continuity between the root and stem

Leaf primordia form in specific patterns on the sides of shoot apical meristems

Stem variations reflect different evolutionary pathways

Some stems have specialized functions in addition to support and conduction

Leaves

A leaf primordium develops into a leaf through cell division, growth, and differentiation

The leaf epidermis provides protection and regulates exchange of gas

Mesophyll, the ground tissue in leaves, carries out photosynthesis

The vascular tissue in leaves is arranged in veins

Leaf shapes and arrangements have environmental significance

Abscission zones form in the petioles of deciduous leaves

Some leaves have specialized functions in addition to photosynthesis and transpiration

As you read in Chapter 3, growth produced by the apical meristems at the tips of shoots and roots results in what is called the *primary plant body*. The roots, stems, leaves, and reproductive structures of plants are all derived originally from apical meristems. Even the lateral meristems that produce secondary growth, making woody trunks and roots thicker, form from cells produced by apical meristems. Here we will focus on the primary growth of roots, stems, and leaves, exploring how these organs in vascular plants develop and function together.

Vascular plants that live only one year or two years, called *annuals* and *biennials,* often have only primary growth. Those that live longer, called *perennials,* add new primary growth every year, lengthening shoots and roots and also replacing damaged or dead tissue. Although many perennial plants such as trees and shrubs have secondary growth, a few trees, such as palms, consist solely of primary growth. That is, they have no lateral, or secondary, meristems.

In a sense, primary growth is about getting from one place to another. Plants cannot move through their environment as animals do, but they can grow through their surroundings to obtain what they need. Roots absorb water and mineral nutrients by growing through the soil to new regions of resources from regions that have been depleted. Meanwhile, stems and leaves acquire the solar energy needed for photosynthesis by growing toward regions of greater illumination.

The growth of roots, stems, and leaves is interrelated. For example, seedlings usually have more roots than shoots because a germinating seed contains a supply of food but needs water for elongation to allow the photosynthetic shoot to develop. As photosynthesis becomes the main source of the plant's food, the root-to-shoot ratio is reduced. Throughout a plant's life, the ratio of shoots and roots changes as necessary so that the light and CO_2 collected by leaves enter the plant in the correct proportions with the water and minerals collected by roots.

Evolutionary changes have resulted in modified roots, stems, and leaves that have contributed to survival in various environments. In some plants, for instance, enlarged roots and stems have evolved that store water, helping the plants to survive droughts, dry seasons, or dry climates. Roots and stems may also store food, producing reserves that can be used when decreases in photosynthesis result from shading or from leaf damage by wind, cold, disease, and predation. Sometimes modified leaves fulfill unusual roles, as in the case of plants like the Venus's flytrap, which "eats" insects to compensate for lack of nitrogen in the soil.

In short, roots, stems, and leaves do not function in isolation but instead work together, not only in producing, transporting, and storing nutrients but also in providing structural support and protection for the plant. As we examine what makes each of these organs unique, we will also explore how they relate to and depend on each other.

The Venus flytrap is a dramatic example of leaf adaptation.

Roots

The main functions of roots are to anchor the plant and to absorb and conduct water and minerals. Roots must transport water and minerals to the stems and leaves, while also receiving organic molecules that come from the stems and leaves. In addition to absorption and conduction, roots produce hormones and other substances that regulate the plant's development and structure. Here we will look more closely at how roots carry out these functions.

Taproot systems usually penetrate more deeply than fibrous root systems

There are two main patterns of root growth: taproot systems and fibrous root systems. Most dicots and gymnosperms have a **taproot system,** which has a large main root known as a *taproot* that functions to "tap" deep sources of water (Figure 4.1a). The taproot develops directly from the radicle (embryonic root) and produces branch roots called **lateral roots.** The lateral roots then develop their own branch roots, resulting in an extensive root system. Taproots generally penetrate deeply and are therefore well suited for plants that become larger each year, such as trees. However, not all taproot systems are deep. Some large trees, such as conifers, have shallow taproot systems, a pattern typically occurring in mountains because the soil is frequently shallow and lies atop rocks. Nor are plants with taproot systems necessarily large. Many small plants have taproot systems, particularly if they must survive long periods without rainfall. The common dandelion, for example, has a single taproot that can be a foot or more in length.

Seedless vascular plants and most monocots, such as grasses, have a **fibrous root system** (Figure 4.1b). Instead of one main root developing from the radicle, the radicle, or embryonic root, soon dies and numerous roots arise from the lower part of the stem. These are called **adventitious** roots because they do not come from the usual location; that is, they do not come from other roots. In a fibrous root system, no single root stands out as the largest. Each adventitious root forms lateral roots, producing a root system that is typically shallower and more horizontal than a

(a)

(b)

Figure 4.1 Taproot systems and fibrous root systems. (**a**) In taproot systems, as in this dandelion, lateral roots branch off from a larger main root called a *taproot*. Taproot systems are typical of most dicots and gymnosperms. (**b**) A fibrous root system has no main root, and the root system is usually shallower. Fibrous root systems are found in most monocots and seedless vascular plants.

taproot system. This generally shallow structure allows roots to quickly obtain water before it evaporates. Fibrous root systems are fairly common in dry regions, where deeper soil layers may have no moisture. They are also often found in plants that do not grow beyond the first growing season, such as corn. Taproot systems and fibrous root systems represent two different strategies for obtaining water, which is scarce in many locations.

Typically, 50% to 90% of a plant's roots occur in the top 30 cm of soil, or about a foot in depth. However, roots can extend significantly deeper in both taproot and fibrous systems. Among cultivated plants, potato roots regularly reach depths of 0.9 meter (3 feet), while the fibrous root systems of wheat, oats, and barley can reach from 0.9 to 1.8 meters (3 to nearly 6 feet). During well drilling and excavating, desert tree roots have been found as deep as 67 meters (220 feet), although these reports are rare. The roots of even a small, herbaceous plant can easily extend to a radius of 0.9 meters (3 feet) around the stem. Indeed, most desert plants have extensive shallow root systems rather than the extremely deep root system mentioned earlier. Also, with extensive branching, the total length of all of a plant's roots can be quite large in comparison with the surface area of the aboveground portion of a plant. For instance, a single corn plant may have almost 457 meters (1,500 linear feet) of roots. The root system of a plant can sometimes weigh as much as the stem and leaves.

Root development occurs near the root tip

Figure 4.2a shows the basic root structure. Whether a root is long or short, its growth—like the growth of a stem—begins with cell division in the apical meristem near its tip. As you know, what makes a meristem a "fountain of youth" is a small supply of dividing cells called *initials*. The initials of root apical meristems are located within a small spherical center of the meristem, typically around 0.1 mm in diameter, and divide at a relatively slow rate. In root apical meristems, this central area is known as the **quiescent center** (from the Latin word meaning "to rest").

When each initial divides, one daughter cell remains as an initial within the apical meristem, and the other becomes a derivative, ready for cell growth and differentiation. If the apical meristem is damaged or destroyed, a few initials and their derivatives can rebuild it. One experiment found that one-twentieth of the apical meristem of a potato plant was enough to regenerate the entire meristem. Every cell of an apical meristem seems to have a developmental "map" enabling it to reproduce the entire structure.

Cell division in a root or shoot apical meristem produces the derivatives that become the primary meristems:

the protoderm, ground meristem, and procambium, all of which occur within a millimeter or two of the apical meristem itself. As you may recall from Chapter 3, cell divisions within these primary meristems produce the cells that will develop into the various tissues. The protoderm, which gives rise to the epidermis, develops from the outside portions of the apical meristem. The ground meristem, which produces the ground tissue, is located to the inside of the protoderm. The procambium, the source of primary vascular tissue, is to the inside of the ground meristem. The derivatives in these three primary meristems divide much more quickly than the initials in the apical meristem. In one study, the cells of the protoderm, ground meristem, and procambium were dividing every 12 hours, while the initials were dividing just once every 180 hours.

In a root, the division, growth, and differentiation of cells can be traced linearly through three overlapping regions known as the *zones of cell division, elongation,* and *maturation* (Figure 4.2c). The **zone of cell division** consists of the root apical meristem and the three primary meristems. The **zone of elongation** is where the derivatives stop dividing and begin to grow in length. This area overlaps with the zone of cell division because some cells are still dividing while others are growing in length. The zone of elongation is where most growth of the root takes place as the cells lengthen, the process that actually extends the root farther into the soil. The zone of elongation overlaps with the **zone of maturation,** where cells begin specializing in structure and function into different cell types, such as epidermal cells and conducting cells. The zone of maturation is also where some epidermal cells form root hairs. Above the zone of maturation, the first lateral roots emerge.

The root cap protects the root apical meristem and helps the root penetrate the soil

A root apical meristem produces a **root cap** consisting of several layers of cells (see Figure 4.2b). The root cap protects cells of the root apical meristem as the root pushes between soil particles. As the root grows, root cap cells are damaged and die, to be sloughed off and replaced by new cells. The outer cells of the root cap produce a slimy polysaccharide known as **mucigel** that lubricates the passage of the root through the soil. All plant cells have the genetic capacity to produce mucigel, but typically only root cap cells express this potential. Within each cell of the root cap, the slime is packaged in the transport vesicles of the Golgi apparatus, which fuse with the cell membrane to release the slime.

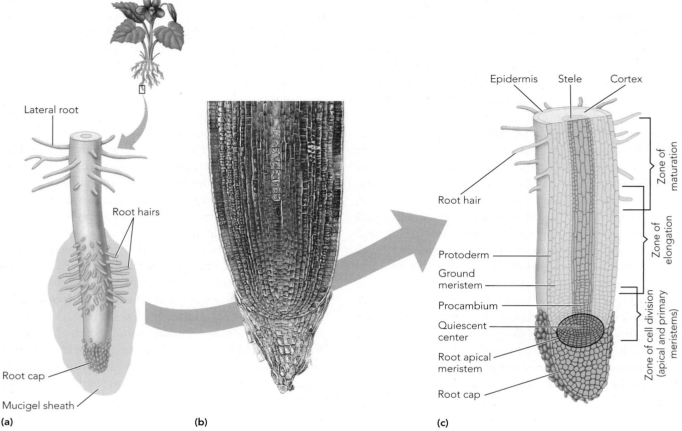

Figure 4.2 The root apical meristem. (a) This drawing of the first several millimeters of a dicot root shows root hairs forming in the region of maturation. Farther up the root, lateral roots begin to form. (b) In this photomicrograph the root cap is clearly visible. (c) The root apical meristem produces the cells of the root itself and cells of the root cap. The root apical meristem consists of a quiescent zone of slowly dividing cells surrounded by a narrow region of more rapidly dividing cells. Just beyond the root apical meristem are regions of increased cell division: the protoderm, procambium, and ground meristem. Above them are regions of cell elongation and cell maturation. Each region gradually blends into the next.

Absorption of water and minerals occurs mainly through the root hairs

It is primarily within the zone of maturation, beyond the zone of elongation, that a root produces the epidermal cells called *root hairs* (Figure 4.2). Specialized to absorb water and minerals from the soil, root hairs typically occur only in the last centimeter or two of the root. As the root continues to grow, older root hairs die and new ones arise in each new zone of maturation.

Most of the significant water and mineral absorption occurs through root hairs, even in large tree roots. In plants with taproots, root hairs may occur a considerable distance underground. In fibrous root systems, roots may not be deep but are often widespread, so that the root hairs are not close to the base of the plant. For these reasons, watering plants for short periods of time where the stems enter the ground is frequently ineffective in getting water into the plant.

The primary structure of roots relates to obtaining water and dissolved minerals

Botanists examine root and stem structure by slicing these organs in various planes and examining thin sections under a microscope. A horizontal cut at a right angle to the long axis is called a **cross section,** also known as a **transverse section.** In examining cross sections of a wide range of vascular plants, botanists have identified typical arrangements of vascular and ground tissues. As you will see later, stems generally have a more complex arrangement of vascular and ground tissues than do roots.

In a three-dimensional view, of course, the tissues form cylinders. The central cylinder of a root or stem, which is surrounded by the cortex, is known as the **stele** (from the Greek word for "pillar"). Most roots have the simplest type of stele—and the earliest to evolve—known as a **protostele** (PRO-toe-steel) (from the Greek *proto,* "before"). In all protosteles, the vascular tissue forms a solid central cylinder that is surrounded by cortex, but the arrangement of vascular tissue can vary. A cross section of the protostele of most dicot and conifer roots reveals solid spokes or lobes of xylem, with phloem occurring between the lobes (Figure 4.3a). In most monocot roots, the stele has parenchyma cells in the center, surrounded by alternating strands of xylem and phloem (Figure 4.3b). Botanists who study the roots of fossil plants have proposed that in monocots the tissue in the center of the stele simply remained as parenchyma cells that failed to develop into conducting tissue. Although these cells are often called *pith* because of their location,

they are not ground tissue. They arise from the procambium rather than from the ground meristem.

In the roots of most seed plants, two important cell layers called the *pericycle* and the *endodermis* surround the stele. The **pericycle** immediately encircles the stele and consists of meristematic cells that give rise to lateral roots, also called *branch roots* (Figure 4.4). As lateral roots arise from the pericycle, they grow through the cortex and epidermis, displacing these tissues, to reach the outside. The xylem and phloem of each lateral root are continuous with the vascular tissue of the parent root and have the same structure.

While a major role of the pericycle is to produce branch roots, the function of the **endodermis** is to regulate the flow of substances between the cortex and the vascular tissue (Figure 4.5a). The endodermis develops from the innermost layer of cortex and consists of a single layer of closely packed cells surrounding the pericycle. Each endodermal cell is surrounded on four of its six sides by a Casparian

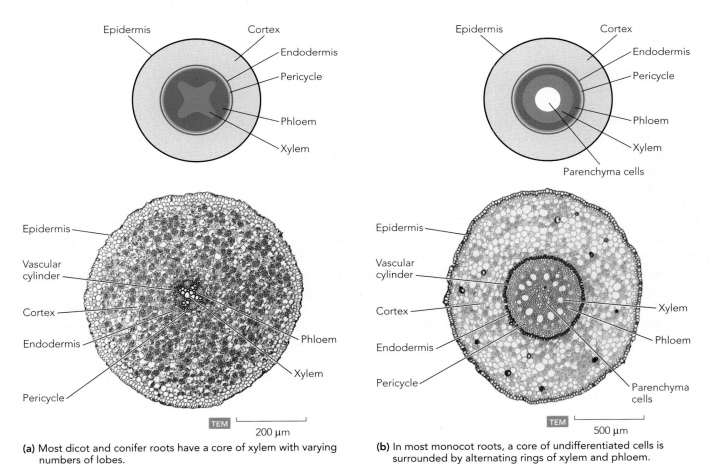

(a) Most dicot and conifer roots have a core of xylem with varying numbers of lobes.

(b) In most monocot roots, a core of undifferentiated cells is surrounded by alternating rings of xylem and phloem.

Figure 4.3 The primary structure of roots. (a) Most roots have a protostele—a solid central cylinder of vascular tissue. In all primary roots, the vascular tissue is surrounded by one or more layers of pericycle and next by one layer of endodermis. **(b)** Monocot roots have a stele with rings of xylem and phloem surrounding a core of parenchyma.

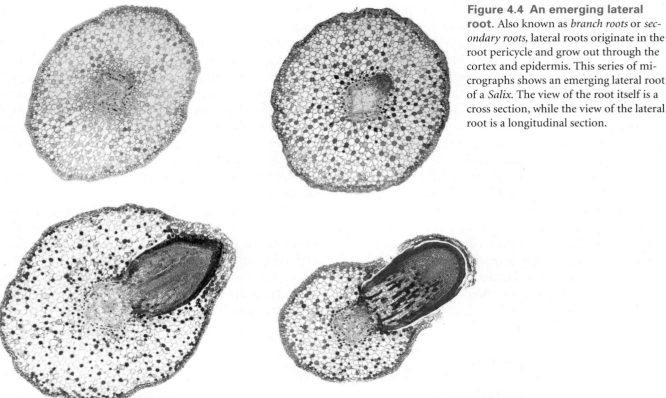

Figure 4.4 An emerging lateral root. Also known as *branch roots* or *secondary roots,* lateral roots originate in the root pericycle and grow out through the cortex and epidermis. This series of micrographs shows an emerging lateral root of a *Salix.* The view of the root itself is a cross section, while the view of the lateral root is a longitudinal section.

strip composed of suberin and sometimes lignin, which impregnate cell walls and seal the space between cells. The Casparian strip is missing from the sides of endodermal cells that face to the outside and inside (Figure 4.5b). Thus, the Casparian strips force water and minerals to pass across cell membranes and through the cytoplasm of the endodermal cells, rather than through or between the cell walls. In this way, the cell membranes of the endodermis control the type and quantity of mineral nutrients that move between the cortex and the vascular tissue. In a growing root, the first endodermis, with its Casparian strips, becomes evident in the root hair region. The endodermis and its Casparian strips function in the root hair regions at the tips of roots, where there is significant water and mineral uptake.

Some roots have specialized functions in addition to anchoring the plant and absorbing water and minerals

As you know, the main functions of roots are to anchor and support the plant and to absorb water and minerals. In many plants, however, modified roots have evolved in ways that serve a variety of plant needs, including reproduction and storage of water or food (Figure 4.6).

Some modified roots provide additional support or anchorage for a plant. Among these are **aerial roots,** adventitious roots that arise from stems. Aerial roots often occur in **epiphytes** (from the Greek *epi-,* "upon," and *phyton,* "plant"), plants that grow on other plants for support but nourish themselves. In orchids, which often grow as epiphytes on trees, aerial roots attach the plant and absorb water and nutrients from rainwater that drips through the leaf canopy above them. Aerial roots can also arise in other types of plants. In corn, aerial roots known as *prop roots* grow out of the stem and into the soil, helping to anchor and support the plant. Many climbing plants such as ivy use aerial roots to anchor themselves to a vertical surface. **Buttress roots** are flared roots that extend from tree trunks, contributing to stability in a fashion similar to the buttresses supporting the walls of medieval cathedrals. Some tropical trees develop huge buttress roots that help stabilize them in the thin soils sometimes found in the tropics. **Contractile roots** in lilies and other plants actually shorten to pull the plant deeper into the soil.

Some modified roots are involved in asexual reproduction, as adventitious buds called *suckers* arise from roots and break through the soil to form new shoots. The production of buds by roots is reasonably common among

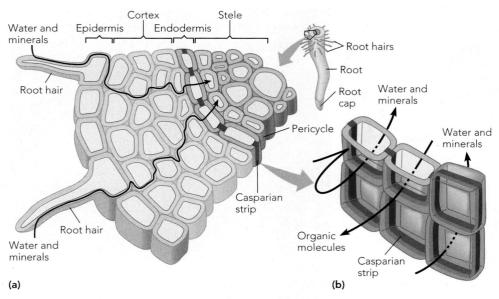

Figure 4.5 The endodermis. (a) Water and minerals can pass between the cells of the epidermis and cortex, but they must pass through the cell membranes of the endodermis because of the Casparian strip. (b) This view of two layers of the endodermal cylinder shows how the Casparian strip forces water and dissolved minerals from the soil to pass through endodermal cells instead of around or between them.

plants. For example, common milkweed (*Asclepias syriaca*) also propagates through root buds, as does the sassafras tree (*Sassafras albidum*) and the weed known as *common leafy spurge* (*Euphorbia escula*).

Pneumatophores, also known as air roots, provide oxygen for plants in areas such as swamps, where the high rate of aerobic decay reduces the oxygen supply in the water. Mangrove and bald cypress trees produce pneumatophores that stick out of the swamp water to take in air to support respiration.

Other types of modified roots store water or food. For example, the buffalo gourd (*Cucurbita foetidissima*) produces large underground water storage roots that can weigh 100 pounds or more. Species that grow in arid regions are most likely to store water in such roots. Many types of roots—including carrots, sweet potatoes, and sugar beets—store starch or sugar as a source of nutrition for the plant between growing seasons and to initiate new growth as a growing season begins. Over the years, plant breeders and farmers have selected individual plants with particularly high storage capacities as food crops. In most cases, roots involved in food storage have a modified type of secondary growth in which additional meristems form in the root.

Some plants, such as mistletoe, have modified parasitic roots known as **haustoria** (singular, *haustorium*), which penetrate stems and roots of other plants to obtain water, minerals, and organic molecules. At least one parasitic plant, *Striga,* has become a major agricultural pest.

All of these root modifications reflect evolutionary changes that have been successful in particular environments. As you will see later in the chapter, many plants also have modified stems or leaves that serve additional roles. The frequently overlapping functions of plant organs underscore the fact that roots, stems, and leaves are closely interrelated in meeting a plant's needs.

Roots have cooperative relationships with other organisms

Roots often form **mutualistic,** or mutually beneficial, associations with other organisms. **Mycorrhizae** (my–kuh–RY–zee) (from the Greek *mykes,* "fungus," and *rhiza,* "root") are mutualistic associations between vascular plant roots and soil fungi, occurring in more than 80% of plant species. The two main types of these associations are endomycorrhizae and ectomycorrhizae. In **endomycorrhizae,** the fungi penetrate plant roots and produce branching structures called *arbuscules*. Parts of the arbuscules press up against the outside of plant cell membranes to obtain nutrients (Figure 4.7a). In **ectomycorrhizae,** the fungi do not penetrate plant roots. Instead, a many-branched fungal net surrounds the root to produce a sheath called a *mantle* (Figure 4.7b). In both of these types of mutualistic associations, the plant gains increased absorption of minerals, such as phosphorus, and consequently does not need to produce as many root hairs. The fungus

(a) Aerial root in epiphytic orchid. In epiphytic orchids, aerial roots absorb water from the air.

(b) Aerial root as prop root. In corn, aerial roots serve as prop roots to stabilize the plant.

(c) Climbing aerial root. Adventitious climbing roots of Boston ivy (*Parthenocissus tricuspidata*) help anchor the plant to vertical surfaces.

(d) Buttress roots. Large buttress roots produced by some tropical trees help stabilize the plant in thin soils.

(e) Pneumatophores (air roots). Pneumatophores, such as those of the white mangrove (*Laguncularia racemosa*), supply oxygen to plants growing in swamps where the water may become deoxygenated.

(f) Storage roots. A number of plants, such as Hamburg parsley (*Petroselinum hortense*), store water or food in their roots.

Figure 4.6 Modified roots.

THE INTRIGUING WORLD OF PLANTS

Parasitic Roots

Ordinarily, the primary growth of plants produces stems to reach the light and roots to reach water and minerals in the soil. However, some plants with primary growth find nutrition in other ways. The genus *Striga* has more than 40 species, around a third of which parasitize crop plants. Each *Striga* plant can produce between 50,000 and 500,000 seeds that are almost microscopic and can germinate even after more than ten years have passed. After germination, the seedlings must attach to a host plant root within a week. The roots of *Striga* seedlings invade the host plant by forming haustoria within the roots of the host. Each haustorium obtains nutrients and water directly from the host. Most of the damage takes place underground, resulting in severe yield reduction and up to a total crop loss.

Three species of *Striga* currently cause major problems with cereals and legumes in Africa and Asia. Overall crop loss in all of Africa and parts of Asia is around 40%, and in some areas surpasses 70%. The parasite infects two thirds of cereal cultivation in Africa, making it the most serious yield-limiting agricultural problem on the continent.

A number of strategies are being implemented to try to eliminate *Striga*. Genes from a wild corn (*Zea diploperennis*) that are resistant to *Striga* have been incorporated into cultivated corn. Scientists are also planting nitrogen-fixing legumes, which cause *Striga* seeds to abort during development. Another strategy is to use host plants with genes that are resistant to herbicides. The host plants can then be sprayed to eliminate the parasite. Scientists are also experimenting with hormones that stimulate premature germination of *Striga* seeds, thus eliminating the natural "seed bank" that builds up in the soil as a result of delayed germination. They have also discovered a fungus (*Fusarium oxysporum*) in northern Africa that kills *Striga* but does not slow the growth of sorghum and other cereals. Sorghum yields increased up to 70% in fields infected with *Striga* that carried the fungus.

Striga, a parasitic crop plant.

Striga is one of about 3,000 species of parasitic flowering plants. Another parasitic plant is a species of the genus *Triphysaria*, which infects the roots of *Arabidopsis thaliana* (common wall cress). Since *Arabidopsis* serves as a model plant for molecular genetics experiments, the interaction between *Triphysaria* and *Arabidopsis* may help unravel the molecular events of plant parasitism. Scientific studies continue in an effort to minimize or perhaps eliminate crop damage caused by parasitic plants.

Haustorium growing into host root.

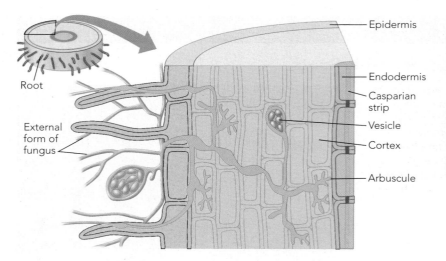

(b) Ectomycorrhizae form a sheath outside the root.

(a) Endomycorrhizae penetrate the root cortex.

Figure 4.7 Mycorrhizae are mutualistic associations of roots and fungi.

(c) The ectomycorrhizae of this pine tree extend for some distance into the soil.

may also help protect the plant from attack by some disease-causing fungi and nematodes (roundworms). Meanwhile, the fungus gains sugar and other organic molecules produced by the plant.

Mycorrhizae (Figure 4.7c) are commonly seen in fossil plants and may have been important in allowing vascular plants to become established on land. Many studies have demonstrated that potted seedlings grow much more rapidly in a mutualistic association with the proper mycorrhizal fungus. Similarly, transplanting works best if the soil in both locations has the correct mycorrhizal fungus.

Some plants also form associations with species of nitrogen-fixing bacteria—bacteria that can convert nitrogen gas from the air into ammonium, which is incorporated into various organic molecules. Plants can then take up fixed nitrogen from these bacteria and incorporate it into amino acids, nucleotides, and other vital nitrogen-containing compounds. This is virtually the only biological pathway by which inorganic nitrogen can enter food chains. Plants in the legume family, Fabaceae, are particularly important for humans because their mutualistic associations with nitrogen-fixing bacteria enrich the soil by adding nitrogen. This enrichment is quite important because harvesting crops results in removal of mineral nutrients, such as nitrate, from the soil. In legumes, nitrogen-fixing bacteria infect the roots, where they cause the plant to form root nodules in which the bacteria live. Nitrate is released from nodules into the soil. Nitrogen-fixing bacteria and their associations with plants will be discussed in more detail in Chapter 10.

Section Review

1. How do taproot and fibrous root systems differ?
2. Describe cell development and maturation through the zones near a root tip.
3. What are the functions of the root cap, mucigel, and root hairs?
4. What are the roles of the pericycle and endodermis?
5. What are some examples of specialized adaptations in roots?
6. What are mycorrhizae, and how are they beneficial?

Stems

Stems and leaves are the organs that plants usually produce above ground and constitute the shoot system. As you read in Chapter 3, together a stem and its leaves are commonly known as a *shoot*. Stems move leaves toward the light and away from the shade of other plants or structures. To support the weight of leaves and to withstand the force of wind, stems must be strong, especially in tall trees. They must also conduct water, minerals, and organic molecules between the roots and the leaves. Here we will look at the primary structure of stems.

As you know, leaves are attached to stems at nodes, and the portion of a stem between nodes is known as an *internode*. In most plants, a dormant axillary bud that has the potential to form a branch is located on the upper surface of the angle, or axil, between the stem and the petiole of a leaf. Nodes, internodes, and axillary buds are some basic features that distinguish stems—including underground stems—from roots and leaves.

Stem growth is more complex than root growth because the stem not only grows in length but also produces leaves and axillary branches, with the shoot apical meristem giving rise to leaf primordia and axillary buds in rapid succession (Figure 4.8). Since the leaf primordia are close together, the internodes are very short at first. Whereas the lengthening of a root occurs in a single zone of elongation near the root tip, a stem typically lengthens simultaneously in several internodes below the shoot apical meristem. Some plants, including grasses such as wheat, have a region of dividing cells at each internode, known as an **intercalary meristem** (in–TER–kuh–lair–ee). These intercalary meristems allow the stem to grow rapidly all along its length.

Botanists have developed zone and cell-layer models to describe stem growth

Botanists have developed two models of how a shoot apical meristem gives rise to the primary meristems: the protoderm, ground meristem, and procambium. One is known as the *zone model*, while the other is called the *cell-layer model*. Both are accurate, although some plants seem to fit one model better than the other.

The **zone model** describes the shoot apical meristem as a dome divided into three regions: the central mother cell zone, peripheral zone, and pith zone (Figure 4.9a). The **central mother cell zone** contains cells that divide infrequently and give rise to the cells of the peripheral and pith zones. The **peripheral zone** is a three-dimensional ring surrounding the central mother zone. The peripheral zone consists of cells that divide rapidly to become leaf primordia and parts of the stem, supplying cells to the protoderm, procambium, and the part of the ground meristem that produces the cortex. Below the central and peripheral zones is the **pith zone,** which produces cells that become the part of the ground meristem that produces pith. Below these zones of shoot apical meristem

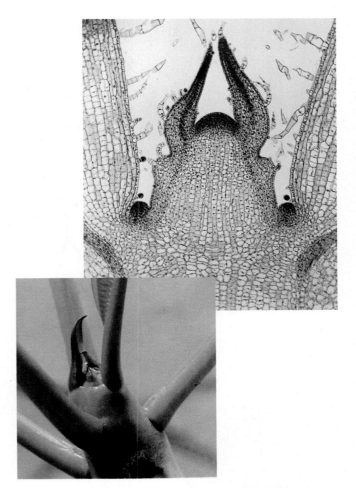

Figure 4.8 The shoot apical meristem. The shoot apical meristem itself consists of a central group of slowly dividing cells surrounded by a narrow region of more rapidly dividing cells. This longitudinal section shows the leaf primordia emerging on both sides of the shoot apical meristem. As the stem grows, new leaf primordia arise along the flanks of the apical meristem. Just beyond and below the apical meristem, the protoderm, procambium, and ground meristem are regions of increased cell division. Cell elongation and cell maturation occur below these regions.

lie the protoderm, procambium, and ground meristem, where cell division continues and where cell growth and differentiation begin. Below these primary meristems, cells continue their elongation and differentiation into mature tissues.

The **cell-layer model,** also called the *tunica-corpus model*, describes the initials of the shoot apical meristem as forming several cell layers, starting from the tip of the apical meristem (Figure 4.9b). The outer layers of initials form the **tunica,** which is equivalent to the outer part of the peripheral zone. Most plants have two layers of tunica, identified as L1 and L2. In the tunica, the initials

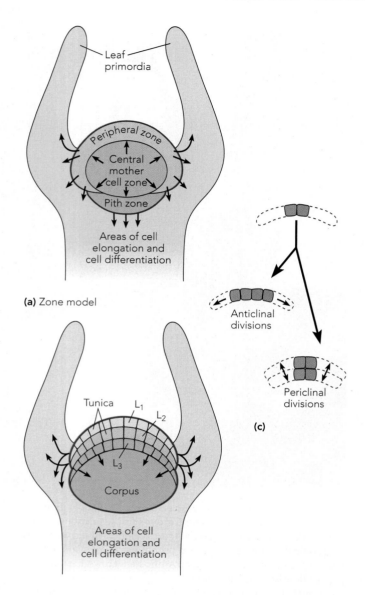

(a) Zone model

(b) Cell layer model (tunica-corpus model)

(c)

Figure 4.9 Organization of the shoot apical meristem into zones and layers. This figure shows two different models used to describe the shoot apical meristem. **(a)** In the zone model, the slowly dividing cells of the central mother zone give rise to rapidly dividing cells in the peripheral and pith zones. Cells from the peripheral zone develop into the leaf primordia and eventually give rise to the dermal and vascular cells of the stem, as well as the cortex. Cells from the pith zone eventually give rise to the ground tissue in the center of the stem. **(b)** The cell layer model, or tunica-corpus model, describes the shoot apical meristem as having three layers of dividing cells. There are typically two outer layers (L1 and L2) that form the tunica, which gives rise to the protoderm. The innermost layer (L3) gives rise to the procambium and ground meristem. **(c)** Cells in the tunica undergo anticlinal divisions (perpendicular to the surface of the apical meristem), while cells in the corpus undergo both anticlinal and periclinal divisions (parallel to the surface of the apical meristem).

typically divide perpendicular to the surface, known as **anticlinal** divisions (Figure 4.9c). The L3 layer and its derivatives form the **corpus,** roughly equivalent to the central mother cell zone, the inner parts of the peripheral zone, and the pith zone. In the initials of the corpus, there are anticlinal divisions and also **periclinal** divisions, which are parallel to the surface. In the cell-layer model, the outermost layer of the tunica produces the protoderm, while the corpus gives rise to the procambium and ground meristem.

In primary growth of most stems, the vascular tissue forms separate bundles

Considerable variation exists in the arrangement of vascular tissue in the primary structure of stems. The stems of a few seedless vascular plants have protosteles, the same primitive pattern found in almost all roots (Figure 4.10a). Protosteles are present in the earliest plants found in the fossil record. The stems of some seedless vascular plants, such as ferns and horsetails, have a **siphonostele** (Figure 4.10b), consisting of a continuous vascular cylinder that surrounds a core of pith (ground tissue). In siphonosteles, which evolved from protosteles, the phloem occurs outside the xylem or on both sides of the xylem. The cylinder is broken by **leaf gaps,** where vascular tissue branches off from the stele to enter leaves.

In the stems of most seed plants—that is, gymnosperms and flowering plants—vascular tissue forms **vascular bundles,** separate strands composed of xylem and phloem. Vascular bundles do not have leaf gaps. In each bundle, the xylem faces toward the center of the stem, and the phloem faces toward the surface of the stem. In most gymnosperm and dicot stems, the vascular bundles form a circle around the pith, an arrangement known as a **eustele.** Like siphonosteles, eusteles evolved from protosteles. There are two types of eusteles. In one type, the bundles form a tight ring, with narrow regions of parenchyma cells between them (Figure 4.10c). In the other type, the bundles form a loose ring, with wider regions of parenchyma cells between them (Figure 4.10d). In contrast with the circular arrangement in eusteles, the vascular bundles in most monocots are scattered throughout the ground tissue (Figure 4.10e). This dispersed arrangement in monocots evolved from eusteles.

As you know, the terms *cortex* and *pith* simply refer to the locations of ground tissue, which consists mainly of parenchyma. These terms are only useful for distinguishing locations of ground tissue in relation to a single cylinder or ring of vascular tissue. Ground tissue to the outside of the vascular cylinder or ring is called *cortex,* while

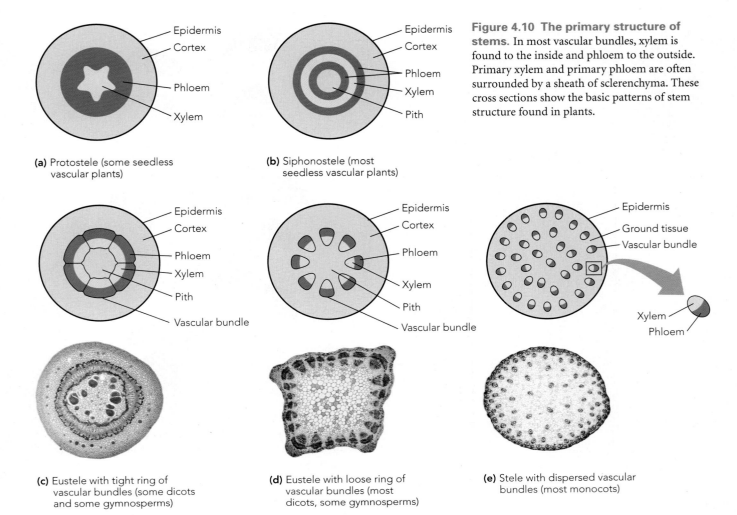

Figure 4.10 The primary structure of stems. In most vascular bundles, xylem is found to the inside and phloem to the outside. Primary xylem and primary phloem are often surrounded by a sheath of sclerenchyma. These cross sections show the basic patterns of stem structure found in plants.

(a) Protostele (some seedless vascular plants)

(b) Siphonostele (most seedless vascular plants)

(c) Eustele with tight ring of vascular bundles (some dicots and some gymnosperms)

(d) Eustele with loose ring of vascular bundles (most dicots, some gymnosperms)

(e) Stele with dispersed vascular bundles (most monocots)

ground tissue to the inside of the vascular tissue is called *pith*. However, if the vascular bundles are scattered, as in most monocot stems, the ground tissue is dispersed between the bundles, so the terms *cortex* and *pith* do not apply. In all stem structures, of course, the dermal tissue surrounds the other tissues. In primary growth, the epidermis is the dermal tissue, which frequently produces a cuticle composed of water-impermeable materials that reduce water loss.

A transition region maintains vascular continuity between the root and stem

Since most roots have protosteles, and most stems have various arrangements of vascular bundles, how do the vascular tissues of the root and stem connect? The answer is that there is a transition zone between stem and root, in which one pattern gradually blends into the other (Figure 4.11). This transition zone forms during the early growth of the seedling and is a few millimeters to a few centimeters in length.

One could ask why the roots and shoots of the same plant have different patterns of vascular tissue. The evolution of roots and stems may suggest an answer. The first vascular plants had no roots, consisting instead of aboveground and underground stems with protosteles. Evidently the protostele structure was retained as underground stems evolved into roots. In aboveground stems, however, the protostele may not have provided sufficient structural support and was gradually replaced in most vascular plants by siphonosteles or eusteles.

Leaf primordia form in specific patterns on the sides of shoot apical meristems

You have seen how the internal structure of stems can vary. There are also variations in leaf arrangement. Leaf primordia form on the sides of shoot apical meristems in an ordered and predictable pattern, or **phyllotaxy** (from Greek words meaning "leaf order") for each species of

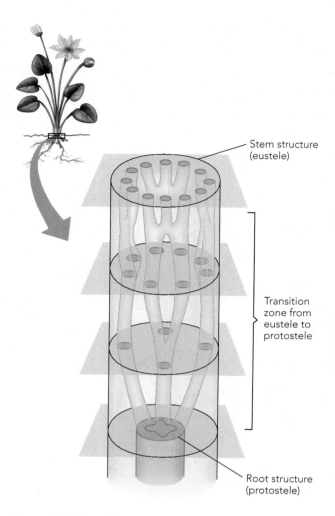

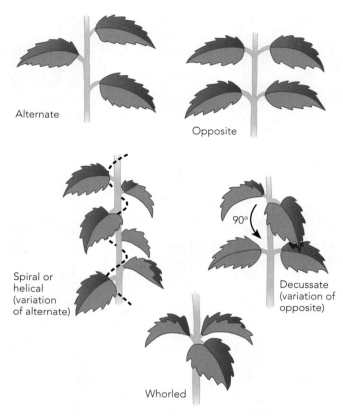

Figure 4.12 Basic patterns of leaf arrangements on a stem. The three most general types of leaf phyllotaxy are alternate, opposite, and whorled.

Figure 4.11 Transition zones between stem and root. In many plants, the stele of the root has a different arrangement of vascular tissues than the stele of the stem. In the transition zone, one arrangement blends into the other, in an often complex pattern that can vary significantly from species to species. The transition usually occurs within an area of only several millimeters to several centimeters. This simplified example shows how part of the vascular tissue of the protostele of the buttercup root (*Ranunculus*) blends into part of the vascular tissue of the eustele of the stem, which has many discrete vascular bundles.

plant. There are three basic patterns, distinguished by the number of leaves per node: alternate, opposite, and whorled (Figure 4.12). An **alternate** arrangement has one leaf per node. In one type of alternate phyllotaxy, each leaf is oriented 180 degrees from the previous leaf. In another common type, the leaves form a spiral or helical pattern, alternating in a spiral that goes around the stem. An **opposite** arrangement consists of two leaves per node. In one variation, each pair of leaves is oriented like the previous pair. In another, known as a decussate pattern,

each pair forms at right angles to the previous pair. A **whorled** arrangement includes three or more leaves per node. Regardless of their phyllotaxy, leaves are generally arranged so that each leaf is exposed well to the light and can easily carry out photosynthesis.

Botanists are looking for clues to why new leaves appear in specific places. Two types of theories attempt to explain the patterns of leaf primordia on the apical meristem: field theories and available-space theories. Field theories, or biochemical theories, point to hormones or other growth substances as the causes. For example, existing leaf primordia might produce a compound that inhibits the formation of new primordia nearby. However, no one has yet identified such hormonal fields. Available-space theories, also known as *biophysical theories*, hold that new primordia arise when space for them is not occupied by existing primordia. One of these theories proposes that forces placed on the apical surface area by primordia cause a sort of spontaneous buckling or bubbling of the apical surface that becomes the next primordium.

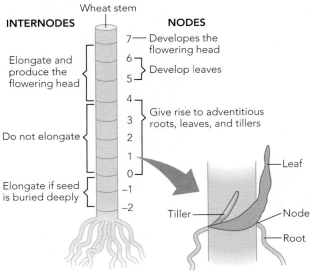

(a) Wheat grows in cooler, temperate regions and grows one to two feet tall. A typical wheat plant consists of up to seven leaves and up to three axillary shoots called tillers, linked to the stem at nodes. Only a few of the internodes elongate during the life cycle of the plant.

Figure 4.13 Comparing nodes and internodes of wheat and of palm plants. A plant shoot consists of stem internodes separated by nodes, where leaves are attached. The principal differences between the stems of wheat and palm are the numbers of internodes and how much elongation occurs.

Stem variations reflect different evolutionary pathways

In addition to variations in phyllotaxy, stems come in many sizes and shapes, reflecting different environmental adaptations. Let us consider two stems—the tall, fibrous stem of a palm tree and the short, slender stem of a wheat plant. Palm trees are the largest plants that have a stem produced entirely by primary growth. Even though palm trees do not have lateral (secondary) meristems, the

trunks still grow considerably in width as a result of continuing cell division and cell growth by derivatives of the apical meristem. In a palm tree, a single apical meristem sits atop a stem that can grow nearly 61 meters (200 feet) high. In contrast, wheat plants are typical grasses with short stems.

Palm trees and wheat plants are good examples of how different environments can result in modification of stems through evolution to serve differing needs. Native to tropical regions with little or no frost, palm trees live for many years and must compete with other plants for sunlight, frequently in lush, tropical forests. By growing tall over a period of years, they receive adequate sunlight, and their leaves and apical meristems are protected from most animals. In contrast, wheat plants are native to the cold mountains and plains of the Middle East and live a single year. They must survive in cold, windy regions with short growing seasons and intermittent browsing by animals. Since its shoot apical meristems are near the ground for most of the growing season, the wheat plant can survive the drying and physically damaging wind, as well as the loss of leaves to grazing animals.

(b) Palms grow in tropical regions and are frequently quite large, with hundreds of leaf bases on the trunk. Palm trees produce many more leaves than do wheat plants, and all internodes elongate. Leaves eventually die and fall off revealing leaf bases.

The stems of wheat plants and palm trees differ significantly in the number of internodes and the patterns of elongation. Each wheat stem produces a main shoot and from one to three axillary shoots called *tillers* (Figure 4.13a). The main shoot develops only about seven leaves, and the tillers produce even fewer. Most of the internodes do not elongate at first, keeping each shoot apical meristem near ground level until the very end of the plant's life cycle. Then, through the action of the intercalary meristems in the internodes, each shoot grows to between a few inches and several feet in length, producing a flowering head that develops into wheat kernels when pollinated. The last and largest leaf, called the *flag leaf,* contributes greatly to the yield of grain. In contrast, palm trees form many leaves, and the internodes gradually elongate to produce a trunk, which elevates the younger leaves near the shoot apical meristem higher and higher (Figure 4.13b).

Old leaves die and eventually fall from the trunk, but their leaf bases remain.

Some stems have specialized functions in addition to support and conduction

You have seen that many plants have modified roots. In addition, a number of specialized stems have evolved.

Some modified stems aid in reproduction. **Stolons,** also known as runners, are horizontal aboveground stems—such as those formed by strawberry plants—that help a plant reproduce (Figure 4.14a). Stolons often originate as axillary buds, forming a stem that grows above the surface of the ground to establish a new plant nearby. Many grasses, like wheat, reproduce through tillers, axillary shoots that extend from nodes at the base of the plant just beneath the apical meristem. Horizontal stems called

rhizomes occur underground (Figure 4.14b). Irises typically have rhizomes that send up new leaves or shoots every year.

Other modified stems mainly store food or water. Food-storage stems become enlarged by accumulating starch in parenchyma cells. **Tubers,** such as those of the sweet potato, are underground stems primarily composed of starch-filled parenchyma cells that form at the tips of either stolons or rhizomes (Figure 4.14c). The eyes of the potato are actually axillary buds arranged in a helical fashion along the surface of the potato. In **bulbs,** such as onions, starch accumulates in thickened, fleshy leaves attached to the stem (Figure 4.14d). **Corms,** such as those in gladiolus, are underground food storage stems that are

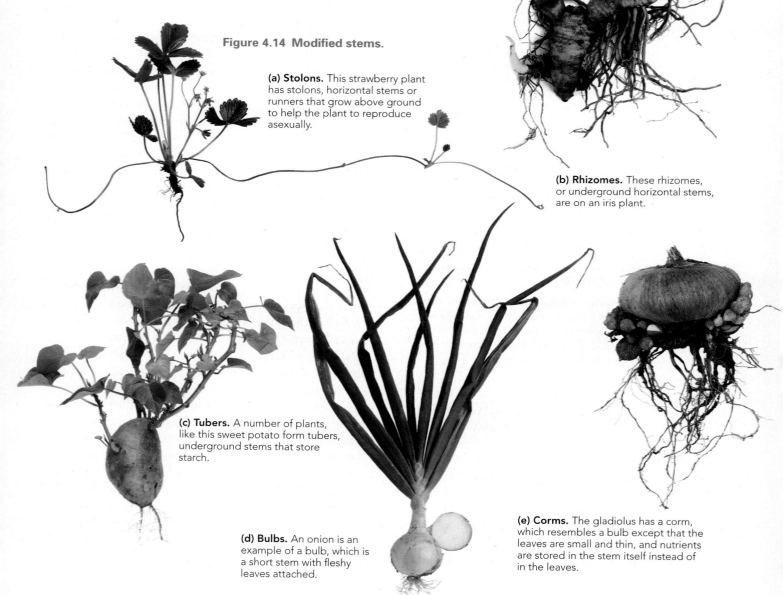

Figure 4.14 Modified stems.

(a) Stolons. This strawberry plant has stolons, horizontal stems or runners that grow above ground to help the plant to reproduce asexually.

(b) Rhizomes. These rhizomes, or underground horizontal stems, are on an iris plant.

(c) Tubers. A number of plants, like this sweet potato form tubers, underground stems that store starch.

(d) Bulbs. An onion is an example of a bulb, which is a short stem with fleshy leaves attached.

(e) Corms. The gladiolus has a corm, which resembles a bulb except that the leaves are small and thin, and nutrients are stored in the stem itself instead of in the leaves.

PLANTS & PEOPLE

Stems and Roots as Foods

Worldwide, cereal grains such as rice, wheat, and corn are the main sources of human food. However, the stem crops, potato and yams, and the root crops, sweet potato and cassava, complete the top six crops that together supply 80% of the calories humans need. Growing root crops has advantages over growing seed crops. A large harvest can easily occur without mechanization, and root crops store successfully without processing or drying. However, their high moisture content makes them heavier than cereal grains and therefore more expensive to ship. Nutritionally, root crops are high in starch and calories but often lack sufficient protein or oil, compared with cereal grain crops.

The most widely grown underground crop is potatoes (*Solanum tuberosum*), a major source of starch in developed countries. Potatoes prefer a cool climate and typically do poorly in the Tropics. Potatoes originated in the Andes Mountains of South America. After Spanish explorers brought potatoes to Europe in the 1500s, they became an important crop in many European countries, particularly Ireland. When a potato blight fungus swept through Europe in the 1840s, Ireland's population dropped 50% after many Irish starved to death or emigrated. Although often called a root crop, potatoes are actually tubers. The buds produced by the stem are called "eyes," and a piece of a potato containing an eye (called a *set*) propagates the plant.

Cassava (*Manihot esculenta*) is the most important root crop in tropical countries, where potatoes will not grow. Also known as *manihot* and *manioc*, cassava was first cultivated by South American Indians. It provides the basic source of calories for more than 300 million people, with both the storage root and the leafy greens being edible. Cassava can be made into granules that are the main ingredient in tapioca pudding.

Cassava

Taro growing on Molokai, Hawaii.

The sweet potato plant (*Ipomoea batatas*), native to South America, has large storage roots that can be boiled, baked, or fried. They can be peeled, sliced, and made into sweet potato fries.

Sometimes sweet potatoes are called yams, but yams are actually plants of the genus *Dioscorea*. Native to the Far East, yams are tubers that probably qualify as the world's largest vegetables, with some weighing as much as 318 kg (700 pounds).

Taro (*Colocasia esculenta* or *Xanthosoma sagittifolim*) produces a corm rich in starch and sugar. Most production occurs in Africa. Traditional Hawaiian poi contains taro roots that are steamed, crushed, and then fermented and perhaps flavored. "One-finger" poi is quite stiff, while "three-finger" poi is runnier.

---

shaped like bulbs but consist primarily of stem tissue rather than thick leaves (Figure 4.14e). Thick stems that store water are also commonly found in some desert plants, particularly cacti, in which the spines are actually modified leaves attached to a fleshy stem. The *Plants & People* box above describes some stems and roots that serve as important food crops.

Section Review

1. How does stem growth differ from root growth?
2. Describe the basic types of steles in stems.
3. What is phyllotaxy?
4. What do the differences between palm trees and wheat plants reveal about stem growth?
5. What are some similarities between modified stems and modified roots?

Leaves

Stems give rise to leaves and to reproductive structures such as flowers and cones, which in terms of their evolution are actually modified shoots. In Chapter 6, we will look at reproductive structures, while here we will focus on the structure and function of leaves.

The first plants were photosynthetic stem systems, and leaves evolved from flattened stems that grew close together. In this way, leaves evolved into the main photosynthetic organs of plants. As you will see, however, leaves also carry out other important functions.

A leaf primordium develops into a leaf through cell division, growth, and differentiation

Earlier in the chapter, you learned that leaves originate on the shoot apical meristem as bumps of tissue called *leaf primordia.* Now we will look more closely at how leaf primordia develop into leaves.

The formation of a bulge, known as a **leaf buttress,** on the flank of a shoot apical meristem provides the first indication that a new leaf will soon appear (Figure 4.15). As the apical meristem grows up and away, the cells in the leaf buttress elongate until a flattened leaf primordium appears. The primordium expands by cell division and growth into a thin petiole and a **blade,** the flattened portion of the leaf. Petioles often have two small leaflike flaps called **stipules** that are attached at the node. Petioles may vary from stemlike to leaflike, so sometimes there is no clear demarcation between the petiole and the leaf, with the petiole becoming progressively more leaflike as it approaches the leaf blade. Some leaves, known as **sessile** leaves, lack petioles altogether and are instead attached directly to the stem (Figure 4.16). Sessile leaves are particularly common in monocots, where the base of the leaf circles around the stem to form a sheath.

In the region of the primordium that will become the leaf blade, longitudinal ridges of dividing cells develop on each side of the primordium. If these two ridges produce new cells uniformly, the edge of the leaf blade, called the *margin,* will be smooth and even. If there are different rates of cell division in different parts of the two ridges, an uneven leaf margin will result. Sometimes these ridges of dividing cells are called *marginal* or *intercalary meristems.*

The leaf epidermis provides protection and regulates exchange of gas

The leaf epidermis, a single layer of cells derived from protoderm, protects the leaf from water loss, abrasions, and

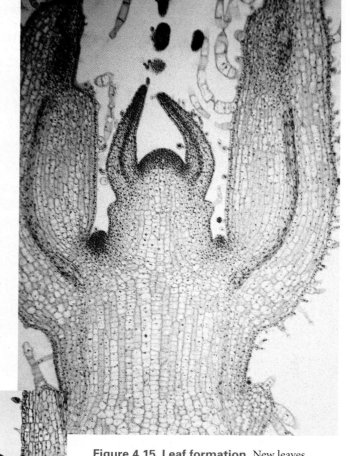

Figure 4.15 Leaf formation. New leaves first appear as small bulges, called *leaf buttresses,* on the flank of a shoot apex. This photomicrograph shows a longitudinal section of the shoot tip of *Coleus blumei.* The two leaf buttresses will develop into opposite leaves, characteristic of this species. The elongating buttress develops into a flattened leaf primordium connected to the stem by a stemlike petiole. The protoderm, procambium, and ground meristem arise from the apical meristem. These primordia already have noticeable procambial strands that extend downward to connect with the stele of the stem itself.

the entry of fungi and bacteria that cause disease. The epidermis also regulates the exchange of gases—such as CO_2, O_2, and water vapor—that are needed by or produced by the leaf. The epidermis is usually non-photosynthetic.

Since leaves exist principally to carry out photosynthesis, the surface area of the leaf maximizes light absorption, which also results in considerable water loss. To counter water loss, the leaf epidermis typically produces an external cuticle made of wax and cutin, a fatty sub-

Figure 4.16 Sessile leaves. Leaves that are joined directly to the stem, without a petiole, are known as *sessile leaves.* Often, sessile leaves form a sheath around the stem.

stance that is impervious to water. Some plants produce a layer of wax outside the cuticle, which gives extra protection against water loss. The cuticle and its external waxy layer also provide a slick surface that discourages the attachment and germination of fungal spores and creates an unsure footing for many insects.

Epidermal cells also sometimes have numerous leaf hairs. Some of these trichomes, which make leaves feel furry or fuzzy, protect against water loss and excessive buildup of heat. Others contain toxic substances that repel insects or other animals that might eat the leaf.

To control the movement of water vapor, CO_2, and O_2, leaves have **stomata** (singular, *stoma;* from the Greek *stoma,* "mouth"), each consisting of a pore regulated by two epidermal cells called **guard cells,** one on each side of the pore. Stomata are typically located in greater numbers on the underside of the leaf, where they are protected from the highest temperatures and accumulate less dust and fewer fungal spores (see Figure 4.17). Through the stomata, CO_2 enters the leaf for incorporation into carbohy-

drates by photosynthesis, and water vapor and oxygen exit the leaf in large quantities. Oxygen, which is produced by photosynthesis and needed for respiration, may exit leaves or enter the leaves, depending on the time of day.

When guard cells take up water, the bands of cellulose in their cell walls cause them to swell and curve so that the stoma opens. When guard cells lose water, the stoma closes. In most plants, the amount of water in the leaf is an important factor controlling whether stomata are open or closed. For example, high temperatures and high winds tend to dry the leaf and cause stomata to close. High concentrations of CO_2 inside the leaf also typically cause stomata to close because they signal the plant that sufficient CO_2 for photosynthesis is available. You will learn more about the factors controlling the opening and closing of stomata in Chapters 8 and 10.

Transpiration, the evaporation of water through the stomata, serves as a suction to pull water and minerals up the stem from the roots. In effect, the plant must lose water from the leaves in order for water to be pulled up from the roots. As long as transpiration supplies more water than the plant needs, the process represents a net gain of water for the plant.

The evaporative process also cools the leaves, which could become quite hot otherwise from direct exposure to the Sun. The surface area of the leaf itself serves as a direct radiator of heat. Heat loss from leaves becomes especially important when the stomata have closed to prevent excess water loss by the plant.

Mesophyll, the ground tissue in leaves, carries out photosynthesis

Photosynthesis in a leaf takes place in chlorenchyma ground tissue called **mesophyll** (from the Greek *mesos,* "middle," and *phyllon,* "leaf"), located between the upper and lower layers of epidermis (Figure 4.18). Mesophyll cells contain chloroplasts and are specialized for photosynthesis. Sometimes mesophyll cells are elongated and lined up underneath the epidermis, an arrangement called **palisade mesophyll,** also known as *palisade parenchyma*

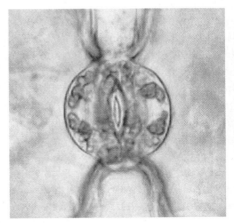

Closed stoma.

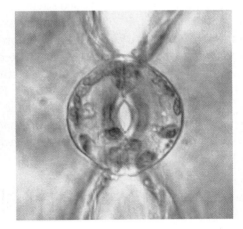

Open stoma.

Figure 4.17 Leaf epidermis. Stomata are pores in the leaf epidermis surrounded by two guard cells, which are photosynthetic and contain numerous chloroplasts. These micrographs show the change in the shape of guard cells, caused by uptake of water, that lead to opening of the stomata.

(from the Latin *palus,* stake). Think of stakes lined up — but not quite touching — to make a picket fence, and you will have a good image of a palisade layer. Palisade layers are usually only one cell thick, although multiple layers

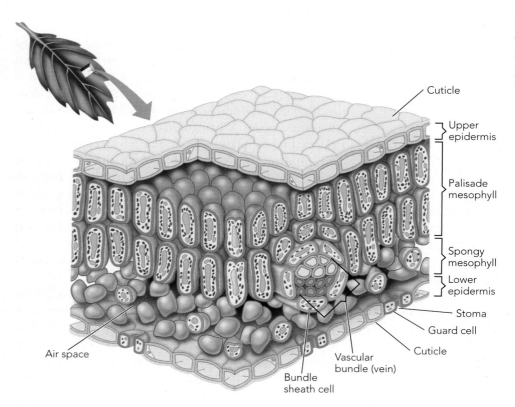

Figure 4.18 Leaf mesophyll. This cross section of a typical leaf shows two layers of palisade mesophyll overlying a layer of spongy mesophyll.

may occur where sunlight is intense. In many dicots, palisade layers occur just beneath the upper epidermis, the side most directly exposed to light. Below the palisade mesophyll is the **spongy mesophyll,** also called *spongy parenchyma.* Spongy mesophyll consists of loosely organized photosynthetic cells that have sufficient space between them to allow diffusion of CO_2 from the stomata to other parts of the leaf. In some plants where the leaf blade is vertical, palisade mesophyll appears on both sides of the leaf, and spongy mesophyll occurs in the center or not at all. Most of the chloroplasts, and therefore most of the photosynthesis in a leaf, typically occurs in the palisade parenchyma.

The vascular tissue in leaves is arranged in veins

The vascular tissue of each leaf merges with the vascular tissue of the stem. At each node on the stem there are typically one or more small vascular bundles called **leaf traces,** which leave the main vascular system of the stem and pass through a connecting petiole to reach the leaf blade. Once inside the petiole and the leaf blade, the vascular bundles are known as **leaf veins** and are continuous with the vascular bundles in the petiole and the stem it-

self (Figure 4.19a). Leaf veins form under the influence of plant hormones, particularly auxin. The portion of the vein that faces the upper surface of the leaf usually consists of xylem, while the bottom portion of the vein usually consists of phloem. In addition to conducting water, minerals, and food, veins provide support to the leaf, sometimes having **bundle sheath cells** around them for added strength and protection.

There are two general patterns of leaf veins, known as netted venation and parallel venation. Most dicots and ferns have **netted venation** (Figure 4.19b), also known as reticulate venation, in which the leaf veins form branching networks. The leaves of most monocots and gymnosperms have **parallel venation** (Figure 4.19c) also called *striate venation,* in which the veins run in long lines parallel with each other and the leaf edges.

Leaf shapes and arrangements have environmental significance

Leaf shape, size, and arrangement help plants carry out photosynthesis and meet other needs. Controlled by genes, the various leaf shapes and structures reflect characteristics that have enabled plants to survive in various environments. Leaves may be large or small, numerous or few,

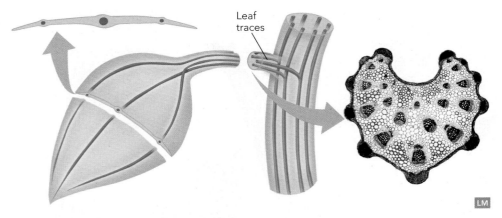

Leaf traces

(a) In the petiole and leaf blade, leaf traces become veins and can branch and merge. As shown in the cross section of a carrot petiole, there are many leaf traces.

(b) Netted venation.

(c) Parallel venation.

Figure 4.19 Leaf traces and patterns of venation.

thick or thin, and persistent or ephemeral, depending on the species, the photosynthetic needs, and the environment. For example, the thick leaves of many desert plants promote their survival by storing water.

In plant identification guides, key traits used to describe leaves include whether the blade is single or divided, the leaf shape, leaf margin characteristics, and venation. For instance, a **simple leaf** consists of a single, undivided blade. Sometimes a simple leaf is deeply lobed. In a **compound leaf,** the blade is divided into leaflets. In palmately compound leaves, the leaflets are all attached at the tip of the petiole, fanning out in a palmlike arrangement. In pinnately compound leaves, the leaflets form rows on opposite sides of an axis, in a featherlike fashion. As you can see in Figure 4.20, these are just a few of the variations in leaf shape.

In general, larger leaves with thin, smooth blades occur in environments with cooler temperatures, lower levels of light, moist conditions, and no wind. The larger surface area compensates for the lower light level. Smaller leaves with variable margins occur in environments with warmer temperatures, high levels of light, dry conditions, and high wind. The small size minimizes buffeting by winds, and the variable margins conduct wind away from the leaf. Leaves that grow well in sunlight differ somewhat in internal structure from those that grow well in shade. The leaves of some plants can orient themselves in relation to the light. The compass plant (*Silphium laciniatum* or *Lactuca biennis*) is so named because the upper surface of the leaves face east and west. At midday, the sun shines down on the edge of the leaves, which thereby avoid overheating during the hottest part of the day.

Plant leaves adapt to wind in various ways. Some plants, like pine trees, have small, stiff leaves that produce minimum drag in the wind. Others simply lose their leaves during windy seasons and regrow them later. Some plants roll their large leaves into a conical shape that avoids drag by pointing into the wind. The leaves of quaking aspen trees can move with the wind because the petioles are flat and set at right angles to the blades, allowing the leaves to rotate. Quaking aspen leaves shake and shimmer in the wind with a characteristic whispering sound.

Abscission zones form in the petioles of deciduous leaves

Plants that lose all their leaves at certain seasons of the year are known as **deciduous** plants (from a Latin word meaning "to fall off"). Examples of deciduous trees in temperate zones are such dicot trees as maple and sycamore. In tropical regions with an alternation of wet and dry seasons, many shrubs and trees lose their leaves in between wet seasons.

In deciduous plants, the areas where the leaves separate from the plant are known as **abscission zones** (from the Latin word meaning "to cut off"). Abscission zones typically form near the node where the petiole attaches to the stem, in response to shorter days and drier, cooler weather in temperate zones (Figure 4.21). These environmental changes trigger the production of hormones that

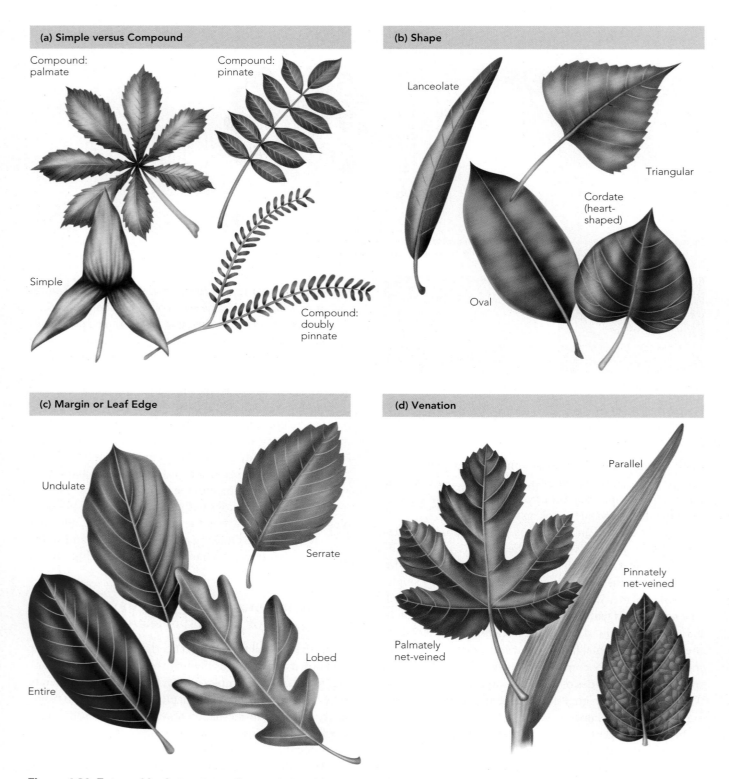

(a) Simple versus Compound

Compound: palmate

Compound: pinnate

Simple

Compound: doubly pinnate

(b) Shape

Lanceolate

Triangular

Cordate (heart-shaped)

Oval

(c) Margin or Leaf Edge

Undulate

Serrate

Entire

Lobed

(d) Venation

Parallel

Pinnately net-veined

Palmately net-veined

Figure 4.20 External leaf structure. Characteristics of the external structure of a leaf include details of whether the blade is single (simple) or divided (compound), the shape of each blade, shape of the margin, or leaf edge, and pattern of venation.

initiate a programmed series of chemical processes. After useful small molecules move from the leaves back into the stem, chemical reactions form the separation layer of the abscission zone, in which the middle lamellas and cell walls are weakened. Eventually the weight of the leaf and the force of wind cause the leaf to separate. Before the separation, however, a protective layer forms on the side of the abscission zone next to the main body of the plant, thereby preventing disease-causing bacteria or fungi from entering. In Chapter 11, you will read more about the chemical processes involved in leaf abscission.

Most pines and other conifers are nondeciduous trees, although they do lose their leaves gradually over a period of years. Pine needles remain photosynthetically active for 1 or 2 years, and sometimes for a dozen years or more, depending on the species and local conditions. You might think that retaining leaves year around would require less energy than growing new leaves each year. However, maintaining leaves through an inhospitable cold or dry season also requires energy and places many restrictions on leaf structure and shape. For example, pine needles have extremely thick cuticles and other structural adaptations that prevent water loss during the dry winter season. Pine needles are thin leaves, not easily damaged by freezing or deep snow, but having very little surface area for photosynthesis.

Some leaves have specialized functions in addition to photosynthesis and transpiration

Leaves have been modified to perform a number of specific functions by plants. Drought-tolerant leaves have adaptations to decrease water loss. Plants that do well in dry, desert environments are known as **xerophytes** (from the Greek *xeros,* "dry," and *phyton,* "plant"). Some desert plants only produce leaves during and following brief rainy seasons. Others form thick, succulent leaves or stems that retain water during dry seasons. Leaves of desert plants frequently have sunken stomata and thick cuticles. The sunken stomata are kept out of the wind, thereby minimizing water loss. Some plants have dense layers of trichomes that give the leaf a white appearance and a wooly feel. These trichomes reduce water loss and also keep the leaf from overheating.

Protective leaves often occur as sharp outgrowths that discourage animals from eating the plant. **Spines** are sharp modified leaves or modified stipules (Figure 4.22a). Sometimes **thorns** are described as modified leaves, but they are actually modified stems that arise from axillary buds found where the leaf joins the stem. **Prickles** are nei-

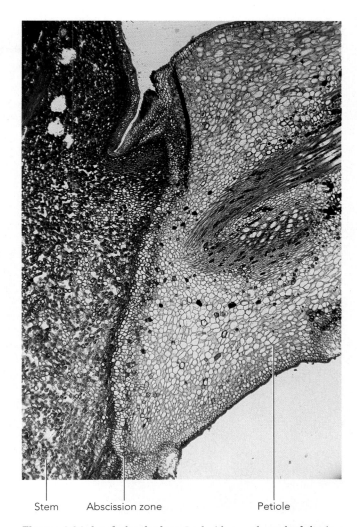

Stem   Abscission zone   Petiole

Figure 4.21 Leaf abscission. In deciduous plants, leaf abscission occurs after formation of an abscission zone in the petiole, typically near the node.

ther modified leaves nor modified stems, but instead develop as outgrowths of cells of the epidermis or cortex.

Some plants form **tendrils**—slender, coiling structures that attach a climbing plant to a supporting structure. Tendrils are typically modified leaves, as in the case of the pea plant (Figure 4.22b), but in some plants stipules or stems can become tendrils, as in the modified stems of grapevines. Tendrils that are stems frequently produce leaves. Tendrils exhibit a growth movement called **thigmotropism** (from the Greek *thigma,* "touch"). When a tendril touches an object, the opposite side of the tendril begins to grow rapidly so that it curves around what it has touched. The amount of growth depends on the degree of stimulation, so continuous contact is necessary for the tendril to wrap tenaciously around an object.

Figure 4.22 Modified leaves.

(a) Spines. When leaves or structures associated with leaves produce sharp projections, they are known as spines. The barberry plant (*Berberis dictophylla*), for example, produces spines from leaves.

(b) Tendrils. Tendrils can be modified leaves, stipules, or stems. In this pea plant (*Pisum sativum*), the end of the leaf has become a tendril while the bottom has remained leaflike. If the tendril doesn't touch an object, it curls around an imaginary central axis.

(c) Window leaves. Window leaves occur in a few plants growing in hot dry regions like this *Haworthia coeperi*. Most of the plant, including the leaf, is buried in soil or sand. The top of the leaf is exposed to the Sun, and its transparent "window" admits light that is transmitted to underground portions of the leaf, where photosynthesis occurs.

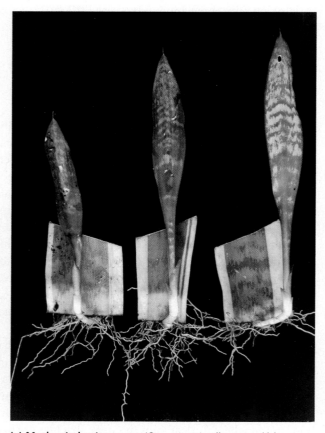

(d) Bracts. The brilliant red color in this poinsettia does not come from the petals of a flower but from modified leaves called *bracts*, found at the base of the flower. The actual poinsettia flower has no petals.

(e) Mother-in-law's tongue (*Sansevieria*) will root and bloom quite readily when leaf cuttings are planted in compost.

EVOLUTION

Leaves That "Eat" Insects

Why do some plants have leaves that "eat" insects? The answer has to do with nitrogen and the environment. Plants typically get nitrogen from the soil, often relying on nitrogen-fixing bacteria. However, swamps and bogs have little nitrogen available in the soil because their acidic environment is unsuitable for the growth of nitrogen-fixing bacteria. Epiphytes, which often grow on other plants high in the rain forest, sometimes lack sufficient nitrogen as well. In more than 200 species of plants, modified leaves have evolved that trap insects as an alternative source of nitrogen. Among these plants are the bladderwort, sundew plant, Venus's flytrap, and pitcher plant.

Bladderwort (*Utricularia vulgaris* and related species) is an aquatic plant that produces tiny bladders with a trap door. The bladders, typically less than a half-

Sundew plant.

centimeter in diameter, open their doors when an aquatic insect brushes against one of four stiff trigger hairs. Water rushes into the bladder, along with the insect. Then the door closes, and enzymes digest the hapless victim.

The leaves of sundew plants (*Drosera rotundifolia* and related species) produce mucilage-coated sticky hairs that attract and trap insects. Once an insect is attached to the hairs, the leaf folds around it. Enzymes released by the leaf and associated bacteria digest the insect and the leaf absorbs the digested molecules. Following digestion, the leaf resumes its normal shape and secretes new mucilage.

Venus's flytraps (*Dionaea muscipula*) have leaves that bend in the middle, with each half looking like half of a trap. When an insect trips two of the three trigger hairs on each half, the leaf snaps shut and enzymes digest the insect. Bits of dirt that might fall into the trap normally do not cause it to close.

Pitcher plants of several genera have leaves shaped like a vase or pitcher that collects water. Glands around their tops secrete nectar that attracts insects. Often insects fall into the trap and drown. Many pitcher plants have bristles that point downward, preventing escape. In Malaysia, several species of tree frogs lay their eggs in the traps, thus assuring sufficient water and food for the offspring. The eggs produce enzymes that prevent the plant's digestive enzymes from acting.

Pitcher plant.

Floating leaves occur in plants such as water lilies, which have leaves with extra air pockets between spongy mesophyll cells that help the leaves float. They also have less vascular tissue because of a reduced need to pull water up from the roots, as well as a reduced need for support. Finally, stomata in leaves of floating plants, such as water lilies, are typically located on the top instead of bottom of the leaves. *Victoria amazonica*, the giant Amazon water lily, has huge floating leaves that can support the weight of a child.

Window leaves are found in succulent plants growing in the Kalahari Desert in southern Africa (Figure 4.22c).

These plants solve the problems of hot, dry desert conditions by burial in the sand, with their leaves barely protruding into the air. The end of each leaf, perhaps a centimeter in diameter, has a transparent waxy cuticle that serves as a "window" to admit light. Light passes through several cell layers of transparent water storage cells before reaching photosynthetic cells, which line the outside of the leaves farther down. You might think that the windows admit light to increase photosynthesis of leaves that remain beneath the soil. However, research indicates that when the windows were blocked with reflective tape, CO_2 uptake by the leaves did not significantly decrease. Therefore, the ecological significance of the windows is not yet understood.

Bracts are modified leaves at the base of flowers but are not flower parts themselves. Poinsettias (*Euphorbia pulcherima*), for example, have red or white bracts (Figure 4.22d). In dogwood, the large white or pink bracts are actually more noticeable than the flowers.

In many plants, leaves can be a means of asexual reproduction. For example, if the leaf or stem of an African violet or a coleus is removed from the plant, it can be rooted either in soil or in water. A root-inducing hormone can be used to increase the speed and intensity of rooting, enabling the rooted leaf, referred to as a "leaf cutting," to be planted in soil.

Rex begonia plants can be reproduced from leaf disks that are punched out of the leaf and placed on moist filter papers, where roots will form from the cut veins in the disk. Another method is to place a leaf on soil and make cuts in the larger veins, where roots will form under moist conditions. With either method, the leaves removed from the parent plant can produce roots and then buds from the cut surfaces of petioles or veins in the blades, particularly if conditions are moist and the leaves are succulent.

Insect-eating leaves trap insects in various ways, digest them into small molecules, and absorb the molecules to supply nitrogen to the plant. The *Evolution* box on page 95 provides information about several types of these modified leaves.

Section Review

1. Describe how a leaf forms from a leaf primordium.
2. What are the functions of the leaf epidermis?
3. What is the difference between palisade mesophyll and spongy mesophyll?
4. What are some examples of how leaves are adapted to particular environments?
5. Describe three types of modified leaves.

SUMMARY

Roots

Taproot systems usually penetrate more deeply than fibrous root systems (pp. 72–73)
Taproot systems have a large main root, while fibrous systems do not.

Root development occurs near the root tip (p. 73)
Initials in the small quiescent center of root apical meristems produce replacement initials and derivative cells that give rise to specialized cell types.

The root cap protects the root apical meristem and helps the root penetrate the soil (p. 73)
The root apical meristem produces a root cap to protect the meristem as the root grows into the soil. Cells of the root cap produce slimy mucigel that eases passage of the root past soil particles.

Absorption of water and minerals occurs mainly through the root hairs (p. 74)
In the maturation zone, near root tips, some epidermal cells elongate to produce root hairs, which absorb water and minerals from the soil.

The primary structure of roots relates to obtaining water and dissolved minerals (pp. 74–76)
Most dicot and conifer roots have a lobed core of xylem, with phloem between the lobes. Most monocot roots have a core of undifferentiated cells surrounded by alternating rings of xylem and phloem. A pericycle, which gives rise to branch roots, surrounds the vascular cylinder. An endodermis surrounds the pericycle. The Casparian strip prevents substances from entering the stele unless they pass through the membranes of endodermal cells.

Some roots have specialized functions in addition to anchoring the plant and absorbing water and minerals (pp. 76–77)
Some modified roots produce adventitious buds, anchor epiphytic plants, provide enhanced stability by buttressing, provide

oxygen to roots of swamp plants, and store water or food. Contractile roots pull the plant deeper into the soil. Parasitic roots, known as *haustoria,* penetrate stems and roots of other plants.

Roots have cooperative relationships with other organisms (pp. 77–80)

Mycorrhizae are mutualistic associations between plant roots and soil fungi. Plants gain increased absorption of minerals and perhaps protection from disease. The fungus gains organic materials from the plant.

Stems

Botanists have developed zone and cell-layer models to describe stem growth (pp. 81–82)

In both models, shoot apical meristems have a central zone of slowly dividing initial cells. Derivatives move into regions of further division, known as the *protoderm, ground meristem,* and *procambium,* and finally into zones of elongation and maturation.

In primary growth of most stems, the vascular tissue forms separate bundles (pp. 82–83)

Transverse sections of stems reveal protosteles in some seedless plants, siphonosteles with leaf gaps in some ferns, eusteles with rings of vascular bundles in most dicots and conifers, and dispersed vascular bundles in most monocots.

A transition region maintains vascular continuity between the root and stem (p. 83)

Where the stem meets the root, the vascular pattern of one transitions into that of the other.

Leaf primordia form in specific patterns on the sides of shoot apical meristems (pp. 83–84)

Leaf primordia form in predictable patterns, or phyllotaxy, on the sides of shoot apical meristems.

Stem variations reflect different evolutionary pathways (pp. 85–86)

The palm produces a stem with many elongated nodes. The wheat plant produces a stem with a few nodes, of which only a few elongate.

Some stems have specialized functions in addition to support and conduction (pp. 86–87)

A number of types of specialized primary stems help in reproduction (stolons and rhizomes) or store starch (tubers, corms, or bulbs).

Leaves

A leaf primordium develops into a leaf through cell division, growth, and differentiation (p. 88)

Leaf buttresses on the apical meristem develop into leaf primordia. A leaf primordium grows into a thin, stemlike petiole and a flattened blade. A ridge of dividing cells on each side of the blade controls leaf shape.

The leaf epidermis provides protection and regulates exchange of gas (pp. 88–89)

Epidermal cells produce a cuticle that reduces water loss and entry of fungi and bacteria. Exchange of CO_2, O_2, and water vapor with the outside air occurs through stomata, pores regulated by guard cells.

Mesophyll, the ground tissue in leaves, carries out photosynthesis (pp. 89–90)

Mesophyll cells can form palisade rows of cells on the sunny side of a leaf. Spongy mesophyll allows movement of gases to all parts of the leaf.

The vascular tissue in leaves is arranged in veins (p. 90)

Leaf vascular bundles called *veins* are continuous with vascular tissue in the stem. Dicots usually have netted venation, while monocots usually have parallel venation.

Leaf shapes and arrangements have environmental significance (pp. 90–91)

Large, thin blades are more common in areas with cooler temperatures, lower light levels, moist conditions, and no wind. Smaller leaves with variable margins are more common with warmer temperatures, high light levels, dry conditions, and high wind.

Abscission zones form in the petioles of deciduous leaves (pp. 91–93)

Deciduous trees lose all their leaves at certain seasons of the year. In abscission zones, the weakening of middle lamellas and cell walls results in the breaking point for the departing leaves.

Some leaves have specialized functions in addition to photosynthesis and transpiration (pp. 93–96)

Some leaves are designed for drought tolerance or protection. Others are modified to grasp structures or float on water. Some form colorful bracts that appear to be part of a flower. Some can produce new plants.

Review Questions

1. What are the principal functions of roots?
2. Why might a tree that normally forms a deep taproot system form a shallow root system that spreads out over a wide area?
3. Describe how a root grows in length.
4. What is the function of root hairs?
5. Why is the Casparian strip important?
6. Describe several types of modified roots.
7. What are mycorrhizae?
8. Why is there a transition zone between stem and root?

9. Describe several types of modified stems.
10. What two functions does transpiration perform for a leaf?
11. How do leaves form?
12. What are some ways in which leaf shapes reflect environmental adaptations?
13. What is the difference between palisade and spongy mesophyll?
14. Describe several types of modified leaves.

Questions for Thought and Discussion

1. Why do you think few plants are parasites on one another?
2. Why do you think some plants store carbohydrates in roots and shoots, while others do not?
3. Considering what you know about the conducting cells of xylem and phloem, why do you think xylem is found to the inside and phloem to the outside in most vascular systems?
4. Why do you think the arrangement of vascular tissue is different in roots and shoots of the same plant?
5. If the CO_2 concentration in the air continues to increase, what changes might you predict in the future evolution of plants?
6. Some plants have very little leaf development, relying mainly on stems for photosynthesis. How would you explain this?
7. Think of a particular type of modified root, stem, or leaf. Then describe some possible intermediate stages of mutations that might have led to that modification and why those mutations might have had selective value for that plant in its environment.
8. Examine Figure 4.11, which shows how the protostele of the root of a buttercup (a dicot) makes the transition into the eustele of the stem. Draw a similar diagram to show how this transition might occur in a monocot.

Evolution Connection

Comparisons between living and fossil vascular plants suggest to biologists that the eustele evolved from a more primitive protostele. If that is so, why do you think that the roots of flowering plants are relatively unchanged in retaining a protostele-type arrangement of tissues whereas the stems of flowering plants are eusteles?

To Learn More

Visit The Botany Place Website at www.thebotanyplace.com for quizzes, exercises, and Web links to new and interesting information related to this chapter.

Bartoletti, C. Susan. *Black Potatoes: The Story of the Great Irish Famine, 1845–1850.* Boston: Houghton Mifflin, 2001. An account of the tragic effects of the potato famine in Ireland.

Bubel, Nancy. *The New Seed Starter's Handbook.* Emmaus, PA: Rodale Press, 1998. An excellent book on growing plants from seeds.

D'Amato, Peter. *The Savage Garden: Cultivating Carnivorous Plants.* Berkeley, CA: Ten Speed Press, 1998. Written by an expert at growing carnivorous plants.

Dekruif, Paul. *Hunger Fighters.* New York: Harcourt, 1967. The story of early research to find improved varieties of wheat and other crops.

Slack, Adrian, and Jane Gate (photographer). *Carnivorous Plants.* Boston: MIT Press, 2000. Beautiful pictures and interesting details about insectivorous plants.

Secondary Growth in Plants

A Dragon Tree (*Dracaena draco*), Canary Islands, Spain.

Secondary Growth: An Overview

Lateral meristems, cylinders of dividing cells, produce secondary vascular and secondary dermal tissue

The vascular cambium produces secondary xylem (wood) and secondary phloem

The cork cambium produces secondary dermal tissue

Bark consists of all the tissues external to the vascular cambium

Growth Patterns in Wood and Bark

The vascular cambium produces secondary xylem, secondary phloem, and ray parenchyma, as well as more vascular cambium

Sapwood conducts water and minerals, but heartwood does not

Growth rings in wood reflect the history of secondary growth in a tree

Dendrochronology is the science of tree ring dating and climate interpretation

Growth patterns in reaction wood counteract leaning

The cork cambium is reformed as the stems and roots enlarge

Lenticels are pathways in the bark for gas exchange

Commercial Uses of Wood and Bark

Wood is used mainly for fuel, paper products, and construction

Wood structure can be studied from three cutting planes

Wood can vary in properties such as hardness and grain

Latex, resin, and maple syrup are some products from wood fluids

Commercial cork comes from the thick outer bark of some trees

Trees are a renewable but limited natural resource

Madagascar, a large island off the eastern coast of Africa, is home to some of the world's most unusual trees, including the giant baobab (*Adansonia grandidieri*). According to an African legend, the baobab was the first tree and became increasingly jealous and vocal as God created the other trees. To shut up the baobab, God finally pulled it out of the ground and put it back upside down. Indeed, when baobab branches lose their leaves, they look as though their roots are high in the air.

The giant baobab is an example of a plant that has secondary growth—wood and bark. However, its cylindrical trunk, which can be up to 25 meters (82 feet) high and about 3 meters (10 feet) in diameter, also contains many parenchyma cells for water storage, enabling the tree to survive in the dry climate of western Madagascar. When hollowed out, giant baobab trunks have been made into huts and even a jail. Meanwhile, the fiber in the bark can be made into rope, paper, and cloth.

The giant baobab is one of eight baobab species, six of which are unique to Madagascar, which has been geographically separated from the east African coast for more than 140 million years. During that time, many species of organisms unique to the island evolved in its many different environments, including mangrove swamps along

These giant baobabs in Madagascar are a rare and endangered species.

the shore, rain forests, dry forests, and deserts. In terms of species diversity, Madagascar is one of the world's richest areas, with approximately 5% of all living species of animals and plants, including about 10,000 plant species. About 80% of those plant species are unique to the island, and a single mountaintop may contain 200 plant species found nowhere else. Many of the island's plant and animal species interrelate in interesting ways. For example, the flowers of the exotic traveler's palm (*Ravenala madagascariensis*) are pollinated only by the black-and-white ruffed lemur, a primate that eats fruits and nectar. Since this lemur is an endangered species, so is the palm. The giant baobab tree, which produces large, fragrant flowers, is pollinated in part by lemurs.

Some woody plants, like the giant baobab, are unusual. Many other woody plants, of course, are quite common. Woody plants consist of all gymnosperms, about 20% of all dicot species, and about 5% of all monocot species. Since so few monocots have secondary growth, this chapter will focus on gymnosperms and woody dicots. Conifers such as pine, fir, and redwood trees are examples of gymnosperms, while woody dicots include such trees as oak, maple, and walnut.

You have already seen in Chapter 4 how primary growth produces roots, stems, and leaves. We will now look at how wood and bark develop, how they make stems and roots thicker and stronger, and how they provide products for human use.

Giant baobab flower.

Secondary Growth: An Overview

Secondary growth is an increase in the girth of a plant initiated by cell divisions in lateral meristems. Primary growth and secondary growth happen simultaneously but in different parts of a woody plant. As primary growth continues, as the result of the activity of the apical meristems at the tips of shoots and roots, secondary growth adds width to older areas of the stems and roots that are no longer growing in length. Typically, stems have much more secondary growth than roots. First we will look in general at how secondary growth occurs in relation to primary growth.

Lateral meristems, cylinders of dividing cells, produce secondary vascular and secondary dermal tissue

In Chapter 4, you saw that apical meristems are areas of dividing cells that give rise to three regions of further cell division: the protoderm, procambium, and ground meristem. The protoderm produces the primary dermal tissue, called the *epidermis;* the procambium produces primary vascular tissues; and the ground meristem produces ground tissue: pith and cortex. Primary growth is growth in length and occurs in regions of cell elongation just behind the meristematic regions of dividing cells. Cell elongation is accompanied or followed by cell differentiation into different types of mature cells.

Making stems and roots thicker rather than longer involves quite a different process. As you saw in Chapter 3, the meristems that produce secondary growth are called *lateral meristems* (also known as *secondary meristems*) and are cylinders rather than clusters of undifferentiated cells. Instead of growth in length, secondary growth is radial—increasing the diameter of a stem or root as dividing cells produce lateral, or sideways, growth. All along a lateral meristem, new cells are added internally, toward the center and toward the surface of the stem or root. Since secondary growth has been going on longer at the base of a trunk than at the top, lateral meristems are actually shaped a bit more like cones than cylinders.

Secondary growth arises in regions of a woody plant where primary growth has ceased, typically during the first or second year of the plant's growth. The process starts when differentiated cells revert to become undifferentiated cells, forming two lateral meristems called the *vascular cambium* and *cork cambium.* The word *cambium* comes from the Latin *cambire,* meaning "to exchange." Cambium cells are cells that have exchanged their previous roles for a new role of dividing repeatedly to produce new growth. As the name indicates, the **vascular cambium** produces vascular tissues, which are called secondary xylem and secondary phloem to distinguish them from the primary xylem and primary phloem formed by the procambium. The vascular cambium itself forms from cells in the cortex and procambium. In roots, pericycle cells are also involved. **Cork cambium,** also known as **phellogen** (from the Greek *phellos,* "cork," and *genos,* "birth"), forms initially from parenchyma cells in the cortex and sometimes in the primary phloem. Cork cambium (plural, *cambia*) produces new dermal tissue, which eventually replaces the epidermis formed by the protoderm. Figure 5.1 provides an overview of the meristems involved in primary and secondary growth.

When we think of woody plants, we usually think of large trees. However, many shrubs are quite woody, and some smaller "herbaceous" plants such as alfalfa form woody stems even during their first year of growth (see the *Plants & People* box on page 102).

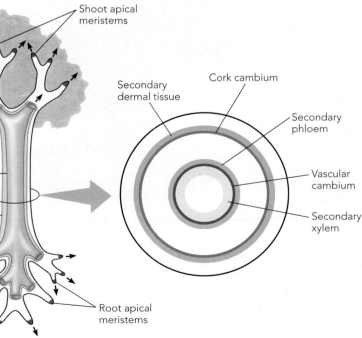

Figure 5.1 Locations of apical and lateral meristems. The apical meristems produce primary growth that lengthens shoots and roots. The lateral meristems—the vascular cambium and cork cambium—are cylindrical and produce secondary growth that thickens stems and roots.

Bonsai Trees

Bonsai is the art of growing trees in a confined space to simulate the environmental conditions that give trees some of their most stunning and beautiful effects. Bonsai is not stunting a tree's growth but causing the plant to grow slowly and in specific directions. The goal is to use a small plant to "suggest" a larger natural scene. The word *bonsai* is translated roughly as "tree in a tray or pot." Bonsai trees are miniature reproductions of trees in their natural state. When properly cared for, these trees can live for many years. Examples exist today of bonsai almost 500 years old.

The art of growing bonsai trees, which originated in China as far back as 200 B.C. and spread to Japan in the tenth century, includes a variety of trunk and planting styles. Some criteria of beauty include the shape and size of roots and the shape and branching of the trunk. Also important is the arrangement of the branches and foliage. Some bonsai even have reproductive structures such as cones, flowers, or fruit. Bonsai trees can be shaped in a wide variety of styles designed to simulate nature, including formal upright, informal upright, slanting, cascade, and windswept.

A bonsai forest of Japanese larch.

Bonsai maple.

Although the art of bonsai can be painstaking and complex, here is a quick and easy way to create these miniature art forms with few resources and minimal time and effort:

- ◆ Select a species. You can use trees, shrubs, or even vines. You will want a relatively small plant, such as one that you can purchase in a small container at a nursery.
  - ◆ Select a shallow, decorative container. A bonsai in a glazed container requires less frequent watering.
    - ◆ Select a small specimen and prune the roots and shoots considerably so that they fit easily in your container. Decide which direction the resulting bonsai will face, and shape accordingly.
- ◆ Add small rocks to the bottom of the container for good drainage. Use potting soil for a growth medium. Plant your bonsai. Cover the soil surface with pebbles or gravel according to your tastes.
- ◆ Supply water frequently, especially if the container is small, and small amounts of slow-release fertilizer occasionally.
  - ◆ Place the bonsai in the level of light it prefers in nature. If the plant is in high light, water frequently.
  - ◆ Wrap tape or wire around individual branches to encourage growth in a particular orientation. Once you have oriented the branch correctly, remove the tape or wire.
- ◆ Be prepared to experiment and learn by doing.

The vascular cambium produces secondary xylem (wood) and secondary phloem

As you saw in Chapter 4, a cross-sectional view of a typical dicot or conifer stem shows the primary xylem and phloem forming either a complete or broken ring of vascular bundles, with xylem facing the center and phloem facing the outside. In a three-dimensional view, these vascular bundles form a cylinder. In dicot and conifer stems, secondary growth begins when vascular cambium cells arise from residual procambium cells between the primary xylem and phloem. If the vascular bundles form a broken ring, as shown in Figure 5.2a, the vascular cambium's cells also arise from parenchyma cells between the bundles. The vascular cambium's cells are existing cells that become meristematic at different times, under the influence of the hormone auxin, until eventually they form a complete cylinder that runs through the middle of each vascular bundle.

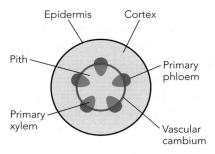

(a) Initial formation of vascular cambium in stem

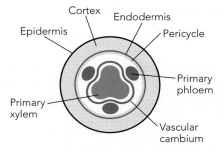

(b) Initial formation of vascular cambium in root

Figure 5.2 Initial formation of the vascular cambium in stems and roots. **(a)** In woody stems, the vascular cambium initially forms between the primary xylem and primary phloem in the vascular bundles and in the parenchyma tissue between bundles. **(b)** In a root, the vascular cambium does not initially form a cylinder because of the lobed arrangement of the primary xylem.

In roots, the arrangement of primary xylem and phloem prevents the vascular cambium from initially forming in a circular configuration (Figure 5.2b). Dicot and conifer roots have a central lobed cylinder of primary xylem, with patches of primary phloem nestled between the lobes. The vascular cambium of roots first forms as separate sections that grow together into an irregular shape, weaving between the primary phloem and the lobes of primary xylem, with pericycle cells becoming vascular cambium at the tips of xylem lobes. Within a year or so, different rates of cell division in the vascular cambium result in the formation of a cylinder.

Secondary xylem expands the plant's capacity to carry water and minerals up from the roots and adds structural support. Secondary phloem increases transport of food from the leaves. Both secondary xylem and secondary phloem contain conducting cells, which replace older cells that no longer conduct. Each year the vascular cambium adds secondary xylem to the inside and secondary phloem to the outside of the vascular cambium (Figure 5.3). In both stems and roots, vascular cambia produce much more xylem than phloem. As the stem or root grows in thickness, the mature primary xylem and primary phloem

tissues are pushed farther apart. Meanwhile, the vascular cambium itself expands in diameter. Later in the chapter, you will see how the cells of the vascular cambium divide to increase the diameter of the vascular cambium and to produce new secondary xylem and secondary phloem.

Secondary xylem is what we commonly call **wood**. In fact, *xylem* comes from the Greek word, *xylon*, or "wood." Like primary xylem, secondary xylem consists largely of dead cells. Only the more recently formed layers of secondary xylem conduct water and minerals, with the primary xylem and older secondary xylem being inactive. Similarly,

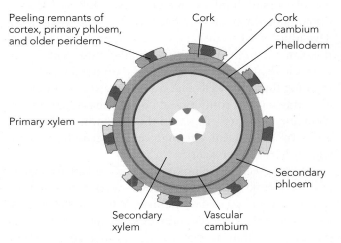

(a) Stem growth after several years

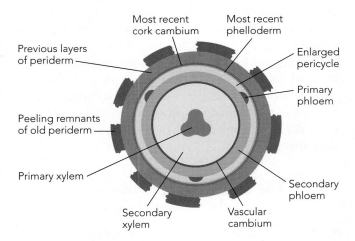

(b) Root growth after several years

Figure 5.3 Stem and root growth after several years.
(a) As a woody stem expands in width, the primary xylem and primary phloem are pushed farther apart as the vascular cambium produces layers of secondary xylem and secondary phloem. Meanwhile, the epidermis ruptures and is eventually sloughed off, as are the older layers of periderm produced by previous cork cambia. **(b)** As the vascular cambium expands in diameter within the root, it eventually becomes circular, and the layers of secondary xylem and phloem look similar to those in a stem. Meanwhile, the layers of periderm have replaced the epidermis as the outer covering of the root.

only the more recent layers of living secondary phloem conduct food because the primary phloem and older secondary phloem cells no longer function as conducting cells. Older phloem cells no longer conduct because they are stretched and broken when new cells produced by the vascular cambium push them outward. Older xylem cells no longer conduct because an increasing number of vessels have broken columns of water and an increasing number of tracheids contain air. These changes will be discussed in Chapter 10.

Secondary growth enhances the two basic functions of vascular tissue: conduction and support. Because of secondary growth, plants can become taller and live longer. The addition of secondary xylem and phloem increases the plant's capacity to conduct water and nutrients, with the basic shapes and structures of the conducting cells remaining the same. Cells that transport water and minerals—tracheids in conifers and both tracheids and vessel elements in dicots—function the same way in primary and secondary xylem, although cells of secondary xylem usually have thicker walls.

Likewise, cells that transport food—sieve cells in conifers and sieve-tube members in dicots—work the same way in primary and secondary phloem. What does change is that secondary xylem cells provide greater support for the plant because they have more lignin, which adds considerable strength to the cellulose backbone of cell walls. Lignin accumulates in and between cell walls, particularly in secondary xylem, and constitutes up to 25% of the dry weight of wood. Lignin is the second most common organic compound on Earth, right behind cellulose itself. Without the additional lignin produced in secondary growth, trees could not grow very tall or withstand strong winds, and roots could not penetrate dense layers of soil. You will see in Chapter 7 how the structure of lignin provides strength.

The cork cambium produces secondary dermal tissue

As you saw in Chapter 3, there are two types of dermal tissue in vascular plants: epidermis and periderm. The epidermis and cortex form during primary growth and are eventually replaced by periderm in plants that have secondary growth. Periderm is produced by the cork cambium and consists of cork, phelloderm, and the cork cambium cells themselves. **Cork,** also known as **phellem** (from the Greek *phellos,* "cork"), forms to the outside of the cork cambium and consists of dead cells when mature. **Phelloderm** (from the Greek *phellos,* "cork," and *derma,* "skin") is a thinner layer of living parenchyma cells that forms to the inside of each of the many cork cambia.

The first layer of cork cambium forms at the same time or somewhat later than the vascular cambium in regions of a stem or root where primary growth has ceased. Unlike the vascular cambium, the cork cambium does not grow in diameter. Every year, or sometimes less frequently, a new cork cambium forms inside the old one, creating another layer of periderm inside the old periderm.

In a stem, the first cork cambium arises from parenchyma cells in the outermost layers of the cortex (Figure 5.4a). Each new cork cambium arises from cortex tissue to the inside, until eventually the cortex is used up in this

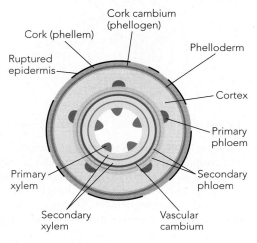

(a) Initial formation of cork cambium in stem

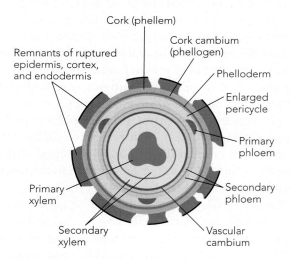

(b) Initial formation of cork cambium in root

Figure 5.4 Initial formation of the cork cambium in stems and roots. (a) In a woody stem, the initial cork cambium arises from the outermost layer of cortex. To the inside, the cork cambium produces a layer of living parenchyma cells called *phelloderm.* To the outside, it produces a layer of dead cork cells. Each successive cork cambium forms to the inside. **(b)** In a woody root, the cork cambium forms from the outermost layer of enlarged pericycle.

manner. As the diameter of the stem expands due to the action of the vascular cambium, the cortex expands. Since no cell division occurs in the cortex, the expansion eventually causes the cortex to break apart and fall off the stem. Subsequent cork cambia then arise from the secondary phloem to the inside.

In a root, the initial cork cambium forms after changes in the endodermis and pericycle, two layers not present in a stem (Figure 5.4b). Since water and minerals are no longer absorbed in root areas undergoing secondary growth, the filtering function of the endodermis is no longer needed, so this tissue layer becomes inactive. Meanwhile, to the inside of the endodermis, the pericycle no longer gives rise to branch roots but instead widens as it is pushed toward the outside while the vascular cambium adds layers of secondary xylem and phloem. From the outer layer of this enlarged pericycle emerges the first cork cambium, which forms a layer of periderm. The outermost layers of the root—the endodermis, cortex, and epidermis—become stretched and eventually rupture and peel off, leaving periderm as the outer covering. As the root continues to expand—generally more slowly than the stem—a new cork cambium and periderm form inside the previous one, inside the older secondary phloem. Later in the chapter, we will look at how cork cambium cells divide to produce the outer layers of a stem or root.

Bark consists of all the tissues external to the vascular cambium

The term *bark* is often misunderstood as simply referring to the protective outer surface of a tree. Botanically speaking, **bark** consists of all tissues outside the vascular cambium—in other words, the part of a stem or root surrounding the wood. Peeling the bark off a tree frequently removes the vascular cambium as well. Bark has two distinct regions: an inner bark and an outer bark (Figure 5.5). The **inner bark** consists of the living secondary phloem, dead phloem between the vascular cambium and the currently active, innermost cork cambium, and any remaining cortex. Each new cork cambium forms inside the previous one, in secondary phloem. There the structure of the inner bark of the stem becomes a mishmash of secondary phloem and periderm. The **outer bark** consists of dead tissue—including dead secondary phloem and all the layers of periderm outside of the most recent cork cambium. As periderm layers build up in the outer

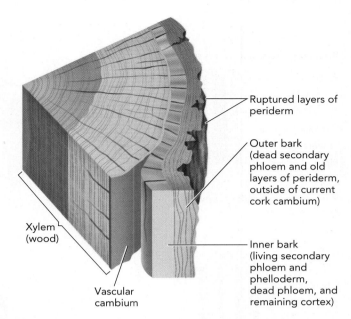

Figure 5.5 Inner and outer bark.

- Ruptured layers of periderm
- Outer bark (dead secondary phloem and old layers of periderm, outside of current cork cambium)
- Inner bark (living secondary phloem and phelloderm, dead phloem, and remaining cortex)
- Xylem (wood)
- Vascular cambium

Figure 5.5 Inner and outer bark. Bark consists of all tissues outside the vascular cambium. These include phloem and periderm. Inner bark consists of living secondary phloem and living phelloderm produced by the most recently formed cork cambium. Outer bark consists of dead secondary phloem plus periderm from earlier, now inactive cork cambia and cork produced by the current cork cambium.

bark, the outermost layers gradually crack and peel off in patterns that vary from species to species (Figure 5.6).

Since periderm does not continue accumulating while secondary xylem does, bark is typically much thinner than the woody portion of a stem or root. Even so, some trees, such as redwood, produce bark up to 30 centimeters thick (1 foot), which protects the trees from fire damage. The

(a) Silver birch (*Betula pendula*)

(b) Birch (*Betula albosinensis*)

(c) Madrone (*Arbutus menziesii*)

Figure 5.6 Some variations in bark. (a) Thick, checkered bark. **(b)** Smooth bark. **(c)** Thin, rough bark.

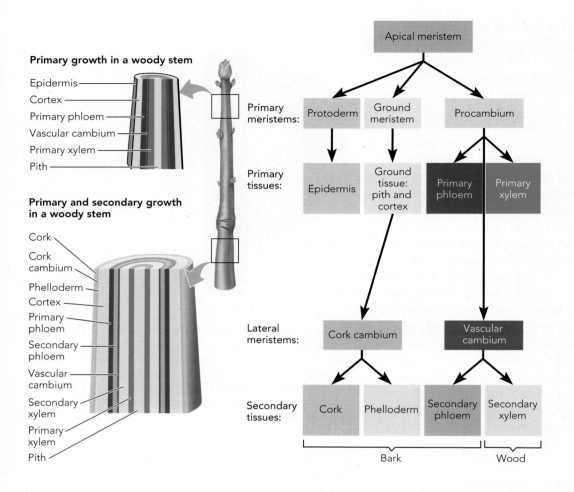

Primary growth in a woody stem

- Epidermis
- Cortex
- Primary phloem
- Vascular cambium
- Primary xylem
- Pith

Primary and secondary growth in a woody stem

- Cork
- Cork cambium
- Phelloderm
- Cortex
- Primary phloem
- Secondary phloem
- Vascular cambium
- Secondary xylem
- Primary xylem
- Pith

Apical meristem

Primary meristems: Protoderm | Ground meristem | Procambium

Primary tissues: Epidermis | Ground tissue: pith and cortex | Primary phloem | Primary xylem

Lateral meristems: Cork cambium | Vascular cambium

Secondary tissues: Cork | Phelloderm | Secondary phloem | Secondary xylem

Bark   Wood

Figure 5.7 Summary of meristems and tissues in primary and secondary growth. Roots and stems have the same primary and lateral meristems and the same layers of vascular and dermal tissues. The diagram shows a typical woody stem. Woody roots generally do not have pith in the center.

splitting of older layers of periderm results in the rough appearance of the outer bark of most trees, but in some trees, such as birches, the bark is relatively smooth because it stretches.

Although bark is relatively thin compared with wood, it is vital to keeping a tree alive. The dead tissue of the outer bark provides protection, while the living secondary phloem of the inner bark carries sugar and other organic molecules from leaves to roots. Since roots generally lack photosynthesis, they need the food transported by the phloem in the bark. If the entire bark is removed in a complete ring around a tree—a process called **girdling**—the transport in the phloem between shoot and root is disrupted, thereby killing the tree, after any food stored in the roots is utilized. If only a narrow strip of the inner bark remains, the tree can still survive because new bark will form from the remaining phloem. Porcupines often kill trees by girdling, for when they eat bark—really their interest is in the sugar-rich phloem of the inner bark—they seem to find it easier to move around a tree than to climb up or down. Since bark includes all of the phloem, it contains—along with the leaves—most of the accumu-

lated nutrients. Therefore, when trees are harvested, the soil fertility can be largely maintained by leaving the bark and leaves on the ground, a practice that fosters the sustainable development of forest resources.

Figure 5.7 provides a summary of primary and secondary growth in a woody stem. As you saw in Figures 5.2, 5.3, and 5.4, secondary growth in woody roots is very similar to secondary growth in woody stems. In both roots and stems, the vascular cambium gives rise to the inner bark and the wood, while the cork cambium produces the outer bark.

Section Review

1. Compare and contrast lateral meristems with apical meristems.
2. What is the role of the vascular cambium?
3. What is the role of the cork cambium, and how does it form differently in roots and stems?
4. Describe the components and basic functions of bark.

Growth Patterns in Wood and Bark

How do dark and light regions form in wood? Why is bark typically rough and uneven? How can a large tree stay alive even after the middle of its trunk has been hollowed out? In this section, we will look at these and other aspects of the growth of wood and bark.

The vascular cambium produces secondary xylem, secondary phloem, and ray parenchyma, as well as more vascular cambium

The wellspring, or continuous source, of wood in a tree is the vascular cambium, which also produces the secondary phloem that conducts food within the inner bark. Two types of meristematic cells form the vascular cambium: fusiform initials and ray initials. Like the initials in apical meristems, these initials can undergo periclinal cell divisions (parallel to the surface of the stem or root) and anticlinal cell divisions (perpendicular to the surface of the stem or root). Periclinal cell divisions add to the thickness of a stem or root, as one daughter cell remains meristematic within the vascular cambium while the other matures either to the inside or to the outside of the vascular cambium (Figure 5.8). Anticlinal cell divisions increase the diameter of the vascular cambium itself, as a fusiform initial or ray initial divides so that both daughter cells remain within the vascular cambium as initials.

Fusiform initials (from Latin *fusus,* "spindle") arise within the vascular bundles and produce new vascular tissue—xylem to the inside and phloem to the outside. The word *fusiform* means "tapered toward each end," referring to the slender shape of these elongated cells, simi-

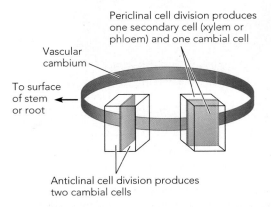

Figure 5.8 Cell division in the vascular cambium. Cambial cells can divide parallel with or perpendicular to the vascular cambium.

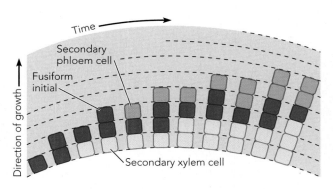

Figure 5.9 Development of the vascular cambium over time. Cells of the vascular cambium divide to produce xylem or phloem.

lar to the conducting cells they produce: the tracheids and vessel elements of the xylem and the sieve cells and sieve-tube members of the phloem. A fusiform initial can divide so that one daughter cell becomes either xylem or phloem, depending on which side of the vascular cambium it appears. Meanwhile, the other daughter cell remains a fusiform initial (Figure 5.9). As you have seen, a fusiform initial can also divide to produce two fusiform initials that remain within the vascular cambium. The fusiform initials, as well as the conducting cells they produce, are aligned lengthwise—parallel to the surface of the stem or root.

The cell divisions of the fusiform initials typically produce much more xylem than phloem, with each year's growth of xylem being visible as a ring. The rings are apparent because the sizes of new cells vary with the season. The small cells of late summer are easily distinguished from the large cells of the following spring. The rings of xylem accumulate year after year, but the innermost rings no longer conduct water and minerals. In the case of phloem, typically only the current year's cells conduct food. The older, outermost layers of phloem cells may still contain sugar molecules attached to cell walls but are no longer living. As the vascular cambium produces new layers of phloem, the oldest layers get pushed into the outer bark and are eventually shed.

Ray initials usually arise between the vascular bundles and are often cubical cells (Figure 5.10). They do not strengthen a tree but instead produce cubical parenchyma cells that serve mainly for storage and also some sideways transport. When a ray initial divides in this manner, one daughter cell remains a ray initial while the other becomes a parenchyma cell. These parenchyma cells usually accumulate in files that radiate out, looking like spokes of a wheel and showing up particularly well in cross sections of the wood of older trunks. Rays in secondary phloem are known as *phloem rays,* and those in secondary xylem are

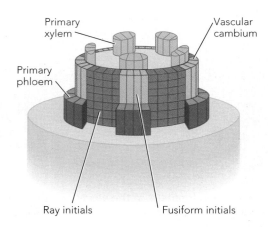

Figure 5.10 Three-dimensional view of vascular tissue in a stem.

known as *xylem rays*. Rays are common in dicots, where they are often more than one cell wide. In conifers and other gymnosperms, they are less common and typically one cell wide. Rays are often shortened if blocked by expansion of xylem or phloem cells on either side or if ray initials stop dividing (Figure 5.11).

Keep in mind, of course, that the vascular cambium is a cylinder of cells in which divisions occur in every cell. When viewed from the side, the layers of fusiform initials may be either lined up in clear rows one on top of an-

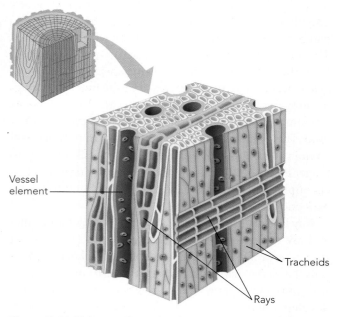

Figure 5.11 Xylem cells and xylem rays in a woody dicot stem. This three-dimensional view of a portion of a woody dicot stem shows the lengthwise orientation of the xylem cells in relation to the ray cells, which run from the center to the surface. Notice that the rays are often interrupted by growth of the conducting cells—in this case the vessel elements and tracheids of the xylem. A three-dimensional view of phloem cells and phloem rays would be similar.

other, an arrangement known as a *storied vascular cambium,* or staggered unevenly in what is called a *non-storied vascular cambium.*

Sapwood conducts water and minerals, but heartwood does not

Although xylem contributes to the strength of a trunk, branch, or root throughout a tree's life, each ring of xylem conducts water and minerals for only a few years before its water columns break, a process called *cavitation*. The older, nonconducting rings of xylem form the center of the trunk or root and are therefore known as **heartwood** (Figure 5.12). Since heartwood no longer transports water and minerals, a large tree can survive even if the center of its trunk is hollowed out. It is the outer rings of xylem, known as **sapwood,** that still transport the xylem sap. Depending on the species, a tree can have anywhere from 1 to 12 xylem rings in its sapwood, with variations also occurring in ratios of heartwood to sapwood and in wood color. Heartwood is usually darker than sapwood, but in some species it is hard to tell the difference. Heartwood and sapwood are generally shades of brown, but species can vary greatly in hue—from the pale gray of ash to the dark chocolate brown of black walnut.

The inactive xylem of the heartwood can endanger a tree by providing a pathway for invading fungi. After cavitation, the entry of fungi into adjacent live cells is often

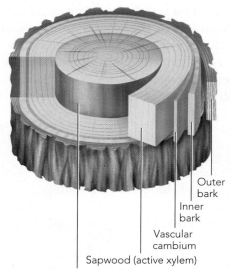

Figure 5.12 Heartwood and sapwood. The newer, outer rings of xylem are actively transporting water and minerals and are known as *sapwood*. The older, inner xylem rings are no longer involved in transport and are known as *heartwood*. Heartwood is often darker in color due to deposits of resins, gums, and other compounds designed to repel insects.

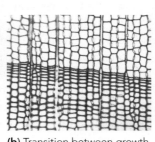

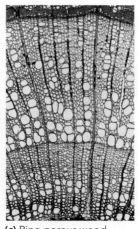

Figure 5.13 Growth rings show the age of a tree. (a) This cross section shows growth rings. **(b)** This micrograph shows the transition between the smaller cells of late-summer growth and the larger cells of spring growth. The contrast is visible as a growth ring. **(c)** This micrograph shows ring-porous wood, in which vessels are contained primarily in early wood.

(a) Growth rings

(b) Transition between growth rings

(c) Ring-porous wood

prevented by the production of tyloses (singular, tylose) that partially or completely plug xylem cells. A tylose is simply a protrusion of cytoplasm from one parenchyma cell into an adjacent conducting cell. Parenchyma cells also produce antibacterial and antifungal substances that make the wood both aromatic and resistant to rotting. For instance, the coast redwood contains significant quantities of preservative compounds, which is why its wood is often used for outdoor furniture and decks. Some trees, like cottonwood, have few or no preservative compounds, so the heartwood rots quickly.

Growth rings in wood reflect the history of secondary growth in a tree

Growth rings are visible because of the distinction between early wood produced in spring and late wood produced in late summer or fall (Figure 5.13a). In spring and early summer, as the days grow longer with abundant moisture and light, the vascular cambium produces large cells with relatively thin secondary cell walls (Figure 5.13b). In late summer, as days grow shorter and cooler, the vascular cambium produces smaller cells with thicker walls. The dividing line between the late-summer wood of one year and the spring wood of the following year is visible as the line between growth rings. In most regions of the world, trees produce one growth ring per calendar year, but a few places have two rainy seasons per year, resulting in two growth rings. Trees in some tropical regions lack clearly visible growth rings because growth continues evenly year-round.

The width of a growth ring may reveal something about the season that produced it. A thick ring results from a season with good growing conditions, while a thin ring indicates the opposite. Variation in ring width due to environmental factors occurs when water, temperature, or nutri-

tion are borderline for the tree's ability to grow. In contrast, abundant water, good soil, and an ideal location result in relatively uniform rings.

Dicot species vary in the distribution patterns of vessels in their annual rings. Some trees, such as oak and sassafras, have ring-porous wood, in which vessels occur principally in spring wood (Figure 5.13c). In other trees, such as aspen and sugar maple, vessels occur throughout each annual ring, a pattern known as *diffuse-porous wood*.

Dendrochronology is the science of tree ring dating and climate interpretation

You already know that a tree's age can be determined by counting the growth rings, and perhaps you have seen large stumps or sections of trees in museums with rings correlated to historical events, such as Columbus's arrival in the Caribbean in 1492. The science of tree-ring dating is called **dendrochronology** (from the Greek *dendron*, "tree," and *chronos*, "time"). Dendrochronology has been used to identify what is perhaps the oldest living organism—a bristlecone pine (*Pinus longaeva*) growing in the rather inhospitable conditions of the White Mountains of California (Figure 5.14). Estimated to be about 4,900 years old, this weathered tree was named Methuselah, after an ancient patriarch of the Old Testament. Since many old bristlecones have many growth rings but are not large trees, 1,000 years of growth rings may be only 6 inches wide.

Growth rings can reveal not only a tree's age but also details about climate changes and human history. For example, a pattern of 20 thin growth rings and 2 fat ones may indicate that a dry spell of 20 years was followed by 2 years of heavy rainfall. Human events and artifacts may also be dated by analyzing growth rings (see *The Intriguing World of Plants* box on page 111). In the American Southwest, dendrochronology has been used to date ancient Native

Figure 5.14 Bristlecone pines—the oldest known living organisms? One of these bristlecone pines (*Pinus longaeva*) in the White Mountains of California is more than 4,000 years old and may be the oldest known living organism, based on studies of its growth rings. Scientists will not reveal which tree, known as Methuselah, is the oldest.

wood may develop as an apparent counterbalance (Figure 5.15). In dicots, reaction wood forms on the upper side of leaning trunks or branches and is called *tension wood* because it "pulls" the trunk or branch toward a vertical position. Tension wood has wider growth rings on top, but the rings are lighter because they contain little or no lignin. In conifers, reaction wood occurs on the lower side of a leaning trunk or branch and is called *compression wood* because it "pushes" the trunk or branch toward a vertical position. Compression wood has wider growth rings on the bottom and is rich in lignin, giving it added strength. Both tension wood and compression wood shrink much more than normal wood when dried. Boards containing reaction wood are not usable for most projects because the wood shrinks at a different rate and to a different degree and also has irregularities in strength.

The cork cambium is reformed as the stems and roots enlarge

As the action of the vascular cambium widens a stem or root, the outermost layers produced by primary growth—the cortex and epidermis—stretch and crack because they are no longer growing. If the cork cambium did not form, the loss of the cortex and epidermis would leave the tree without a protective outer covering. The rate of water loss would increase, and therefore roots and stems would dry out. They would also be open to infection by disease and

American cliff dwellings. For example, if a roof is split by the growth of a living tree, we know that the dwelling itself is older than the tree. If a tree trunk is incorporated into the ruin, we can tell approximately when the construction occurred. Scientists do not have to cut down a tree to examine its growth rings because a core drill can remove a small cylinder of wood that reveals the growth ring pattern without harming the tree. The growth ring patterns of living trees can be overlapped with those in dead trees to extend the ability to date back thousands of years. Dendrochronology has also been used to date pieces of wood from the remains of ancient campfires as far back as 9,000 years.

Growth patterns in reaction wood counteract leaning

The growth pattern of rings can sometimes be irregular as a result of a tree's response to the forces of wind and gravity. In trunks or branches that are leaning, **reaction**

Figure 5.15 Reaction wood. This cross section of a conifer limb has compression wood on the lower side. The thicker growth rings help support the horizontal weight of the limb.

THE INTRIGUING WORLD OF PLANTS

Tree Clues to a Colonial Mystery

In May of 1587, more than 100 British settlers arrived on Roanoke Island, one of North Carolina's Outer Bank islands. Things did not go well and in August they sent their governor, John White, back to England for badly needed supplies. However, war broke out between England and Spain and White was unable to return for three years. When he finally returned in 1590, all trace of the colonists had disappeared except for the letters *CRO* carved on one tree and the word *Croatoan*, carved on another. *Croatoan* was both the name

"James Fort Construction," 1607.

Growth rings on an ancient bald cypress near the Roanoke colony site.

of an Indian tribe and the name of a nearby island, suggesting that the colonists may have moved there. However, the colonists were never found.

A second colony at Jamestown, Virginia, nearly failed as well, with only 38 of the original 104 settlers surviving the first year. Writings show that the colonists complained of hunger and thirst. Captain John Smith, of *Pocahontas* fame, reported that the Indians

Graph of ring width data.

refused to share food because of concerns about their own supply.

In 1998, scientists reported on an investigation of the tree rings of ancient bald cypress trees growing in the area of these colonies. As you know, the width of growth rings correlates with growing conditions for a particular year. Wide rings mean good conditions, while narrow rings indicate poor conditions. It turns out that the English colonists were stranded on Roanoke during the area's worst 3-year drought in 800 years. The Jamestown residents suffered through the driest 7-year period between 1215 and 1998. These clues from dendrochronology may indicate the cause of the failure and near failure of these two colonies.

---

to predation by insects and other organisms attracted to the exposed sugar-rich phloem. The formation of the cork cambium avoids this scenario. As the cork cells produced by the cork cambium enlarge and mature, their cell walls become coated and impregnated with a fatty substance called **suberin,** which makes them waterproof and prevents the sugar-rich phloem sap from seeping through to the surface. In this way, the suberin makes stems and roots

less attractive to animals. After the cork cells die, they still function as an "armor" around the vascular tissues.

Cork cambium cells, which are often cube-shaped, divide to produce either cork cells to the outside or phelloderm cells to the inside. Unlike the cells of the vascular cambium, cork cambium cells do not divide to produce more cambium cells. That is why the cork cambium cannot grow in diameter and eventually becomes too small for the expanding stem or root. As the stem or root expands in width, the cork cambium and phelloderm stretch and break apart, becoming part of the outer bark. Then a new cork cambium forms to the inside of the old one, creating another layer of periderm. The cork cambium and the living phelloderm cells it produces are relatively isolated from both water and minerals in the xylem and carbohydrates in living phloem. This may be another reason why a new cork cambium reforms inside the old one from time to time.

The first cork cambium generally forms during the first or second year of a tree's life and may remain active for a year or for up to two dozen years if the tree is not increasing in width rapidly. Although successive cork cambia create multiple layers of periderm, in most trees the outer bark does not become very thick because some of the older bark splits and peels off each year as a result of the force of new growth underneath.

Lenticels are pathways in the bark for gas exchange

The suberin in the cell walls of cork cells blocks the passage of oxygen into the stem or root. However, stems and roots have **lenticels,** small openings in the outer bark where the cork layer is thin and there is enough space between cells to allow for exchange of gases (Figure 5.16). As new cork cambium arises, new lenticels develop that are aligned with the outer lenticels, providing a continuous pathway for

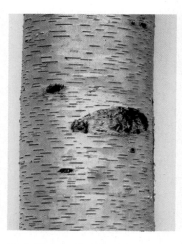 

Birch (*Betula albosinensis*)

Figure 5.16 Lenticels. Gas exchange into and out of roots and stems occurs through the lenticels, which are formed by the rapid division and growth of cork cells that split the epidermis and narrow the cork layer. Gases—oxygen in particular—diffuse through intercellular spaces of the remaining cork cells because the cells are rounded and do not fit tightly together.

oxygen. In trees with smooth bark, lenticels are easily observable, usually appearing as short streaks, slits, or raised dots on the surface of twigs, branches, trunks, and roots. If the bark is thicker and textured, lenticels occur at the base of cracks and are not readily visible. In addition to appearing in stems and roots, lenticels can be seen as spots and streaks on the surfaces of some fruits, such as apples and pears. Lenticels are also visible as some of the markings on commercial cork products. In the next section, we will look at products made from wood and bark.

Section Review

1. Compare the structure and functions of the two types of meristematic cells in the vascular cambium.
2. How do heartwood and sapwood differ?
3. What can growth rings reveal about the life of a tree?
4. How does the development of bark differ from that of wood?

Commercial Uses of Wood and Bark

So far we have examined how secondary growth provides additional support, conduction, and protection for a woody plant. At the same time, variations in growth patterns and other physical characteristics of wood and bark can have important implications for human uses. In this section, we will look in general at how the qualities of wood and bark lend themselves to various products.

Wood is used mainly for fuel, paper products, and construction

Worldwide, the two main uses of wood are for fuel and paper. While most developed nations use electricity and fossil fuels such as gas for heating and cooking, the majority of the world's population uses wood—approaching half of the wood harvested. About the same amount is used to make paper, a thin film of cellulose made from wood pulp—a solution of water and crushed wood (Figure 5.17). Before the mid-1800s, people in the Western world used paper typically made of linen, but the invention of machines to make paper from wood pulp made trees the more economical source. More than half of the yearly tree harvest in the United States goes to make paper and paper products, with white spruce (*Picea glauca*) being

Figure 5.17 Making paper. This highly automated machine makes paper from a liquid solution of ground-up trees.

the major source of paper for newspapers. Worldwide, the average person uses 15 kilograms (33 pounds) of paper per year, while the average American uses 333 kilograms (732 pounds) per year. An average tree makes 200 copies of the want ads in a typical metropolitan Sunday paper, while one edition of a college newspaper requires the pulp from several acres of trees.

About a third of the paper used throughout the world—and nearly half of the paper in the United States—is recycled. However, the cutting of forests for paper production continues to increase because of the growing world population and also the use of printers, copy machines, and personal computers. According to one estimate, personal computers and printers alone have increased paper use by over 100 billion sheets per year, with the need for paper pulp expected to double in the next 50 years.

Only about one-fourth of paper pulp production worldwide comes from forests managed for year-to-year pulp production. Worldwide, the current need for paper and paper products could be supplied by an area about the size of California, if the land were correctly managed for continuous production. However, the use of forest resources for paper production must compete with other potential land uses. If nonforest crops provided a source of pulp, this area could be considerably reduced.

Paper can be made from many plants that grow more rapidly than trees. One promising source is kenaf (*Hibiscus cannabinus*), a drought-resistant annual crop that grows to 15 feet in a five-month season and requires few or no pesticides or herbicides (Figure 5.18). Although native to central Africa, kenaf can be cultivated in many southern U.S. states. One of its advantages is that the whole plant can be used. About one-third of the plant consists of phloem fibers, which can be used for paper,

rope, twine, rugs, sacks, and even clothing. The other two-thirds—the inner fibers—can be used for animal bedding, potting soil, oil-absorbent materials, and filler in plastics. Kenaf requires less energy to produce pulp than trees and is also stronger. Kenaf and other alternative sources of pulp will undoubtedly play an important role in future paper production.

In worldwide use of wood, construction is a distant third behind fuel and paper. However, wood is still the main construction material in developing countries, where most people cannot afford other materials. Even in the United States, 94% of houses have at least a wood framework, commonly made of Douglas fir or pine. Western red cedar is the most desirable wood for making roof shingles, while oak is commonly used for hardwood floors, boats, and railroad ties.

Wood structure can be studied from three cutting planes

Wood in a lumberyard can have various appearances depending on how it is cut (Figure 5.19). A **transverse** cut gives a circular cross section. A **radial** cut is longitudinal and passes through the center of the stem. A **tangential** cut is also longitudinal but crosses the radius at a right angle instead of passing through the center of the stem.

Figure 5.18 Kenaf, a nonwoody alternative source of paper pulp. The kenaf plant can be cultivated in the field in warm regions. It grows tall and fast and produces a stem that can be almost entirely used to make paper.

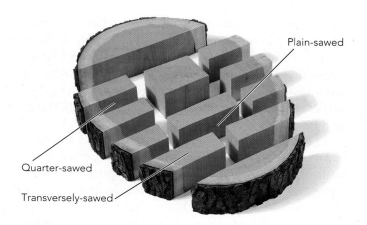

Figure 5.19 Lumber cuts. At lumberyards, wood can be plain-sawed, quarter-sawed, or transversely sawed. Most long boards at lumber yards are plain-sawed. Quarter-sawed wood has a much more uniform grain and is stronger, but there are fewer cuts of it per log. Transversely-sawed wood is the cut made to make firewood.

In terms of lumber, a tangential cut can be trimmed to yield a rectangular **plain-sawed** board, also known as a *flat-cut board*. In plain-sawed lumber, growth rings are variable, roughly parallel wavy lines. Since such boards frequently warp, they are rarely used in construction projects. A radial cut can be trimmed to produce a rectangular **quarter-sawed** board, in which growth rings appear as parallel lines running the length of the board. Such uniform boards do not warp, but the sawing wastes lumber and must be done exactly.

Veneers, peels from large logs, consist of a continuous tangential cut, which is angled in slightly so that the cut does not stop. Usually no more than 1 or 2 millimeters thick, veneers provide a pleasing outer layer when glued to inexpensive, less beautiful wood or composites such as particleboard in furniture or panel production. Plywood results from gluing together several thicker veneers for strength. Usually, each layer is placed in a different orientation for added strength. Particleboard is made of small pieces of wood or wood chips glued together.

Wood can vary in properties such as hardness and grain

As you might imagine, the features of wood have been thoroughly studied from a practical as well as a botanical point of view. Some properties of wood are hardness, density, durability, grain, texture, and water content. Depending on their properties, some types of wood are used mainly for one purpose, while others have diverse

uses. Red spruce, for instance, is a major source of wood pulp but is also used to make sounding boards for pianos as well as boat paddles. Hickory, an important source of handles for tools, is also used in smoking meats. Slash pine is used not only for wood pulp and turpentine but also for railroad ties and large timber pieces in construction.

The wood of dicot trees—such as hickory, maple, and oak—is commonly called **hardwood** because it usually contains many fibers, making it difficult to nail, saw, or damage. These qualities make hardwood an excellent component for many durable products (Fig. 5.20). Hardwood generally also has a much higher energy content per unit weight than the wood of gymnosperms such as conifers. Therefore, it makes excellent firewood and charcoal because it burns a long time. Charcoal is made by burning hardwood in an atmosphere with low oxygen, resulting in most of the carbon being left behind while the liquid components of the wood evaporate.

In contrast, the wood of conifers is commonly called **softwood** because it typically contains fewer fibers and no vessels, making it softer than dicot wood. In general, the wood of conifers is less dense and therefore typically floats easily but burns more quickly. It is usually easier to saw but is also more easily damaged. Douglas fir, the most common tree in the forests of the western United States, provides most of the softwood used for construction.

The terms *hardwood* and *softwood* are generalizations because not all dicots have hardwood, and not all conifers have softwood. For instance, balsa wood is an extremely

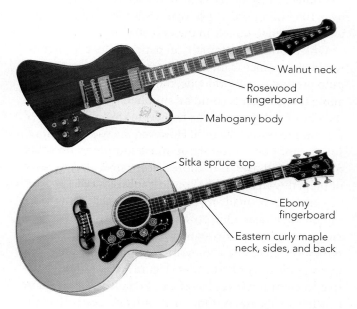

Figure 5.20 Guitars are made from a variety of hardwoods.

soft, light wood that comes from a tropical dicot, while some pine species have harder wood than some dicot trees.

The actual hardness of wood relates to its **density**—the amount of matter per a unit of volume. A related characteristic is **specific gravity,** the ratio of a wood's weight to the weight of an equal volume of liquid water at room temperature. The difference in a wood's specific gravity is attributed to the proportion of cell wall substance to the lumen, the space once occupied by cell contents. Fibers may also cause differences in specific gravity. If the fibers have thick walls and narrow lumens, the specific gravity is usually high. On the other hand, if the fibers are thin-walled with wide lumens, the specific gravity is usually low. Basically, specific gravity tells you whether something floats or sinks. Most woods float, but some hardwoods, such as ironwoods, will actually sink.

Durability is the extent to which wood is resistant to breakdown and decay by fungi, bacteria, and insects. Durable woods such as cedar and black locust contain tannic acid, also known as *tannin,* which repels many of the organisms that cause wood to decay.

Grain is the overall alignment of the conducting cells of xylem. In *straight-grained wood,* the conducting cells are aligned parallel to the longitudinal axis of the piece of wood. In *cross-grained wood,* the conducting cells are not parallel to the long axis of the wood. In *spiral-grained wood,* the arrangement of the conducting cells makes a gradual spiral up and down the trunk. In *interlocking-grained woods,* the orientation of the grain changes during the life of the tree. An example of straight-grained wood is English elm (*Ulmus procera*), while sycamore (*Platanus*) has an interlocking grain.

Texture refers to the sizes of the cells in the xylem and phloem and to the sizes of growth rings. Coarse texture is caused by the presence of many large vessels and wide rays. Fine texture is caused by the presence of small vessels and narrow rays. Uneven texture is caused by large size differences between the cells of early (spring) wood and late (late summer or fall) wood within a growth ring.

Water content is the percentage of water by weight. In the forest, wood is about 75% water. Dried (cured, seasoned) wood is around 10% to 20% water. Dry wood contains around 55% to 75% cellulose, 15% to 25% lignin, and 20% or less oils, tannins, resins, and other ingredients. The tannin-to-cellulose ratio can vary considerably. Dry wood is used to make furniture and houses because it does not shrink, and therefore the wood is less likely to split.

In addition to these properties, knots can affect the quality and appearance of wood. Knots are simply regions where branches intersect the trunk. While a branch is living, its vascular system connects to that of the trunk. When the branch dies, the trunk continues to expand and grow around the branch, but no further vascular connections occur. When the wood is cut, the parts of the branch with vascular connections form tight knots, but loose knots form where there are no vascular connections to the trunk. Since branches often form reaction wood where they merge with the trunk, they often consist of dense wood that is difficult to saw. Wood that is relatively free of knots comes from very tall, wide trees that have most of their branches near the top. However, knots can be considered a positive feature in wood, as in the knotty pine used in paneling and furniture.

Latex, resin, and maple syrup are some products from wood fluids

Many flowering plants produce various forms of latex, a milky substance that blocks the entry of disease-causing organisms and insects through wound sites and frequently contains compounds that prevent the growth of fungi and bacteria. Plant latex is extremely useful to humans as the basis of the rubber industry and as a source of pharmaceuticals (see the *Plants & People* box on page 116).

Pines and many other nonflowering seed plants produce resin, a sticky substance that flows through canals in the secondary xylem and phloem, the periderm, and the leaves (Figure 5.21). The word *resin* is derived from both the Greek and Latin words for "pine." Also known as

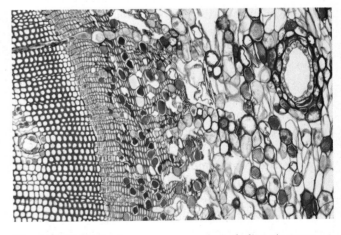

Figure 5.21 Resin. Some gymnosperms, including pine trees, produce resin that seals wounds to protect them from further damage and from disease. The resin flows through ducts in the xylem, phloem, and periderm.

PLANTS & PEOPLE

Different Roads to Producing Rubber

The elastic latex used to make rubber originally came from the rubber tree (*Hevea brasiliensis*), which is native to Central and South America. As long ago as 1,600 B.C., the Mayans collected latex from the trees and made it into large rubber balls and binding material to secure ax heads to handles. The Mayans added juice from morning glory vines to the latex to prevent the rubber from becoming brittle. In 1839, Charles Goodyear accidentally discovered a variation of this process when he dropped a mixture of rubber, lead, and sulfur onto a hot stove. This process of heating rubber, known as *vulcanization,* results in long-lasting, elastic rubber products.

Latex is collected from rubber trees by "tappers" who gash the bark of each tree in the cool morning and collect the latex in a container. The latex—which contains rubber, water, and proteins—must be rapidly processed to avoid spoilage. A single tree can be a useful producer for 25 or more years. Now, 80% of natural rubber is

A woman collects sap from a rubber tree in Yunnan Province, China.

collected from plantations in Southeast Asia, especially in Malaysia and Thailand.

During World War II, when the Japanese cut off the supply of natural rubber to the United States, chemists rapidly developed a method for making artificial rubber from petroleum. Now, more than two-thirds of the world's rubber is synthetic. However, the high costs of making artificial rubber from petroleum and of transporting natural rubber from the tropical countries have prompted scientists to study the ability of other plants to produce latex. For example, sunflowers (*Helianthus*) species and the guayule bush (*Parthenium argentatum*) are possible sources of rubber. Sunflowers grow throughout the United States, while guayule is native to the southwestern United States and northern Mexico. Guayule rubber contains only 20% of the protein of rubber from *Hevea brasiliensis* trees. As a result, rubber allergies from products made of guayule rubber—for example, from latex gloves—are much less common.

pitch, resin is apparently involved in sealing injuries and preventing the entry of disease organisms. A damaged tree will secrete often copious quantities of resin onto the surface of the trunk. Turpentine, a semifluid component of the resin, evaporates and leaves behind a waxy compound known as *rosin* that seals the wound. Some insects, such as the sequoia pitch moth, are attracted to pitch and lay eggs in it. The larvae enter the tree, causing further damage and more release of pitch.

When soft wood from conifers is burned, the turpentine causes the wood to burn rapidly, with the popping and cracking resulting from the explosion of resin pockets. Hard woods from flowering trees typically lack resin and therefore usually burn more smoothly and slowly. Turpentine extracted from softwoods such as pine is used to make paints and paint thinners, inks, lacquers, soaps, and polishes, among other compounds. Rosin also has a variety of uses, including sealants and adhesives. During the days of wooden sailing ships, rosin was used to make them watertight. Today it is used to make varnishes and waxes for finishing wood. Baseball pitchers and batters use rosin to get a better grip on the ball or bat, while dancers use it for im-

proved traction. Rosin also strengthens the degree and crispness of contact between a violin bow and strings.

Extinct conifers near the Baltic Sea and in a few other regions produced huge quantities of resin, which collected in lumps weighing from a few ounces to 100 pounds. This fossilized resin, known as *amber,* is millions of years old. Ancient insects and bacteria often became trapped in this resin and were fossilized along with it (Figure 5.22). Bacteria that are more than 30 million years old have been revived from amber. In general, however, DNA from amber is quite degraded. The idea that dinosaur blood from fossilized mosquitoes could be the source of genes to recreate dinosaurs is simply the source of ideas for book and movie plots. Still, fossilized amber serves as a window into the prehistoric past and has also been literally mined as material for jewelry.

Another product from wood fluids is the sweet sap collected after tapping into the outer xylem of maple trees in early spring. After sufficient heating and evaporation, the sap becomes maple syrup and then maple sugar. People often think that the sap comes from the phloem that transports sugars from leaves to roots, but it actually

Figure 5.22 Insect in amber. Amber is fossilized resin formed through natural polymerization of organic compounds.

Figure 5.23 Commercial cork. Outer bark of the cork oak (*Quercus suber*) is stripped and used commercially to make corks for wine bottles and other commercial products. The tree lives and repairs itself by establishing a new cork cambium and producing more cork from the secondary phloem that was left behind.

arises from starch stored in root xylem during winter, when leaves are not present. During warm spring mornings, the sap expands and is pushed up through the xylem in the trunk. In a process called *sugaring*, the sap is collected by drilling holes into the trunk and inserting taps, which drain the sap into buckets or into tubes that conduct it to a central location, where the water is boiled away to make maple syrup.

Commercial cork comes from the thick outer bark of some trees

The cork cells of the outer bark provide flexibility and water tightness for the trunk, so cork makes excellent removable seals or "corks" for wine bottles and other containers. Cork's flexibility allows it to return to its original shape after being deformed, making it useful for bulletin boards. The most important source is the cork oak (*Quercus suber*), which has thick outer bark that consists principally of cork cells (Figure 5.23). A harmless harvest of cork oaks, which are native to the Mediterranean region, occurs about every 10 years. In recent years, some wineries have replaced natural corks with those made of synthetic plastic for economic reasons.

Trees are a renewable but limited natural resource

Before humans existed, forests covered much more of Earth's surface than today, and the amount of forested land remained relatively constant. When trees died from disease or burned in fires caused by lightning, they were replaced by new tree growth. With the advent of human civilization around 10,000 years ago, however, the extent of Earth's forests began to decrease. Hunters sometimes started forest fires to drive game animals out of hiding, and farmers burned or cut down trees to clear land for crops. When the useful wood was gone and the soil was depleted of nutrients, people simply moved on. This technique, often called *slash-and-burn,* remains popular today in tropical rain forests.

Throughout human history, wood has been used for heating, cooking, and construction. Forests have been replaced with farms and cities or simply harvested and not replenished. Today only about half of the world's original forested land remains, and only a small percentage of those forests are "old growth," unaffected by human use. The Pacific Northwest and Canada, for instance, have large managed forests but very few areas of native forests. In the continental United States, 98% of the forests have been logged at least once.

Deforestation, the process of clearing forests, is reducing the supply of trees available for cutting. Each second an area of tropical rain forest at least the size of a football field is cut or burned. At current rates of deforestation, native temperate and tropical rain forests will have disappeared by the year 2030, with only parks and managed forests remaining. Despite the fact that seedlings are planted in managed forests, trees are being used at a much greater rate than they can be replaced, a trend that new forestry practices seek to reverse (see the *Conservation Biology* box on page 118).

CONSERVATION BIOLOGY

Sustainable Use of Wood Resources

In recent years, sustainable forestry, sometimes called *new forestry*, has sought to develop methods whereby timber can be extracted from forests while biodiversity and long-term ecosystem health are maintained. An important component of new forestry has been implementation of management strategies to ensure continued production of harvestable trees. Sustainable forestry seeks to maintain timber production for an identified geographic area at a constant rate over time. Basically, this means that new trees must be planted at the same rate they are cut and allowed to grow to a similar size as the ones they replace. Until recently, sustainable forestry has succeeded in only a few isolated locations. Even when replanting rates equal the harvest rates, new trees seldom reach the size of those originally cut. The days of harvesting timber from old growth trees are over in the United States because the old growth trees are gone, except for protected forests. Experts realize that achieving sustainable forestry may require a long time and involve

Selective tree harvesting.

Tree being processed by debarking machine.

ferred method of timber harvest because a new forest is most rapidly reestablished by replanting with small seedlings. For species that do best in partial shade, such as many types of spruce, selective cutting is usually the best harvesting method. In forests of mixed species, a combination of harvesting methods may be suitable.

The problem with farming of any kind is that the harvested crop carries with it nutrients from the soil. After generations of farming, the soil becomes nutrient-poor and the yield and the nutritional quality of the crop declines. Many agricultural regions suffer from this problem. In fact some nutritionists believe the increase of certain diseases in modern populations can be directly linked to nutrient-poor soils.

In the case of timber harvesting, the second crop can take decades to produce. Once virgin timber is cut, noticeably poorer second and third crops follow. This is usually the result of each crop's removing nutrients from the soil, which becomes nutrient-poor. One solution to the

several crop rotations. Converting a nonsustainable forest to a sustainable forest is part of a process known as *forest stewardship,* in which forests are viewed as communities of living organisms that must be understood and respected if they are to continue to provide harvests of useful wood to humans.

An important principle of sustainable forestry is to harvest trees in a manner that leads to the most rapid reestablishment of a similar forest. Some species, such as Douglas fir or lodgepole pine, have seedlings that grow best in full sunlight and naturally revegetate areas cleaned by storms or forest fires. For such species, clear cutting is the pre-

problem of soil nutrient depletion in forestry involves removing bark from the trees on site instead of at some other location. Alternatively, the bark can be removed at the mill and returned to the harvest site. The bark, which contains phloem, has part of the soil nutrients in trees. While valuable ions from the soil travel up the tree in the xylem, they are incorporated into organic molecules in the leaves. When cut, the usable wood contains very few dissolved minerals. Removing the bark at the harvest site involves changes in procedure and equipment. The economic payoff is years down the line, but sound management dictates that this practice should be followed.

Deforestation, particularly in tropical rain forests, reduces not only the quantity of trees but also the diversity of tree species. There are between 80,000 and 100,000 species of trees in the world. The United States currently has about 1,000 species, more than three times as many as in all of Europe. The Appalachian region alone has about 300 tree species. However, those numbers pale in comparison with the tree diversity in tropical rain forests, even though those regions now cover less than 2% of the land on Earth. A study of a 10.4-hectare (25-acre) plot in a Malaysian rain forest identified 780 tree species, which is approaching the number of tree species native to the United States and Canada. In 1996 a scientist found 476 tree species in just one hectare (2.4 acres) of the Atlantic rain forest region of Brazil, including 104 species previously unknown to the region and 5 species previously unknown to scientists. By comparison, all temperate forests typically have between 2 and 20 tree species per hectare. Unfortunately, at least 20% of the world's tree species face extinction in the near future, including about 250 species in the United States. The World Resources Institute estimates that deforestation in tropical rain forests results in the loss of 100 species of all types of organisms per day.

Section Review

1. Why might alternative sources of paper products be important in the future?
2. What are the meanings of the terms *hardwood* and *softwood*?
3. Describe the following characteristics of wood: density, durability, grain, texture, water content.
4. What are some ways in which wood fluids are commercially useful?
5. What does it mean to say that trees are a renewable but limited natural resource?

SUMMARY

Secondary Growth: An Overview

Lateral meristems, cylinders of dividing cells, produce secondary vascular and secondary dermal tissue (p. 101)
The lateral meristems—the vascular cambium and cork cambium (phellogen)—increase the diameter of a stem or root. The vascular cambium produces secondary xylem and secondary phloem, while the cork cambium produces secondary dermal tissue.

The vascular cambium produces secondary xylem (wood) and secondary phloem (pp. 102–104)
The vascular cambium forms between the primary xylem and phloem and then grows in diameter as it adds secondary xylem to the inside and secondary phloem to the outside. Secondary xylem and phloem increase conduction, and secondary xylem contains added lignin that provides greater support.

The cork cambium produces secondary dermal tissue (pp. 104–105)
Cork cambium produces periderm, which replaces the epidermis and cortex. Periderm consists of the cork cambium, cork (phellem), and phelloderm. Since a cork cambium does not grow in diameter, it must be replaced. Outer periderm layers peel off as the outer layer of bark.

Bark consists of all the tissues external to the vascular cambium (pp. 105–106)
The inner bark is mainly secondary phloem, while the outer bark is all the layers of periderm outside the most recent cork cambium.

Growth Patterns in Wood and Bark

The vascular cambium produces secondary xylem, secondary phloem, and ray parenchyma, as well as more vascular cambium (pp. 107–108)
Fusiform initials produce xylem and phloem, and ray initials produce files of parenchyma cells called *rays* for storage and sideways transport. Fusiform and ray initials can divide to increase the vascular cambium's diameter.

Sapwood conducts water and minerals, but heartwood does not (pp. 108–109)
Heartwood consists of inactive inner rings of xylem, usually darker than the conducting outer rings called *sapwood*. Trees produce antibacterial and antifungal substances that help protect the heartwood from rot and fungi.

Growth rings in wood reflect the history of secondary growth in a tree (p. 109)
Growth rings are typically dividing lines between late-summer growth of one year and spring growth of the next. Thick rings indicate good growing conditions.

Dendrochronology is the science of tree ring dating and climate interpretation (pp. 109–110)

Growth rings can reveal tree age and growing conditions and can help determine the dates of human events and artifacts.

Growth patterns in reaction wood counteract leaning (p. 110)

Reaction wood forms in the upper or lower sides of leaning trunks and branches as a counterbalance. In dicots, it appears on the upper side and in conifers on the lower side.

The cork cambium is reformed as the stems and roots enlarge (pp. 110–112)

Cork cells become impregnated with suberin, making them waterproof and preventing phloem from seeping to the surface. Cork cambium breaks apart and a new cork cambium and periderm form underneath. Since outer layers of periderm peel off, bark does not accumulate as much as wood.

Lenticels are pathways in the bark for gas exchange (p.112)

Lenticels are typically small slits or streaks on the surface of stems, allowing passage of gases through the bark.

Commercial Uses of Wood and Bark

Wood is used mainly for fuel, paper products, and construction (pp. 112–113)

Worldwide, the two primary uses of wood are for fuel and paper, in about equal measure, with construction a distant third. Growing demand for paper is prompting research into nonforest sources. Wood is still the main construction material in developing countries.

Wood structure can be studied from three cutting planes (pp. 113–114)

A transverse, or plain-sawed, cut is a circular cross section. A radial, or quarter-sawed, cut is longitudinal cut through the center of a stem. A tangential cut is a longitudinal cut that does not pass through the center of the stem.

Wood can vary in properties such as hardness and grain (pp. 114–115)

Dicots produce hardwood, which is usually difficult to cut. Conifers produce softwood, which is generally easier to cut. Hardness relates to density and specific gravity. Durability is resistance to decay. Grain is alignment of xylem cells. Texture refers to sizes of cells and growth rings. Water content is the percentage of water by weight. Knots are regions where branches intersect the trunk.

Latex, resin, and maple syrup are some products from wood fluids (pp. 115–117)

Latex is used to make rubber and pharmaceuticals. Turpentine and rosin are extracted from resin. Some turpentine products are paint thinners, inks, and lacquers, while rosin is used in sealants and adhesives. Fossilized resin, called *amber,* is used in jewelry. Maple syrup and sugar come from maple xylem sap.

Commercial cork comes from the thick outer bark of some trees (p. 117)

Water tightness and flexibility make cork a good sealing material. Most commercial cork comes from the cork oak.

Trees are a renewable but limited natural resource (pp. 117–119)

Deforestation has greatly reduced forested land and reduced tree diversity, prompting the need for new forestry practices.

Review Questions

1. In terms of the life of the plant, what are the purposes of lateral meristems?
2. Name the two lateral meristems and describe the types of cells produced by each meristem.
3. How does the outer bark differ from the inner bark in structure and purpose?
4. Describe the difference between fusiform initials and ray initials.
5. What is the function of rays?
6. Explain the difference between heartwood and sapwood.
7. Explain how tree rings can be used to date human events and artifacts.
8. Why is the cork cambium essential for a tree's health?
9. How is the outer bark both protective and porous?
10. Why are nonforest sources of paper pulp being considered?
11. How can you tell if a board was cut radially or tangentially?
12. Why is dicot wood called *hardwood* and conifer wood called *softwood*?
13. Choose two general characteristics used to describe wood (such as grain and durability) and explain why they are relevant to potential wood uses.
14. Explain why bark is commercially useful.
15. Why is slash-and-burn agriculture a problem?

Questions for Thought and Discussion

1. In what ways do you think secondary growth can help a plant compete for resources? In what ways might secondary growth make a plant vulnerable?
2. A family builds a wooden home in the middle of a beautiful forest. While sitting on the redwood deck one evening, they notice that the forests surrounding their home are being cleared and the view is rapidly deteriorating. Discuss the inherent moral issues of this situation.
3. Why do you think the vascular cambium is permanent, while the cork cambium reforms periodically?

4. Some extinct seedless plants produced very large trees. Their vascular cambia produced xylem and phloem but did not produce more cambium cells to expand with the tree. What eventually happened to trees with this pattern of growth?

5. Are the words *xylem* and *wood* synonymous? Explain.

6. Some parks have "tunnel trees," in which the center of the trunk has been cut out at the base but the tree is still able to live. What is your opinion about this practice?

7. If trees have the potential to be immortal, why do you think each tree species has a characteristic life span?

8. What effect do you think the use of computers, the Internet, and e-mail will eventually have on the use of paper?

9. Draw a small section of the vascular cambium of a woody dicot stem as seen in cross-sectional view; 2 or 3 cells should suffice. Now, imagine that these cells each contribute one new phloem element followed by one new xylem element, then divide once so that the circumference of the vascular cambium increases and keeps pace with the increasing diameter of the stem. Using the cells you've drawn as a starting point, diagram a sequence of cell divisions and subsequent cell enlargements that would accomplish the above developmental processes. Now diagram these same processes as they would appear in both radial and tangential longitudinal sections.

Evolution Connection

Imagine the first subpopulation of herbaceous plants in which a vascular cambium had become active and was producing secondary xylem and phloem. Why was this meristematic activity so advantageous that it became a major evolutionary development in land plants? Why was the activation of a more superficial cork cambium an adaptive co-development?

To Learn More

Visit The Botany Place Website at www.thebotanyplace.com for quizzes, exercises, and Web links to new and interesting information related to this chapter.

Davis, Wade. *One River: Explorations and Discoveries in the Amazon Rain Forest.* New York: Simon & Schuster, 1996. The author is an ethnobotanist who studies native uses of plants and tells fascinating stories

Eastman, John. Hansen, Amelia (Illustrator). *The Book of Forest and Thicket: Trees, Shrubs, and Wildflowers of Eastern North America.* Mechanicsburg, PA: Stackpole Books, 1992. This book includes interesting stories about plants, bugs, and their habitats.

Maclean, Norman. *Young Men & Fire.* Chicago: The University of Chicago Press, 1992. The author pieces together the obscure events surrounding the tragedy of the 1949 Mann Gulch forest fire in Montana, where a smoke-jumper crew of 15 was reduced to 3 men in 2 hours.

Pakenham, Thomas. *Remarkable Trees of the World.* New York: W. W. Norton, 2002. This book features photographs and unusual facts about 60 of the world's most fascinating trees.

Watts, T. May. *Tree Finder: A Manual for the Identification of Trees by Their Leaves.* Rochester, NY: Nature Study Guild, 1991. Easy identification of 161 species of deciduous trees, including "finders" guides for Pacific coast, Rocky Mountain, and Desert trees.

Life Cycles and Reproductive Structures

Cones, flowers, fruits, and seeds are all reproductive structures of plants.

Plant Reproduction: An Overview

Asexual reproduction occurs through mitosis and results in offspring that are genetically identical to each other and the parent

Sexual reproduction results in genetic variation

Meiosis and Alternation of Generations

Daughter nuclei produced by meiosis have one copy of each chromosome

Plant sexual life cycles feature both haploid and diploid multicellular forms

Cone and Flower Structure

In gymnosperms, some apical meristems produce cones

In angiosperms, some apical meristems produce flowers

A flower can consist of up to four types of modified leaves

The number and symmetry of flower parts can vary

Flowers can vary in the position of their ovaries

Flower structures are examples of how natural selection modifies what is already present

Seed Structure

Seeds form from ovules on bracts of cones or in carpels of flowers

Seeds nourish and protect developing embryos

In seed germination, first the embryonic root grows through the seed coat, and then seedling formation begins

Fruit Structure

During seed development in a flowering plant, the ovary expands to become part or all of a fruit

Fruits can be categorized as simple, aggregate, or multiple

A number of mechanisms disperse seeds and fruits to new locations

Like all other organisms, plants reproduce their own kind. As you learned in Chapter 1, plant reproduction is more varied and complex than human reproduction. Unlike humans, who reproduce only sexually and by one method of sexual reproduction, plants have a variety of ways of reproducing both asexually and sexually. As you know, asexual reproduction inolves only one parent and produces offspring that are genetically identical to the parent. In sexual reproduction, the genetic material of two parents combines to produce offspring that are a genetic combination of both parents. A few examples will illustrate some of the variations in both asexual and sexual reproduction that occur in plants.

A strawberry plant, for instance, reproduces asexually by producing horizontal stems called *runners* or *stolons,* which form new plants at their ends. During a growing season, one strawberry plant that actively produces stolons can spread many yards and produce a dozen or more new plants, all exactly like the parent. To start a strawberry bed, a gardener needs only a few plants and will soon have plants to give away. Horticulturalists know many different methods of reproducing plants by asexual means. As you will learn later, some of these methods require human intervention.

Many plants, such as the garden marigold (various species of the genus *Tagetes*), reproduce exclusively by sexual reproduction. Sperm and egg combine in fertilization to produce a zygote that grows and divides to produce an embryo inside a seed. In such a sexual process, the genetic material of male and female plants combines to produce offspring that are similar but also different from either parent. Sexual reproduction in most plants involves seeds. However, seeds do not always arise from sexual reproduction. Some plants, such as the common dandelion and the strawberry, can produce seeds by either a sexual or an asexual process.

In Chapter 1, you saw that plants have two adult forms that alternate in producing each other. In this chapter, we will look at how those two adult forms vary among plants. We will also examine the structures involved in the sexual reproduction of seed plants: cones, flowers, seeds, and fruits.

Plant Reproduction: An Overview

When you look forward to a wonderful meal, it may seem like you are "living to eat." Biologically speaking, of course, you are "eating to live." Needless to say, reproduction also has a biological context. The relative reproductive success of different species is the basis of their evolutionary success or failure. For example, if Ponderosa pine trees do not reproduce at least as successfully as other plants competing for their growing space, they will eventually become extinct.

When you give someone flowers, you do not usually think of them as being beautiful structures designed to facilitate sexual reproduction in flowering plants. Nevertheless, it is true. Flowers are structures that attract insects and other organisms, which in turn help distribute pollen. In a sense, almost everything about a plant or any other organism improves its reproductive success. Structures, processes, and behaviors that do not improve reproductive success tend not to occur because they represent an investment of energy that can be more efficiently used.

Reproduction can be described in terms of life cycles. A **life cycle** of a species is a sequence of stages leading from the adults of one generation to the adults of the next generation. As you have seen, a life cycle can be either sexual or asexual. In this section we will look in general at the differences between asexual and sexual reproduction in plants.

Asexual reproduction occurs through mitosis and results in offspring that are genetically identical to each other and the parent

Asexual reproduction, also known as *vegetative reproduction,* involves one parent, and the offspring are genetically identical to the single parent. Such offspring are often referred to as **clones** of the parent. Through asexual reproduction, a plant that is well adapted to a stable environment can quickly produce offspring that will be equally successful in that environment.

Asexual reproduction involves cell division by mitosis, nuclear division that produces two daughter nuclei identical to the parent nucleus (Chapter 2). As you can see in Figure 6.1, plants reproduce asexually in a variety of ways. One method is through adventitious shoots, also called "suckers," that emerge from roots of some species, such as aspen trees. In some plants, offspring can arise asexually from leaves, as in the cases of the *Kalanchoe* plant and the "walking fern," which produces a new plant whenever one of its leaves touches the soil. In the greenhouse and

lab, many plants can be propagated by rooting stem cuttings, by other similar techniques, or by tissue culture methods. Students majoring in horticulture often take classes in plant propagation to train them to use these various methods. Each plant that results from asexual reproduction is a genetically identical, naturally produced clone of the parent.

Plants rely more frequently on asexual reproduction than animals do. One reason could be that the structure of plants, in which primary growth is restricted to apical meristems, lends itself easily to asexual reproduction. Therefore, if cells of a root, stem, or leaf can develop into a shoot apical meristem, a new stem can form, and the stem can readily form roots at its base. It is also possible that plants rely more frequently on asexual reproduction because they compete with other plants for growing space, which asexual reproduction enables them to rapidly obtain.

Sexual reproduction results in genetic variation

Sexual reproduction involves one parent of each sex, leading to offspring that are genetically different from both parents and from one another, having a mixture of genes from each parent. In this way, sexual reproduction results in new genetic combinations. Although sexual reproduction occurs in all environments, it is particularly important for plants in environments that are changing or for plants that grow in a variety of different environments. Since the offspring of sexual reproduction are genetically varied and different from the parents, one or more offspring might be better adapted to a particular environment, thereby giving the species a competitive advantage.

Sexual reproduction in plants involves three types of reproductive cells: spores, sperm, and eggs. A spore can develop into an organism without fusing with another reproductive cell. In contrast, a sperm (male reproductive cell) fertilizes an egg (female reproductive cell), creating a zygote that becomes an embryo, which develops into an adult organism. Sperm and eggs are therefore called **gametes** (from the Greek *gamein,* "to marry"). In gymnosperms and angiosperms, the embryo is contained in a seed, a structure that includes the embryo and a store of food, packaged together within a protective coat.

Sexual reproduction involves some risk because egg cells or sperm cells can be damaged or destroyed, preventing fertilization from taking place. Indeed, some scientists think that one purpose of sexual reproduction might be to rid populations of harmful genes by forcing each organism through a single-celled stage when certain genes must work properly if the cell is to survive.

(a)

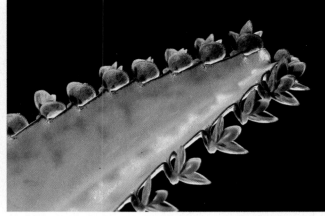

(b)

(d)

(c)

(e)

Figure 6.1 Asexual reproduction. Plants have many different methods of asexual reproduction. (**a**) In some plants, like quaking aspen trees, adventitious shoots arise from buds that form on roots. The entire group of trees is referred to as a *clone*. In autumn, the different clones are sometimes visible because the leaves change colors at a different rate. (**b**) Some succulents, such as *Kalanchoe daigremontiana*, produce adventitious plantlets at the edges of leaves. The plantlets fall off the leaf and readily root in the soil.

(**c**) The water hyacinth (*Eichhornia crassipes*) produces many new plants on short stems that break to release them. The water hyacinth reproduces so rapidly that it clogs waterways in tropical and semi-tropical regions like Florida. (**d**) Cholla cacti (various *Opuntia* species) produce stem segments that easily break off to fall to the ground and start new plants. Segments frequently hitch rides on clothes or animal fur, thereby distributing the species. (**e**) The piggyback plant (*Tolmiea menziesii*) starts new plants at the base of every leaf.

Sexual reproduction is also energy-intensive. First, sperm and eggs must be produced. Then, a large input of energy is required for the development of a zygote into an embryo and finally into an adult multicellular organism. Given these considerations, why does sexual reproduction occur? Apparently, the genetic diversity generated by sexual reproduction provides a tremendous selective advantage to many organisms. If offspring have genetic variability, the species has an increased potential of surviving in a particular environment, especially if some parameter of that environment is changing.

Plants such as dandelion and violet that reproduce both sexually and asexually provide scientists with a means for testing the selective value of asexual and sexual reproduction in a variety of different changing and unchanging environments. Several laboratory groups are working on such experiments.

Section Review

1. **How does asexual reproduction differ from sexual reproduction?**
2. **What are some methods of asexual reproduction in plants?**
3. **What might be some reasons why plants reproduce both sexually and asexually?**

Meiosis and Alternation of Generations

At some point during their sexual life cycles, all multicellular organisms return to two single cells, the sperm and the egg, which combine to become a zygote that develops into a new multicellular organism. A sperm and an egg must each have half the chromosome number of the cells of a multicellular adult. If the chromosome number were not halved, each generation would have twice the number of chromosomes in each cell as the previous generation. Within a few generations, cells would be crammed full of chromosomes, with no room for anything else. In this section you will learn about **meiosis**, a type of nuclear division that is involved only in sexual reproduction and results in daughter cells with half the original number of chromosomes.

Daughter nuclei produced by meiosis have one copy of each chromosome

The two types of nuclear division—mitosis and meiosis—have different effects on the number of chromosomes in the daughter cells. In mitosis, the original nucleus divides to produce *two* daughter nuclei, each with the *same* number of chromosomes as the original nucleus. In this way, mitosis and cytokinesis—division of the cytoplasm—produce two daughter cells with chromosomes that are identical to those of the original cell, as long as there are no mutations. In meiosis, the nucleus divides *twice* to produce *four* daughter nuclei, each with *half* the number of chromosomes as the original nucleus.

The number of chromosomes in a cell can vary from species to species. A typical body cell of an organism—any cell other than a reproductive cell (spore, sperm cell, or egg cell)—is called a *somatic cell* (from the Greek *soma*, "body"). In plants, a somatic cell typically contains either one set of chromosomes or two sets of chromosomes. In a cell with two sets of chromosomes, each pair of chromosomes consists of one chromosome originally derived from the egg and one from the sperm. Each pair of chromosomes is called a pair of **homologous chromosomes** because they both contain genes that control the same characteristics. A cell with two sets of chromosomes is said to be **diploid** (from the Greek *diplous*, "double"). A cell with a single set of chromosomes is said to be **haploid** (from the Greek *haplous*, "single"). The diploid chromosome number is referred to as $2n$, and the haploid number is referred to as n. In humans, for example, the diploid number ($2n$) is 46, and the haploid number (n) is 23. In the evening primrose (*Oenothera lamarckiana*), the diploid number ($2n$) is 14, and the haploid number (n) is 7. Plant cells, and in fact entire species of plants, can be **polyploid**, which means having more than the diploid ($2n$) number. As you will see in later chapters, polyploid plants can have significant effects on speciation.

Meiosis I

In plants, mitosis can occur in haploid, diploid, or polyploid cells. For example, if the original cell is haploid, then each daughter cell will be haploid, having just one set of chromosomes. If the original cell is diploid, then each daughter cell will be diploid.

In contrast, meiosis can only occur within a diploid cell or within a polyploid cell that has an even number of chromosomes—$4n$, $6n$, and so on. If the chromosome number were halved in a haploid cell, neither daughter cell would have all of the genes necessary for the organism. As you will see, meiosis involves two stages. We will look at the example of meiosis in a diploid cell, resulting in the original diploid cell being divided into four haploid cells. Each of these daughter cells has only half as many chromosomes as the original cell.

The preparation for meiosis involves an S phase in which chromosomes are replicated, just as they are in preparation for mitosis. As in mitosis, the stages of meiosis include prophase, metaphase, anaphase, and telophase (Figure 6.2). However, rather than one nuclear division, meiosis involves two nuclear divisions, distinguished by Roman numerals as meiosis I and meiosis II. During **meiosis I**, homologous chromosomes separate. During **meiosis II**, the sister chromatids separate. The phases of each of these nuclear divisions are also identified by Roman numerals. For example, the first phase of meiosis I is called prophase I.

Prophase I

Prophase I is the first—and most complex—stage of meiosis. Prophase I starts after interphase and begins with events that also occur in mitosis: the formation of the spindle, the breaking up of the nuclear envelope, and the disappearance of the nucleolus. Unlike prophase of mitosis, however, prophase I of meiosis involves homologous chromosomes. In prophase of mitosis, each chromosome, which is composed of two chromatids, aligns individually and without regard to other chromosomes. In prophase I of meiosis, homologous chromosomes—each having the same genes and consisting of two chromatids—form pairs. This pairing, called **synapsis**, results in the formation of a **tetrad**, a structure consisting of four chromatids. By chance, the chromatids may have overlapped during interphase. During prophase I, such an overlap may result in an exchange of chromosome segments, a process

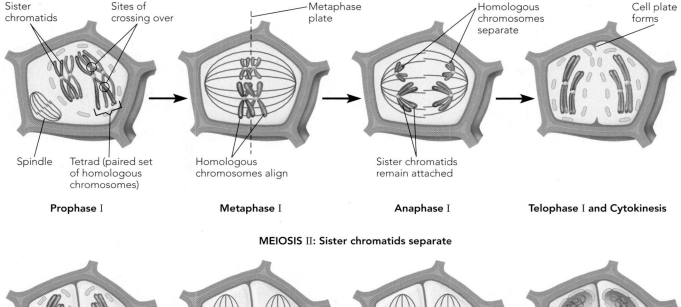

MEIOSIS I: Homologous chromosomes separate

Sister chromatids · Sites of crossing over · Metaphase plate · Homologous chromosomes separate · Cell plate forms

Spindle · Tetrad (paired set of homologous chromosomes) · Homologous chromosomes align · Sister chromatids remain attached

Prophase I   **Metaphase** I   **Anaphase** I   **Telophase** I **and Cytokinesis**

MEIOSIS II: Sister chromatids separate

Sister chromatids separate

Chromosomes condense   Chromosomes align   Sister chromatids separate   Haploid daughter cells

Prophase II   **Metaphase** II   **Anaphase** II   **Telophase** II **and Cytokinesis**

Figure 6.2 Meiosis. The two divisions of meiosis are shown in diagrammatic form. The principal event in prophase I is synapsis, the pairing of homologous chromosomes. The homologous pairs of chromosomes line up on the metaphase plate in metaphase I. Separation of homologous chromosomes occurs in anaphase I. Sometimes reformation of the nuclear envelopes and cytokinesis occur during and after telophase I. Meiosis II is quite similar to mitosis in that each replicated chromosome behaves as an independent unit.

known as **crossing over.** This source of genetic variation is discussed in more detail in Chapter 15.

Metaphase I

Metaphase I of meiosis is similar to metaphase of mitosis, except that tetrads of homologous chromosomes, instead of single chromosomes, move onto the metaphase plate. As in mitosis, chromosome movement is under the control of spindle microtubules. Spindle microtubules from one pole of the spindle apparatus are attached to one chromosome of each pair. Microtubules from the opposite pole are attached to the other chromosome.

Anaphase I

In **anaphase I,** homologous chromosomes are separated and move to opposite poles. As in anaphase of mitosis, motor proteins "walk" the chromosomes toward the opposite poles. Unlike mitosis, however, the sister chro-

matids of a chromosome remain attached to their centromeres and therefore move together as one unit toward the same pole. Meanwhile, the homologous chromosome moves to the opposite pole. Thus, each pole gains one set of chromosomes, and the overall chromosome number is halved. In mitosis, by contrast, the chromatids of one chromosome separate during anaphase, so the overall chromosome number remains the same.

Telophase I and cytokinesis

During **telophase I,** a variety of events occur, depending on the species. In general the cell returns, at least briefly, to its pre-meiotic state before entering meiosis II. The chromosomes continue their movement of anaphase I, ending up close to the opposite poles. In many cases, the nucleoli reappear, the nuclear envelope reforms, and chromosomes uncoil. In other cases, cells proceed immediately into meiosis II. Cytokinesis—division of the cytoplasm—usually occurs

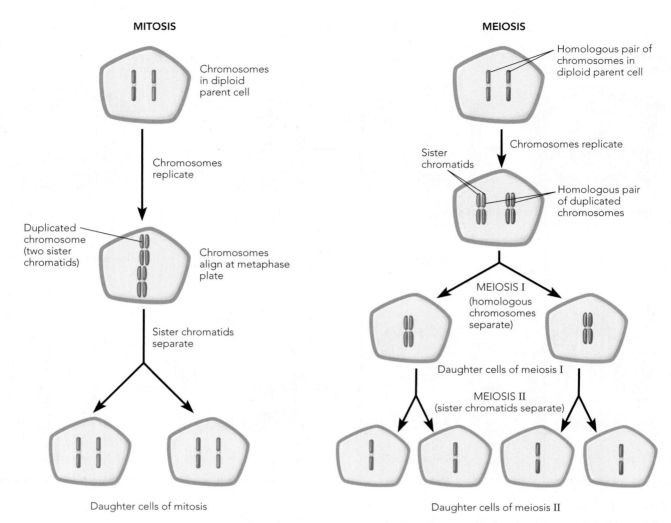

(a) In mitosis, the chromosomes shorten during prophase, line up along the cell's equator during metaphase, and separate during anaphase. Each chromosome, with its two chromatids, behaves independently. The result is two daughter cells with the same number of chromosomes as the parent cell.

(b) In meiosis, chromosomes shorten during prophase, and similar chromosomes pair and line up on the cell's equator. The chromosome pairs separate during anaphase of the first meiotic division, thus reducing the chromosome number of the cell in half. The second meiotic division is similar to mitosis.

Figure 6.3 A comparison of mitosis and meiosis. In both types of nuclear division, the chromosome has already replicated in the S phase prior to both the G$_2$ phase and the start of nuclear division.

during telophase I. Meiosis I is called the *reduction division* because it is here that the number of chromosomes is cut in half.

Meiosis II

During **meiosis II** the sister chromatids separate in a process that is basically the same as mitosis. The main difference is that meiosis II begins with a haploid cell, whereas mitosis can begin with a haploid, diploid, or polyploid cell. After the stages of prophase II, metaphase II, anaphase II, and telophase II are completed, the result is four nuclei, each with half the number of chromosomes as the original. Cytokinesis follows. In the case of animals,

the cells become sperm and eggs. In the case of plants, the cells become spores. Figure 6.3 summarizes the differences between mitosis and meiosis.

Plant sexual life cycles feature both haploid and diploid multicellular forms

The sexual life cycles of plants differ greatly from the life cycles of higher animals, such as humans. In humans and other higher animals, the multicellular form is diploid, and the only haploid cells are either sperm or eggs. Plant sexual life cycles are more complex because two multicellular forms of each plant exist. One form is called the

sporophyte (from the Greek words meaning "spore-producing plant") and consists of diploid cells. The other form is called the **gametophyte** (from the Greek words meaning "gamete-producing plant") and consists of haploid cells. Sexual life cycles in plants involve **alternation of generations,** in which these two adult forms alternate in producing, or generating, each other. A typical sexual life cycle of a plant involves five steps (Figure 6.4):

1. Some cells of a multicellular diploid sporophyte undergo meiosis to produce haploid spores.
2. The spore undergoes mitosis to produce a multicellular, haploid gametophyte.
3. One or more cells of the gametophyte undergo mitosis to produce haploid sperm or egg.
4. Sperm and egg combine in **fertilization** to produce a diploid zygote.
5. The zygote undergoes mitosis to produce a multicellular, diploid sporophyte.

In plants, meiosis only produces haploid spores, whereas in animals it only produces gametes. In fungi, spores can arise by either meiosis or mitosis and can be either haploid or diploid.

Plant species vary in the relative sizes of sporophytes and gametophytes, and in whether each form can live independently (Figure 6.5). In most bryophytes, the gametophyte is larger than the sporophyte. When you look at a carpet of moss, for instance, you mostly see gametophytes, with the sporophytes appearing as stalks attached to them. In vascular plants, however, the gametophyte is much smaller than the sporophyte. In most seedless vascular plants, such as ferns, the gametophytes are typically separate, free-living plants. The fern sporophyte is the familiar fern plant, whereas the fern gametophyte is a heart-shaped leafy structure only a few millimeters in diameter. In seed plants, such as conifers and flowering plants, the gametophyte is microscopic in relation to the sporophyte and depends on the sporophyte for nutrition. When you look at a conifer tree or a flowering plant, you are observing the sporophyte. Gymnosperm gametophytes are found in the cones, and angiosperm gametophytes are located within the flowers.

There is also sexual variation in the types of spores produced and therefore in the types of gametophytes arising from those spores. In ferns and most other seedless vascular plants, a single type of spore gives rise to a bisexual gametophyte that produces both eggs and sperm. Most seed plants, however, have two types of spores: a **megaspore** that produces a female gametophyte and a **microspore** that produces a male gametophyte.

Botanists have debated why plant sexual life cycles involve both sporophytes and gametophytes. The most

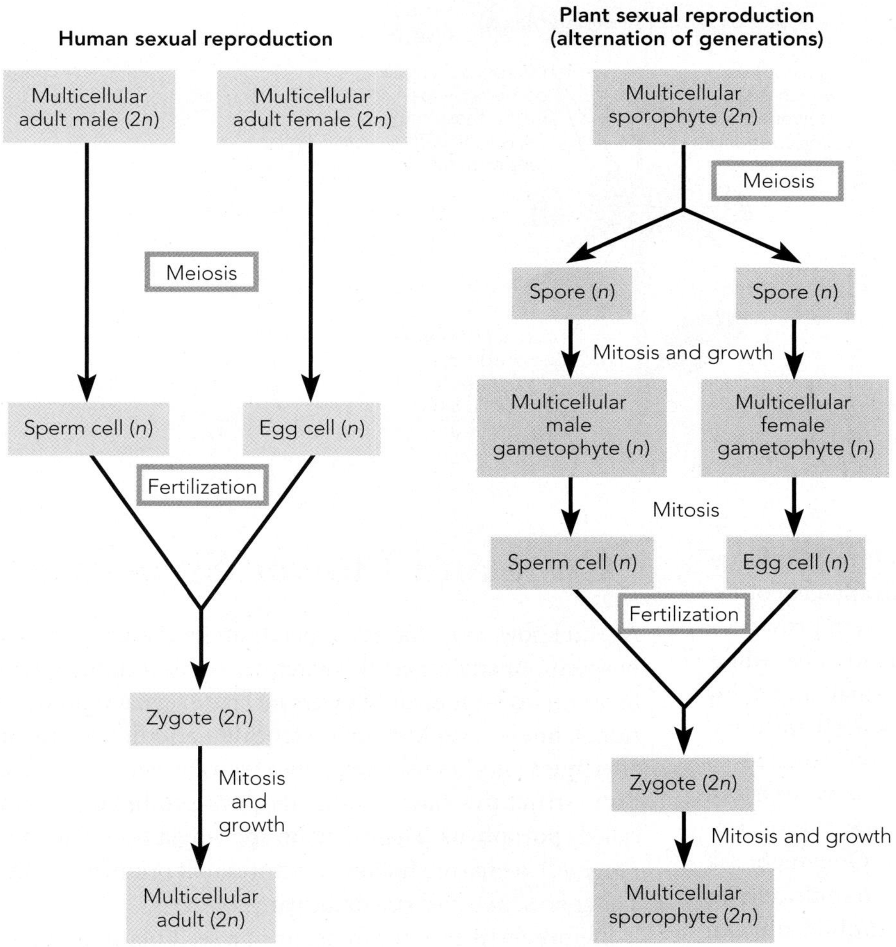

Figure 6.4 Alternation of generations. As you can see, plant sexual reproduction is more complex than human reproduction, involving additional steps between the key stages of meiosis and fertilization. Instead of directly producing sperm and eggs, meiosis in plants produces spores that develop into gametophytes, which then produce the sperm and eggs. In this way, the diploid sporophytes and the haploid gametophytes alternate in producing each other. In some plant species, sporophytes produce spores that give rise to bisexual gametophytes that produce both sperm and eggs.

In most bryophytes, such as mosses:

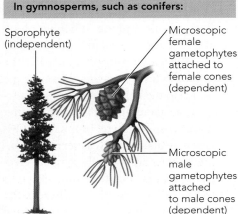

Sporophytes (dependent)

Gametophytes (independent)

In most seedless vascular plants, such as ferns:

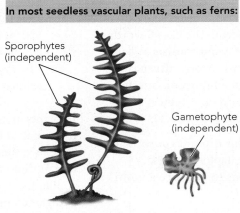

Sporophytes (independent)

Gametophyte (independent)

Figure 6.5 Reproduction in typical sporophytes and gametophytes. Among plants, there are variations in the relative sizes of sporophytes and gametophytes, as well as the physical relationships between the two multicellular plant forms. The gametophyte is the dominant, independent form in bryophytes, but in vascular plants the gametophytes are typically microscopic and the sporophyte is the familiar plant.

In gymnosperms, such as conifers:

Sporophyte (independent)

Microscopic female gametophytes attached to female cones (dependent)

Microscopic male gametophytes attached to male cones (dependent)

In angiosperms (flowering plants):

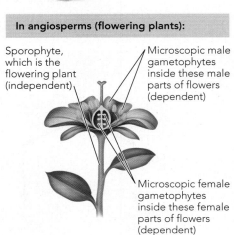

Sporophyte, which is the flowering plant (independent)

Microscopic male gametophytes inside these male parts of flowers (dependent)

Microscopic female gametophytes inside these female parts of flowers (dependent)

prominent theory is that each of these multicellular forms provides the organism with a selective advantage. For example, in seedless plants the sporophyte often provides a significant advantage in growing on dry land. Meanwhile, the free-living gametophytes are usually small and allow sexual reproduction to occur quickly with a minimal amount of water—frequently just a drop will suffice—to allow the sperm to swim to the egg. That is why ferns can live far from water but still obtain from dew, fog, or limited rain the water necessary to reproduce. One might ask why amphibians like toads do not have two multicellular life-forms. The answer is that toads are mobile and can return to the water for fertilization of their eggs.

Section Review

1. **What is the principal difference between mitosis and meiosis?**
2. **Describe pairing of homologous chromosomes.**
3. **What is alternation of generations?**

Cone and Flower Structure

As you know, most species of plants are seed plants—gymnosperms or angiosperms. Therefore, in this section we will focus on how cones and flowers facilitate sexual reproduction. Cones—also known as **strobili**—and flowers form after apical meristems develop into reproductive meristems. Both structures have modified spore-producing leaves called **sporophylls.** These contain **sporangia** (singular, *sporangium*), which are hollow structures that produce spores. The spores then give rise to the gametophytes.

Before fertilization can occur in a seed plant, pollination must take place. In seed plants, the male gametophytes are **pollen grains,** collectively known as **pollen.** The process of transferring pollen from the "male" part of a plant to the "female" part of a plant is known as **pollination.** Keep in mind, though, that pollination does not guarantee fertilization. In order for fertilization to take place, a sperm produced by a pollen grain must unite with an egg in the female part of a seed plant. Each egg is contained in a structure called an **ovule** (Latin *ovulum*, "little egg"). Fertilization does not immediately follow pollination and might not take place until months later, if at all. If fertilization does occur, the ovule will develop into a seed.

Some species of seed plants are capable of **self-pollination** because the male and female gametophytes are on the same plant. In most gymnosperm species, for instance, the male and female gametophytes are on different cones on the same plant. In most angiosperm species, male and female gametophytes are not only on the same plant but also on the same flower. In some angiosperm species, the male and female gametophytes are on different flowers of the same plant. Such species are said to be **monoecious** (from the Greek words for "one house") because each plant has both male and female flowers. Pumpkins and corn are examples. By contrast, in some angiosperm species the male and female flowers are on different plants. Such a species is called **dioecious** (from the Greek words for "two houses"). In a dioecious species, pollination can only occur through **cross-pollination** between separate plants. Marijuana (*Cannabis sativa*) and willows (species of the genus *Salix*) are examples of dioecious species.

In gymnosperms, some apical meristems produce cones

Gymnosperms are known for their characteristic cones that bear exposed seeds. Apical meristems

that become reproductive develop into either pollen cones (sometimes called *simple cones* or *male cones*) or ovulate cones (sometimes called *compound cones* or *female cones*) (Figure 6.6). In pine trees, for example, a pollen cone forms from a meristem that produces a stem with attached leaves. The stem becomes the cone axis, whereas the leaves are modified into papery sporophylls containing sporangia that form spores by meiosis.

The "female" cones are called *ovulate* cones because they contain the egg-bearing ovules. Just as in pollen cones, the meristem forms a stem that becomes the cone axis. However, the leaves become modified into woody bracts. An axillary bud at the base of each bract forms sporophylls, which are fused to form a scale that typically bears two ovules. Wind or insects transfer male gametophytes (the pollen grains) from the pollen cones to the female gametophytes attached to the ovulate cones.

In angiosperms, some apical meristems produce flowers

Angiosperms are known for their characteristic flowers that bear enclosed seeds. Apical meristems that become reproductive develop into male, female, or bisexual flowers, depending on the species. Wind, insects, birds, and mammals such as bats transfer pollen grains to female parts of flowers. Pollinators are attracted to the

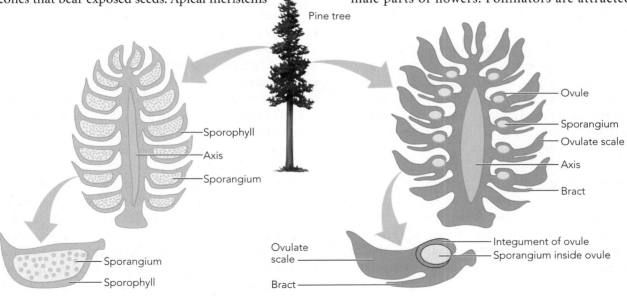

(a) The pollen cones form each season and have papery sporophylls containing sporangia that release pollen in the spring.

(b) The ovulate cones take several seasons to mature and have woody ovulate scales typically bearing two ovules that become seeds. Each ovulate scale is formed from fused sporophylls derived from a highly modified axillary bud forming in the axil between the bract (a modified leaf) and the cone axis. (In this cross section, only one ovule is visible on each scale.)

Figure 6.6 Cones in gymnosperms. In a common gymnosperm, such as a pine tree, pollen cones ("male cones") and ovulate cones ("female cones") occur on the same tree.

flowers by colors, sugary nectar, and other attractants and inadvertently carry sperm-producing pollen from one plant to the other.

As you know, flowers appeared relatively recently in the evolutionary development of plants, arising from modifications to apical meristems and the leaves they produce. The developmental relationship between leaves and flowers can be easily seen in some flowers and not in others. As will be discussed in more detail in Chapter 12, signals produced in the leaves and transmitted to an apical meristem initiate the formation of a flower. Seasonal change—specifically the length of nights—controls production of these signals from leaves. The identity of these signals is unknown.

When the induction signal arrives from the leaves, the apical meristem begins to enlarge and eventually produces flower parts from primordia that would otherwise have become leaves. All flower parts are modified leaves. As you will see, some of these modified leaves are sporophylls. The role of sporophylls in reproduction will be discussed in more detail in Chapters 21 and 22.

A flower can consist of up to four types of modified leaves

Because angiosperms include nearly 258,000 species and have existed for more than 140 million years, great diversity occurs in flower structure among living species. Before discussing the many variations in floral structure, let us examine the components of a typical flower (Figure 6.7). This generalized flower sits atop a stem called a **peduncle.** At the top of the peduncle is a swollen tip called a **receptacle,** which bears the parts of the flower. A receptacle can form up to four types of modified leaves: sepals, petals, stamens, and carpels. Sepals and petals are sterile, while the stamens and carpels are the sporophylls—the fertile, reproductive modified leaves.

Sepals (from the Latin *sepalum,* "covering") protect the flower bud before it opens. They are frequently green and form first and lowest on the receptacle. The sepals are together known as the **calyx** and can be fused.

Petals (from the Latin *petalum,* "to spread out") are the colorful modified leaves that attract pollinators. Petals form second and inside or above the calyx on the receptacle. The petals are together known as the **corolla** and can be fused. The calyx and corolla—consisting of the two sterile types of modified leaves—are known as the **perianth,** meaning "around the flower."

Stamens are the pollen-producing, or "male," parts of a flower. The stamens are together known as the **androecium** (an–DREE–she–em; from Greek words meaning "the house of man"). Stamens form third and inside or above the peri-

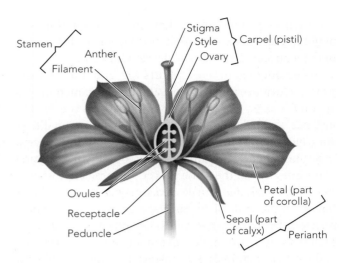

Figure 6.7 Generalized flower structure. A flower sits atop a floral stem called a *peduncle.* The tip of the peduncle is swollen and becomes a receptacle. Attached to the receptacle are up to four types of modified leaves: sepals, petals, stamens, and carpels (also called *pistils*). Sepals and petals, which are sterile, are known collectively as the *calyx* and the *corolla,* respectively. The stamens and carpels are fertile and are known collectively as the *androecium* and *gynoecium*. This diagram is generalized in the sense that many flowers do not include all of these parts.

anth. Each stamen has a long filament with an **anther** composed of two lobes and four pollen sacs. The pollen grains—the male gametophytes—form within these pollen sacs.

Carpels (from the Greek *karpos,* "fruit") are the ovule-bearing, or "female," part of the flower, known collectively as the **gynoecium** (jeh–NEE–she–em; from Greek words meaning "the house of woman"). Carpels form last and inside or above the androecium. The gynoecium can consist of one or more carpels, and carpels can be either separate or fused. The term **pistil** is sometimes used to refer to an individual carpel, which is known as a simple pistil or simple carpel, or to a group of fused carpels, known as a compound pistil or compound carpel. Each carpel or group of fused carpels (pistil) consists of a stigma, style, and ovary. Located at the top of the carpel, the **stigma** provides a sticky surface for pollen. The **style** is the middle section connecting the stigma to the ovary and can be short or long. The **ovary** contains one or more ovules and will eventually swell to become part or all of the fruit. A pollen grain deposited on the stigma germinates to produce a pollen tube that grows down through the style to the ovary. Upon penetrating the female gametophype, one sperm fertilizes the egg which then becomes the zygote. After fertilization the ovule becomes a seed. A second sperm from the pollen grain combines with two or more other nuclei produced by the female gametophyte to form a polyploid endosperm nucleus that will divide to produce food for the developing embryo. This so-called **double fertilization** is one of the defining features of flowering plants. In flowering plants, most endosperm nuclei are $3n$ and can divide by mitosis only.

A peduncle can bear a single flower or a group of flowers called an **inflorescence.** Within an inflorescence, each flower sits atop a stalk called a **pedicel.**

The number and symmetry of flower parts can vary

Considerable diversity occurs in flower structure. For instance, flowers can vary in the types of structure present. A flower is said to be **complete** if it contains all four types of modified leaves: sepals, petals, stamens, and carpels, as in Figure 6.8. A flower that lacks one or more of these types is called **incomplete.** Flowers can also be classified according to whether or not they have both of the fertile types of modified leaves. A **bisexual** flower, also known as a **perfect** flower, has both stamens and carpels. A **unisexual** flower, also known as an **imperfect** flower, has either stamens or carpels but not both. A male (staminate) flower is an imperfect flower that has only stamens. A female (carpellate or pistillate) flower is an imperfect flower that has only carpels. A bisexual, or perfect, flower can be either complete or incomplete, depending on whether it has both sepals and petals. A unisexual flower, of course, is always incomplete. That is, even if it includes both sepals and petals, it still lacks either carpels or stamens.

Flowers also vary in their symmetry. Those with radial symmetry are said to be **regular** flowers. They are also known as **actinomorphic,** or ray-shaped, flowers (from the Greek *aktis,* "ray"). In radial symmetry, the floral parts radiate out from a center (Figure 6.9a). Examples are the flowers of apples and tulips. Some flowers have bilateral symmetry, which means they can be divided only along a single imaginary line into two equal parts that mirror each other (Figure 6.9b). Such flowers are said to be **irregular,** or **zygomorphic** (from the Greek *zygon,* "yoke" or "pair"). Some examples are snapdragons and many orchids.

Natural selection frequently modifies floral structure to reflect the needs or structure of pollinating organisms. Some flowers produce sugary nectar that attracts pollinators. Pollen then sticks to the animals, which carry it to flowers on other plants, causing cross-pollination. The structures of many flowers, shaped by natural selection, attract particular pollinating organisms that will visit other flowers produced by the same plant species. Similarly, the structure of animals frequently reflects characteristics that facilitate obtaining food from a particular type of plant. For example, flowers pollinated by hummingbirds often

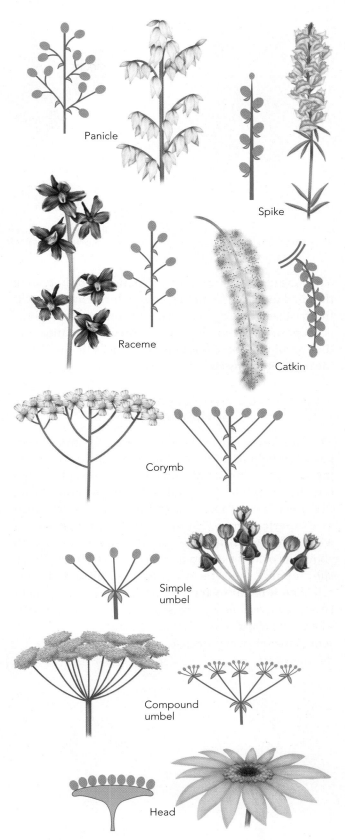

Figure 6.8 Inflorescences. Inflorescences bear a group of flowers on a single peduncle.

(a) Regular, or actinomorphic, flower (radial symmetry)

(b) Irregular, or zygomorphic, flower (bilateral symmetry)

Figure 6.9 Types of flower symmetry. Flowers in which the members of each whorl have radial symmetry are known as *regular,* or *actinomorphic,* flowers. The lance-leaved coreopsis is an example. Flowers that have bilateral symmetry are called *irregular,* or *zygomorphic,* flowers. The calypso orchid is an example.

Flower structures are examples of how natural selection modifies what is already present

We have observed that leaves become modified to serve various other functions, such as spines and flower parts. This evolutionary pattern, in which a modified structure serves new functions, occurs frequently in all life-forms because mutations can only modify existing structures. If the modifications increase a plant's ability to survive and produce offspring, they are retained in subsequent generations. For example, petals have bright colors that attract insects, which distribute pollen more widely than the wind. In this case, a mutation that changed the color of a green leaf can increase the number of offspring and their geographic distribution. Subsequent mutations might have changed the petal shape in such a way that the flower resembled a desirable insect mating partner. Eventually, the shape and color of a petal can be so modified that its evolutionary origin as a leaf is not obvious to an unknowing observer. In Chapter 23, you will see how stamens and carpels may have evolved from leaves.

have long necks with sweet nectar located at the bottom of the floral tube. The hummingbirds, in turn, have long bills that facilitate reaching the nectar. The anthers of such flowers deposit pollen on the birds' feathers. Pollination and the coevolution of plants and the animals that pollinate them will be discussed in more detail in Chapter 23.

Flowers can vary in the position of their ovaries

The location of the ovary with respect to other flower parts can help in flower classification. If the fused stamens, sepals, and petals attach to the receptacle below the ovary, they are said to be hypogynous (from the Greek *hypo-,* "under"), and the ovary is described as a **superior ovary** (Figure 6.10). An arrangement in which the fused stamens, sepals, and petals attach above the ovary is said to be epigynous (from the Greek *epi-,* "upon"). Since the ovary is below these parts, it is called an **inferior ovary.** In a perigynous arrangement (from the Greek *peri-,* "around"), the parts attach halfway up the ovary, which is then called a **semi-inferior ovary,** or half-inferior ovary.

Section Review

1. Describe how cones function as reproductive structures in gymnosperms.
2. Describe the functions of the four types of modified leaves in a flower.
3. How does a complete flower differ from a perfect flower?

Superior ovary

Semi-inferior ovary

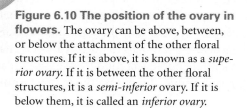

Inferior ovary

Figure 6.10 The position of the ovary in flowers. The ovary can be above, between, or below the attachment of the other floral structures. If it is above, it is known as a *superior ovary.* If it is between the other floral structures, it is a *semi-inferior* ovary. If it is below them, it is called an *inferior ovary.*

Seed Structure

Seeds get plants through the hard times—through the seasons when extreme temperatures or lack of moisture can make growth, and even life, impossible. Seeds might never have evolved if the climate were conducive to plant growth year-round. In fact, ferns and other seedless plants ruled the botanical world when the continents, as a result of the constantly moving plates of the Earth's crust, were clustered near the equator. In response to the present seasonable environments, however, animals hibernate, store food in burrows, migrate to warmer climates, or build houses to weather unpleasant seasons. Gymnosperms and angiosperms produce seeds—tiny dormant copies of themselves that germinate when favorable conditions return. Overproduction of seeds is largely responsible for feeding the animal world.

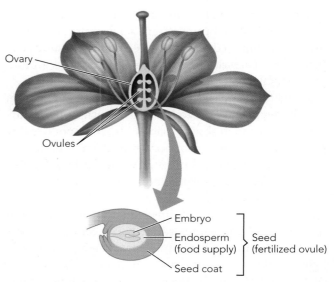

Figure 6.11 The ovule in flowering plants. Each carpel in a flowering plant contains one or more ovules within an ovary, located at the base of the carpel. Each fertilized ovule becomes a seed.

Seeds form from ovules on bracts of cones or in carpels of flowers

Gymnosperms produce seeds on or near bracts, exposed leaves found on cones, whereas angiosperms produce seeds inside the ovaries of flowers. Seeds protect plant embryos from drying out, facilitating the survival of embryos on dry land. As you have seen, seeds develop from sporangia attached to modified leaves—the sporophylls of a cone or carpels of a flower. The ovule forms as an attachment to a modified leaf and develops into a seed after fertilization. The outer layers of the ovule, known as *integuments,* will harden to become the seed coat. An ovule typically has one or two integuments. As you saw in Figure 6.6, in the ovulate cone of gymnosperms such as pine, each scale typically has two ovules. Each fertilized ovule develops into a seed sitting atop the scale. In angiosperms, seeds develop as fertilized ovules inside the ovary, which forms the base of the carpel (Figure 6.11). The ovary eventually matures into a fruit. You will study the formation of embryos in seeds in Chapters 22 and 23.

Seeds nourish and protect developing embryos

If you were to explain to a young child what a seed is, you might say it is "a baby plant in a box with its lunch." The baby plant is, of course, the embryo. The walls of the box are the seed coat. The food consists of molecules of starch, protein, and fat that surround the embryo within the seed.

A gymnosperm seed contains an embryo with several cotyledons attached to a short stem and an embryonic root. The embryo is surrounded by varying amounts of tissue used as nutrition by the developing embryo and by a seed coat.

In Chapter 3, you saw the basic structures of dicot and monocot seeds (see Figure 3.11). The seeds of dicots and monocots differ not only in how many cotyledons they have but also in their structure. Dicot seeds contain an embryo with fleshy, prominent cotyledons that contain starch, protein, and lipids, which are broken down and used for energy and as carbon skeletons by the germinating seed. The cotyledons attach to a short stem that culminates in the embryonic root called a *radicle.* The embryo can curl back on itself, depending on how much the embryo grows for each particular species. In dicot seeds, the embryo is surrounded by varying amounts of nourishing endosperm, which in turn is surrounded by a seed coat.

Monocot seeds, such as cereals and other grasses, contain a small embryo on one side and a large starchy endosperm that fills most of the seed. The embryo has one cotyledon known as a **scutellum,** which is attached to an embryonic axis containing shoot and root meristems. A layer of protein called the *aleurone layer* surrounds the endosperm and responds to signals from the scutellum by producing enzymes to break down starch (Chapter 11). A seed coat surrounds the aleurone layer and the embryo. In response to hormonal cues after embryo formation is complete, most seeds lose considerable water and become dormant.

In seed germination, first the embryonic root grows through the seed coat, and then seedling formation begins

When mature seeds are shed from fruits or cones, they contain between 5% and 20% water by weight. Seed germination begins with **imbibition,** a passive process in which the dry seed takes up water like a sponge. In most seeds, germination begins within a few hours after imbibition is completed. For example, in Grand Rapids lettuce seeds (*Lactuca sativa*), germination begins around 16 hours following the start of water uptake.

Many seeds are dormant following their formation, even if they are allowed to imbibe. Frequently, these dormant seeds contain abscisic acid or other compounds that prevent germination from occurring. After abscisic acid gradually deteriorates during the winter and reaches low levels the following spring, the seeds can germinate. In this way, abscisic acid serves as a kind of clock to prevent seeds from germinating during winter warm spells when the young plants would soon freeze. Sometimes the breakage of dormancy requires specific temperatures or light conditions related to the production of hormones, such as gibberellins, by the embryos. Some types of lettuce germinate only when they are in sun or shade, depending on the preference of the adult plant.

The first visible sign of germination is growth of the radicle through the seed coats. The root apical meristem in the radicle becomes active and begins, by cell division and elongation, to produce the seedling's root. Food molecules in the seedling, and particularly in the cotyledons, break down to supply the energy and structural molecules needed for root growth. Sometime after root growth begins, the embryonic shoot, called the plumule, begins to grow, and the complete seedling forms.

In some dicots and most monocots, the hypocotyl barely develops, and the cotyledon(s) remain on or below the soil surface. Since the cotyledons remain underground, this type of germination is called *hypogeous* (from the Greek words meaning "under earth"). Corn is an example, as you can see in Figure 6.12a. However, in most dicots and some monocots, the hypocotyl grows and pushes the cotyledons above the surface of the soil. Since the cotyledon or cotyledons are aboveground, this is known as *epigeous germination* (from the Greek words meaning "above earth") (Figure 6.12b). In epigeous germination the cotyledons also function as photosynthetic organs. However, they are more exposed to harsh weather conditions, which can occur in the spring.

Although seeds are generally the result of sexual reproduction, many plants produce asexual seeds through a process known as **apomixis** (from Greek words meaning "away from the act of mixing"). Apomictic seeds form without the "mixing" or joining of sperm and eggs (see the *Biotechnology* box on page 137). In dandelions, some plants produce sexual seeds while others produce apomictic seeds. The species is thus poised to take advantage of either a changing or an unchanging environment. In Chapter 11, we will look in more detail at the process of seed germination.

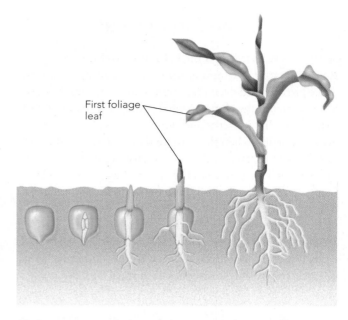

(a) Germination in which cotyledon or cotyledons are below ground (hypogeous germination). Example: corn.

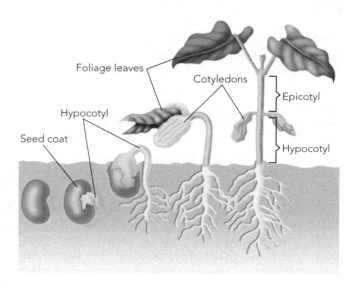

(b) Germination in which cotyledon or cotyledons are above ground (epigeous germination). Example: garden bean.

Figure 6.12 Seed germination. **(a)** In some plants, such as corn, the hypocotyl does not elongate, and so the cotyledons remain belowground. **(b)** In other plants, such as the garden bean, an elongated hypocotyl forms between the cotyledon(s) and the roots. The hypocotyl carries the shoot, the epicotyl, above the ground. In beans, the hypocotyl is bent to form a hook that gradually straightens.

Section Review

1. Describe how seeds originate.
2. What are the functions of seeds?
3. Describe the process of germination.

BIOTECHNOLOGY

Apomixis in Agriculture

As you know, by a process known as *apomixis,* some plants produce asexual seeds that germinate to produce offspring exactly like the parent. Scientists worldwide are interested in the genetics of this process. For example, researchers studying dandelions have discovered three genes that control apomixis.

Plant breeders are extremely interested in the possibility of adding the ability to produce apomictic seeds to the genetic repertoire of a number of crop plants. Here's why: In 1908 a corn breeder named G. H. Shull discovered that if he crossed two pure-breeding corn lines, the resulting hybrid produced four times as much corn. This so-called *hybrid vigor* holds true for many species of plants, including the major cereals that feed the world. The trouble is that hybrid seeds must be recreated each year. Scientists must interbreed two varieties—with one serving as the female parent and the other serving as the male parent—to produce hybrid seed. This costly yearly process uses resources such as land and the time of workers. If a hybrid plant produced seed by apomixis, scientists or farmers could simply harvest the hybrid seeds directly and use them the following year. The high yields of hybrids can be passed to future generations of plants easily and at low cost.

Apomictic seed would make it more difficult for a private company to protect a patent on hybrid seed. Anyone who grew the crop could save back seed for the next year, although they would be subject to legal action. Despite patent infringement problems, companies would be able to devote many of the resources currently devoted to hybrid seed production to producing new, useful products. In 1997 two scientists—one American and one Russian—received the first patent for an apomictic plant. They produced apomictic corn by crossing corn to eastern gamagrass (*Tripsacum dactyloides*), a wild grass.

At the International Wheat and Maize Center in Mexico, scientists have continued to work on transferring genes for apomixis from a wild grass to corn. Although more than 300 species of plants can produce apomictic seeds, most of them, except for citrus and a few others, are not cultivated crops. Wild relatives of sorghum, beets, strawberries, and mangos produce seeds by apomixis. In the coming years, you will no doubt be hearing more about the promise of apomixis for increasing the quantity and quality of our food.

A blackberry produces seed by apomixis.

Gamagrass, an apomictic wild grass, is distantly related to corn.

Fruit Structure

Before the evolution of flowers, plants either did not produce seeds or produced them exposed on the bracts of cones. In flowering plants, seeds are enclosed in an ovary, produced as part of the flower. After fertilization and seed development, the ovary and sometimes other parts of the flower will expand to produce a fruit. Fruits can serve various purposes, depending on the species. They protect the developing embryo (seeds) from drying out and to some extent protect from disease and herbivores. They promote seed distribution by animals that eat the fruit. In addition, they provide ready-made fertilizer for the germinating seed.

In the United States, most people are familiar with fruits grown in temperate climates, such as apples and oranges, as well as some tropical fruits that are transported without spoiling quickly, such as pineapples. Many delicious tropical fruits, however, do not ship well and are usually unavailable in American stores (see *The Intriguing World of Plants* box on page 138).

During seed development in a flowering plant, the ovary expands to become part or all of a fruit

Botanically, a typical fruit consists of a mature ovary or ovaries, which include seeds. In common usage, however,

THE INTRIGUING WORLD OF PLANTS

Tropical Fruits

Your familiarity with tropical fruits depends on where you live and on how much you have traveled. People in temperate regions such as most of Europe and the United States have eaten bananas, citrus fruits, and pineapples and, less frequently, avocados, coconuts, and dates. The few varieties of tropical fruits that ship well and have a long shelf life have become important export crops for many tropical countries. Many other varieties occur in local markets. A few other fresh, tropical fruits such as guavas and mangos appear fresh in stores in temperate latitudes from time to time and in various conditions of edibility.

Tropical fruits play an important role in providing both nutrition and calories for at least half of the world's people. Since many grow wild and sometimes year-round in tropical and sub-tropical countries, they provide plentiful food that is either inexpensive or free. Often they do not keep well after harvest and so do not constitute an important part of an export economy. For many tropical fruits, cost and unfamiliarity, as well as short shelf life, have limited their inclusion in U.S. and European diets.

In terms of human nutrition, bananas and plantains, members of the genus *Musa*, are the world's fourth most important crop in gross value of production. Generally speaking, people eat bananas raw and cook the more starchy plantains. Bananas, the largest selling fruit in the United States, serve as the second most important crop in the diet of many of the world's people, after rice or corn. Despite their popularity and importance, bananas

Red banana

Papaya

Carambola

Durian

and plantains remain a poorly studied crop. Hundreds of varieties exist, many of which grow best in very specific eco-geographic regions. In developing countries, bananas grow year-round as a weed and provide food in between the harvests of other crops. Some other important tropical fruits include the following:

- Durians (*Durio zebethinus*) are large, spiny fruits that are very popular in Southeast Asia. Their strong, cheese-like odor is quite different from that of any other fruit and offends some people while delighting others. The flesh of the fruit tastes sweet and buttery.
- Mangosteens (*Garcinia mangostana*) have a sweet, slightly acidic flavor.
- Lychees (*Litchi chinensis*) are a popular dessert fruit.
- Mangos (*Mangifera indica*) are considered by some to be the queen of tropical fruits because of their exotic taste.
- Guava (*Psidium guayava*). To many people, guavas taste like a mixture of banana and pineapple.
- Jackfruit (*Artocarpus heterophyllus*). This sweet fruit, related to breadfruit, can weigh as much as 50 kilograms (110 pounds) and be nearly 1 meter in length. Jackfruit hangs from the trunk, which is known as a cauliflorous arrangement.
- Breadfruit (*Artocarpus communis*). These starchy fruits, native to Polynesia, are often served like potatoes and squash.
- Carambola (*Averrhoa carambola*), or starfruit, is becoming more common in grocery stores. The crisp taste has been described as being between that of apples and grapes. Carambola is also a cauliflorous fruit.
- Papaya (*Carica papaya*). Papayas are cauliflorous fruits that seasonally arrive in grocery stores. Their taste reminds some people of a peachy apricot.

most people use the term *fruit* to refer only to juicy varieties that are either sweet or tart, such as apples, oranges, and lemons. Meanwhile, a number of fruits that are not sweet—such as tomatoes, squash, green beans, and eggplant—are called "vegetables." In fact, the term *vegetable* has become a catchall label for a wide variety of edible

parts of plants that are not sweet, such as tubers (for example, potatoes), modified roots (sweet potatoes), leaves (lettuce), and unopened flowers and peduncles (broccoli).

The primary function of fruits is to distribute seeds to new areas where the plants might grow. They are sweet and inviting to animals, which carry them off directly or

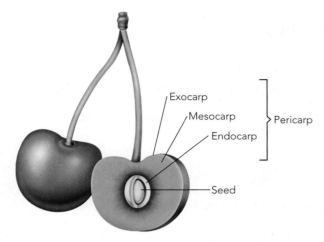

Figure 6.13 The ovary wall in a fruit is swollen and is called the *pericarp*. The ovary wall consists of an inner layer (endocarp), a middle layer (mesocarp), and an outer layer (exocarp). In various types of fruits, one or more of these layers may be thick or thin and fleshy or dry.

deposit the seeds in feces. In nature, fleshy fruits can also help nourish germinating seeds. After all, seeds do not find themselves naturally placed under the ground in good garden soil. Rotting fruit often provides enough of a "starter soil" to enable a seedling to become established so that its roots can penetrate into the available soil.

Fruits can be either fleshy or dry. The ovary wall, called the **pericarp,** consists of three parts: an outer part called the **exocarp** (often the skin), a middle part known as the **mesocarp,** and inner parts named the **endocarp** (Figure 6.13). In Chapter 11, we will look at how hormones control fruit development and ripening. Here we will focus on the basic varieties of fruits.

Fruits can be categorized as simple, aggregate, or multiple

There are three general types of fruits: simple, aggregate, and multiple. Most fruits are **simple fruits,** which develop from one carpel or from several fused carpels. **Aggregate fruits** originate from one flower that has many separate carpels. Each carpel becomes a tiny fruit that is gathered, or aggregated, with other similar "fruitlets" on a single receptacle. Some examples are blackberries, strawberries, and magnolias. **Multiple fruits** develop from the carpels of more than one flower in a single inflorescence, as in pineapples and figs. Multiple fruits can be fleshy or dry at maturity. Table 6.1 provides a brief summary of common edible fruits. Keep in mind, though, that all species of flowering plants produce fruits, whether we eat them or not.

Simple fleshy fruits
In simple fruits that are fleshy, one or more layers of the pericarp become soft during ripening. Basic types include berries, hesperidia, pepos, drupes, and pomes.

- In **berries,** all three pericarp layers soften to varying degrees as the fruit ripens. Berries may originate from either one carpel or several carpels, and each carpel may have one seed or multiple seeds. Some examples are tomatoes, grapes, and bananas. In many plants that produce berries, the flower has an inferior or semi-inferior ovary. Therefore, several parts of the flower may contribute to the fruit. Fruits in which other flower parts besides the ovary become part of the fruit are sometimes referred to as **accessory fruits.** Berries usually have shriveled flower parts at the tip, as in cranberries and bananas. Some fruits that are commonly called *berries* are, botanically speaking, aggregate fruits instead. The strawberry, for example, is a fleshy receptacle with many individual achenes (described below) on its surface.
- **Hesperidia** are botanically similar to berries, but with a leathery skin or exocarp that produces pungent, acidic oils. The inner ovary wall produces sacs that fill with juice. Hesperidia include all citrus fruits such as grapefruit, kumquats, lemons, limes, and tangerines.
- **Pepos** are similar to berries but with a thick rind or exocarp. They consist of members of the pumpkin family (Cucurbitaceae), such as watermelons, pumpkins, and cantaloupes. The mesocarp and endocarp may not be distinct.
- **Drupes** develop from flowers with superior ovaries and one ovule. The simple and usually fleshy fruit resembles a berry but has a hard endocarp often known as a *pit*, which is typically attached to the single seed. Some examples of drupes are olives and coconuts. In the coconut, the fibrous husk, consisting of the mesocarp and exocarp, is removed before what we commonly know as the coconut is sold. The fibrous husk can be used to make brushes and matting. The outside of the hard coconut we normally see in the grocery store, which is about the size of a softball, is the endocarp. Inside is coconut meat, which is a cellular endosperm, and coconut water, which is an endosperm composed only of nuclei, not of complete cells. The embryo is cylindrical and embedded in the cellular endosperm. Coconut milk is a commercial product made by blending the two endosperms.
- **Pomes** resemble berries, but the bulk of the fleshy fruit comes from an enlarged receptacle that forms from the swollen end of the peduncle or pedicel. Since the calyx and corolla attach here, their base becomes part of the

Table 6.1 Fruit types

Type	Description		Examples
Aggregate Fruits	Fruits forming from one flower with several carpels		blackberries, strawberries, raspberries, magnolias
Multiple Fruits	Fruits forming from more than one flower on one peduncle		pineapples, mulberries, figs, breadfruit
Simple Fruits—Fleshy			
Berries	Fruits having one to many seeds and a pericarp that becomes soft and often sweet and slimy as it matures		grapes, dates, eggplant, tomatoes, green peppers, blueberries, gooseberries, mangosteens, guavas, bananas, persimmons
Hesperidia	Fruits similar to berries but with a leathery pericarp that produces fragrant oils		all citrus fruits, such as oranges, lemons, grapefruit
Pepos	Fruits similar to berries but with a thick rind (exocarp)		pumpkins, cucumbers, squashes, cantaloupes, watermelons
Drupes	Single seed surrounded by a hard endocarp that forms what is popularly known as the pit. The mesocarp and exocarp are fleshy or fibrous		olives, peaches, almonds, coconuts
Pomes	Bulk of the fruit is formed from a swollen receptacle		pears, apples
Simple Fruits—Dehiscent Dry			
Follicles	Fruits that open along one seam when the seeds are to be released		milkweed, columbines, peonies, magnolia
Legumes	Fruits that split into two seed-bearing halves. Legumes belong to the large plant family that includes peas and beans. Seeds can be born in a common ovary or in separate compartments		garden peas, beans, mesquite, peanuts
Siliques	Dry fruits in which the seeds reside on a partition between halves of the ovary		shepherd's purse, cabbage, watercress, radish
Capsules	Fruits of two or more carpels, split along seams or forming caps or pores		poppies, irises, snapdragons, orchids, yucca
Simple Fruits—Indehiscent Dry			
Nuts	Dry fruits with a hard, thick pericarp and a basal cup		acorns, hickory nuts, chestnuts, hazelnuts
Schizocarps	Have a hard, thin pericarp that splits into two halves		members of the parsley family, such as carrots
Achenes	Have thin pericarps. The single seeds connect to the pericarp only at their base		sunflowers, buttercups
Samaras	Have thin pericarps. The seeds occur in pairs and have wings that allow dispersal by the wind		maple, elm, ash
Caryopses or grains	Grass seed in which the fruit has a hard pericarp, which is fastened to the embryo all the way around		all members of the grass family, such as corn and rice

fruit, and their remnants can often be seen at the end of the fruit opposite the stem. Pomes are therefore also examples of accessory fruits. Some familiar pomes are apples and pears.

Simple dry fruits

Simple fruits that are dry can be categorized as being either dehiscent or indehiscent. **Dehiscent** dry fruits split open at maturity to shed seeds. The word *dehiscent* comes from the Latin word meaning "to split open."

- **Follicles** have single carpels that split along one seam to release seeds. Examples are milkweed, columbines, and magnolias. Magnolia fruits are multiple fruits consisting of individual follicles that each contains one seed.
- **Legumes** resemble follicles except that, at maturity, legumes have two seams that split the fruit into two halves. Each half bears seeds. Members of the legume family (Fabaceae) include such plants as beans, peanuts, and peas. Legumes produce fruits that are also known as *legumes*.

◆ **Siliques,** produced by species in the mustard family (Brassicaceae), are dry fruits consisting of two carpels that split into two halves, with the seeds found on a central partition between the halves. Examples are shepherd's purse and cabbage.

◆ **Capsules** split in several different ways, depending on the species that produces them. All capsules develop from at least two carpels. Some split along the seams between the carpels; others split within carpels; still others form caps or pores at the top of the ovary. Poppies, irises, and orchids are examples of plants that produce capsules.

Indehiscent dry fruits remain closed at maturity. Examples are nuts, schizocarps, achenes, samaras, and caryopses.

◆ **Nuts** have stony pericarps as shells and originate from compound carpels. Some examples are acorns and hazelnuts. Some "nuts" are not actually nuts in the botanical sense. For instance, almonds and walnuts are drupes with the exocarp and mesocarp removed. A Brazil nut is actually a seed that comes from a capsule, while pistachios are seeds found inside drupes.

◆ **Schizocarps** occur in members of the parsley family (Apiaceae) such as parsley, carrots, dill, and celery, as well as in maples (Aceraceae). Schizocarps have a hard, thin pericarp, which opens into two or more parts, each of which contains a seed. In maple seeds, two seeds have wings attached.

◆ **Achenes** resemble small nuts with their hard, thin pericarps and single seed. Achenes attach to the pericarp at the root end. Thus, the seeds can be easily removed from the fruits. Achenes form from single carpels. Sunflowers and buttercups produce achenes. As you have already seen, a strawberry is an aggregate fruit containing many achenes.

◆ **Samaras,** from ash and elm trees, are like achenes with the addition of hard, thin, elongated pericarps that produce wings around a single seed.

◆ **Caryopses** (singular, *caryopsis*) or grains occur in all plants of the grass family (Poaceae), which includes rice and corn. The grain is a dry, achene-like fruit with a hard pericarp. In contrast to the achene, the pericarp and seed coat of a grain fuse all around the embryo, so the fruit does not split open when mature. What most people call a corn or rice seed is really a fruit, a caryopsis. The embryo and endosperm are surrounded by a pericarp. The integuments disappear during the maturation of the fruit.

A number of mechanisms disperse seeds and fruits to new locations

Seed distribution follows many different patterns, depending on whether the plant produces edible fruits, fruits or seeds that attach to animals, or fruits or seeds carried by the wind or by water. Sometimes the seed itself is distributed; other times the fruit is distributed, as in the case of fleshy fruits; or the plant itself may be distributed. In the tumbleweed (*Salsola*), for example, the entire plant is blown along by the wind and distributes seeds as it goes. This effective seed dispersal method has resulted in the tumbling tumbleweed becoming a signature plant of western North America. The tumbleweed is actually a native of Russia. Russian immigrants accidentally brought tumbleweed seeds with them to the United States, mixed in with crop seeds they had packed for planting.

Some seeds have plumes of fluff (such as dandelion seeds, Figure 6.14a) or wings (such as maple seeds, Figure 6.14b) to assist in being carried aloft by the wind. Others are round, like poppy seeds or tobacco seeds, and therefore can roll along by the action of wind. Still others, like orchids or petunias, have tiny seeds that blow through the air like dust.

Self-induced seed dispersal through the air occurs in several different species but in a number of different ways. An unusual method of dispersal through the air is found in fruits of dwarf mistletoes, which are forcibly and rapidly shot into the air when triggered by the heat of a passing animal. In dwarf mistletoe, discharge occurs after the buildup of water pressure in the fruit. In other fruits, such as those of witch hazel, discharge is promoted by drying of the fruit. In some members of the pumpkin family, heat and fermentation within the fruit result in an explosive discharge of bubbly goo containing seeds wherever the ovary wall first becomes weak enough to succumb to the internal pressure generated by CO_2.

Some fruits and seeds, like coconut, disperse quite successfully by floating in water. In ocean waters, these fruits need thick hard or spongy outer layers to keep seawater out. Thus, the coconut is adapted to survive for long periods of time floating in the ocean, and currents have carried these fruits to many new locations.

Other plants produce seeds or fruits with pockets of trapped air to promote flotation. Sedges, for example, are marsh and bog plants that produce seeds covered by membrane-covered sacs that help to float the seeds. Water-repellant waxes coat some fruits and seeds so that they can survive water dispersal. During floods, entire plants may be ripped from their moorings and sent on their way to a new location. In capsules filled with small seeds, raindrops may splash the seeds like dry dust to new locations several feet away.

Some seeds rely on dispersion by animals by serving as food and by hitching a ride. In some cases the seed passes through the digestive system to germinate, within a pile of ready-made fertilizer, at a new location (Figure 6.14c). Fruit

Figure 6.14 Seed dispersal. (**a**) Seeds and fruit can be dispersed by the wind. Some seeds, such as dandelion, have the calyx modified into a plume, which allows for wind dispersal. Seeds of maple and pine (**b**) have wings. In the case of tumbleweed, the whole plant is uprooted and blown along the ground. Passage of some seeds such as these raspberries (**c**) through an animal's digestive system dissolves away part of the seed coat, aiding germination. The animal also serves to distribute the seed. (**d**) Many seeds have barbs or hooks on their seed coats or fruit coats that stick to the fur of animals. Most often, seeds or fruits are able to hitch a ride because they grab onto fur (or clothing). Cocklebur (*xanthium*) fruits are well known to hikers. Devil's claw (*proboscidea*) (**e**) is a fruit that catches on the legs of animals and eventually splits open to release seeds. (**f**) Some seeds produce elaiosomes, shown here as the artificially colored components on the ends of the seeds. These are oil bodies that attract ants, which disperse the seeds while eating the eliaosome as food. The ants carry the seed and its eliaosome underground. The food benefits the ants, and the underground destination benefits the seeds.

ripening involves changes to red, yellow, or orange coloration that attracts animals. Also, ripe fruits produce pleasant or at least interesting odors and are sugary sweet when eaten. Again, these features of fleshy fruits attract animal dispersal agents. A change to the bright color often occurs at the time sugar content of the fruit is increasing. Green fruits, with immature seeds, are often sour or cause the mouth to pucker, which encourages animals to leave them alone. For example, some types of persimmon fruits are particularly rich in astringent tannins before they ripen. Tannin is the same compound that sometimes makes tea leaves bitter.

In other cases, the seed or fruit produces hooks, barbs, and stickers that attach to animal fur or skin for a journey to new horizons (Figure 6.14d and e). Many species of plants—up to a third of the species in some ecosystems—are transported to new locations by ants. Some seeds produce white appendages called elaiosomes, which are used as food by the ants (Figure 6.14f). In North America, these plants include bleeding heart, trillium, Dutchman's breeches, and some violets.

Humans also transport plants, seeds, and fruits willingly and accidentally. This concerns farmers since new and virulent plant diseases and noxious plants can be rapidly spread around the world by human contact.

Section Review

1. **What are the functions of a fruit?**
2. **Describe the basic structure of a fruit.**
3. **What is the difference between simple, aggregate, and multiple fruits?**

SUMMARY

Plant Reproduction: An Overview

The structural, functional, and biochemical features of an organism are all designed to ensure success of the individual and the production of successful offspring.

Asexual reproduction occurs through mitosis and results in offspring that are genetically identical to each other and the parent (p. 124).
Plants tend to use asexual reproduction in stable environments. Some methods include adventitious shoots from roots and plantlets arising from leaves.

Sexual reproduction results in genetic variation (pp.124–125).
In sexual reproduction, offspring are a combination of the genetic traits of each of two parents. Sexual reproduction is more common in varied or changing environments, where a high degree of variation in traits of offspring could be useful to ensure survival of the species.

Meiosis and Alternation of Generations

Meiosis produces nuclei that have half of the original chromosome number. Such cells are necessary in sexual reproduction to keep chromosome numbers constant.

Daughter nuclei produced by meiosis have one copy of each chromosome (pp. 126–128).
In meiosis, homologous chromosomes pair in prophase I so that as a result of meiosis I, the chromosome number is effectively cut in half. Meiosis II resembles mitosis. The end result of meiosis of a diploid cell is four haploid cells, which in plants are spores.

Plant sexual life cycles feature both haploid and diploid multicellular forms (pp. 128–130).
In plant sexual life cycles a multicellular diploid ($2n$) sporophyte alternates with a multicellular haploid (n) gametophyte. In plant life cycles, meiosis produces haploid spores. Sperm and eggs are produced by mitosis from structures arising from these spores. In most plants, gametophytes are less conspicuous than sporophytes.

Cone and Flower Structure

In seed plants, fertilization first requires pollination. In most gymnosperm species, male and female gametophytes are on different cones of the same plant. Most angiosperm species produce flowers that contain both male and female parts. Some angiosperm species are monoecious, with each plant having male and female flowers. Others are dioecious, with each plant being either male or female.

In gymnosperms, some apical meristems produce cones (p. 131).
Cones are reproductive meristems that develop from vegetative meristems. The stem becomes the central cone axis. In pine, the leaves have been modified to become sporophylls in male cones and bracts in female cones. In female cones, sporophylls develop from axillary buds.

In angiosperms, some apical meristems produce flowers (pp. 131–132).
Prompted by hormonal signals, apical meristems can give rise to male, female, or bisexual flowers. All flower parts are modified leaves.

A flower can consist of up to four types of modified leaves (pp. 132–133).
The four types of modified leaves that form parts of the flowers are sepals, petals, stamens, and carpels.

The number and symmetry of flower parts can vary (pp. 133–134).
Complete flowers have all four types of modified leaves. Incomplete flowers lack one or more types. Perfect flowers have stamens and carpels, while imperfect flowers have either stamens or carpels. Regular flowers have radial symmetry, while irregular flowers have bilateral symmetry. The structure of flowers has often evolved in conjunction with the evolving structure and habits of pollinating animals.

Flowers vary in the position of their ovaries (p. 134).
Sepals, petals, and stamens can be attached below, above, or in the middle of the ovary.

Flower structures are examples of how natural selection modifies what is already present (p. 134).
Flowers evolved as a result of leaf mutations that increased the ability of plants to survive.

Seed Structure

Seeds are the product of plant reproduction on dry land. They exist to carry plants through seasons of the year that are inhospitable for plant growth.

Seeds form from ovules on bracts of cones or in carpels of flowers (p. 135).
In gymnosperms, seeds form from ovules on the upper surface of the bracts of cones. In angiosperms, seeds form from ovules inside carpels of fruits. Bracts, cone scales, carpels, and fruits are formed from modified leaves.

Seeds nourish and protect developing embryos (p. 135).
A seed is an embryo surrounded by varying amounts of nutritious tissue and then by a seed coat. The nutritious tissue is used by the embryo as it develops inside the ovule. The integuments, or layers, surrounding the embryo become the seed coat.

In seed germination, first the embryonic root grows through the seed coat, and then seedling formation begins (pp. 135–136).
Seeds contain little water. A period of water uptake, called *imbibition*, precedes germination, which begins as the radicle pushes through the seed coat to contact the soil. Many seeds contain abscisic acid, which prevents germination for a period of months after the seed is formed. Seed dormancy prevents seeds from germinating when environmental conditions are inappropriate for survival.

Fruit Structure

During seed development in a flowering plant, the ovary expands to become part or all of a fruit (pp. 137–139).
Botanically, a fruit consists of a mature ovary or ovaries. Fruits can be either fleshy or dry. The outer part of the ovary wall is called the *exocarp*, the middle part the *mesocarp*, and the inner part the *endocarp*.

Fruits can be categorized as simple, aggregate, or multiple (pp. 139–141).
Fruits can be simple, aggregate (having more than one carpel in a single flower), or multiple (forming from more than one flower). Simple fruits can be fleshy or dry. Dry fruits can be dehiscent (splitting open at maturity) or indehiscent (remaining closed at maturity).

A number of mechanisms disperse seeds and fruits to new locations (pp. 141–143).
Seeds can be dispersed by wind and by floating on water. Some fruits have bright colors and sugary tastes when they are ripe and thus attract animals that aid in seed dispersal. A few fruits explosively discharge their seeds.

Review Questions

1. Compare and contrast asexual and sexual reproduction.
2. Distinguish between haploid and diploid.
3. What is the usefulness of meiosis?
4. How would you describe homologous chromosomes to a friend who has not taken a biology course?
5. What happens in homologous pairing?
6. Why is meiosis I known as the reduction division?
7. Distinguish between a gametophyte and a sporophyte.
8. Outline the basic sexual life cycle of a plant, showing where mitosis and meiosis occur.
9. Describe the basic variations in gametophytes and sporophytes.
10. What is the difference between a monoecious and a dioecious species?
11. Identify the four types of modified leaves in a flower and briefly describe their functions.
12. Is a complete flower perfect? Is a perfect flower complete? Explain.
13. Describe the basic structure of a seed and explain how germination takes place.
14. What are the differences between simple, multiple, and aggregate fruits?
15. What are the basic methods of seed dispersal?

Questions for Thought and Discussion

1. Suppose a plant breeder and a genetic engineer worked together and produced a plant that grew wild in most any climate and produced abundant and tasty fruit and seeds with perfect nutrition for humans. What do you think would be the biological, sociological, economic, and political consequences?
2. Suppose you were transported to the Carboniferous Period, 320 million years ago. Seed plants do not exist, so

fruit and seeds are not part of your diet, yet you need adequate nutrition to survive. By the way, you are a vegetarian. What is for dinner?

3. Why might it be dangerous for a plant to rely on a single species of animal as a pollinator?

4. Some flowering plants are self-pollinating. Given the fact that they are not pollinated by organisms, why do you think these plants still have flowers?

5. Some fruits are poisonous, while others are edible. How do both of these characteristics facilitate seed dispersal?

6. Draw fully labeled diagrams to illustrate and compare the gametophytes of a moss, a fern, a gymnosperm, and an angiosperm.

Evolution Connection

Biologists believe that, in the history of life on earth, asexual reproduction evolved first and sexual reproduction evolved later. The latter is clearly a successful strategy that has great adaptive value because most eukaryotes have life cycles that include sexual reproduction. Moreover, in many eukaryotes, sexual reproduction is the only reproductive process in the life cycle. Why is sexual reproduction so advantageous?

To Learn More

Visit The Botany Place Website at www.thebotanyplace.com for quizzes, exercises, and Web links to new and interesting information related to this chapter.

Bubel, Nancy. *The New Seed Starter's Handbook.* Emmaus, PA: Rodale Press, 1988. Lots of basic botanical information and useful techniques for growing plants from seeds.

Hutton, Wendy, and Heinz Von Holzen. *Tropical Fruits of Asia.* Boston: Periplus Editions, 1996. A beautifully illustrated book with lots of details about tropical fruits of Thailand, Malaysia, and Indonesia.

Klein, Maggie Blyth. *All About Citrus and Subtropical Fruits.* New York: Ortho Books, 1985. Pictures and growing information on 50 citrus varieties and 16 exotic fruits.

Schneider, Elizabeth. *Uncommon Fruits & Vegetables: A Commonsense Guide.* New York: William Morrow, 1998. Interesting information about buying, storing, and eating 80 unusual fruits and vegetables.

Susser, Allen, and Greg Schneider. *The Great Mango Book.* Berkeley, CA: Ten Speed Press, 2001. Mangos started as an important ingredient in Indian cooking. Now more than 50 varieties are used all over the world in many different drinks and dishes.

The Functions of Plants

Basic Plant Biochemistry

Coffee, eucalyptus, ginger, basil, and cloves all produce biochemicals used by humans.

The Molecular Components of Living Organisms

Carbohydrates, which supply and store energy and serve as structural building blocks, include sugars and polymers of sugars

Proteins, which catalyze reactions and are structural building blocks, are polymers of amino acids

The nucleic acids DNA and RNA, which code and express genetic information, are polymers of nucleotides

Lipids are membrane components consisting mainly of carbon and hydrogen atoms derived from acetates and other molecules

Secondary metabolites such as phenolics, alkaloids, and terpenoids often protect or strengthen plants

Energy and Chemical Reactions

Energy can be stored and can move or change matter

Chemical reactions involve either a net input or a net output of free energy

Redox reactions release energy as a result of movement of electrons between atoms or molecules

The terminal phosphate bond in ATP releases energy when broken

NADH, NADPH, and $FADH_2$ are universal carriers of energy-rich electrons in living organisms

Chemical Reactions and Enzymes

Collision theory describes product formation by reactions in gases or liquids

Enzymes position reactants, allowing reactions to occur with minimal activation energy or increase in temperature

Cofactors such as coenzymes interact with enzymes to assist reactions

Competitive and noncompetitive inhibition can slow or stop enzymatic reactions and pathways

Enzymatic reactions are linked together into metabolic pathways

One of the realities of the universe is that large objects are made from many smaller components. Chemically speaking, everything is made of the small pieces of matter called *molecules,* and molecules themselves are made from atoms. Cars, made from pieces of metal and plastic, ultimately consist of small molecules. Cake ingredients such as flour, sugar, and oil all have a molecular structure. Living organisms are no exception. Just as small interlocking pieces fit together to make a jigsaw puzzle or a child's Lego® structure, an organism can be broken down into components consisting of trillions of molecules.

Biochemistry is the study of how organic molecules, based on frameworks of carbon atoms, form the basic structure of organisms. Although biochemistry is a complex subject, the biochemist's overall view of a living organism is deceptively simple, consisting of three categories of components:

◆ **Organic "building block" molecules.** Photosynthetic organisms, such as plants, make a number of different types of small organic molecules. Using CO_2 from the air and H_2O from the soil, photosynthesis produces three-carbon sugar-phosphates, which in turn are used—sometimes with minerals from the soil—to make other types of molecules. Other life-forms obtain the molecular building blocks of life either directly or indirectly from plants and other photosynthetic organisms. For instance, animals either eat plants or eat animals that ate plants. What we refer to as *food* consists biochemically of organic molecules that supply energy and that our bodies reorganize to make the organic molecules we need to produce and sustain ourselves.

◆ **Enzymes.** Enzymes are proteins that assist with chemical reactions in cells. They modify small organic molecules and assemble and reassemble them into larger, smaller, or different molecules. A living cell has thousands of different enzymes, each carrying out a particular modification or assembly of organic molecules. In short, enzymes are the tools that produce cellular components, cells, and finally organisms.

◆ **A plan for producing all the various types of molecules.** The DNA in the chromosomes contains the plan for the organism—specifically, the structure of enzymes. Once the enzymes are produced, they begin the process of modifying organic molecules into cellular and organismal components.

In a general sense, biochemistry involves examining how all these components fit together and interact. In this chapter, we will look first at the basic types of molecules that form the building blocks of plants and other organisms. Then we will examine the roles of energy and enzymes in chemical reactions, providing a foundation for exploring photosynthesis and respiration in the next two chapters. If you need to review any basic chemistry, you can refer to Appendix A.

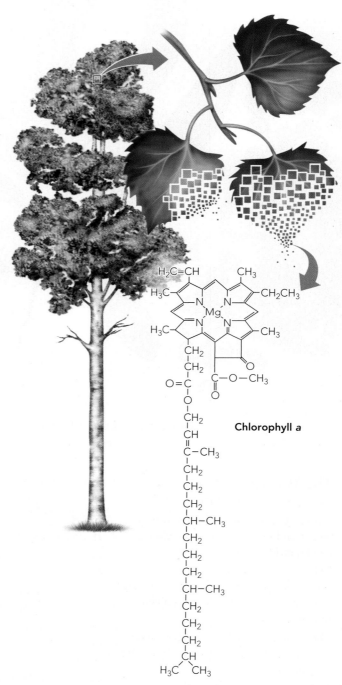

Chlorophyll *a*

This chlorophyll *a* structure represents one of millions of leaf molecules.

The Molecular Components of Living Organisms

Most large molecules in plants and other organisms are made from only a few types of smaller molecules. A typical macromolecule is a **polymer,** a long molecule composed of repeating structural units called **monomers.** You might think of monomers as being similar building blocks. Most carbohydrates, for instance, are polymers made of linked sugar molecules. As you will see, various combinations of just a few types of monomers can result in a wide variety of polymers. Monomers are typically linked to form polymers through a chemical reaction known as a **dehydration synthesis,** so named because the bond is formed by removing a molecule of water (Figure 7.1a). Dehydration synthesis is also called a *dehydration reaction* or *condensation reaction.*

In addition to forming macromolecules, cells also have to break them down into smaller molecules. As you might expect, breaking a bond formed by dehydration synthesis involves adding a molecule of water, a process known as **hydrolysis.** Basically, hydrolysis is the reverse of dehydration synthesis (Figure 7.1b).

As you saw in Chapter 2, there are four main types of macromolecules in living organisms: carbohydrates, proteins, nucleic acids, and lipids. These macromolecules are called **primary metabolites** because they are essential products of the metabolism, or chemical reactions, involved in the growth and development of every plant cell and, indeed, the cells of all organisms. We will now focus on the structures of these macromolecules. In addition, we will look at some molecules that are known as **secondary metabolites** because they are not essential for basic plant growth and development. Secondary metabolites are not found in all plant cells, or even in all plant species, but they do play a variety of important roles, such as providing structural support and protecting many plants from herbivores and disease.

Carbohydrates, which supply and store energy and serve as structural building blocks, include sugars and polymers of sugars

Carbohydrates, which usually have names ending in *-ose,* consist of all sugars and their polymers. Every carbohydrate contains carbon, hydrogen, and oxygen and can be classified as either a monosaccharide, a disaccharide, or a polysaccharide.

Monosaccharides are the simplest type of carbohydrate, with a molecular formula that is usually a multiple of CH_2O. Monosaccharides are also called simple *sugars* or *simple carbohydrates.* The most common monosaccharide in plants is glucose, a combination of two sugar phosphate compounds that are immediate products of photosynthesis (Figure 7.2a). Plants use glucose as their primary source of energy. Another common monosaccharide is fructose, a six-carbon sugar produced by almost all fruits and some vegetables (Figure 7.2b). Two five-carbon sugars, ribose and deoxyribose, are parts of the nucleic acids RNA and DNA, respectively.

Two monosaccharides can be linked to form a **disaccharide.** The most common disaccharide is sucrose ($C_{12}H_{22}O_{11}$), or table sugar, which is formed when

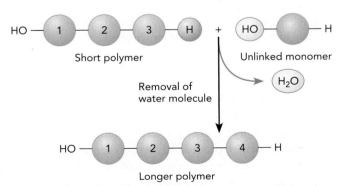

(a) Dehydration synthesis of a polymer, the process by which monomers are linked into polymers. The formation and removal of a water molecule results in linking a monomer to another monomer.

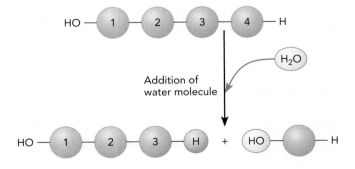

(b) Hydrolysis of a polymer, the reverse process by which a monomer is separated from other monomers by the addition of a water molecule.

Figure 7.1 Dehydration synthesis and hydrolysis. Carbohydrates, proteins, nucleic acids, and lipids are all important classes of polymers found in cells.

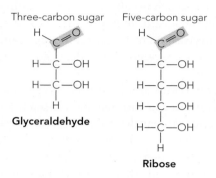

(a) Structure of glucose, a monosaccharide

Linear form

Ring form

Abbreviated ring form

Or

(b) Examples of monosaccharides

Three-carbon sugar

Five-carbon sugar

Glyceraldehyde

Ribose

Six-carbon sugars

Glucose

Galactose

Fructose

Figure 7.2 Some monosaccharide structures. **(a)** Glucose, a six-carbon sugar, exists as either a linear chain or a ring. In this diagram, the transition between the two forms is shown as well as an abbreviated structural formula. **(b)** Sugars can have different numbers of carbon atoms, but they typically have from four to seven. Three-carbon sugars do not have ring forms. Larger sugars may be either chains or rings. Notice in comparing glucose and galactose that variation around asymmetric carbons, shown in gray, gives rise to different sugars. Sugars can also vary in the location of the carbonyl group (shown in pink).

glucose is linked to fructose (Figure 7.3a). Linking two glucose units forms maltose (Figure 7.3b). By bonding together different sugar monomers, many different types of disaccharides can be produced. Glucose and glucose derivatives link together as the result of a dehydration synthesis.

Dehydration synthesis reactions can also link hundreds to thousands of monosaccharides to form polymers called **polysaccharides,** which usually store energy or provide structural support. Some examples of polysaccharides are starch and cellulose. Starch stores energy when photosynthesis produces more glucose than can be immediately used by the plant. Plant starch is amylose, while animals

make glycogen, an energy-storing polysaccharide. Plants store large amounts of starch in seeds for use during germination to supply energy for a developing plant.

Cellulose, which provides structural support in the cell walls of plants and algae, is structurally similar to starch in that the glucoses are both linked by the first carbon of one to the fourth carbon of the other. The structural difference between starch and cellulose is that in cellulose every other glucose is turned upside down (Figure 7.4a). Technically, the glucoses in starch link by alpha-alpha (α-α) bonds, whereas those in cellulose are linked by beta-beta (β-β) bonds (see Figure 7.4b,c). Mammalian digestive enzymes

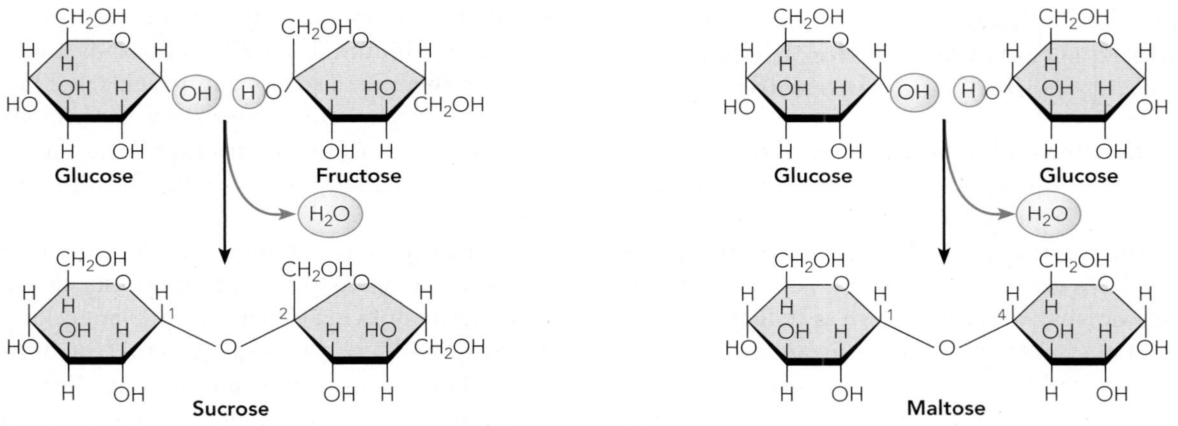

(a) Glucose and fructose can be linked to form the disaccharide sucrose, or *table sugar*.

(b) Two glucose molecules can be linked to form maltose. This linkage involves joining the number 1 carbon of one glucose to the number 4 carbon of the other. A diffferent linkage of the two glucoses would produce a different disaccharide.

Figure 7.3 Dehydration synthesis of disaccharides.

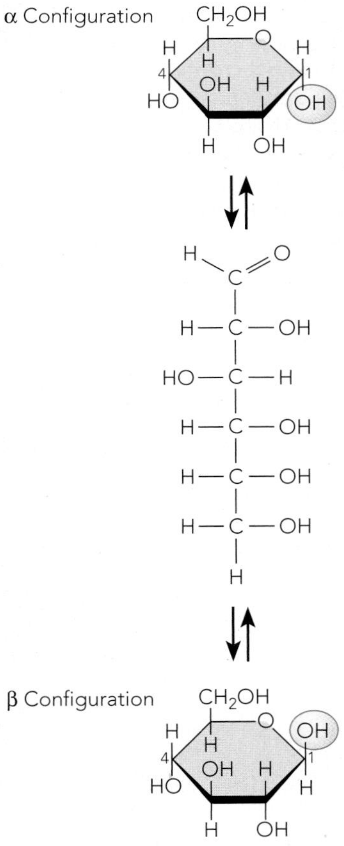

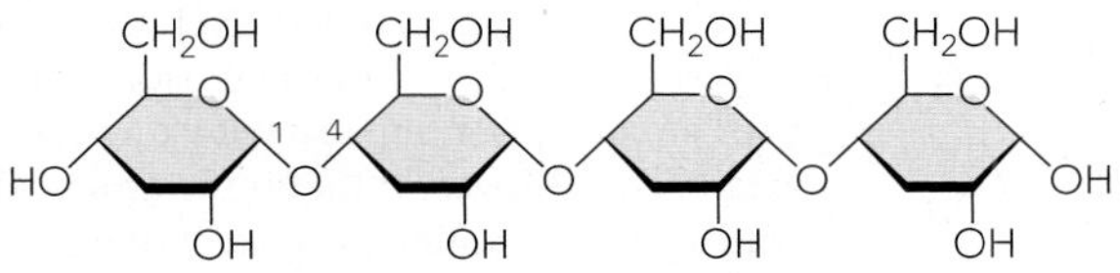

(b) Starch: 1–4 linkage of α glucose

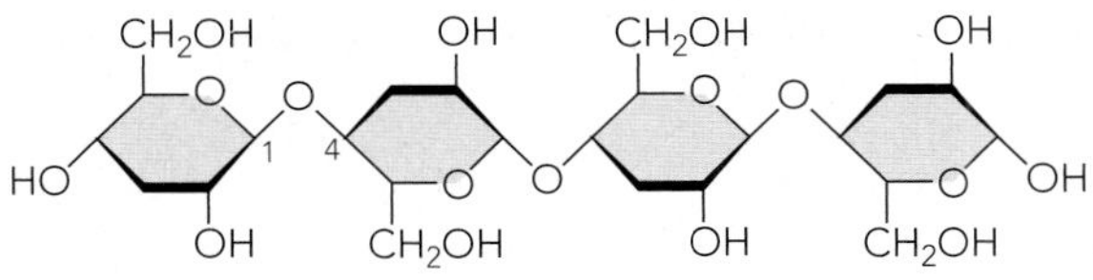

(c) Cellulose: 1–4 linkage of β glucose

(a) α and β glucose ring structures

Figure 7.4 Starch and cellulose: two polysaccharides. (a) Glucose can form alpha (α) and beta (β) rings, which differ in the location of the hydroxyl group (–OH) attached to the number 1 carbon. This difference distinguishes two of the polymers of glucose: starch and cellulose. (b) The α-ring glucose is the monomer for starch, which stores energy. (c) The β-ring glucose is the monomer for cellulose, the principal component of plant cell walls. The bond angles make every other glucose "upside down."

can break the alpha-alpha bonds, enabling starch to be used as a food source. However, mammals cannot digest cellulose, which has β-β bonds. Mammals such as cows and horses that "eat" cellulose material in grass and wood do not actually break down the cellulose. Instead, the microorganisms that inhabit their gut produce cellulase, an enzyme that digests the cellulose.

Plants can convert sugars into many other compounds besides disaccharides and polysaccharides. In a general sense, the carbons in all molecules in plants, and in all other life-forms, originate as simple three-carbon sugar phosphates in photosynthesis. As we study the many molecules and biochemical reactions that occur in organisms, we must not lose sight of their complete dependence on the carbohydrates produced in photosynthesis.

Proteins, which catalyze reactions and are structural building blocks, are polymers of amino acids

There are 20 amino acids that cells use in various combinations to form thousands of different proteins. Each amino acid has the same basic structure, consisting of a central carbon atom to which are bonded an amino group ($-NH_2$), a carboxyl group ($-COOH$), a hydrogen atom, and a variable side chain known as an R group (Figure 7.5a). At neutral pH inside a cell, an amino acid usually has an ionized form, in which the carboxyl group loses a proton (H^+) by ionization, while the amino group gains a proton (Figure 7.5b). It is the R group that distinguishes each amino acid from the others and determines the properties of that amino acid (Figure 7.5c). For instance, some amino acids are water soluble, while others are not. In addition to the 20 amino acids used to make proteins, there are also free amino acids that have other functions, such as supplying energy and providing part of the structure of hormones such as auxin. However, amino acids usually occur as the building blocks of proteins.

Some of the most abundant molecules on earth are proteins, making up to 50% or more of the dry weight in the majority of living organisms. In plant cells, proteins are the second most common general type of molecule after carbohydrates, comprising about 10–15% of the dry weight of a typical cell. Often, the largest concentration of proteins in plants is found in its seeds, in certain seeds as much as 40% of the dry weight may come from protein. Most proteins in living cells are enzymes, which help

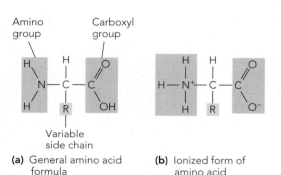

(a) General amino acid formula

(b) Ionized form of amino acid

Figure 7.5 Structure of amino acids. (a) This diagram shows the general structure of an amino acid: a central carbon atom bonded to an amino group ($-NH_2$), a carboxyl group ($-COOH$), a hydrogen atom, and a variable side chain known as an R group. **(b)** The carboxyl group can release a proton (H^+) and is therefore acidic. The nitrogen of the amino group will accept a proton, which gives the amino group a positive charge. These changes result in an ionized form of the amino acid, as shown here. The ionized form is the normal structure at a neutral pH within a cell. **(c)** These are examples of some of the 20 amino acids that can be used to make proteins, showing the variation in R groups.

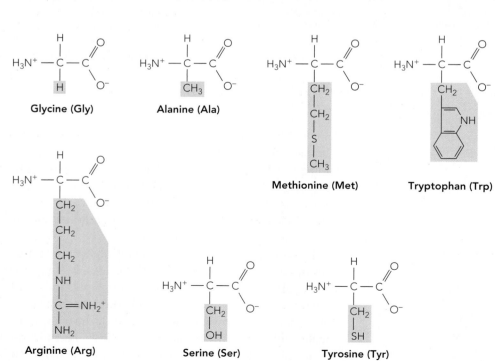

(c) Some examples of amino acids

Forests Made of Carbohydrates

Photosynthetic organisms convert simple molecules and ions into a collection of organic molecules used to supply both molecular building blocks and energy for all organisms. Every carbon, nitrogen, phosphorous, and sulfur atom in your body has been chemically incorporated into organic molecules by the enzymes of photosynthetic organisms such as plants.

Plants and animals have somewhere between 30–50% of their genes in common. Yet photosynthetic organisms such as plants differ significantly from non-photosynthetic organisms such as animals. If we study a large animal such as a human and a large plant such as a tree, we find that water comprises 60–70% of their weight. In humans, however, the next most common compound is protein, making up 15–20% of a person's weight, mainly in the form of muscle. In contrast, the second most common compound in plants is cellulose, which is 20–30% of a

plant's weight. When you walk in a forest, you are surrounded by huge amounts of carbohydrates linked into cellulose. In other words, large animals are composed primarily of water and polymerized amino acids, while large plants are composed mainly of water and polymerized sugar.

The biochemical facts that large animals are meaty while trees are woody tell us something about the respective behaviors of these organisms. In a general sense, it tells us that animals move, for much of animal protein is in the form of muscles, while plants form permanent, often woody, stems that support their leaves. Animals move to reproduce and to obtain food. To some extent, plant reproduction involves movement, as many plants have evolved mechanisms that use animals to disperse seeds. However, plants remain in one location and "reach" with roots and leaves for the food resources they need.

speed up chemical reactions. As you saw in Chapter 2, structural proteins such as actin and tubulin form important parts of the cytoskeleton. Storage proteins supply free amino acids to germinating seeds.

Amino acids are linked to form proteins in dehydration synthesis reactions that involve peptide bonds, which is why a polymer of amino acids is called a **polypeptide** (Figure 7.6). Most proteins consist of a single polypeptide, but some have more than one. A protein may have hundreds or even thousands of amino acids. The sequence of the amino acids in a protein, known as the **primary structure** (Figure 7.7a), can be very diverse because theoretically each position in a protein can be occupied by any of 20 different amino acids. As a result of this diversity, proteins have a more variable structure when

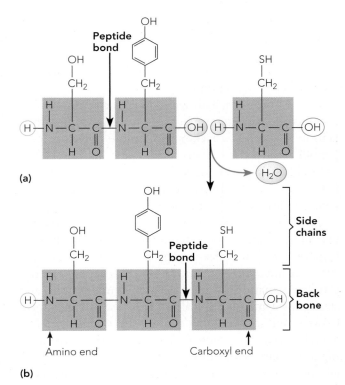

Figure 7.6 The formation of a polypeptide. (a) Through dehydration synthesis, amino acid monomers are linked by peptide bonds to form a chain. **(b)** Since they are formed by multiple peptide bonds, proteins are also known as *polypeptides*. In a peptide bond, the carbon of the carboxyl group bonds with the nitrogen of the amino group of the next amino acid. Peptide bonds form one at a time, beginning with the amino acid at the amino end of the polypeptide, until the protein is synthesized.

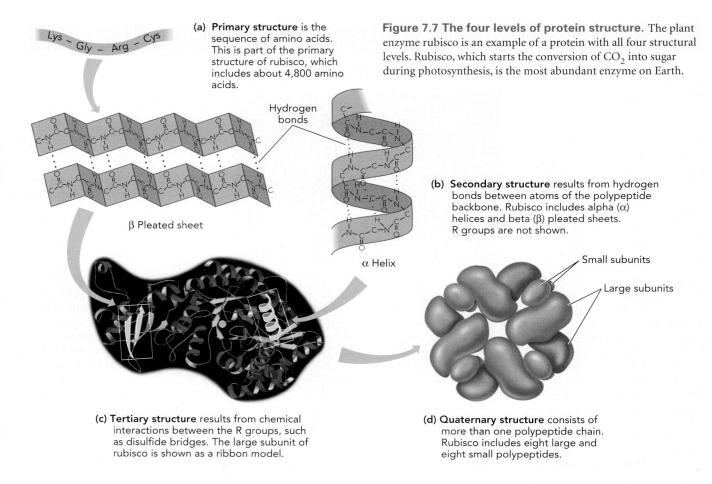

(a) **Primary structure** is the sequence of amino acids. This is part of the primary structure of rubisco, which includes about 4,800 amino acids.

Figure 7.7 The four levels of protein structure. The plant enzyme rubisco is an example of a protein with all four structural levels. Rubisco, which starts the conversion of CO_2 into sugar during photosynthesis, is the most abundant enzyme on Earth.

Hydrogen bonds

β Pleated sheet

(b) **Secondary structure** results from hydrogen bonds between atoms of the polypeptide backbone. Rubisco includes alpha (α) helices and beta (β) pleated sheets. R groups are not shown.

α Helix

Small subunits

Large subunits

(c) **Tertiary structure** results from chemical interactions between the R groups, such as disulfide bridges. The large subunit of rubisco is shown as a ribbon model.

(d) **Quaternary structure** consists of more than one polypeptide chain. Rubisco includes eight large and eight small polypeptides.

compared with sugar polymers, which are often made of a single, repeating subunit.

In most proteins, hydrogen bonds form between hydrogens and oxygens and between hydrogens and nitrogens in the protein's backbone of amino acids joined by peptide bonds. These interactions result in several types of local twistings or foldings known as **secondary structure** (Figure 7.7b). In particular, alpha (α) helices and beta (β) pleated sheets often form.

The R groups interact to form **tertiary structure,** the overall three-dimensional pattern of folding in a protein (Figure 7.7c). The helices and pleated sheets of secondary structure are imbedded in the tertiary structure, which is stabilized mainly by charge-charge interactions and strong covalent bonds called *disulfide bridges,* which form between sulfur-containing amino acids such as cysteine. So-called *chaperone proteins* frequently assist in the folding of protein chains into their final configuration. Usually, the primary sequence of amino acids in a protein will form one preferred secondary and tertiary configuration that is energetically stable. Heat can disrupt the tertiary structure of a protein, in a process called **denaturation,** as when a clear egg white turns opaque upon cooking.

Quaternary structure occurs when a protein consists of more than one polypeptide chain (Figure 7.7d), as in the case of the plant enzyme rubisco, which starts the process of converting CO_2 into sugar during photosynthesis. Composed of eight large polypeptides and eight small polypeptides, rubisco is a protein complex with a molecular weight about 500,000 times that of a hydrogen atom. (The name *rubisco* is actually an abbreviation. We will look at the origin of the name when discussing enzymes later in this chapter.)

The genes of each organism provide the instructions for synthesizing proteins from amino acids (see the *Biotechnology* box on page 157). Adult humans cannot make eight of the 20 amino acids that are needed to synthesize proteins. In addition, infants cannot make a ninth, histidine (Figure 7.8). The amino acids that the human body cannot make are called **essential amino acids** because they must be obtained from the diet. Since most plant proteins lack one or more of these essential amino acids, vegetarian diets should be carefully monitored to assure that they are supplying all necessary amino acids. Latin American cultures, for example, frequently use both corn and beans in their diet. Many Native American cul-

BIOTECHNOLOGY

Weapons Against Weeds

Some commercial herbicides inhibit the synthesis of specific amino acids, thereby killing weeds. Weeds can be defined as plants that grow well where humans do not want them. Anyone who has ever tended a garden understands this definition. Weeds compete with crop plants for sunlight, fertilizer, and water. Removing weeds mechanically from agricultural fields increases crop production but requires time and money. Herbicides can be sprayed on fields to eliminate weeds if they do not eliminate crop plants at the same time. For this reason, herbicide-resistant crop plants are potentially very valuable and useful in agriculture.

Specific herbicides have different biochemical mechanisms of action. Some inhibit photosynthesis, and others interfere with hormonal regulation of growth. Roundup® is the commercial name of a herbicide containing glyphosate. This compound kills plants by inhibiting the synthesis of the amino acids phenylalanine and tryptophan. Specifically, glyphosate inhibits the action of an enzyme known as EPSP synthetase, an enzyme that plants need to make these amino acids. While there is still much research to be done, herbicide-resistant plants may be engineered to have additional copies of the gene that codes for EPSP synthetase. By overproducing the enzyme, plants can survive even when enzyme action is partially inhibited. Herbicide-resistant plants may also be engineered to have a bacterial gene that does not respond to glyphosate, thereby allowing the plant to manufacture amino acids in the presence of this herbicide. Testing is still being conducted, however, scientists hypothesize that any residual Roundup® on crops produced by glyphosate-resistant plants would probably not harm humans because the human body does not synthesize either phenylalanine or tryptophan, which are acquired instead through the diet.

A herbicide-resistant crop in a field infested with weeds, before and after the application of the herbicide.

---

tures use beans and squash. In both cases, the combined diet supplies all essential amino acids. Dietitians recommend that a typical human eat 50–100 grams of protein a day, although the amount can be less if the protein is of high quality. This means that the amino acids must be well balanced in terms of what humans need.

The nucleic acids DNA and RNA, which code and express genetic information, are polymers of nucleotides

The nucleic acids—DNA (deoxyribonucleic acid) and RNA (ribonucleic acid)—play important roles in coding and expressing genetic information. DNA stores the hereditary information in the nucleus and in the mitochondria and chloroplasts. RNA participates in decoding the information in DNA into the structure of proteins.

Essential amino acids

Corn

Methionine
Valine
(Histidine)
Threonine
Phenylalanine
Leucine
Isoleucine
Tryptophan
Lysine

Beans and other legumes

Figure 7.8 Essential amino acids. Vegetarians must eat complementary vegetables to ensure sufficient levels of all essential amino acids. For example, a combination of beans and corn will supply an adult all of the essential amino acids.

Nucleic acids are polymers made of nucleotides. A **nucleotide** consists of three parts: base, sugar, and phosphate group (Figure 7.9). A base is a nitrogen-containing compound and can be either a double-ringed structure called a *purine* or a single-ringed structure called a *pyrimidine*. The specific types of purines are adenine (A) and guanine (G), and the pyrimidines are thymine (T), cytosine (C), and uracil (U). The sugar found in nucleotides is either ribose (in RNA) or deoxyribose (in DNA). A phosphate group consists of a phosphorous atom with covalent bonds to four oxygen atoms. Although nucleotides can occur in other parts of a cell, they are most commonly found in DNA and RNA, as well as in a modified form in ATP, which living organisms use as a source of energy.

DNA nucleotides all have the sugar deoxyribose and differ only in their bases. Each DNA nucleotide has one of the following four bases: adenine, thymine, cytosine, or guanine. The DNA structure is a **double helix,** in which two strands of nucleotides wrap around each other and are attached to each other by hydrogen bonds between the bases (Figure 7.10). Guanine always bonds to cytosine with three hydrogen bonds, and adenine always bonds to thymine with two hydrogen bonds. Accordingly, the base sequence of one strand can be predicted from the known sequence of the other strand. The nucleotides of each individual strand are joined by covalent bonds between the sugar and the phosphate of adjoining nucleotides.

RNA differs from DNA in several ways. First, RNA nucleotides contain the sugar ribose instead of deoxyribose. Secondly, one of the four RNA nucleotides has the base uracil instead of thymine. Finally, RNA is single-stranded instead of double-stranded, although the single strand sometimes loops around and bonds to itself.

Lipids are membrane components consisting mainly of carbon and hydrogen atoms derived from acetates and other molecules

The fourth main category of macromolecules consists of lipids, a group of molecules composed mainly of carbon and hydrogen atoms. Unlike carbohydrates, proteins, and nucleic acids, lipids are not simple polymers. Instead, they are diverse molecules that are grouped together in the same category because they are generally **hydrophobic** ("water-fearing"), meaning that they are insoluble in water. Many lipids are large molecules that are formed by dehydration synthesis and are made principally of modi-

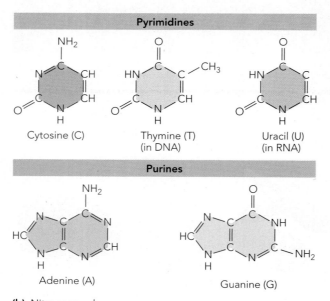

(a) Nucleotides

(b) Nitrogenous bases

Figure 7.9 Nucleotides. DNA and RNA are composed of monomers called nucleotides. **(a)** Each nucleotide consists of a nitrogenous base, a sugar, and a phosphate group. **(b)** Nitrogenous bases are either single-ringed structures called pyrimidines or double-ringed structures called purines. In DNA, the bases are adenine (A), guanine (G), thymine (T), and cytosine (C). In RNA, uracil (U) replaces thymine, and ribose is the sugar instead of deoxyribose.

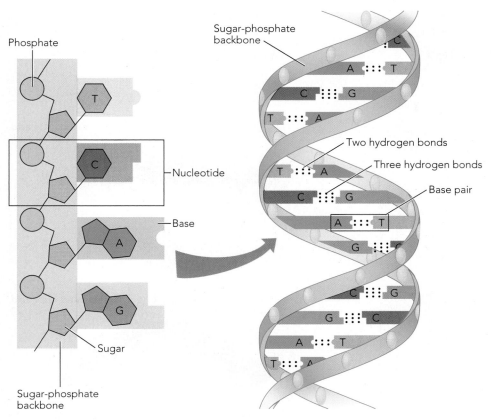

Figure 7.10 DNA structure. The structure of DNA consists of two polynucleotide strands that twist around each other to form a double helix. The two strands are held together by hydrogen bonds between the base pairs, with adenine pairing with thymine and cytosine pairing with guanine. In contrast, RNA consists of a single strand of nucleotides joined only by sugar-phosphate bonds, which form the backbone of the molecule. The bases of the nucleotides are usually not paired, unlike the structure of DNA.

Polynucleotide strand

DNA double helix

fied two-carbon acetate fragments linked together. The most common lipids are fats, phospholipids, and steroids.

Fats consist of glycerol, a three-carbon molecule derived from sugars, and fatty acid chains, which are built from acetate (Figure 7.11a). Fatty acids are long chains of carbons and hydrogens bonded to glycerol by dehydration synthesis (Figure 7.11b). Fatty acids and fats can be either saturated or unsaturated (Figure 7.11c). In saturated fatty acids, the carbons are all connected by single covalent bonds. In unsaturated fatty acids, one or more double bonds occur between carbons in a chain.

Animal fats are usually saturated. Saturated fats are solid at room temperature—think of butter or bacon grease. Plant fats are usually unsaturated fats, which are liquid at room temperature—think of corn or olive oil. Oils can be made into saturated fats by bubbling hydrogen through them, a process known as hydrogenation. Examples of hydrogenated vegetable oils include margarine and peanut butter. Many seeds store unsaturated fats, or oils, which provide nourishment for germinating seeds.

Diets high in saturated fats increase fat deposits or plaque on the inside of blood vessels, leading to reduced blood flow, less elastic vessels, and cardiovascular disease. Unsaturated fats do not have this effect, but scientists are not sure why. One hypothesis contends that unsaturated fats do not cause heart disease because they remain oily rather than solid at body temperature, making plaque deposits less likely. Animal fats, in addition to being saturated, also have higher levels of cholesterol than the solid vegetable fats. Research continues on the effects of hydrogenated vegetable fats.

Like fats, phospholipids contain glycerol, but they have only two fatty acids instead of three. Also, they include a phosphate molecule that is linked to the third carbon on glycerol (Figure 7.12). This phosphate "head" is water-soluble, or **hydrophilic** ("water-loving"), while the fatty acid "tails" are hydrophobic. Phospholipids are the main constituents of many membranes, with the hydrophilic heads facing toward the outside of the membrane, where they can absorb water (Figure 7.13).

Steroids are structurally distinct from other lipids and consist of four rings of interconnected carbons to which various smaller side groups are attached. You have no doubt heard of the steroid cholesterol, which is common in animal cells but not an important constituent of plant

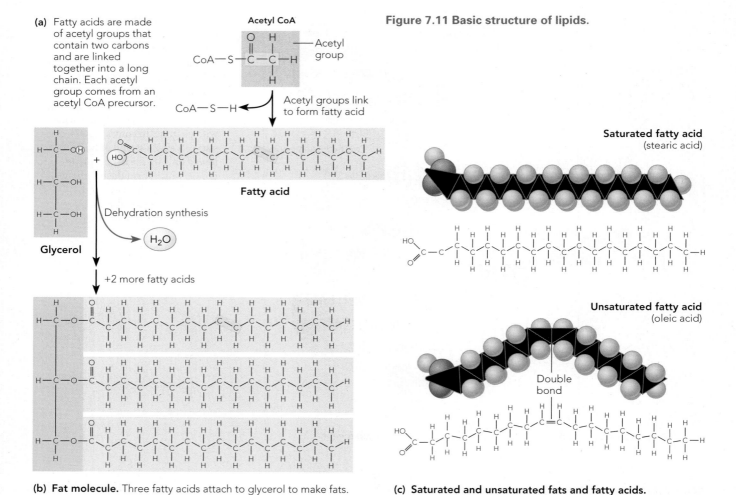

(a) Fatty acids are made of acetyl groups that contain two carbons and are linked together into a long chain. Each acetyl group comes from an acetyl CoA precursor.

Acetyl CoA

Acetyl group

Acetyl groups link to form fatty acid

Fatty acid

Glycerol

Dehydration synthesis

H_2O

+2 more fatty acids

(b) Fat molecule. Three fatty acids attach to glycerol to make fats.

Figure 7.11 Basic structure of lipids.

Saturated fatty acid (stearic acid)

Unsaturated fatty acid (oleic acid)

Double bond

(c) Saturated and unsaturated fats and fatty acids. Saturated animal fats contain no double bonds, while unsaturated plant fats contain one or more double bonds.

cells, so plants are only a minor source of cholesterol in our diets. In plants, steroids are the structural basis of brassinosteroids, a recently discovered class of plant hormones involved in cell division and elongation, which you will read about in Chapter 11. Plant growth hormones called *gibberellins* are structurally related to steroids. Steroids also serve to stabilize plant membrane structure.

Secondary metabolites such as phenolics, alkaloids, and terpenoids often protect or strengthen plants

Unlike carbohydrates, proteins, nucleic acids, and lipids, secondary metabolites are not essential for basic plant growth and development, but they do play important roles in helping many plants survive, most notably in providing protection from herbivores and disease. There are

three main categories of secondary metabolites: phenolics, alkaloids, and terpenoids.

Phenolics, made chiefly from the amino acids phenylalanine and tyrosine, are a group of many different ringed hydrocarbons that do not have nitrogen in their structure. For the most part, phenolics strengthen plants or protect them from various threats. In many cases, they have become essential enough so that the plant produces large quantities of them. Approximately 40% of the carbon circulating in the biosphere is in the form of phenolics, which are typically found in the cell walls and vacuoles of cells that produce them. The major types of phenolics are lignins, flavonoids, and allelopathics.

◆ Lignins are complex phenolic molecules that strengthen cell walls and repel herbivores (Figure 7.14a). Trees could not grow tall without the presence of lignins in their cell walls. Lignins are the second most common specific or-

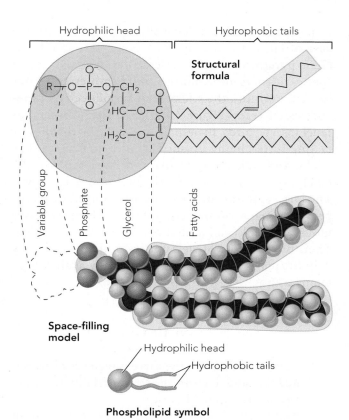

Hydrophilic head   Hydrophobic tails

Structural formula

Variable group | Phosphate | Glycerol | Fatty acids

Space-filling model

Hydrophilic head
Hydrophobic tails

Phospholipid symbol

Figure 7.12 Phospholipids. Phospholipids have two fatty acid "tails" attached to glycerol. The third glycerol linkage attaches to a water-soluble (hydrophilic) phosphate "head." Variable small molecules are then added to the phsophate.

ganic molecule after cellulose, making up as much as 30% of plant tissue. Their presence in cell walls has been described as "encrusting." In some cells, when lignin synthesis begins, cells have already entered the final stages of programmed cell death to become tracheids or vessel elements in woody plants. In fact, the word *lignin* is derived from *lignum,* the Latin word for wood.

♦ Flavonoids include thousands of water-soluble molecules and are commonly found in fruits and vegetables. Some deter herbivores and prevent bacterial decay, as in the case of the brown, bitter compounds called *tannins,* which can be used to preserve leather. Many flavonoids—such as lycopene in tomatoes and procyanidins in apples, grapes, and strawberries—are used

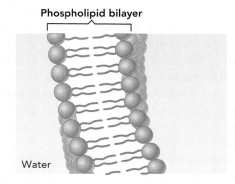

Phospholipid bilayer

Water

Figure 7.13 Phospholipid bilayers. Phospholipids form the prinicipal molecular constituents of cell membranes, with the charged phosphate "heads" facing outward, where they can attract water.

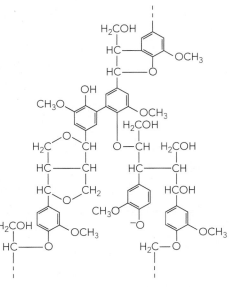

(a) Part of the complex structure of lignin.

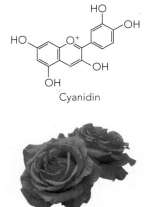

Cyanidin

(b) Cyanidin, a color-producing flavonoid in roses and other flowers.

Figure 7.14 Phenolics. Plant phenolics, which are derived from the amino acid phenylalanine, provide structural support and either attract or repel insects. **(a)** Lignins are molecules that strengthen cell walls and repel herbivores and fungi. The complex, weblike structure of lignin impregnates and coats cell walls to make wood hard and sturdy. **(b)** Flavonoids serve a variety of functions, including repelling herbivores and, as in the case of cyanidin, attracting pollinators.

in medicine to help control and prevent cancer and cardiovascular disease and as antiviral agents. Others function as flavoring agents and fragrances, as in black pepper, cloves, ginger, vanilla, cinnamon, and maple syrup. Flavonoids called *anthocyanins* cause the red, blue, and purple colors of some flowers that attract pollinating insects and other organisms (Figure 7.14b).

◆ Allelopathics are phenolic compounds secreted by plant roots or leached from leaves by rain and fog. They inhibit neighboring plants and can therefore decrease competition for light and minerals.

Alkaloids, made from several amino acids, serve primarily to protect plants against herbivores. Alkaloids are ringed compounds containing nitrogen as a part of at least one of the rings. They have a highly variable structure, made largely from the amino acids tryptophan, tyrosine, phenylalanine, lysine, and arginine. There are more than 12,000 known types of alkaloids, which are produced by 20 percent of flowering plants. They deter herbivory by insects and frequently affect the neurological system of animals. Many alkaloids are useful medically because of their neurological effects and effects on cell division. Caffeine, heroin, quinine, nicotine, vinblastine, ephedrine, and cocaine are all alkaloids.

Terpenoids, also known as terpenes, protect plants from herbivores and disease (Figure 7.15a). Three acetates combine to form a five-carbon isoprene subunit plus one molecule of carbon dioxide. Isoprene subunits are then hooked together to make the classes of terpenoids, which can have 10, 15, 20, 30, or (in the case of latex) thousands of carbons. Terpenoids include pyrethrum (an insecticide first isolated from a species of chrysanthemum), a number of essential oils like peppermint, the sticky resins produced by pines and related trees (Figure 7.15b), and latex. Terpenoids repel insects, birds, mammals, and other herbivores by tasting bitter, being poisonous, or being sticky. The sticky and poisonous resins and latexes physically block wounds in plants, thereby preventing infections. The large amounts of terpenoids produced by plants provide a major component of the blue haze seen over mountains, hills, and farm fields during warm weather. Their function is not well understood, but they may help the plant to protect itself internally against high temperatures.

Section Review

1. Name and describe the three categories of carbohydrates and describe the structural difference between starch and cellulose.
2. Name and describe the four levels of protein structure.
3. How do DNA and RNA differ in structure?
4. What is the structural difference between a fat and a phospholipid?
5. How do secondary metabolites protect and strengthen plants?

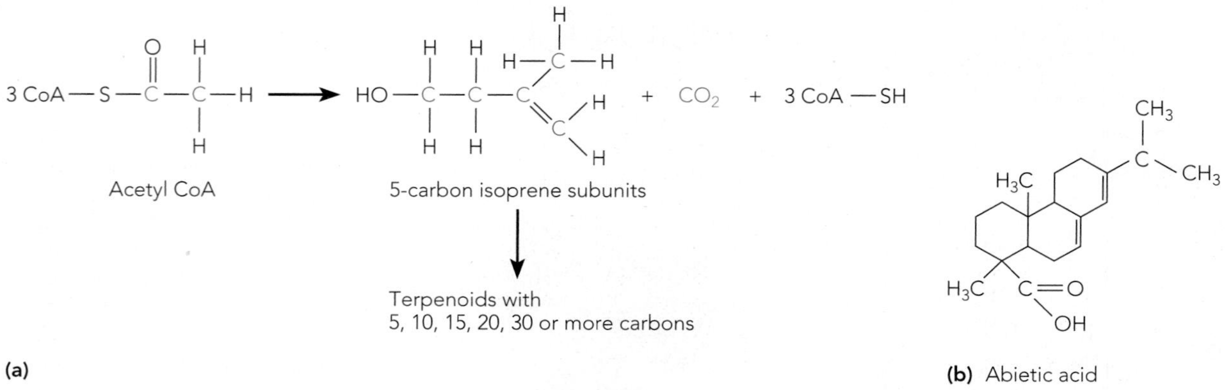

(a)

(b) Abietic acid

Figure 7.15 Terpenoids, or terpenes. (a) Plants produce terpenoids, also known as *terpenes*, by linking together five-carbon isoprene subunits. Isoprene subunits form from three acetates in a series of reactions. The acetates enter the reaction linked to a large molecule called coenzyme A (CoA). A carbon dioxide is lost as three acetates link. Terpenoids often repel insects. **(b)** A typical terpenoid is abietic acid, which is a resin in conifers that seals wounds and fossilizes as amber.

Energy and Chemical Reactions

Energy is the capacity to perform work. We cannot see energy, but we can see its effects on matter. Energy moves matter, changes its shape, and causes chemical reactions, as in the movement of chromosomes in mitosis or the growth of roots through the soil. This section will provide an overview of the role of energy in chemical reactions.

Energy can be stored and can move or change matter

Energy exists in two general forms, potential energy and kinetic energy. **Potential energy** is stored energy. A charged battery, for example, contains potential energy. **Kinetic energy** is energy exhibited by motion, for example, of a battery powering a toy. Either form of energy can be transformed into the other, but is never lost (Figure 7.16). A snowboarder at the top of a hill has potential energy that becomes transformed into kinetic energy when moving down the hill. Kinetic energy is then required to return to the top of the hill, restoring a state of high potential energy. In the examples of both the toy and the snowboarder, the kinetic energy appears as heat and motion that can then interact in other ways with matter in the environment.

The transformation of energy, known as *thermodynamics,* is governed by two basic laws. According to the **first law of thermodynamics,** energy can be harnessed and transformed, but it cannot be created or destroyed. According to the **second law of thermodynamics,** every transformation of energy increases the disorder of matter in the universe. As reflected in the examples of the battery-powered toy and the snowboarder, there are differ-

ent sources of energy. Electrical energy in a battery, for example, is one source of potential energy. In this section, we will be looking in general at how plant cells produce and use another type of potential energy called *chemical energy.*

The potential energy stored in matter can be measured in terms of calories (cal), or more often kilocalories (Cal or kcal = 1,000 cal). Calories with a small *c* are sometimes called small calories; calories with a capital *C* are called big calories. One small calorie is the amount of heat required to raise one cubic centimeter of water one degree Celsius (centigrade). When people talk about calories in food, for example, a candy bar with 300 calories, they are really talking about big Calories, or kilocalories. The energy contained in foods and molecules is determined by literally burning them in a machine called a *calorimeter* and measuring how much heat is produced.

Chemical reactions involve either a net input or a net output of free energy

Chemical reactions can be classified according to whether they have a net output or a net input of free energy—the amount of energy available to do work. If a chemical reaction has a net release of free energy into the surroundings, it is said to be an **exergonic** ("energy outward") reaction (Figure 7.17a). In exergonic reactions, the potential energy of the products is less than the potential energy of the reactants because there has been a net output of free energy. If a chemical reaction requires a net input of free energy, it is said to be an **endergonic** ("energy inward") reaction (Figure 7.17b). In endergonic reactions, the potential energy of the products is greater than the potential energy of the reactants as a result of the net input of free energy. Whether exergonic or endergonic, any chemical reaction requires some initial input

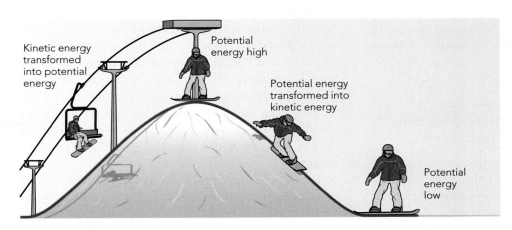

Kinetic energy transformed into potential energy

Potential energy high

Potential energy transformed into kinetic energy

Potential energy low

Figure 7.16 Kinetic and potential energy. As a snowboarder goes down a hill, potential energy is transformed into kinetic energy, released as motion and as heat. As the snowboarder goes back up to the top of the hill, kinetic energy supplied by the ski lift is transformed into potential energy.

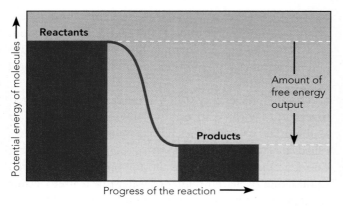

(a) Exergonic reaction (net output of free energy)

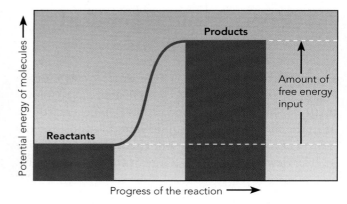

(b) Endergonic reaction (net input of free energy)

Figure 7.17 Exergonic and endergonic reactions. (a) In an exergonic ("energy outward") reaction, there is a net output of free energy because the potential energy of the products is less than the potential energy of the reactants. **(b)** In an endergonic ("energy inward") reaction, there is a net input of free energy because the potential energy of the products is greater than the potential energy of the reactants.

of energy, known as **activation energy,** before the reaction will occur.

In chemical reactions, two different types of energy must be considered: energy that is expressed as heat and the energy that is required to maintain order. If a reaction releases energy as heat, the change in heat is negative. That is, the reaction system loses heat to the surrounding environment. If a reaction takes up energy as heat, the change in heat is positive. In other words, the reaction system gains heat from the environment.

The degree of disorder in a piece of matter is known as its **entropy.** A greater level of disorder means higher entropy. For example, a decaying log has higher entropy than a living tree. Living organisms expend considerable energy maintaining a highly ordered molecular system with low entropy. For most organisms, including plants and animals, the energy to maintain this order comes ultimately from the sun through photosynthesis, a process that is overall heavily endergonic. When organisms die, the entropy of their molecular components soon increases markedly as they decay.

Redox reactions release energy as a result of movement of electrons between atoms or molecules

The chemical bonds of molecules that hold atoms together contain potential energy. As chemical bonds are broken or formed in reactions, energy can be released or absorbed. **Oxidation** means a loss or partial loss of one or more electrons, and **reduction** means a gain or partial

gain of one or more electrons (Figure 7.18). (The gain of one or more electrons is called *reduction* because it reduces the amount of positive charge.) In living organisms, oxidation and reduction are coupled together in oxidation-reduction reactions, known as **redox reactions.** In a redox reaction, electrons are either lost from one atom or molecule (oxidation) and added to another atom or molecule (reduction), or else they move away from the nucleus of one atom (partial oxidation) and toward the nucleus of another atom (partial reduction). The movement of an electron or electrons from one atom or molecule to another releases energy. This situation is somewhat similar to an object, such as a meteor, releasing energy as it is attracted to Earth by the force of gravity.

Redox reactions release energy when electrons move closer to particular nuclei. Such a movement of electrons occurs when electrons in a covalent bond move toward one of the bonding atoms and away from another. For example, consider a carbon atom bonded to another carbon atom, which is the case in sugars. A carbon-to-carbon bond is symmetrical because the electrons forming it are usually situated an equal distance from each atom. If the same carbons are then bonded to oxygens, which is the case in CO_2, the electrons typically spend more time near oxygen. Then the bonds are asymmetrical, and energy is released.

To understand why conversion of a symmetrical bond to an asymmetrical bond releases energy, consider the analogy of a satellite orbiting and then falling back to Earth. There is a kinetic release of energy as heat and light and perhaps as physical force if chunks hit the ground. The closer the satellite comes to Earth, the more energy it releases. Similarly, electrons orbit, in a sense, around nuclei. The positively charged protons in certain nuclei attract the negatively charged electrons, which release energy as they move closer to the nucleus. In the next two

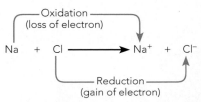

(a) This is the gerneral equation for a redox reaction for substances X and Y, reflecting the loss of an electron from substance X and a gain of electron in substance Y.

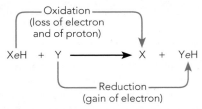

(b) In this example, sodium (Na) is oxidized while chlorine (Cl) is reduced. Na is the electron donor, known also as the reducing agent, and Cl is the electron acceptor, also called the oxidizing agent.

Oxidation
(loss of electron
and of proton)

$$XeH + Y \longrightarrow X + YeH$$

Reduction
(gain of electron)

(c) Sometimes a proton (H^+) is attracted, neutralizing the negative charge caused by an excess electron.

chapters, you will see how redox reactions are involved in the processes of photosynthesis and respiration.

The terminal phosphate bond in ATP releases energy when broken

As you read in Chapter 2, the main source of energy in a cell is ATP (adenosine triphosphate). ATP is made of three parts: adenine (a base), ribose (a sugar), and three phosphates (Figure 7.19). In Chapter 9, we will look at how ATP is synthesized. Here we will focus on how the breakdown of ATP releases energy.

The two covalent bonds between phosphates in ATP link oxygen and phosphorus. Breaking those bonds releases the energy that was used to make them. For instance, when the terminal phosphate is removed, ATP becomes ADP (adenosine diphosphate) plus P_i (inorganic phosphate), and energy is released. The actual amount of energy released depends on the concentration of reactants and products and the concentration of other ions.

ATP frequently supplies activation energy for both endergonic and exergonic reactions. For an endergonic reaction to occur in a cell, it must be paired with an exergonic reaction, a process known as **energy coupling** (Figure 7.20). In other words, the energy produced by an exergonic reaction is used to power an endergonic reaction. Usually ATP participates in reactions by being converted to ADP plus P_i and then adding the phosphate to a reactant of the endergonic reaction. The transfer of a phosphate to another molecule is known as **phosphorylation.** Removing the phosphate from the endergonic reactant then releases energy that drives the endergonic reaction.

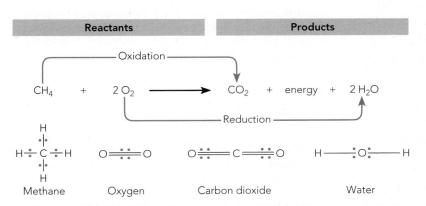

(d) Oxidation can involve either the loss of an electron or the movement of an electron away from a nucleus. Some atoms attract electrons more strongly than others. In this example, the electrons in the carbon-hydrogen bonds of methane (CH_4) are shared more or less equally by the hydrogen and carbon atoms. In an oxygen (O_2) molecule, the electrons are shared equally by the two oxygen atoms. However, in the case of carbon dioxide (CO_2) and water (H_2O) molecules, the oxygen atoms are electrophilic (electron-loving) and therefore attract the electrons much more strongly than do the hydrogen or carbon atoms. Therefore, oxygen is partially reduced, while carbon and hydrogen are each partially oxidized.

Figure 7.18 Oxidation and reduction. Oxidation involves a loss of electrons, and reduction involves a gain of electrons. Oxidation and reduction reactions occur together, in what are known as *redox reactions*. Each loss of electrons from one atom or molecule must be combined with a gain of electrons in another atom or molecule.

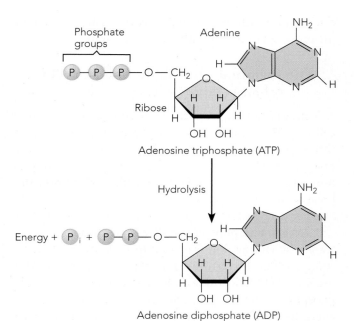

ATP (adenosine triphosphate) is made of three parts: adenine (nitrogenous base), ribose (sugar), and three phosphates. ATP can be broken down to ADP and a phosphate group, releasing energy.

Figure 7.19 ATP structure and breakdown.

ATP powers three types of work in cells: chemical, transport, and mechanical. Chemical work involves supplying activation energy for reactions and powering endergonic reactions. Transport work moves molecules across membranes. Mechanical work involves the movement of cells and cellular parts.

Breakdown of ATP releases energy for several reasons. First, energy is released because the phosphate removed from ATP or ADP can assume a number of different structures that were not all available before hydrolysis. In other words, entropy increases. Also, energy is released when ADP and the phosphate repel each other because both are surrounded by negative charges. Finally, energy is released because in phosphate-oxygen-phosphate bonds, the phosphate and the oxygen share electrons equally, whereas in phosphate-oxygen-hydrogen bonds, the electrons from hydrogen move toward the phosphate.

To understand how the formation of ATP stores energy released by the breakdown of ATP, think of the plastic latches on backpacks. You insert one piece of plastic into the other and use energy to push until the two parts of the latch lock together with a click. When you release the latch with your fingers, the two parts are pushed apart, releasing energy. In the case of molecules, ADP and P_i are the two parts of the latch. You must add energy to push them together—overcoming the repulsion of the electrons surrounding each molecule—until they lock by

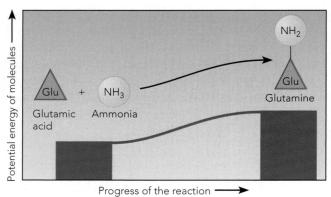

Endergonic reaction (requires input of free energy).

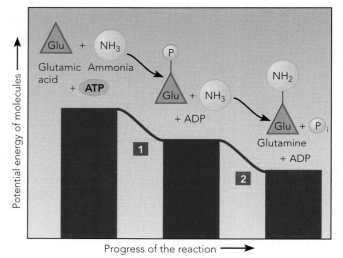

Paired reactions: overall process exergonic

Figure 7.20 Energy coupling. For an endergonic reaction to occur in a cell, it must be paired with an exergonic reaction, a process known as energy coupling. The conversion of glutamic acid to glutamine is endergonic, as shown in the top graph. If the change of glutamic acid to glutamine is coupled with the exergonic breakdown of ATP to ADP, the overall process is exergonic (bottom graph). The reaction proceeds in two steps. (1) First, the breakdown of ATP adds a phosphate group to glutamic acid. This primes the glutamic acid by adding an energy-rich phosphate bond. (2) Ammonia displaces the phosphate group, and energy is released as glutamine is formed.

covalently bonding to each other. When ATP is broken down into ADP plus P_i, the energy is released.

NADH, NADPH, and FADH$_2$ are universal carriers of energy-rich electrons in living organisms

Many important reactions in cells need both energized electrons and energy to proceed. NADH, NADPH, and FADH$_2$ are electron carriers that supply both. NADH is formed by the addition of 2 electrons and 1 proton (H^+) to NAD^+ (Figure 7.21). NADPH is formed by the addi-

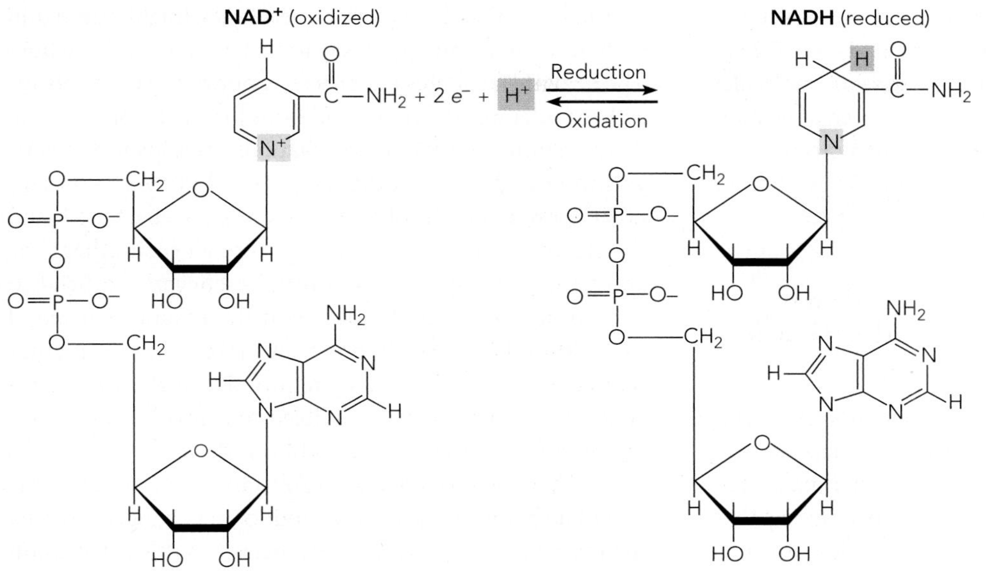

NAD$^+$ (oxidized)

NADH (reduced)

Figure 7.21 NADH. Cells use compounds such as NADH to carry energy-rich electrons, which are donated to enzymatic reactions that need them. NAD$^+$ combines with 2 electrons and 1 proton (H$^+$) to form NADH. NADP$^+$ combines in a similar fashion to produce NADPH.

tion of 2 electrons and 1 proton to NADP$^+$. FADH$_2$ is formed by adding 2 electrons and 2 protons to FAD.

NADH, NADPH, and FADH$_2$ supply energy-rich electrons to enzymatic reactions throughout the cell. For example, electrons are released from the breakdown of glucose in cells before the synthesis of ATP uses the energy. NADH, NADPH, and FADH$_2$ carry these electrons to the sites where ATP is made. The energy from the electrons is used to synthesize ATP. In this way, the energy from glucose is eventually stored in ATP. This process will be discussed in more detail in Chapters 8 and 9.

Section Review

1. **What is the difference between kinetic and potential energy?**
2. **What are the first and second laws of thermodynamics?**
3. **Why is it necessary to have both exergonic and endergonic reactions in living cells?**
4. **What is the difference between oxidation and reduction?**
5. **Describe the three types of work that ATP powers in cells.**

Chemical Reactions and Enzymes

Living cells are biochemical factories that produce the molecules and larger structures that give the cell form and allow it to function. We will continue our study of plant biochemistry by considering the enzymes that facilitate chemical reactions in living cells.

Collision theory describes product formation by reactions in gases or liquids

The many different types of molecules that make up a living cell are involved in numerous chemical reactions to meet the structural and physiological needs of the cell. Thus, understanding how these reactions occur provides basic knowledge about how living cells work.

Consider a typical chemical reaction: A + B → C. In gases and liquids, collision theory adequately describes the interactions and reactions of molecules, such as A and B. Molecules have mass, m, and move about with a velocity, v. The velocity of molecules and their kinetic energy increase with temperature. As molecules A and B move faster, their more energetic interactions increase the likelihood of interactions that will cause the formation of covalent bonds resulting in product C.

Given typical temperatures, many necessary biochemical reactions in living cells occur too slowly to maintain normal metabolism. Heat increases the rate of reactions described by collision theory. While this might be appropriate in chemistry lab, heat is not a useful way of increasing the rate of reactions in living organisms. First, high temperatures damage complex organic molecules. Second, while plants could conceivably produce enough heat to substantially increase reaction rates, the production would require the expenditure of considerable metabolic energy.

The shape or electron configurations of the molecules make certain collisions more likely than others (Figure 7.22a). The speed of collision is also a factor because as the molecules move faster, collisions are more frequent. However, if the molecules are oriented correctly, the speed of the molecules can be considerably less and a successful reaction will still occur (Figure 7.22b). This is where enzymes come into play.

Enzymes position reactants, allowing reactions to occur with minimal activation energy or increase in temperature

An enzyme is the chemical version of a matchmaker. Imagine two people who might hit it off if they ever happened to bump into each other in the right setting. A matchmaker acts to facilitate the process of bringing the two people together under the right conditions. Similarly, the participants in a

(a) In a solution or a gas, molecles move at random. Some collisions will result in a reaction between molecules, while other collisions will not.

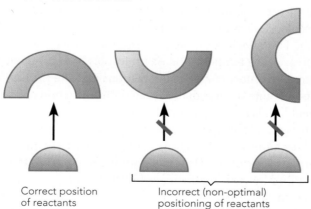

Correct position of reactants

Incorrect (non-optimal) positioning of reactants

(b) Reactants must be positioned correctly in order for a reaction to occur. If reactants are positioned correctly, minimal energy is required for a reaction to occur. Raising the temperature increases reaction rates by increasing both the frequency and the intensity of collisions.

Figure 7.22 Collision theory.

chemical reaction, known as the **reactants,** might bump into each other in just the right way to make a chemical reaction. The enzyme facilitates the process by bringing reactants together under just the right conditions for a reaction to occur.

In chemical terms, we say that enzymes lower the activation energy of reactants (Figure 7.23a). As two reactants approach each other, the electrons of reactant A and the electrons of reactant B begin to repel each other. The activation energy is the amount of energy required to overcome this initial tendency of the reactants to repel each other. By correctly positioning reactants, enzymes reduce the activation energy required in order for the reaction to begin. The reaction itself may involve the breaking of bonds or the formation of bonds.

An enzyme is a **catalyst**—a substance that increases the rate of a chemical reaction without being changed itself by the reaction (Figure 7.23b). A reactant that is acted upon by an enzyme is known as a **substrate** of that enzyme. The tertiary structure of each enzyme has projections, ridges, grooves, and cavities. This shape provides a binding site, known as the **active site,** for the substrate (S) to bind with the enzyme (E) to produce an **enzyme-substrate complex** (ES). The product or products (P) are formed by the chemical reaction and then separate from the enzyme. For example, the enzyme sucrase breaks the substrate sucrose into the products of glucose and fructose. Each binding site positions a substrate so that covalent bonds can be broken or formed, thereby converting the substrate into products. After the products detach from the enzyme, the enzyme is free to participate in the reaction again. The course of an enzyme-catalyzed reaction can be summarized as follows:

$$E + S \rightarrow ES \text{ complex} \rightarrow E + P$$

One of the important functions of enzymes in living cells is to provide binding sites so that endergonic reactions can be coupled to exergonic reactions. Enzymes work remarkably rapidly, often carrying out thousands or even millions of reactions in a few seconds. Each substrate molecule may be bound to the active site for a millisecond or less before the product must be gone and the next substrate molecule has moved in. Since enzymes, substrates, and products are all of microscopic size, the distances they must move are only a few nanometers. Also, the cytoskeleton often delivers molecules, in transport vesicles, to locations where enzymes will act on them. Thousands to billions of molecules may be processed by a particular type of enzyme during the "life" of a single cell.

The binding of a substrate to an enzyme can occur in a variety of ways and is a complex phenomenon. Not only does the shape of the substrate fit that of the binding site on the enzyme, but also various types of bonds stabilize the interaction. These include covalent bonds, ionic bonds, and hydrophobic interactions. These bonds facilitate binding at the active site and may also initiate electron flow in the direction required for the reaction to occur. Sometimes the

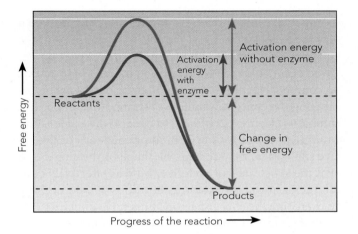

(a) By correctly positioning reactants, enzymes lower the activation energy, making it possible for reactions to occur without relying mainly on heat or speed of molecule movement.

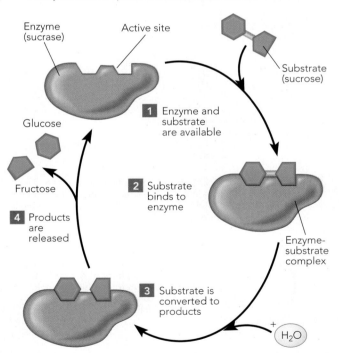

(b) Reactants resemble small puzzle pieces, while enzymes resemble large ones. This diagram shows how the enzyme sucrase breaks sucrose apart by hydrolysis into glucose and fructose. The substrate and enzyme come together to form an enzyme-substrate complex that breaks down into the two product molecules, glucose and fructose. The enzyme is then free to bind another substrate, and the process repeats.

Figure 7.23 Enzymes provide binding sites for reactants.

enzyme and substrate or substrates fit together like puzzle pieces (Figure 7.24a). Binding of a substrate or substrates can also change the shape of the enzyme—usually only slightly—thereby enhancing both binding and the rate of the subsequent chemical reaction. Binding that alters the shape of the active site is known as **induced fit** (Figure 7.24b). For example, after the enzyme hexokinase changes shape following the binding of a glucose, a phosphate can also bind to form glucose-phosphate, the product.

The names of enzymes, which typically end in -*ase*, usually relate to what each enzyme does. Earlier in the chapter, you looked at the structure of the plant enzyme rubisco. That name is an abbreviation for *ribu*lose 1,5-*bis*phosphate *c*arboxylase/*o*xygenase. The full name reflects the fact that rubisco can add either carbon dioxide or oxygen to a compound. An enzyme that catalyzes a reduction is called a *reductase,* and an enzyme that synthesizes a compound is called a *synthase.* For example, the enzyme that makes ATP from ADP and inorganic phosphate is called *ATP synthase.*

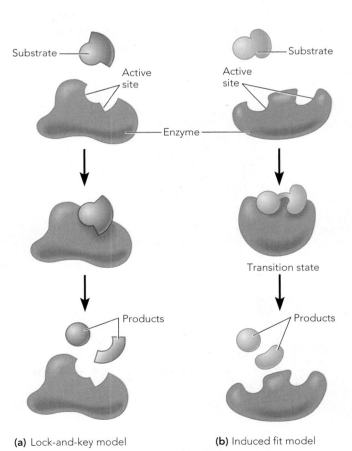

(a) Lock-and-key model   (b) Induced fit model

Figure 7.24 Two models for enzyme-substrate interaction. (a) The lock-and-key model shows substrates binding to enzymes like jigsaw puzzle pieces. (b) In the induced fit model, enzymes and their substrates may sometimes modify their shapes during binding.

Cofactors such as coenzymes interact with enzymes to assist reactions

Some small nonprotein molecules, known as **cofactors,** bind to enzymes or substrates and assist reactions by providing energy, supplying electrons or protons, or facilitating the reaction in other ways. Cofactors can be either inorganic mineral ions, such as magnesium (Mg^{++}) and calcium (Ca^{++}), or organic compounds such as vitamins. Cofactors that are nonprotein organic compounds are also known as **coenzymes.** Some cofactors bind temporarily to an enzyme or substrate, while others bind permanently. Some cofactors can be used repeatedly in reactions catalyzed by enzymes, but others must be partially resynthesized before reacting again, as in the case of ATP.

Many of the small molecules that we associate with living organisms act as cofactors. Some examples are hormones; vitamins; positively charged ions (cations); negatively charged ions (anions); ATP; carriers of energy-rich electrons such as NADH and NADPH; and many common molecules such as sugars, amino acids, nucleic acids, and acetate. The binding of such molecules to enzymes or to reactants facilitates the flow of electrons necessary to complete the reaction

or even to supply electrons or energy needed in the reaction (see the *Plants & People* box on page 171).

Competitive and noncompetitive inhibition can slow or stop enzymatic reactions and pathways

Competitive inhibitors bind to the active site in such a way as to block binding of the substrate and prevent the catalyzed reaction from occurring (Figure 7.25a). Competition occurs between substrate and inhibitor molecules, so competitive inhibition can often be overcome by increasing the concentration of substrate. Noncompetitive inhibitors (Figure 7.25b) bind to the enzyme at a location other than the active site or bind permanently at the active site. Often, they change the shape of the enzyme so that the substrate can no longer bind as effectively or at all. Increasing substrate concentrations will do nothing to overcome the effect of a noncompetitive inhibitor.

End-product inhibitor or feedback inhibition (Figure 7.25c) occurs when the product of a series of enzymatic reactions inhibits one of the enzymes responsible for the product's formation. By this mechanism, organisms make

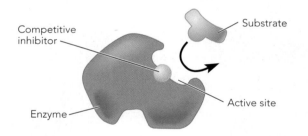

(a) Competitive inhibitors compete with the substrate and block the active site of the enzyme. By increasing the substrate concentration, the substrate can overcome the competition with the inhibitor and the reaction can still occur.

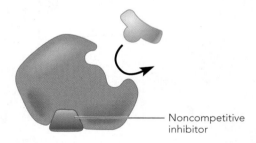

(b) Non-competitive inhibitors are more effective inhibitors. They bind to a site other than the active site. Binding in this manner changes the shape of the enzyme and the reaction is less likely to occur or does not occur at all.

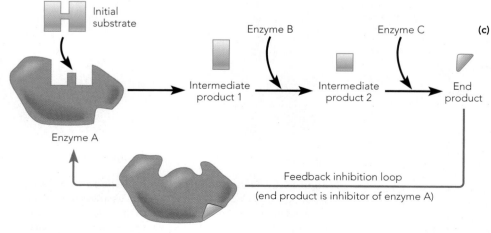

(c) Feedback inhibition occurs when the end product of a series of reactions inhibits the enzyme at the beginning of the reactions, thereby preventing overproduction of the end product.

Figure 7.25 Enzymatic inhibitors.

PLANTS & PEOPLE

Take Your Cofactors Every Day

Humans require at least 13 essential vitamins and perhaps as many as 60 minerals. These vitamins and minerals often function as cofactors assisting enzymatic reactions. Plants produce these vitamins and obtain minerals from the soil. Humans get them from plants, directly or indirectly, or from nutritional supplements such as vitamin and mineral tablets. Minerals for supplements come from the soil or sometimes from seawater or from organisms, as in the case of calcium from shells. Vitamins come from other living organisms such as bacteria or plants or are synthetically produced. Since these compounds are used and reused by enzymes, they are required in small amounts. However, their presence is essential. Deficiencies in essential vitamins and minerals can have devastating results (see the table below).

Many sailors in the old days died of scurvy, caused by a deficiency of vitamin C. Once the cause of scurvy was identified, British ships carried supplements of lime juice for the sailors. Lime juice proved useful because it provided vitamin C and didn't spoil without refrigeration. Thus, British sailors came to be known as "limeys."

We think of such deficiencies as being mainly of historical interest. However, agricultural soils in the United States and throughout the world have sometimes become depleted of minor minerals because most fertilization programs do not replace them. Modern humans, therefore, experience some risk of developing mineral deficiencies.

Selenium deficiencies, which are common in soils and therefore in diets, are associated with an increased risk of certain types of cancers. Studies by the National Cancer Institute indicated that increasing selenium in the diets of people deficient in the mineral causes highly significant decreases in the rates of stomach, lung, colorectal, and prostate cancer but no decrease in the rates of skin cancer. While some soils in the world have a selenium deficiency that can cause human health problems, other soils have excess selenium and can cause poisoning.

Copper deficiencies in soils may cause weakened hearts and death as a result of ruptured aortas. Copper is a required cofactor for important enzymes involved in the synthesis of collagen and other molecules that strengthen the walls of veins and arteries. The relationship of insufficient dietary copper to an increase in deaths caused by aortic aneurysms was recognized by veterinarians in animals many years ago when adequate levels of copper were not being supplied in animal feeds. Since some animals, such as poultry, eat only factory-produced pellets, adequate amounts of essential vitamins and minerals must be supplied.

Plants require a number of trace minerals, but they make their own vitamins. Humans require several minerals not required by plants. As agricultural soils become further depleted of essential minerals, mineral deficiency diseases will become increasingly well known and studied and mineral supplementation will become a more important part of human diets.

Some Vitamin Deficiency Diseases in Humans

Deficiency	Symptoms/disease	Prevention, in addition to vitamin supplements
Vitamin A	Xerophthalmia is the major cause of blindness among young children in most developing countries. The condition is most common in southern and eastern Asia, where polished rice, devoid of carotene, is the staple food.	A diet that includes green leafy vegetables and yellow fruits
Vitamin C	Scurvy is characterized by hemorrhagic manifestations and abnormal bone and tooth formation. This condition was common among sailors while on long sea voyages.	A diet that includes fresh fruits and vegetables
Vitamin D	Osteomalacia is the adult form of vitamin D deficiency and is characterized by softening of the bones, resulting in deformation. Rickets is the childhood form and results in improper bone formation and in bone deformation.	Adequate levels of sunlight or vitamin D–enriched foods such as milk.
Vitamin K	Failure of blood to clot; fatal hemorrhaging	A diet including fresh fruits and vegetables
Vitamin B12 (Cobalamin)	Pernicious anemia is a fatal disease of elderly people resulting from a low level of red blood cells.	Large-scale consumption of raw liver is a specific cure; consumption of meats in general
Vitamin B1 (Thiamine)	Beriberi is a potentially fatal disease resulting in neurological problems and heart failure.	Avoid polished rice (rice with fruit and seed coat layers and embryo removed) as a major constituent of diet.
Nicotinamide	Pellegra is a disease characterized by skin problems, diarrhea, and mental deterioration.	Nicotinamide is made from the amino acid tryptophan. Diets rich in corn frequently cause a deficiency in this nutrient.

sure that overproduction of products does not occur. End-product inhibition can be either competitive or non-competitive but is typically competitive.

Enzymatic reactions are linked together into metabolic pathways

The 25,000 to 50,000 enzymatically catalyzed reactions in living cells are linked together into various metabolic pathways. For example, as you will learn in Chapters 8 and 9, the reactions of photosynthesis and respiration are grouped into joined reactions that produce particular products. Groups of enzymatic reactions can be either linear or circular and can have branch points by which specific compounds enter or leave the pathway. Linear reactions can generate products that serve as feedback inhibitors, and circular pathways always regenerate the starting compound.

Section Review

1. Explain how enzymes facilitate chemical reactions.
2. Describe the role of cofactors.
3. What are competitive and noncompetitive inhibitors?

SUMMARY

The Molecular Components of Living Organisms

Most large organic molecules are polymers—long molecules made of repeating subunits called *monomers*. Monomers are linked into polymers by dehydration synthesis. Hydrolysis separates monomers from polymers. Carbohydrates, proteins, nucleic acids, and lipids are primary metabolites found in all plant cells. Secondary metabolites are not found in all plant cells.

Carbohydrates, which supply and store energy and serve as structural building blocks, include sugars and polymers of sugars (pp. 151–154)
Carbohydrates include monosaccharides, disaccharides, and polymers. Sugars are linked together via a dehydration synthesis. Some important carbohydrates in plants are starch, used for energy storage, and cellulose, the chief component of plant cell walls.

Proteins, which catalyze reactions and are structural building blocks, are polymers of amino acids (pp. 154–157)
Amino acids have the same basic structure with a variable R group. In living organisms, 20 types of amino acids are linked by peptide bonds to form proteins.

The nucleic acids DNA and RNA, which code and express genetic information, are polymers of nucleotides (pp. 157–158)
Nucleotides, composed of bases, sugar, and phosphates, serve as the building blocks for DNA and RNA, as hormones, and as sources of energy. DNA consists of two strands of nucleotides. The energy-carrying molecule ATP is a modified nucleotide.

Lipids are membrane components consisting mainly of carbon and hydrogen atoms derived from acetates and other molecules (pp. 158–160)
Lipids include fats, phospholipids, steroids, and terpenoids. Fats are made of fatty acid chains that are derived from acetate and attached to glycerol. Phospholipids contain glycerol, two fatty acid chains, and a phosphate molecule. Phospholipids are the principal molecular component of membranes. Steroids are multiringed compounds that occur in membranes and can function as plant hormones.

Secondary metabolites such as phenolics, alkaloids, and terpenoids often protect or strengthen plants (pp. 160–162)
Major types of phenolics are lignins, flavonoids, and allelopathics. Alkaloids, which are made from amino acids, and terpenoids, which are made of isoprene subunits, often deter herbivores.

Energy and Chemical Reactions

Energy can be stored and can move or change matter (p. 163)
Potential energy is stored energy, while kinetic energy is energy having to do with motion. The first law of thermodynamics states energy can be harnessed and transformed but not created or destroyed. The second law of thermodynamics states that every transfer of energy increases the entropy (disorder) of matter in the universe.

Chemical reactions involve either a net input or a net output of free energy (pp. 163–164)
Exergonic reactions have a net output of energy, while endergonic reactions have a net input. Both require activation energy. In chemical reactions, energy can be expressed as heat or as entropy, a measure of disorder in matter.

Redox reactions release energy as a result of movement of electrons between atoms or molecules (pp. 164–165)

In chemical reactions, oxidation and reduction are paired in redox reactions so that one molecular component is oxidized while the other is reduced.

The terminal phosphate bond in ATP releases energy when broken (pp. 165–166)

ATP is universally used by living organisms to supply energy to reactions. The covalent bonds linking phosphates in ATP require considerable energy to form. When the bonds are broken, energy is released. ATP supplies activation energy for both endergonic and exergonic reactions, both of which are needed in living cells.

NADH, NADPH, and FADH2 are universal carriers of energy-rich electrons in living organisms (pp. 166–167)

NADH, NADPH, and $FADH_2$ are molecules that carry energy-rich electrons and donate them to reactions catalyzed by enzymes that require both energy and electrons.

Chemical Reactions and Enzymes

Collision theory describes product formation by reactions in gases or liquids (pp. 167–168)

According to collision theory, molecules that can interact to form covalent bonds do so if they collide with sufficient energy. The velocity and energy of collisions are increased by raising the temperature. In living organisms, many necessary reactions would proceed too slowly at existing temperatures.

Enzymes position reactants, allowing reactions to occur with minimal activation energy or increase in temperature (pp. 168–170)

Reactants and enzymes form an enzyme-substrate complex, which breaks down to the enzyme and products. Cofactors interact with enzymes, altering the reactions they catalyze. Both competitive and noncompetitive inhibition can stop enzyme action by blocking or altering the three-dimensional shape of the active site.

Cofactors such as coenzymes interact with enzymes to assist reactions (p. 170)

Some common cofactors include vitamins, hormones, ions, and ATP. Coenzymes are organic cofactors.

Competitive and noncompetitive inhibition can slow or stop enzymatic reactions and pathways (pp. 170–172)

Both competitive and noncompetitive inhibition can slow or stop enzyme action by blocking or altering the shape of the active site.

Enzymatic reactions are linked together into metabolic pathways (p. 172)

Pathways may be either linear or circular and can have branch points where specific compounds enter or leave.

Review Questions

1. Describe the general structure of carbohydrates.
2. Describe the structural differences between starch and cellulose.
3. Describe the four levels of protein structure.
4. What is the difference between primary and secondary metabolites? What are some functions of secondary metabolites?
5. Why is the tertiary structure of an enzyme important in terms of reactions catalyzed by the enzyme?
6. What is the difference between kinetic and potential energy?
7. How do endergonic reactions differ from exergonic reactions?
8. What are redox reactions?
9. What is the source of ATP, and how does ATP supply energy?
10. Why are electron carriers like NADH important for chemical reactions?
11. Why are enzymes essential for the metabolism of a cell?
12. Explain how enzymes work.
13. What are cofactors, and why do people take them in pills?
14. Describe how inhibition can slow or stop enzyme action.

Questions for Thought and Discussion

1. How might genetic engineers produce plants with adequate quantities of all the essential amino acids?
2. Why do you think the entropy of the universe is gradually increasing?
3. If activation energy were not required for reactions to occur, what would be the effect on cells?
4. Scientists think that movement, nervous systems, brains, and the ability to learn developed because animals move around to obtain food. What might be the nature and characteristics of a plant that could move around like an animal?
5. Use the phospholipid symbol shown in this chapter to draw diagrams showing how the presence of a mixture of phospholipids, some with saturated and some with unsaturated fatty acid components, prevents the close packing of phospholipids that occurs in membranes consisting of phospholipids with saturated fatty acids only. (Recall that it's this lack of close packing that causes plant fats, which contain a significant proportion of unsaturated fatty acids, to be liquid at room temperature whereas animal fats, which contain mainly saturated fatty acids that can pack together tightly, are solid at this temperature.)

Evolution Connection

An important difference between animals and plants in terms of biochemistry is that plants produce a much more diverse array of secondary metabolites than do animals. Several classes of these metabolites have been introduced in this chapter. Suggest reasons why this biosynthetic capacity has adaptive value in plants but not, apparently, in animals.

To Learn More

Visit The Botany Place Website at www.thebotanyplace.com for quizzes, exercises, and Web links to new and interesting information related to this chapter.

Gilbert, F. Hiram, ed. *Basic Concepts in Biochemistry: A Student's Survival Guide.* New York: McGraw-Hill Professional, 1999. This book stresses the mastering of fundamental concepts by using simple, jargon-free language, algorithms, mnemonics, and clinical examples.

Buchanan, B. Bob, Wilhelm Gruissem, L. Russell Jones. *Biochemistry and Molecular Biology of Plants.* New York: John Wiley & Sons, 2002. This book is current, quite technical, and has it all.

Edelson, Edward. *Francis Crick & James Watson and the Building Blocks of Life (Oxford Portraits in Science).* New York: Oxford University Press, 2000. This intriguing biography of Watson and Crick explores the structure of DNA, the molecule that carries our genes and determines everything from the color of our eyes to the shape of our fingernails.

Watson, D. James, and Lawrence Bragg. *The Double Helix: A Personal Account of the Discovery of the Structure of DNA.* New York: Touchstone Books, 1991. This book recounts both the scientific and personal aspects of the discovery.

Watson, D. James. *Genes, Girls, and Gamow: After the Double Helix.* New York: Alfred A. Knopf, 2002. The story of the discovery of the double helix continues.

8

Photosynthesis

Photosynthesis uses light energy to produce organic molecules.

An Overview of Photosynthesis

Photosynthesis produces food, molecular building blocks, and O_2, which support almost all life on Earth

Photosynthesis uses light energy to convert CO_2 and H_2O into sugars

The processes of photosynthesis and respiration are interdependent

Converting Light Energy to Chemical Energy: The Light Reactions

Chlorophyll is the principal light-absorbing molecule of photosynthesis

Light energy enters photosynthesis at locations called *photosystems*

The light reactions produce O_2, ATP, and NADPH

In the light reactions, ATP is synthesized using energy from chemiosmosis

Converting CO_2 to Sugars: The Calvin Cycle

The Calvin cycle uses ATP and NADPH from the light reactions to make sugar phosphates from CO_2

The Calvin cycle is relatively inefficient at converting CO_2 into sugars

The enzyme rubisco also functions as an oxygenase, resulting in photorespiration

The C_4 pathway limits the loss of carbon from photorespiration

CAM plants store CO_2 in a C_4 acid at night for use in the Calvin cycle during the day

Sugarcane is a perennial, tropical plant that grows up to 5 meters (16.4 feet) in height. It has been cultivated in Asia and in the Americas as a source of sugar for thousands of years. In 510 B.C., the Persian King Darius I saw it growing on the banks of the Indus River. Later, Alexander the Great (356–332 B.C.) brought plants back to Greece from India. The Persians and Greeks were both surprised to find a plant that produced sweetener without the need for bees to make honey. The Chinese had also been extracting sugar from sugarcane for millennia before Marco Polo found large sugar mills in China when he visited in the 1200s.

The European explorations of the Americas in the 1500s and 1600s introduced many Europeans to sugar produced from sugarcane. Christopher Columbus himself brought the plant back from the Caribbean and established a sugarcane plantation there. Sugar became a wildly popular and expensive replacement for honey and provided a sure sign of affluence.

Unfortunately, slavery in the Americas got an economic impetus from the need for workers on sugar plantations in Central America, the Caribbean, and the southern British colonies that would become the United States. An insatiable sweet tooth made Europe the prime market for sugar. During the 1600s and 1700s, sugar was a key element in the so-called *triangular trade* that carried cane sugar and rum (made from sugarcane) from the Americas to Europe; tools, weapons, and other trade goods from Europe to Africa; and ultimately more than 11 million enslaved Africans to the Americas. It is tragic

Sugar beets are primary sources of processed sugar.

to note that the sweetness of sugar played a role in changing the course of history and the lives of so many enslaved Africans.

Eventually, a source of cheaper sugar undermined the market for cane sugar. In 1744, a German chemist named Andreas Marggraf discovered sugar in the juice of certain types of beets (*Beta vulgaris*). By mostly haphazard selection, growers eventually produced sugar beets with markedly increased concentrations of sugar. Sugar-beet production in continental Europe increased slowly over the next half century. By 1802, the first large factory for the extraction of beet sugar was built in Germany. By this time, the production of sugar from sugar beets in several European countries began to reduce the market for more expensive cane sugar brought by boat from the Americas and played a role in ending slavery in the Americas.

The importance of sugar in human history reminds us of the vital role played by plants in supplying our food. As you know, humans and other animals ultimately depend on photosynthesis for all of their food. Sugar in particular is produced as a direct result of photosynthesis and serves as a source of both energy and organic molecules for all living organisms. Understanding photosynthesis and protecting plants and other organisms that carry it out are therefore of vital interest.

Sugarcane being harvested on St. Kitts in the Caribbean.

An Overview of Photosynthesis

Photosynthesis makes the basic organic molecules a plant needs to survive, prosper, and reproduce. In general, photosynthetic organisms make the lives of nonphotosynthetic organisms possible.

Photosynthesis produces food, molecular building blocks, and O_2, which support almost all life on Earth

Like it or not, your molecular ingredients are remarkably similar to those of broccoli and earthworms. All organisms use carbon-based molecules as building blocks for assembling and maintaining themselves. In almost all cases, pho-

Figure 8.1 Photosynthetic organisms. Photosynthetic species—which include almost all plants, algae, and some bacteria—harvest solar energy to produce organic molecules.

tosynthesis is the ultimate source of those molecules. Plants, algae, and photosynthetic bacteria, of course, rely on photosynthesis directly and are known as *autotrophs* because they make their own food (Figure 8.1). Since they get their energy through photosynthesis, they are known more specifically as **photoautotrophs.** Most nonphotosynthetic life-forms, such as animals and fungi, are heterotrophs, depending totally on other organisms for organic molecules to build their bodies, energy to function, and oxygen. You can consume all the carbon dioxide gas (CO_2) you want—that is what makes carbonated drinks bubbly—but as an animal you cannot use CO_2 to produce organic molecules. Through photosynthesis, however, plants can convert CO_2 and H_2O into sugars that form the basis of thousands of other organic molecules that make up all living organisms.

Since most heterotrophs rely either directly or indirectly on photosynthetic organisms for nutrition, photosynthetic organisms underlie almost every food chain. In terrestrial food chains, land animals either eat plants or eat animals that have eaten plants. For example, cows eat grass, and humans eat cows. Meanwhile, fungi absorb energy-rich compounds from the remains of organisms whose organic carbon molecules were originally produced by plants and other photosynthetic organisms. In aquatic food chains, animals eat algae or eat animals that have eaten algae. For example, sea urchins eat algae, and sea otters eat sea urchins. The only organisms that do not rely either directly or indirectly on photosynthesis are bacteria known as **chemoautotrophs,** which get their carbon from CO_2 and their energy from inorganic chemicals (see *The Intriguing World of Plants* box on page 178).

Every carbon atom in your body has existed at some previous point in a photosynthetic organism and has been processed by that organism from CO_2 and the energy of sunlight. The carbon-containing molecules produced by photosynthesis are responsible for more than 94% of the dry weight of living organisms. They are combined with minerals from the soil to produce the many different kinds of molecules found in living organisms. When you die, your body will eventually become CO_2, water, and a few minerals. Those substances will again be used in photosynthesis.

Photosynthesis also sustains life by releasing oxygen (O_2). In 1771, Joseph Priestley, an English clergyman and scientist, used closed-container experiments to show that plants "restore" air, thereby allowing a candle to burn and a mouse to survive. Later he discovered oxygen. A citation associated with a medal he received for this discovery said, "For these discoveries we are assured that no vegetable grows in vain . . . but cleanses and purifies our atmospheres." Later, in 1779, Dutch physician Jan Ingenhousz repeated and extended Priestley's observations, showing that oxygen was restored by plants only in the presence of sunlight and only by green plant parts.

THE INTRIGUING WORLD OF PLANTS

Nonphotosynthetic Plants

Indian pipe (*Monotropa uniflora*).

Some plants are actually nonphotosynthetic, parasitic entirely. Indian pipe (*Monotropa uniflora*) also called ghost flower is a parasite flowering plant. Only the nonphotosynthetic flowering stalks appear above-ground. The remainder of the plant is underground, absorbing nutrients from the fungal hyphae of mycorrhizae of nearby living plants. A Cherokee legend says that after a week of watching quarreling chiefs of several tribes smoke the peace pipe, the Great Spirit changed the chiefs into the flowers called Indian Pipes to remind everyone that smoking the peace pipe was supposed to end quarreling. According to the legend, the flowers grow wherever people have quarreled. In the western United States, the snow plant (*Sarcodes sanguinea*) is a saprophytic member of the family that includes huckleberries and blueberries.

Snow plant (*Sarcodes sanguinea*).

O_2 in the atmosphere is split by ultraviolet (UV) light into molecular oxygen, which combines with O_2 to produce ozone (O_3). The ozone layer in the atmosphere absorbs harmful UV light from the Sun and thereby helps to make life on land possible.

Photosynthesis uses light energy to convert CO_2 and H_2O into sugars

Photosynthesis in plants and algae takes place in microscopic organelles called *chloroplasts* (Chapter 2). Typically around 3–5 μm in diameter, chloroplasts can be circular or elongated and are most common in leaf tissue, where a typical cell may contain between 5 and 50 chloroplasts.

The process of using light energy to convert CO_2 and H_2O into sugars consists of two series of reactions: the light reactions and the Calvin cycle (Figure 8.2). The **light**

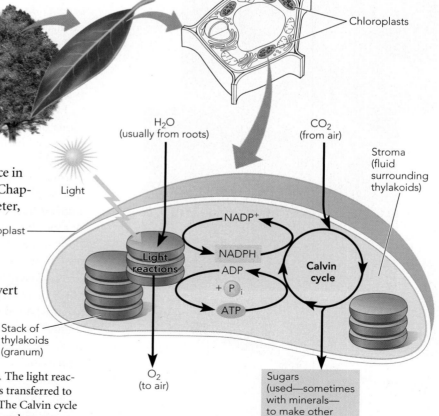

Figure 8.2 An overview of photosynthesis. The light reactions use chlorophyll to capture light energy that is transferred to ATP and NADPH, in electrons supplied by water. The Calvin cycle uses ATP, NADPH, and CO_2 to make simple three-carbon sugar phosphates, which become building blocks for more complex molecules.

reactions, which occur within the thylakoid membranes, are the *photo* part of photosynthesis because they capture light energy. They use light energy and H_2O to generate chemical energy in the form of ATP and NADPH, and they release O_2 as a by-product. The **Calvin cycle,** the *synthesis* part of photosynthesis, assembles (synthesizes) simple three-carbon sugars, using ATP and NADPH from the light reactions and CO_2 from the air. The Calvin cycle takes place in the stroma, the fluid-filled region surrounding the thylakoids.

The simple three-carbon sugars produced by the Calvin cycle become building blocks for complex molecules such as glucose ($C_6H_{12}O_6$). Photosynthesis can be summarized in terms of the amount of CO_2 and H_2O needed to produce one molecule of glucose:

$$6CO_2 + 12H_2\mathbf{O} + \text{light energy} \rightarrow C_6H_{12}O_6 + 6\mathbf{O_2} + 6H_2O$$

The use of bold in this equation indicates that oxygen atoms from H_2O combine to form molecules of O_2. We can simplify the formula by noting only the net consumption of H_2O:

$$6CO_2 + 6H_2O + \text{light energy} \rightarrow C_6H_{12}O_6 + 6O_2$$

This formula shows that CO_2, water, and light energy are used to make sugar and oxygen. We can further simplify it by dividing the previous formula by 6. Now we have the basic formula for making sugars and other carbohydrates (molecules with the basic structure CH_2O) one carbon at a time:

$$CO_2 + H_2O + \text{light energy} \rightarrow CH_2O + O_2$$

In summary, photosynthesis captures light energy from the Sun and uses it to assemble CO_2 into sugars. We will look at specific reactions of the process later in the chapter.

The processes of photosynthesis and respiration are interdependent

Before we look at the light reactions and the Calvin cycle, it is worth keeping in mind that photosynthesis by itself does not sustain life. Photosynthesis produces food, but all organisms—whether or not they are photosynthetic—must then extract energy from that food in a process known as *respiration.* In **respiration,** which takes place in mitochondria, organisms break down organic molecules in the presence of oxygen and convert the stored energy into the form of ATP. Cells then use the energy in ATP to do work. In other words, respiration harvests energy from the food produced by photosynthesis. Each process relies on products of the other, as respiration uses sugars and O_2 in producing CO_2, H_2O, and ATP, while photosynthesis uses CO_2 and H_2O in producing sugars and O_2 (Figure 8.3).

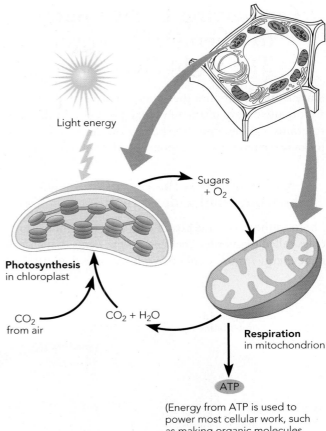

Figure 8.3 The interdependence of photosynthesis and respiration. The products of photosynthesis—sugars and O_2—are used in respiration to produce the ATP that powers most cellular work. The by-products of respiration—CO_2 and H_2O—are used in photosynthesis.

Respiration is an exergonic process, resulting in a net output of free energy in the form of ATP, while photosynthesis is an endergonic process, resulting in a net input of free energy. In Chapter 9, we will be looking at how respiration yields energy from organic molecules produced by photosynthesis.

Section Review

1. **How do nonphotosynthetic organisms obtain the carbon molecules necessary for life?**
2. **How do heterotrophs and autotrophs differ?**
3. **What is the source of the oxygen produced during photosynthesis?**
4. **Briefly summarize how photosynthesis relates to respiration.**

Converting Light Energy to Chemical Energy: The Light Reactions

Since sunlight supplies the energy to drive photosynthesis, organisms are ultimately solar-powered. In the light reactions, light energy absorbed by chlorophyll is used to make two energy-rich compounds: ATP and NADPH.

Chlorophyll is the principal light-absorbing molecule of photosynthesis

Photosynthesis is made possible by light-absorbing molecules called **pigments.** The types of pigments that absorb light energy for use in photosynthesis are either attached to or part of the thylakoid membranes of chloroplasts. The pigment that is directly involved in the light reactions is the green pigment **chlorophyll** (Figure 8.4).

Each type of photosynthetic pigment absorbs light energy at certain wavelengths. Visible light and all other forms of electromagnetic energy move through space as packets of energy called **photons,** which vary in their energy content, depending on their wavelengths. A photon with a shorter wavelength has more energy than a photon with a longer wavelength (Figure 8.5a). For instance, photons that are visible as blue light contain more energy than photons visible as red light. Chlorophyll absorbs photons from the red and blue portions of the visible spectrum but either transmits or reflects photons from the green portion (Figure 8.5b). In other words, the green color is what is visible after chlorophyll has absorbed the

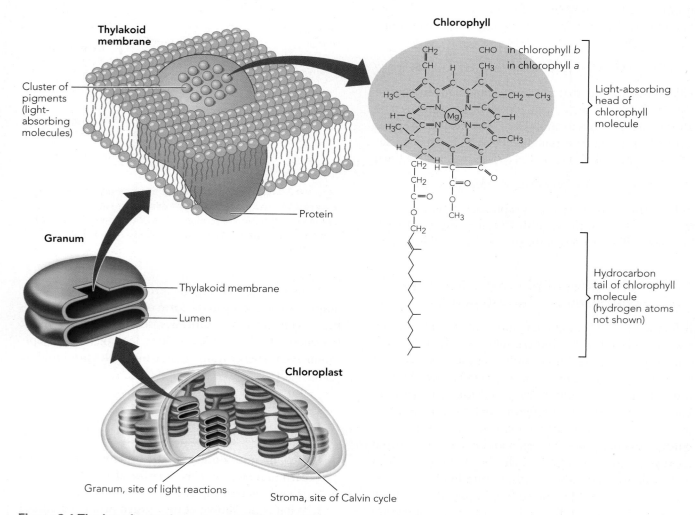

Figure 8.4 The location and structure of chlorophyll. Photosynthesis in plants and algae occurs within organelles called *chloroplasts,* which in plants are found in cells in leaves and some stems. Within the chloroplast, membrane-bound structures called *thylakoids* are stacked together to form grana. The thylakoid membranes absorb light through clusters of pigments, with the green pigment chlorophyll *a* being directly involved in the light reactions.

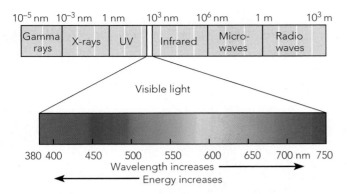

(a) Electromagnetic spectrum

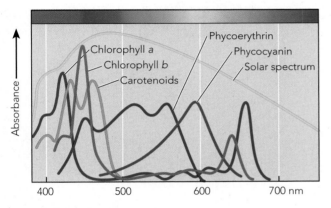

(b) Absorption of pigments

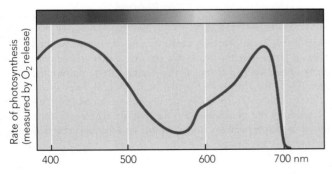

(c) Action spectrum

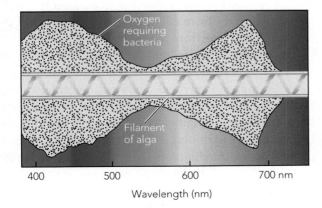

(d) Engelmann's experiment

light used in the light reactions of photosynthesis. As a result of the thylakoid membranes reflecting green light, the photosynthetic parts of plants—leaves and some stems—typically appear green.

There are two main types of chlorophyll in plants and green algae, known as chlorophyll *a* and chlorophyll *b*. In plants, **chlorophyll *a*** is the only pigment directly involved in the light reactions. It primarily absorbs light from the blue-violet and red ranges of the spectrum and appears dark green because it mainly reflects green light. In photosynthesis, a chlorophyll *a* electron that has absorbed a photon from the blue portion of the spectrum loses the extra energy as heat and ends up with the same energy as an electron that has been energized by a photon from the red portion of the spectrum. In other words, blue light is not used directly in photosynthesis in plants. **Chlorophyll *b*** does not take part directly in the light reactions but instead transmits absorbed energy to those chlorophyll *a* molecules that are directly involved. Chlorophyll *b* is therefore known as an **accessory pigment.** Some other accessory pigments, known as *carotenoids,* primarily absorb blue-green light and reflect yellow or yellow-orange light. In plants, these accessory pigments are typically not visible until chlorophyll breaks down as when leaves of deciduous plants change color. It is the carotenoids that cause autumn coloration after short days and cold temperatures have slowed photosynthesis and chlorophyll has broken down.

Measuring O_2 production as a function of wavelength reveals the **action spectrum** for photosynthesis—that is, a profile of how effectively different wavelengths of light promote photosynthesis. The action spectrum for photosynthesis has peaks in the blue and red regions of the spectrum, corresponding closely to chlorophyll's **absorption spectrum**—the range of a pigment's ability to absorb wavelengths of light. This correlation indicates that chlorophyll is the primary pigment involved in photosynthesis (Figure 8.5b and c). The action spectrum of photosynthesis can also be demonstrated by placing oxygen-requiring

Figure 8.5 Chlorophyll absorbs light. (**a**) The electromagnetic spectrum contains a relatively narrow band of visible light. (**b**) Chlorophyll absorbs in both the blue and red regions of the absorption spectrum but transmits green. (**c**) The action spectrum of photosynthesis corresponds to the absorption spectrum of chlorophyll and accessory pigments. (**d**) In 1883, Thomas Engelmann placed single-celled oxygen-requiring bacteria along a filament of photosynthetic algae. The bacteria congregated around the regions of the algal cells receiving blue and red light from a prism. In this way, he demonstrated which wavelengths of light were causing oxygen production by promoting photosynthesis. As you can see, the distribution of the bacteria correlates with the action spectrum of photosynthesis.

bacteria next to a strand of photosynthetic algae and then exposing the algae to different wavelengths of light. The bacteria cluster where O_2 is being released as a by-product of photosynthesis—the areas receiving blue and red light (Figure 8.5d).

One could reasonably ask why photosynthesis does not use all available light, in which case plants would appear black. One possible answer might be that in Earth's early days, more than 2.5 billion years ago, some other life-form absorbed green light first, making it unavailable for photosynthesis. Perhaps this life-form floated on the ocean's surface, where it was first to get the light. For example, the prokaryote *Halobacterium* absorbs green light. Some photosynthetic prokaryotes called *cyanobacteria* absorb light primarily in the green but also in the blue regions of the spectrum.

Light energy enters photosynthesis at locations called photosystems

Within the thylakoid membranes, chlorophyll *a*, chlorophyll *b,* and other pigments such as carotenoids form clusters of pigments, with each cluster consisting of about 200 to 300 pigment molecules, along with associated protein molecules. Experiments indicate that light reactions in each cluster are triggered by one chlorophyll *a* molecule, which absorbs the energy of a photon and ejects an electron that is then absorbed by a molecule known as a *primary electron acceptor.* Together, this chlorophyll *a* molecule and the primary electron acceptor are called a **reaction center.** The reaction center and accessory pigments in each cluster work together as a light-harvesting unit known as a **photosystem** (Figure 8.6). Each photosystem absorbs light energy on the outside—the stroma side—of the thylakoid membrane.

There are two types of photosystems, known as photosystem I and photosystem II. The numbers designate the order in which they were discovered. Photosystem I has little chlorophyll *b,* whereas photosystem II has more chlorophyll *b,* in almost equal amounts with chlorophyll *a.* These photosystems occur repeatedly throughout the thylakoid membranes. The accessory pigments are critical components. The chlorophyll *a* molecule that triggers the light reactions is not going to get hit directly by photons very often because it represents less than 1% of the pigments in a photosystem. Accessory pigments, however, funnel the energy from the photon to the chlorophyll *a* in the reaction center. These accessory pigments, along with chlorophyll *a* molecules that transfer energy to the chlorophyll *a* molecule in the reaction center, are sometimes called *antenna pigment molecules* because they

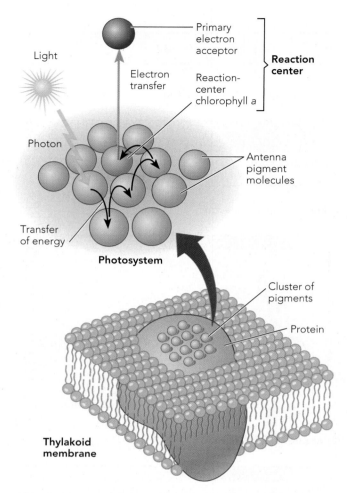

Figure 8.6 Each photosystem is a light-harvesting complex. A photosystem consists of a reaction center and antenna pigment molecules that transfer light energy to the chlorophyll *a* molecule in the reaction center. The reaction center also includes a primary electron acceptor. Light energy causes an electron to be ejected from the chlorophyll *a* in the reaction center and transferred to the primary electron acceptor. This process is carried out repeatedly in each photosystem.

function like an antenna in receiving and transmitting energy, somewhat like a satellite dish. Energy can be transferred in the form of energized electrons, or the energy itself can move from molecule to molecule.

The chlorophyll *a* molecule in the reaction center absorbs energy at slightly longer wavelengths (lower energy) than chlorophyll in general. The reaction center chlorophyll *a* molecule in photosystem II is called *P680,* with the *P* standing for pigment and the number indicating that it absorbs light best at a wavelength of 680 nm. In photosystem I, the chlorophyll *a* molecule in the reaction center is called *P700* because it absorbs light best at 700 nm. As a result of the energy transfer from the antenna pigment molecules, much more light energy can reach the

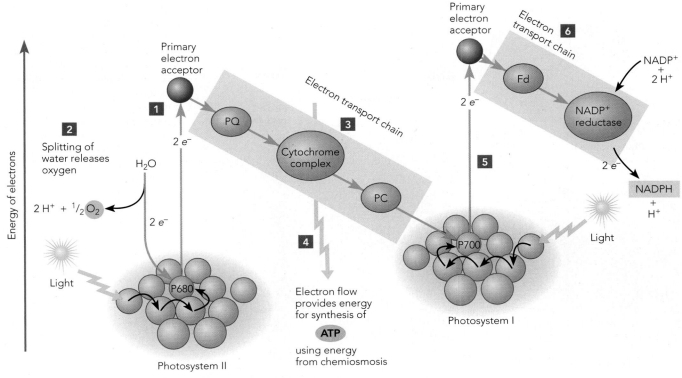

Figure 8.7 Movement of electrons in the light reactions. This zigzag diagram, known as the *Z scheme*, provides an overview of the flow of energy in the light reactions. As each electron flows through the electron carriers—plastoquinone (PQ), a cytochrome protein complex, plastocyanin (PC), ferredoxin (Fd), and NADP⁺ reductase—each carrier attracts the electron more strongly than the previous carrier. The numbered steps are described in the text.

chlorophyll *a* molecule in the reaction center than it would have absorbed on its own. Each time the reaction center chlorophyll *a* molecule is activated, an energized electron is transferred to the primary electron acceptor.

The light reactions produce O₂, ATP, and NADPH

The two light-absorbing photosystems are linked together in a zigzag pattern sometimes called a *Z-scheme* (Figure 8.7). The Z-scheme consists of a series of protein-based electron carriers that form a pathway for the movement of electrons. Keep in mind that this pathway is repeated thousands of times within a typical thylakoid membrane. The movement of electrons in each Z-scheme constitutes the light reactions.

We will now follow the flow of energy to see how the light reactions produce O₂, ATP, and NADPH. The steps correlate with Figure 8.7. Although each electron passes through the light reactions one at a time, notice that the figure shows two electrons, the number needed at the end of the light reactions in order to transform one molecule of NADP⁺ into one molecule of NADPH.

1 When light energy reaches the chlorophyll *a* in the reaction center, it energizes an electron in the chlorophyll *a* molecule. This electron is transferred to the primary electron acceptor.

2 Each ejected electron is quickly replaced by an electron from H₂O, after an enzyme splits an H₂O molecule into 2 electrons, 2 hydrogen ions (H⁺), and one oxygen atom. (A hydrogen ion is a solitary proton.) The chlorophyll *a* is positively charged because it lost an electron, so it attracts the negatively charged electron from water. In addition to replacing the electron in the chlorophyll *a* molecule, the splitting of water produces oxygen gas, releasing one molecule of O₂ for every two molecules of H₂O that are split.

3 Each electron ejected from the chlorophyll *a* molecule passes from the primary electron acceptor and gradually loses energy by transfer through a series of electron carriers known as an **electron transport chain.** The movement from carrier to carrier occurs in a series of oxidation-reduction (redox) reactions.

4 The energy that is released by the flow of electrons is indirectly used to power the synthesis of ATP, a process we will look at shortly.

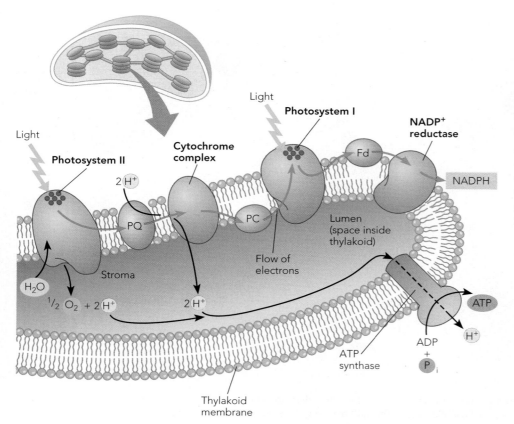

Figure 8.8 Chemiosmosis and ATP synthesis in the light reactions. This diagram shows the flow of hydrogen ions (H⁺) through ATP synthase, a process called *chemiosmosis*. Some of the hydrogen ions come from the splitting of water, while others are pumped through the membrane by the energy released through the flow of electrons in the electron transport chain. ATP synthase uses the energy of chemiosmosis to add an inorganic phosphate (P_i) to ADP to form ATP.

5 Each electron that passes through the electron transport chain neutralizes the positively charged chlorophyll *a* molecule in the reaction center of photosystem I. This chlorophyll *a* molecule is positively charged because the absorption of a photon has already ejected a reenergized electron that is transferred to a primary electron acceptor.

6 Each electron ejected from photosystem I passes through another electron transport chain. The last electron carrier in this chain is an enzyme that transforms NADP⁺ to NADPH. As noted previously, 2 electrons must pass through the light reactions before one NADPH can be synthesized.

So, the light reactions use the energized electrons ejected from photosystem II in producing energy-rich ATP and use the reenergized electrons ejected from photosystem I in producing NADPH. The light reactions capture 32% of the sunlight absorbed by chlorophyll, making them more efficient than any energy-capturing device made by humans. For example, solar panels typically capture, as electricity or heat, around 5% of the solar energy they absorb. As you will see, the ATP and NADPH produced by the light reactions will be used in the Calvin cycle to make CO_2 into simple sugar phosphates.

In the light reactions, ATP is synthesized using energy from chemiosmosis

As electrons move through the electron transport chain between photosystem II and photosystem I, they lose energy. This energy is used to pump protons from the stroma into the thylakoid compartment, causing a concentration difference in H⁺ ions across the thylakoid membrane (Figure 8.8). The charge difference is also a pH difference, creating a kind of battery that stores energy for doing work. In a process known as **chemiosmosis** (from the Greek *chemi-*, referring to "chemical," and *osmos*, meaning "a push"), H⁺ ions move back across the membrane, resulting in a release of energy. Many of these H⁺ ions flow through an enzyme called **ATP synthase,** which uses the energy to add an inorganic phosphate (P_i) to ADP, thereby forming ATP. This phosphorylation is known as **photophosphorylation** because the energy to carry it out comes originally from light.

Electrons typically move from photosystem II to photosystem I, in what is called *noncyclic electron flow*. During the light reactions as they occur naturally in plants, the synthesis of ATP relies on this noncyclic electron flow and is therefore known as *noncyclic photophosphorylation*. In some bacteria and in laboratory experiments with plants, it is possible to achieve what is called *cyclic photophosphorylation,* which only involves photosystem I. Electrons flow in a cycle from the reaction center of photosystem I to the electron transport chain and back to the same reaction center, indirectly producing ATP but no NADPH. However, plant physiologists continue to debate whether cyclic photophosphorylation occurs in plants in nature.

Section Review

1. Describe the role of chlorophyll in photosynthesis.
2. How does light energy enter the light reactions?
3. Identify the products of the light reactions and explain how they are formed.

Converting CO_2 to Sugars: The Calvin Cycle

As you have seen, the light reactions use light energy and H_2O to yield chemical energy in the form of ATP and NADPH. These products fuel the second part of photosynthesis—the Calvin cycle, which makes simple sugar phosphates. The Calvin cycle is named after Melvin Calvin, who—together with his graduate student Andrew Benson and later James Bassham—in 1953 determined the pathway by which plants convert CO_2 into sugars. In 1961, Calvin won the Nobel Prize for this discovery, which resulted from experiments involving exposure of photosynthetic algae to radioactive CO_2 for shorter and shorter lengths of time (Figure 8.9). After 5 seconds of exposure to the radioactive CO_2, the principal radioactive compound in the algae was a three-carbon molecule known as 3-phosphoglycerate (PGA). The rest of the Calvin cycle was worked out by similar experiments. Since the first product has three carbons, the Calvin cycle is sometimes known as the C_3 pathway.

The Calvin cycle uses ATP and NADPH from the light reactions to make sugar phosphates from CO_2

The reactions of the Calvin cycle are sometimes called the *dark reactions* or the *light-independent reactions* because they can take place in the dark, as long as the products of the light reactions—ATP and NADPH—are supplied. However, these terms can be misleading because they

Figure 8.9 Calvin's experiment. Melvin Calvin, working with Andrew Benson and other colleagues, conducted an experiment in which they traced the process of photosynthesis in green algae, using radioactive CO_2, in this "lollipop" apparatus. After varying lengths of time, the lollipop contents were emptied into boiling alcohol, thereby killing the algae so that the progress of the radioactivity through various compounds could be followed. In this way, Calvin and his colleagues were able to identify how CO_2 fixation occurs in photosynthesis.

imply that the Calvin cycle can continue indefinitely in the dark, which is not the case. Cellular stockpiles of ATP and NADPH last only a few seconds or minutes at most. Cells do not store large quantities of either ATP or NADPH, so the Calvin cycle must rely on these molecules being resupplied by the light reactions.

As you have seen, in plants and algae the Calvin cycle occurs outside the thylakoids, in the stroma of chloroplasts. In the case of plants, CO_2 enters through the pores called *stomata* in the leaf epidermis and then diffuses into mesophyll cells, where photosynthesis takes place. The Calvin cycle uses the energy-rich products of the light reactions—ATP and NADPH—to incorporate three CO_2 molecules (one at a time) into a three-carbon sugar phosphate. You might imagine that synthesis of sugars occurs by linking together molecules of CO_2 while adding needed electrons and hydrogens. However, this is not the case. In fact, the Calvin cycle adds one CO_2 to a five-carbon compound. After the cycle repeats three times, enough carbon has been added to account for one molecule of a three-carbon sugar phosphate called *glyceraldehyde-3-phosphate* (*G3P*), also known as *3-phosphoglyceraldehyde* (*PGAL*). Outside of the Calvin cycle, molecules of G3P are used to produce molecules of several types of six-carbon sugars, including fructose and glucose, which can be combined to form sucrose, composed of 12 carbons. Sucrose is the principal sugar used in translocation of carbohydrates from the leaves to other parts of the plant.

The Calvin cycle incorporates CO_2 and uses ATP and NADPH from the light reactions to create building blocks of life. The carbon atoms fixed into sugar by the Calvin cycle will ultimately become the carbons of all of the organic molecules found in plants, animals, and almost all other life-forms. Figure 8.10 provides an overview of one turn of the Calvin cycle:

1 One ATP powers addition of a phosphate to a five-carbon sugar phosphate to make a sugar phosphate molecule with two phosphates. The addition of this second phosphate energizes the five-carbon molecule. Specifically, one ATP is used to make ribulose-1,5-bisphosphate (RuBP) from ribulose-5-phosphate (Ru5P).

2 Carbon dioxide is added to the five-carbon sugar phosphate. Specifically, the enzyme **rubisco** adds CO_2 to RuBP. As you may recall from Chapter 7, the name *rubisco* is the abbreviation for ribulose 1,5-bisphosphate carboxylase/oxygenase. Rubisco is called a *carboxylase* because it can add carbon from CO_2 to another molecule. In this case, the resulting short-lived six-carbon compound immediately breaks down into two molecules of a three-carbon organic acid, 3-phosphoglycerate (PGA). This is called **carbon fixation** because the carbon from CO_2 is incorporated ("fixed" or bound) into a nongaseous, more complex molecule.

3 Two molecules of ATP add phosphates to the three-carbon organic acids. The addition of phosphates energizes the three-carbon organic acids. Specifically, two molecules of ATP are used to convert two molecules of PGA into two molecules of 1,3-bisphosphoglycerate (BPG).

4 Two molecules of NADPH add electrons to the three-carbon organic acid phosphates, reducing each BPG into one molecule of glyceraldehyde-3-phosphate (G3P). The result is two G3P molecules, representing a total of six carbons.

5 After three turns of the Calvin cycle, enough carbon has been fixed to allow one molecule of G3P to leave the cycle and become available for making other sugars, while still leaving enough carbons to regenerate Ru5P to complete the cycle.

6 Most G3P continues through the rest of the Calvin cycle. Other reactions of the cycle make four-carbon, six-carbon, and seven-carbon sugar phosphates. Eventually, the cycle regenerates the five-carbon Ru5P molecule, which initiates another turn of the cycle.

Here is the overall equation reflecting the products of three turns of the Calvin cycle:

$$3CO_2 + 6NADPH + 9ATP + 6H^+ \rightarrow 1G3P + 6NADP^+ + 9ADP + 8P_i + 3H_2O$$

This equation reflects the three CO_2 molecules required for the sugar, the nine ATP molecules, and the six NADPH molecules used for three turns of the Calvin cycle. The products $NADP^+$, ADP, and P_i return to the light reactions as reactants.

Glucose, an important sugar of living cells, is produced indirectly from two molecules of G3P formed in the Calvin cycle. The metabolic fates of G3P include:

◆ Conversion in respiration to CO_2 and H_2O, with energy stored in ATP.

◆ Conversion to intermediate compounds in respiration that are synthesized into amino acids and other compounds.

◆ Conversion to fructose 6-phosphate (F6P) and fructose bisphosphate.

◆ Conversion of F6P into glucose 6-phosphate (G6P) and glucose 1-phosphate (G1P).

◆ Use of G1P to form cellulose in cell walls and starch for energy storage.

◆ Use of G1P and F6P to make sucrose for transport throughout the plant.

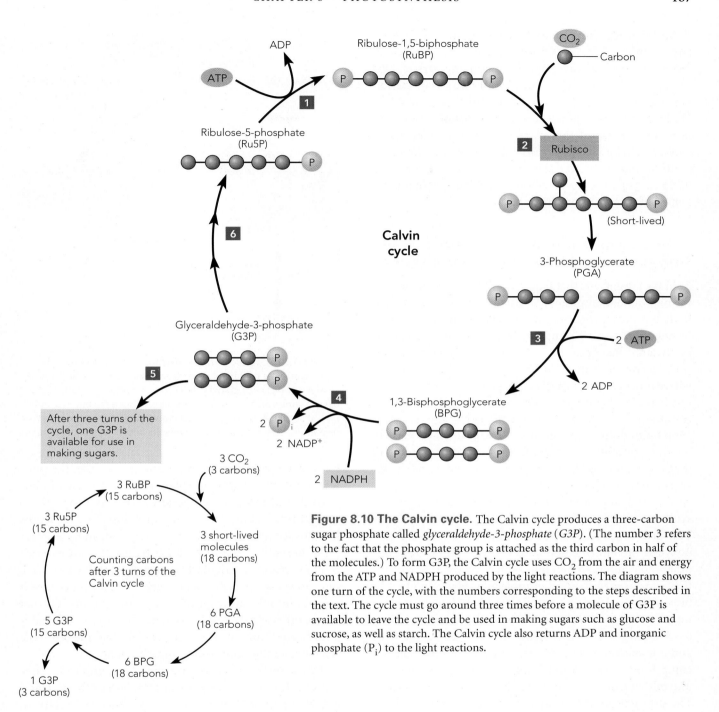

Figure 8.10 The Calvin cycle. The Calvin cycle produces a three-carbon sugar phosphate called *glyceraldehyde-3-phosphate* (*G3P*). (The number 3 refers to the fact that the phosphate group is attached as the third carbon in half of the molecules.) To form G3P, the Calvin cycle uses CO_2 from the air and energy from the ATP and NADPH produced by the light reactions. The diagram shows one turn of the cycle, with the numbers corresponding to the steps described in the text. The cycle must go around three times before a molecule of G3P is available to leave the cycle and be used in making sugars such as glucose and sucrose, as well as starch. The Calvin cycle also returns ADP and inorganic phosphate (P_i) to the light reactions.

The Calvin cycle is relatively inefficient at converting CO_2 into sugars

The efficiency of the Calvin cycle is the amount of chemical energy it actually uses to convert CO_2 compared to the amount of light energy received by the light reactions. The Calvin cycle's efficiency can be measured by comparing how much energy the cycle uses to fix CO_2 and how much light energy is needed to generate the NADPH that enters the cycle. The theoretical maximum for photosynthetic efficiency has been calculated to be around 35%. In the real world, however, most plants and algae achieve only between 1% and 4%. This reduced efficiency results in part from the fact that the Calvin cycle may waste up to half of the carbon it fixes. This occurs in a process known as *photorespiration,* which we will look at shortly.

Since production totals for photosynthesis involve large numbers that are difficult to imagine, let us consider a summary that involves one plant and one person. One corn plant fixes 0.23 kg (about a half pound) of carbon per season. A 45-kg (100-lb.) person contains about 6.8 kg (15 lbs.) of fixed carbon. Corn kernels that are eaten contain 10% or less of that carbon. A typical large ear of cooked corn has 100 kilocalories (kcal), so to maintain an adult human would require 25 ears of corn per day—or 9,125 ears of corn per year—equivalent to approximately a third of an acre of production.

The enzyme rubisco also functions as an oxygenase, resulting in photorespiration

Rubisco, the enzyme that fixes carbon in the Calvin cycle, is the most abundant protein on Earth. Every carbon of your body has been processed by rubisco because the carbon comes directly or indirectly from plants. In the summer, rubisco can cause up to a 15% decrease in daytime atmospheric concentrations of CO_2. Within plant canopies, where most photosynthesis takes place, the decrease is as much as 25%. Plants grow more rapidly if the CO_2 concentration of their atmosphere is artificially increased, as is sometimes done in greenhouses. The gradual increase in the CO_2 concentration of Earth's atmosphere as a result of the burning of fossil fuels by humans may be slowed by increased photosynthesis by plants.

The oxygen produced by the light reactions of photosynthesis actually inhibits net carbon fixation by rubisco. This is because rubisco, in addition to its role as a carboxylase (an enzyme that adds carbon from CO_2 to another molecule), can function as an oxygenase, an enzyme that adds oxygen to another molecule (Figure 8.11). Rubisco does not bind CO_2 strongly, and at higher temperatures and lower CO_2 concentrations rubisco is just as likely to bind oxygen. In this case, no carbon is fixed. One PGA and one molecule of 2-phosphoglycolate (a two-carbon compound) result from ribulose 1,5-bisphosphate. The phosphoglycolate is eventually broken down into CO_2. The production of CO_2 as a result of rubisco activity is known as **photorespiration** because it occurs in the light, produces CO_2, and uses O_2. Unlike respiration (see Chapter 9), photorespiration does not produce ATP. Photorespiration begins with rubisco activity and involves chloroplasts, peroxisomes, and mitochondria. At low CO_2 concentrations and high O_2 concentrations, the oxygenase function of rubisco predominates.

Overall, on bright, sunny days, when the temperature is around 25°C, O_2 may reduce the rate of CO_2 fixation

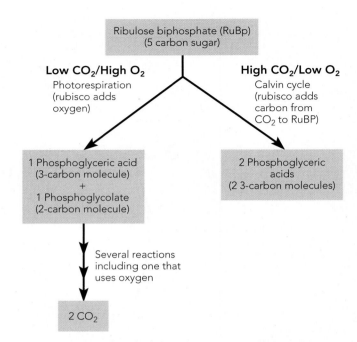

Figure 8.11 Rubisco and photorespiration. Rubisco, the fixation enzyme of the Calvin cycle, can function as a carboxylase (fixing carbon) or as an oxygenase (fixing oxygen), depending on oxygen concentration. As a carboxylase, rubisco facilitates production of sugars. As an oxygenase, it converts two carbons of ribulose bisphosphate back into CO_2, the process known as *photorespiration*.

by rubisco by 33%. As temperatures increase, the photorespiration rate of many plants equals the rate of photosynthesis. At higher temperatures, plants also begin to close their stomata, preventing water loss. Consequently, less CO_2 enters the leaf, less O_2 is released, and photorespiration results. Photorespiration reduces the growth rate of many plants, particularly on bright, hot days. Overall, photorespiration reduces net carbon fixation because it releases into the atmosphere a large amount of CO_2 that would otherwise be fixed in sugars. The situation is analogous to collecting sugar in a box, except that the box has a hole in it so that some of the sugar escapes.

By converting some fixed carbon back into CO_2, photorespiration wastes a significant fraction of the ATP and NADPH produced by the light reactions. Indirectly, in terms of productivity, it wastes minerals, water, light, and any other resources needed by plants to live and reproduce. In an environment in which one or more necessary resources are scarce, a plant that wastes up to 50% of fixable CO_2 may have difficulty surviving on its own and will be easily outcompeted for necessary resources by a more efficient plant.

Rubisco has been around at least 2 billion years and became the carbon fixation enzyme at a time when the at-

EVOLUTION

Evolution and O₂ Concentration

If you think about it, the O_2 produced in photosynthesis should be used up by respiration in both photosynthetic and nonphotosynthetic organisms. So, why does our current atmosphere have 21% O_2 when the Earth started out with none? In general, the idea that O_2 concentrations have gradually increased from zero to the current 21% is far too simple. The course and causes of changes in atmospheric O_2 concentration are areas of very active research.

After the Earth cooled enough so that gravity could retain an atmosphere, that atmosphere had a high concentration of CO_2—maybe up to 80%—and no O_2. When photosynthesis evolved in bacteria around 3.5 billion years ago, plenty of CO_2 was available. For the first 2 billion years or so, the O_2 produced by photosynthesis was taken up by iron deposits on the ocean floor. Then O_2 began to gradually appear in the atmosphere. Photosynthetic organisms evolved into a number of different forms, and so did nonphotosynthetic bacteria, thanks to the biomass and O_2 production of their photosynthetic kin.

Around 2.5 to 1.9 billion years ago, there was a surge in the amount of atmospheric O_2. Some investigators think the cause was an increased runoff and erosion from the continents, resulting in a buildup of ocean sediments that buried and killed many of the nonphotosynthetic bacteria living on the ocean floor. Thus, photosynthetic organisms, which absorbed light near the surface, predominated, causing the rapid increases in O_2 concentrations. Increased levels in O_2 may have made possible the evolution of eukaryotic cells around 2.2 billion years ago.

A second surge in O_2 concentrations occurred just before the Cambrian period, around 600 million years ago, and quite possibly for the same geologic reason. This increase in O_2 concentration may have fueled the huge adaptive radiation in invertebrate animals around that time. By the late Cambrian period, O_2 concentration had reached at least 2%, enough to allow the survival of land-based eukaryotes. Starting around 430 million years ago, the evolution and rapid spread of land plants caused the O_2 concentration to increase to a high of around 35% in the Carboniferous period, around 370 million years ago. This may account for the giant, two-foot dragonflies and other huge insects that existed then. The inefficient oxygenation system of insects puts an upper limit on insect size for any given O_2 concentration.

By the end of the Permian period, around 250 million years ago, the percentage of O_2 had crashed to 15%. The decline may have been caused in part by a large die-off of organisms, followed by their decay. Following these mass extinctions, photosynthetic organisms again predominated, and O_2 levels again increased.

---

mosphere had little or no free oxygen. The fact that the enzyme has an oxygenase function is a historical accident that only became significant when oxygen concentrations gradually increased as a result of the actions of photosynthetic organisms. Evidently, no single mutation can change the enzyme to eliminate photorespiration while retaining carbon fixation, because in the long history of competition and natural selection, such a mutation would surely have occurred. The resulting plant would have had a tremendous selective advantage and would have rapidly taken over most environments (see the *Evolution* box on this page).

In the future, it might be possible for genetic engineering to design a rubisco that has little or no oxygenase function and therefore up to twice the photosynthetic efficiency in plants. A computer that could simulate the effect of particular mutations on enzyme structure and function might demonstrate what specific amino acid changes would produce rubisco with no oxygenase function and a strong carboxylase function.

The C₄ pathway limits the loss of carbon from photorespiration

Remember that photosynthesis first evolved among bacteria and later algae in the water, where moderate light levels and temperature prevailed. On land, plants may have to contend with lack of water, higher light levels, and more extreme temperatures. These environmental conditions lead to closure of stomata, which helps plants by preventing water loss but also hinders photosynthesis by keeping CO_2 from entering the leaves. When the stomata close, the CO_2 concentration in leaves falls as a result of Calvin cycle activity, while the O_2

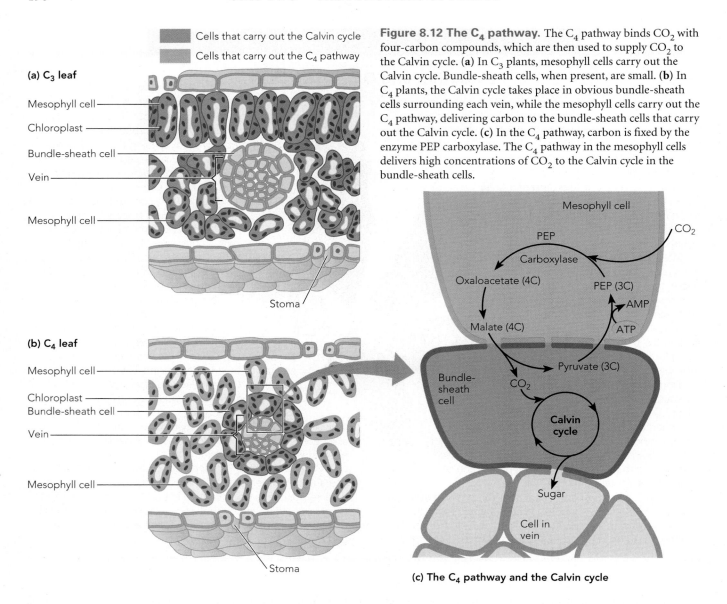

Figure 8.12 The C₄ pathway. The C_4 pathway binds CO_2 with four-carbon compounds, which are then used to supply CO_2 to the Calvin cycle. (**a**) In C_3 plants, mesophyll cells carry out the Calvin cycle. Bundle-sheath cells, when present, are small. (**b**) In C_4 plants, the Calvin cycle takes place in obvious bundle-sheath cells surrounding each vein, while the mesophyll cells carry out the C_4 pathway, delivering carbon to the bundle-sheath cells that carry out the Calvin cycle. (**c**) In the C_4 pathway, carbon is fixed by the enzyme PEP carboxylase. The C_4 pathway in the mesophyll cells delivers high concentrations of CO_2 to the Calvin cycle in the bundle-sheath cells.

(c) The C₄ pathway and the Calvin cycle

concentration increases as a result of the light reactions. Under these conditions, rubisco is likely to add O_2 to other molecules instead of adding CO_2, leading to increased photorespiration and reduced photosynthesis.

Plants that waste energy in photorespiration also waste resources such as water and mineral nutrients. Natural selection will favor any change that aids photosynthesis in hot, sunny conditions. Among flowering plants, a number of tropical monocots and some dicots have an add-on to the Calvin cycle called the C_4 *pathway*. The **C₄ pathway** binds CO_2 into four-carbon compounds, which are then used to supply an increased concentration of CO_2 to the Calvin cycle. Scientists discovered the C_4 pathway in the 1960s, when they noticed that in sugarcane the first product of carbon fixation is a four-carbon molecule, hence the term C_4 (Figure 8.12).

The C_4 pathway either prevents or limits photorespiration because its carbon-fixing enzyme, known as PEP carboxylase, only binds CO_2 and not O_2. Unlike rubisco, PEP carboxylase keeps binding CO_2 to carbon compounds even when the CO_2 concentration is low within the leaf.

Plants that have the C_4 pathway are known as **C₄ plants** and are more common in the Tropics, in arid regions, and in hot, dry, sunny environments. Plants that only carry out the Calvin cycle for carbon fixation are known as **C₃ plants.** Most C_4 plants have a different leaf anatomy than C_3 plants, a difference that is crucial to the operation of the C_4 pathway. In C_3 plant leaves, the Calvin cycle occurs in all photosynthetic cells, but in leaves of C_4 plants it typically occurs only in bundle-sheath cells, which appear in a prominent single or double layer surrounding each leaf vein (Figure 8.12a and b).

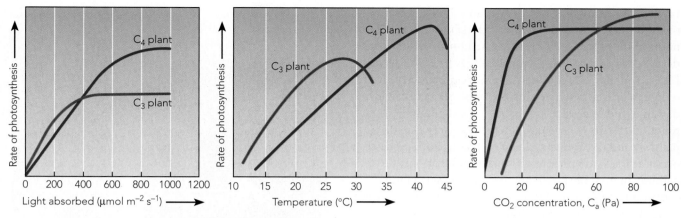

Figure 8.13 C$_4$ plants are more efficient than C$_3$ plants. When light intensity or temperature is high, or when CO$_2$ concentration is low, C$_4$ plants are much more efficient at photosynthesis, and therefore at utilization of water and minerals, than are C$_3$ plants. These conditions are often found in deserts and temperate grasslands.

This ringlike arrangement is often called *Kranz anatomy,* from the German word *kranz,* meaning "wreath or halo." All of the mesophyll cells in a C$_4$ leaf have only the C$_4$ pathway. These mesophyll cells feed CO$_2$ trapped in organic compounds to the bundle-sheath cells, where it is released and refixed by the Calvin cycle. Thus, the bundle-sheath cells have high concentrations of CO$_2$, allowing rubisco to fix CO$_2$ rather than O$_2$.

Apparently a C$_4$ pathway evolved several times in plants. It occurs in more than 19 families of flowering plants. Many cereals and other grasses are C$_4$ plants, but the pathway exists in some dicots as well. As you can see in Figure 8.12c, the enzyme PEP carboxylase fixes CO$_2$ in mesophyll cells of C$_4$ plants. The more efficient carbon fixation by PEP carboxylase is especially notable at low CO$_2$ concentrations, when rubisco would tend to bind O$_2$ to other molecules. PEP carboxylase attaches bicarbonate to PEP to make oxaloacetate. This four-carbon acid is typically converted to malate, a process that uses NADPH from the light reactions. Malate—or, in some plants, aspartate—moves into the bundle-sheath cells through plasmodesmata. The bundle-sheath cells break it down to pyruvate, a common metabolite in cells, regenerating both CO$_2$ and NADPH. Pyruvate—or, in some plants, alanine—is then moved back to the mesophyll cells, where PEP is regenerated by enzyme action.

The C$_4$ pathway would seem to be a relatively energy-inefficient process because ATP is needed to convert pyruvate to PEP, in addition to the three molecules of ATP used in the Calvin cycle. Despite this inefficiency, the C$_4$ pathway combined with the Calvin cycle will outperform the Calvin cycle alone on hot, sunny days, when photosynthesis is rapid and leaf CO$_2$ concentrations may drop (Figure 8.13). When temperatures are cooler and as the CO$_2$ concentration increases, the Calvin cycle alone—the C$_3$ pathway—is more energy-efficient because it requires less ATP.

The relative efficiency of C$_4$ plants can be demonstrated by a competition experiment. Wheat (*Triticum aestivum*), a C$_3$ plant, is put into competition in a closed container with corn (*Zea mays*), a C$_4$ plant. PEP carboxylase, the C$_4$ enzyme, is much more efficient at fixing carbon than is rubisco. Therefore the C$_4$ plant takes up most of the CO$_2$ in the air, the CO$_2$ produced by photorespiration in the C$_3$ plant, and the CO$_2$ produced by respiration in the C$_3$ plant. Before long, the C$_4$ plant is flourishing and the C$_3$ plant dies. If the plants occupy separate closed containers, each does well. Another common example of a C$_4$ plant outcompeting C$_3$ plants is found when the C$_4$ plant crabgrass (*Digitaria sanguinalis*) overgrows more desirable C$_3$ lawn grasses, such as Kentucky bluegrass (*Poa pratensis*), during hot, dry summer days.

Earth's atmosphere has a current CO$_2$ concentration of 365 parts per million (ppm), or 0.0365%. C$_4$ plants achieve maximum photosynthesis rates at around 50 ppm (0.005%) CO$_2$. Even on hot days with high light intensities, photosynthesis is occurring at maximum rates in the leaves of these plants. C$_3$ plants usually increase their photosynthesis rates as CO$_2$ concentrations increase up to 500 ppm (0.05%) and in some cases at even higher concentrations. These increases relate to the fact that increased CO$_2$ concentration in the leaves lowers the rate of photorespiration.

CAM plants store CO$_2$ in a C$_4$ acid at night for use in the Calvin cycle during the day

Some plants have a variation of the C$_4$ pathway called **crassulacean acid metabolism** (**CAM**), in which they take up CO_2 at night using the C_4 pathway and then carry out the Calvin cycle during the day (Figure 8.14). The name comes from the Crassulaceae family of succulent desert plants, in which the process was first discovered. Like C_4 plants, **CAM plants** live in regions where high temperatures necessitate closing the stomata during the day to avoid excessive water loss. CAM plants and C_4 plants differ in where and when they carry out the C_4 pathway and the Calvin cycle. In C_4 plants, both processes occur at the same time but in different locations: the C_4 pathway in mesophyll cells and the Calvin cycle in bundle-sheath cells (Figure 8.14a). In contrast, CAM plants carry out both processes in the mesophyll cells, but at different times: the C_4 pathway at night and the Calvin cycle during the day. At night, when it is cool, mesophyll cells use the C_4 pathway to bind CO_2 temporarily in malate within the vacuoles (Figure 8.14b). During the day, malate is transferred to the chloroplasts and converted to pyruvate and CO_2, which enters the Calvin cycle. During the day, CAM plants rapidly use the CO_2 stored during the night, so their overall photosynthetic production is less than that of other plants.

Like the C_4 pathway, CAM has also arisen several times in the evolution of plants. A typical environment favoring CAM is one with high temperatures during the day, high light intensity, and low water availability. Currently CAM exists in 18 or more plant families, mostly dicots. Some nonsucculent examples are pineapple and several seedless vascular plants, including some ferns.

Section Review

1. What are the main reactions of the Calvin cycle?
2. Describe the difference between the C_3 and C_4 pathways.
3. Compare and contrast C_4 plants and CAM plants.

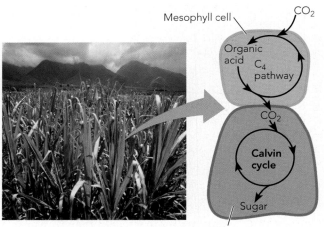

Sugarcane (C_4 plant)

(a) C_4 plants, such as sugarcane, run the C_4 pathway and the Calvin cycle simultaneously during the day. The C_4 pathway is carried out in mesophyll cells, while the Calvin cycle is carried out in bundle-sheath cells.

Pineapple (CAM plant)

(b) CAM plants, such as pineapple, run the C_4 pathway at night, allowing organic acids to accumulate in the vacuole. During the day, these organic acids are used to supply CO_2 to the Calvin cycle. The CAM pathway allows plants to keep stomata closed during hot, dry days while using CO_2 accumulated the previous night for photosynthesis.

Figure 8.14 Comparing C_4 and CAM plants.

SUMMARY

An Overview of Photosynthesis

Photosynthesis produces food, molecular building blocks, and O_2, which support almost all life on Earth (pp. 177–178)
Plants, algae, and photosynthetic bacteria are photoautotrophs, getting all their energy from photosynthesis. Most nonphoto-synthetic organisms are heterotrophs, depending totally on other organisms for organic molecules. Photosynthesis provides the building blocks of life and sustains life by supplying O_2.

Photosynthesis uses light energy to convert CO_2 and H_2O into sugars (pp. 178–179)
The light reactions take electrons from water and use sunlight absorbed by chlorophyll to energize them. Solar energy is used first to produce ATP and then to deposit energy-rich electrons

in NADPH. Chlorophyll and other light-absorbing molecules are located in the thylakoids of chloroplasts. The Calvin cycle uses ATP and NADPH from the light reactions, as well as CO_2, to produce three-carbon sugars.

The processes of photosynthesis and respiration are interdependent (p. 179)

In photosynthesis, electrons from H_2O are energized by the Sun and linked with CO_2 to form sugars. In respiration, the carbons in sugars are used to make CO_2 while the energy in sugar is transferred to ATP and the de-energized electrons are added to O_2 to produce H_2O.

Converting Light Energy to Chemical Energy: The Light Reactions

Chlorophyll is the principal light-absorbing molecule of photosynthesis (pp. 180–182)

Chlorophyll and other light-absorbing pigments are located inside the thylakoid membranes of chloroplasts. Chlorophyll uses blue and red light while transmitting green light.

Light energy enters photosynthesis at locations called *photosystems* (pp. 182–183)

Solar energy energizes electrons in chlorophyll within light-harvesting units called photosystem I and photosystem II. These photosystems occur repeatedly throughout the thylakoid membranes. The light reactions capture about a third of the sunlight absorbed by chlorophyll.

The light reactions produce O_2, ATP, and NADPH (pp. 183–184)

In the light reactions, electron flow first leads from water to a chlorophyll *a* molecule in photosystem II, where the electron is energized and transferred to an accepter. The electron gradually loses energy, which is transferred to ATP. The de-energized electron is transferred to a chlorophyll in photosystem I, where the electron is energized and transferred to NADPH.

In the light reactions, ATP is synthesized using energy from chemiosmosis (pp. 184–185)

Between photosystem II and photosystem I, an electron transport chain uses light energy to pump H^+ ions across the thylakoid membrane. Some H^+ ions release energy as they move back across the membrane and through the enzyme ATP synthase, a process called *chemiosmosis*. ATP synthase uses the energy from chemiosmosis to produce ATP by phosphorylation.

Converting CO_2 to Sugars: The Calvin Cycle

The Calvin cycle uses ATP and NADPH from the light reactions to make sugar phosphates from CO_2 (pp. 185–187)

The Calvin cycle, which occurs in the stroma of chloroplasts, uses CO_2 from the air and ATP and NADPH from the light reactions to produce sugars. After three turns of the cycle, enough

CO_2 has been added to produce one molecule of G3P, which is used to make other sugars.

The Calvin cycle is relatively inefficient at converting CO_2 into sugars (pp. 187–188)

The theoretical maximum for photosynthetic efficiency in synthesizing sugars is about 35%, but actual efficiency is between 1% and 4%, in part because the Calvin cycle may waste up to half of the carbon it fixes.

The enzyme rubisco also functions as an oxygenase, resulting in photorespiration (pp. 188–189)

Rubisco, the carbon-fixing enzyme in the Calvin cycle, is the most abundant protein on Earth. At higher temperatures and lower CO_2 concentrations, rubisco binds O_2. When oxygen is bound, no carbon is fixed and two carbons are eventually released as CO_2, a process known as *photorespiration*. On bright, hot days, rubisco can be involved in carbon loss as frequently as in carbon fixation.

The C_4 pathway limits the loss of carbon from photorespiration (pp. 189–191)

The C_4 pathway adds CO_2 to a three-carbon compound to produce a four-carbon oxaloacetate. C_4 plant anatomy features the C_4 pathway in mesophyll cells and the C_3 pathway in bundle-sheath cells surrounding vascular bundles in leaves. Malate from C_4 cells moves into bundle-sheath cells, where it is broken down into pyruvate and CO_2. Thus, high concentrations of CO_2 are supplied to the Calvin cycle, reducing photorespiration. C_4 plants are particularly efficient on hot, sunny days when leaf CO_2 concentrations are low and O_2 concentrations are high.

CAM plants store CO_2 in a C_4 acid at night for use in the Calvin cycle during the day (pp. 191–192)

Some succulent, desert plants have a variation of the C_4 pathway called *crassulacean acid metabolism (CAM)*. By taking up CO_2 at night through the C_4 pathway and using it in the Calvin cycle during the day, they can keep stomata closed during hot days.

Review Questions

1. What is the importance of photosynthesis for life on the planet?
2. Explain how photosynthesis and respiration are interdependent.
3. Describe the function of chlorophyll in photosynthesis.
4. Describe how a photosystem captures light energy.
5. What are the products of the light reactions and the Calvin cycle?
6. How do the light reactions and the Calvin cycle depend on each other?

7. Trace the pathway of an electron through the light reactions. Where does each electron begin, and where does it end?

8. Explain how ATP is synthesized during the light reactions.

9. How does the Calvin cycle provide building blocks for making more complex molecules?

10. Explain how the process of making sugar phosphates is a cycle.

11. Explain how rubisco is involved in photorespiration.

12. What is the difference between an oxygenase and a carboxylase?

13. How does the anatomy of a C_3 leaf differ from that of a C_4 leaf?

14. What types of environments favor C_4 plants? Why?

15. Compare and contrast C_4 plants and CAM plants.

Questions for Thought and Discussion

1. What kinds of life-forms do you think would exist after 6 billion years on Earth if photosynthesis had never developed? What life-forms would you expect if photosynthetic organisms were the only ones to evolve?

2. Suppose the Earth becomes warmer and has ever-higher concentrations of CO_2 over the next 500 million years. What types of plants might evolve?

3. If an asteroid strikes Earth and kicks up a thick cloud of dust that lowers the rate of photosynthesis by 90%, what immediate and long-term effects would you expect to find in animal and plant populations?

4. Suppose you own a greenhouse that produces salad greens and vegetables. Would it be worth your while to artificially increase the CO_2 concentration in your greenhouse's atmosphere? Explain.

5. Some scientific studies have shown that spraying methanol (CH_2OH) on plants increases carbon fixation by photosynthesis. What is happening?

6. From considerations of input and output of CO_2, O_2, H_2O, ATP, NADPH, and glucose, draw a flow diagram to illustrate the interdependence of respiration, the light reactions of photosynthesis, and the carbon-fixation (Calvin cycle) reactions of photosynthesis. It is suggested that you start by drawing three rectangular boxes side by side and label these, from left, "light reactions," "Calvin cycle," and "respiration." You can then indicate processes that occur within each box and the chemicals that flow from one box to another and interconnect the processes.

Evolution Connection

Biologists believe that the first living organisms on Earth, some 3.8 to 4 billion years ago, were chemoheterotrophs that utilized organic molecules formed in the Earth's atmosphere and oceans by nonbiological chemical processes, and that photosynthesis evolved later, once the supply of these organic compounds became depleted. From your knowledge of the processes involved, do you think that non-cyclic photophosphorylation evolved first and cyclic photophosphorylation evolved later, or was it more likely the other way around? Defend your answer.

To Learn More

Visit The Botany Place Website at www.thebotanyplace.com for quizzes, exercises, and Web links to new and interesting information related to this chapter.

Hessayon, G. D. *The House Plant Expert: The World's Best-Selling Book on House Plants.* London: Sterling Publications, 1992. Correct lighting is a key to having happy house plants.

Hobhouse, Henry. *Seeds of Change: Five Plants That Transformed Mankind.* New York: HarperCollins Publishers, 1999. Sugarcane is one of these five influential plants.

9
Respiration

Skunk cabbage (*Lysichiton americanum*).

An Overview of Nutrition

All living organisms need sources of energy and carbon

Plants use photosynthesis to store light energy in sugars and use respiration to transfer the energy from sugars to ATP

The breakdown of sugar to release energy can occur with or without oxygen

Respiration

Glycolysis splits six-carbon sugars into two molecules of pyruvate

The Krebs cycle generates CO_2, NADH, $FADH_2$, and ATP

The electron transport chain and oxidative phosphorylation transfer energy from the energy-rich electrons of NADH and $FADH_2$ to ATP

The energy yield from respiration is high

In some plants, the electron transport chain can generate excess heat

Plants, unlike animals, can make fatty acids into glucose

Fermentation

In the absence of oxygen, pyruvate produced by glycolysis is converted to ethanol or lactate

Some important industries rely on fermentation

Fermentation has a low energy yield compared to that of respiration

Energy is vital for carrying out the biochemical and physiological functions necessary for life, both to construct organisms and to maintain them. In fact, living organisms are islands of order in a universe that is overall becoming more disordered. Living organisms use ATP and electron carriers such as NADH, NADP, and FADH$_2$ to facilitate chemical reactions. Nonphotosynthetic organisms rely on photosynthetic organisms for organic molecules that can be broken down to obtain energy.

In photosynthesis, plants make ATP in the light reactions and use all of it in the Calvin cycle. Therefore, even though plants and other photosynthetic organisms can make their own food, they still must break down that food to produce the ATP and electron carriers necessary to build and maintain themselves. Unlike photosynthesis, which requires light and therefore occurs during the day, the breakdown of sugar and related molecules to obtain metabolic energy can occur around the clock.

An Indonesian corpse flower (*Amorphophallus titanum*) in Kew Gardens.

Environment can considerably influence how much metabolic energy an organism must use to survive. As a result of their biochemistry, living organisms can tolerate only a limited range of temperatures. Many environments have seasonal temperature extremes below or above those optimal for life. A number of different structures, physiological mechanisms, and behaviors have evolved that enable organisms to survive extreme temperatures.

As you will learn in this chapter, the process of breaking down sugar to transfer the energy to ATP is not totally efficient. Some energy is always lost as heat. Some animals, such as mammals and birds, maintain a constant body temperature by trapping this heat, using fur, feathers, and body fat as insulating materials. When conditions

are too cold, they produce more body heat by breaking down more sugar and then breaking down the resulting ATP. In some cases, they use an alternate pathway to break down sugar, while releasing all of the energy as heat.

In contrast to mammals and birds, the temperatures of plants—as well as those of reptiles, amphibians, and fish—are usually close to that of the outside environment. Such organisms are much less active in using metabolic energy directly to control temperatures. Plants, for example, cease photosynthesis and respiration when temperatures are too cold. They may lose their leaves and may also enter a dormant state in which metabolism is slower or suspended until suitable temperatures return.

While plants do not maintain a constant body temperature, a few plants can maintain temperatures considerably warmer than the surrounding air by generating heat instead of producing ATP. In some plant species, the heat is used to melt snow and ice, enabling the plant to take advantage of sunny but cold days in early spring. In other plants, such as the corpse flower, the heat evaporates fragrant molecules from the flowers, attracting particular pollinating organisms.

The corpse flower and other "hot plants" are unusual examples of how plants use energy produced through respiration. In this chapter, we will examine how plants use respiration to obtain ATP and heat from sugars produced in photosynthesis and from other organic compounds. During respiration, these organic molecules are broken down in the presence of oxygen into CO_2 and H_2O, releasing energy that is either transferred to ATP or given off as heat. We will also look at an alternative metabolic pathway, known as *fermentation,* which occurs sometimes in plants and other organisms when oxygen is absent.

An Overview of Nutrition

The processes by which an organism takes in and uses food substances are known as **nutrition.** Once food is produced, it must be broken down by a series of biochemical reactions in order to release the energy it contains. In plants, animals, and fungi, respiration is the process that breaks down sugars, in the presence of O_2, into CO_2 and H_2O while using the energy released to make ATP and heat.

All living organisms need sources of energy and carbon

Organisms need carbon and energy in order to create organic compounds, which form the structural and energetic basis of life as we know it. Based on their source of carbon, organisms can be classified as either autotrophs or heterotrophs (Table 9.1). Plants are examples of autotrophs, organisms that obtain carbon from CO_2 and use it to make their own organic compounds. Animals are examples of heterotrophs, organisms that must obtain carbon by consuming organic compounds from other organisms.

Both autotrophs and heterotrophs can be further classified based on their energy source. Plants and most other autotrophs are photosynthetic organisms and are known as *photoautotrophs* because they get energy from light. Nonphotosynthetic autotrophs, consisting of a few types of prokaryotes, are known as *chemoautotrophs* because they get energy from inorganic chemical compounds rather than light. Most heterotrophs, including humans, get both their energy and their carbon from organic compounds, which means that you qualify as a **chemoheterotroph.** Some heterotrophs, consisting of a few types of prokaryotes, are **photoheterotrophs,** getting their energy from light but their carbon from organic compounds.

In addition to carbon, most organisms—whether they are autotrophs or heterotrophs—need mineral nutrients and specific organic molecules such as vitamins. However, only plants and other autotrophs can make their own organic molecules.

Plants use photosynthesis to store light energy in sugars and use respiration to transfer the energy from sugars to ATP

Like all organisms, plants and other photosynthetic organisms carry out respiration. Figure 9.1 provides an overview of the relationship between photosynthesis and respiration. The overall relationship between the two processes can be described as follows:

- Plants and other photosynthetic organisms collect solar energy to make ATP and NADPH. These reactions constitute the light reactions of photosynthesis.
- They use the energy of ATP and the energy-rich electrons of NADPH to convert CO_2 to sugars. These reactions constitute the Calvin cycle reactions of photosynthesis.
- The sugars produced as a result of photosynthesis are combined with minerals from the soil to make a host of different organic molecules, which are used as a source of energy and as a source of structural components such as carbon skeletons.
- If oxygen is available, the process of respiration converts some of the sugars produced by photosynthesis to CO_2 and H_2O, while the energy is released as heat or transferred to ATP.

The net result of the processes of photosynthesis and respiration is the transfer of light energy to chemical energy in ATP and various organic molecules.

As in photosynthesis, the synthesis of ATP during respiration involves phosphorylation—the addition of a phosphate group to a molecule. In the case of ATP synthesis, the phosphorylation is the addition of an inorganic phosphate (P_i) to an ADP molecule to make ATP.

Table 9.1 Sources of Energy and Carbon for Organisms

Type of nutrition	Energy source	Carbon source	Types of organisms
Autotroph			
Photoautotroph	Light	CO_2	Photosynthetic prokaryotes, plants, algae
Chemoautotroph	Inorganic compounds	CO_2	Some prokaryotes
Heterotroph			
Photoheterotroph	Light	Organic compounds	Some prokaryotes
Chemoheterotroph	Organic compounds	Organic compounds	Many prokaryotes and protists, fungi, animals, some parasitic plants

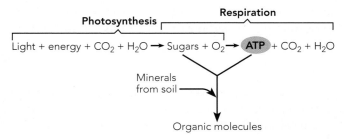

(a) Summary of the relationship of respiration to photosynthesis. Photosynthesis begins with carbon dioxide and water, and respiration ends with the same compounds.

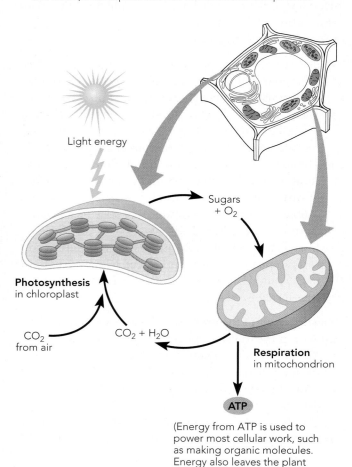

(b) Locations of respiration and photosynthesis in plant cells. In plants, photosynthesis occurs in chloroplasts, while respiration occurs in mitochondria.

Figure 9.1 Respiration and photosynthesis: An overview.

ATP synthesis can be carried out in several ways. The synthesis of ATP during photosynthesis is called *photophosphorylation* because it is powered by light energy. That is, light energy stimulates the electron flow through an electron transport chain that causes hydrogen ions (H^+) to move across the membrane—the movement known as *chemios-*

mosis. The enzyme ATP synthase then uses chemiosmotic energy to make ATP. In respiration, ATP is synthesized by two other types of phosphorylation: substrate-level phosphorylation and oxidative phosphorylation.

In **substrate-level phosphorylation,** an enzyme transfers a phosphate group from one phosphate-containing organic molecule to ADP, producing ATP (Figure 9.2a). This type of phosphorylation is so named because it involves an enzyme acting on two substrates: an ADP molecule and another phosphate-containing molecule. Chemiosmosis and ATP synthase are not involved, and the phosphorylation can occur with or without oxygen.

Oxidative phosphorylation is quite similar to photophosphorylation because it involves an electron transport chain, chemiosmosis, ATP synthase, and oxygen (Figure 9.2b). However, in oxidative phosphorylation it is energy from NADH—rather than light energy—that stimulates the electron flow for chemiosmotic synthesis of ATP. The process is called *oxidative phosphorylation* because it begins with oxidation, the loss of electrons. Specifically, NADH loses electrons to the electron transport chain, starting the flow of energy that ultimately powers the synthesis of ATP.

The breakdown of sugar to release energy can occur with or without oxygen

In plants, as well as in all other eukaryotes and some prokaryotes, the breaking down of sugars to obtain energy in the form of ATP follows one of two general pathways. One pathway is **aerobic,** which means it uses oxygen. The other is **anaerobic,** which means it does not use oxygen. Both pathways begin with a series of anaerobic enzymatic reactions known collectively as **glycolysis,** which takes place in the cytosol, the fluid part of the cell's cytoplasm. Glycolysis splits a six-carbon sugar into two molecules of pyruvate and also produces ATP and NADH.

Respiration is the aerobic pathway, in which cells ultimately require oxygen when breaking down organic molecules and converting the energy into the form of ATP. This process within cells is often called *cellular respiration,* to distinguish it from the use of the word *respiration* to refer to supplying oxygen to cells, as in the case of breathing in animals. In scientific usage, however, the term *respiration* means cellular respiration.

Respiration begins with glycolysis in the cytosol of the cell. The pyruvate produced by glycolysis then enters the mitochondrion, where it is broken down to form a compound called *acetyl coenzyme A,* or *acetyl CoA* (Figure 9.3). Next, acetyl CoA is broken down to supply two-carbon fragments that enter the phase of respiration known as the **Krebs cycle.** The Krebs cycle generates ATP by substrate-

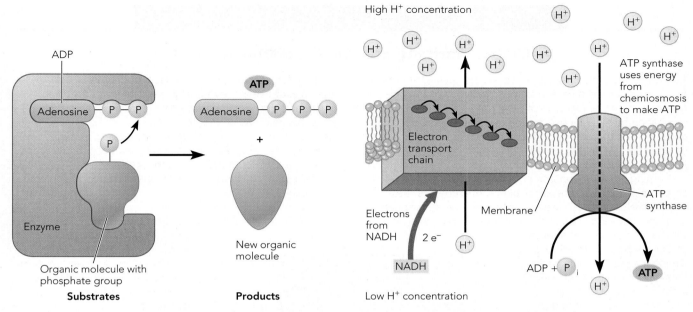

(a) Substrate-level phosphorylation. In synthesis of ATP through substrate-level phosphorylation, an enzyme transfers a phosphate from one substrate molecule to ADP, thereby forming ATP. Since this method relies only on enzyme action, it can occur with or without oxygen.

(b) Oxidative phosphorylation. In oxidative phosphorylation, ATP synthase produces ATP by using energy from chemiosmosis—the flow of hydrogen ions (H⁺) from a region of high concentration to low concentration.

Figure 9.2 ATP synthesis by substrate-level phosphorylation and oxidative phosphorylation.

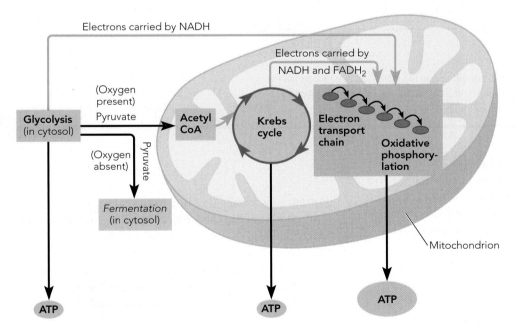

Figure 9.3 An overview of ATP production in respiration and related processes. Glycolysis takes place in the cytosol of the cell and produces a small amount of ATP. If oxygen is present, respiration occurs. First, the pyruvate from glycolysis proceeds to the mitochondria. Within the mitochondria, the following phases of respiration take place: the conversion of pyruvate to acetyl CoA, the Krebs cycle, the electron transport chain, and oxidative phosphorylation. The Krebs cycle produces a small amount of ATP. The greatest yield of ATP comes from oxidative phosphorylation, which is powered by chemiosmosis. If oxygen is absent, the pyruvate from glycolysis goes through the process of fermentation, and the only ATP produced is the small amount that came from glycolysis.

level phosphorylation. It also supplies the electron carriers NADH and FADH$_2$ to the last phase of respiration, consisting of an electron transport chain and synthesis of large amounts of ATP by oxidative phosphorylation. By definition, the term *respiration* refers to an energy-yielding process that uses oxygen. Actually, only the electron trans-

port chain directly requires oxygen. However, the Krebs cycle cannot occur unless oxygen is available for the electron transport chain.

If oxygen is not present, organic molecules are broken down through the anaerobic pathway, which is known as **fermentation** and takes place completely within the

CONSERVATION BIOLOGY

Global Warming and the Greenhouse Effect

Living organisms are carbon based, consisting of organic molecules, with carbon frameworks that were originally made by plants in photosynthesis. As you know, respiration breaks glucose and other sugars down to CO_2 and transfers the energy to ATP. Indeed, when any organically based material is burned, either metabolically or in a fire, the carbon is converted into CO_2.

Burning of large amounts of fossil fuels releases large amounts of CO_2 into the atmosphere. Even before the advent of civilization, CO_2 was released into the atmosphere as a result of volcanoes and forest fires. However, civilization has increased the release of CO_2 by burning fossil fuels. During the past century, scientists have monitored increasing atmospheric concentrations of CO_2 and noted increasing average temperatures.

Scientists theorize that temperatures might be increasing because of what is known as the greenhouse effect. This theory suggests that the gases that accumulate in the atmosphere, like CO_2, prevent heat from radiating into space. Instead, this heat is reflected back onto the earth's surface, raising temperatures, similar to the way a greenhouse traps heat. Scientists worry about the prospect of continued global warming as a result. They note

that, in this scenario, polar ice caps would melt, increasing the level of oceans, eventually inundating coastal cities.

In fact, the consequences of the greenhouse effect and of global warming are complex. Aside from human contributions to global warming, some scientists believe that warm temperatures may alternate naturally with cool temperatures in a cycle of hundreds or thousands of years. Consider the following scenario, which has been proposed by some researchers:

◆ As CO_2 concentrations and temperatures increase, so does photosynthesis. Rubisco performs very well at high CO_2 concentrations, and warmer temperatures encourage plant growth in temperate and even in subpolar regions.

◆ As a result of increased global photosynthesis, CO_2 concentrations in the atmosphere drop, causing temperatures to fall and global photosynthesis to decline. Polar ice caps once again begin to increase in size, and climates worldwide become colder.

◆ The colder temperatures cause plants to die off again. The plants are degraded by bacteria, releasing considerable CO_2 back into the atmosphere through respiration. This, of course, causes a renewed greenhouse effect and the whole cycle begins again.

This graph shows the steady increase of atmospheric CO_2 (blue line) and a general warming trend (red line) since 1958.

---

cytosol. In fermentation, pyruvate is converted into either ethanol or lactate, depending on the organism. The process of fermentation is carried out by enzymes without an electron transport chain, and the only ATP produced is the small amount that came from glycolysis (see the *Conservation Biology* box on this page).

Section Review

1. Describe how organisms differ in their modes of nutrition.
2. Describe the relationship between photosynthesis and respiration.
3. How does substrate-level phosphorylation differ from oxidative phosphorylation?
4. How do respiration and fermentation differ?

Respiration

Respiration is usually described as including glycolysis, the Krebs cycle, and the electron transport chain with associated oxidative phosphorylation. We will describe glycolysis in connection with respiration because in plants and most other types of organisms respiration occurs more often than fermentation. Keep in mind, though, that glycolysis is necessary for both respiration and fermentation.

Glycolysis splits six-carbon sugars into two molecules of pyruvate

The term *glycolysis* (from the Greek *glyco*, "sweet" or "sugar," and *lysis*, "splitting") reflects the fact that the process involves splitting six-carbon sugars into two molecules of pyruvate, a three-carbon molecule (Figure 9.4).

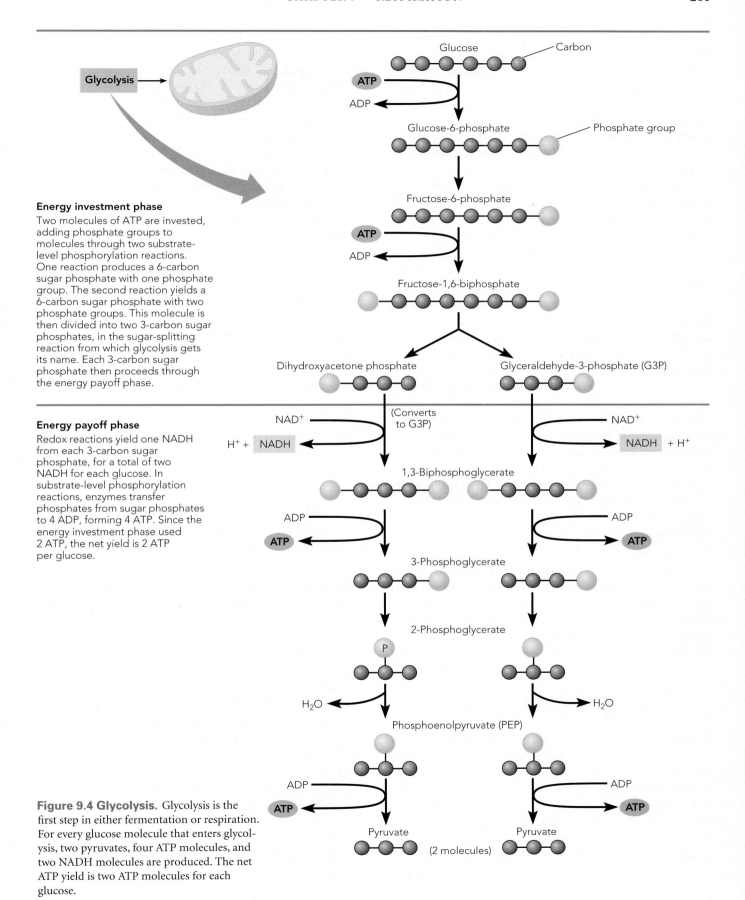

Glycolysis

Energy investment phase
Two molecules of ATP are invested, adding phosphate groups to molecules through two substrate-level phosphorylation reactions. One reaction produces a 6-carbon sugar phosphate with one phosphate group. The second reaction yields a 6-carbon sugar phosphate with two phosphate groups. This molecule is then divided into two 3-carbon sugar phosphates, in the sugar-splitting reaction from which glycolysis gets its name. Each 3-carbon sugar phosphate then proceeds through the energy payoff phase.

Energy payoff phase
Redox reactions yield one NADH from each 3-carbon sugar phosphate, for a total of two NADH for each glucose. In substrate-level phosphorylation reactions, enzymes transfer phosphates from sugar phosphates to 4 ADP, forming 4 ATP. Since the energy investment phase used 2 ATP, the net yield is 2 ATP per glucose.

Glucose — Carbon

Glucose-6-phosphate — Phosphate group

Fructose-6-phosphate

Fructose-1,6-biphosphate

Dihydroxyacetone phosphate   Glyceraldehyde-3-phosphate (G3P)

NAD^+ (Converts to G3P) NAD^+
H^+ + NADH   NADH + H^+

1,3-Biphosphoglycerate

3-Phosphoglycerate

2-Phosphoglycerate

H_2O H_2O

Phosphoenolpyruvate (PEP)

Pyruvate   Pyruvate
(2 molecules)

Figure 9.4 Glycolysis. Glycolysis is the first step in either fermentation or respiration. For every glucose molecule that enters glycolysis, two pyruvates, four ATP molecules, and two NADH molecules are produced. The net ATP yield is two ATP molecules for each glucose.

Glycolysis occurs in a series of ten reactions, each catalyzed by a specific enzyme.

The reactions of glycolysis resemble an assembly line, with the enzymes serving as metabolic control points. If the activity of one enzyme slows or stops as a result of being inhibited, so does the entire assembly line. A good example is phosphofructokinase, the enzyme that catalyzes the conversion of fructose-6-phosphate into fructose-1,6-bisphosphate. A molecule of ATP breaks down to supply the energy and the phosphate gained in this reaction. Inhibitors of phosphofructokinase include molecules such as ATP that indicate the cell has a good energy supply. Activators of phosphofructokinase include molecules such as ADP that indicate the cell may not have enough ATP.

For each molecule of glucose that enters glycolysis, two molecules of ATP are used in carrying out reactions and four molecules of ATP are produced, for a net gain of two ATP molecules. Meanwhile, two molecules of NADH are formed. The fact that glycolysis produces ATP and NADH tells us that two pyruvates contain fewer calories than one glucose, which can be confirmed with a calorimeter. Glycolysis generates an apparently meager amount of ATP and NADH for such a lengthy series of reactions. The reactions do, however, also produce intermediate compounds that are important sources of organic molecules for various cell processes. Glycolysis supplies sugars to make sucrose, the principal form of sugar transported from leaves to other parts of the plant. In addition, polysaccharides that help form cell walls originate from glycolysis. Glycolysis also supplies carbon frameworks for the synthesis of nucleic acids, some amino acids, and lignin, as well as glycerol used in the synthesis of lipids.

Scientists believe that the earliest organisms, which were prokaryotes that first evolved around 3.5 billion years ago, may have produced ATP solely through glycolysis. Respiration probably did not evolve until after significant oxygen accumulated in the atmosphere around 2.7 billion years ago (see the *Plants & People* box on page 204).

The Krebs cycle generates CO_2, NADH, $FADH_2$, and ATP

If oxygen is present, each pyruvate produced by glycolysis enters the mitochondrion and is transformed into a compound called *acetyl coenzyme A*, known more commonly as *acetyl CoA* (Figure 9.5). To form acetyl CoA, first a carbon is removed as CO_2 from pyruvate. The remaining two-carbon fragment is converted to become acetate, a process that generates one NADH. The acetate is then linked to a large cofactor called *coenzyme A*, forming acetyl CoA. Then acetyl CoA is broken down, with the

coenzyme A (CoA) being recycled for use with another pyruvate, while the two-carbon fragment enters the Krebs cycle. Accordingly, this conversion process forms the link between glycolysis and the Krebs cycle.

The Krebs cycle takes place in the mitochondrial matrix, the part of the mitochondrion that is interior to both mitochondrial membranes. The conversion of pyruvate to acetyl CoA and the Krebs cycle itself generate all of the CO_2 produced by respiration. Meanwhile, each turn of the Krebs cycle involves considerable energy transfer, producing one ATP, one $FADH_2$, and three NADH molecules. The cycle begins when a four-carbon molecule, oxaloacetate, combines with a two-carbon fragment from acetyl CoA to make citrate. Since citrate is the first compound formed, the Krebs cycle is sometimes called the *citric acid cycle*. The resulting six-carbon citrate compound is converted to isocitrate. In each of the next two conversions, a carbon leaves the cycle in the form of CO_2 and one NADH is generated. The remaining reactions involve a series of four-carbon compounds and yield 1 ATP, 1 $FADH_2$, and 1 NADH. The cycle is completed with the regeneration of oxaloacetate, which can accept another two-carbon fragment from an acetyl CoA to begin the cycle again. The NADH and $FADH_2$ molecules produced by the Krebs cycle supply energy-rich electrons to the next phase of respiration: the electron transport chain and oxidative phosphorylation.

The electron transport chain and oxidative phosphorylation transfer energy from the energy-rich electrons of NADH and $FADH_2$ to ATP

The synthesis of ATP in the inner mitochondrial membrane depends on the electron transport chain. The oxidative phosphorylation of ADP to ATP is powered by the energy supplied by chemiosmosis—the flow of hydrogen ions (H^+) across the membrane. These ions have been pumped out by the electron transport chain. Some of them move back across the membrane in association with the enzyme ATP synthase, which uses this chemiosmotic movement of hydrogen ions as a source of energy to synthesize ATP. The process is quite similar to ATP synthesis in the light reactions of photosynthesis, which also involve an electron transport chain, chemiosmosis, and ATP synthase. As noted previously, though, in respiration the flow of electrons is caused by the oxidation of NADH (the removal of electrons from NADH) rather than by light energy. That is why the ATP synthesis during this phase of respiration is called *oxidative phosphorylation*.

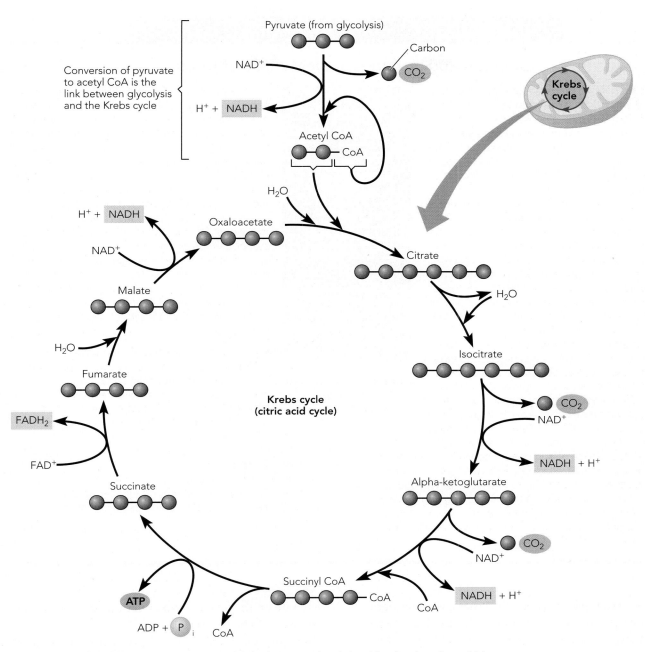

Figure 9.5 The Krebs cycle. The Krebs cycle, also known as the *citric acid cycle,* takes place within the mitochondrial matrix, the interior region of each mitochondrion. The conversion of pyruvate to acetyl CoA provides the link between glycolysis and the Krebs cycle. As acetyl CoA is broken down, a two-carbon fragment enters the Krebs cycle, combining with a four-carbon compound to produce citrate. Although the Krebs cycle generates only a small amount of ATP, it plays a key role in supplying the electron carriers NADH and $FADH_2$ to the electron transport chain, which makes possible the high yield of ATP from oxidative phosphorylation. Each turn of the Krebs cycle produces one ATP, one $FADH_2$, and three NADH molecules. Since glycolysis breaks each glucose into two pyruvates, the yield of the Krebs cycle per molecule of glucose is two ATP, two $FADH_2$, and six NADH molecules.

As in the light reactions, electrons pass from one electron carrier to another. Most of these electron carriers are protein complexes, with each electron carrier attracting the electron more strongly than the previous carrier. In this way, oxidation/reduction (redox) reactions move electrons along the electron transport chain while energy is released and transferred. The energy released by the electron transport chain pumps hydrogen ions into the

PLANTS & PEOPLE

Sucrose and Fructose: Sweeteners of Choice

When it comes to sweeteners, sucrose and fructose are used far more often in foods than glucose. After all, we sweeten coffee and almost everything else with sucrose (a disaccharide composed of fructose linked to glucose), and the most common sweetener in prepared foods is high-fructose corn syrup.

Sucrose, or table sugar, is so common partly because it tastes quite sweet as sugars go. The main reason, though, is that sucrose is the transport form of sugar in plants. Glucose is made in the chloroplasts and synthesized into sucrose for transport in the phloem. Sucrose is isolated from plants such as sugarcane and sugar beets, where it is found in high concentrations. Selection by plant breeders has resulted in varieties of the plants with high levels of sugar.

High-fructose corn syrup is frequently found in prepared foods because it is produced at low cost from corn kernels, which are rich in starch. The starch is converted by enzymes to glucose, which is then made into fructose. To our taste, fructose is considerably sweeter than glucose. On a weight basis, HFCS is 75% sweeter than sucrose. A 12-ounce soda sweetened with HFCS would require 10 teaspoons of sucrose for equivalent sweetness. HFCS consists of 14% fructose, 43% dextrose (glucose), 31% disaccharides, and 12% other products. The choice between sucrose and HFCS as a sweetening agent is frequently driven by sweetness considerations as well as by the market price and availability of corn, sugarcane, and sugar beets.

High-fructose corn syrup is the major sweetener in most processed foods.

In general, humans and other animals should consume only moderate levels of these sweeteners. All natural sweeteners add significant numbers of calories to a diet that contains them. Dangers of diets high in sweeteners such as sucrose or HFCS include diabetes, heart problems, and cholesterol buildup. In a United States Department of Agriculture (USDA) study, rats fed diets high in fructose and low in copper begin dying in five weeks instead of living a normal two years. Human studies on related diets were stopped when some subjects developed heart abnormalities. In general, diets high in sweeteners require relatively high levels of minerals such as magnesium, chromium, and copper to prevent known health risks.

Artificial, non-caloric sweeteners have come into use during the last century. They include

- aspartame, marketed as Nutrasweet, which is a dipeptide of two amino acids. Aspartame is 200 times sweeter than sugar.
- saccharin, which was discovered accidentally in 1879 in research designed to find food preservatives. Saccharin is 300 times sweeter than sugar.
- sorbitol, which is used mainly in pharmaceuticals, is 50% as sweet as sugar.
- sucralose, marketed as Splenda, is a modified form of glucose. Sucralose is 600 times as sweet as sugar.
- Acesulfame potassium is 130 times as sweet as sugar.

All artificial sweeteners are associated with health risks for some people who consume them.

---

intermembrane space between the inner and outer mitochondrial membranes (see Figure 9.6). The charge separation between the hydrogen ions outside the inner membrane and the electrons in the electron transport chain forms a potential energy gradient—a kind of battery that is measured as a pH difference between the solutions on each side of the membrane. The movement of the hydrogen ions back across the membrane and through ATP synthase then powers ATP synthesis. For every three hydrogen ions that move through ATP synthase, one ATP is synthesized.

The last electron carrier in an electron transport chain is known as the *terminal electron acceptor*. In the case of respiration, the terminal electron acceptor is an oxygen atom from the air, which is why organisms that carry out respiration require oxygen. The electrons and hydrogen

ions unite on the inside of the inner membrane with O_2 from the air to become H_2O.

In theory, the energy in each NADH gives rise to three molecules of ATP, while the energy of each $FADH_2$ gives rise to two molecules of ATP because the electrons in $FADH_2$ carry less energy than those in NADH. Actually, the number of ATP molecules for each NADH could be higher or lower, depending on whether the NADH comes from glycolysis or from the Krebs cycle and also on how much ATP is already present in the cell.

ATP synthase has been called "a molecular machine" and is composed of three parts: a cylindrical rotor, a rod or "stalk," and a knob, with each part consisting of protein subunits. The cylindrical rotor spans the membrane and surrounds a channel through which the hydrogen ions flow. In the center of the channel is a rod connecting the rotor to the knob. The knob, which protrudes into the mitochondrial matrix, contains sites where inorganic phosphate (P_i) is joined to ADP to make ATP. The most interesting feature of ATP synthase is that both the cylindrical rotor and the rod spin, activating sites in the knob where ATP is synthesized.

The energy yield from respiration is high

Figure 9.7 summarizes the estimated energy yield from glycolysis, the Krebs cycle, and oxidative phosphorylation for one molecule of glucose: 36 molecules of ATP. This is an ideal value based on the assumption that chemiosmotic pumping of hydrogen ions from one NADH and associated hydrogen ions will yield three molecules of ATP, and that each $FADH_2$ will yield 2 molecules of ATP. As noted

earlier, the actual value can be higher or lower and would be expected to vary from one type of cell to the other.

The total yield of ATP from one glucose is sometimes calculated as 38 molecules. However, that does not account for the fact that two ATP molecules must be used in shuttling electrons across the mitochondrial membrane—specifically the electrons from the NADH produced in glycolysis. This shuttling is necessary because the inner mitochondrial membrane is impermeable to NADH. Subtracting these two ATP molecules gives the net ATP yield of 36.

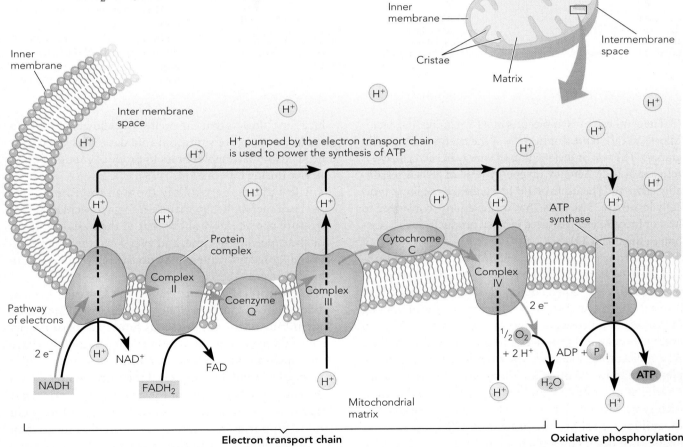

Figure 9.6 The electron transport chain and oxidative phosphorylation. While the Krebs cycle occurs in the mitochondrial matrix, the electron transport chain and oxidative phosphorylation occur within the inner mitochondrial membrane. Actually, many copies of the chain occur in the inner membrane, made possible by the increased surface area provided by fingerlike projections called *cristae*. NADH and $FADH_2$ enter each chain, which consists mainly of electron-carrying protein complexes. Complex I removes high-energy electrons and associated protons from NADH. Complex II removes high-energy electrons and associated protons from $FADH_2$. Complex III transfers the electrons to Complex IV, where they are combined with oxygen to make water. At complexes I, III, and IV, the energy released from the electrons pumps hydrogen ions into the intermembrane space. Chemiosmosis—the flow of hydrogen ions back across the membrane and through ATP synthase—provides the energy to synthesize ATP by oxidative phosphorylation.

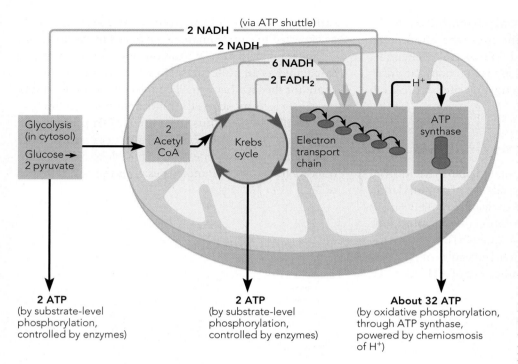

2 NADH (via ATP shuttle)
2 NADH
6 NADH
2 FADH₂

Glycolysis (in cytosol)
Glucose → 2 pyruvate

2 Acetyl CoA

Krebs cycle

Electron transport chain

H⁺

ATP synthase

2 ATP
(by substrate-level phosphorylation, controlled by enzymes)

2 ATP
(by substrate-level phosphorylation, controlled by enzymes)

About 32 ATP
(by oxidative phosphorylation, through ATP synthase, powered by chemiosmosis of H⁺)

Figure 9.7 Summary of estimated maximum ATP production in respiration. The numbers reflect the estimated maximum ATP yield per molecule of glucose. The estimates are based on the energy in each NADH being converted to 3 ATP and the energy in each $FADH_2$ being converted to 2 ATP. Since each glucose is converted into two pyruvates, two turns of the Krebs cycle are involved. Glycolysis produces 2 ATP, the Krebs cycle produces another 2 ATP, and the yield from the electron transport chain and oxidative phosphorylation is 32 ATP. At first it may appear that the latter figure should be 34 ATP. After all, the diagram shows 10 NADH and 2 $FADH_2$ entering the electron transport chain. However, we need to subtract the 2 ATP used in shuttling electrons from the NADH molecules produced in glycolysis. Therefore, the net maximum yield is 36 ATP.

The synthesis of 36 molecules of ATP requires 262.8 kilocalories (kcal), representing 38% of the energy contained in glucose. The rest of the 686 kcal of energy in each glucose is released as heat. Actually, an energy yield of 38% is reasonably efficient. The useful yield in a gasoline engine is typically less than 25%, with 75% of the energy converted to heat or incompletely oxidized exhaust products, such as carbon monoxide (CO).

The synthesis and use of ATP in living cells is an undertaking of considerable magnitude. The average person who is neither a couch potato nor a lumberjack uses around 2,000 kcal per day, equivalent to about 0.45 kg (1 pound) of glucose per day. Calculations reveal that an average human cell produces and uses around 10 million ATP molecules per second. The overall metabolic rates of plants are 10 to 100 times less than those of most animals, but the number of ATP molecules produced and used in each typical living plant cell every second is still often in the millions. In short, the process of ATP synthesis and breakdown occurs on a phenomenal scale in living cells.

In some plants, the electron transport chain can generate excess heat

In some plants, an enzyme called *alternative oxidase* moves electrons from NADH to O_2 without leading to oxidative phosphorylation. When this alternative oxidase moves electrons, no ATP results, and the energy is all released as heat. This type of mechanism is used by plants to pro-

duce "hot" flowers that can melt snow, allowing the plants to take advantage of sunny but cold days. This is the same mechanism that enables bears to produce enough heat to survive during hibernation.

A few plants, particularly those in the Family Araceae, can metabolically maintain their flowers at a temperature considerably above that of their environment, for short periods of time, and even keep their temperatures at constant values. Why do they devote energy to this endeavor?

Many tropical plants in the Family Araceae—such as philodendrons, caladiums, elephant ears, dieffenbachias, and anthuriums—are grown as houseplants or in gardens in warm regions. These plants frequently have foul-smelling flowers that attract insects such as flies and beetles. A notable example is the corpse flower, which you saw at the beginning of the chapter. This extremely large Indonesian flower grows up to 3.7 m (12 feet) tall and is supported by a fleshy root weighing more than 46 kg (90 pounds). The corpse flower's common name comes from its aroma. The heating of the flower parts causes a large amount of aromatic molecules to evaporate into the air, thereby more effectively attracting pollinators. Presumably, plants with stronger odors attracted more flies and eventually produced more seeds, initiating the next generation. In this way, selection for flowers that were warmer and had a stronger odor occurred over successive generations (see *The Intriguing World of Plants* box on page 207).

THE INTRIGUING WORLD OF PLANTS

Skunk Cabbage

Skunk cabbage (*Lysichiton americanum*)
melting surrounding snow.

Among the examples of "hot plants" in North America are several species of skunk cabbage, such as *Symplocarpus foetidus,* found in the eastern United States, and *Lysichiton americanum,* found in the west. These members of the Araceae family bloom in January or February, and the heat produced by the flower bud can raise its temperature as high as 16°C (60°F). The bud frequently melts surrounding snow and readily survives many nights with temperatures considerably below freezing. The heat released also activates fragrant molecules from the flowers, giving the plant its distinctive name. Skunk cabbage maintains its high floral temperatures by converting starch, stored in a large fleshy root, into glucose or CO_2. The advantage of a skunk cabbage's high metabolism continues to be the subject of debate. Not many pollinating insects are out and about in January and February. On the other hand, pollinating insects are often found in the swamps of eastern and western Canada and the United States, where skunk cabbage grows. Any "early bird" flies would profit from using skunk cabbage as a food source and as a source of life-sustaining warmth, while the plants would profit by having pollinating organisms early in the season. Beginning growth so early in the season also gives the plant weeks of direct sunlight and, therefore, of photosynthesis without being shaded by other plants.

Some have suggested that skunk cabbage simply retains an adaptation that was useful in the Tropics (where the plant's strong odor would increase its chances of being pollinated) but has no use in temperate regions. This seems unlikely because the various species of skunk cabbage devote considerable energy to producing heat. A variety that saved the energy probably would have rapidly multiplied to become the dominant form in the population.

Plants, unlike animals, can make fatty acids into glucose

Animals can obtain energy from several sources. Starches and other carbohydrates break down or change into glucose that is metabolized by respiration. Fats break down to acetyl CoA that enters the Krebs cycle. Proteins break down to amino acids that enter the Krebs cycle at various places. Most organisms, including humans, can metabolize fats to glycerol and acetyl CoA units that can enter glycolysis and the Krebs cycle to produce energy (Figure 9.8). Hibernating animals, for example, have a sophisticated hormonal control system to regulate this process. Thus, organisms can store energy as fat, when extra food is available, and use fat to obtain ATP when food is in short supply. However, most animals—including all mammals—cannot convert fatty acids into glucose.

In contrast, plants and some bacteria can break down fatty acids into acetyl CoA, which is then used to make glucose. In this way, plants are more versatile than animals because they can use fatty acids as either a source of energy or as a source of glucose, which is water-soluble and readily convertible into forms that can be moved throughout the plant. The ability of plants to use fatty acids for energy or for structural molecules may explain why so many plants have oil as a storage compound used to nourish germinating seeds.

Plants can convert fatty acids into sugars by virtue of the glyoxylate cycle, which occurs partly in microbodies called *glyoxysomes* (see Chapter 2) and partly in mitochondria. Basically, the glyoxylate cycle is nothing more than the Krebs cycle with two additional enzymes that bypass the two steps of the Krebs cycle that release some carbon as CO_2. Since these carbons are not lost, they are available to synthesize glucose.

Section Review

1. Describe the relationship between glycolysis and the Krebs cycle.
2. Explain the roles of the electron transport chain and ATP synthase in the production of ATP.
3. Summarize the products of glycolysis, the Krebs cycle, the electron transport chain, and oxidative phosphorylation.

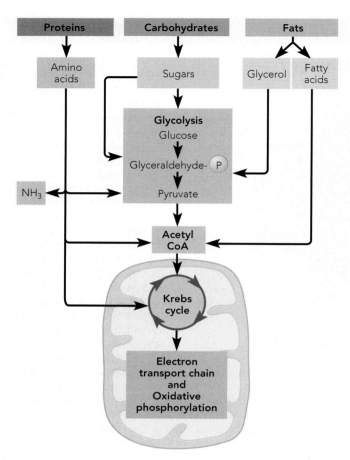

Figure 9.8 Substrates other than glucose can be used in respiration. Proteins, carbohydrates, and fats all feed into respiration at various locations.

Fermentation

Before photosynthesis evolved, respiration was not possible because of lack of oxygen. In today's world, anaerobic environments still occur where oxygen is excluded or when oxygen is used more rapidly than it can be replaced. Under such conditions, fermentation can occur. Some microorganisms, known as *obligate anaerobes,* require anaerobic conditions to survive. Others, known as facultative anaerobes, have the ability (faculty) to carry out respiration if oxygen is present or to carry out fermentation if oxygen is absent.

In the absence of oxygen, pyruvate produced by glycolysis is converted to ethanol or lactate

Fermentation converts pyruvate to other organic molecules, such as ethanol or lactate, while transferring electrons to NAD^+ (Figure 9.9). Because the NAD^+ concentration in living cells is very low, it must be rapidly regenerated so that glycolysis can continue and the cell can thereby obtain ATP. In the absence of O_2, the electron transport chain produces no NAD^+, so the regeneration of NADH to NAD^+ becomes the purpose of fermentation. In the early days of life on Earth, before photosynthesis had evolved, the atmosphere contained very little if any free oxygen, so glycolysis combined with fermentation was the only source of ATP. Living cells used primitive forms of glycolysis and fermentation to break down sugars and other molecules produced spontaneously in shallow, ancient oceans. Today, fermentation is restricted to certain environments where specialized bacteria live and to certain times in the life of all cells, but it plays a role in physiology, commercial uses, and disease. For example, yeasts are fungi that are used in the production of beer and wine by fermentation. (The term *fermentation* comes from the Latin word for yeast, *fermentum.*) Anaerobic bacteria of the genus *Clostridium* cause diseases such as gangrene and tetanus.

Most plant cells produce ethanol if deprived of oxygen; however, some species produce lactate, malate, glycerol, or both ethanol and lactate. Plants encounter oxygen deprivation when their roots are flooded because oxygen diffuses three million times more slowly in pure water than in air. In bogs and swamps, where many organisms compete for a limited supply of oxygen, seeds sometimes undergo oxygen deprivation during the early phases of germination. A lack of oxygen promotes grass seed germination, probably by stimulating synthesis of the plant hormone ethylene.

Animal cells typically cannot carry out alcoholic fermentation. If they could, then humans could get drunk simply by holding their breath. Instead, when oxygen is scarce, a process called *lactic acid fermentation* converts pyruvate to lactate. Typically, lactic acid fermentation occurs when the animal uses ATP to move muscles, which may become sore as a result of the buildup of lactate. If the circulatory system cannot keep up by supplying enough oxygen for oxidative phosphorylation, respiration is inhibited by lack of oxygen, but the organism still continues to produce pyruvate and ATP by glycolysis. In fact, both respiration and fermentation can occur in an organism at the same time. Unlike alcohol fermentation, lactic acid fermentation is reversible. When oxygen is again available, lactate converts to pyruvate, and respiration proceeds.

Some important industries rely on fermentation

The ability of yeast, a facultative anaerobe, to metabolize pyruvate into ethanol gave rise to the brewing and baking industries (Figure 9.10). In winemaking, sugary fruit juice, mixed with yeast cells, ferments until the alcohol concentration reaches 12%. At this point, the yeast cells die as a result of the alcohol they have produced. Any alcoholic beverage with a higher concentration of ethanol is fortified, meaning that alcohol that is concentrated by distillation was added to make the final product. If oxygen enters the process before completion, bacteria from the air rapidly convert ethanol to acetic acid. A solution of 9% acetic acid is vinegar. To make beer, wheat or some other starch-containing grain is germinated long enough to break down some of its starch to maltose, which can serve as food for yeast. When yeast is added, alcohol fermentation begins.

The fermentation process that produces ethanol also yields CO_2, which in turn causes the solution to bubble and appear active. In winemaking the CO_2 usually dissipates, whereas in the brewing of beer the CO_2 remains.

In the baking industry, yeast is mixed with a starchy, sugary dough that provides an anaerobic environment. The CO_2 produced by glycolysis and fermentation of sugar causes the dough to rise, and the alcohol that is produced evaporates during the baking process. People who say they love to be in the kitchen when bread bakes may be responding to the alcohol-enhanced aroma of the bread.

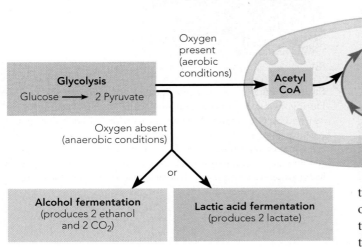

(a)

Fermentation has a low energy yield compared to that of respiration

Keep in mind that in fermentation, the conversion of pyruvate to ethanol or to lactate does not produce any additional ATP. The only ATP comes from glycolysis, which produces two ATP molecules per glucose. Each ATP has 7.3 kcal of energy, while glucose has 686 kcal. The energy yield of glycolysis plus fermentation is therefore 14.6/686, or just over 2%.

In contrast, respiration can generate a maximum of about 38 ATPs per glucose, for an energy yield of approximately 40%. One of the reasons you do not find anaerobic organisms dancing or playing basketball is that they do not have the energy and could not get it without consuming massive amounts of glucose or other foods.

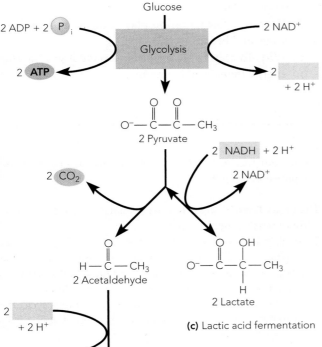

(c) Lactic acid fermentation

Figure 9.9 Fermentation. (**a**) In the absence of oxygen, the Krebs cycle and the electron transport chain cannot function. Instead, pyruvate is converted to ethanol or to lactate in the cytosol. The production of both ethanol and lactate in fermentation serves to regenerate NAD^+, enabling the limited ATP production of glycolysis to continue. (**b**) Alcohol fermentation occurs in yeasts, most plant cells, and some bacteria. (**c**) Lactic acid fermentation occurs in a variety of cells in many types of organisms, including animal muscle cells. Lactic acid fermentation by some fungi and bacteria is used to make cheese and yogurt.

(b) Alcohol fermentation

(a)

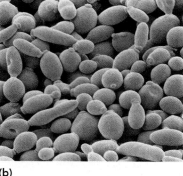

(b)

Figure 9.10 Some commercial uses of fermentation. During the making of beer and wine, yeast converts sugar to pyruvate and then to ethanol. In wine, the CO_2 produced is allowed to escape, whereas in beer it is retained in the final product. **(a)** Modern wineries and breweries such as this microbrewery often use stainless steel containers. **(b)** This scanning electron microscope image of yeast *(Saccharomyces cerevisiae)* shows yeast in the process of "budding" or reproducing.

Section Review

1. What is fermentation, and how does it differ from respiration?
2. How does fermentation play a role in making beer, wine, and bread?
3. Why do anaerobic organisms have less available energy than aerobic organisms?

SUMMARY

An Overview of Nutrition

All living organisms need sources of energy and carbon (p. 197)

Most autotrophs are photoautotrophs, obtaining energy from light and carbon from CO_2. A few are chemoautotrophs, obtaining energy from inorganic compounds. Most heterotrophs are chemoheterotrophs, obtaining both energy and carbon from organic compounds, but a few are photoheterotrophs, obtaining energy from light instead.

Plants use photosynthesis to store light energy in sugars and use respiration to transfer the energy from sugars to ATP (pp. 197–198)

In respiration, organisms break down sugars and other organic compounds to make ATP. During respiration, ATP syn-

thesis occurs by substrate-level phosphorylation and oxidative phosphorylation.

The breakdown of sugar to release energy can occur with or without oxygen (pp. 198–200)

Through either respiration or fermentation, all living cells break down glucose to CO_2 and H_2O, yielding ATP. Both fermentation and respiration use glycolysis to break glucose into pyruvate. Under aerobic conditions, respiration occurs, which involves the breakdown of pyruvate to acetyl CoA, the Krebs cycle, the electron transport chain, and oxidative phosphorylation.

Respiration

Glycolysis splits six-carbon sugars into two molecules of pyruvate (pp. 200–202)

Glycolysis consists of ten reactions that convert a six-carbon sugar into two molecules of pyruvate. From one glucose, glycolysis produces two molecules of ATP and two molecules of

NADH. Intermediate compounds serve as reactants to form various compounds.

The Krebs cycle generates CO₂, NADH, FADH₂, and ATP (p. 202)

As pyruvate leaves glycolysis, it is converted to two molecules of acetyl CoA and two molecules of CO_2. In the Krebs cycle, the acetyl groups are converted to CO_2. In two turns of the cycle, energy from one glucose is transferred to two molecules of ATP, while energy-rich electrons and accompanying hydrogens are incorporated into six molecules of NADH and two molecules of $FADH_2$.

The electron transport chain and oxidative phosphorylation transfer energy from the energy-rich electrons of NADH and FADH₂ to ATP (pp. 202–205)

Energy released from the electron transport chain moves hydrogen ions across a membrane. This chemiosmotic coupling creates a charge difference and pH difference across the membrane, which functions as a battery to power oxidative phosphorylation by ATP synthase. The electrons from the electron transport chain, along with associated hydrogen ions, are combined with oxygen to produce water.

The energy yield from respiration is high (pp. 205–206)

The net maximum energy yield from one glucose is 36 molecules of ATP, representing about 40% of the energy in glucose. The remaining energy is released as heat.

In some plants, the electron transport chain can generate excess heat (pp. 206–207)

Using an alternative oxidase, electrons can bypass the electron transport chain, resulting in almost all of the stored energy being released as heat.

Plants, unlike animals, can make fatty acids into glucose (p. 207)

Plants and animals can convert fatty acids into acetyl CoA, which is metabolized to CO_2 in the Krebs cycle. Plants can also break down fatty acids into acetyl CoA, which is used to make glucose, without production of CO_2.

Fermentation

In the absence of oxygen, pyruvate produced by glycolysis is converted to ethanol or lactate (p. 208)

Fermentation converts pyruvate to other organic molecules, such as ethanol or lactate, while transferring electrons from NAD^+ to NADH.

Some important industries rely on fermentation (pp. 208–209)

The baking, brewing, and winemaking industries are based on the ability of yeast to ferment sugars into ethanol and CO_2.

Fermentation has a low energy yield compared to that of respiration (pp. 209–210)

The ATP yield of fermentation per molecule of glucose consists only of the two ATP molecules produced by glycolysis, or about 2% of the energy in glucose.

Review Questions

1. What is the difference between autotrophs and heterotrophs?
2. What is the net result of the processes of photosynthesis and respiration?
3. Distinguish between the three types of ATP synthesis.
4. What is the function of ATP and of NADH in cells?
5. What are the end products of glycolysis?
6. What enters the Krebs cycle and what are the end products?
7. Compare and contrast glycolysis and the Krebs cycle.
8. Explain how oxidative phosphorylation is both separate from and dependent upon the electron transport chain.
9. Describe in general how glycolysis, the Krebs cycle, the electron transport chain, and oxidative phosphorylation are related.
10. Describe ATP synthase and what it does.
11. How does fermentation differ from respiration in terms of process and the amount of ATP yielded?
12. What can plants do with fat that animals cannot do? Explain why.

Questions for Thought and Discussion

1. If you hold your breath, what happens to the glucose molecules in your cells that are being used for energy?
2. Which do you think evolved first, photosynthesis or respiration? Explain.
3. Why do most eukaryotes die if oxygen is cut off? Why are they unable to survive using fermentation?
4. Plants produce ATP in photosynthesis, so why do they need to carry out respiration?
5. When ATP is broken down some energy is released as heat. Does this mean that the temperature of a plant will always be somewhat higher than the outside temperature? Explain.
6. Make a series of diagrams to illustrate the process of aerobic respiration in a plant. Your diagrams should be, in order: (a) an entire plant; (b) an individual plant cell; (c) a close-up of a portion of the cytoplasm of a plant cell, showing an individual mitochondrion; and (d) a close-up of a portion of a mitochondrion. In each diagram, draw and label the individual processes and reactions to a degree of detail appropriate to the diagram.

Evolution Connection

Biologists believe that the reactions of glycolysis and fermentation evolved early on in the history of life on Earth and that the Krebs cycle was added later. Explain why this hypothesis is reasonable. Is there any evidence to support it?

To Learn More

Visit The Botany Place Website at www.thebotanyplace.com for quizzes, exercises, and Web links to new and interesting information related to this chapter.

Gardenway Staff and P. Hobson. *Making Cheese, Butter, and Yogurt.* North Adams, MA: Storey Books, 1997. This book is packed with information about cheese and has a variety of recipes.

Mathews, C. K., Van Holde, K. E., and K. G. Ahern. *Biochemistry.* San Francisco: Benjamin Cummings, 2000. This excellent text contains detailed information on respiration.

Robbins, Louise. *Louis Pasteur: And the Hidden World of Microbes.* New York: Oxford Portraits in Science, 2001. This book examines Pasteur's experiments of microbes in fermentation and various diseases, as well as the changes in medicine and public perception of disease that resulted from his work.

Long-stemmed water lilies.

Molecular Movement Across Membranes

Diffusion is the spontaneous movement of molecules down a concentration gradient

Facilitated diffusion and active transport use proteins to assist in movement across membranes

Exocytosis and endocytosis transport large molecules

Osmosis is the movement of water across a selectively permeable membrane

In plant cell growth, the osmotic potential inside the cell interacts with pressure generated by the cell wall

Movement and Uptake of Water and Solutes in Plants

Water evaporation from leaves pulls water through the xylem from the roots

Stomata control gas exchange and water loss for the plant

Sugars and other organic molecules move from leaves to roots in the phloem

Soil, Minerals, and Plant Nutrition

Soil is made of ground-up particles of rocks surrounded by negative charges that bind water and minerals

Plants require 17 essential elements, most of which are obtained from soil

Soil particles bind water and mineral ions

Bacteria in the soil make nitrogen available to plants

What are plants made of? Today we tend to answer that question by mentioning molecules such as DNA and enzymes, sugars and amino acids, and hormones that a plant makes out of simpler inorganic components. In any case, our answer involves the system of chemistry we have studied. Human understanding of chemistry used to be much simpler, back when everything was thought to consist of only four elements. The Greek philosopher Empedocles (around 450 B.C.) and later Aristotle (384–322 B.C.) believed that everything in the universe consisted of various combinations of earth, air, fire, and water. Some Greek philosophers added a fifth element, called *quintessence,* which characterized the celestial as opposed to the terrestrial realm.

In truth, some reasonably good science was done under the earth, air, fire, and water system of what we will call the "old" chemistry. Around 1600, the Belgian chemist Jan Baptista van Helmont conducted an experiment to determine the relative contributions of earth and water to plant growth. He planted a 2.3 kilogram (5 pound) willow tree in a tub containing 90.9 kg (200 pounds) of oven-baked, dry soil. For five years, he watered and cared for the tree. At the end of that time, the tree weighed 76.9 kg (167 pounds), but the soil had lost only 57 mg (2 ounces). Observing that the tree had absorbed large quantities of water but only a small amount of soil, van Helmont concluded that it was made almost completely of water. Indeed, even modern chemistry would concede that water is the most common molecule found in a plant, accounting for perhaps 60% of its weight.

In 1699, an Englishman named John Woodward carried out an experiment in London using spearmint plants, arriving at a conclusion distinctly different from van Helmont's. He placed plants in four water sources: rainwater, Thames River water, Hyde Park sewer water, and Hyde Park sewer water plus garden soil. After 77 days he collected the following data on the weight gain of the four groups of plants:

Water source	Weight gain [in grains, 1 grain = 64.8 milligrams (mg) = 0.002 ounce (oz)]
Rain	17.5
Thames River	26.0
Hyde Park sewer	139.0
Hyde Park sewer and garden soil	284.0

Woodward observed that plant growth increased in proportion to the amount of soil or silt in each water source. He concluded that plants are made primarily from earth. We now know that minerals from the soil are essential for plant growth, but actually they contribute only a small percentage of a plant's weight.

Farmers have known for centuries that plant growth improved when animal manure was added to the soil. In the 1700s, they began to take note of the fact that various naturally occurring mineral deposits worked as well in the fields as manure. For example, marl—which we know as lime or calcium carbonate ($CaCO_3$)—was known to be useful if worked into the soil.

Farmers recognized also that saltpeter (potassium nitrate, KNO_3) from decaying plant and animal remains could help plants grow. Around 1731, an English agriculturalist named Jethro Tull asserted that saltpeter was a fifth element in plants. Tull was probably the first investigator to propose that the old four-element chemistry could not adequately describe the composition of plants. He also believed that plant roots had tiny mouths that they used to eat the earth, and that plowing the earth into bite-sized pieces would make it easier for plants to consume.

Around this same time, scientists began to identify the chemical elements recognized by chemists today. For example, in 1771, Joseph Priestley determined that plants produced something that enabled candles to burn and animals to survive. He had discovered oxygen. Scientists continued to define individual elements of the new chemistry, and in 1866, Demitri Mendeleev published the first periodic table, listing some 46 elements. The old chemistry was officially dead.

We now know that plants require at least 17 elements to compose their biochemical structures. Carbon comes from the air as CO_2, plants can obtain the oxygen and hydrogen they need by splitting water molecules, and other elements come from the soil. Since water and mineral absorption occurs in the roots and photosynthesis takes place in the leaves, plants need a system of transport to move molecules to where they are needed. In this chapter, we will look at how inorganic and organic molecules are transported between cells and throughout the plant as a whole.

Molecular Movement Across Membranes

Plant cells have several ways to import and export molecules that are important for cellular growth and development. These molecules include both water and various **solutes,** molecules that dissolve in water. Some solutes used by plants are mineral ions, such as potassium and phosphorus, found in the soil. Others are organic molecules such as sugars that are synthesized by plants in particular cells and needed by cells throughout the plant.

Molecules can move through the interior of cells or within cell walls. Movement through the interior of cells is known as **symplastic transport** (from the Greek *sym,* "with") because molecules move within the cytoplasm. The continuum of cytoplasm between cells, joined by the channels called *plasmodesmata,* is known as the *symplast* of a plant. The plasma membrane is selectively permeable, governing entry of molecules into the cytoplasm of each cell, often restricting movement of some molecules while enhancing movement of others.

The continuum of cell walls throughout the plant is known as the *apoplast* (from the Greek *apo,* "away from").

The movement of molecules within the cell walls is called **apoplastic transport,** in which molecules pass around ("away from") the cytoplasm of cells. Apoplastic transport can be rapid because the molecules are not being filtered through the plasma membrane and cytoplasm of cells, but the cells have no control over the types of molecules being transported.

The movement of a molecule through a plant typically involves both apoplastic transport and symplastic transport. We will now look at the types of symplastic transport across plasma membranes: diffusion, facilitated diffusion, active transport, the movement of large molecules by exocytosis and endocytosis, and osmosis.

Diffusion is the spontaneous movement of molecules down a concentration gradient

If you place a drop of red food coloring in one end of a filled bathtub and a drop of blue food coloring in the other end, the molecules in each drop will spontaneously spread out until the concentration of each food coloring is uniform throughout the tub. The tendency of molecules to spread out spontaneously within the available space is called **diffusion** (Figure 10.1a). In diffusion,

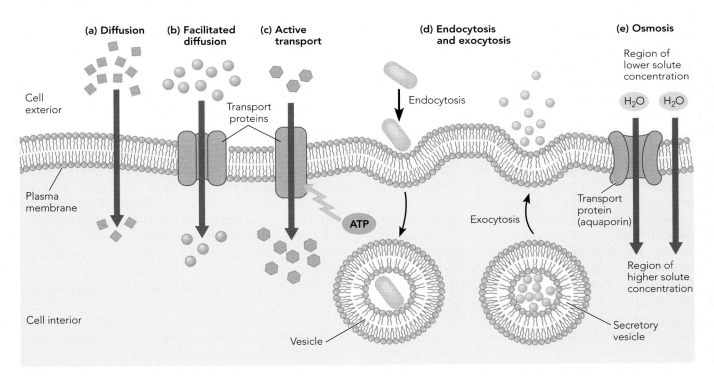

Figure 10.1 Transport of molecules across membranes.
(a) In diffusion, a solute moves spontaneously to a region of lower solute concentration. (b) In facilitated diffusion, transport proteins help solutes diffuse more rapidly through the membrane. (c) Unlike diffusion and facilitated diffusion, active transport requires energy, as transport proteins move solutes "uphill" to a region of higher solute concentration. (d) Vesicles move large molecules into a cell (endocytosis) or out of a cell (exocytosis). (e) Water movement across a membrane, called *osmosis,* occurs with or without transport proteins. Water moves to a region of higher solute concentration (lower water concentration).

solutes gradually move through a **concentration gradient,** a transition between regions of higher and lower concentrations. In diffusion, the movement is *down* a concentration gradient, from a region of higher concentration to a region of lower concentration. Such movement leads to an **equilibrium**—a random, equal distribution. Diffusion is passive transport because it does not require energy. It can occur in open solutions or between two solutions separated by a membrane, particularly for lipid-soluble substances that move readily through membranes.

Facilitated diffusion and active transport use proteins to assist in movement across membranes

For many water-soluble molecules, transport proteins assist diffusion through the plasma membrane, a process called **facilitated diffusion** (Figure 10.1b). The transport proteins are typically embedded in the plasma membrane. When a transport protein binds with a solute, the protein changes shape in a manner that delivers the solute to the other side of the membrane. Facilitated diffusion is similar to regular diffusion because the solute moves from a region of higher solute concentration to a region of lower solute concentration. Also, as with all types of diffusion, the transport is passive, which means that it does not involve any input of energy.

Some transport proteins appear to act alone. Others associate to form channels in the plasma membrane and can be shaped to either block or open a so-called *gated channel,* thereby regulating the transport of solutes. The diameter of the channel regulates the size of molecules that can move through from one side of the membrane to the other. Specific binding sites also control whether specific solutes can enter the channel. The channels can be opened or closed by specific solutes to be transported or by other molecules that control diffusion.

Sometimes transport across a membrane moves *up* a concentration gradient, from a region of lower concentration to a region of higher concentration. Since this transport requires energy to move up the gradient, it is not passive, like diffusion, but is instead called **active transport** (Figure 10.1c). The energy for most active transport is supplied by ATP or by the release of energy in electron transport chains. Active transport can involve a single protein or two proteins. For example, a transport protein in the plasma membrane of many plant cells uses the energy from ATP to pump hydrogen (H^+) ions to the outside of the cell. A second protein, called a *cotransporter protein,* then allows the H^+ ions to flow back across the membrane if accompanied by a sucrose molecule. Chemiosmosis is a form of active transport that uses energy from an electron transport chain to pump H^+ ions across the plasma membrane (see Chapters 8 and 9). The battery-like charge differential across the membrane is then used as a source of energy to synthesize ATP as the H^+ ions flow back across the membrane through the enzyme ATP synthase.

Exocytosis and endocytosis transport large molecules

Large molecules and multimolecular components often leave plant cells by **exocytosis,** a process in which small membrane-bound vesicles containing specific molecules fuse with the plasma membrane to release their contents from the cell (Figure 10.1d). Mucigel secretion from the root cap, placement of cell wall components, and the release of digestive enzymes of carnivorous plants are all examples of exocytosis in plants. Plant cells can also take up large molecules—a process known as **endocytosis,** in which the plasma membrane surrounds a large molecule and pinches off, enclosing the molecule within a vesicle inside the cell. In other words, endocytosis is the reverse of exocytosis. Since plant cells have cell walls, endocytosis is not as important a process as in animal cells. Many single-celled algae are photosynthetic and can absorb organic molecules as well. Individual molecules are taken up by facilitated diffusion, while larger fragments of many molecules can sometimes be taken up by endocytosis.

Osmosis is the movement of water across a selectively permeable membrane

The term **osmosis** (from the Greek *osmos,* meaning "a push") refers to the movement of water or any other solvent across a selectively permeable membrane. In the case of cells, of course, the solvent is always water. Water flows spontaneously from a region of lower solute concentration (higher water concentration) to a region of higher solute concentration (lower water concentration) (Figure 10.1e). Although water can pass directly through the membrane, transport proteins called *aquaporins* usually facilitate osmosis by forming channels that specifically admit water.

The idea that water moves spontaneously to a region of *higher* solute concentration may not be intuitive. After all, diffusion of a solute involves spontaneous movement "downhill" to a region of *lower* solute concentration. However, keep in mind that water is the solvent, not a solute. Its movement is actually "downhill" to a region of lower concentration as well. It is just that water moves to a region of lower *water* concentration. In a region of higher solute concentration, some water molecules are bound to the solute molecules, so there are fewer water molecules that are free to move, resulting in a lower water concentration. In an area of lower solute concentration, there are fewer solute molecules, so there are more unbound water molecules that are free to move. Therefore, water moves to an area of lower water concentration (higher solute concentration). Osmosis is similar to the diffusion of solutes in the sense that each substance moves spontaneously toward a region where it is less concentrated. Like other substances that move across a membrane, water tends to flow to equalize its concentration.

In plant cell growth, the osmotic potential inside the cell interacts with pressure generated by the cell wall

Living plant cells contain about 70–80% water. Since water takes up space, a cell taking up additional water must increase in size. Recall that plant cells have rigid cell walls that resist expansion. Therefore, cell enlargement requires both increased water and a weakening of the cell wall. Growth of a plant cell resembles enlargement of a water balloon that is surrounded by a cardboard box. To make the balloon larger, you can increase the internal pressure by putting more water in the balloon, but you must also weaken the walls of the box or increase its size.

The cell contents take up water as a result of a force called **osmotic potential,** the measurement of water's tendency to move across a membrane as a result of solute concentrations. Osmotic potential is also called *solute potential.* Since water moves to a region of higher solute concentration, the direction of its movement depends on the solute concentrations inside and outside of the cell. Frequently, osmotic potential is demonstrated by placing a membrane-bound bag of sugar solution into a container of pure water. The sugar molecules are large solutes that cannot cross the selectively permeable membrane, whereas the smaller water molecules can pass through. The sugar solution inside the bag has a higher solute concentration than the solution outside, and it is therefore described as being **hypertonic** (from the Greek *hyper,* "above") with respect to the solution outside. The solution with the lower solute concentration is said to be **hypotonic** (from the Greek *hypo,* "below"). Under these conditions, water flows into the bag, expanding the bag. The osmotic flow is from a region of lower solute concentration (higher water concentration) to a region of higher solute concentration (lower water concentration). If the two solutions were to have equal solute concentrations, they would be called **isotonic** solutions (from the

Greek *isos,* "equal"), characterized by equilibrium, with no net flow of water in either direction.

The bag of sugar solution can be compared to a protoplast—the contents of a plant cell minus the cell wall. The solute concentration of a plant cell—which includes minerals and organic molecules such as sugars and amino acids—is typically higher than that of the cell's surroundings. Like the bag of sugar solution, the typical cell is surrounded by a hypotonic solution, resulting in a net flow of water into the cell (Figure 10.2a). The protoplast spontaneously takes up water from the surroundings until the pressure of the surrounding cell wall, known as **pressure potential,** prevents further expansion of the protoplast. Under these conditions, the plasma membrane is pressed against the cell wall, making the cell turgid, or stiff, which is the normal, desirable state for a plant cell. If the solute concentrations inside and outside of the cell are isotonic, or equal, the protoplast is flaccid, or limp (Figure 10.2b). If many of a plant's cells become flaccid, stems and leaves may droop. If the solute concentration outside the cell exceeds the solute concentration inside the cell, there is a net flow of water out of the cell, causing the plasma membrane to shrink away from the cell wall, a condition known as **plasmolysis** (Figure 10.2c). When plasmolysis occurs, plants wilt and cell-to-cell cytoplasmic connections are broken, so transport through the phloem is restricted. In extreme cases of plasmolysis, the plant dies.

The ideal state of a plant cell differs from that of an animal cell. Since animal cells do not have cell walls, they expand or shrink as water moves in or out of the cell, potentially bursting or shriveling. In the normal animal cell, the solute concentrations inside and outside the cell are isotonic. In contrast, the desirable state for a plant cell is turgidity, in which the cell has a higher solute concentration than its surroundings.

The term **water potential** is used to refer to a measurement that predicts which way water will tend to flow between a plant cell and its surroundings or between different parts of a plant, such as roots and leaves. Water potential is defined as the combination of the osmotic potential (the effect of solute concentrations) and the pressure potential (the effect of cell wall pressure). These potentials are measured by the same units, represented by the Greek letter *psi* (ψ). Water potential is identified as ψ_w, pressure potential as ψ_P, and osmotic potential (solute potential) usually as ψ_O. The equation for water potential is $\psi_w = \psi_P + \psi_O$. Osmotic potential is always zero or a negative number, while pressure potential is always a positive number. Water potential can be positive, zero, or negative, depending on whether the cell is shrinking, at equilibrium, or expanding. If the osmotic potential

Outside of cell:
- Lower solute concentration
- Higher water potential

Inside of cell:
- Higher solute concentration
- Lower water potential

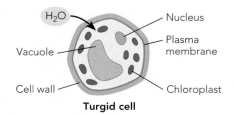

Turgid cell

(a) Plant cell surrounded by hypotonic solution. If the solution outside the cell is hypotonic, there is a net flow of water into the cell. This is the normal condition for a plant cell – being turgid. The expanded protoplast presses the plasma membrane against the cell wall.

Outside and inside of cell:
- Equal solute concentrations
- Equal water potentials

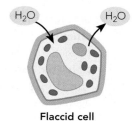

Flaccid cell

(b) Plant cell in isotonic conditions. If the solutions inside and outside have equal solute concentrations, there is equilibrium. The plant cell is flaccid, and the loss of turgor may cause stems and leaves to droop.

Outside of cell:
- Higher solute concentration
- Lower water potential

Inside of cell:
- Lower solute concentration
- Higher water potential

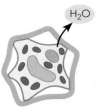

Cell in state of partial plasmolysis

(c) Plant cell surrounded by hypertonic solution. If the outside solution is hypertonic, there is a net flow of water out of the cell. This loss of water can result in plasmolysis.

Figure 10.2 Osmosis and regulation of water balance. In osmosis, water moves from an area of higher water potential (lower solute concentration) to an area of lower water potential (higher solute concentration). In a hypotonic environment, the cell wall prevents a plant cell from taking in too much water and bursting. However, the cell wall cannot prevent a cell from losing water in a hypertonic environment, which can result in plasmolysis.

and pressure potential balance each other, the water potential is zero and the cell neither expands nor shrinks. If the osmotic potential is more negative—that is, stronger—than the pressure potential, then the water potential is negative and the cell expands by taking up water.

For most living plant cells, the water potential is either zero or negative, indicating that if the cell wall were not there, the protoplast would take up water. Since we are dealing with negative numbers, "higher" and "lower" water potentials mean "less negative" and "more negative," which can be confusing. Mathematically, a lower water potential is indeed a lower (more negative) number. Physically, however, a cell or plant organ with a lower (more negative) water potential has *more* capacity to take up water. With water potential, therefore, you might say that "less is more." Just remember that a more negative water potential means greater capacity to absorb water. Water

moves from a region of higher water potential to a region of lower water potential—that is, from where the water potential is zero or negative to where it is more negative. In terms of the overall flow of water in a plant, water potential becomes increasingly negative—that is, increasingly strong—as water moves from the roots to the leaves. Leaf cells have more capacity to take up water than do root cells.

The pressure of water potential can be expressed mathematically in a variety of ways, such as atmospheres, pounds per square inch, millimeters (mm) of mercury, and kilopascals (KPa). For example, if you are at sea level, the pressure of all the atmospheric gases can be expressed as 1 atmosphere, 14.7 pounds per square inch, 760 mm of mercury, or 101.3 KPa, or 0.101 megapascal (MPa). By comparison, growing shoots and roots generate water potentials in the range of 33 to 165 pounds per square inch.

The Power of Plants

The force of water potential by which the cell expands and takes up water is frequently in the range of 30 to 100 pounds per square inch. Germinating seeds use the pressure generated by water potential to push their way up through the soil, leaves, and other materials that have covered them. Meanwhile, as roots expand, they generate substantial pressures that help in penetrating the soil. Think about the effort needed to dig a hole with a shovel in dry, hard ground, and yet plant roots can penetrate this dense material.

Sidewalk damage caused by roots.

One obvious effect of the power of plants is on city sidewalks, which are frequently pushed out of place, raised, and even broken by the growth of plant roots. Less obviously, many home sewer systems have been damaged by roots growing toward a source of water.

In nature, tree seedlings sometimes germinate in crevices on the top of large boulders and end up splitting enormous boulders entirely in two. Also, houseplants sometimes split open their pots as a result of root growth. Even individual small seedlings sometimes lift rocks many times their size during germination.

You can see why growing roots can raise sections of sidewalk and topple retaining walls (see *The Intriguing World of Plants* box on this page).

The water potential generated by a plant cell or organ can be measured in different ways. The cell or organ can be put in competition for available water with an external solution containing a solute that does not enter the plant cell. The minimum concentration of the solution that stops expansion of the cell or plant organ is equal to the water potential. Alternatively, plant tissue can be allowed to absorb water in a closed chamber from a small water supply, with the water temperature closely monitored. As water evaporates, the temperature of the remaining water cools, indicating the rate of evaporation. The rate of evaporation is monitored electronically to measure water movement into the plant cell or organ.

Section Review

1. **Explain the difference between symplastic transport and apoplastic transport.**
2. **How does facilitated diffusion differ from diffusion?**
3. **Compare and contrast osmosis and diffusion of solutes.**
4. **How do variations in solute concentrations affect a plant cell?**
5. **What is water potential?**

Movement and Uptake of Water and Solutes in Plants

Having explored transport at the cellular level, we will now look at the overall movement of water and solutes in a plant. Plants derive water and minerals from the soil and use the xylem to transport them from the roots to the rest of the plant. Leaves need both water and minerals to carry out photosynthesis and to synthesize the many types of molecules used by plants. In the leaves, photosynthesis and other biochemical processes make sugar and other organic molecules, which are then transported throughout the plant by the phloem (Figure 10.3).

Water evaporation from leaves pulls water through the xylem from the roots

The vascular system transports water, minerals, and organic molecules throughout the plant. Xylem is composed of tracheids and, in the case of flowering plants, vessel elements (see Chapter 4). These dead cells handle transport of water and minerals from the roots to stems and leaves, where water evaporates through the stomata—the process called *transpiration*.

A large tree in a forest can transpire between 700 and 3,500 liters (185 and 925 gallons) a day during the summer. By comparison, typical crop plants transpire much

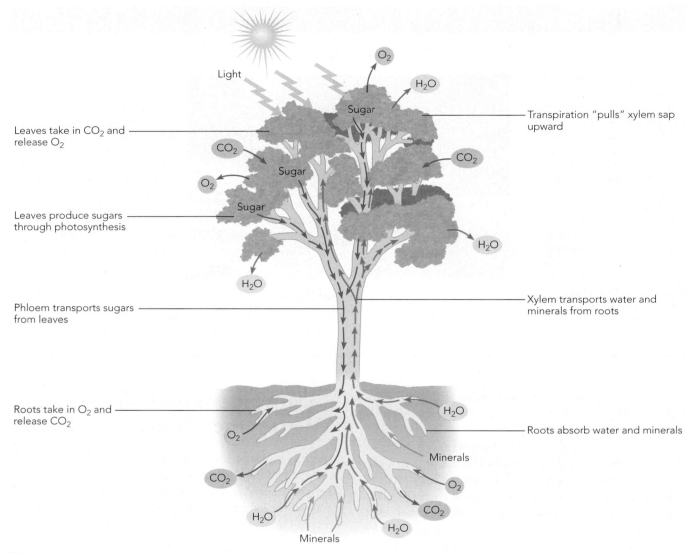

Figure 10.3 Overview of water and solute transport in plants.

less. Corn, for instance, transpires approximately 2 liters (0.53 gallon) a day per plant. However, that amount is still significant, equating to a daily transpiration of 60,000 liters (15,850 gallons) per acre, an area roughly the size of a football field. During a growing season, an acre of corn will use 6 million liters (about 1.6 million gallons). If applied all at once, this amount of water would cover the acre about 5 feet deep. Plant breeders are therefore interested in developing crop plants that require less water (see the *Biotechnology* box on page 221). The plant needs water for cell growth and photosynthesis and to provide minerals for the biosynthesis of proteins, nucleotides, and other molecules. However, small amounts of water would supply these needs.

Transpiration, which appears to waste water, actually serves two necessary functions. First, it cools the leaves, which are considerably heated by the sunlight absorbed

in photosynthesis. Second, it serves as the pump to pull water and water-soluble minerals up from the roots. The fact that a plant pumps water from the top was puzzling to plant physiologists, considering how mechanical pumps, such as a well pump, work. A pump located at the top of a water well can only pump water 10.36 m (34 feet) or less because water columns pulled higher than 10.36 m (34 feet) will break as a result of their weight. For this reason, most wells have pumps at the bottom, where this limitation does not apply. How, then, is water pumped from the top of a plant?

Plants, and in particular tall trees, are able to "pump" from the top of the water column because of the design of xylem tissue and the characteristics of water. Water is a **polar molecule**—a molecule with positively charged and negatively charged ends. Therefore, the positively charged end of

BIOTECHNOLOGY

Water-Efficient Crops

Can scientists develop plants that use less water and are more efficient at transpiration? Indirectly, *any* increased efficiency of the plant translates into water saved. If the plant has a shorter time to maturity, or decreased photorespiration, or an architecture that produces more grain on each plant, the net effect saves water. For example, the following traits help wheat tolerate drought:

◆ Large seed size provides more food for germination in dry, hard soil.

◆ Long coleoptiles (the sheath covering the first leaves that push up through the soil) allow for deep sowing where the soil is cooler and has more moisture.

◆ Thinner, wider leaves shade the soil and radiate heat more rapidly.

◆ Prostrate (horizontal) growth habit shades the soil.

◆ High photosynthetic capacity of flowering spike ensures rapid grain development.

◆ Rapid adjustment to osmotic stress means that the cells produce extra solute molecules to prevent water loss.

◆ Accumulation of abscisic acid controls the closing of stomata and thus regulates water loss.

◆ Hairy, waxy leaves diffuse sunlight and prevent water loss through epidermal cells.

◆ Heat tolerance offers protection from high temperatures that often accompany drought.

Wheat with these traits can be obtained by traditional breeding and by genetic engineering if specific genes have been identified.

The use of genetic engineering to improve drought tolerance is attracting increasing interest and following many experimental routes. Here are three examples that provide an introduction.

Scientists are interested in producing genetically engineered plants that convert table sugar, sucrose, into sugar polymers called *short-chain fructans*, produced naturally by some plants, such as the onion. These molecules taste sweeter than sucrose but provide no calories to humans. Currently short-chain fructan synthesis involves an expensive, industrial process. An agronomic side benefit of this research has been the discovery that plants, genetically transformed to produce short-chain fructans, also have increased drought tolerance. They may increase drought tolerance because of interactions with membranes or because the molecules serve as an osmotic protectant that is not translocated out of the cells.

The successful introduction of the C_4 photosynthetic pathway into rice (see Chapter 14) should markedly increase production under the high temperature, high light intensities in which rice grows.

Irrigated fields in Oregon.

Scientists at the University of California produced *Arabidopsis* that are hypersensitive to abscisic acid and close their stomata more rapidly in response to increasing stress. When plants were not watered for 12 days, the engineered plants looked healthy, whereas normal plants had wilted and shriveled.

While these plants are not yet found in farmer's fields, such fascinating results imply that further research is warranted.

---

one water molecule attracts the negatively charged end of another. This characteristic helps explain three behaviors of water molecules: adhesion, cohesion, and tension.

◆ **Adhesion** is the attraction between different kinds of molecules. In plants, there is adhesion between water molecules and the cell wall molecules. Water moves to the top of plants in a continuous stream sometimes thought of as a column of water. Actually, the column passes through millions of narrow xylem cells, where

cellulose walls adhere to water molecules, binding and supporting tiny segments of the column. Therefore, the column is in no danger of breaking as a result of its own weight. Paper towels, made from wood pulp, provide a good demonstration of adhesion of cellulose to water molecules.

◆ Water columns in xylem also exhibit **cohesion**—the attraction between molecules of the same kind. Since they are highly polar, water molecules bind to each other, which helps support the water column.

◆ Water columns in xylem experience **tension,** the negative pressure on water or solutions. In the xylem, tension is caused by transpiration through stomata. Water evaporating from the stomata into the air "pulls" the water column up, in much the same way as a person sucking on a straw "pulls" the fluid up through the straw. In plants, tension is transmitted down the stem or trunk. In fact, the diameter of tree trunks actually shrinks during transpiration, as in the way a straw starts to collapse when you suck hard.

Most physiologists use the **tension-cohesion theory** to explain transport in the xylem (Figure 10.4). Actually, while both tension and cohesion are important, adhesion is crucial. A water well that is 15.2 m (50 feet) deep cannot be successfully pumped from the top, but through transpiration a 15.2 m (50-foot) tree can "pull" water from the bottom of the tree to the top. Two important differences explain the failure of the well and the success of the tree. In both cases, the column of water has cohesion between water molecules and tension created by the pump. One difference is that the pipe carrying water up from the well does not shrink in response to tension, so the tension created by the pump is not transmitted down the column, in contrast with the tree. The other difference is that the well pipe provides little or no adhesive component.

Studies using radioactive isotopes dissolved in water have indicated clearly that xylem transports water. Sometimes the tension of the water column in a stem causes the water column to break and air bubbles to form. In some cases, plants can repair the damage by redissolving the air bubble, usually at night. This occurs because the pressure of surrounding tissues drives water back into the cell with the air bubble, reducing the size of the bubble and finally eliminating it. Since there are many single cells in the xylem, air bubbles are usually confined to just a few tracheids, and the water stream simply flows around them. In the case of vessels, the breaking of a water column interrupts transport in the entire vessel, not just one vessel element. Taller trees have more danger of broken water columns. Higher rates of transpiration increase tension on the columns and increase the danger of breakage. The tallest trees, coast redwoods, live in moist habitats with considerable fog and low or moderate levels of transpiration.

As a result of transpiration, water potential becomes more and more negative as water moves from the soil to the leaves. For example, the water potential of soil might be −0.3 MPa whereas the water potential of root hairs is −0.6 MPa. Thus, water will flow from the soil into the root. Midway up a tree trunk the water potential would be −0.7 MPa, whereas in the leaf it would drop to −3.0

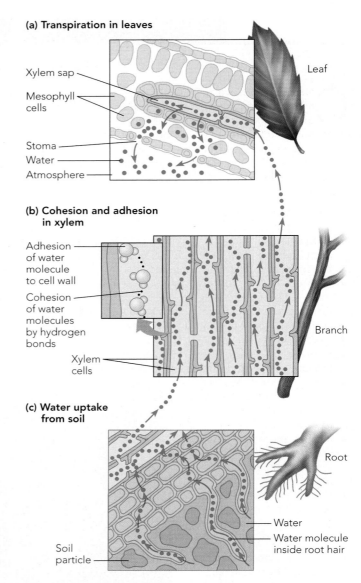

(a) Transpiration in leaves

Xylem sap

Mesophyll cells

Stoma
Water
Atmosphere

Leaf

(b) Cohesion and adhesion in xylem

Adhesion of water molecule to cell wall

Cohesion of water molecules by hydrogen bonds

Xylem cells

Branch

(c) Water uptake from soil

Root

Water
Water molecule inside root hair

Soil particle

Figure 10.4 Water and dissolved solutes flow from roots to shoots. **(a)** The evaporation of water through stomata creates a gradient of water potential that is most negative at the top of the tree. **(b)** Cohesion of water molecules to each other and adhesion of water molecules to the cellulose walls of tracheids keep the column of sap intact. **(c)** Roots take up water from soil.

MPa. The water potential of air outside the leaf could be between −5.0 and −100.0 MPa. Thus, increasingly negative water potential keeps water flowing from the soil, into and up the stem (trunk), and out the leaves into the air.

Water absorption in roots occurs through elongated epidermal cells known as *root hairs,* which develop just above the root apical meristem (see Chapter 4). The water potential of root hairs reflects whether the plant needs water or not. The root hairs also compete directly with soil particles for water and can either win or lose the competition, depending on how dry the soil is. Between the root

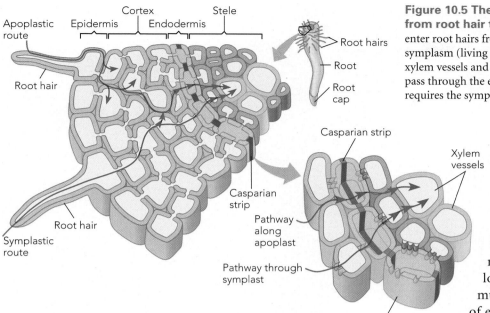

Figure 10.5 The pathway of minerals and water from root hair to xylem. Water and mineral ions enter root hairs from the soil and flow through the symplasm (living cells) or apoplasm (cell walls) to xylem vessels and tracheids. As water and mineral ions pass through the endodermis, the Casparian strip requires the symplastic route.

hairs and the endodermis, water can flow between cells—apoplastic transport—or through the cytoplasm of cells—symplastic transport (Figure 10.5). However, when water reaches the endodermis, the Casparian strip ensures that water and dissolved minerals must be filtered through the endodermal cells, thereby giving the membrane the opportunity to control the uptake of ions (see Chapter 5).

The negative water potential of root cells causes enough water uptake to generate root pressure. If a stem is removed, roots continue to push water up the stem. The water pushed into the stem by root pressure may end up leaving the leaves as droplets through specialized epidermal regions, in a process called **guttation.** Early investigators thought root pressure was responsible for the movement of water to the tops of tall trees, operating much like a pump at the bottom of a well. However, root pressure can only move water a few feet and is at its lowest during the day, when maximal transpiration occurs. The water potential of plant organs can be measured by applying pressure within an airtight container. For instance, a stem can be placed in such a container, with a cut end of the stem sticking through a hole at the top. Pressure is applied until water appears at the protruding end of the cut stem or in the stomata of leaves. At this point, the applied pressure equals the water potential of the stem.

Stomata control gas exchange and water loss for the plant

Plants must maintain enough water in tissues to prevent loss of turgor and consequent wilting, which occurs when the plasma membrane is not pushing against its cell wall. A loss of turgor interrupts cell-to-cell communication

and disrupts the supply of both nutrients and hormones necessary to maintain the plant and control its functioning. As a result of the high rate of transpiration necessary for leaf cooling and for pumping water from the roots, dangerous amounts of water loss can occur quickly. The plant must be able to respond to a variety of environmental stimuli to control water balance, which occurs when the water potential of a cell or a tissue is zero.

The waxy layer, called the cuticle, on the outside of most epidermal cells of leaves allows very little water loss. Ninety percent of water loss by a plant occurs through the stomata, the pores surrounded by guard cells. Stomata occur in the epidermis of all aboveground parts of the plant, with most occurring on the undersides of leaves, where the temperature is lower and where they are less likely to become clogged by dust deposited by the air. Up to 10,000 stomata can occur per square centimeter of leaf surface.

When sufficient water is available, the photosynthetic guard cells of stomata take up water and become curved, like overinflated balloons, to open a pore that allows gas exchange with the air spaces that make up 15–40% of the inside of a leaf (Figure 10.6). Stomata open in response to decreasing internal concentrations of CO_2 and to light from the blue portion of the visible spectrum. They close in response to increasing internal concentrations of CO_2, high temperatures, wind, low humidity, and the hormone **abscisic acid** (**ABA**). On a typical summer day, stomata are closed at daybreak. As sunlight stimulates photosynthesis, the CO_2 concentration in the leaf drops and the stomata open in response to the CO_2 level and to blue light, remaining open until nightfall unless conditions that promote water loss occur.

Abscisic acid controls the opening and closing of the pores of stomata by guard cells. When high levels of ABA occur in guard cells, they lose water, and the pore closes. Abscisic acid, produced in roots in response to dry soil, is transported to leaves and provides advance warning of drought. When low levels of ABA occur in guard cells, they take up water, and the pore opens.

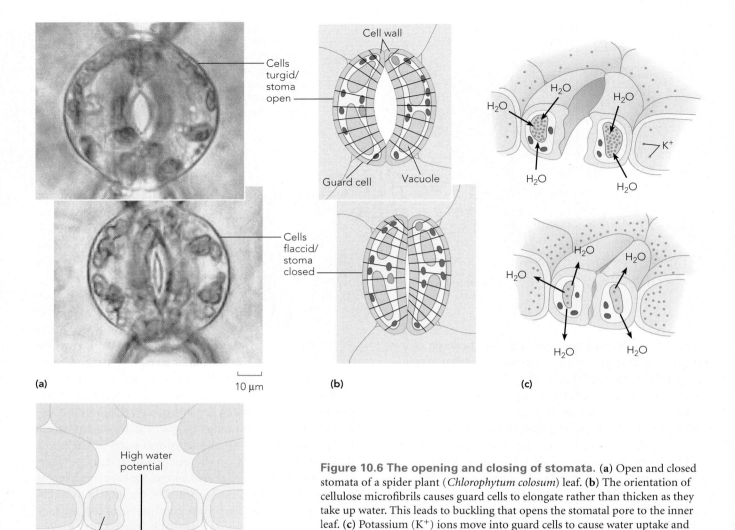

Figure 10.6 The opening and closing of stomata. (a) Open and closed stomata of a spider plant (*Chlorophytum colosum*) leaf. **(b)** The orientation of cellulose microfibrils causes guard cells to elongate rather than thicken as they take up water. This leads to buckling that opens the stomatal pore to the inner leaf. **(c)** Potassium (K^+) ions move into guard cells to cause water uptake and opening of the stomatal pore. **(d)** Inside the leaf, water vapors are abundant, humidity is high, and water potential is negative. Outside the leaf, water vapor is less abundant, humidity is lower, and water potential is very negative.

By controlling the diameter of the stomata, the plant can regulate the rate of water loss caused by transpiration. On hot, dry, windy days, stomata will be closed. Of course, closing the stomata saves water but also reduces the uptake of CO_2 required for photosynthesis. Under these conditions the plant will lose carbon through photorespiration (see Chapter 8). Plants are constructed so that stomata close, resulting in lower photosynthesis, before photorespiration markedly increases.

Sugars and other organic molecules move from leaves to roots in the phloem

In plants, transport of sugar and other organic molecules takes place in the phloem. In the phloem of flowering plants, organic molecules are transported through sieve-tube members and their companion cells (see Chapter 3). Phloem moves sap from a sugar source to a sugar sink. A **sugar source** is a part of a plant—usually leaves and also green stems—that produces sugar. A **sugar sink** is a part of a plant that mainly consumes or stores sugar, such as roots, stems, and fruits. Sugar transport is driven by water uptake by osmosis and therefore requires the selectively permeable plasma membrane of a living cell.

As with water and minerals, sugar and other organic molecules can be moved by either symplastic transport or apoplastic transport. In the case of sugar, the sugar synthesized in leaf mesophyll cells must be transported to the cells of the phloem (Figure 10.7a). Symplastic transport is most common in plants from warm environ-

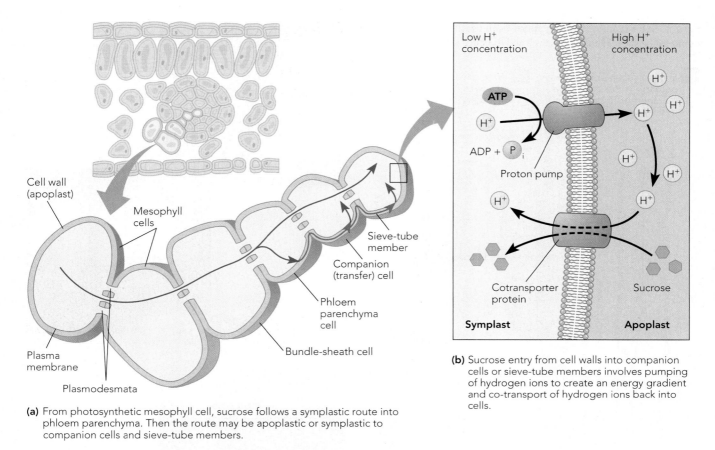

(a) From photosynthetic mesophyll cell, sucrose follows a symplastic route into phloem parenchyma. Then the route may be apoplastic or symplastic to companion cells and sieve-tube members.

(b) Sucrose entry from cell walls into companion cells or sieve-tube members involves pumping of hydrogen ions to create an energy gradient and co-transport of hydrogen ions back into cells.

Figure 10.7 Loading sucrose into the phloem.

ments, with molecules remaining inside cells to pass through plasmodesmata (the channels between cells) from mesophyll cells into phloem cells. Apoplastic transport is most common in plants from temperate or cold environments, with molecules following a pathway outside the plasma membrane as they move from mesophyll cells to the phloem. Frequently, these plants store sugar in the cell walls of cells near the phloem. Companion cells take up the sugar and pass it to sieve-tube members through plasmodesmata. Some companion cells have cellular protrusions and ingrowths that increase the surface area between them and sieve-tube members. Such modified companion cells are known as *transfer cells.*

Energy is required when molecules transported in the apoplastic pathway enter sieve-tube members, as ATP is used to pump H^+ ions out of the cells (Figure 10.7b). Then the H^+ ions and sugar molecules enter the cell together, with the help of a cotransporter protein. The mechanism of phloem transport differs from the transpiration-driven movement of water and minerals in the xylem. In the phloem, the sugar that enters sieve-tube members generates osmotic potential and water uptake. The turgor pres-

sure developed by water uptake moves water and sugar down the phloem until the sugar is downloaded into root cells and other cells requiring energy. Open pores at each end of a sieve-tube member allow direct connections between cells so the sugary solution can move easily through the phloem. Building high pressure at the leaf end (sugar source) and reducing it at the root end (sugar sink) keeps the phloem sap moving. When sugar reaches sugar sinks, such as roots, water then leaves the sieve-tube members, along with solutes like sugar. The mechanism for phloem transport, first proposed by Ernst Munch in 1927, is known as the **pressure-flow hypothesis** (Figure 10.8). Although living cells are required, the actual transport process, driven by osmosis, is passive.

Phloem-feeding aphids have provided useful information about phloem transport. Phloem sap contains between 10–20% sugar and small percentages of other organic molecules such as amino acids. A typical aphid feeds by sticking its pointed, straw-like stylet through leaf or stem tissue and into the sugary phloem. The turgor pressure of sieve-tube members then pushes phloem sap through the aphid's gut to emerge as drops of "honeydew"

Figure 10.8 Pressure flow in sieve-tube members. In this example, the sugar source is a leaf cell and the sugar sink is a root cell.

1 Entry of sugar reduces water potential in the sieve-tube members, causing the tube to take up water.

2 Water pressure forces sap to flow through sieve-tube members.

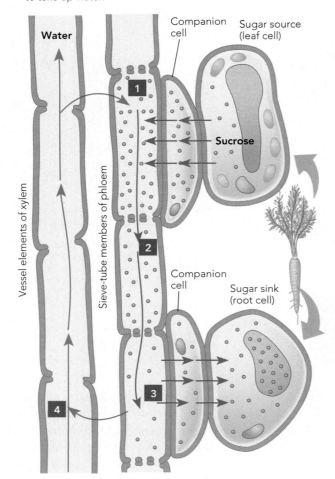

3 As sugar is unloaded at the sink, the pressure in the sieve tubes drops, creating a pressure gradient. Most of the water then diffuses back to the xylem.

4 Xylem recycles water from the sugar sink (root) to the sugar source (leaf).

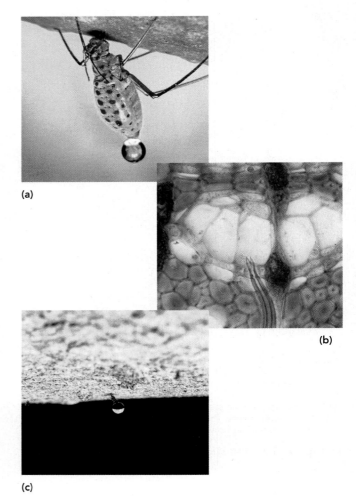

Figure 10.9 Using aphids to study flow of phloem sap. (**a**) The sieve tube pressure pushes phloem sap into the aphid. (**b**) The aphid sticks its stylet directly into a phloem sieve-tube member. (**c**) If the insect is removed, phloem sap can be collected from the stylet to measure flow.

on the tip of the aphid's abdomen. If the aphid is anesthetized to prevent stylet withdrawal, and the rest of its body is removed, the stylet exudes pure phloem sap for several hours, serving as a tap that botanists can use to measure the flow (Figure 10.9). Phloem sap moves at speeds of up to 1.0 m per hour. Neither diffusion nor cytoplasmic streaming can account for such high speeds. The pressure developed in phloem cells of leaves by osmotic uptake of water accounts for the observed transport rates.

Section Review

1. Define transpiration and explain how water gets to the top of tall trees.
2. How do stomata control gas exchange and water loss?
3. How does sugar get from the leaves to the roots?

Soil, Minerals, and Plant Nutrition

You already know that plants obtain the minerals they need from the soil. The uptake of mineral ions occurs through root hairs at the same time as water uptake, and transport of the solution occurs through the xylem. In this section, you will learn about the structure of soil and how it binds solute molecules and water.

Soil is made of ground-up particles of rocks surrounded by negative charges that bind water and minerals

Rocks, weathered by wind and rain and fractured by water expanding into ice, break down to produce stones and gravel and eventually become soil. Bacteria, algae, fungi, lichens (associations of algae and fungi), mosses, and plant roots all secrete acids that contribute to the breakdown of rocks into soil. Classified by size, soil particles include **sand,** which contains particles 0.02 to 2 mm (0.0008–0.08 inch) in diameter; **silt,** which contains particles 0.002 to 0.02 mm (0.00008–0.0008 inch) in diameter; and **clay,** which contains particles smaller than 0.002 mm (0.00008 inch) in diameter.

Soil occurs in layers called **horizons** (Figure 10.10). In a simplified view, the uppermost horizon, or A horizon, consists of **topsoil,** which contains the smallest soil particles and is most suited to support plant growth. Topsoil ranges in depth from a few millimeters to several feet and generally contains soil particles of the three basic sizes (sand, silt, and clay); decaying organic material called *humus;* and various organisms such as bacteria, fungi, nematodes, and earthworms. Ideal garden topsoil is loam, which contains roughly equal amounts of sand, silt, and clay. Loam is most suited for planting because the soil particles are small enough to permit the growth of roots between them and because the soil has enough surface area to bind sufficient water and minerals to support growth. Plant roots can penetrate sand easily, but the particles are large and the total surface area of sandy soil is insufficient to support most plant growth. The second horizon, the B horizon, contains larger and less-weathered particles of sand and rocks, as well as less organic matter. The deepest horizon, the C horizon, is quite rocky but supplies raw materials for soil produced in the upper horizons. Groundwater, in the form of underground deposits known as **aquifers,** occurs in various locations in or below the C horizon. Wells tap into groundwater, replenished by rain, which percolates through the soil horizons. Below the C horizon is rock, the Earth's crust or bedrock, which can extend 40 kilometers (15.4 miles).

Plants require 17 essential elements, most of which are obtained from soil

By the mid-1800s, scientists realized that plants depended on soil to supply water and minerals. This conclusion was supported by experiments in supplying plants with solu-

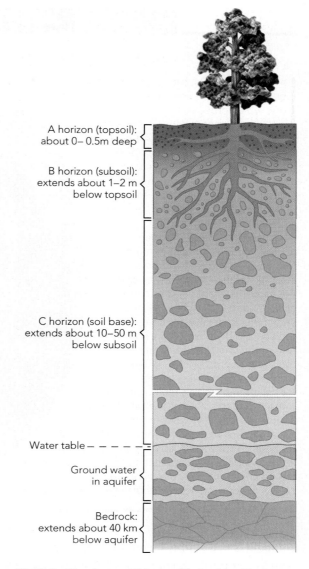

A horizon (topsoil): about 0– 0.5m deep

B horizon (subsoil): extends about 1–2 m below topsoil

C horizon (soil base): extends about 10–50 m below subsoil

Water table – – – –

Ground water in aquifer

Bedrock: extends about 40 km below aquifer

Figure 10.10 Soil horizons. This simplified soil profile shows the A, B, and C horizons. In general, soil becomes rockier the deeper you go.

tions of water and minerals in a lab and also by the invention of **hydroponics** (from the Greek *hydor,* "water," and *ponos,* "labor or toil"), or soil-less gardening. In hydroponics, the mineral nutrients normally supplied by the soil are mixed into a liquid solution, which is used to water the plant roots. The first hydroponics recipe—containing KNO_3, $Ca(NO_3)_2$, KH_2PO_4, $MgSO_4$, and $FeSO_4$—supported the growth of many kinds of plants in liquid culture or in sand. It seemed that plant mineral nutrition was totally understood (see the *Plants & People* box on page 228). By the 1900s, however, plants were no longer growing consistently with this mineral recipe. Had the laws of plant nutrition changed? As it turns out, chemical

PLANTS & PEOPLE

Justus von Liebig—A Father of Modern Agriculture

In the early 1800s, most agricultural scientists believed in some variation of the humus theory, which proposed that all major components of plants consisted of soil and water taken in through the roots. After all, people had known for thousands of years that soil fertilized with organic matter such as manure grew much better crops than unfertilized soil. The contribution of air to plant growth was not widely recognized.

In 1840, however, German chemist Justus von Liebig put the humus theory to rest. He presented strong evidence that most or all of the carbon in plants comes from atmospheric CO_2, while the necessary minerals and water come from the soil. He developed a Law of the Minimum, which states that plant growth is limited by the necessary mineral present in the smallest amount. He called this limiting factor the "minimum." Liebeg's Law of the Minimum has been supported by experiments and has governed soil testing and fertilizer application for nearly 150 years. Liebig invented the first artificial fertilizer, a combination of chemical elements known to promote plant growth. Unfortunately, several ingredients formed a concrete-like

Justus von Liebig.

substance, and his venture into the marketplace failed. Two British scientists, J. B. Lawes and J. H. Gilbert, developed the first commercially successful artificial fertilizers in 1843. In 1862, German scientist W. Knop published a list of five chemicals that allowed for hydroponics, or gardening without soil. The nutritional contribution of soil to plants had been defined in chemical terms.

Through his views and the students he trained, Justus von Liebig played an indirect but important role in improving the productivity of American agriculture. When Abraham Lincoln created the Department of Agriculture in 1860, he appointed one of Justus Liebig's students as the department's first scientist. The passage of the Homestead Act in 1862 furthered the expansion of American agriculture by giving 1.29 square km (0.25 square mile) of free land for farming to any head of a family or anyone at least 21 years old. The Land Grant College Act, passed that same year, led to the establishment of agricultural colleges in each state. Meanwhile, Liebig and his students influenced the establishment of agricultural experimental stations throughout Europe and the United States.

---

companies had simply begun to manufacture chemicals with higher purity, leaving out many impurities that were actually essential for plant growth.

Plants contain 60 or more chemical elements, but only 17 of them are currently believed to be essential. These are classified as either macronutrients or micronutrients (Table 10.1). **Macronutrients** are used in large amounts for producing the body of the plant and for carrying out essential physiological processes. The air supplies oxygen and carbon, while the other macronutrients come from the soil. **Micronutrients** are usually necessary cofactors for enzymes and are therefore recycled by the plant. Plants exhibit characteristic deficiency symptoms when inadequately supplied with one or more essential nutrients. Some minerals may be required by plants at such low concentrations that adequate amounts are supplied by dust, and deficiencies would be very difficult to demonstrate. Plants with inadequate mineral nutrition pass on these inadequacies to animals that consume

them. Humans and other animals require several minerals (selenium, chromium, and fluoride) that are usually found in plants but typically not required by them.

When soils in a region lack sufficient quantities of a micronutrient, plants and animals may develop deficiency symptoms. The United Nations estimates that micronutrient malnutrition afflicts more than 40% of the world's people. For example, soil in parts of China commonly lacks sufficient selenium. Common human signs of the deficiency are heart and bone defects. A soil study revealed that in regions with low levels of soil selenium, three times as many people died from cancer as in the regions with high levels of soil selenium. Studies in which people received selenium supplements showed dramatic decreases in the incidence of many types of cancer.

Agriculture removes nutrients from the soil and reduces soil fertility. Nutrient depletion is particularly a problem where farming has occurred for thousands of years. Soil testing can determine soil fertility with respect

Table 10.1 Essential Nutrients for Most Vascular Plants

Element	Chemical symbol	Form available to plants	Importance to plants
Macronutrients			
Carbon	C	CO_2	Major element in organic compounds
Oxygen	O	CO_2	Major element in organic compounds
Hydrogen	H	H_2O	Major element in organic compounds
Nitrogen	N	NO_3^-, NH_4^+	Elements in nucleotides, nucleic acids, amino acids, proteins, coenzymes, and hormones
Sulfur	S	SO_4^{2-}	Elements in proteins, coenzymes, and amino acids
Phosphorus	P	$H_2PO_4^-$, HPO_4^{2-}	Elements in ATP and ADP, several coenzymes, nucleic acids, and phospholipids
Potassium	K	K^+	Cofactor in osmosis and ionic balance, action of stomata, and protein synthesis
Calcium	Ca	Ca^{2+}	Essential for stability of cell walls, in maintaining membrane structure and permeability, acting as enzyme cofactor, and regulating some stimulus responses
Magnesium	Mg	Mg^{2+}	Enzyme activator and a component of chlorophyll
Micronutrients			
Chlorine	Cl	Cl^-	Essential in water-splitting step of photosynthesis that produces oxygen and functions in osmosis and ionic balance
Iron	Fe	Fe^{3+}, Fe^{2+}	Activator of some enzymes, forms parts of cytochromes and nitrogenase, and is required for chlorophyll synthesis
Boron	B	$H_2BO_3^-$	Required for chlorophyll synthesis; might be involved in nucleic acid synthesis, carbohydrate transport, and membrane integrity
Manganese	Mn	Mn^{2+}	Activator of some enzymes, active in the formation of amino acids, required in water-splitting step of photosynthesis, and involved in the integrity of chloroplast membrane
Zinc	Zn	Zn^{2+}	Activator of some enzymes, involved in formation of chlorophyll
Copper	Cu	Cu^{2+}, Cu^+	Activator of some enzymes involved in oxidation/reduction reactions, component of lignin-biosynthetic enzymes
Molybdenum	Mo	MoO_4^{2-}	Involved in nitrogen fixation and nitrate reduction
Nickel	Ni	Ni^{2+}	Cofactor for an enzyme that functions in nitrogen metabolism

to specific nutrients. Application of the appropriate fertilizer to soil can remedy specific problems.

Soil particles bind water and mineral ions

Approximately 93% of the Earth's crust consists of **silicates** (SiO_4^{-4}). Therefore, soil particles display negative charges on their outside layers. Being polar molecules, water molecules have a positively charged side and a negatively charged side, so rings of water form around each soil particle. In the water, some mineral ions dissolve as cations (positively charged ions) and others as anions (negatively charged ions). The first ring of water surrounding a soil particle has cations, the next has anions, and so on. Some cations dissolve in the water, while others bind directly to soil particles.

Water, dissolved mineral ions, and dissolved O_2—together known as the **soil solution**—occupy around 50% of soil volume and provide the source of these nutrients for plants (Figure 10.11a). Soil binds water molecules with a force called the **matrix potential,** which is a negative number. Think of all the soil as consisting of a matrix of particles of various sizes, which are separated by

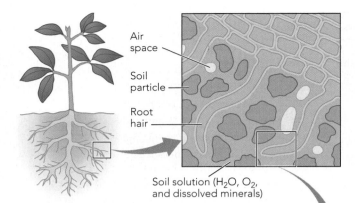

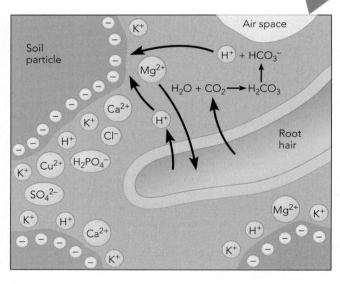

(a) Soil particles and soil solution. Root hairs cannot absorb minerals directly from soil particles. Instead, they absorb soil solution that includes water, dissolved oxygen, and dissolved minerals, which are present as either positively charged ions (cations) or negatively charged ions (anions).

(b) Cation exchange. Since cations are positively charged, they bind closely to the negatively charged soil particles, but they can be displaced by H^+ ions, the process known as cation exchange. When dislodged, they are available for absorption. The diagram shows a magnesium ion (Mg^{2+}) displaced by two H^+ ions. Root hairs supply the H^+ ions directly by secreting them and indirectly by producing CO_2, causing a chemical reaction that yields H^+. Negatively charged ions usually do not bind as closely to soil particles, and are therefore more readily absorbed but also more easily drained from the soil.

Figure 10.11 Absorption of minerals by root hairs.

water and air. For a root hair to absorb water, its water potential must be more negative than the matrix potential of the soil.

Ions bind to soil particles in a preferential order, depending on the relative strength of their positive or negative charges. For example, cations bind to soil particles according to three rules: Cations with more positive

charges bind first, smaller ions bind before larger ions, and ions at high concentrations bind before ions at low concentrations. For example, according to the first two rules, Ca^{2+} binds before Na^+. High concentrations of Na^+ bind before low concentrations of Ca^{2+}, according to the third rule.

These rules become important for plants when toxic ions occur in the soil solution. For example, in regions with salty soil, the concentration of sodium ions (Na^+) is high and Na^+ displaces ions that the plant needs from soil particles. The sodium ions remain in the soil, while the useful ions end up in the groundwater, where they are largely unavailable to plants. For that reason, salty soil is nutrient-poor soil. Large regions of the southwestern United States, which was an ocean millions of years ago, suffer from this problem.

The order in which ions bind to soil particles also plays an important role in acidic soils, which occur in regions of high rainfall. In rain, CO_2 dissolves in water to produce H^+ ions according to the following reaction: $CO_2 + H_2O \rightarrow H_2CO_3$ (carbonic acid) $\rightarrow H^+ + HCO_3^-$ (bicarbonate). Hydrogen binds tightly to soil particles, displacing other cations, including those important to plants. Also, acidic soil brings previously insoluble and highly toxic aluminum ions into the soil solution. Acidic soil is therefore nutrient-poor and often contains toxic aluminum.

The displacement of the mineral cations by H^+ ions plays a role in normal mineral uptake by roots (Figure 10.11b). Both water molecules and mineral ions bind directly to soil particles. Since anions are not found in the first binding layer, directly next to the soil particles, they are more easily drained away from the soil and "lost" to plants in the groundwater, a process known as *leaching*. As roots penetrate the soil, they release CO_2 produced in respiration, which combines with water to become bicarbonate and H^+ ions. Roots can also secrete H^+ ions directly. The CO_2 dissolves to produce H^+ ions, which replace the mineral cations bound to the soil, a process known as **cation exchange.** In this way, the minerals are released from the soil and made available in the soil solution for uptake by roots.

Bacteria in the soil make nitrogen available to plants

In Chapter 4 you read about mycorrhizae, the mutualistic associations between plant roots and soil fungi that increase plant absorption of soil minerals. Some plants also form associations with bacteria. Plants need nitrogen, but they cannot absorb nitrogen gas (N_2) from the air. They must absorb it from nitrogen compounds in the soil, primarily as nitrate (NO_3^-) but also as ammonium (NH_4^+). Some soil bacteria carry out **nitrogen fixation**—the conversion of nitrogen gas into nitrate or ammonium (Figure 10.12). In some soils, nitrifying bacteria convert ammonium to nitrite (NO_2^-) and then into nitrate. In addition to nitrogen-fixing bacteria, there are ammonifying bacteria that release ammonium by breaking down organic matter called *humus* and denitrifying bacteria that convert nitrate back to N_2.

Nitrogen-fixing bacteria first convert nitrogen gas into ammonia (NH_3), through the action of the enzyme nitrogenase. Ammonia then picks up an H^+ ion from the soil solution to become ammonium NH_4^+. Some nitrogen-fixing bacteria are free-living, but most form mutualistic associations with certain plants, particularly legumes such as alfalfa, peas, beans, and clovers. Some non-legumes such as alder trees and a genus of water ferns (*Azolla*) form similar associations. Plants that have associations with nitrogen-fixing bacteria take less nitrogen from the soil than other plants and actually add nitrogen back to the soil. Therefore, farmers often rotate legumes with other crops to enrich the soil. In a single growing season, a legume crop can add 300 kg (660 pounds) of fertilizer per 1 hectare (2.47 acres). The bacteria associated with

one legume plant yield between 1 and 3 grams (0.035–0.11 ounce) of fixed nitrogen. The crop is then plowed back into the soil to release additional nitrogen by normal decay. *Azolla* ferns floating in rice paddies supply nitrogen to the rice after they die as a result of shading by the rice and absence of water in the paddy.

Nitrogen fixation by bacteria is a complex process. In legumes, bacteria of the genus *Rhizobium* enter roots through a modified root hair called an *infection thread*. In response to the infection thread, the plant produces a flavonoid (see Chapter 7) that activates a series of chemical signals, resulting in the plant forming **root nodules** where the bacteria will live (Figure 10.13). Once inside the nodules, the bacteria change into an enlarged form called *bacteroids,* which live in vesicles inside root cells.

Nitrogen can also be fixed commercially, but the process is energy-intensive and expensive. Nevertheless, farmers frequently use commercially produced nitrogen fertilizer because it eliminates the need for crop rotation with legumes and is easily applied along with other fertilizers. In developing countries, where artificial fertilizer is often prohibitively expensive, farmers sometimes grow legumes side-by-side with a non-nitrogen-fixing crop. Once the latter is harvested, the legume crop is plowed

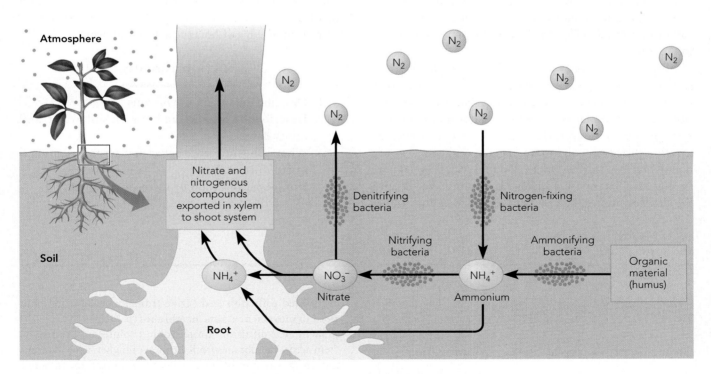

Figure 10.12 Soil bacteria regulate the soil nitrogen available to plants. Plants can absorb either nitrate or ammonium from the soil. Most soil nitrogen is nitrate because of the presence of nitrifying bacteria that convert ammonium, from decaying organic matter and from nitrogen-fixing bacteria, to nitrate.

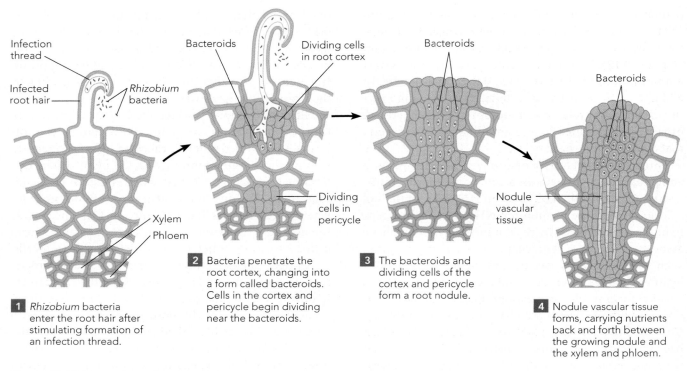

Infection thread

Infected root hair

Rhizobium **bacteria**

Xylem

Phloem

1 *Rhizobium* bacteria enter the root hair after stimulating formation of an infection thread.

Bacteroids

Dividing cells in root cortex

Dividing cells in pericycle

2 Bacteria penetrate the root cortex, changing into a form called bacteroids. Cells in the cortex and pericycle begin dividing near the bacteroids.

Bacteroids

3 The bacteroids and dividing cells of the cortex and pericycle form a root nodule.

Bacteroids

Nodule vascular tissue

4 Nodule vascular tissue forms, carrying nutrients back and forth between the growing nodule and the xylem and phloem.

Figure 10.13 Formation of root nodules.

under to provide more nitrogen through the decay process. In the case of legumes like peas and beans, a crop can be harvested before the rest of the plant is returned to the soil.

Since commercially produced nitrogen fertilizer is expensive, scientists have long dreamed of transferring the ability to form nitrogen-fixing bacterial associations to all crop plants. Although one primary enzyme, nitrogenase, carries out the fixation, the bacterial colonization process involves several genes and complex signaling between the host plant and bacteria. Each legume species associates with a specific bacterium species, and the signaling molecules vary with each association. Trans-

ferring the nitrogen-fixing process to a new plant would involve introducing a number of genes and an overall greater knowledge of how the association works.

Section Review

1. **Describe the three soil horizons.**
2. **Describe the interaction between soil, water, and mineral ions.**
3. **What is the difference between macronutrients and micronutrients?**
4. **How do bacteria supply nitrogen to plants?**

SUMMARY

Molecular Movement Across Membranes

Diffusion is the spontaneous movement of molecules down a concentration gradient (pp. 215–216)
Solute molecules move spontaneously from a region of higher solute concentration to a region of lower solute concentration.

Facilitated diffusion and active transport use proteins to assist in movement across membranes (p. 216)
In facilitated diffusion, proteins bind solutes and transport them across membranes from a region of higher solute concentration to a region of lower solute concentration. In active transport, solutes move up a concentration gradient with the help of energy supplied by the cell.

Exocytosis and endocytosis transport large molecules (p. 216)
Molecules leave the cell by exocytosis and enter by endocytosis. In exocytosis, molecules are packaged in membrane-bound vesicles, which fuse with the plasma membrane. In endocytosis, the plasma membrane forms pockets around molecules, which become vesicles.

Osmosis is the movement of water across a selectively permeable membrane (pp. 216–217)
In osmosis, water moves from a region of lower solute concentration (higher water concentration) to a region of higher solute concentration (lower water concentration). Osmosis can occur with or without transport proteins, called *aquaporins*.

In plant cell growth, the osmotic potential inside the cell interacts with pressure generated by the cell wall (pp. 217–219)
Cells grow when their water potential is negative. Water potential is the sum of the osmotic potential and pressure potential. Osmotic potential is generated by the cell's solute concentration. Pressure potential is generated by cell wall resistance to protoplast expansion.

Movement and Uptake of Water and Solutes in Plants

Water evaporation from leaves pulls water through the xylem from the roots (pp. 219–223)
Transpiration through stomata creates tension in water columns of xylem. Cohesion of water molecules to each other and adhesion to cell walls help support the water in tracheids and vessel elements. Transpiration is regulated by the opening and closing of stomata in response to light, CO_2, and abscisic acid.

Stomata control gas exchange and water loss for the plant (pp. 223–224)
A full 90% of a plant's water loss occurs through stomata, which open in response to lowered CO_2 concentrations or to blue light and close due to increases in abscisic acid.

Sugars and other organic molecules move from leaves to roots in the phloem (pp. 224–226)
According to the pressure-flow hypothesis, osmotic pressure caused by production of sugars in photosynthesis pushes a sugar solution from its source in leaves to sinks in stems, roots, and fruits.

Soil, Minerals, and Plant Nutrition

Soil is made of ground-up particles of rocks surrounded by negative charges that bind water and minerals (p. 227)
Soil occurs in layers called *horizons* and consists of a mixture of sand, silt, and clay mixed with organic matter.

Plants require 17 essential elements, most of which are obtained from soil (pp. 227–229)
Plants require nine macronutrients and at least eight micronutrients. Mineral deficiencies in soils are passed onto the food chain to nonphotosynthetic organisms.

Soil particles bind water and mineral ions (pp. 229–230)
The soil solution consists of water and dissolved minerals bound to negatively charged soil particles by a matrix potential. Cations bind to soil particles, depending on the relative charges and their concentrations.

Bacteria in the soil make nitrogen available to plants (pp. 230–232)
Plants absorb nitrogen as nitrate or ammonium ion from the soil. All nitrogen in the soil is secured by nitrogen-fixing bacteria, which convert ammonium (NH_4^+) to nitrate (NO_3).

Review Questions

1. How do symplastic transport and apoplastic transport differ?
2. Explain the difference between diffusion, facilitated diffusion, and active transport.
3. Name some cellular processes in plants that use exocytosis.
4. If a selectively permeable membrane separates a solution and fresh water, in which direction will water flow? Why?
5. Why is a plant cell like a water balloon inside a cardboard box?
6. How does water potential affect the flow of water in a plant?
7. Describe plasmolysis and how it occurs.
8. How does transpiration relate to transport?
9. How do the properties of water facilitate transport in plants?
10. Why are pumps placed at the bottom of water wells?
11. Explain how stomata function and why they are important.
12. How does rainfall produce acidic soils?
13. Give three examples of macronutrients and three examples of micronutrients, explaining the importance of each.
14. Describe the conversion of nitrogen gas to nitrate by bacteria.
15. Why does farming result in nutrient-poor soils?

Questions for Thought and Discussion

1. During droughts, trees lining irrigation ditches have sometimes been cut to save water. Is this a sound idea? Explain.
2. As the relative humidity approaches 100%, transpiration decreases. Does this result in plants being nutrient-poor? Explain.

3. Describe some creative ways to maintain soil fertility on agricultural land.

4. The presence of cell walls in plants is thought by some scientists to indicate that plants evolved in fresh water rather than from oceanic algae. Why are cell walls useful to plant cells surrounded by fresh water?

5. Why do cells that are carrying out photosynthesis take up water?

6. Diagram the pathway taken by a single water molecule that, by chance, is carried from the soil to a leaf of a tall tree and then returns from the leaf to the roots of the same plant. Include in your diagram the key cells and tissues through which this movement occurs.

Evolution Connection

As you have read in this chapter, agronomists hope to one day transfer the ability to fix atmospheric nitrogen from certain microorganisms to crop plants. Why do you think this ability has apparently never evolved naturally in plants.

To Learn More

Visit The Botany Place Website at www.thebotanyplace.com for quizzes, exercises, and Web links to new and interesting information related to this chapter.

Brady, Nyle and Ray Weil. *The Nature and Properties of Soils.* Upper Saddle River, NJ: Prentice Hall, 2001. The physical, chemical, and biological properties of soil are explored.

Gleick, Peter H. *The World's Water 2002–2003: The Biennial Report on Freshwater Resources.* Washington, D.C.: Island Press, 2002. An interesting general view of all aspects of water availability, conflicts, and sanitation.

Postel, Sandra. *Pillar of Sand: Can the Irrigation Miracle Last?* New York: W. W. Norton, 1999. She discusses the role of irrigation in human history; the present status of insufficient resources; and how the future of irrigated agriculture might be improved.

Taiz, Lincoln, and E. Zieger. *Plant Physiology.* Sunderland, MA: Sinauer Associates, 2002. This excellent text explores all areas of plant physiology including mineral nutrition and water relations of cells and whole plants.

Plant Responses to Hormones and Environmental Stimuli

A lesser celandine (*Ranunculus ficaria*) follows the sun.

Effects of Hormones

Auxin plays a central role in cell enlargement and formation of new tissue

Cytokinins control cell division and differentiation and also delay aging

Gibberellins interact with auxins to regulate cell enlargement and stimulate seed germination

Abscisic acid causes seed dormancy and regulates plant responses to drought

Ethylene allows the plant to respond to mechanical stress and controls fruit ripening and leaf abscission

Brassinosteroids are a newly discovered group of plant hormones that act like auxin

Additional compounds may play a role as plant hormones

Responses to Light

Blue light absorption controls the growth of stems toward the light and the opening of stomata

Absorption of red and far-red light determines when seed germination, stem and root growth, and flowering occur

Photoperiodism regulates flowering and other seasonal responses

Plants respond to repeating cycles of day and night

Responses to Other Environmental Stimuli

Roots and shoots respond to gravity

Plants respond to mechanical stimuli, such as touch and wind

Plants prepare for environmental conditions that prevent normal metabolism and growth

Plants react to environmental stresses such as drought

Plants deter herbivores and pathogens

In Chapter 1 you learned that Charles Darwin and his son Francis studied the bending of plants toward the light, with the aim of discovering the "eye" of the plants. Of course, plants do not have an eye, but they do grow toward the light and in some cases follow the movement of the Sun across the sky. Responses to internal and external stimuli help plants survive and thrive in a changing, sometimes hostile environment. For organisms with no nervous system that are literally rooted to the ground, plants have a remarkable variety of developmental responses. Through the use of translocated hormones, one part of the plant can "communicate" with another. For example, when the shoot apical meristem is damaged or lost, hormonal signals stimulate axillary buds to grow. Plants also sense the outside environment. For example, some seeds are biochemically programmed to germinate only when light conditions are similar to those favored by the mature plant. In other plants, metabolic energy heats their flowers, melting the surrounding snow. Still others shed seeds only when singed by fire that clears out the vegetation to allow sunlight needed by the young trees.

Forest fires stimulate some cones to open and release seeds.

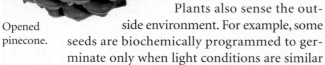

Opened pinecone.

Developmental responses of plants occur throughout their life cycles and are initiated by changes in environmental variables such as temperature, light intensity, and daylength. Developmental responses of plants vary with the season and even with time of day, maximizing short-term and long-term survival. For example, the century plant (*Agave parryi*) grows for around 25 years before it is developmentally ready to respond to environmental cues of a particular summer and produce a large group of flowers before dying.

Developmental responses are typically initiated by internal changes and external environmental stimuli. One major external stimulus is light, which triggers responses by hormones and photoreceptors. A photoreceptor consists of a large protein and a chemical compound that absorbs light of a particular wavelength. Hormones are also involved in responses to other environmental stimuli, such as gravity or touch.

Because photoreceptors and hormones are products of enzyme-catalyzed reactions, their synthesis is subject to genetic control and therefore to the variation induced by mutation. Plants with superior developmental adaptations produce more offspring and so are more successful at transferring their genes to future generations. Therefore, adaptive developmental patterns become fixed in populations by evolution.

In this chapter, we will consider the hormones and photoreceptors that provide the plant with information about the internal and external environment. They regulate responses throughout development from seeds into seedlings and then into mature plants. They also regulate responses to seasonal changes and to daily variations in light and temperature.

Century plants in bloom.

Effects of Hormones

Hormones in plants, as well as in animals and other organisms, direct growth and development, as well as responses to environmental stimuli. Although plants do not have a nervous system or a brain to assist in responding to stimuli, they do use hormones in remarkable ways.

The term *hormone* comes from the Greek *hormon,* meaning "to stir up," because scientists initially observed that hormones stimulated responses. However, they discovered later that hormones also inhibit responses. For example, some suppress the growth of axillary buds, while others promote such growth. The effect of a particular hormone may vary, depending on its concentration or on the types and locations of the cells upon which it acts. Also, a particular aspect of plant growth or development may involve multiple hormones, frequently in response to external stimuli such as light. For instance, three hormones and two photoreceptors are involved in regulating stem elongation.

A hormone is a small molecule that carries information from the cell where it was produced to particular target cells, causing a change in response to internal needs or external stimuli. Some plant hormones, for example, indicate that leaves need water from the roots or that the roots require sugar from the shoot. Some are released in response to changes in the outside environment, such as temperature, daylength, light, wind, or to herbivores and disease-causing organisms. Hormones are therefore key players in a plant's communication system, both internally and in reacting to the external environment.

Hormones almost always act by binding to a protein, thereby initiating a **signal-transduction pathway,** a series of events that ultimately stimulates or inhibits a cellular response (Figure 11.1). A signal-transduction pathway consists of interactions between molecules as they relay the signal. Most of these molecules are proteins, but some are small water-soluble molecules or ions called *second messengers,* which relay the information from the primary messenger—a hormone or environmental stimulus. Most developmental or growth responses in plants integrate the activity of several hormones and photoreceptors, each acting through a signal-transduction pathway. Each hormone or photoreceptor may affect a particular developmental event, such as germination or flowering, differently. In this section, we will look at six main types of plant hormones: auxins, cytokinins, gibberellins, abscisic acid, ethylene, and brassinosteroids (Table 11.1).

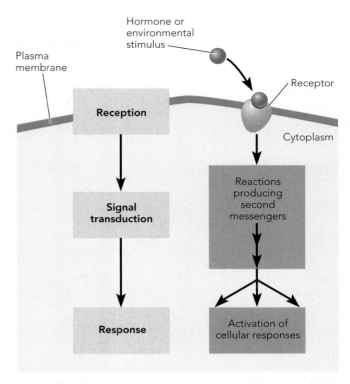

Figure 11.1 Signal-transduction pathway. Hormones bind to proteins associated with the membranes of the cells they will affect. An environmental stimulus or the binding of a hormone initiates a series of events called a *signal-transduction pathway,* which leads to activation or inactivation of particular enzymes or proteins.

Auxin plays a central role in cell enlargement and formation of new tissue

Auxin was the first plant hormone discovered, with experiments by Charles Darwin and his son Francis in 1880 laying the groundwork. The Darwins observed that a grass seedling bends toward the light only if the tip of the coleoptile—the sheath covering the young grass shoot—is present. If the tip is removed or covered with an opaque cap, bending does not occur (Figure 11.2a). The Darwins proposed that the coleoptile tip receives light and sends a signal down the coleoptile to the elongating, curving region. In 1913, Danish botanist Peter Boysen-Jensen discovered that the signal is mobile and can pass through permeable agar but not through impermeable mica (Figure 11.2b). In 1926, a Dutch graduate student named Fritz Went removed coleoptile tips, placed them on agar blocks, and found that a substance accumulates in the agar, causing growth when the blocks are placed on coleoptiles with the tips removed (Figure 11.2c). He called this substance *auxin* (from the Greek *auxein,* "to increase"). Furthermore, Went observed that placing the

Table 11.1 Plant Hormones

Hormone		Where synthesized in plants	Major functions
Auxins (Example shown: IAA)		Embryos, meristems, buds, young leaves	Stimulates stem and root growth; promotes cell differentiation in tissue culture and in procambium; regulates development of fruit; apical dominance; causes phototropism and gravitropism
Cytokinins (Example shown: zeatin)		Roots, seeds, fruits, leaves	Promotes root growth and differentiation; stimulates cell division and growth in tissue culture; stimulates germination; retards aging
Gibberellins (Example shown: GA$_3$)		Meristems, young leaves, embryos	Promotes seed germination and bud growth; promotes stem elongation and leaf growth; stimulates flowering and fruit development
Abscisic acid (ABA)		Leaves, stems, roots, fruits	Inhibits growth; closes stomata during water stress; promotes dormancy
Ethylene		Ripening fruits, aging leaves and flowers	Promotes ripening of some fruits and thickening of stems and roots
Brassinosteroids (Example shown: brassinolide)		Seeds, fruits, shoots, leaves, and flower buds	Auxin-like effects; inhibits root growth; retards leaf abscission; promotes xylem differentiation

block on half of the cut surface of the tip would cause the coleoptile to grow only on that side. Later, in 1931, the structural formula of auxin was determined to be that of indoleacetic acid (IAA).

There are some synthetic auxins and several natural auxins, but the most common natural form of auxin in plants is indoleacetic acid, the auxin referred to throughout this chapter. This auxin, a modified amino acid derived from tryptophan, is produced mainly in shoot apical meristems, young leaves, and embryos. It is transported through phloem parenchyma, rather than in the conducting cells of either phloem or xylem.

The major short-term effect of auxin is to stimulate cell growth. The proposed explanation of this process is the acid growth hypothesis, which was developed over the past 40 years and states that auxin stimulates certain proteins to pump hydrogen ions (H$^+$) into the cell wall. These ions activate enzymes called *expansins*, which weaken the cell wall by breaking links between cellulose microfibrils, enabling the cell to expand. In terms of water transport, expansins reduce the resistance of the cell wall, allowing water to flow into the cell through osmosis, thereby expanding the cell. Recent evidence suggests that auxin may activate genes involved in cell wall expansion.

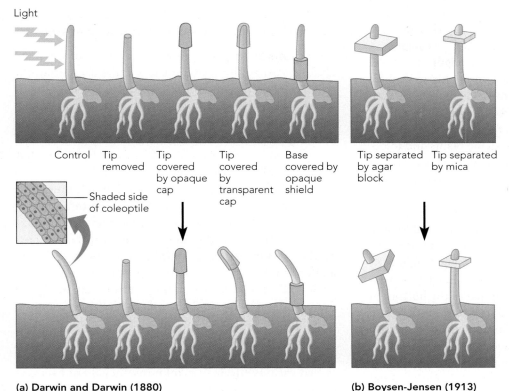

(a) Darwin and Darwin (1880)

(b) Boysen-Jensen (1913)

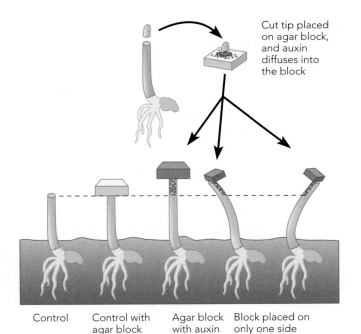

(c) Went (1926)

Figure 11.2 Experiments with auxin. (a) Charles Darwin and his son Francis demonstrated that only the tip of the coleoptile perceives light. As shown in the closeup view, the cells on the shaded side elongate. **(b)** Experiments by Peter Boysen-Jensen showed that the tip of the coleoptile produces a stimulus that will pass through agar but not mica. **(c)** Fritz Went demonstrated that the tip secreted auxin into agar blocks, which could be applied to coleoptiles with no tips to stimulate growth. Agar blocks applied to half of the coleoptile stimulated growth on only that side.

Experiments with grass seedlings and other tissues reveal that growth response varies with the auxin concentration. Growth increases until the auxin concentration reaches a peak and then falls as high auxin levels promote synthesis of ethylene. The optimal concentration of auxin for cell elongation is higher for shoot cells than for root cells. During transport, the concentration decreases as a result of an auxin-destroying enzyme called IAA-oxidase.

Although mainly associated with cell enlargement, auxin has many developmental effects, some occurring in minutes and others taking hours or longer. There is still much to learn about how auxin works, and its many effects probably involve a variety of signal-transduction pathways. Studies of tissue cultures and seedlings reveal that auxin in apical meristems stimulates formation of vascular tissue, as well as lateral and adventitious roots.

Auxin is also responsible for the development of lateral meristems—the vascular cambium and cork cambium.

Auxin is involved in apical dominance (see Chapter 3), the suppression of axillary bud growth (Figure 11.3). Removal of the shoot apical meristem terminates apical dominance and stimulates the growth of axillary buds. If a shoot apex is damaged or destroyed, one of the axillary buds quickly becomes the leader shoot. Gardeners frequently pinch off shoot apices to shape plants and to develop more fruit-bearing buds. Cytokinins, which you will read about shortly, promote growth of axillary buds, thereby countering the effects of auxin.

Synthetic auxins such as naphthaleneacetic acid (NAA), which are not degraded by IAA-oxidase, are used to both stimulate and inhibit growth. Some, such as indolebutyric acid, are used commercially to propagate plants by inducing root formation on cuttings. A powder or paste containing a synthetic auxin is applied to the bottom of the cutting, which is then placed in water, sand, or soil to allow for root formation. Some synthetic auxins, such as 2,4 = dichlorophenoxyacetic acid (2,4 = D) are used today as weed killers, causing the plants to produce high concentrations of ethylene, which triggers aging. Broad-leaved plants are killed more readily than grasses, probably because of greater absorption of the auxin. Agent Orange, a chemical agent used for defoliation during the Vietnam War, was a mixture of synthetic auxins. The health problems caused by Agent Orange were not the result of the auxins themselves, which are not toxic to humans, but instead were traced to a contaminating chemical used in the production of synthetic auxins.

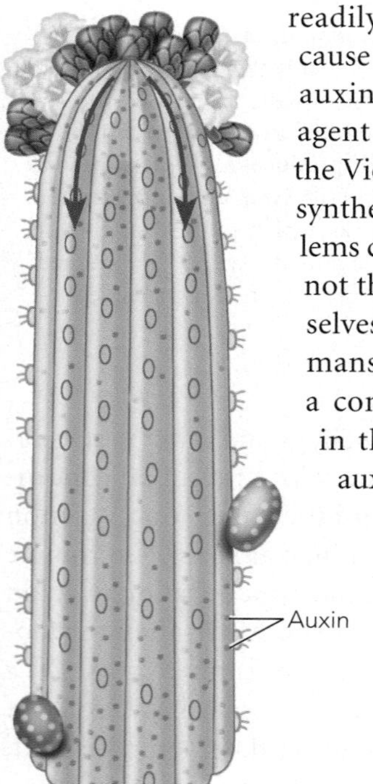

Figure 11.3 Apical dominance is caused by auxin. Cacti have strong apical dominance. When the auxin concentration drops below a critical level, an axillary bud located at a spine cluster begins growth. Sometimes the bud lowest on the stem is not the first to begin growth, indicating that hormones other than auxin produced by the apical meristem may be involved.

Cytokinins control cell division and differentiation and also delay aging

Cytokinins affect plant growth in a variety of ways, including controlling cell division and differentiation, counteracting apical dominance, and delaying aging of leaves. The name *cytokinins* refers to their role in cell division, or cytokinesis. They are usually modified forms of adenine and were originally discovered as a result of experiments on tobacco plants to find chemicals that stimulate cell growth. Cytokinins are synthesized in the roots and transported through the xylem to other plant organs, where overall they foster a more youthful state of development. In the stem, for example, they promote axillary bud growth. If the shoot apical meristem is damaged or removed, the cytokinin-to-auxin ratio in the axillary buds increases, thereby promoting faster bud growth. Direct application of cytokinins can also promote bud growth even if the apical meristem is intact. Cytokinins delay leaf aging and increase leaf longevity in various ways, including attracting amino acids from other parts of the plant. Although scientists have observed various effects of cytokinins, they do not yet understand the signal-transduction pathway for these hormones.

In plant tissue cultures, cytokinins are associated both with cell division and with differentiation leading to the production of shoot buds. By themselves, cytokinins have few effects on cultured cells, but when they are supplied along with auxin, cultured cells begin to divide and differentiate. These effects depend greatly on the hormone concentrations. If only auxin is supplied, the cultured cells elongate but do not divide. If a cytokinin is also added, the effects depend on the auxin-to-cytokinin ratio. If the cytokinin concentration is low compared to auxin, the cells grow, divide, and differentiate into roots. At moderate cytokinin concentrations, they grow and divide rapidly to produce a mass of undifferentiated cells, called a callus, but do not differentiate. At high cytokinin concentrations, they grow, divide, and differentiate into shoot buds. The effects of cytokinin and auxin in tissues have been found to apply to many types of plants (see the *Biotechnology* box on page 241).

Gibberellins interact with auxins to regulate cell enlargement and stimulate seed germination

Gibberellins are a class of hormones affecting a wide variety of developmental phenomena in plants, including cell elongation and seed germination. The name is de-

B I O T E C H N O L O G Y

Effects of Auxin and Cytokinins on Cultured Plant Cells

The ability to regenerate plants from tissue culture is important not only in cloning useful plants but also in regenerating cells that contain new genetic material from experiments in genetic engineering. In 1941, Johannes van Overbeek found that the liquid endosperm of the coconut (*Cocos nucifera*), called coconut water, promoted growth, cell division, and survival of plant tissue cultures. A decade later, Folke Skoog and Carlos Miller found similar results using coconut water in tissue cultures of tobacco. By chance, they discovered that "old," chemically degraded DNA was also useful in maintaining successful tissue cultures. Some chemical detective work led to the discovery and structural identification of the first cytokinin, called kinetin.

In tobacco tissue cultures a nutrient medium containing macronutrients, micronutrients, vitamins, 0.18 milligram per liter (mg/l) of auxin, and no kinetin produced roots. A medium containing 1.08 mg/l of auxin and 0.2 mg/l of kinetin only produced a callus of growing and dividing cells, while a medium containing 0.03 mg/l of auxin and 1.0 mg/l of kinetin produced shoots and eventually entire plants. For a large number of dicots, varying the auxin-to-cytokinin ratio led to the ability to control the type of differentiation arising from cultured cells. High ratios led to roots; medium, to callus; and low, to shoots.

For monocots, however, the situation was more difficult. While cultured cells grew well as callus on tissue culture medium, no amount of variation in the auxin-to-cytokinin ratio produced regenerated plants. In the 1970s, it was discovered that many monocots produce two different types of cells in culture. Large, highly vacuolated cells constitute the majority of cells that form, while a few small, nonvacuolated cells also occur. The larger, faster growing cells do not produce regenerated plants on any known tissue culture medium. The smaller, slower growing cells respond quite well to the addition of cytokinins by regenerating large numbers of embryos that can be induced to form plants on media with high cytokinin-to-auxin ratios. Embryo-producing callus forms readily when the seed leaf, or cotyledon, is placed on a tissue culture medium containing auxin. Transfer of callus to a medium containing auxin and high levels of cytokinins results in embryo and plant formation.

Today, the challenges are to develop tissue culture methods for regenerating so-called difficult plants, which include many forest trees. Also, investigators are looking for individual genes that influence the regeneration process and can be moved from one species to the other by genetic engineering.

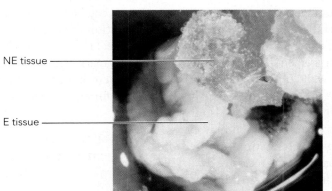

Embryogenic and non-embryogenic tissue. The crystalline mass of tissue is 1.0 centimeter (0.39 inch) across and contains around 50,000 non-embryogenic (NE) cells, which cannot readily be induced to form plants. The creamy mass of tissue contains an equal number of embryogenic (E) cells, which can be induced to form asexual embryos by hormones included in the tissue culture medium.

NE tissue

E tissue

(a)

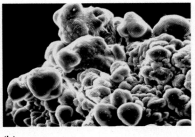

(b)

(c)

Embryos forming on rice tissue cultures. (a) A light micrograph showing asexual embryos forming as bumps on the tissue culture. (b) A scanning electron micrograph showing asexual embryos. (c) This clump of asexual embryos from tissue culture shows several that have begun to germinate.

rived from a fungus of the genus *Gibberella*, discovered by Japanese scientists to secrete a chemical that caused infected rice stems to grow too tall and then topple. The chemical was given the name *gibberellin* and was later found to be present naturally in plants in regulated amounts and in a variety of forms. There are more than 110 different gibberellins, but in any given plant species only a few kinds are biologically active. Like auxin, gibberellins are synthesized in apical meristems, young leaves, and embryos. While auxins and cytokinins are made from amino acids and bases, gibberellins are made by linking five-carbon isoprenoid units together to make a characteristic structure containing four rings.

Gibberellins are one of several types of hormones involved in stem elongation. You already know that auxins are thought to promote cell enlargement by activating expansin proteins that serve as wall-loosening enzymes. Gibberellins may facilitate the movement of expansins into the correct positions in the cell wall. They also increase cellular concentrations of auxin, which may explain their dramatic effect on cell elongation. Applied gibberellins can reverse dwarfism in many recessive dwarf mutants with low gibberellin levels. By examining mutants in which cell elongation is inhibited, researchers are investigating how various hormones and photoreceptors participate and interact in cell elongation.

Gibberellins play roles in both embryo growth and seed germination. In one signal-transduction pathway in germinating barley seeds, they stimulate the production of the enzyme alpha-amylase, which breaks down starch to provide glucose for the seedling. In another pathway, also in barley, they activate the secretion of this enzyme. Gibberellins also promote seed germination. The hormone abscisic acid, which you will read about shortly, prolongs seed dormancy, which is characterized by high concentrations of abscisic acid and low concentrations of gibberellins in the embryo. With time, the abscisic acid decays and gibberellin synthesis increases. The processes that allow seeds to germinate after a passage of time following their formation are known generally as "after ripening." After imbibition, the passive uptake of water by the seed, gibberellins released from the embryo signal that it is time for the seed to end its dormancy and to germinate.

Gibberellins are also involved in promoting flowering of some plants, including plants that normally require cold treatments, usually provided by winter and plants that "bolt" to form a tall inflorescence in their second year of growth. In agriculture, and in experiments using native plants for land restoration, flowering can be accelerated by storing seeds or plants at near-freezing temperatures before planting them, thereby substituting for the effect of a long winter. The practice of using cold treatment to hasten flowering is known as **vernalization** (from the Latin *vernus*, "spring") because it reduces the dormancy period before spring. Botanists have found that treating plants with gibberellins has the same effect as vernalization, resulting in a shorter growing season and more rapid flowering. Such treatments are often used in temperate regions with short growing seasons because accelerated flowering may spell the difference between a successful crop and a failed crop.

Gibberellins contribute to fruit formation as well, leading to useful commercial applications. When applied to developing bunches of grapes, for example, they promote elongation of stem internodes and increased grape size (Figure 11.4). The resulting bunch of large grapes has greater market value, as well as fewer problems with fungi and bacteria because there is more space between the grapes for air circulation.

Abscisic acid causes seed dormancy and regulates plant responses to drought

Abscisic acid (ABA), a terpenoid hormone synthesized in leaves, stems, roots, and green fruit, establishes dormancy in seeds and other plant organs and helps the plant adapt to water stress. Originally, abscisic acid was mistakenly thought to play a major role in leaf abscission, hence the name (see Chapter 4). However, this role now appears to be minor.

In terms of its role in seed dormancy, ABA promotes the production of storage proteins in seeds and increases in concentration as the seed matures. Then it prevents germination during a period of cold temperatures, until increasing synthesis of gibberellins allows seeds to germinate. In some plants, single-gene mutants either reduce the concentration of ABA or reduce its effect on germinating seeds, resulting in premature germination.

During periods of water stress, ABA promotes the closing of stomata, preventing additional water loss. The ac-

Figure 11.4 Gibberellin causes stem elongation and fruit growth. When sprayed on grapes, gibberellin promotes elongation of stems bearing the fruit and increases the size of the fruits themselves.

tion of ABA on guard cells is not well understood but appears to involve at least three signal-transduction pathways. Stomatal opening and closing respond to a number of environmental signals, which may explain the complexity of ABA's mechanism of action.

Ethylene allows the plant to respond to mechanical stress and controls fruit ripening and leaf abscission

Ethylene is a gas that acts like a hormone by causing responses to mechanical stress and also by stimulating aging responses such as fruit ripening and leaf abscission. Ethylene synthesis is initiated by high concentrations of auxin, stress, and various developmental phenomena.

Ethylene keeps plants from becoming too spindly in the face of wind and other stresses that might damage them. The gas actually represses growth in length while stimulating expansion in width. Therefore, plants stimulated by auxin become tall, while plants stimulated by ethylene become stout. Ethylene production increases when the plant is touched, buffeted by wind, or in any way damaged (Figure 11.5).

Ethylene also enables the plant to adapt successfully to the hazards of growth underground. When an underground shoot or root encounters an obstacle, the pressure induces ethylene synthesis. Ethylene then initiates a growth maneuver called the **triple response** that allows the shoot or root to push aside or grow around the obstacle (Figure 11.6). The triple response includes (1) a slowing of stem or root elongation, (2) a thickening of the stem or root, and (3) a curving to grow horizontally. The second and third parts of the response enable the stem or root to successfully go around the obstacles.

During the 1800s, ethylene was used as an illuminating gas on city streets, where leaks in the gas lines would occasionally lead to the defoliation of trees, a sign of ethylene's role in leaf abscission. Ethylene action in the abscission of leaves and of fruits has been particularly well studied. The ethylene receptor is a transmembrane protein. The components of the signal-transduction pathway, which are involved after ethylene binds to this protein, include other proteins and a gene that apparently codes for a protein that controls the opening and closing of a membrane channel or pore.

In some fruits, known as *climacteric fruits,* ethylene regulates fruit ripening. Examples are apples, avocados, bananas, cantaloupes, figs, mangos, peaches, plums, and tomatoes. Climacteric fruits have a large, rapid increase in ethylene production that precedes a sharp increase in CO_2 production during ripening (Figure 11.7). In non-

Figure 11.5 Touch or other physical stress induces ethylene production. The shorter *Arabidopsis* plant was touched twice a day, stimulating a release of ethylene that represses growth in length. The taller plant was untouched.

Ethylene concentration (parts per million)

Figure 11.6 The triple response to ethylene. In the triple response to ethylene, plants cease growth in length, become thicker, and grow horizontally. The extent of the response varies with ethylene concentration.

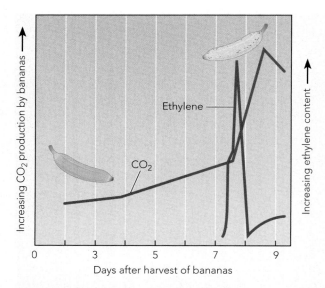

Figure 11.7 Ethylene and fruit ripening. In climacteric fruits such as bananas, a rapid rise in ethylene production precedes the increase in respiration, indicating that ethylene is involved in ripening.

climacteric fruits—such as citrus fruits, grapes, peppers, pineapples, strawberries, and watermelons—CO_2 production decreases gradually as ripening occurs. Tomatoes with less ethylene than normal will ripen more slowly. Such fruits grow and lose chlorophyll normally, but show less reddening and high resistance to overripening and shriveling, thereby remaining fresh and tasty for several weeks after purchase. As you can imagine, growers and grocers are eager to bring such tomatoes to the marketplace. Even tropical fruits with very short shelf lives can be successfully marketed if ethylene-deficient mutants behave like these tomatoes.

Currently, ripening of climacteric fruits can be delayed by storing them at low temperatures in atmospheres of low O_2 and high CO_2. The low temperature and low O_2 level prevent any new synthesis of ethylene, while the high CO_2 level prevents action of any ethylene that was already produced. Before leaving the warehouse, each fruit shipment receives a carefully calculated dose of ethephon, a water-soluble chemical that gradually releases ethylene. In this way, stores can carry ripe, climacteric fruits year-round.

Brassinosteroids are a newly discovered group of plant hormones that act like auxin

Steroid hormones are common in animals but until recently were not identified in plants. Plant steroid hor-

mones were first discovered in the genus *Brassica,* which includes cabbages, and were accordingly named **brassinosteroids.** They bind to a receptor protein in the plasma membrane and do not enter the cell, in contrast to animal steroids. Brassinosteroids stimulate cell division and elongation in stems, cause differentiation of xylem cells, promote pollen tube growth, slow root growth, and delay leaf abscission, while enhancing ethylene synthesis. In general, these effects are very similar to those of auxin, but mutants that do not synthesize brassinosteroids usually respond to applied auxins, so apparently the two types of hormones act through different pathways.

Additional compounds may play a role as plant hormones

Plant physiologists recognize at least two additional classes of compounds as being potential plant hormones: polyamines and jasmonic acid. Polyamines, so named because they are synthesized from amino acids, promote cell division and synthesis of DNA, RNA, and proteins. Bacterial and animal cells appear to use polyamines as hormonal substances. In plants, polyamines promote root initiation and tuber formation and are involved in the development of embryos, flowers, and fruits. Jasmonic acid, synthesized from fatty acids, inhibits growth of seeds, pollen, and roots, while promoting accumulation of proteins during seed development. Jasmonic acid also stimulates formation of flowers, fruits, and seeds and is active in plant defenses against pathogenic organisms. Some experimental evidence indicates that a related compound, methyl-jasmonate, may serve as an early warning, signaling plants downwind that pathogens or herbivores are damaging plants in the vicinity. When sagebrush plants in the field were injured by clipping, they released methyl-jasmonate, which caused nearby wild tobacco plants (*Nicotiana attenuata*) to produce a defensive enzyme. Subsequent damage to the tobacco leaves from grasshoppers and cutworms was greatly reduced.

Section Review

1. **What is a hormone?**
2. **How are cytokinins similar to and different from auxin?**
3. **How do gibberellins and abscisic acid interact?**
4. **Describe several roles of ethylene.**

Responses to Light

Hormones are often involved in growth responses of plant organs that result in a turning toward or a turning away of stems, leaves, or roots from external stimuli. Some of these growth responses are called **tropisms** (from the Greek *tropos,* "turn") because they are movements in response to external stimuli. One major type of response to light is **phototropism,** growth toward or away from light. Growth toward light is known as *positive phototropism,* while growth away from light is called *negative phototropism.* Later in the chapter, we will look at some other types of tropisms.

Several photoreceptors are involved in phototropism. Although plants have neither eyes nor brains, they obtain a surprising amount of information from their exposure to light. Many developmental phenomena, from flowering to the growth of stems toward the light, are mediated by particular wavelengths of light, all of which are components of sunlight.

Blue light absorption controls the growth of stems toward the light and the opening of stomata

Blue light regulates a number of developmental phenomena in plants, including opening of stomata, inhibition of hypocotyl elongation, and phototropism. The bending of the stem toward blue light, an example of phototropism, is caused by the movement of auxin to the darker side of the stem. In phototropism, shoots grow toward light, while roots grow away from it. A recently discovered flavin-containing protein, named *phototropin,* absorbs blue light and initiates a signal-transduction pathway leading to movement of auxin to the dark side of the stem. This movement stimulates growth on the dark side, causing movement of the stem toward the light (Figure 11.8). Winslow Briggs and his colleagues at Stanford University discovered phototropin and have begun to understand its actions. When blue light is absorbed, phosphate is added to phototropin, which initiates signal transduction. Phototropin is similar to other proteins that help organisms detect light, oxygen, and voltage.

Earlier work by Briggs and coworkers supported the hypothesis that light causes auxin to move to the dark side of the coleoptile. Auxin is not simply destroyed on the lighted side, as had been supposed. Roots respond to blue light by growing away from it, a response that occurs in nature during the germination of seeds on the ground or near the soil surface. Roots react differently than

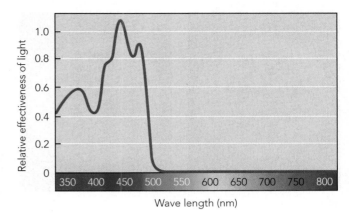

(a) Action spectrum for phototropism stimulated by blue light

(b) Coleoptiles before light exposure

(c) Coleoptiles after 90-minute exposure to the indicated colors of light

Figure 11.8 Blue light causes phototropism.

shoots because they have a greater sensitivity to auxin than do shoots, so concentrations that promote shoot growth actually inhibit root growth.

Recent studies have shown that stomatal opening is promoted by blue light and reversed by green light. Identification of the photoreceptor is an area of current research, with the carotenoid zeaxanthin and the flavin-containing protein phototropin having been proposed as photoreceptor molecules.

Absorption of red and far-red light determines when seed germination, stem and root growth, and flowering occur

Light controls many developmental phenomena, from flowering to stem elongation. The effects of light on plant growth and development are known as *photomorphogenesis* (from Greek words that essentially translate as "being shaped by light"). In the case of red and far-red (bordering on infrared) light, the photoreceptor that absorbs light and causes developmental effects is called **phytochrome.** This photoreceptor is like an on-off switch.

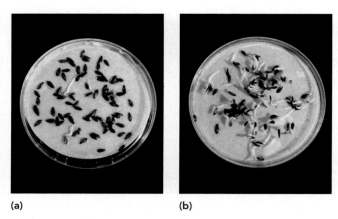

(a)   **(b)**

Figure 11.9 Photodormancy of seeds. Lettuce seeds have served as a model for studying photodormancy and the effect of phytochrome. (**a**) Lettuce seeds do not germinate if kept in the dark or if they receive a series of exposures to red and far-red light, ending with far-red light. (**b**) Lettuce seeds germinate if exposed to red light or a series of exposures to red and far-red light, ending with red light.

When exposed to red light (660nm), phytochrome is converted to a form that absorbs far-red light (720nm). This form of phytochrome is abbreviated P_{FR} (for phytochrome far-red). When exposed to far-red light, the same phytochrome is converted to a form that absorbs red light. This form of phytochrome is known as P_R (for phytochrome red). For most effects, P_{FR} is the on switch; that is to say, red light causes the developmental effect.

Phytochrome causes short-term and long-term effects, all of which are modulated by signal-transduction pathways. It controls many genes, including those involved in greening of leaves, and also the expression of several key proteins in photosynthesis.

Phytochrome was discovered as a result of studies on seed germination. Some plants produce seeds that are **photodormant,** meaning that they require activation by light. Photodormancy in seeds of plants that grow in the sun, such as some types of lettuce (*Lactuca sativa*), usually responds to red light (Figure 11.9). Sunlight contains both red and far-red light, but overall is more of a red-light source. In seeds of plants that grow in the shade, such as *Phacelia,* photodormancy breakage requires far-red light. Sunlight filtered through leaves has a greater far-red component. Therefore, photodormancy typically blocks germination until conditions are suitable for growth of the seedling.

Some varieties of lettuce with photodormant seeds have served as a model system for studying seed dormancy. In these seeds, dormancy can be broken by activation of the phytochrome system, by applied gibberellins, by stratification (cold treatment), or by removing the seed

coat and endosperm layers surrounding the embryo (see *The Intriguing World of Plants* box on page 247).

Phytochrome and cold treatment appear to stimulate the synthesis or action of gibberellins that promote germination. The actual uptake of water that causes the radicle to elongate is promoted by the breakdown of storage compounds such as fats, starch, and proteins in the embryo.

In general, phytochrome alerts sun-loving plants when they are shaded. Stimulated by the red-light component of sunlight, phytochrome inhibits cell elongation, probably by interfering with gibberellin production and action. Elongation resumes if the plant is shaded and receives primarily far-red light. In this way, the phytochrome response keeps stems short unless the plant faces competition for light. In the case of shade-loving plants, stem growth is promoted about equally by red and far-red light, but leaves appear healthier and greener in shade.

Photoperiodism regulates flowering and other seasonal responses

In temperate regions of the Earth, plants grow and flower in spring, summer, and fall but not in winter. To determine what time of year it is, plants in temperate regions integrate many environmental cues but usually rely mainly on detecting changes in the photoperiod, the relative lengths of night and day. This response to the relative lengths of night and day is known as **photoperiodism.**

In the 1920s, W. W. Garner and H. A. Allard studied a variety of tobacco, called *Maryland Mammoth,* which did not flower during the summer in Maryland. With the coming of autumn, they took plants into the greenhouse and observed that the plants finally flowered in December. On investigation, they found that the plants flowered when the day length is shorter than 14 hours. They coined the term **short-day plant (SDP)** to describe the tobacco and other plants that flower only when the days are shorter than a critical length (Figure 11.10). Short-day plants flower in late summer and early fall. Some examples are poinsettias, cockleburr, soy bean, violets, and some strawberries. Subsequently, long-day plants and day-neutral plants were discovered. A **long-day plant (LDP)** is one that flowers only when the daylength is longer than a critical length. Some examples are clover, petunias, and wheat. Long-day plants flower in late spring and early summer. A **day-neutral plant** flowers regardless of the daylength. Some examples are impatiens, corn, and holly. During the 1940s, researchers discovered that plants actually measured the night length, not the daylength. Therefore, it would really be more accurate to call a long-day plant (LDP) a short-night plant and to call a short-

THE INTRIGUING WORLD OF PLANTS

Studying Photodormant Seeds

The mechanism of action by which phytochrome breaks dormancy and promotes germination of some seeds is still not completely understood. Lettuce seeds (*Lactuca sativa,* Grand Rapids variety) have frequently served as a model system for studying photodormancy. If seeds are imbibed for several hours and then given a red light treatment, the radicle begins to protrude through the seed coat about 16 hours later. The promoting effect of red light can still be reversed by far-red light up to eight hours after red light. Whatever phytochrome (P_{FR}) does—its signal-transduction pathway—occurs during the eight hours preceding the start of radicle protrusion.

Removal of the lettuce seed's outer layers, consisting of the seed coat and endosperm, produces so-called *naked embryos*, which germinate in the dark with no red light

Phytochrome and seed dormancy.

treatment. Naked embryos treated with red light take up water and germinate more rapidly than embryos receiving a far-red treatment or no light at all. Photodormancy can be restored to these naked embryos by surrounding them with solutes, which compete with the embryo in binding with water molecules. The solutes mimic the seed coat and endosperm layers by providing a physical force preventing water uptake.

Red light induces a more negative water potential in embryos, which means a greater capacity to take up water. Either solute concentrations increase in cells of the radicle, where cell elongation occurs first in germination, or the cell walls weaken to allow cell expansion. These are the only two factors that can decrease the water potential to promote germination.

Graph: y-axis H_2O *Content in embryo (ml/g dry weight)* ranging 0.6 to 3.4; *x-axis* Hours after red light or dark treatment, 8 to 44. Two curves labeled "Red light" and "Dark".

---

day plant a long-night plant. However, the earlier terminology had already become well established.

Some plants, such as *Xanthium,* need just one exposure to the correct photoperiod to induce flowering; others such as soybean require several to many exposures. The phytochrome system plays a role in measuring night length. If a short-day (long-night) plant receives even a brief red-light interruption during an otherwise inductive long night, it will not flower. If a long-day (short-night) plant is kept on long nights, which do not induce flowering, a light flash during such non-inductive nights will restore flowering. Only red light causes this night-interruption effect (Figure 11.11). The effect of red can be reversed with far-red light, indicating that phytochrome is involved. However, phytochrome itself does not measure the night length like some kind of clock. Within a few hours, and well before the end of the long night, the form of phytochrome that absorbs far-red light, which predominates at the end of a sunny day, has all decayed back to the red-absorbing form. Scientists do not yet know what does measure the night length, although a subpopulation of phytochrome molecules may be involved.

Commercial growers manipulate the photoperiod of various plants to trigger flowering. For example, poinsettias are short-day plants that must be ready to sell in December.

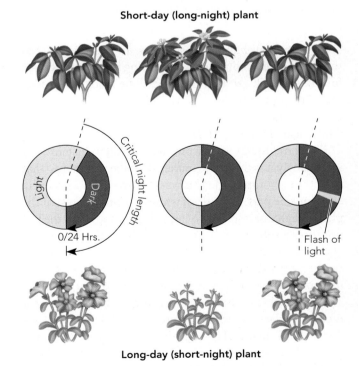

Short-day (long-night) plant

Long-day (short-night) plant

Figure 11.10 Short-day and long-day plants. Short-day (long-night) plants flower when the night exceeds a critical length. Long-day (short-night) plants flower when the night is shorter than a critical length.

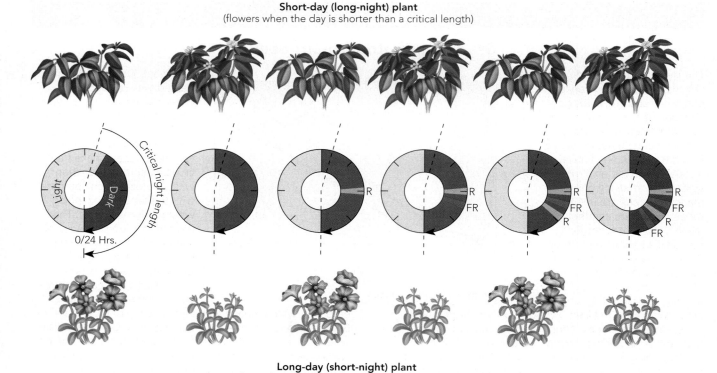

Figure 11.11 Night interruptions of flowering. In a short-day plant, a red-light interruption during a flower-inducing long night prevents flowering. If the red light is followed by far-red light, flowering occurs. In a long-day plant kept on a nonflower-inducing long night, a red-light interruption during the night restores flowering.

Started from cuttings in late summer, the plants are placed under short-day conditions beginning in September, for four to six weeks, so that flowers will be ready for Christmas. The poinsettia flowers, actually quite small, are produced along with colorful modified leaves called *bracts,* which also require short days.

Floral meristems form from shoot apical meristems, but surprisingly it is the leaves, rather than the meristems, that detect the photoperiod. This was discovered by selectively covering one part of the plant while exposing the rest of the plant to light. In response to inductive photoperiod, the leaf produces a substance that travels up the stem to the shoot apical meristem and promotes flowering. An induced leaf can be removed from a plant and grafted, for a few days at a time, onto a series of plants that were not induced to flower. The induced leaf continues to promote flowering for a number of days and is able to cause flowering in several additional host plants. The hypothetical substance that promotes flowering has not been identified but has been named **florigen.** Florigen does not appear to be any particular known plant hormone, although it may be a mixture of hormones. Despite many attempts, no compound acting like florigen has been isolated, possibly indicating that florigen is easily broken down during extraction or active only in extremely small quantities. The economic value of isolated florigen would be immense. The compound could be sprayed on plants to induce earlier flowering, flowering at a particular date, or uniform flowering of all plants in a field.

The prominent plant physiologist, Anton Lang, discovered that some long-day plants and biennials can be induced to flower by applied gibberellin, even when grown on short days. The hormone causes rapid elongation of a long flowering stalk in a process known as **bolting** (Figure 11.12). Lang and his colleagues also discovered that long-day plants produced inhibitors of flowering, which can be translocated through grafts into other plants. They found no evidence for flowering inhibitors in the leaves of short-day plants. Therefore, florigen, or one aspect of it in some plants, may be related to the disappearance of an inhibitor when an inductive photoperiod occurs.

A few plants, such as pineapple, flower in response to high concentrations of applied auxin, causing the synthesis of ethylene. Surprisingly, such plants can be induced to flower by turning them on their sides, where they produce

Figure 11.12 Some plants bolt to flower. Plants such as skunk cabbage will bolt, producing a long flowering stalk. This process occurs naturally but can be induced with gibberellins.

an increase in auxin on the lower side, promoting ethylene synthesis. Tipping plants over would work well for plants grown in containers, but this is, of course, impractical for field cultivation. Pineapple plants in fields are sprayed with an artificial auxin that induces them to flower and produce ripe fruit at the same time. In this way, harvesting the fruits is mechanized and considerably simplified.

Plants respond to repeating cycles of day and night

Plants have biological cycles of about 24 hours, known as **circadian rhythms** (from the Latin *circa*, "about," and *dies*, "day"). Many legumes close leaflets at night in response to decreasing light and temperature (Figure 11.13a). Many biochemical and physiological features of plants and other organisms ebb and flow on 24-hour cycles. Rates of transpiration and the activity of many enzymes oscillate on a circadian rhythm. Organisms other than plants also follow these cyclic rhythms. For example, the single-celled marine algae *Gonyaulax polyedra* photosynthesizes maximally in the day and produces its own light, by bioluminescence, maximally in the night.

Circadian rhythms continue on an approximate 24-hour cycle, even if plants are maintained in an unchanging environment of light or darkness. In nature, the phytochrome system plays an important role in maintaining and in resetting the circadian clock each day, as morning sunlight rapidly changes most of the phytochrome into the red-absorbing form.

Scientists who study circadian rhythms look for mutants in which the process has been altered, searching for one or more genes that are the master controlling elements. In one series of experiments, they studied the gene coding for a protein that binds chlorophylls in the photosystems of photosynthesis. Synthesis of this chlorophyll-binding protein occurs daily, beginning at first light. By attaching the firefly luciferase gene to the DNA coding for this protein, scientists cultivated an *Arabidopsis* plant that produced luminescence, like a firefly, whenever the gene was turned on. Mutants with an altered circadian

(a) In beans and many other plants, the leaves move from a horizontal, light-catching position in the day to a vertical position that minimizes heat loss to the atmosphere

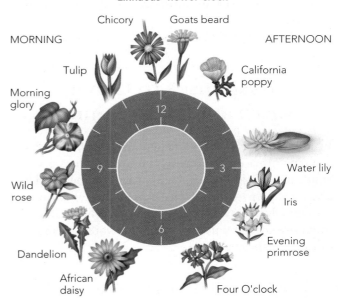

(b) The Swedish botanist Carolus Linnaeus planned a garden that could theoretically function as a "floral clock" because different plants open and close their flowers at different times of the day. By observing the status of the flowers, a person might determine the approximate time.

Figure 11.13 Sleep movements.

rhythm could then be selected by looking for plants that do not display luminescence at daybreak.

Many legumes position their leaves horizontally in the day and vertically at night. Darwin's proposed explanation is that these "sleep" movements reduced heat loss during the night by limiting the leaf surface area exposed to the night sky. The sleep movements of plants vary considerably among species. The flowers of many plants also open and close at specific times during the day or night. Flower opening is frequently correlated with the activity of particular pollinators such as insects, bats, or birds. Swedish botanist Carolus Linnaeus (1707–1778) designed a flower garden consisting of pie-shaped segments corresponding to different times of day (Figure 11.13b). Each segment would contain a plant that opened its flowers at that time of the day. Theoretically, a person could determine the time of day by observing which plants were opening or closing their flowers. It is not known, however, whether Linnaeus successfully planted such a "floral clock."

Section Review

1. Describe what happens in phototropism.
2. What are the effects of blue light and red light on plants?
3. Describe the effects of phytochrome.
4. What is meant by short-day plants and long-day plants?
5. How do cycles of day and night affect plants?

Responses to Other Environmental Stimuli

In addition to hormones and light, plant cells are subject to many other environmental stimuli, such as gravity, wind, and physical touch. In many environments, plants also must prepare for seasonal change and for periods of drought, flooding, and excessive temperatures. In order to survive, a plant must also deter herbivores and disease-causing organisms.

Roots and shoots respond to gravity

Growth toward or away from gravity is called **gravitropism.** When roots grow toward gravity, their growth is known as *positive gravitropism.* When shoots grow away from gravity, their growth is called *negative gravitropism.* Two current hypotheses attempt to explain how plants re-

spond to gravity. One hypothesis attributes gravitropism to specialized plastids called **statoliths,** which are full of dense starch grains and occur in root cap cells. If these cells die, gravitropism markedly decreases. Because of their density, statoliths settle to the bottom of cells, thereby identifying the direction of gravity for the plant (Figure 11.14). Auxin is transported to the gravitationally "down" side of roots, where it represses growth, allowing the "up" side of the root to grow resulting in downward curvature. In stems, statoliths in cells of the cortex and epidermis sense gravity and respond in a similar fashion, promoting cell elongation on the "down" side resulting in upward curvature. However, some plants that lack statoliths, such as certain *Arabidopsis* mutants, still have a reduced but significant response to gravity. Accordingly, statoliths may not be the only mechanism involved in gravitropism.

A second hypothesis, known as the gravitational pressure hypothesis, asserts that gravity stretches proteins on the upper plasma membrane and adds pressure to proteins on the lower plasma membranes to allow the plant to distinguish the upper and lower membranes. For example, proteins called *integrins,* found in both plants and animals,

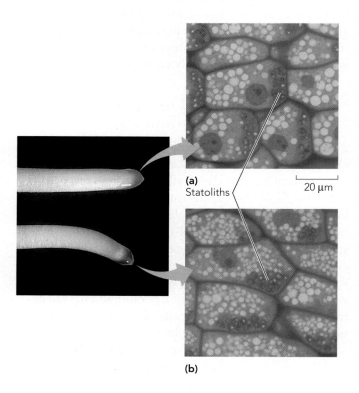

Figure 11.14 Statoliths may cause gravitropism. (a) This micrograph shows a root immediately after being placed on its side. (b) Notice that in the gravitropic response the statoliths accumulate at the bottom of cells.

Figure 11.15 Thigmotropism. Tendrils provide a good demonstration of thigmotropism, which is growth response to touch.

link the outside of the cell membrane to the cytoskeleton inside. Integrins have been recently discovered in the membranes of statoliths and may occur in plasma membranes as well. Perhaps the statolith hypothesis and the gravitational pressure hypothesis will turn out to have much in common.

Roots and shoots respond simultaneously to several stimuli, including gravity, light, and the presence of water. Therefore, the actual direction of shoot and root growth depends on the interaction of these stimuli. For example, roots may modify a purely vertical gravitational response to grow horizontally toward a source of water.

Plants respond to mechanical stimuli, such as touch and wind

Thigmotropism (from the Greek *thigma*, "touch") is a growth response to touch (see Chapter 4). Touch affects plants in several different ways. One involves ethylene production. You have already learned that ethylene switches plants from growth in length to growth in width. In fact, if a plant stem is touched or rubbed, which probably simulates the action of wind, it will respond by producing ethylene and reducing growth in length. Tendrils curl around objects they encounter because of a touch response. The side of the tendril that touches a foreign object produces ethylene, which inhibits growth on that side, while growth on the other side causes the tendril to curve (Figure 11.15).

Three other observable tropisms require further study: hydrotropism, heliotropism, and chemotropism. **Hydrotropism,** growth toward or away from water, occurs when

roots grow toward water or moist soil. The root cap contains the sensor, and the signal-transduction pathway involves calcium. **Heliotropism,** or sun tracking, relates to flowers or leaves that either follow the Sun, as in the case of sunflowers, or avoid the Sun throughout the day. Heliotropism may be an ongoing phototropism, although it seems to occur on mature flower stalks or leaves that no longer grow and may involve turgor loss on the shaded side of the stem. **Chemotropism,** growth toward or away from a chemical stimulus, is suspected in directing growth of the pollen tube to the female gametophyte of seed plants. In flowering plants, the pollen tube must grow down the length of the style, which can be several centimeters. In the lily (*Lilium longiflorum*), a small protein is suspected of being the chemical attractant.

Other sensitivities to touch translate mechanical stimuli into rapid action. For example, the leaves of the so-called sensitive plant (*Mimosa pudica*) close by folding vertically when touched (Figure 11.16). In the case of *Mimosa,* cells in specialized organs, called pulvini (singular, pulvinus), at the joints of leaves and leaflets lose potassium and then lose water in response to touch. As a result, they become flaccid instead of turgid, and the leaflets fold together. A slow electrical impulse transfers the stimulation from one leaflet to the next. The impulse is much slower than in animal nerves because no direct pattern of conduction, such as a neuron, exists.

The Venus flytrap (*Dionaea muscipula*), a plant living in bogs with inadequate nitrogen supplies, catches insects in modified leaves, which fold to produce a "trap" within half a second after an insect trips the trigger hairs. Three trigger hairs are located inside on each half of the leaf. If any two are touched, in close succession, or if one is touched twice, the trap will close. Following mechanical stimulation of trigger hairs, the stimulus must be transduced to an electrical signal and then propagated across the leaf. Closure involves uptake of water by mesophyll cells in the "hinge" area of the trap, which swell to mechanically close the trap. Stimulation of trigger hairs depolarizes the membrane. The mechanism of electrical propagation remains mysterious.

Plants prepare for environmental conditions that prevent normal metabolism and growth

Although the shoot and root apices of plants allow continual growth, many plants temporarily cease growth or die at predictable times of the year. At the end of each growing season, annual plants die, while deciduous plants

(a)

(b)

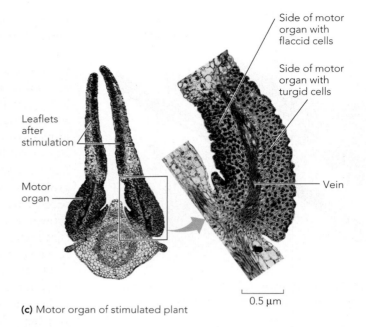

(c) Motor organ of stimulated plant

Figure 11.16 The "sensitive plant," Mimosa pudica. Even a light touch causes leaflets to close in a second or two. (**a**) Unstimulated plant. (**b**) Stimulated plant. (**c**) Light micrograph of motor organ of stimulated plant. Flaccid cells on the inside have lost water, while turgid cells on the outside have retained water.

lose their leaves and enter a dormant state. In temperate zones, shorter days and lower temperatures signal this transition. In tropical and desert regions, the beginning of the dry season serves as the signal.

Bud dormancy involves a series of anatomical and biochemical changes involving two plant hormones: abscisic acid (ABA) and gibberellins. In temperate zones, shorter days (which actually begin after June's summer solstice) will eventually signal shoot apices to prepare for inclement environmental conditions. In tropical zones with an alternation of wet and dry seasons, lack of water signals the plant to begin similar preparations. The preparations, known as **acclimation,** begin in temperate zones before autumn and involve the production of bud scales (Figure 11.17) and the accumulation of ABA. Bud scales help insulate buds and protect them from drying out; abscisic acid induces dormancy and prevents growth during temporary warm or wet periods during an otherwise inclement season. In many cases, cold or drought helps break dormancy by leading to the gradual destruction of ABA. The synthesis of gibberellins, in response to longer, warmer spring days or wetter days of the rainy season, also helps break bud dormancy.

Many trees lose their leaves prior to or with the arrival of inclement environmental conditions such as cold or drought. Preceding abscission, an abscission zone devel-

ops near the junction of a leaf petiole with the larger stem. This zone contains two layers of cells. The separation layer closest to the leaf blade consists of thin cells with weak cell walls. Prior to leaf drop, these cells swell and may divide, and cell walls break down to become gelatinous. The separation layer closest to the stem has several cell layers containing cell walls impregnated with fatty suberin, which will seal the break when leaf drop occurs.

Prior to the loss of leaves, usable molecules are translocated from leaf blades and return to stems. Proteins and

Figure 11.17 Bud dormancy. In this micrograph of a longitudinal section of an axillary bud, bud scales surround the dormant shoot apical meristem, protecting and insulating it during winter months.

EVOLUTION

The Arms Race Between Plants and Herbivores

As you know, plants can produce compounds that repel herbivores and disease organisms. An example is *Datura wrightii*, more commonly known as jimsonweed. Named after *dhatura*, the ancient Hindu word for plant, jimsonweed protects itself from insects by producing potent alkaloids, such as atropine and hyoscine. Alkaloids from *Datura* have medical uses in heart disease and can also be dangerous drugs.

Jimsonweed can also produce either sticky or velvety trichomes (leaf hairs). Sticky jimsonweed plants produce glandular trichomes full of a sticky substance composed of sugars and water, which traps some herbivores. Velvety jimsonweed plants produce non-glandular trichomes. In jimsonweed, a single dominant gene determines whether a plant will have sticky trichomes or velvety trichomes.

Sticky jimsonweed plants are widely resistant to insects that devour velvety plants. However, insects can produce chemicals that deter

Jimsonweed (*Datura wrightii*).

the action of plant compounds and one type of mirid bug, *Tupiocoris notatus,* has anatomically adapted to sticky jimsonweed plants and constitutes their principal pest.

Production of a sugary solution that produces a sticky leaf surface and repels predators has a cost in terms of the total energy and resources available to the plant. Sticky jimsonweed plants require more energy and water to counter the evaporation from the sticky substance, thereby reducing the amount available for seed production. If herbivory is removed as a factor in reducing seed production, sticky jimsonweed plants produce 45 percent fewer seeds than do velvety jimsonweed plants under dry conditions. This reduction may be related to the amount of metabolic energy the sticky plants use to produce glandular trichomes. The sticky, glandular trichomes may represent an unduly expensive adaptation, which may reduce the likelihood of survival for sticky plants.

---

starch break down into movable subunits of amino acids and glucose. Even vitamins and mineral ions are "recycled" by the plant. A decrease in cytokinin production by roots leads to the reclamation of these molecules. Applications of cytokinins can prevent the loss of molecules and keep leaves green well beyond the time when normal preparations for winter would begin.

The beautiful autumn leaf colors of red, yellow, and orange actually have two sources. Red anthocyanin pigments are newly synthesized in autumn, while yellow and orange carotenoids, already present in leaves, become visible with chlorophyll breakdown. The synthesis of anthocyanins in leaves at the end of their growing season has puzzled scientists. Why would plants "waste" the energy? Researchers at Harvard University have hypothesized that the layer of red anthocyanins, which accumulates in the mesophyll cells of some plants as leaves begin to die, protects leaves from damage while nutrients are being removed. Ordinarily, chlorophyll absorbs damaging ultraviolet-to-blue wavelengths. However, as leaves age,

chlorophyll breaks down, so the red pigments appear to take over the job. Anthocyanins frequently accumulate in plants suffering from drought or cold temperature stress where they may serve as osmotically active protective molecules.

Ethylene promotes leaf abscission, while auxin and cytokinins suppress it. Ethylene promotes synthesis and activity of enzymes that cause development of the abscission zone. Commercially, ethylene is used in promoting fruit drop in cherries, grapes, and many berries. The induction of a uniform drop time makes mechanical harvesting easier.

Plants react to environmental stresses such as drought

The general mechanism by which plants respond to environmental stresses involves receiving and identifying the environmental signal, communicating the signal throughout the plant, and altering gene expression and

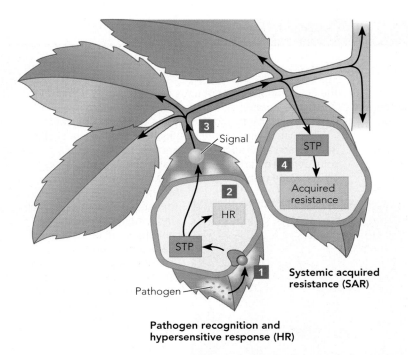

1. The binding of a pathogen to the plasma membrane induces a signal transduction pathway (STP).

2. The STP causes a hypersensitive response (HR), which kills infected plant cells. Before they die, they release antimicrobial molecules.

3. Dying cells release salicylic acid, which is transported throughout the plant.

4. In healthy cells throughout the plant, salicylic acid induces a STP that produces antimicrobial molecules, preventing further infection. This response is known as systemic acquired resistance (SAR).

Figure 11.18 Plant cell responses to infection by fungi, bacteria, or viruses. As a result of an initial infection and a subsequent signal-transduction pathway, plants produce a variety of antimicrobial molecules .

metabolism to counter the stress. For example, in the case of drought stress, the root hairs are the first cells affected. When these cells no longer have sufficient water potential to remove water from soil particles, transpiration slows and the message about insufficient water is thereby communicated to the entire plant. As a result, abscisic acid synthesis increases in the roots and leaves, and the stomata close.

The cells of plants enduring drought also accumulate one or more specific proteins, such as dehydrins, which can account for up to 12% of total protein. The function of these proteins remains unclear, but genetic engineering to increase production by adding additional genes that code for the proteins sometimes results in plants with increased drought tolerance.

In many plants, a process of osmotic adjustment begins, in which cells begin synthesizing solute molecules that increase the ability of the cells to attract and retain water. Such molecules, known as *osmolytes*, generally accumulate in the cytoplasm, while the vacuole accumulates charged ions and solutes that would interfere with metabolism in the cytoplasm itself.

Plants deter herbivores and pathogens

Herbivores and disease organisms are major destroyers of plants. About 50% of mammalian and insect species are herbivores. Fungal, bacterial, and viral diseases destroy about 30% of cultivated crops worldwide, causing both starvation and an economic loss in the trillions of dollars. Plants in nature are no more immune. Any mutation that brunts the attack of these predators would give a plant a huge evolutionary advantage (see the *Evolution* box on page 253).

In many plants, secondary metabolites such as alkaloids, phenolic compounds, and terpenes (see Chapter 7) defend against both herbivores and disease-inducing bacteria and fungi. They may increase in concentration during the summer season as the number of insects increases. For example, the concentration of the phenolic compound tannin in oak leaves increases from 0.7% dry weight in April to 5.5% dry weight in September. Secondary metabolites are frequently contained in trichomes, the leaf hairs that are usually the first part of the plant encountered by the mouth of an herbivore. Many plants increase production of secondary metabolites in response to an attack, so a second successful attack is less likely. Of course, energy is required to produce secondary metabolites so protection comes at a cost.

The response of a plant infected with a disease-causing organism is complex. For example, plants have several induced responses, which occur only as the result of interaction with an herbivore or infection by a pathogen.

Attack by herbivores such as insects activates an induced resistance (IR) pathway that produces jasmonic acid and other compounds that contribute to an overall effort to rid the plant of herbivores and pathogens. The IR pathway makes the plant less tasty and also attracts predatory insects that feed on herbivorous insects. Tomato plants treated with jasmonic acid, to simulate IR, attract twice the number of parasitic wasp larvae as untreated plants. The larvae parasitize and kill larvae of a butterfly species that eats tomato leaves.

Infection by fungi, bacteria, or viruses induces resistance in plants through a second pathway (Figure 11.18). Molecules from a pathogen bind to specific receptor proteins in the plasma membrane of the plant. This binding leads to a localized reaction called a *hypersensitive response* (*HR*), in which plant cells seal off the area of infection and produce compounds to destroy the pathogen or retard its growth and cell division. By programmed cell death, the HR produces a noticeable lesion on the leaf.

Cells dying from the initial infection around the region of the HR release salicylic acid (a modified form of the active ingredient in aspirin) and nitric oxide gas (NO), which initiate a signal transduction in many parts of the plant. This signal transduction leads to a widespread reaction known as *systemic acquired resistance* (*SAR*), which is a general resistance acquired as a result of infection. This SAR results in the production of compounds that help deter pathogens.

Section Review

1. **How are statoliths involved in gravitropism?**
2. **Give an example of thigmotropism.**
3. **What is the role of leaf abscission, and how does it take place?**
4. **Describe some ways in which plants respond to herbivores and pathogens.**

SUMMARY

Effects of Hormones

Auxin plays a central role in cell enlargement and formation of new tissue (pp. 237–240)

Auxins were the first plant hormones to be discovered and characterized. The principal plant auxin is IAA, which promotes cell enlargement, differentiation of vascular tissue, bud suppression, and ethylene synthesis.

Cytokinins control cell division and differentiation and also delay aging (p. 240)

Cytokinins induce cell division and keep plants in a youthful state of development. In tissue cultures, the auxin-to-cytokinin ratio controls whether undifferentiated callus cells are produced or whether cells differentiate to form roots or shoots.

Gibberellins interact with auxins to regulate cell enlargement and stimulate seed germination (pp. 240–242)

Gibberellins initiate the breakdown of starch in the endosperm of grass seeds. They also break seed dormancy, promote flowering of plants, and promote cell elongation in cooperation with auxin.

Abscisic acid causes seed dormancy and regulates plant responses to drought (pp. 242–243)

Abscisic acid promotes closure of stomata during periods of water stress. It also fosters production of storage proteins in developing seeds and then increases in concentration to prevent germination and maintain seed dormancy.

Ethylene allows the plant to respond to mechanical stress and controls fruit ripening and leaf abscission (pp. 243–244)

Ethylene inhibits longitudinal elongation of shoots and roots and promotes increase in width. It initiates the triple response of shoots and roots, and plays a role in ripening fruit.

Brassinosteroids are a newly discovered group of plant hormones that act like auxin (p. 244)

Brassinosteroids are steroid hormones that influence many of the same developmental systems as auxin.

Additional compounds may play a role as plant hormones (p. 244)

Two classes of compounds that potentially function as plant hormones are polyamines and jasmonic acid.

Responses to Light

Blue light absorption controls the growth of stems toward the light and the opening of stomata (p. 245)

Blue light absorption causes auxin to move to the dark side of the stem and root in phototropism and promotes opening of the stomata.

Absorption of red and far-red light determines when seed germination, stem and root growth, and flowering occur (pp. 245–246)

Phytochrome is a photoreceptor converted from one form to the other by red or far-red light. Phytochrome causes germination of photodormant seeds.

Photoperiodism regulates flowering and other seasonal responses (pp. 246–249)

Plant response to photoperiod determines the timing of flowering. Measurement of night length is controlled by phytochrome for a few hours, but phytochrome itself is not the clock. The flowering stimulus, called *florigen* but as yet unidentified, moves from leaves to meristems. Gibberellin stimulates flowering in long-day plants, while ethylene promotes flowering in some plants, such as pineapple.

Plants respond to repeating cycles of day and night (pp. 249–250)

Circadian rhythms occur with a repeating 24-hour period and affect the sleep movements of leaves, the opening of flowers, and some enzyme activity. Phytochrome helps plants maintain the rhythms.

Responses to Other Environmental Stimuli

Roots and shoots respond to gravity (pp. 250–251)

One hypothesis attributes gravitropism to statoliths, starchy plastids that settle to the bottom on cells. Another states that gravity adds pressure to proteins on the lower plasma membrane. The extent of gravitropism depends on the simultaneous responses of root and shoots to other stimuli, such as light and the presence of water.

Plants respond to mechanical stimuli, such as touch and wind (p. 251)

In thigmotropism, a growth response to touch, ethylene inhibits growth when the plant is touched by a physical object or by wind. In an unrelated response, Venus flytraps can translate mechanical stimulus into rapid folding of the flytrap leaf.

Plants prepare for environmental conditions that prevent normal metabolism and growth (pp. 251–253)

Bud dormancy is caused by shorter days of late summer, as are preparations for leaf abscission, triggered by ethylene.

Plants react to environmental stresses such as drought (pp. 253–254)

Plants alter gene expression and metabolism to counter stress. For example, to counter drought stress, plants produce abscisic acid in the leaves to stimulate closing of the stomata.

Plants deter herbivores and pathogens (pp. 254–255)

Plants produce secondary metabolites that repel herbivores and disease-causing organisms. Herbivores induce plants to produce jasmonic acid in the induced resistance pathway. Bacteria, fungi, and viruses induce plants to produce salicylic acid and nitric oxide, which lead to the production of compounds that deter pathogens.

Review Questions

1. Explain why the origin of the term *hormone* ("to stir up") is misleading.
2. How do hormones carry out their actions within plants?
3. How was auxin discovered?
4. How does the origin of the term *cytokinins* relate to their role?
5. What are some effects of gibberellins?
6. Describe some ways in which different hormones have opposite effects.
7. Describe some commercial applications of plant hormones.
8. Explain why the name *abscisic acid* is misleading.
9. What is the triple response to ethylene?
10. Compare the effects of blue light and red light.
11. How does light affect dormancy in plants?
12. In what way are the terms *long-day plant* and *short-day plant* inappropriate?
13. What evidence indicates that plants have a biological clock?
14. Describe the main types of tropisms.
15. How do plants prepare for winter?
16. What is the function of secondary metabolites?

Questions for Thought and Discussion

1. Why do some plants flower in response to short-days (long nights), while others flower in response to long days (short-nights)? What ecological advantage do plants gain?
2. Why do you suppose that leaves measure the night length, even though the shoot apical meristem is the actual site of flower formation?
3. Why do you suppose that the identity of the flowering stimulus remains a mystery?

4. Why do some trees keep their leaves year-round? Why do you think so many plants are deciduous? After all, growing new leaves each year requires a lot of energy.

5. Why do plants have specific mechanisms that cause fruit ripening in summer and fall? Why do you think they do not save the energy and just let fruit fall when winter comes?

6. Draw a graph to show the probable germination response of seeds (*Y*-axis) to a series of treatments (*X*-axis) ranging from a high concentration of gibberellins and no abscisic acid at one extreme to no gibberellins and a high concentration of abscisic acid at the other extreme.

Evolution Connection

Explain why the interrelationships between plants and herbivores can justifiably be viewed as an evoutionary "arms race" between the two populations.

To Learn More

Visit The Botany Place Website at www.thebotanyplace.com for quizzes, exercises, and Web links to new and interesting information related to this chapter.

Darwin, Charles. *The Power of Movement by Plants.* New York: Da Capo Press, 1966. This book details Darwin's experiments on phototropism.

Went, Fritz W. *Phytohormones.* City: Universe Books, 1937. This book discusses Went's classical experiments.

Genetics and Gene Expression

The garden pea (*Pisum sativum*).

Mendel's Experiments on Inheritance

Making sense of Mendel's experiments requires a basic understanding of genes and chromosomes

Monohybrid crosses involve individuals that have different alleles for a specific gene

Segregation of alleles occurs during anaphase I of meiosis

A testcross demonstrates the genotype of an individual with a dominant phenotype

Dihybrid crosses involve individuals that have different alleles for two specific genes

Beyond Mendel's Work

Mendel's laws also apply to crosses that involve more than two traits

Some characters are not controlled by one dominant and one recessive allele

The locations of genes affect inheritance patterns

Genes interact with each other and with the environment

Mendel's gene for height in peas controls the production of a growth-promoting hormone

In all living things, offspring resemble their parents in some ways but not in others. This fact has always been familiar to people who observed their own families and who raised domesticated animals and crop plants. Farmers selected the seeds of plants that had high yields, flavorful fruits, or disease resistance and sowed those seeds to produce the next generation. They knew such selection processes resulted in gradual crop improvement. For example, people have gradually improved corn through selective breeding for at least 7,000 years. Modern corn (*Zea mays,* subspecies *mays*) originated as a wild grass known as *teosinte* (*Zea mays,* subspecies *palviglumis*), which still grows in southern Mexico. In teosinte, the exocarp completely surrounds each kernel, making it difficult to eat. In corn, the kernels remain exposed. The kernels of teosinte are also few, small, and not fused to an ear as in modern corn. The differences between teosinte and corn result from variations in as few as five genes. The two plants can still breed with each other, enabling scientists to study the inheritance of their individual traits.

By the nineteenth century, scientists had proposed several explanations for how desirable features are passed from one generation to the next. One explanation held that each part of a plant contributed a tiny replica or particle of itself to its offspring. The proponents of this explanation in-cluded ancient Greek philosophers and Charles Darwin. Another popular idea likened inheritance to the blending of fluids. According to this idea, breeding a tall plant with a short plant caused the fluid for tallness to blend with the fluid for shortness. We now know that these explanations are incorrect.

It took the careful observations of an obscure Austrian monk, Gregor Mendel (see the *Plants & People* box on page 263), to uncover the laws of inheritance that apply to all living things. Mendel was well acquainted with the scientific method and had access to many varieties of garden peas (*Pisum sativum*) with visually distinct features. He knew that, in general, offspring frequently resemble parents and grandparents. Mendel observed that breeding different varieties of pea plants produced offspring with certain features in predictable ratios. His training in applied horticulture and appreciation for the usefulness of mathematics in science gave him a unique background and led to his discoveries.

Knowing that plants contain some sort of independently acting units that carry heredity information, Mendel developed the term *elemente* for what we call *genes*. In just the last half century, scientists investigating the biochemistry of genes and gene expression have shown exactly what causes the various plant features Mendel studied.

Teosinte

Male tassels

Cob

Teosinte

Corn x teosinte hybrid

Male tassels

Cob

Modern corn

PLANTS & PEOPLE

A Brief Biography of Gregor Mendel

Gregor Johann Mendel was born in 1822 in the town of Hyneiee, which was then in Austria but is now in the Czech Republic. As a child, he studied agricultural as well as academic subjects at the local school, and he learned about the practical aspects of plant reproduction by helping his father graft fruit trees in the family orchard. Although Mendel lacked both the money and the good health to attend high school, he overcame these obstacles and, in 1840, graduated from the Olmutz Philosophical Institute, which was equivalent to a high school. In 1843, he continued his education at the Augustinian Monastery of St. Thomas in Brünn (now called Brno). After his ordination as a priest in 1847, Mendel became a temporary teacher, but this position ended in 1850 when he failed the state certification exam to teach natural sciences. His failure was due to poor answers in natural history and physics.

Gregor Mendel.

Undaunted, Mendel attended the University of Vienna from 1851 to1853. There, he became interested in plant variations and in the power of the scientific method. He had extensive training in the use of mathematics to test scientific hypotheses.

After leaving the university, Mendel returned to the monastery and, while there, also served as a teacher at the Brünn Modern School. The monks had been involved in agriculture for many years and had worked extensively with peas. Thus, it is not surprising that Mendel turned to these plants and, drawing on his university studies, began the research that ultimately made him famous. He carried out his experiments between 1856 and 1863, in the monastery garden, which was 7 meters (m) by 35 m (23 feet by 115 feet).

Mendel presented the results of his research in a public lecture in 1865 and wrote a paper on his findings titled "Experiments on Plant Hybrids." He tried to publish his paper in Germany, the scientific center of Europe, but his results did not impress the scientific establishment, who rejected the paper with unenthusiastic remarks. Mendel finally got his paper published in an obscure, local journal, *The Proceedings of the Brünn Society of Natural History,* in 1866. It was sent to only 120 libraries. Mendel then moved into the administrative post of abbot at the monastery, and his scientific career ended. He died in 1884 at the age of 61.

Not long before his death Mendel commented on his scientific obscurity by saying, "My scientific work has brought me a great deal of satisfaction, and I am convinced that it will be appreciated before long by the whole world." His words were prophetic: In 1900 his results were rediscovered independently by three researchers in plant genetics—Hugh de Vries of Holland, Carl Correns of Germany, and Erich von Tschermak of Austria. To their credit, they gave Mendel the recognition he justly deserved.

In this chapter, we will examine the principles of inheritance revealed by Mendel's experiments and relate those principles to the events that occur during meiosis. We will then consider several aspects of inheritance that go beyond what Mendel observed.

Mendel's Experiments on Inheritance

Mendel could only speculate about the carriers of inheritance. Although he never saw chromosomes, he deduced that something like chromosomes must exist. Later, microscopy revealed that chromosomes are the particulate carriers of genetic information and demonstrated that the results of genetics experiments are tightly linked to the mechanics of meiosis. During meiosis, the movement of genes into daughter cells portrays Mendel's laws of inheritance.

Making sense of Mendel's experiments requires a basic understanding of genes and chromosomes

As you know from your reading of earlier chapters, chromosomes contain genetic information that is encoded by genes. While sometimes represented as beads on a chromosomal string, genes actually consist of a series of nucleotides on the strands of DNA's double helix. A typical eukaryote, such as a pea plant, has 25,000 to 50,000 genes spread out on various numbers of chromosomes, depending on the species. Peas have 14 chromosomes, 7 provided by a sperm from the male parent and 7 provided by an egg from the female parent. Thus, an adult pea plant is diploid, and each gene is represented twice in every somatic (nonsexual) cell. This means that, on average, each pea chromosome contains 3,600 to 7,200 genes ($2 \times 25{,}000/14$ to $2 \times 50{,}000/14$).

Genes carry the information about an organism's features, such as seed color and plant height. We refer to these features as **characters.** Each character may appear in two or more forms, called **traits.** In peas, for example, the character of seed color may appear as the trait of green seeds or yellow seeds, while the character of plant height may appear as the trait of shortness or tallness. Variant forms of a gene, known as **alleles,** code for each trait. Some genes, such as the genes that control flower color in certain plants, have several alleles. However, no more than two alleles for a gene can be present in a diploid cell. One allele came from each parent, and the alleles can be the same or different, depending on the genetic makeup of the parents.

Monohybrid crosses involve individuals that have different alleles for a specific gene

Mendel worked with varieties of garden peas that had two different traits for each of seven characters. Each variety was pure breeding, which means that when two plants with the same trait were bred with each other, all of their offspring had that trait. In his first experiment, Mendel crossbred plants that had different traits for one character. For example, he bred plants that had purple flowers with plants that had white flowers. Mendel then bred the offspring of this cross with each other. The offspring were hybrids of their parents, so this second breeding is called a **monohybrid cross.** Most importantly, he counted the numbers of offspring that showed each trait after every cross.

Figure 12.1 illustrates the procedure Mendel used to cross purple-flowered and white-flowered pea plants. First, **1** Mendel removed the stamens of all the flowers on a purple-flowered plant, thus preventing the plant from pollinating itself. Then, **2** he collected pollen from the stamens of a white-flowered plant and brushed it on the carpels of the purple flowers. This allowed the sperm produced by the white-flowered plant to fertilize the eggs in the purple-flowered plant. **3** Mendel then allowed the fruits (the pea pods) to develop. The hybrid peas he harvested from those pods were the offspring of each cross. He referred to the plants that were crossed as the *parental generation* and the offspring as the **first filial (F_1) generation.** Filial comes from the Latin word *filius,* meaning "son." Finally, **4** Mendel planted the hybrid peas and all of them produced plants with purple flowers. There were no plants with white flowers at all. Although it isn't shown in Figure 12.1, Mendel also performed the reciprocal cross,

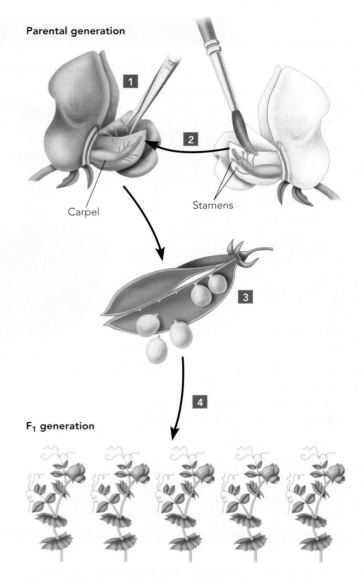

Figure 12.1 A monohybrid cross. Mendel crossed purebreeding varieties of pea plants that had two different traits for one character—in this example, purple flowers and white flowers. Mendel used the same procedure to study the inheritance of six other characters.

brushing pollen from the stamens of purple flowers on the carpels of white flowers. The result was the same: All of the F_1 plants had purple flowers.

Finally, Mendel allowed the F_1 plants to interbreed. Since peas are self-pollinating, this amounted to leaving the plants alone. The offspring of this cross he called the **second filial (F_2) generation.** When he planted the F_2 peas, he found that 75% of the resulting plants had purple flowers and 25% had white flowers. Mendel interpreted his results by suggesting that the white-flower trait was present in the F_1 plants but was masked somehow by the visible, purple-flower trait. He called the visible trait **dominant** and the masked trait **recessive.** Crossing the F₁ plants allowed the recessive trait to reappear in the F₂ generation.

Today we know that dominant and recessive traits are controlled by two alleles of a specific gene. The allele that controls the dominant trait is represented by a capital letter, and the allele that controls the recessive trait is represented by the lowercase form of the same letter. Thus, for flower color, *P* represents the dominant (purple-flower) allele and *p* the recessive (white-flower) allele. Because each cell in a pea plant has two alleles for each gene, the cells could have either two *P* alleles, two *p* alleles, or one of each. A plant that has two copies of the same allele for a gene (either *PP* or *pp* in the case of flower color) is said to be **homozygous** for that allele, while a plant that has two different alleles (*Pp*) is **heterozygous.** A homozygous individual is called a *homozygote,* and a heterozygous individual is called a *heterozygote.*

The combination of alleles that a plant has (such as *PP*, *pp*, or *Pp*) is known as the plant's **genotype.** In contrast, the physical appearance of the plant (purple flowers or white flowers, in this case) is called the plant's **phenotype.** It takes only one dominant allele to cause a plant to show the dominant phenotype, but two recessive alleles are needed for the recessive phenotype to be revealed. Thus, plants that have either the *PP* or *Pp* genotype will produce purple flowers, but only *pp* plants will produce white flowers.

The results of each cross in Mendel's first experiment can be represented by using a Punnett square (Figure 12.2). This visual device was invented in 1905 by British geneticist Reginald Crundall Punnett. The gametes produced by each parent are listed on two adjoining sides of the square. The filled-in boxes represent possible combinations of alleles that can occur as the result of fertilization. In the first cross, both parent plants were homozygous. The purple-flowered parent had the genotype *PP*, so all of its gametes carried the *P* allele. The white-flowered parent had the genotype *pp*, so all of its gametes carried the *p* allele. As a result, all of the offspring in the F₁ generation received a *P* allele from one parent and a *p* allele from the other parent and were heterozygous.

Because the F₁ plants had both alleles, half of their gametes carried one allele and half carried the other. Therefore, when the F₁ plants were allowed to interbreed in the second cross, 25% of their offspring (the F₂ generation) received two *P* alleles, 50% received a *P* allele from one parent and a *p* allele from the other, and 25% received two *p* alleles (Figure 12.2). In other words, the ratio of *PP*, *Pp*, and *pp* genotypes in the F₂ generation was 1:2:1. Since plants with either the *PP* or *Pp* genotype produce purple flowers, 75% of the F₂ offspring had the dominant phenotype, while 25% had the recessive phenotype. Thus, the

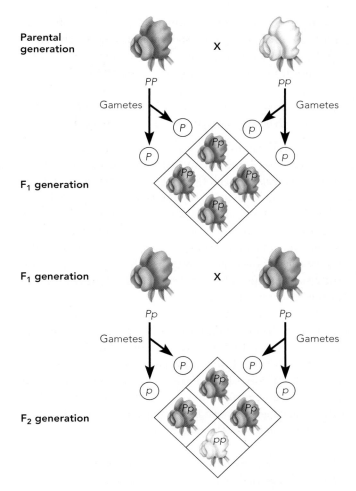

Figure 12.2 Mendel's law of segregation. You can use a Punnett square to show what Mendel found in each cross in his first experiment. Write the alleles of the gametes produced by each parent on adjacent sides of the square. Then fill in each box with the alleles corresponding to the row and column of that square. The pairs of alleles in the boxes represent the possible genotypes of the offspring. In a cross between two pure-breeding parents (*PP* and *pp*), all of the F₁ offspring have the genotype *Pp* and produce purple flowers. When two F₁ plants are crossed, the F₂ offspring occur in a genotypic ratio of 1 *PP*: 2 *Pp*: 1 *pp* and a phenotypic ratio of 3 purple-flowered: 1 white-flowered.

ratio of dominant to recessive phenotypes in the F₂ generation was 3:1.

Mendel obtained similar results in crosses involving all seven of the characters he investigated. On the basis of these results, he formulated the first of his laws of inheritance, the **law of segregation.** In modern terms, the law of segregation states that alleles segregate during meiosis and then come together randomly during fertilization. Because fertilization is random, in a cross between two heterozygotes an egg has an equal chance of being fertilized by a sperm carrying a dominant allele or a recessive allele. In that sense, crossing two heterozygotes is like tossing two coins simultaneously. The laws of probability affect each coin separately, so each has an equal chance of landing heads up or tails up. What happens to one coin does not influence what happens to the other.

Segregation of alleles occurs during anaphase I of meiosis

A key to understanding the segregation of alleles is to remember what happens to homologous chromosomes during meiosis I. Recall from Chapter 6 that homologous chromosomes pair up to form tetrads during prophase I of meiosis. Thus, seven tetrads form in the reproductive cells of a pea plant. The two homologous chromosomes in each tetrad contain the same genes, although the alleles for each of those genes may be different. The genes on the chromosomes of one tetrad are different from those of the other tetrads.

During anaphase I of meiosis, the alleles begin their separate journeys to end up in different gametes. Figure 12.3 illustrates this process for a pea plant that is heterozygous for the flower-color gene. When the reproductive cells of this plant undergo meiosis, two types of gametes will be produced: One half will have the *P* allele, while the other half will have the *p* allele. This process is the biological basis of segregation.

A testcross demonstrates the genotype of an individual with a dominant phenotype

Suppose you had a pea plant that exhibited the dominant phenotype for some character. For example, the character of seed shape has a dominant phenotype (smooth seeds) and a recessive phenotype (wrinkled seeds). A plant that produces smooth seeds could be either homozygous for the dominant allele (*SS*) or heterozygous (*Ss*). You couldn't deduce the plant's genotype simply by looking at the plant or its seeds. However, you can determine the genotype of a plant that has a dominant phenotype by performing a **testcross,** also called a *backcross.* In a testcross, a plant whose genotype is unknown is crossed with a plant that exhibits the recessive phenotype for the character in question—in this example, wrinkled seeds. Since two recessive alleles are required for the recessive phenotype to be revealed, you know that a plant that produces wrinkled seeds must be homozygous (*ss*). As Figure 12.4 shows, there are two possible outcomes of a testcross, depending on the genotype of the parent plant with the dominant phenotype. If that plant is homozygous, then all of the offspring will have the dominant phenotype. On the other hand, if the parent plant with the dominant phenotype is heterozygous, then half the offspring will have the dominant phenotype and half will have the recessive phenotype.

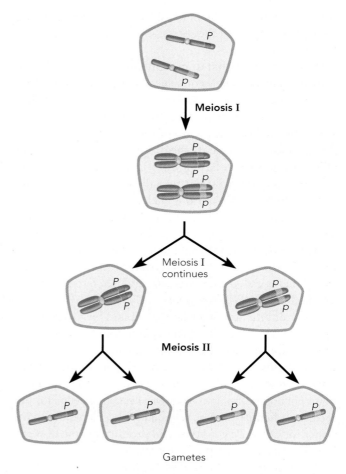

Figure 12.3 Segregation of alleles during meiosis. A pair of homologous chromosomes is shown in this diploid reproductive cell of a pea plant. (For simplicity, the other six pairs of homologous chromosomes are not shown.) One of the chromosomes has the *P* allele for purple flowers, and the other chromosome has the *p* allele for white flowers. Segregation of alleles occurs during meiosis I, as the chromosome that has the *P* allele moves to one pole of the cell, while the chromosome that has the *p* allele moves to the opposite pole. As a result, these chromosomes end up in different cells at the end of meiosis I. These cells form four haploid gametes during meiosis II, two gametes with each allele.

Dihybrid crosses involve individuals that have different alleles for two specific genes

Having performed numerous monohybrid crosses in which he considered only *one* character, Mendel began to cross pure-breeding plants that differed in *two* characters. In one of these crosses, for example, he crossed a plant that produced smooth, yellow seeds (two dominant phenotypes) with a plant that produced wrinkled, green seeds (two recessive phenotypes). Because the plants were pure breeding, we know that their genotypes were *SSYY* and *ssyy,* respectively. All of the F$_1$ plants had the dominant phenotype for both characters: They produced smooth, yellow seeds (Figure 12.5). The F_1 plants received two dominant alleles (*S* and *Y*) from one parent and two recessive alleles (*s* and *y*) from the other parent, so they had the genotype *SsYy*. Since the F_1 plants were heterozygous for two characters, they are referred to as *di*hybrids.

When Mendel allowed the F_1 plants to interbreed in a **dihybrid cross,** he obtained an F_2 generation that consisted of plants with four phenotypes in a ratio of 9 smooth, yellow: 3 wrinkled, yellow: 3 smooth, green: 1 wrinkled, green (Figure 12.5). Notice that a Punnett square for a dihybrid cross is a 4 × 4 matrix. As in a monohybrid cross, the gametes produced by each parent are listed on adjoining sides of the Punnett square, and the boxes are filled in with the genotypes of the F_2 individuals produced.

To understand how a 9:3:3:1 ratio arises in a dihybrid cross, think of such a cross as two independent, monohybrid crosses performed at once. Remember that a monohybrid cross results in an F_2 generation with a 3:1 ratio of dominant to recessive phenotypes. By multiplying (3:1) × (3:1), you can predict the phenotypic ratio in the F_2 generation of a dihybrid cross. To do this, multiply the first 3 by the second 3, the first 3 by the second 1, the first 1 by

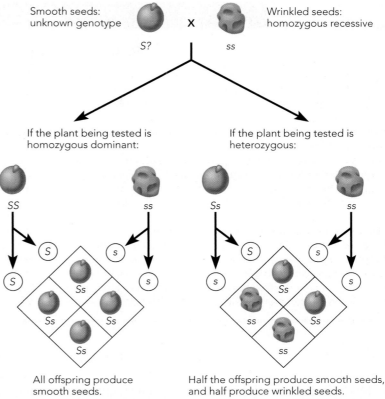

Figure 12.4 A testcross. To discover the genotype of a plant that has a dominant phenotype, such as smooth seeds, the plant is crossed with one that has the recessive phenotype (wrinkled seeds). If all the offspring have the dominant phenotype, the parent with the unknown genotype must be homozygous for the dominant allele. If the dominant and recessive phenotypes occur in a 1:1 ratio in the offspring, the parent with the unknown genotype must be heterozygous.

Figure 12.5 Mendel's law of independent assortment. When a pure-breeding plant that has smooth, yellow seeds (*SSYY*) is crossed with a pure-breeding plant that has wrinkled, green seeds (*ssyy*), all the F_1 plants have smooth, yellow seeds and are heterozygous (*SsYy*). The F_1 plants produce four types of gametes (*SY, sY, Sy,* and *sy*). When two F_1 plants are crossed, the F_2 generation exhibits a phenotypic ratio of 9 smooth, yellow: 3 wrinkled, yellow: 3 smooth, green: 1 wrinkled, green.

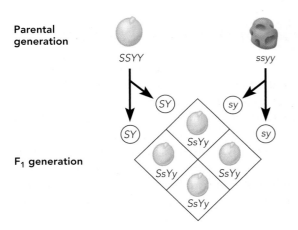

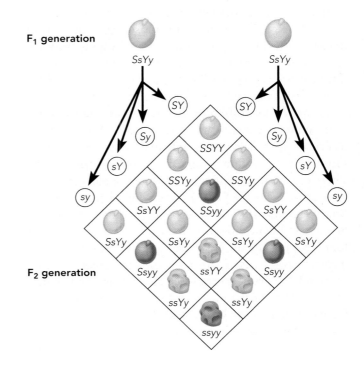

PLANTS & PEOPLE

Genetics Before Mendel

Many workers before Mendel had wondered about and investigated heredity in both animals and plants. In the 1760s Josef Kölreuter found that hybrid plants in tobacco either had the phenotype of just one parent or were intermediate in phenotype between both parents. Kölreuter also noted that phenotypes could be absent in one generation but reappear in subsequent generations. In the 1790s an Englishman named T. A. Knight crossed a pea plant that had purple flowers with one that had white flowers. Like Mendel, he noticed that the offspring all had purple flowers and that white-flowered plants reappeared in subsequent generations. Unfortunately, Knight did not quantify his results, noting simply that purple flowers had a "stronger tendency" to reappear than white flowers. In the early 1800s a German, Karl Friedrich von Gärtner, noticed dominant and recessive traits in the garden pea and other plants. Von Gärtner carried out thousands of crosses and established that pollen transmits traits to the female plant.

In his 1866 paper, Mendel cited the work of these and other investigators, saying that they worked with "inexhaustible perseverance." He also commented that "so far, no generally applicable law governing the formation and development of hybrids has been successfully formulated. . . .Those who survey the work done in this department will arrive at the conviction that among all the numerous experiments made, not one has been carried out to such an extent and in such a way as to make it possible to determine the number of different forms under which the offspring of the hybrids appear, or to arrange these forms with certainty according to their separate generations, or definitely to ascertain their statistical relations." In other words, before Mendel no one else had both collected and analyzed quantitative data about the results of genetic crosses.

---

the second 3, and the first 1 by the second 1. The result is 9:3:3:1.

Mendel tried several dihybrid crosses involving pairwise combinations of different characters and found similar results. The results of these crosses formed the basis for Mendel's second law of inheritance, the **law of independent assortment.** This law simply states that each pair of alleles segregates independently during meiosis. For this reason, the law of independent assortment could also be called the *law of independent segregation.*

Independent assortment of alleles occurs because the segregation of homologous chromosomes in one tetrad does not affect the segregation of homologous chromosomes in any other tetrad. Consider the F_1 pea plants in the dihybrid cross described above. These plants are heterozygous for both the seed-shape gene and the seed-color gene, which are on two different chromosomes (Figure 12.6). During meiosis, a chromosome that contains either the *S* allele or the *s* allele may end up in a gamete with a chromosome that contains either the *Y* allele or the *y* allele. Therefore, meiosis produces four types of gametes: One-quarter will have the *S* and *Y* alleles, one-quarter will have *S* and *y*, one-quarter will have *s* and *Y*, and one-quarter will have *s* and *y*. Note, however, that while the plant produces four types of gametes, any given reproductive cell can produce only two types of gametes when it undergoes meiosis.

Mendel's laws of segregation and independent assortment explain the simple rules by which traits are passed from one generation to the next. Mendel's contribution to the science of genetics was enormous, but he did not develop his laws in an intellectual vacuum. In fact, scientists before Mendel had crossed plants that differed in various traits and had observed similar results (see the *Plants & People* box on this page). Unlike these earlier scientists, however, Mendel kept track of the numbers of offspring that showed different traits. The numbers gave him the insight that led to his discovery of the laws of inheritance.

Section Review

1. **How many different genes occur in a typical eukaryote like peas or humans?**
2. **How did Mendel obtain a 3:1 phenotypic ratio in the F_2 generation of his monohybrid crosses?**
3. **When does segregation of alleles occur in meiosis?**
4. **Explain the difference between a monohybrid cross and a dihybrid cross.**

Beyond Mendel's Work

In the 150 years since Mendel began his experiments with garden peas, biologists have investigated the inheritance of many genes in numerous organisms. In general, scientists have turned to organisms that have shorter life cycles than peas, including the fruit fly *Drosophila melanogaster* and, more recently, a small flowering plant, *Arabidopsis thaliana* (see *The Intriguing World of Plants* box on page 270). While peas require more than two months to complete a life cycle, fruit flies require two weeks and *Arabi-*

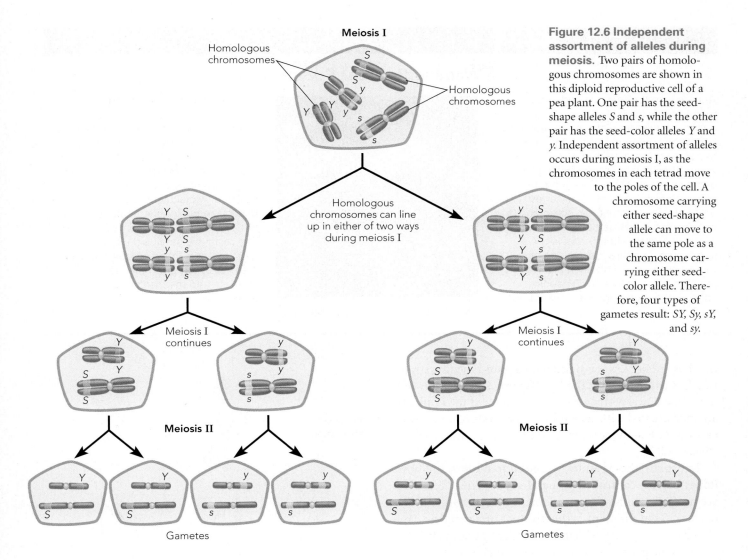

Meiosis I

Homologous chromosomes

Homologous chromosomes

Homologous chromosomes can line up in either of two ways during meiosis I

Meiosis I continues

Meiosis I continues

Meiosis II

Meiosis II

Gametes

Gametes

Figure 12.6 Independent assortment of alleles during meiosis. Two pairs of homologous chromosomes are shown in this diploid reproductive cell of a pea plant. One pair has the seed-shape alleles *S* and *s*, while the other pair has the seed-color alleles *Y* and *y*. Independent assortment of alleles occurs during meiosis I, as the chromosomes in each tetrad move to the poles of the cell. A chromosome carrying either seed-shape allele can move to the same pole as a chromosome carrying either seed-color allele. Therefore, four types of gametes result: *SY, Sy, sY,* and *sy.*

dopsis requires six weeks or less. The research carried out since Mendel has shown that his first and second laws explain the mechanics of inheritance for the vast majority of eukaryotic traits. However, some patterns of inheritance are more complex than the ones that Mendel studied. We'll investigate some of those patterns next. Then we'll take a brief look at the molecular basis of one of the characters Mendel studied—plant height.

Mendel's laws also apply to crosses that involve more than two traits

Mendel didn't report on trihybrid crosses, which involve plants that differ in three characters, but such crosses are easily analyzed using his methods. As an example, consider crossing a pure-breeding plant that is tall and has smooth, yellow seeds (*SSTTYY*) with a plant that is short and has wrinkled, green seeds (*ssttyy*). The F$_1$ plants (*SsTtYy*) will be heterozygous and phenotypically domi-

nant for all three characters. Such an F$_1$ plant can produce eight different types of gametes.

Crossing two F$_1$ plants will produce an F$_2$ generation with a rather complex phenotypic ratio of 27:9:9:9:3:3:3:1. The "27" in this ratio represents plants that have all three dominant traits, each "9" represents plants that have two dominant traits and one recessive trait, each "3" represents plants that have one dominant trait and two recessive traits, and the "1" represents plants that have all three recessive traits. You can verify this ratio yourself either by drawing an 8 × 8 Punnett square or by multiplying (3:1) × (3:1) × (3:1).

Some characters are not controlled by one dominant and one recessive allele

If Mendel had performed his experiments on snapdragons (*Antirrhinum majus*) instead of peas, he would have had much more difficulty discovering the rules that govern inheritance. When a pure-breeding strain of snapdragons

THE INTRIGUING WORLD OF PLANTS

A Weed with Great Potential

Arabidopsis thaliana, a weed in the mustard family (*Brassicaceae*), has become a popular plant for genetic experiments in recent years. *Arabidopsis* has several features that make it a favorite of scientists interested in plant genetics, growth, and development. First, the plant is small—only a few centimeters high—and hardy, so botanists can easily grow large numbers of the plants indoors. Second, *Arabidopsis* completes its entire life cycle in only four to six weeks, and each plant can produce more than 10,000 seeds. Third, like the garden pea, *Arabidopsis*

Mouse-ear cress (*Arabidopsis thaliana*).

is self-pollinating, which simplifies F_1 crosses. Finally, *Arabidopsis* has only five pairs of chromosomes and 26,000 genes, a relatively small number of genes for a eukaryote and fewer than in any other flowering plant in which the genes have been identified and counted. Molecular biologists determined the nucleotide sequence of all of the plant's chromosomes in 2000. The relatively small amount of genetic material in *Arabidopsis* makes it easier to establish the identity, location, and mechanism of action of specific genes.

---

that has red flowers is crossed with a pure-breeding strain that has white flowers, the F_1 offspring have neither red nor white flowers—they have pink flowers! If the F_1 plants are then crossed with each other, they produce an F_2 generation with a phenotypic ratio of 1 red: 2 pink: 1 white (Figure 12.7).

To explain these results, you need to understand how flower color is controlled in snapdragons. The red flowers are red because they have a pigment that the white flowers lack. That pigment is the product of a reaction catalyzed by an enzyme, which is coded for by one allele (C^R) of a specific gene. Another allele of that gene (C^W) codes for a form of the enzyme that does not catalyze the reaction, so no pigment is produced. A plant that is homozygous for the C^R allele produces a lot of the pigment and therefore has red flowers. One that is homozygous for the C^W allele produces no pigment and therefore has white flowers. A heterozygote ($C^R C^W$) produces some pigment, but only enough to turn the flowers pink. The flower color of the heterozygote is intermediate between that of both homozygotes because neither allele exhibits complete dominance over the other. Consequently, this type of inheritance is known as **incomplete dominance.** Since this pattern of inheritance does not involve one dominant and one recessive allele, the alleles are represented

Figure 12.7 Incomplete dominance in snapdragons. If a pure-breeding snapdragon with red flowers is crossed with a pure-breeding snapdragon with white flowers, the F_1 generation will have pink flowers. If the F_1 plants are crossed, the F_2 generation will have a 1:2:1 phenotypic ratio of red-, pink-, and white-flowering plants.

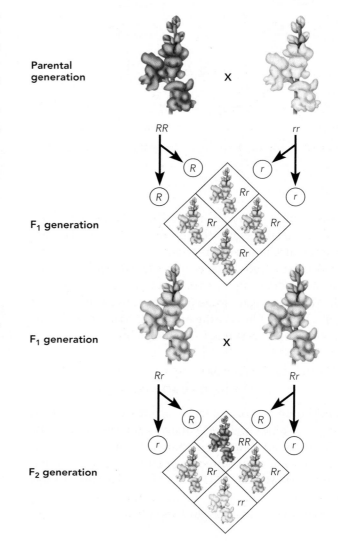

Parental generation

RR   X   rr

F_1 generation

Rr   Rr   Rr   Rr

F_1 generation

Rr   X   Rr

F_2 generation

RR   Rr   Rr   rr

by capital letters with superscripts instead of a capital letter and a lowercase letter.

Another variation on the one dominant allele–one recessive allele system occurs with genes that have more than two alleles. A good example of such a gene is the one that controls the appearance of clover leaves (Figure 12.8). That gene has at least seven alleles, any two of which can be present in a particular clover plant. Each pairwise combination of alleles results in leaves with a specific size, shape, and pattern of markings. The pollen-incompatibility gene, which causes self-sterility in some plants, is another example of a gene that has many alleles. If two plants have the same allele for this gene, pollen from one plant will not germinate when it lands on a stigma of the other plant's flowers. In certain plants, flower color also appears to be determined by genes with multiple alleles.

While some characters are controlled by genes with more than one allele, other characters are controlled by more than one gene. Inheritance in such cases is referred to as **polygenic inheritance.** The patterns of polygenic inheritance are usually very different from those of the characters that Mendel studied, where there were only two phenotypes for each character—for example, purple flowers and white flowers, or smooth seeds and wrinkled seeds. In polygenic inheritance, the phenotypes often exhibit a continuum of values. In wheat, for example, two genes control the color of the kernels. When a plant that has dark red kernels is crossed with one that has white kernels, the F_1 plants all have medium-red kernels. Crossing the F_1 plants produces an F_2 generation with a phenotypic ratio of 15 red:1 white. There are four degrees of redness in the F_2 generation, ranging from dark red to light red. Polygenic inheritance is also involved in the determination of sex in annual mercury (*Mercurialis annua*) and cob length in some varieties of corn.

The converse of polygenic inheritance is **pleiotropy,** where a single gene controls more than one character. In fact, three of the characters that Mendel investigated—flower color, seed color, and the presence of a colored spot on the leaf axil—are controlled by one gene. Mendel observed that purple flowers, brown seeds, and a brown axil spot always occurred together, as did white flowers, light seeds, and no axil spot. In tobacco (*Nicotiana tabacum*), a single gene affects at least five plant characters (Figure 12.9). In general, the dominant allele for this gene confers a long, narrow phenotype, whereas the recessive allele (when homozygous) confers a broader, shorter phenotype. Wheat breeders have found a gene that causes

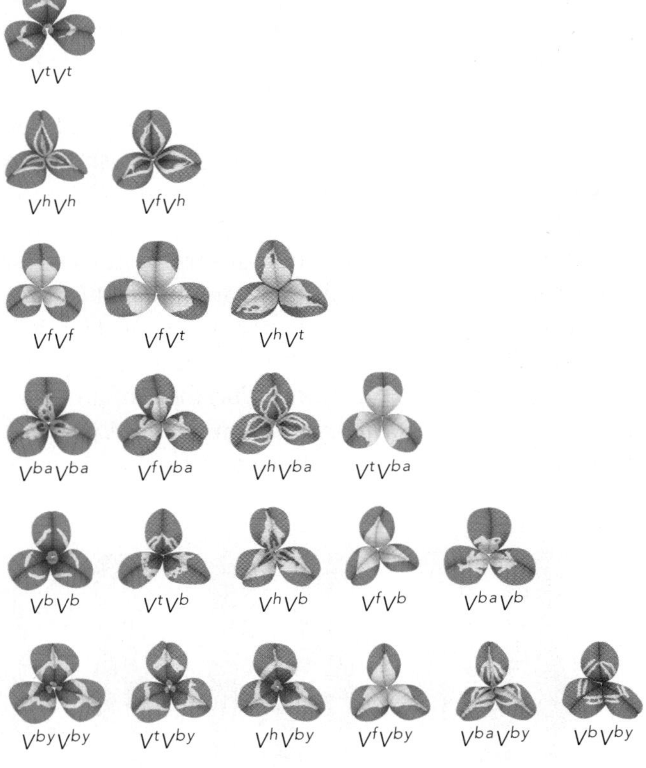

Figure 12.8 Inheritance involving multiple alleles. The chevron (V or inverted-V) on clover leaves is determined by a gene with multiple alleles. The phenotypes associated with various combinations of six of those alleles (V^t, V^h, V^f, V^{ba}, V^b, and V^{by}) are shown here.

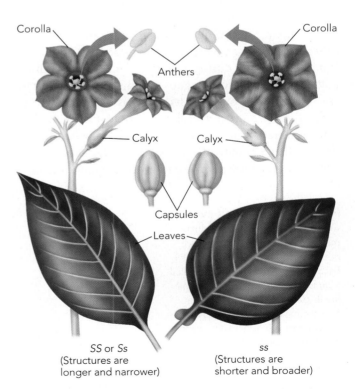

Figure 12.9 Pleiotropy in tobacco. Pleiotropy occurs when a gene has more than one effect on an organism's phenotype. In tobacco plants, the *S* gene affects the corolla, anther, calyx, leaf, and capsule.

both high yields and the presence of awns (bristles on the flower clusters). By looking for awns in the offspring of crosses, breeders can predict the yields of the offspring even before the offspring reproduce.

The locations of genes affect inheritance patterns

As you learned earlier in this chapter, each chromosome in a typical eukaryote contains thousands of genes. Therefore, there is a tendency for the genes on a chromosome to segregate as a unit during meiosis, causing the traits specified by those genes to be inherited together. Such genes are known as **linked genes.** When linked genes are involved in a genetic cross, the law of independent assortment does not apply because the alleles in question are physically incapable of segregating independently.

Since the seven characters that Mendel studied are inherited independently and peas have seven pairs of homologous chromosomes, you might assume that each character is controlled by a gene on a different chromosome. In fact, three of those characters (plant height, flower position, and pod shape) are controlled by genes on chromosome 4, and two other characters (seed color and flower color) are controlled by genes on chromosome 1. Why, then, don't those genes that are on the same chromosome behave as though they are linked?

The answer to this question lies in the phenomenon called *crossing over,* which was discussed in Chapter 6. When tetrads form during prophase of meiosis I, the chromatids of the two homologous chromosomes in a tetrad can sometimes be observed lying across each other, creating an X-shaped structure called a *chiasma* (Figure 12.10a). The chromatids involved in a chiasma may break and then exchange fragments, in the process exchanging any alleles that are located on the fragments (Figure 12.10b).

If the exchanged alleles are different (for example, one coding for purple flowers and one for white flowers), each chromatid will acquire a new combination of alleles. When two genes are very close together on a chromosome, it is likely that they will be included in the same exchanged fragment, so they will remain linked even after crossing over takes place. The farther apart they are, the more likely it is that crossing over will involve one gene but not the other. When that happens, the linkage between the genes is broken. The three genes on chromosome 4 whose inheritance Mendel studied are so far apart that they behave as though they were on separate chromosomes. The same is true for the seed-color and flower-color genes on chromosome 1. As far as we know, Mendel did not work with any characters that appear to be controlled by linked genes.

Location is also important for genes on sex chromosomes, the pair of homologous chromosomes that determine sex in some species. In dioecious plants, which have separate sexes, sex chromosomes are common. For example, in asparagus (*Asparagus officinalis*), female plants have two X chromosomes with identical genes—but not necessarily the same *alleles* for those genes—while male plants have one X chromosome and one Y chromosome. Some of the genes on the Y chromosome are different from those on the X chromosome. (The same pattern of sex determination occurs in humans.) Therefore, if a male has a recessive allele for a gene on the X chromosome, the trait specified by that allele is always expressed. Females must have two copies of the recessive allele to show the recessive trait. In plants, genes on the X and Y chromosomes that are involved in sex determination most often control the production and action of various hormones.

Inheritance patterns are further complicated by genes that are not located on chromosomes in the nucleus. Recall from Chapter 2 that mitochondria and chloroplasts have small, circular DNA molecules like those of prokaryotes. Characters coded for by genes in the DNA of these organelles are passed from parents to offspring through

Figure 12.10 Crossing over during meiosis. (a) Two chromatids of a tetrad cross over each other during prophase of meiosis I. Each point of crossing over is called a *chiasma.* **(b)** The chromatids may break and exchange fragments at a chiasma. When that happens, the alleles on each fragment are swapped between the chromatids.

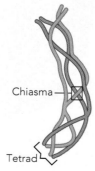

(a) Two chromatids of a tetrad cross over each other during prophase of meiosis I. Each point of crossing over is called a chiasma.

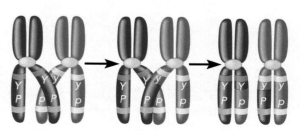

(b) The chromatids may break and exchange fragments at a chiasma. When that happens, the alleles on each fragment are swapped between the chromatids.

Figure 12.11 Cytoplasmic inheritance. The yellow patches on the leaves of this *Pelargonium* are areas where none of the chloroplasts are capable of producing chlorophyll. This trait is controlled by an allele of a gene in the chloroplasts' DNA. When cells divide, chloroplasts that have this allele are distributed randomly into daughter cells. Cells that receive these chloroplasts instead of normally functioning chloroplasts give rise to the yellow patches.

cytoplasmic inheritance. During cytokinesis, the existing mitochondria and chloroplasts randomly segregate into daughter cells. Inheritance of this type is also known as *maternal inheritance* because the egg carries cytoplasm with organelles to the next generation, but the sperm does not.

Cytoplasmic inheritance is responsible for the appearance of white or yellow patches on the leaves of certain plants (Figure 12.11). In many instances, these patches consist of cells with white chloroplasts because they have an allele that fails to code for the production of chlorophyll. A patch develops when a cell that has a mix of green and white chloroplasts divides and one of the daughter cells, by chance, receives only white chloroplasts. The patch grows as that daughter cell and its progeny divide. Cytoplasmic inheritance also occurs with mitochondrial genes that cause male sterility in plants. Plant breeders value male sterility because self-pollination does not occur in sterile male flowers, so stamens need not be removed before crosses are performed.

Genes interact with each other and with the environment

Genes do not act in isolation to affect an organism's phenotype. We have already discussed one example illustrating this fact in polygenic inheritance, where a single character is controlled by more than one gene. Another example is a phenomenon known as **epistasis,** which occurs when one gene alters the effect of another. In some sweet peas (*Lathyrus* spp.) , for instance, two genes interact to determine flower color. Each gene codes for an enzyme in the biochemical pathway of pigment synthesis, so

a plant must have at least one dominant allele (*C* and *P*) of both genes to produce purple flowers. A plant that is homozygous for one or both of the recessive alleles will have white flowers. Thus, if a white-flowered plant with the genotype *CCpp* is crossed with a white-flowered plant with the genotype *ccPP*, the F₁ offspring will be heterozygous for both alleles (*CcPp*) and have purple flowers (Figure 12.12). However, crossing two of

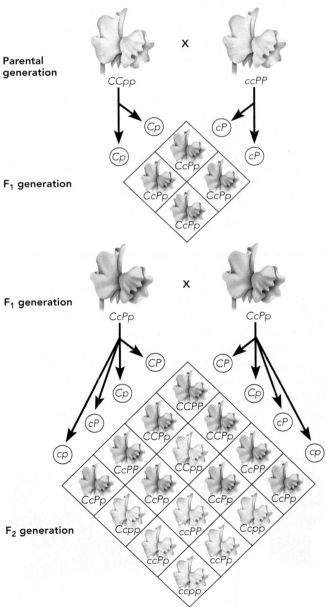

Figure 12.12 Epistasis. Two genes interact to control flower color in sweet peas. A cross between two white-flowered plants with different genotypes (*CCpp* and *ccPP*) produces only purple-flowered plants in the F₁ generation. Crossing two F₁ plants results in an F₂ generation with a 9 purple: 7 white ratio.

Table 12.1 Epistasis in Plants

Variations in the classic F_2 phenotypic ratio of 9:3:3:1 can occur when two genes interact with each other.

Phenotypic Ratio in F_2	Type of Gene Interaction	Example
9:3:3:1	Two non-linked genes	Mendel's crosses
9:3:4	Homozygous recessive of one gene prevents all color	Color of onion bulbs
9:6:1	Each gene has the same effect if the dominant allele is present	Fruit shape in squash
9:7	Dominant allele for both genes is required for a trait to be expressed	Flower color in sweet pea
13:3	One gene is the dominant suppressor of the other gene	Malvidin (floral pigment) synthesis in *Primula*
12:3:1	Dominant allele of one gene replaces the effect of both alleles of the other gene with a new effect	Fruit color in squash

these F_1 plants yields a phenotypic ratio of 9 purple: 7 white in the F_2 generation. Table 12.1 presents some additional instances of epistasis in plants.

While genes provide the plans according to which an organism is built and operates, the actual interpretation of those plans often depends on environmental factors. Hydrangeas (*Hydrangea macrophylla*) provide an example of environmental influence that is familiar to many

Figure 12.13 An environmental effect on phenotype. These two hydrangeas have the same genotype. The plant with blue flowers is growing in acidic soil, while the one with pink flowers is growing in neutral or alkaline soil.

home gardeners (Figure 12.13). Their flowers can show a range of colors depending on the pH of the soil. In acidic soil, a hydrangea typically has blue flowers. In neutral or alkaline soil, the same plant will grow pink flowers. In other plants, temperature affects flower color. Leaves of the water buttercup (*Ranunculus peltatus*) look very different when they grow in air than when they grow in water. Large, lobed leaves develop above the water, while thin, rootlike leaves develop on submerged parts of the plant. These and many other examples show quite clearly that an organism's phenotype reflects a combination of hereditary instructions and environmental forces.

Mendel's gene for height in peas controls the production of a growth-promoting hormone

Mendel's experiments made clear that inherited traits pass from one generation to the next via discrete but unknown units that behave independently. The nature of Mendel's units of heredity remained unknown for nearly a century. From the 1920s through the 1950s, a series of experiments with bacteria and viruses established clearly that DNA, and sometimes RNA, behaves as the molecule that carries genetic information from one generation to the next. In 1953, Watson and Crick's analysis revealed the structure of DNA. During the 1960s, the mechanism by which DNA determines the characteristics of organisms became clear, at least in general terms (see Chapter 13). Indeed, by the 1960s, the very genes that Mendel himself had investigated were again being studied to determine their biochemical effects.

The gene for height in Mendel's tall and short pea plants is now known as the *Le* gene. Tall plants contain one or two *Le* alleles, and short plants contain two *le* alleles. (In this case, each allele is represented by *two* letters. For the dominant allele, the first letter is capitalized. For the recessive allele, both letters are lowercase.) A good deal of research has been done on the effects of the alleles of these genes. For example, in studies at the University of Tasmania in Australia, tall pea plants were found to contain 10 to 18 times as much of the growth-promoting hormone gibberellin #1 (GA1) as dwarf pea plants. However, dwarf plants contain three to five times as much of another hormone, gibberellin #20 (GA20) as tall plants. GA20 does not promote growth. Subsequent work has confirmed that the recessive allele, *le*, codes for an enzyme that is inefficient at converting GA20 to GA1. As a result, plants with two *le* alleles cannot produce enough GA1 to grow tall. Studies such as these reveal the directions that biologists are taking today to explain mechanisms of heredity.

Section Review

1. Give an example of incomplete dominance in plants.
2. Explain the statement, "Linked genes segregate together."
3. Discuss an example of cytoplasmic inheritance.
4. What biochemical process is affected by Mendel's gene for plant height?

SUMMARY

Mendel's Experiments on Inheritance

Making sense of Mendel's experiments requires a basic understanding of genes and chromosomes (pp. 263–264)
An organism's features, called *characters,* may appear in two or more forms known as *traits.* Each trait is coded for by an allele, a variant form of a gene. A diploid cell contains two alleles of every gene, one from each parent.

Monohybrid crosses involve individuals that have different alleles for a specific gene (pp. 264–266)
Gregor Mendel crossed pure-breeding varieties of garden peas that had different traits for the same character. For the characters that Mendel studied, all the offspring of this cross (the first filial or F_1 generation) had the trait of just one of the parents. He called this trait the *dominant trait.* When Mendel crossed two F_1 plants with each other (a monohybrid cross), 75% of their offspring (the second filial or F_2 generation) had the dominant trait and 25% had the other trait, which he termed *recessive.* Individuals that show a recessive trait have two recessive alleles for a specific gene and are said to be homozygous for that allele. Individuals that show a dominant trait have either two dominant alleles or one dominant and one recessive allele for the same gene; the latter condition is described as *heterozygous.* An individual's combination of alleles is its genotype, which controls its physical appearance, or phenotype. The results of a cross can be represented by a Punnett square. Based on the results of his monohybrid crosses, Mendel formulated the law of segregation, which states that alleles segregate during meiosis and then come together randomly during fertilization.

Segregation of alleles occurs during anaphase I of meiosis (p. 266)
Alleles segregate when an organism's reproductive cells undergo meiosis. If the individual is heterozygous, each reproductive cell will produce two types of gametes, half with the dominant allele and half with the recessive allele.

A testcross demonstrates the genotype of an individual with a dominant phenotype (p. 266)
To determine whether an individual with a dominant phenotype is homozygous or heterozygous, it is crossed with an individual that has the recessive phenotype for that character. The parent with the dominant phenotype must be homozygous if all the offspring have the dominant phenotype. It must be heterozygous if there is a 1:1 ratio of dominant to recessive phenotypes in the offspring.

Dihybrid crosses involve individuals that have different alleles for two specific genes (pp. 266–268)
Mendel also crossed pure-breeding varieties of garden peas that had different traits for each of two characters. For the characters that Mendel studied, all F_1 plants had the dominant trait for both characters. When he crossed the F_1 plants (a dihybrid cross), 9/16 of the F_2 plants had both dominant traits, 3/16 had the dominant trait for the first character and the recessive trait for the second character, 3/16 had the recessive trait for the first character and the dominant trait for the second character, and 1/16 had both recessive traits. These results exemplify Mendel's law of independent assortment, which states that each pair of alleles segregates independently during meiosis. Independent assortment occurs because tetrads segregate independently.

Beyond Mendel's Work

Mendel's laws also apply to crosses that involve more than two traits (p. 269)
In trihybrid crosses, which involve plants that are heterozygous for three characters, the offspring have a phenotypic ratio of 27:9:9:9:3:3:3:1.

Some characters are not controlled by one dominant and one recessive allele (pp. 269–272)
In incomplete dominance, neither allele exhibits complete dominance over the other, so heterozygotes have a phenotype that is intermediate between the dominant and recessive phenotypes. Some genes have more than two alleles, and each pairwise combination of alleles results in a different phenotype. In polygenic inheritance, a character is controlled by more than one gene, and phenotypes often exhibit a continuum of values. Pleiotropy is a situation in which a single gene controls more than one character.

The locations of genes affect inheritance patterns (pp. 272–273)
Genes that are located on the same chromosome and segregate as a unit during meiosis are known as *linked genes.* However, crossing over between homologous chromosomes during meio-

sis can allow the alleles of two genes on the same chromosome to segregate independently. Independent segregation for such genes is more likely the farther apart they are. Cytoplasmic inheritance involves characters coded for by genes in the DNA of mitochondria and chloroplasts.

Genes interact with each other and with the environment (pp. 273–274)

Various phenotypic ratios in an F_2 generation can result from epistasis, in which one gene alters the effect of another. Environmental factors can influence an organism's phenotype.

Mendel's gene for height in peas controls the production of a growth-promoting hormone (p. 274)

The *Le* gene controls height in garden peas. Plants with two recessive *le* alleles produce insufficient amounts of the growth-promoting hormone gibberellin #1 and have a dwarf phenotype.

Review Questions

1. Define *gene* and *allele* in terms of characters and traits. Provide an example to illustrate your definitions.
2. Explain the difference between *phenotype* and *genotype*.
3. How did Mendel determine which of a gene's alleles was dominant?
4. Follow the cross $TT \times tt$ through two generations. Draw a Punnett square that shows the phenotypic ratio in the F_2 generation.
5. What can you learn by performing a testcross?
6. Draw a Punnett square for the cross $SsYy \times SsYy$. What letters belong on adjoining sides of the square, and what do those letters represent? What is the phenotypic ratio of the offspring produced by this cross?
7. Carry out the testcross $AaBb \times aabb$. Draw a Punnett square and determine the phenotypic ratio of the offspring.
8. Give an example of polygenic inheritance in plants.
9. How do linked genes differ from nonlinked genes?
10. Give an example of epistasis.

Questions for Thought and Discussion

1. In plants and other multicellular organisms, each cell contains all the genes of the organism. Why, then, do different parts of a plant look different?
2. Do you think that tallness is dominant over shortness for all plants as it is for garden peas?
3. If an organism has ten genes and each gene has two different alleles, what is the probability that a particular reproductive cell will produce a pair of identical gametes?

4. Suppose you had a population of pea plants consisting of equal numbers of tall plants (TT) and short plants (tt). Can you think of an environment that would favor only the tall or the short plants? If you raised many generations of plants in that environment by allowing random pollination between the plants, how would the percentage of T and t alleles in the population change?
5. Suppose you observe a population of plants containing individuals of many different heights. How could Mendel's laws account for this continuous variation in a character?
6. A plant breeder performs numerous crosses and observes that whenever self-pollination occurs, the flowers never produce seeds. How could you explain this observation?
7. Consider a homologous pair of chromosomes, each with three gene loci *A*, *B*, and *C*, listed in order from the centromere towards the end of one of the chromosome arms. One member of the pair has the three dominant alleles (*A*, *B*, and *C*) while its homologue has the three recessive alleles (*a*, *b*, and *c*). Diagram the outcome of each of the following crossover events between non-sister chromatids: (1) between the centromere and the *A* locus; (2) between the *A* and *B* loci; (3) between the *A* and *B* and between the *B* and *C* loci; (4) two crossovers between the *A* and *B* loci.

Evolution Connection

What selective forces would most likely have influenced the evolutionary development of two different leaf shapes in submerged versus aerial leaves of the water buttercup, *Ranunculus peltatus*?

To Learn More

Visit The Botany Place Website at www.thebotanyplace.com for quizzes, exercises, and Web links to new and interesting information related to this chapter.

Gonick, Larry and Mark Wheelis. *The Cartoon Guide to Genetics*. Markham: Harper Perennial, 1991. A cartoon view of Mendelian genetics and genetic engineering.

Henig, M. Robin. *The Monk in the Garden: The Lost and Found Genius of Gregor Mendel, the Father of Genetics*. Wilmington: Mariner Books, 2001. The history of naturalist Gregor Mendel.

Tagliaferro, Linda, and Mark Bloom. *The Complete Idiot's Guide to Decoding Your Genes*. Madison: Alpha Books, 1999. Explains the world of genetics.

Gene Expression and Activation

Arabidopsis flowers, wild type and mutant.

Gene Expression

During replication, DNA is copied

DNA codes for the structure of proteins

During transcription, RNA is made from DNA

During translation, a protein is made from messenger RNA

Mutations can cause changes in gene expression

Differential Gene Expression

Gene expression is controlled at various levels

Regulatory proteins control transcription

Hormones and light can trigger the activation of transcription factors

Identifying Genes That Affect Development

Experiments on *Arabidopsis* illustrate the use of mutations to understand plant development

Transposons can be used to locate genes that affect development

Homeotic genes control development in plants and animals

As a young plant develops from a zygote, it forms roots, stems, and leaves. Each organ has tissues composed of many types of differentiated cells. As a plant matures, it continues to produce shoots and roots as well as various kinds of reproductive structures, including flowers, cones, fruits, and seeds.

Almost every cell in a plant contains the full complement of 15,000 to 50,000 genes that are characteristic of the species. Cells in an embryo or a meristem have the potential to express any of those genes and to give rise to any part of the plant. Such cells are said to be "totipotent." In a developing or mature plant, however, the totipotency of cells is normally lost as each cell differentiates into a particular cell type.

Differentiated cells express only a fraction of the genetic information they contain. Genes that are necessary for life itself, such as those that code for enzymes used to synthesize ATP, are expressed by all cells. Other genes are expressed only in particular types of cells. For example, genes coding for proteins involved in the Calvin cycle of photosynthesis are expressed by photosynthetically active chlorenchyma cells in leaves but not by root cells. Conversely, genes that cause root hairs to develop from epidermal cells are expressed only by root cells.

Even within one part of a plant, different genes and alleles are expressed by different individuals in a population. For example, some of the pea plants Mendel studied were tall while others were short. In the common garden plant, scarlet sage (*Salvia splendens*), at least seven genes control the beautiful red, rose, salmon, pink, white, purple, and burgundy colors of the flowers.

In this chapter, you'll learn how cells transfer information in the genetic code of DNA into the structure of proteins that control every aspect of plant development and functioning. You'll also learn how gene activity is regulated—that is, how genes are "turned on" and "turned off."

Mature tree

In a mature tree, all living cells continue to use many enzymes, such as those that produce and utilize ATP. In addition, new leaves and reproductive structures are produced, and the tree utilizes different sets of enzymes to adapt to different seasons and even to different times of day.

4–5 Year old tree

As the tree grows older, it produces secondary meristems and may become reproductive. Specialized enzymes are required to produce bark, flowers, fruits, and seeds.

Year old tree

As the seedling grows into a young plant, enzymes continue to direct formation of specialized cells in the stem, roots, and leaves. Each type of cell contains a number of maintenance enzymes in common as well as unique enzymes to direct unique cellular functions.

Seedling

As the seedling produces leaves, the enzymes and other proteins associated with photosynthesis and respiration are synthesized and begin to act.

Seed

As the seed begins to germinate, enzymes are produced that break down the storage compounds, such as fats, starch, and protein, to provide nourishment to the seedling.

Gene Expression

An evolutionary biologist views life as populations of individuals competing for reproductive success. To a cell biologist, life consists of organisms composed of one or more cells. Not surprisingly, a biochemist tends to view life as an organized collection of interacting molecules. To move from a population view to an organismal view or from a cellular view to a molecular view is to be a reductionist—to reduce something to smaller pieces in order to understand it. A reductionist will say that many of an organism's characteristics are determined by its proteins and that its genes provide the code for making those proteins.

How does a genetic code made of nucleotides lead to proteins made of amino acids? As you will learn in this section, the conversion from genetic code to protein involves two basic steps: transcription and translation. Transcription changes one language into another form of the same language. You might transcribe spoken English into printed English. In cells, transcription changes the nucleotide-based language of deoxyribonucleic acid (DNA) into another form of nucleotide-based language, that of ribonucleic acid (RNA). Translation, on the other hand, changes one language to another—English to Chinese, for example. In cells, translation involves going from the nucleotide-based language of RNA to the amino acid-based language of proteins. The common manner in which all living things transfer information from genes to proteins is sometimes referred to as the *central dogma of molecular biology.*

During replication, DNA is copied

When a cell divides, it passes on all of its genetic information to each of its daughter cells. For that to happen, the dividing cell must replicate, or copy, its DNA. You learned in Chapter 7 that DNA is a polymer composed of four different nucleotides. Two strands of DNA, each millions of nucleotides in length, twist around each other to form a double helix stabilized by hydrogen bonds. Recall from Chapter 2 that DNA replication in eukaryotes occurs before nuclear division, during the S phase of the cell cycle. The double-stranded nature of DNA plays a crucial role in the replication process (Figure 13.1).

Figure 13.1 DNA replication—an overall view. (a) During replication, the parental DNA double helix is separated, and a daughter strand is synthesized along each parental strand. **(b)** The nucleotides in the daughter strands contain bases that are complementary to those in the parental strands.

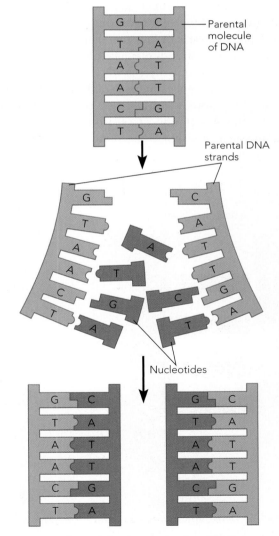

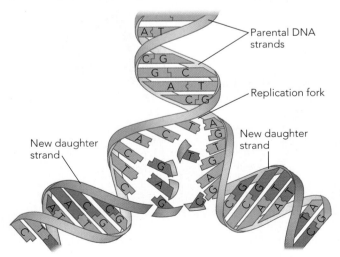

(a)

(b)

Identical daughter molecules of DNA

If you know the nucleotide sequence of one DNA strand, you can determine the sequence of the other strand. That's because nucleotides containing the base adenine (A) always pair with nucleotides containing the base thymine (T), and nucleotides containing the base guanine (G) always pair with nucleotides containing the base cytosine (C). The pairing between A and T and between G and C is called *complementary pairing* and is due to the formation of hydrogen bonds between the nucleotide bases.

The overall process of DNA replication is simple: The DNA double helix separates into two strands, and a new complementary strand is formed for each of the preexisting, or parental, strands. Replication is said to be *semiconservative* because it produces two "daughter" DNA molecules, each consisting of one parental strand and one new strand. The Y-shaped region where the parental double helix unwinds and the new strands are synthesized is called a *replication fork*. In eukaryotes, replication occurs simultaneously at multiple locations on a chromosome.

DNA polymerase is the enzyme that catalyzes the synthesis of new DNA strands during replication. In fact, two molecules of DNA polymerase work at each replication fork, one molecule replicating each of the parental strands. As Figure 13.2 shows, the mechanism of replication is different for the two strands. The reason for this difference is that the sugar-phosphate backbones of the two strands run in opposite directions (Figure 13.2a), but DNA polymerase can add nucleotides in only one direction. Therefore, one new strand is synthesized continuously by a DNA polymerase that moves along a parental strand *toward* the replication fork. The other new strand is synthesized as a series of short pieces (Okazaki fragments) by a DNA polymerase that moves along the other parental strand *away* from the replication fork.

Several other enzymes assist the DNA polymerases during replication. One of these enzymes is helicase, which unwinds the parental DNA strands at the replication fork. Another is DNA ligase, which links together the short pieces of DNA produced by one of the DNA polymerases (Figure 13.2b).

DNA codes for the structure of proteins

The idea that DNA carries the genetic code developed gradually during the first half of the twentieth century. Transplantation experiments in the alga *Acetabularia* (Figure 13.3) demonstrated that the hereditary material in eukaryotes resides in the nucleus. Cytological studies of mitosis and meiosis identified chromosomes as the units of heredity. However, chromosomes contain both proteins and DNA. Proteins, composed of 20 different amino acids,

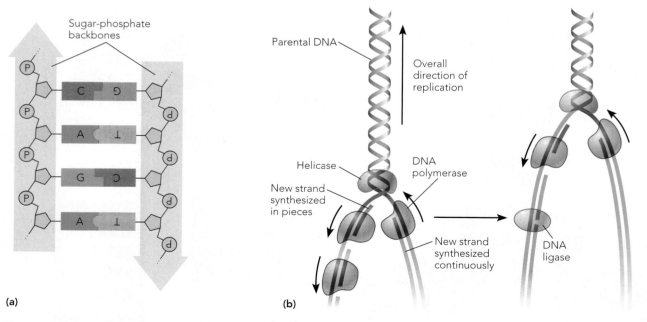

(a)

(b)

Figure 13.2 DNA replication—a closer look. (a) In a DNA molecule, the sugar-phosphate backbones of the two strands are oriented in opposite directions. **(b)** Because of the oppositely oriented backbones, DNA polymerases replicate one strand continuously and the other strand in short pieces. DNA ligase joins the short pieces together. Helicase unwinds the parental DNA, allowing the replication fork to advance.

seemed to have the complexity demanded of the coding material. DNA, with only four different nucleotides, did not seem complex enough.

Several critical experiments proved that the genetic material is DNA, not protein. In 1928, Frederick Griffith found that the contents of bacterial cells contain genetic material but that the protein-rich cell coats of bacteria do not. Oswald Avery and his colleagues identified the genetic material of these bacteria as DNA in 1944. Because of the relative simplicity of DNA, their results were met with considerable skepticism. In 1952, Alfred Hershey and Martha Chase discovered that DNA is the genetically active material of a bacterial virus. A year later, James Watson and Francis Crick demonstrated the double-helical structure of DNA. At that point, investigators began to think that the structure of DNA might be able to accommodate the genetic code, although the exact nature of the code remained a mystery until the 1960s.

A British physician, Archibald Garrod, was the first to suggest that genes code for enzymes. In 1909, he suggested that each genetic disease was caused by a defect in a specific enzyme. Since the identity of the genetic material was unknown at that time, his idea received little publicity.

The real breakthrough in linking genes with enzymes came in the 1940s with the experiments of George Beadle and Edward Tatum, who studied the biosynthetic pathways of amino acids in the bread mold *Neurospora crassa* (Figure 13.4). For example, the amino acid arginine is made in a three-step pathway: In step 1, a precursor substance is converted to ornithine; in step 2, ornithine becomes citrulline; and in step 3, citrulline is changed to arginine. The wild-type, or natural, strain of *Neurospora* can carry out all three steps in the pathway. However, Beadle and Tatum found other strains in which the pathway was blocked at one of the steps. A strain with a block at step 1 could make arginine only when ornithine or citrulline was added to the growth medium. A strain with a block at step 2 could use citrulline but couldn't use ornithine. One with a block at step 3 could not make arginine at all. These observations led the researchers to propose that a different gene codes for each enzyme in a biosynthetic pathway. They called their proposal the "one gene-one enzyme hypothesis." Since not all proteins are enzymes, and many proteins contain more than one polypeptide chain, the Beadle-Tatum hypothesis is usually restated as the "one gene-one polypeptide hypothesis." Over the years, numerous experiments in all kinds of organisms have demonstrated the correctness of this hypothesis.

In the late 1950s, investigators started to think in earnest about how a sequence of nucleotides in DNA could code for a sequence of amino acids in a protein. Clearly, one nucleotide couldn't code for one amino acid because there

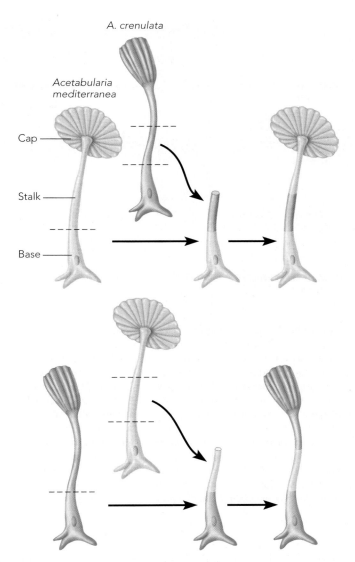

Figure 13.3 Locating the hereditary material. *Acetabularia* is a single-celled marine alga that is much larger than most other cells—up to several centimeters in length. Each cell consists of a cap, a stalk, and a base, which contains the nucleus. Two species of the alga, *A. crenulata* and *A. mediterranea,* have caps with different shapes. If an *A. crenulata* stalk is grafted onto an *A. mediterranea* base, an *A. mediterranea* cap will form. Conversely, if an *A. mediterranea* stalk is grafted onto an *A. crenulata* base, an *A. crenulata* cap will form. Thus, information in the nucleus-containing base determines the shape of the cap.

are 20 kinds of amino acids but only four kinds of nucleotides. Even two nucleotides wouldn't be enough—there are 4^2 or 16 ways to arrange pairs of nucleotides, so only 16 amino acids could be coded for. However, three nucleotides would be sufficient, as there are 4^3 or 64 ways to arrange triplets of nucleotides. Research eventually confirmed that the code words in DNA consist of triplets of nucleotides, which are called **codons.** As you'll learn in the next section, each DNA codon specifies a three-nucleotide

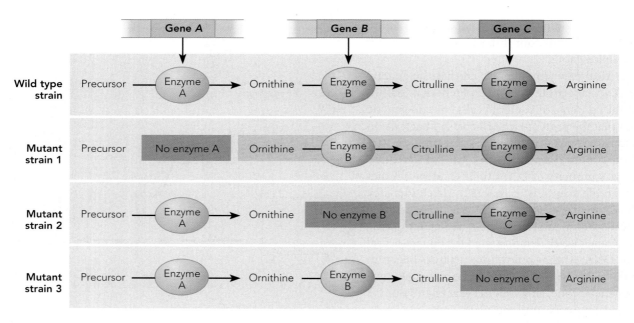

Figure 13.4 Support for the "one gene-one enzyme hypothesis." Beadle and Tatum raised four strains of the bread mold *Neurospora crassa* in a simple growth medium. The wild-type strain could make the amino acid arginine from a precursor substance because it had all three enzymes (A, B, and C) that catalyze the steps in the biosynthetic pathway. Mutant strain I lacked enzyme A; it could make arginine only if ornithine or citrulline was added to the growth medium. Mutant strain II lacked enzyme B; it needed citrulline to make arginine. Mutant strain III lacked enzyme C; it could not make arginine under any condition. Beadle and Tatum concluded that each enzyme was specified by a different gene and that each mutant strain had a defect in one of those genes, leaving the strain unable to produce one enzyme.

codon of RNA. It is the RNA codons that actually code for specific amino acids.

In 1961, Marshall Nirenberg began to decipher the genetic code by using an artificial protein-synthesis system containing ribosomes, amino acids, and other ingredients. When he added a long strand of RNA containing only uracil (U) nucleotides, the system produced a protein that contained just one kind of amino acid: phenylalanine. That result indicated that the RNA codon UUU codes for phenylalanine. A similar approach identified the amino acids coded for by the codons AAA, CCC, and GGG. More complex procedures were needed to decipher codons that had more than one type of nucleotide, but within four years the entire code had been determined. As you can see in Figure 13.5, 61 of the 64 codons specify amino acids. One of these 61 (AUG) codes for the amino acid methionine and also acts as a "start" signal, directing ribosomes to start making a protein at that codon. The other three codons are "stop" signals marking the end of a protein.

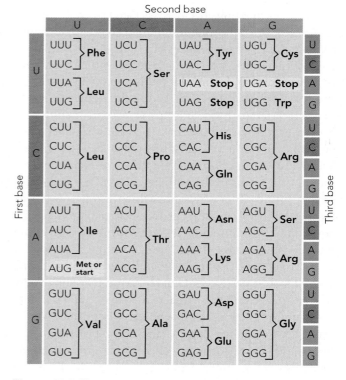

Figure 13.5 The genetic code. Three-nucleotide combinations in RNA, called *codons,* code for the 20 amino acids found in proteins. Note that the order of nucleotides in a codon is important. For example, ACG codes for threonine, whereas GCA codes for alanine. One codon (AUG) codes for methionine and signals ribosomes to begin making a protein. Codons UAA, UAG, and UGA do not code for amino acids. Instead, they signal ribosomes to stop adding amino acids to a protein.

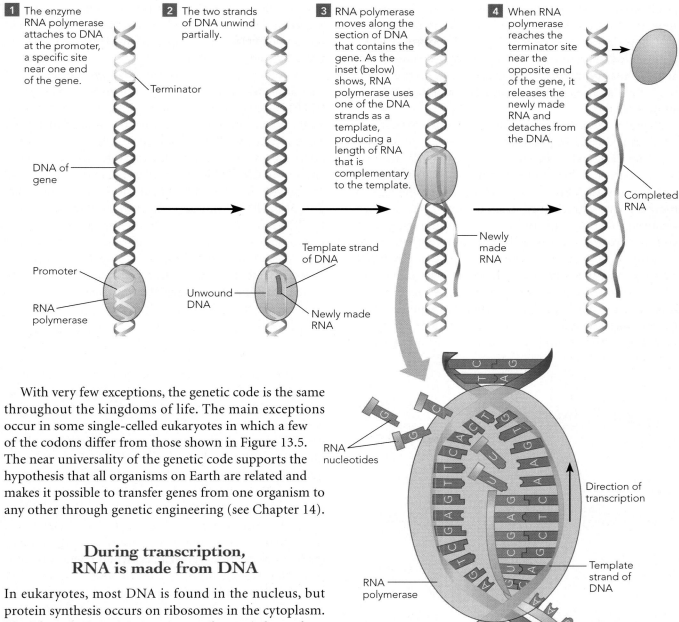

1 The enzyme RNA polymerase attaches to DNA at the promoter, a specific site near one end of the gene.

Terminator

DNA of gene

Promoter

RNA polymerase

2 The two strands of DNA unwind partially.

Unwound DNA

Template strand of DNA

Newly made RNA

3 RNA polymerase moves along the section of DNA that contains the gene. As the inset (below) shows, RNA polymerase uses one of the DNA strands as a template, producing a length of RNA that is complementary to the template.

Newly made RNA

4 When RNA polymerase reaches the terminator site near the opposite end of the gene, it releases the newly made RNA and detaches from the DNA.

Completed RNA

RNA nucleotides

RNA polymerase

Direction of transcription

Template strand of DNA

Newly made RNA

Figure 13.6 Gene transcription. Transcription of a gene involves four steps.

With very few exceptions, the genetic code is the same throughout the kingdoms of life. The main exceptions occur in some single-celled eukaryotes in which a few of the codons differ from those shown in Figure 13.5. The near universality of the genetic code supports the hypothesis that all organisms on Earth are related and makes it possible to transfer genes from one organism to any other through genetic engineering (see Chapter 14).

During transcription, RNA is made from DNA

In eukaryotes, most DNA is found in the nucleus, but protein synthesis occurs on ribosomes in the cytoplasm. Therefore, the genetic message must be carried somehow from the nucleus to the cytoplasm. The agent that carries the genetic message is a particular type of RNA known as **messenger RNA (mRNA),** which is made by **transcription.**

Transcription of a gene begins when an enzyme, RNA polymerase, attaches to the DNA at a specific site called a **promoter,** a sequence of several dozen nucleotide pairs located at one end of a gene (Figure 13.6). The two strands of DNA then unwind partially, and RNA polymerase starts to move along the section of DNA where the gene is located. As RNA polymerase moves, it synthesizes a length of RNA using one of the DNA strands as a template. The nucleotides that are added to the RNA are complementary to the nucleotides in the DNA template strand. That means

the RNA has C where the template has G, G where the template has C, A where the template has T, and U where the template has A. Thus, the specific sequence of nucleotides that makes up a gene is transcribed into a complementary sequence of nucleotides in the RNA. (Poisons called *amatoxins* prevent the movement of RNA polymerase along DNA, thereby blocking transcription. Amatoxins are produced by mushrooms in the genus *Amanita* and are responsible for most lethal cases of mushroom poisoning.)

Transcription ends when RNA polymerase reaches a site called the *terminator* near the opposite end of the gene. At that point, RNA polymerase releases the newly made RNA and detaches from the DNA, and the two DNA strands rewind.

In prokaryotes, the newly made RNA is ready to be translated immediately. In plants and other eukaryotes, however, the RNA (called *heterogeneous nuclear RNA*) must be processed and then moved out of the nucleus before it can be translated. Processing includes removing one or more sections of nucleotides from the RNA. Why are they removed? In eukaryotes, most genes contain regions that code for protein interrupted by regions that do not code for protein (Figure 13.7). The coding regions are called **exons** because they will eventually be *ex*pressed in protein. The noncoding regions are called **introns** because they *int*errupt the coding regions. When RNA polymerase moves along a gene, it transcribes the entire gene into RNA, including introns as well as exons. While the newly made

RNA is still in the nucleus, the introns in the RNA are removed and the exons are spliced together. The removal of introns takes place in a complex of small nuclear RNAs and proteins called a *spliceosome*.

Two other events occur during RNA processing. First, a chemically modified molecule of guanosine triphosphate (GTP) is added to one end of the RNA. (GTP is similar to ATP in structure and in many of its cellular functions.) The modified GTP that is added is called a *cap*. Second, 50 to 200 adenine nucleotides are added to the opposite end of the RNA. This sequence of nucleotides is known as a poly(A) tail. The poly(A) tail may promote the transport of the RNA from the nucleus to the cytoplasm. After the RNA enters the cytoplasm, the cap and poly(A) tail help bind the RNA to the ribosome and protect the RNA from digestion by enzymes. Only after the RNA is fully processed and has entered the cytoplasm is it called *messenger RNA*. It is then ready to be translated.

During translation, a protein is made from messenger RNA

The second step in converting the genetic code to protein is **translation.** During translation, the sequence of nucleotides in mRNA is converted into a sequence of amino acids in a protein. Translation requires the participation of ribosomes, which are composed of proteins and another type of RNA called *ribosomal RNA (rRNA)*. Also required for translation are mRNA and a third type of RNA, **transfer RNA (tRNA).** As Figure 13.8 shows, tRNA molecules consist of 70 to 80 nucleotides. Base pairing between some of the nucleotides folds each tRNA molecule into a shape that has three loops. The middle loop contains a triplet of nucleotides called an **anticodon,** which is complementary to one of the codons in mRNA. At the other end of a tRNA molecule is a site where an amino acid can attach. Enzymes in the cytoplasm link each of the 20 amino acids with tRNA molecules that have specific anticodons. For example, tRNA molecules that have the anticodon UAC are linked to the amino acid methionine. As its name suggests, tRNA functions to transfer specific amino acids to the ribosomes as proteins are being made.

Translation includes three stages: initiation, elongation, and termination (see Figure 13.9). In initiation, **1** an mRNA molecule binds to a fragment of ribosome known as the *small ribosomal subunit*. A tRNA molecule with the anticodon UAC then binds to the "start" codon (AUG) on the mRNA. As we noted above, this tRNA molecule carries the amino acid methionine. Next, **2** another fragment of ribosome, the large ribosomal subunit, attaches to the small subunit, forming a complete ribosome. The mRNA molecule lies in a groove between the two ribosomal subunits.

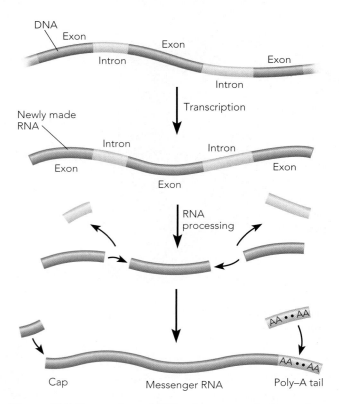

Figure 13.7 Processing of RNA after transcription. When a eukaryotic gene is transcribed, both the coding regions (exons) and noncoding regions (introns) in the gene are transcribed into RNA. The newly made RNA is then processed: The introns are removed, the exons are spliced together, a modified-GTP cap is added to one end, and a poly(A) tail is added to the other end. The completely processed molecule is messenger RNA.

Figure 13.8 Transfer RNA. A molecule of transfer RNA consists of 70 to 80 nucleotides arranged to form a structure with three loops. In the middle loop, a sequence of three nucleotides called an *anticodon* is complementary to a codon in mRNA. The opposite end of the tRNA molecule has a site that can bind to an amino acid. Each of the 20 amino acids binds only to tRNA molecules that have specific anticodons.

The elongation stage begins when a second tRNA molecule moves into place next to the first **3**. If the second tRNA molecule has an anticodon that is complementary to the second codon on the mRNA, hydrogen bonds will form between the codon and anticodon. For example, if the second codon is AAG, the complementary anticodon is UUC. As soon as a second tRNA molecule with a complementary anticodon settles into place, an enzyme creates a peptide bond between the amino acids carried by the first and second tRNA molecules. **4** When that happens, the first tRNA molecule unbinds from the start codon, leaving its amino acid (methionine) attached to the second amino acid. The ribosome then advances along the mRNA molecule to the third codon, and the elongation process is repeated. **5** Elongation continues until an amino acid has been added for every codon in the mRNA molecule. Addition of each amino acid takes only about 60 milliseconds.

If the protein being synthesized will be used in the cytoplasm or in mitochondria, chloroplasts, or the nucleus, the ribosome remains free in the cytoplasm during the elongation stage. However, if the protein is destined to be exported from the cell or inserted into cellular membranes, the ribosome will move to the endoplasmic reticulum (ER) as translation proceeds. A sequence of about

20 amino acids at the start of such proteins forms a *signal peptide* that directs the ribosome to the ER. Recall from Chapter 2 that the ER serves as a synthesis and assembly site for these proteins.

The termination stage of translation occurs when the ribosome encounters a "stop" codon (UAA, UAG, or UGA) on the mRNA molecule. **6** There are no tRNA molecules for the stop codons; instead, a protein known as a *release factor* binds to the stop codon. The release factor severs the link between the last tRNA molecule and the chain of amino acids—now a protein. **7** The newly synthesized protein and the mRNA molecule are released from the ribosome, which separates into its small and large subunits, and translation is complete.

Mutations can cause changes in gene expression

Transcription and translation convert information encoded in the order of nucleotides in DNA into information encoded in the order of amino acids in proteins. It should be obvious, then, that any change in the order or structure of the DNA in a cell could lead to a change in the proteins that the cell makes. Changes in the order or structure of DNA are called **mutations.** Because most proteins are enzymes, mutations often affect the structure of enzymes and their ability to catalyze specific reactions. On a higher level, the genetic differences between individuals in populations arise originally because of mutations.

The simplest and most common type of mutation, a **point mutation,** is a change in one nucleotide in the DNA. Point mutations are also called **single nucleotide polymorphisms** (**SNPs,** pronounced "snips"). Point mutations include substitutions, insertions, and deletions. A *substitution* occurs when one nucleotide is replaced by another nucleotide, such as when C is replaced by A. When a gene that has a substitution is transcribed, the RNA that is produced will also have a substitution. For example, replacing C with A in a gene results in an RNA molecule that has a U where a G would normally be (Figure 13.10). This change in the RNA may cause the codon that contains the substitution to code for a different amino acid. For example, changing the codon CGA, which codes for arginine, to CUA will yield a codon that now codes for leucine. When the mutant RNA is translated, the resulting protein will have leucine at the location where the normal protein has arginine. If the protein is an enzyme, the amino acid substitution may affect the shape of the active site and the catalytic activity of the enzyme. The altered enzyme may be unable to catalyze a reaction or may do so less rapidly or specifically than the normal enzyme. Alternatively, the change in amino acid composition might

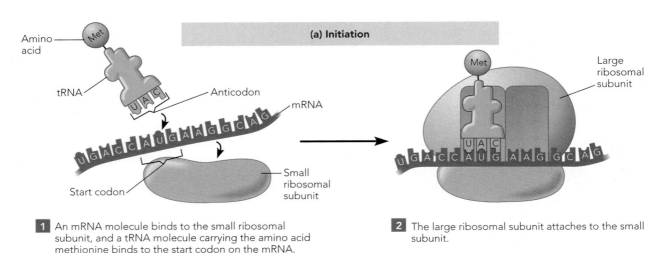

(a) Initiation

1 An mRNA molecule binds to the small ribosomal subunit, and a tRNA molecule carrying the amino acid methionine binds to the start codon on the mRNA.

2 The large ribosomal subunit attaches to the small subunit.

(b) Elongation

3 A second tRNA molecule binds to the second codon on the mRNA, and the amino acid carried by that tRNA molecule is linked to the first amino acid.

4 The first tRNA molecule is released from the ribosome, which advances to the third codon on the mRNA.

5 A third tRNA molecule binds to the third codon on the mRNA, and its amino acid is linked to the second amino acid. The elongation stage continues until the ribosome reaches the last codon.

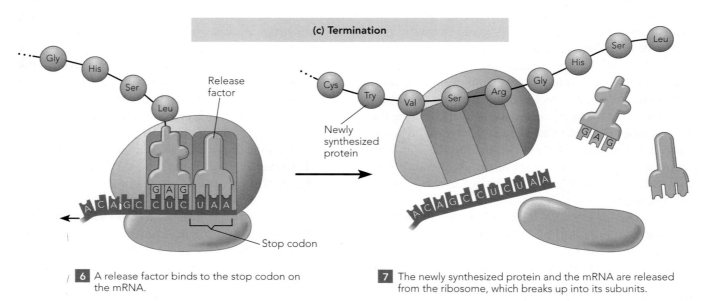

(c) Termination

6 A release factor binds to the stop codon on the mRNA.

7 The newly synthesized protein and the mRNA are released from the ribosome, which breaks up into its subunits.

Figure 13.9 Translation. There are three stages in the translation of mRNA into protein: initiation, elongation, and termination.

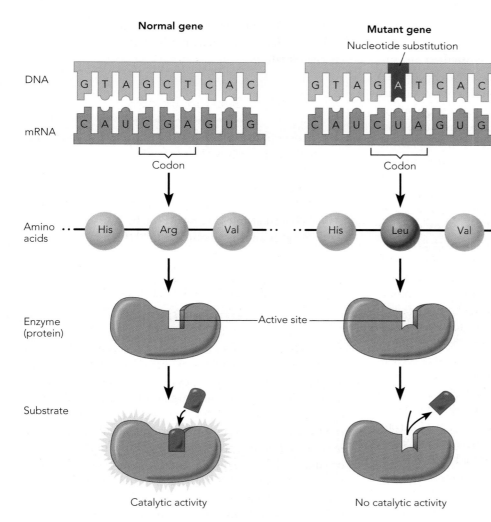

Normal gene

Mutant gene

Nucleotide substitution

DNA

mRNA

Codon

Codon

Amino acids

His Arg Val ·· ·· His Leu Val ··

Enzyme (protein)

Active site

Substrate

Catalytic activity

No catalytic activity

Figure 13.10 Alteration of protein function by a nucleotide substitution. When a single nucleotide in a gene is changed from C to A, transcription produces an mRNA molecule that has U in place of G. As a result, an mRNA codon that normally codes for arginine now codes for leucine. When the mutant mRNA is translated, this single amino-acid change may result in an enzyme that has an abnormal shape near its active site. The enzyme may lack catalytic activity because the substrate cannot bind to the active site.

make the enzyme *more* effective as a catalyst. In either case, the physiology or anatomy of the organism may be altered as a result.

Not all nucleotide substitutions lead to changes in protein function, however. For one thing, most amino acids are coded for by more than one codon, so changing one nucleotide in a codon does not always cause the codon to code for a different amino acid. Changing the RNA codon CGG to CGU, for example, will have no effect on the protein that is synthesized because both CGG and CGU code for arginine. SNPs of this sort are known as silent mutations. Even when a nucleotide substitution results in a protein with a different amino acid, the function of that protein may not be affected. Such would be the case if the substituted amino acid and the one it replaced have similar properties, or if the amino acid change occurs far from a critical part of the protein, such as an active site. Mutations that do not affect protein function are referred to as neutral mutations.

Insertions and *deletions* occur when a nucleotide is added to or removed from DNA, respectively. When they

occur in an exon, these point mutations almost always have a significant effect on the protein that is produced. To understand why, consider a DNA sequence consisting of four codons: GTG-TCG-CAT-TTG. The complementary RNA sequence would be CAC-AGC-GUA-AAC, which would be translated into the amino acid sequence histidine-serine-valine-asparagine. Now suppose a C nucleotide is inserted after the T in the first DNA codon. This insertion will cause the position of the codons to shift by one nucleotide: GT**C**-GTC-GCA-TTT-G. . . . The codons in the RNA will then be CAG-CAG-CGU-AAA-C . . ., which will code for a very different amino acid sequence: glutamine-glutamine-arginine-lysine. A similar shifting of codons occurs when a nucleotide is deleted from a DNA sequence. Because insertions and deletions cause codons to shift, these point mutations are known as **frameshift mutations.**

Thousands of point mutations have been identified in prokaryotes and eukaryotes. If you consider that a typical eukaryote has from 15,000 to 50,000 genes, each containing hundreds to thousands of nucleotides, you can easily calculate that a staggering number of point mutations could occur.

Chromosomal mutations involve more than one nucleotide. They include changes in as few as two nucleotides but often involve hundreds of nucleotides or even entire chromosomes. Chromosomal mutations may produce four types of alterations in chromosome structure: deletions, where all or part of a chromosome is missing; duplications, where there is an extra copy of a chromosome or one of its parts; inversions, where part of a chromosome is removed and then reattached in a reverse orientation; and translocations, where part of a chromosome is attached to a nonhomologous chromosome. Depending on their length, these alterations may affect one or more genes.

Mutations that affect entire chromosomes may result in cells that have too few or too many copies of particular chromosomes, a condition called **aneuploidy.** Aneuploidy usually results from **nondisjunction,** the failure of sister chromatids or homologous chromosomes to separate during mitosis or meiosis (see Chapters 2 and 6). Nondisjunction produces one daughter cell that has an extra copy of a chromosome and one daughter cell that has none. Aneuploidy that arises from nondisjunction during meiosis is often lethal, blocking development at an early stage. However, botanists have been able to produce lines of certain plants, such as wheat, that have zero, one, or three copies of specific chromosomes. These aneuploid lines are useful in efforts to identify the genes that are located on each chromosome.

Polyploidy occurs when cells have more than two complete sets of chromosomes. Cells with three sets are said to be triploid ($3n$), while those with four sets are tetraploid ($4n$). Tetraploidy can be induced in some organisms by the drug colchicine, obtained from the autumn crocus (*Colchicum autumnale*). Colchicine prevents spindle formation during mitosis or meiosis. Without a functional spindle, the chromatids cannot separate and all end up in one daughter cell, which then has twice the normal chromosome number.

Chemicals, radiation, and mistakes made by enzymes during DNA replication can all cause mutations. Chemicals interact directly with DNA to cause changes in nucleotide bases, which are then misread during transcription. Radiation in the form of relatively low-energy ultraviolet (UV) light causes bonds to form between the bases of adjacent nucleotides on a DNA strand, leading RNA polymerase to misread the sequence of nucleotides. In contrast, high-energy radiation (gamma rays and X rays) breaks chromosomes into pieces, and pieces without centromeres are often lost during cell division. The ozone layer in Earth's atmosphere filters out a good deal of the UV light

from the Sun, thereby reducing the mutation rate considerably. However, the ozone layer does not block high-energy radiation.

Section Review

1. How did the research of Beadle and Tatum specifically link genes with enzymes?
2. How many nucleotides are used to code for one amino acid?
3. What molecules are produced by transcription? What molecules are produced by translation?
4. What is another name for an SNP?

Differential Gene Expression

The development of a plant from a fertilized egg follows a specific pattern of cell division and differentiation known as *embryogenesis.* Many types of cells develop during the formation of a multicellular organism such as a plant. Each cell type requires a specific set of proteins determined by the functions it carries out. Moreover, cells of one type may need different proteins during different stages of the organism's life cycle or even at different times of the day as environmental conditions vary. Although each cell type contains all of the organism's genes, it doesn't express every gene. Instead, cells express only those genes that code for the proteins they need. This selectivity for certain parts of an organism's genetic information is called *differential gene expression.* It likely evolved because cells that had it were energetically much more efficient than cells that produced proteins they did not need as well as ones they did.

Gene expression is controlled at various levels

Cells exert differential gene expression at every step between the reading of the genetic code and the use of proteins. Transcription is prevented, or turned off, for some genes and activated, or turned on, for others. The processing of newly made RNA into mRNA can be facilitated or inhibited. Once mRNA enters the cytoplasm, it may be translated, ignored, or destroyed by enzymes. Often, proteins synthesized during translation are not functional or active as they detach from the ribosome. They may have to be chemically modified, split into one or more pieces, or transported to a specific part of the cell. Generally, genes that code for proteins the cell may need at a moment's notice are transcribed and their mRNA is translated, but the resulting proteins are left in an inactive state until they are needed. Genes coding for enzymes involved in ATP synthesis or some other aspect of energy supply typically fall into this category. In contrast, genes whose products are needed with less urgency are generally controlled at the level of transcription. Good examples of such genes in plants are ones involved in flowering and other long-term developmental events.

Regulatory proteins control transcription

You learned earlier that transcription of a eukaryotic gene begins when RNA polymerase attaches to the promoter, a site on the DNA near the gene. The attachment of RNA polymerase to the promoter cannot occur without the assistance of a number of other proteins called **transcription factors.** Some transcription factors bind to the DNA, some bind to RNA polymerase, and some bind to each other. Together, the transcription factors and RNA polymerase form a complex of proteins that makes transcription possible. In eukaryotes, most transcription factors stimulate transcription, but some inhibit transcription, thereby reducing the expression of certain genes.

In addition to the promoter for each gene, eukaryotic chromosomes have other specific segments of DNA where transcription factors can bind. These segments are known as **control elements** because the binding of transcription factors to them exerts control over the expression of one or more genes. Control elements are not always located close to the genes they control. In fact, they may be thousands of nucleotides from the promoter on either side of a gene.

How can a control element that is located far from a gene have any influence over the transcription of that gene? This question is currently a topic of considerable research. The model illustrated in Figure 13.11 presents one possible explanation. According to this model, the bending of DNA brings distant control elements close to the promoter, allowing transcription factors attached to those control elements to interact with RNA polymerase and other transcription factors at the promoter.

In prokaryotes, genes that relate to common physiological processes are located next to each other, along with promoters and other control elements, on the single, circular chromosome and are known as an *operon.* The genes in an operon are transcribed as a group. These groups of genes code for enzymes that are all needed at the same time. Control elements near each group determine whether or not the genes will be transcribed and how rapid transcription will be if it does occur. With rare exceptions, similar groups of genes do not occur in eukaryotic chromo-

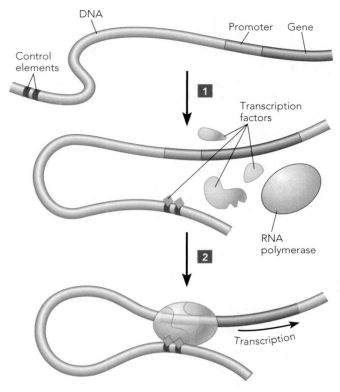

Figure 13.11 How control elements may work. Control elements are segments of DNA involved in controlling the transcription of genes, which may be thousands of nucleotides away. According to the model shown here, control elements are brought close to a gene's promoter when DNA bends (Step 1). Transcription factors can then bind to the control elements, the promoter, RNA polymerase, and each other, allowing the gene to be transcribed (Step 2).

somes. Instead, eukaryotic genes involved in a particular physiological process are typically scattered among several chromosomes. Such genes can still be expressed as a group, however, if they are controlled by the same type of control element. The binding of transcription factors specific for that type of control element will cause all of the genes to be transcribed simultaneously.

Hormones and light can trigger the activation of transcription factors

Environmental factors such as light, temperature, gravity, and wind frequently affect plant growth and development by causing production or movement of a hormone that then activates enzymes or genes. Any protein that binds to a control element is, by definition, a transcription factor. In eukaryotes, hundreds of transcription factors have been identified. They include proteins that are activated when a hormone or light interacts with a cell. A hormone is a substance produced by cells in one part of multicellular organism that affects the function of cells in another part of the organism. In Chapter 3 you encountered one plant hormone, auxin, which is produced in or near apical meristems and suppresses the growth and auxin production of axillary buds. Auxin also promotes the formation of vascular tissue in roots and shoots. You read about other hormones that control other developmental processes in Chapter 11. In plants, at least six hormones and three colors of light affect growth and development in various ways.

In general, hormones and light can activate transcription factors by a common general mechanism. Either a hormone binds to a receptor protein or light is absorbed by a protein. In both situations, the protein is located in the plasma membrane of the target cell. This interaction between a hormone or light and a protein on the outside surface of a cell sets in motion a chain of events that occur inside the cell, ultimately leading to some change in the activity of the cell. The series of steps that link receptor binding to a change in activity is called a **signal-transduction pathway (STP)** because it transduces, or changes the form of, the signal carried by the hormone or light into a cellular response. Photosystems that activate STPs include flavins or carotenoids that absorb blue light and phytochrome, a pigment-protein complex that absorbs red or far-red (infrared) light.

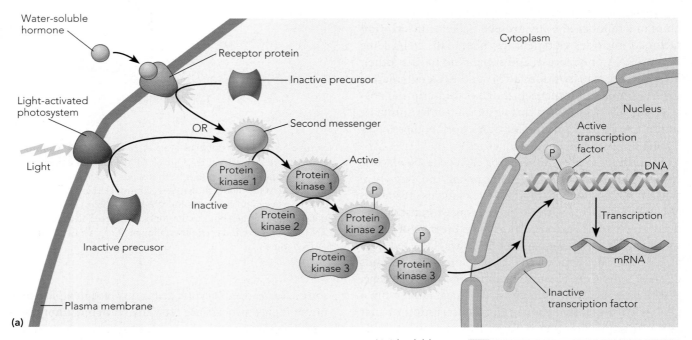

(a)

Figure 13.12 Activation of transcription factors by hormones and light—two mechanisms. (a) Most water-soluble hormones bind to receptor proteins in the plasma membrane. The signal carried by these hormones is relayed to the nucleus via a signal-transduction pathway. Binding of the hormone to its receptor activates a cytoplasmic second messenger, which then activates a protein kinase. The activated protein kinase adds a phosphate group (Ⓟ) to a second protein kinase, activating it. The second protein kinase does the same to a third protein kinase, which enters the nucleus and activates a transcription factor. The activated transcription factor binds to a control element, promoting transcription. Signal-transduction pathways are also activated when light is absorbed by specific membrane proteins. **(b)** Most lipid-soluble hormones bind to receptor proteins in the cytoplasm of their target cells. The hormone-receptor complex enters the nucleus and acts as a transcription factor, promoting transcription.

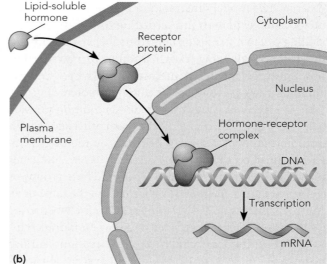

(b)

THE INTRIGUING WORLD OF PLANTS

The Closing of Stomata in Response to Drought Is a Typical STP

In recent years, research scientists have learned much about STPs in plants. As an example, consider the closure of stomata by plants in response to drought. When a plant experiences significant water loss by transpiration through the stomata, abscisic acid, which is produced in roots and in leaves, is transported to the guard cells surrounding stomatal openings:

1. ABA binds to cell receptor proteins in membranes of guard cells surrounding stomata and initiates at least two STPs.
2. Components of signal-transduction pathways include protein kinases, G proteins, and IP3. The exact order is under investigation, and additional components are involved.
3. One STP leads to opening of Ca^{+2} channels in the cell membrane, which allow Ca^{+2} to enter cytoplasm.
4. Another STP leads to opening of Ca^{+2} channels in the tonoplast, which allows Ca^{+2} to enter cytoplasm.
5. Ca^{+2} causes Cl^- to leave cell and prevents K^+ from entering cell.
6. Ca^{+2} causes H^+ to decrease in cell.

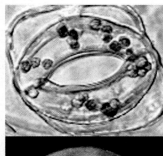
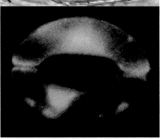

Visualization of calcium in guard cells.

7. Decreased H^+ in cell causes K^+ to exit cell.
8. New loss of K^+ and Cl^- from cell lowers osmotic concentration so water leaves cell.
9. Cells change shape with loss of water resulting in closure of stomata.

Scientists investigating the action of ABA in guard cells are able to make use of several interesting techniques to help their research. In one technique, the increase of Ca^{+2} in guard cells is visualized by using compounds that become colored when calcium is present. Another technique uses so-called caged molecules. These are molecules that scientists believe might be involved in one of the STPs. They are caged because they are bound to carrier molecules that allow them to enter cells but block their activity. The carrier molecules are removed by various means, such as a flash of UV light. Then scientists look to see if the molecule of interest has an effect on, for example, Ca^{+2} concentration or stomatal closure. So, if a guard cell is treated with caged IP3, when the cage is removed, the calcium content of the cells rapidly increases.

The steps in one common STP are diagrammed in Figure 13.12a. At the beginning of the pathway, hormone binding or light absorption leads to the production of a specific substance in the cytoplasm. The hormone or light serves as the "first messenger," and the cytoplasmic substance that is produced is called a **second messenger.** Second messengers include cyclic AMP, diacylglycerol, inositol triphosphate (IP3), and calcium ions. The second messenger in turn activates a type of enzyme called a **protein kinase,** which phosphorylates other proteins (transfers phosphate groups from ATP to them). There are many varieties of protein kinases, each specific for a certain substrate. In STPs, the substrate of the protein kinase that is activated by the second messenger is usually another protein kinase. Phosphorylation of the second protein kinase typically activates that enzyme, which then phosphorylates and activates a third protein kinase, and so on. There may be several protein kinases in an STP. The final protein kinase in the pathway may directly activate an enzyme or may enter the nucleus and activate a transcription factor,

which can then bind to a control element and regulate the transcription of a gene.

Many variations occur in STPs. For example, a G protein may relay the message to the enzyme that produces the second messenger. Calcium ions may activate a regulatory protein called *calmodulin.* Sometimes, hormones or light may initiate more than one signal-transduction pathway at once. For example, in germinating grass seeds, the hormone gibberellin initiates an STP to synthesize the enzyme α-amylase, which breaks down starch into sugar, which the embryo uses for energy. Gibberellin also initiates a second STP to promote secretion of the enzyme from the cells where it is produced. As an example of a typical STP in plants, consider the effect of drought on stomata. When the plant experiences excessive water loss through the stomata, an increase in concentrations of the hormone abscisic acid in leaves causes the guard cells to lose water and the stomata to close. The STPs for the action of abscisic acid in this system involve a number of steps, resulting in loss of water from guard cells (see *The Intriguing World of Plants* box on this page).

Section Review

1. Name several specific steps at which gene activity can be controlled.
2. What is the difference between a control element and a transcription factor?
3. Explain the functions of second messengers and protein kinases in signal-transduction pathways.

Identifying Genes That Affect Development

Genes control all activities of a plant, including growth and development, sexual reproduction, and the responses of the plant to environmental stimuli. One way to identify genes that affect development is to find plants with mutations that alter some aspect of development. You have already come across some of these mutations in earlier chapters. For example, mutations occur that affect trichome formation or the formation of tracheids in xylem tissue. Mutations also affect the number and pattern of stomata on leaves. Scientists compare the proteins of plants that have developmental mutations with the proteins of normal plants, looking for differences in specific proteins. Once they identify such a protein, they study its interactions with other molecules and try to piece together a sequence of molecular events that explains exactly how the protein and the gene that codes for it affect development.

Scientists use at least two other methods to find genes that are involved in development. The first method, called *gene trapping*, is discussed below. The second involves using DNA microarrays and is detailed in the *Biotechnology* box on page 293.

Experiments on *Arabidopsis* illustrate the use of mutations to understand plant development

A small weed in the mustard family, *Arabidopsis thaliana* (see Chapter 12), has become the preferred plant among many botanists for studying the genetic control of development. The plant's short life cycle, hardiness, high seed production, and relatively small number of genes are among the characteristics that make it well suited for genetics experiments.

Figure 13.13 summarizes a typical experimental approach that has been used on *Arabidopsis* to produce and identify developmental mutants. Large numbers of seeds

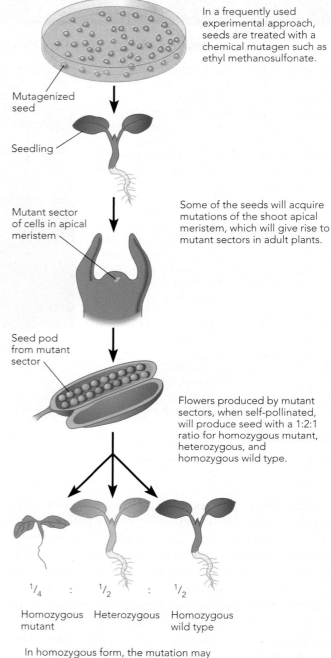

In a frequently used experimental approach, seeds are treated with a chemical mutagen such as ethyl methanosulfonate.

Mutagenized seed

Seedling

Some of the seeds will acquire mutations of the shoot apical meristem, which will give rise to mutant sectors in adult plants.

Mutant sector of cells in apical meristem

Seed pod from mutant sector

Flowers produced by mutant sectors, when self-pollinated, will produce seed with a 1:2:1 ratio for homozygous mutant, heterozygous, and homozygous wild type.

$\frac{1}{4}$: $\frac{1}{2}$: $\frac{1}{2}$

Homozygous mutant   Heterozygous   Homozygous wild type

In homozygous form, the mutation may disrupt the development of the plant.

Figure 13.13 Identifying mutants in *Arabidopsis*.

are treated with a chemical that causes mutations at random locations in the DNA of the embryo inside the seed. Some of the mutations will affect cells of the shoot apical meristem, leading to mutant sectors on the adult plant. When flowers produced by these sectors self-pollinate, one-quarter of their seeds will be homozygous for the

BIOTECHNOLOGY

DNA Microarrays

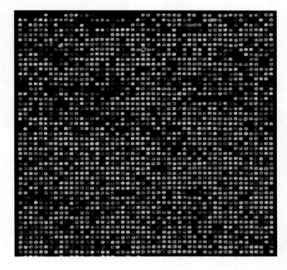

An enlarged view of a DNA microarray.

DNA microarrays are a new method for identifying the genes that are expressed in a particular cell type or tissue. The advantage of this method is that it can test thousands of genes at the same time. The following steps describe how a DNA microarray is prepared and how it can be used to investigate gene expression in plants.

1 A tiny amount of single-stranded DNA from each gene in a plant is deposited as a microscopic spot on a glass slide. Typically, the DNA in each spot consists of a short, unique sequence of nucleotides found in one gene. Thousands of spots, each representing a different gene, are arranged in a grid on the slide. The grid, called a *DNA microarray,* is no larger than a dime.

2 Messenger RNA molecules are isolated from a particular tissue in the plant. This step takes advantage of a simple fact—for a gene to be expressed, it must be transcribed into mRNA. Therefore, the mRNA molecules that are isolated reflect the selective transcription of genes that are expressed in the tissue.

3 The mRNA is treated with reverse transcriptase, a viral enzyme that synthesizes single-stranded DNA using RNA as a template. (The enzyme is so named because the process it catalyzes is the reverse of transcription.) The re-

sulting DNA molecules, called *complementary DNA (cDNA),* are synthesized from nucleotides that have been chemically modified to contain a fluorescent dye.

4 The cDNA is applied to the DNA microarray. Since both the cDNA and the DNA in the microarray are single stranded, they will bind to each other wherever their nucleotide sequences are complementary.

5 Unbound cDNA is washed off the DNA microarray. The microarray is then observed for fluorescence. Spots where binding has occurred fluoresce with the color of the cDNA dye. Such spots represent genes that were transcribed into RNA.

The pattern of fluorescent spots on a DNA microarray indicates the different sets of genes that are expressed by different tissues or under different conditions. A leaf from a plant experiencing drought, for example, will show a different pattern of active genes than a leaf from a well-watered plant. Furthermore, different fluorescent dyes can be used on the same microarray to compare the patterns of genes expressed before and after a developmental or physiological change. For example, genes that are active before flowering can be labeled with green fluorescent cDNA, while those that are active after flowering can be labeled with red fluorescent cDNA.

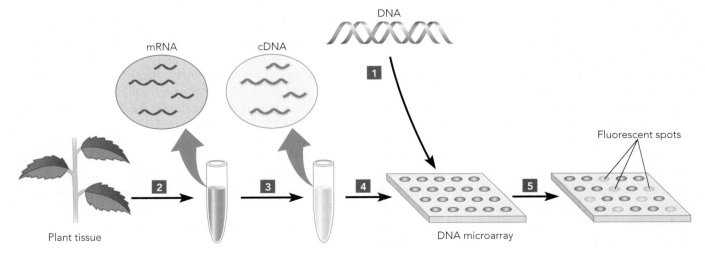

Using a DNA microarray to identify expressed genes.

(a) Flowers

(b) Leaves

Figure 13.14 Developmental mutants in *Arabidopsis thaliana*. **(a)** Flowers: wild type. The wuschel mutant has one stamen instead of six. The agamous mutant is a series of flowers within flowers producing a cabbage-like appearance. The clavata mutant is the opposite phenotype to wuschel with too many central organs producing a massive club of a gynoecium in the center. **(b)** Leaves: Scientists know much about vein anatomy and function but almost nothing is known about how vein patterns are established.

mutation, one-half will be heterozygous, and one-quarter will lack the mutation. If the mutation is recessive, its effect on the plants that develop from those seeds will be visible in plants that are homozygous for the mutation. Although the mutation may be lethal when homozygous, it can be maintained in the heterozygous plants.

Hundreds of genes that govern embryonic development in *Arabidopsis* have been identified through experiments such as the one just described. Some of these genes affect the apical-basal pattern of the seedling. Another group of genes regulates the transition of a vegetative apex into a floral apex in relation to daylength. At least one gene controls the number of flower parts in each series of modified leaves that forms the flower. (Recall from Chapter 6 that flowers are composed of several series of modified leaves called *sepals, petals, stamens,* and *carpels.*) Three more genes, known as *floral organ identity genes,* code for transcription factors involved in determining the identity of flower parts. Mutations in the floral organ identity genes can cause one flower part to develop where another part usually develops (Figure 13.14a). Leaves are the primary site for photosynthesis, making the distribution and transport functions of veins a very important area of study (Figure 13.14b). Mutations in vascular distribution systems are studied in cotyledons, the leaf-like organs with simple vein patterns formed in embryogenisis.

Transposons can be used to locate genes that affect development

Chromosomes in both prokaryotes and eukaryotes contain pieces of DNA called **transposons** that can move from one location to another or can produce copies of themselves that move to other locations. Originally called "jumping genes," transposons were first characterized in the 1940s by corn geneticist Barbara McClintock (Figure 13.15). The simplest transposons contain a single gene that codes for transposase, an enzyme required for the transposon to move. Other transposons contain additional genes and have complex modes of movement involving both DNA and RNA synthesis. Transposons are quite common. In corn as well as humans; for example, they make up approximately half of the genetic material in the nucleus.

Researchers can use so-called designer transposons to inactivate or "knock out" genes affecting development. In this approach, known as **gene trapping** or *transposon tagging,* two homozygous plants are crossed. One plant has a transposon that includes the transposase gene. However, this transposon also has an unrelated defect that prevents it from moving. The other plant has a transposon that lacks the transposase gene but contains a bacterial gene, *GUS,* that codes for the enzyme β-glucuronidase. When exposed to a specific substrate, this enzyme produces a blue product.

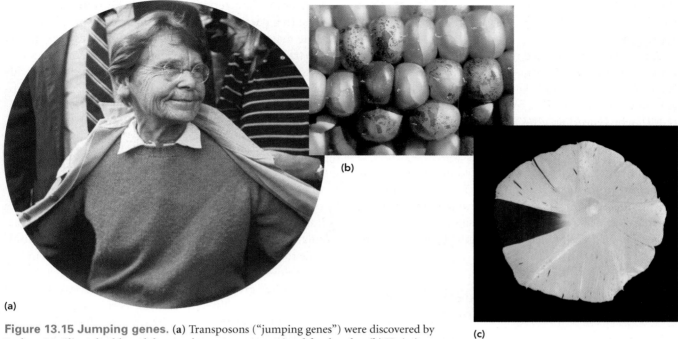

(a)

(b)

(c)

Figure 13.15 Jumping genes. **(a)** Transposons ("jumping genes") were discovered by Barbara McClintock, although her results were not appreciated for decades. **(b)** Variations in the color of corn kernels can happen when a transposon inserts itself into a pigment gene in a cell, thereby inactivating the gene and blocking pigment formation in that cell and its progeny. Conversely, a transposon may jump out of a pigment gene in another cell, restoring pigmentation to that cell and its progeny. Thus, each kernel can develop different spots of color. **(c)** The white parts of this morning glory flower are the result of transposons disrupting the gene for pigment production.

In the F_1 offspring of this cross, the transposon with the *GUS* gene will move around and insert itself at various locations in the chromosomes, using the transposase produced by the other transposon. Since it has no promoter, the *GUS*-containing transposon will not be transcribed unless it inserts itself near an active promoter. Let's suppose it lands in a gene that controls meristem development. That gene will be split apart and inactivated by the insertion of the transposon, but the gene's promoter will activate the *GUS* gene. If this event occurs early enough in the plant's development, the plant will have a defective meristem that stains blue when exposed to the *GUS* substrate (Figure 13.16a). Thus, the *GUS* gene is acting as a "reporter gene," since it informs the researcher that another gene is functioning in a particular developmental event.

Of course, in many of the offspring of such a cross, the *GUS*-containing transposon will land in a gene that was not being expressed. In these individuals, the *GUS* gene will be turned off as well. To avoid having to examine vast numbers of offspring to find ones with an activated *GUS* gene, investigators can insert a gene conferring resistance to a specific toxin into the *GUS*-containing transposon. When the F_1 seedlings germinate, the investigator adds

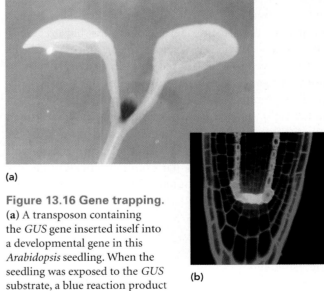

(a)

(b)

Figure 13.16 Gene trapping. **(a)** A transposon containing the *GUS* gene inserted itself into a developmental gene in this *Arabidopsis* seedling. When the seedling was exposed to the *GUS* substrate, a blue reaction product was formed only in the shoot apex, indicating that the developmental gene was active only in the shoot meristem. **(b)** The green endodermis of this root tip demonstrates that a particular gene is active only in that cell layer. A transposon has inserted the gene for green fluorescent protein into a gene that is normally active in the endodermis.

the toxin. The only plants that survive will be those in which the *GUS*-containing transposon has landed in an active gene.

Another commonly used reporter gene in plants comes from *Aequorea victoria,* a species of jellyfish found in the Pacific Ocean that emits bright green flashes of light when agitated. The gene coding for green fluorescent protein (GFP), which produces the light, can be introduced into chromosomes by transposons in much the same manner as the *GUS* gene (Figure 13.16b).

A different kind of knockout procedure involves using the bacterium *Agrobacterium tumefaciens.* When this bacterium infects plants, it introduces a small piece of DNA into the chromosomes of the plant's cells (see Chapter 14). If the piece of DNA is incorporated into a gene, that gene will be knocked out.

Homeotic genes control development in plants and animals

Once a gene that controls a particular aspect of development is identified, one of the next steps is to determine the sequence of nucleotides in the gene. You might think that quite different genes would control the unique developmental patterns of plants and of animals. On the contrary, several families of genes with similar nucleotide sequences occur in plants and animals, even though they control different processes. Genes that have regions with a similar nucleotide sequence in a wide variety of organisms have what are termed "conservative" structures. This usually means that the proteins they code for must have very specific sequences of amino acids for the organism to survive.

Conserved nucleotide sequences usually occur in **homeotic genes,** genes that control the body plan of multicellular organisms by directing certain organs to form in the right places during development. The floral organ identity genes mentioned earlier are homeotic genes. Mutations of homeotic genes can cause inappropriate organs to form in certain places. For example, a petal might develop where a stamen should be. The homeotic genes that

have been characterized so far code for transcription factors. In a particular region of the organism, a homeotic gene may activate a number of other genes that specify the traits of structures that form in that region.

One group of homeotic genes contain a 180-nucleotide sequence called the *homeobox.* (Many conserved nucleotide sequences are referred to as "boxes.") Genes that include the homeobox are very active in animal development, and a few have been identified that affect plant development. One homeobox-containing gene in plants, called *KN1* or *KNOTTED,* seems to be involved in determining where and when the shoot apical meristem appears. A mutation of the *KNOTTED* gene results in leaves that have a bumpy appearance due to abnormal cell divisions in vascular bundles. Therefore, the normal gene may play multiple roles in the formation of various plant tissues in and just below the shoot apical meristem.

Another group of homeotic genes that are active in plants contain a different 180-nucleotide sequence called the *MADS box* (named after the first letters of four transcription factors coded for by these genes). MADS box-containing genes are also found in fungi, animals, and bacteria. In plants, these genes control the kinds and locations of floral structures, among other things. Additional boxes have been discovered in plants and frequently occur in other organisms as well. The evolutionary history of these conserved regions of DNA that have controlled various phases of development since the most ancient times will likely be very interesting indeed as the details are worked out.

Section Review

1. Describe the action of a floral organ identity gene.
2. What is the importance of transposase in the action of transposons?
3. Describe how reporter genes can be used to pinpoint genes active in specific phases of development.
4. What are homeotic genes?

SUMMARY

Gene Expression

During replication, DNA is copied (pp. 279–280)
DNA is replicated during the S phase of the cell cycle by a complex of molecules, including DNA polymerase, helicase, and DNA ligase. The double helix is separated, and a complementary copy of each strand is made.

DNA codes for the structure of proteins (pp. 280–283)
Experiments carried out during the first half of the twentieth century demonstrated that DNA carries the genetic code and that genes code for proteins. The code words in DNA consist of triplets of nucleotides called *codons*, each of which specifies a particular amino acid and/or a start or stop signal during protein synthesis.

During transcription, RNA is made from DNA (pp. 283–284)
Transcription, the synthesis of RNA using DNA as a template, begins when RNA polymerase attaches to a promoter, a site on the DNA at one end of a gene. The strands of DNA unwind, and RNA polymerase creates a length of RNA containing nucleotides complementary to those on one of the DNA strands in the gene. In eukaryotes, the newly made RNA is modified in three ways: noncoding regions (introns) are removed and the remaining regions (exons) are spliced together; a cap consisting of chemically modified GTP is added to one end; and a poly(A) tail is added to the other end. The resulting molecule is messenger RNA (mRNA).

During translation, a protein is made from messenger RNA (pp. 284–285)
Translation, the synthesis of protein using mRNA as a template, requires ribosomes and transfer RNA (tRNA) in addition to mRNA. One end of each tRNA molecule contains an anticodon complementary to one of the codons in mRNA; the other end carries the amino acid specified by that codon. As the ribosome moves along the mRNA, each mRNA codon binds briefly to a tRNA molecule with the correct anticodon, and the amino acid carried by the tRNA is linked to the amino acid carried by the previous tRNA. A protein is formed as amino acids continue to be added. When the ribosome encounters a stop codon, the newly synthesized protein is released.

Mutations can cause changes in gene expression (pp. 285–288)
A mutation is any change in the order or structure of DNA. Point mutations include substitutions, insertions, and deletions. Substitutions often cause one codon to specify a different amino acid, while insertions and deletions cause codons to shift, affecting many codons. Chromosomal mutations, which involve two to hundreds of nucleotides or entire chromosomes, may result in cells that have an abnormal number of particular chromosomes (aneuploidy) or more than two complete sets of chromosomes (polyploidy). Chemicals, radiation, and mistakes made by enzymes during DNA replication can cause mutations.

Differential Gene Expression

Gene expression is controlled at various levels (pp. 288–289)
Cells exert differential gene expression at the level of transcription, processing of RNA, translation, and processing of proteins.

Regulatory proteins control transcription (p. 289)
Proteins called *transcription factors* control gene expression by binding to specific sites on DNA called *control elements*. Some transcription factors stimulate transcription while others inhibit it.

Hormones and light can trigger the activation of transcription factors (pp. 289–291)
Hormones that bind to a receptor in the plasma membrane typically initiate a series of reactions in their target cells called a *signal-transduction pathway*. The final step in the pathway may involve the activation of a transcription factor. Some hormones bind to a receptor in the cytoplasm or nucleus. This binding produces a hormone-receptor complex that functions as a transcription factor.

Identifying Genes That Affect Development

Experiments on *Arabidopsis* illustrate the use of mutations to understand plant development (pp. 292–294)
Attempts to produce mutants in *Arabidopsis* often involve treating large numbers of seeds with a chemical that causes mutations, allowing the plants that grow from those seeds to self-pollinate, and looking for mutant individuals in their offspring. Such procedures have identified many genes involved in development.

Transposons can be used to locate genes that affect development (pp. 294–296)
Transposons are pieces of DNA that can move from one location to another in chromosomes. A transposon that lands in an active gene usually inactivates the gene. This principle is used in gene trapping experiments to identify genes that have particular functions.

Homeotic genes control development in plants and animals (p. 296)
Homeotic genes control the formation of organs in specific places during development. Mutations of these genes cause inappropriate organs to form in certain places. Homeotic genes contain conserved nucleotide sequences that are found in many kinds of organisms.

Review Questions

1. Explain what is meant by the *central dogma of molecular biology.*
2. What is the meaning of *semiconservative replication* when the term is applied to DNA?
3. How did Beadle and Tatum demonstrate that genes code for enzymes?
4. Define the terms *codon* and *anticodon.*
5. Describe the process of transcription.
6. If the nucleotide sequence of a DNA strand is C-G-G-T-A-C-T-G-A, what would be the sequence of complementary RNA? What would be the sequence of amino acids following translation?
7. Distinguish between frameshift, insertion, and deletion mutations.
8. What is the name for the sites on DNA where RNA polymerase attaches?
9. Explain the role of tRNA in translation.
10. What is the difference between a silent mutation and a neutral mutation?
11. What is aneuploidy?
12. Explain how transcription factors work.
13. How do hormones trigger signal-transduction pathways?
14. What is the role of protein kinases in signal-transduction pathways?
15. How are the *GUS* and *GFP* genes used as reporter genes?
16. How do transposons knock out genes?
17. What is the MADS box?

Questions for Thought and Discussion

1. Would it make more sense for each cell in a plant to have only the genes it needs instead of all the plant's genes?
2. Why do you suppose eukaryotic chromosomes have a large amount of "junk DNA" in the form of introns? How did it get there? Could it have an unknown function?
3. Considering that transposons can inactivate genes by landing in the middle of them, what use can transposons possibly have in an organism?
4. Scientists believe that RNA appeared before DNA in the evolution of life. How could this be?

5. Do you think that transposons could cause infectious diseases? How might this occur? Do you think a transposon could become something like a virus?
6. Why do you think a long series of events, a STP, separates arrival of a hormone at a cell with the specific action of the hormone?
7. A section of a single strand of DNA has nucleotide base sequence 3′–TAAGAACCGTAAGCG–5′ Replication of the DNA double helix, of which this strand is one component, is about to proceed from left to right. Diagram the replication process for both strands and show the final sequence of DNA bases that would be synthesized. Make sure to indicate the 5′ and 3′ ends of each parent and daughter strand.

Evolution Connection

What evidence for the unity of plants with members of the other kingdoms of organisms comes from our understanding of (a) the genetic code, (b) post-transcriptional RNA processing, and (c) homeotic genes?

To Learn More

Visit The Botany Place Website at www.thebotanyplace.com for quizzes, exercises, and Web links to new and interesting information related to this chapter.

Echols, Harrison, Carol Gross, and Arthur Kornberg. *Operators and Promoters: The Story of Molecular Biology and Its Creators.* Berkeley: University of California Press, 2001. Interesting personal account of the people who invented molecular biology.

Jacob, Francois, and Betty Spillman. *The Logic of Life.* Princeton, NJ: Princeton University Press, 1993. Jacob is a co-discoverer of mRNA and of operons in bacteria. His view of living organisms is interesting and worth reading.

Levin, A. Donald. *The Origin, Expansion, and Demise of Plant Species (Oxford Series in Ecology and Evolution).* London: Oxford University Press, 2000. Fascinating stories and examples of why particular plant species become extinct.

14

Plant Biotechnology

A plant biologist working with rice tissue cultures.

The Methods of Plant Biotechnology

Genes can be transferred between species through genetic engineering

Plasmids often serve as vectors for gene transfer in plants

Restriction enzymes and DNA ligase are used to make recombinant DNA

Cloning produces multiple copies of recombinant DNA

The polymerase chain reaction clones DNA without using cells

Several methods can be used to insert cloned genes into plant cells

In tissue culture, whole plants are grown from isolated cells or tissues

The Accomplishments and Opportunities of Plant Biotechnology

Genetic engineering has made plants that are more resistant to pests and harsh soil conditions as well as more productive

Transgenic plants contribute to human health and nutrition

Genetically engineered crops require extensive field and market testing before they are released

Genetically engineered crops must be safe for the environment and for consumers

The future holds many opportunities for plant biotechnology

Genomics and proteomics will provide information needed for future efforts in genetic engineering

As long as humans have been planting and cultivating crops for food, we have been competing with other animals—chiefly insects—for that food. The negative effects of insects on agricultural production are significant. For example, the Colorado potato beetle costs U.S. potato growers between $20 million and $40 million a year in pesticide application and reduced yields. Another insect pest, the European corn borer, can destroy entire fields of corn by tunneling through the leaves and eating the kernels. Corn borers also infect corn plants with the spores of pathogenic fungi, forcing the plants to contend with both an insect and a fungal infestation. According to the United Nations Food and Agricultural Organization, as much as half of the world's cultivated crops are lost each year to insects, disease, and weeds. The greatest losses occur in developing countries, which generally cannot afford insecticides, herbicides, or resistant varieties of crop plants.

In 1911, a soil bacterium was discovered in the German province of Thuringia. This bacterium, *Bacillus thuringiensis,* has insecticidal properties. Several of its genes, called *Bt* genes, code for proteins that are converted into toxins in the gut of many insects. The toxins create pores in the gut that allow the bacterium to enter and rapidly colonize the insect. *B. thuringiensis* lives in association with the roots of plants, and the toxins apparently protect the plants from being destroyed by insects. In the 1930s, the bacterium became the source of a commercial insecticide used against the larvae of moths, butterflies, flies, and mosquitoes. The insecticide is sprayed on plants and often kills susceptible larvae, but frequent reapplications are necessary.

To achieve longer-lasting protection against such insect pests, scientists in the 1990s took a more direct approach: They produced plants that carry *Bt* genes in their own chromosomes. To do this, they incorporated the genes into

Colorado potato beetle.

plant cells and then regenerated whole plants from those cells. The resulting plants contain *Bt* genes in every cell and are resistant to susceptible insect pests. Since 1996, several dozen species of plants containing *Bt* genes have been produced, including potato, cotton, corn, sweet potato, tomato, and rice. Some plants have been given multiple *Bt* genes and thus are able to make more than one toxin. Growers pay a premium for seeds containing *Bt* genes because the plants' inherent resistance to insects can markedly increase harvests and reduce the need for pesticide use, thus lowering costs once the seeds are planted.

Corn containing *Bt* genes is now planted widely in the United States with excellent results. However, this new agricultural tool is not without problems. For example, insects can develop resistance to *Bt* toxins. Although the development of resistance can be delayed significantly if around 5–10% of a field is planted with corn that lacks *Bt* genes, the useful lifetime of *Bt* crops may be limited. Another potential problem is the unintended killing of harmless insects by *Bt* toxins. Initial experiments indicated that the larvae of monarch butterflies can be killed by pollen from *Bt* corn if the pollen falls on the leaves of milkweed plants, the preferred food of monarch larvae. However, other experiments have shown that monarch larvae generally avoid leaves covered with pollen. Moreover, larval feeding usually does not coincide with the release of corn pollen. Since *Bt* corn reduces the need for pesticides that may drift to surrounding areas, it may even have beneficial effects on monarchs and other harmless insects in those areas.

The development of *Bt* plants is just one example of how biotechnology is being applied in plants to increase food production and improve human health. In this chapter, we'll investigate the methods of plant biotechnology, many of which are rapidly developing and currently in the news. Then we'll examine some of biotechnology's major accomplishments in plants and its possibilities for the future.

European corn borer.

The Methods of Plant Biotechnology

Biotechnology is simply the application of scientific methods to manipulate living cells or organisms for practical uses. In Chapter 13, you learned how DNA is replicated and transcribed and how RNA is translated to make proteins, including enzymes that catalyze the chemical reactions of organisms. If you understand these processes, you have the foundation you need to understand most of the methods of biotechnology.

Genes can be transferred between species through genetic engineering

Because all living things store genetic information in DNA, and because DNA has the same structure in every organism, genes can be transferred from one organism to any other organism. Genetic engineering includes methods for identifying and isolating genes and for moving them rapidly from one organism to another using molecular techniques. The result of such a gene transfer is a **transgenic organism,** meaning one that contains a gene from a different kind of organism. A transferred gene, if expressed, will produce the same protein in the transgenic organism that it produces in the organism from which it was transferred. Thus, transgenic plants containing *Bt* genes produce *Bt* toxins, just as *B. thuringiensis* bacteria do. Using genetic engineering, it is possible to put an elephant gene in a cabbage or a firefly gene in a tobacco plant (Figure 14.1).

Of course, plant breeders have been moving alleles between individuals of the same or closely related species for thousands of years through traditional methods of plant breeding. For example, they might cross a strain of wheat that makes good flour but lacks drought tolerance with a strain that makes poor flour but has drought tolerance. The goal is to produce a new strain that has the best traits of both parental strains: making good flour and having drought tolerance. The F_1 offspring of such a cross will be heterozygous—that is, they will carry one copy of the alleles for both useful traits.

When F_1 plants that are heterozygous for two traits are crossed, the F_2 plants will have four (2^2) different phenotypes. The number of different F_2 phenotypes increases exponentially as the number of traits increases. For instance, if the F_1 plants differ in 10 traits, then 2^{10} or 1,024 different phenotypes would result from the independent assortment of tetrads during meiosis. Crossing over would raise that number even higher. Therefore, although traditional breeding is quite successful at moving alleles from one organism to another, it may take five to ten years of field testing and breeding to obtain stable, homozygous plants that have certain desired traits and pass them on in a consistent fashion.

In contrast, genetic engineering can rapidly move specific alleles or even entirely new genes into plants. It has the potential to shorten by years the time needed to introduce new, useful varieties of plants. Furthermore, as we noted above, genetic engineering can incorporate genes into plants from many different kinds of organisms, simulating crosses that could never occur in nature.

Figure 14.2 outlines a commonly used procedure for introducing new genes into plants. Most genetic engineering in plants is done with the Ti plasmid and involves four basic steps. Step **1** Ti plasmids are isolated from

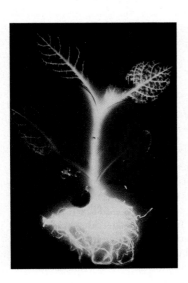

Figure 14.1 An animal gene in a plant. The gene coding for the enzyme luciferase, found in fireflies, was inserted into the chromosomes of a tobacco plant, *Nicotiana tabacum.* When the enzyme's substrates—luciferin, ATP, and oxygen—are present, the tobacco plant glows with the greenish-yellow light of a firefly.

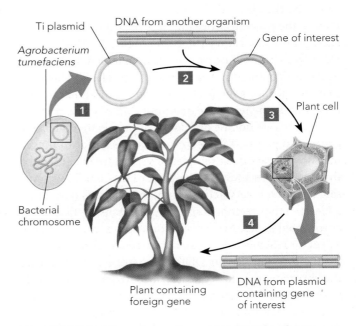

Figure 14.2 Outline of genetic engineering in plants.

Agrobacterium tumefaciens. Step **2** The plasmids are mixed with DNA from another organism, and a gene from the other organism's DNA is inserted into the plasmids. Step **3** A plasmid containing the gene of interest is introduced into a plant cell. DNA from the plasmid, including the foreign gene, is incorporated into the cell's chromosomes. Step **4** A whole plant is grown from the cell. All of the plant's cells contain the foreign gene. Throughout the rest of this half of the chapter, we'll examine the steps in this procedure in more depth. We'll also delve briefly into a few variations on this approach.

Plasmids often serve as vectors for gene transfer in plants

Genetic engineering usually begins with a **vector,** which is an agent that carries a gene from one organism to another. The vector that is used most often in plants is a plasmid found in the soil bacterium *Agrobacterium tumefaciens,* which causes crown gall disease. **Plasmids** are self-replicating, circular DNA molecules in bacteria that are separate from and smaller than the bacterial chromosome. The plasmid that occurs in *A. tumefaciens* is known as the Ti (tumor-inducing) plasmid because it plays a key role in the tumors that are produced when *A. tumefaciens* infects plants.

When a plant is infected with *A. tumefaciens,* a segment of DNA from the Ti plasmid is inserted into the DNA of the plant's chromosomes. If another piece of DNA containing a gene of interest is added to that segment of the plasmid, the gene will also be transferred to the plant's chromosomes by the plasmid. To accomplish this, scientists first remove the plasmids from the bacteria, which can be done easily in the laboratory (Step 1 in Figure 14.2). Pieces of DNA from another organism are then added to the plasmids, as explained in the next section. The Ti plasmids that are used for genetic engineering are modified so that they transfer genes to the plant but do not cause disease.

Plasmids are not the only kinds of vectors that can carry genes into plants. Tobacco mosaic virus and cauliflower mosaic virus are two examples of viruses that are used for this purpose. In addition, the DNA of the cauliflower mosaic virus has a very active promoter, which is frequently added to other vectors to speed up the transcription of incorporated genes. Artificial chromosomes can also serve as vectors. Scientists construct artificial chromosomes from a binding site for DNA polymerase (which allows the chromosomes to replicate), a centromere (which allows them to participate in cell division), and a foreign gene to be transferred. Both bacterial artificial chromosomes (BACs) and yeast artificial chromosomes (YACs) have been produced.

Restriction enzymes and DNA ligase are used to make recombinant DNA

To put foreign DNA into a plasmid, genetic engineers use a **restriction enzyme** to cut DNA into fragments. Restriction enzymes bind to restriction sites on DNA, specific sequences consisting of four to eight nucleotides. Most restriction sites are palindromic (reading the same way forward on one DNA strand and backward on the other strand). For example, the restriction site for the enzyme *Eco*R1, obtained from the bacterium *Escherichia coli,* consists of the complementary nucleotide sequences GAATTC and CTTAAG (Figure 14.3). *Eco*R1 breaks the bonds between the G and A nucleotides on each DNA strand, wherever the sequences GAATTC and CTTAAG occur in DNA. Hundreds of other restriction enzymes exist, and each

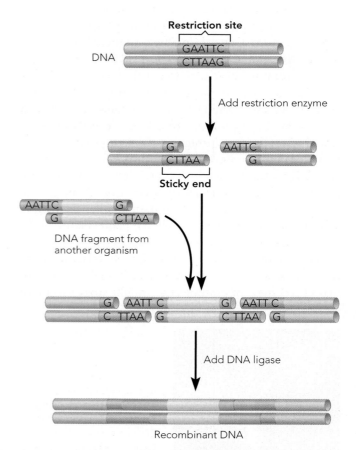

Figure 14.3 Making recombinant DNA. A restriction enzyme binds to restriction sites on DNA, places where the DNA has a specific sequence of nucleotides (in this example, GAATTC or CTTAAG). The enzyme cuts the DNA between two nucleotides in the restriction site. If the same enzyme is used to cut DNA from two different organisms, DNA fragments from each organism will be produced that have complementary sticky ends. Adding DNA ligase causes the fragments to bind together permanently at their sticky ends. When fragments from different sources are combined in this way, the result is recombinant DNA.

binds to a specific restriction site. A long DNA molecule usually contains many restriction sites for each enzyme, so it will be cut into numerous fragments by any restriction enzyme.

Notice in Figure 14.3 that the double-stranded DNA fragments produced by most restriction enzymes have a short, single-stranded sequence at each end. These single-stranded sequences are called **sticky ends** because they readily bind to complementary sequences on other DNA fragments that have been produced by the same enzyme. To insert a foreign gene into a plasmid, scientists combine DNA fragments cut from a plasmid with fragments cut from the DNA of the organism that has that gene. Both sets of fragments were prepared with the same restriction enzyme, so all of the fragments have complementary sticky ends and can bind with each other. Among the many combinations of fragments that are formed, some will consist of plasmid DNA bound to DNA from the other organism. The bonds between the sticky ends are weak hydrogen bonds, but strong, covalent bonds can be formed by adding **DNA ligase,** the enzyme that links DNA fragments to form the lagging strand during replication. When the resulting DNA is a combination of DNA from different sources, it is called **recombinant DNA.**

Cloning produces multiple copies of recombinant DNA

Making a transgenic organism requires many copies of the gene that is to be transferred. Therefore, after a DNA fragment containing a gene of interest has been incorporated into a vector, multiple copies of the recombinant DNA are made through a process called **gene cloning.** When plasmids are used as vectors, they are usually re-inserted into bacteria, and the bacteria are allowed to reproduce many times. The reproduction of each bacterial cell and its progeny produces a clone of genetically identical cells. If the original cell contains a recombinant plasmid, all of the cells in the clone will have a copy of the plasmid and its foreign gene. Under optimal conditions, bacteria reproduce very rapidly—within 12 hours, a single cell can produce a clone of more than 10 million cells.

However, not every clone contains a plasmid with the specific gene that an investigator may want to transfer into a plant. That's because many different combinations of DNA fragments can form when DNA is cut with a restriction enzyme and the fragments are joined with DNA ligase. For example, a plasmid can be resealed without incorporating any foreign DNA, or several pieces of foreign DNA may be joined together without a plasmid. Plasmids that do incorporate foreign DNA may contain any of the 25,000 to 50,000 genes typically found in a eukaryote. How,

then, can researchers know which clones contain a recombinant plasmid with the specific gene they're interested in transferring?

To eliminate bacteria that have not taken up any plasmids, researchers start with plasmids that have been modified to include another gene that codes for resistance to a certain antibiotic. Bacteria that lack the plasmids will be killed when that antibiotic is added to the culture medium. Bacteria that have taken up a plasmid with the resistance gene will survive and produce clones.

That still leaves thousands of clones containing plasmids with different segments of the foreign DNA. These clones constitute a **gene library.** Just as a library of books is a storehouse of written information, a gene library is a storehouse for the genetic information of the organism whose DNA was incorporated into the plasmids. A researcher can save the library and search it at any time for the particular DNA sequences that designate specific genes. Figure 14.4 illustrates one way to do that. DNA from each clone is heated gently or treated with a chemical to separate its two strands, a process called *denaturation.* The denatured DNA is then exposed to a **nucleic acid probe,** a short piece of RNA or single-stranded DNA that is complementary to a DNA sequence in the gene of interest. The probe, which has been labeled with a fluorescent molecule or radioactive isotope, will bind to any cloned DNA that contains the gene of interest. The label on the probe identifies clones where this binding has occurred.

The polymerase chain reaction clones DNA without using cells

The **polymerase chain reaction (PCR)** enables scientists to prepare large amounts of DNA from specific DNA fragments without using plasmids or bacteria. PCR takes place in a small tube that contains the sample DNA, DNA polymerase, all four nucleotides, and short pieces of single-stranded DNA called *primers* (Figure 14.5). The primers are complementary to the ends of the segment of sample DNA that is to be copied. PCR is a cyclical process that begins when the mixture in the tube is heated to denature the sample DNA. The mixture is then cooled to allow the primers to bind to the denatured DNA. Next, DNA polymerase adds complementary nucleotides to each strand of the sample DNA, using the primers as starting points. (The DNA polymerase that is used in PCR was obtained originally from bacteria that live in hot springs, so it is not destroyed by the heating that begins each cycle.) Thus, the first cycle of heating and cooling produces two double-stranded DNA molecules for each molecule of sample DNA that was present in the tube.

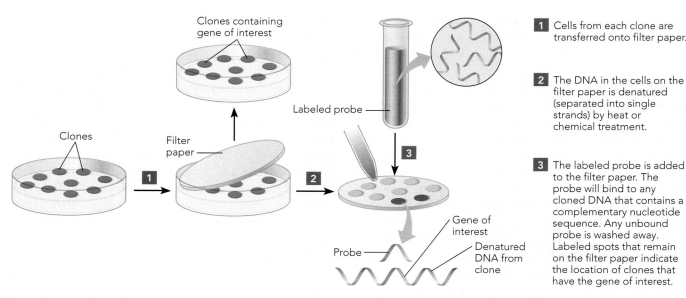

Figure 14.4 Identifying a cloned gene. To determine which clones among thousands contain a specific gene, the clones are exposed to a labeled probe consisting of RNA or single-stranded DNA complementary to part of the gene.

The heating and cooling are repeated through many cycles, each lasting about 2–3 minutes, and in every successive cycle the number of DNA molecules doubles. Billions of copies can be made from a single DNA molecule in a few hours. A thermal cycling machine automates the entire process.

Developed in 1985, PCR has become an indispensable tool in many areas of biotechnology because of its ability to copy DNA fragments quickly and accurately. It is also the preferred method for cloning DNA fragments that are available in very small quantities. For example, it could be used to clone DNA from a rare, mutant plant. Forensic investigators use PCR to copy the miniscule traces of DNA that are sometimes found at crime scenes.

Several methods can be used to insert cloned genes into plant cells

After a gene is cloned, the next step in making a transgenic plant is to insert the gene into cells of the plant. When the Ti plasmid is used as a vector, this step can be accomplished in either of two ways. One way is to return the recombinant plasmid to *A. tumefaciens,* the bacterium that normally carries the plasmid, and allow the bacterium to infect plant cells. The other way is to introduce the recombinant plasmid directly into the plant cells. In either case, the plasmid will insert part of itself, including the gene of interest, into the plant cells' chromosomes (see step 3 in Figure 14.2). The Ti plasmid works well with dicots, which are naturally susceptible to infection by

A. tumefaciens, but less well with monocots. Although modified strains of the bacterium have been developed recently that can infect monocots with some success, there are a variety of other methods that can also be used to introduce foreign genes into plants.

Gene guns

Transgenic plants are sometimes produced by actually firing a gun at plant cells (Figure 14.6a). The targets are often cultured cells (see below) or protoplasts, which are cells that have had their cell walls removed by treatment with various enzymes. Cells in the apical meristems of growing plants are also targeted by this technique. The gene gun is aimed at the cells and fired, but a shield in the gun prevents the casing from leaving the gun. Pellets coated with the gene of interest are ejected from the casing, pass through a hole in the shield, and enter the cells. Occasionally, if enough copies of the gene are introduced into the cells, one or more copies will be incorporated into a chromosome by crossing over. The use of gene guns is, literally and figuratively, a shotgun approach, meaning that it uses relatively unsophisticated technology and is not always successful.

Electroporation

In **electroporation,** a brief pulse of electric current is applied to a solution containing plant cells and copies of the gene of interest. The current creates small pores in the cells' plasma membranes, through which some of the gene copies can enter the cells. If one of these copies enters the

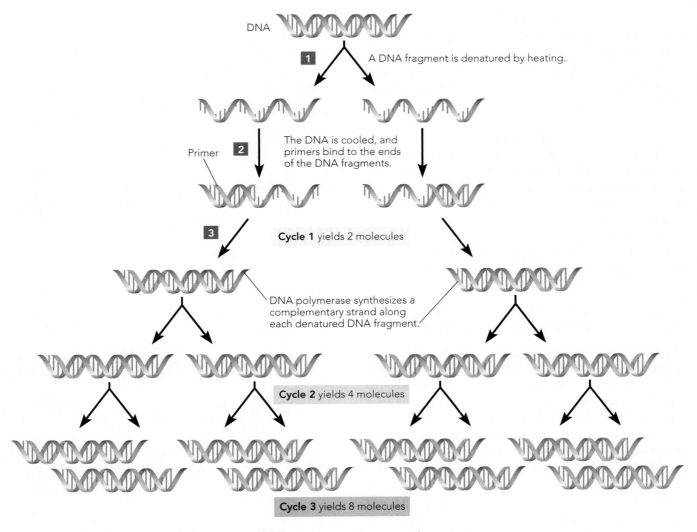

DNA

1 A DNA fragment is denatured by heating.

Primer **2** The DNA is cooled, and primers bind to the ends of the DNA fragments.

3 Cycle 1 yields 2 molecules

DNA polymerase synthesizes a complementary strand along each denatured DNA fragment.

Cycle 2 yields 4 molecules

Cycle 3 yields 8 molecules

Figure 14.5 The polymerase chain reaction (PCR). PCR is a rapid, automated process for making a vast number of copies of a specific DNA fragment in a tube. The process consists of a three-step cycle that is repeated approximately every 2–3 minutes.

nucleus, it may be incorporated into a chromosome by recombination. Electroporation is another shotgun approach and works because large numbers of cells can be treated at once. If a gene conferring herbicide resistance is incorporated along with the gene of interest, adding herbicide will eliminate the vast majority of cells in which gene incorporation has not occurred.

Figure 14.6 Two alternative methods for introducing foreign genes into plant cells. (a) A gene gun fires DNA-coated pellets at plant cells. (b) A fine (right) needle injects genes directly into the cytoplasm of a protoplast, while a suction micropipette (left) stabilizes the protoplast. With both methods, genes that enter the cytoplasm occasionally make their way into the nucleus and may be incorporated into chromosomes.

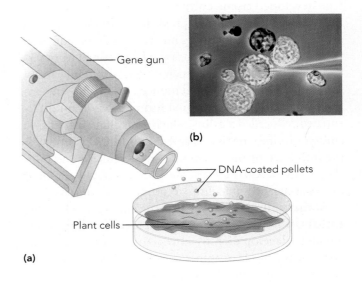

Gene gun

(b)

DNA-coated pellets

Plant cells

(a)

Microinjection

Genes can be injected directly into protoplasts and even nuclei through a very fine needle under a microscope (Figure 14.6b). A micropipette applying light suction is generally used to secure the protoplast so that it isn't pushed away by the needle.

Liposomes

Liposomes are small spheres made of lipid molecules that can fuse easily with plasma membranes. The liposomes are loaded with multiple copies of the gene of interest and placed in close contact with protoplasts. Fusion of the liposomes and protoplasts can be promoted by adding polyethylene glycol or other chemicals. Again, delivery of the genes to the cytoplasm does not ensure that they will either enter the nucleus or be incorporated into a chromosome.

In tissue culture, whole plants are grown from isolated cells or tissues

The final step in creating a transgenic plant is to grow an entire plant from a cell that has incorporated a foreign gene into one of its chromosomes. Plants are multicellular organisms composed of many different types of cells. Each cell in a plant typically contains all of the plant's genes and is totipotent—it has the potential, just like a zygote, to express any of those genes and to produce an entire plant. If an individual cell gives rise to an entire plant, that cell is expressing its totipotency.

In 1905, the German botanist Gottlieb Haberlandt recognized the totipotency of plant cells and carried out experiments in which he isolated single cells from plants and tried to encourage their development into new plants. Haberlandt was unsuccessful because he used highly differentiated cells. Beginning in the 1960s, however, scientists developed **tissue culture** methods that encourage individual plant cells to express their totipotency in an artificial medium containing nutrients and hormones. In tissue culture, cells and tissues can be obtained from virtually any part of a plant—including leaves, stems, roots, shoot tips, and flowers—and induced to develop into entire plants. Often, the cells first form a mass of undifferentiated cells called a **callus,** which then differentiates into embryos or into the tissues and organs of the plant under the influence of the hormones in the culture medium (Figure 14.7). Alternatively, the cell walls can be removed from selected cells in culture to produce protoplasts.

Anther culture is a form of tissue culture in which the anthers of flowers are placed on a medium that causes pollen to develop directly into plants without fertilization. The usual goal of anther culture is to produce haploid plants, in which all alleles, whether dominant or reces-

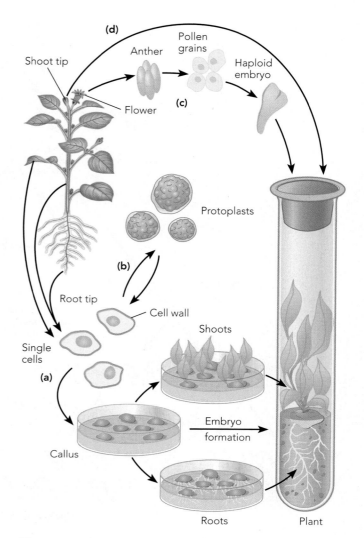

Figure 14.7 Plant tissue culture. An entire plant, or individual plant organs such as stems or roots, can be grown from isolated cells or tissues on an artificial medium that contains nutrients and hormones.

sive, are expressed. Haploid plants grow normally but are sterile because pairing of homologous chromosomes cannot occur during meiosis. However, such plants can be treated with colchicine to obtain homozygous diploid plants, which reproduce normally. (Recall from Chapter 13 that colchicine disrupts spindle formation during mitosis, leading to a doubling of chromosome number.) The homozygous diploid plants will be true breeding for all of the traits they express, a condition that would require many generations of selection with traditional plant breeding. Since each pollen grain produces a phenotypically unique plant, plant breeders use anther culture to produce plants with many types of useful variation.

In **meristem culture** or shoot tip culture, the top few millimeters of a shoot apex are cultured on a medium that encourages axillary buds at the base of each leaf or leaf primordium to develop into a whole plant. Because cells around shoot meristems divide frequently, they are usually free of viruses that may infect the rest of the plant. Meristem culture is therefore an efficient way to produce large numbers of virus-free plants, including many houseplants and crop plants such as bananas.

For the purposes of genetic engineering, tissue culture provides a way to produce any number of genetically identical transgenic plants from a single plant cell that has acquired a foreign gene. If the gene was inserted into a protoplast, the protoplast can be induced to form cell walls and the resulting cells to regenerate plants. Plants regenerated from these cells contain the foreign gene in every cell.

Section Review

1. **What is a plasmid, and how can it serve as a vector?**
2. **Explain how a restriction enzyme works.**
3. **How does the polymerase chain reaction clone fragments of DNA?**
4. **Explain how gene guns are used to produce transgenic plants.**

The Accomplishments and Opportunities of Plant Biotechnology

The American Association for the Advancement of Science identified the ability to move genes from one organism to another as one of four major scientific revolutions of the twentieth century. The other three were understanding the structure of the atom, escaping Earth's gravity, and the development of sophisticated computers.

Table 14.1 shows some of the historical milestones that have occurred in the development of plant biotechnology. Genetically engineered plants, animals, and bacteria began to appear in the 1980s. By 2000, more than half the world's soybeans and a third of its corn came from genetically modified (GM) plants. Products from these plants appear in hundreds of food products, including animal feed, cereals, cooking oil, syrups, and soft drinks.

Genetic engineering in plants has a simple goal: to transfer genes from a variety of organisms into plants, where those genes can be expressed to produce useful traits not normally found in the recipient plants. In this half of the chapter, we'll take a look at several examples of transgenic plants that have been produced so far. Then we'll consider some of the problems regarding the pro-

duction of transgenic plants and a few of the future opportunities for plant biotechnology.

Genetic engineering has made plants that are more resistant to pests and harsh soil conditions as well as more productive

Plants have been engineered for resistance to insects, fungi, and viruses. The production of insect resistance via introduction of the *Bt* gene, which was discussed at the beginning of the chapter, is a good example of this sort of change. A more direct technique produces plants with resistance to devastating viral diseases. A gene coding for a viral coat protein has been transferred to tomato plants, for example, resulting in plants that are virus resistant. Resistance occurs because the plants produce the protein, which either prevents the virus from binding to plant cells or blocks its replication inside the cells. This technology has been applied to many other crop plants and viral diseases. For instance, papayas resistant to the devastating ring spot virus were produced by introducing a viral coat protein gene into a papaya chromosome (Figure 14.8).

A coupling of animal physiology and plant biotechnology has led to plants that use antibodies to avoid disease. Mouse cells are induced to make antibodies against specific toxins released by organisms that cause diseases in plants. The genes coding for these antibodies are isolated and incorporated into plant chromosomes, resulting in transgenic plants that can produce the antibodies (called "plantibodies") themselves. The antibodies bind to and disable the toxins, making the plants resistant to those diseases.

Figure 14.8 Virus-resistant papayas. The papaya plant (*Carica papaya*) on the left was given a gene coding for the coat protein of the ring spot virus and thus is resistant to the virus. The plant on the right lacks the gene and is susceptible to infection by the virus.

Table 14.1 A Few Important Events in the Development of Plant Genetic Engineering

Year	Event
1866	Gregor Mendel determines the basic laws of heredity.
1882	Walther Fleming observes chromosomes.
1944	Oswald Avery, Colin MacLeod, and Maclyn McCarty prove that DNA is the genetic material.
1944	Frederick Sanger uses chromatography to determine the amino acid sequence of insulin.
1947	Gene transfer by plasmids is discovered.
1947	Barbara McClintock reports on transposons.
1953	James Watson and Francis Crick determine the structure of DNA.
1957	Francis Crick and George Gamov propose the central dogma of molecular biology, explaining how genes code for proteins.
1961	Marshall Nirenberg deciphers the first codon.
1964	Charles Yanofsky and colleagues prove that nucleotide sequences in DNA correspond to amino acid sequences in proteins.
1965	Restriction enzymes are discovered.
1969	The first gene is isolated from bacteria.
1972	Paul Berg uses restriction enzymes and DNA ligase to make the first recombinant DNA molecule.
1973	Stanley Cohen, Annie Chang, and Herbert Boyer make the first transgenic organism, a bacterium with a viral gene.
1976	The first genetic engineering company, Genentech, is founded in California.
1977	Frederick Sanger announces his chain termination method of DNA sequencing.
1978	Genentech and the City of Hope National Medical Center announce the lab production of a gene for human insulin.
1980	Human insulin becomes the first useful product made by transgenic bacteria.
1980	The first patent is issued for genetically engineered bacteria.
1983	Eli Lilly receives a license to make human insulin.
1985	Kary Mullis develops the polymerase chain reaction.
1985	The first automated DNA sequencer is invented.
1985	Disease-resistant transgenic plants are field-tested for the first time.
1985	The first genetically engineered crop, a tobacco, is approved for release by the EPA.
1986	The first field test is carried out on a genetically engineered plant.
1987	Genetic engineering patents are applied to plants and animals.
1987	Advanced Genetic Sciences conducts a field trial of bacteria to prevent frost formation on strawberries.
1987	Calgene receives a patent to extend the shelf life of tomatoes by producing antisense RNA that silences the polygalacturonase gene.
1990	Transformation of corn with a gene gun is announced.
1994	Calgene wins FDA approval for the Flavr Savr® tomato.
2000	The genome of *Arabidopsis thaliana* is completely sequenced.
2002	The genome of rice (*Oryza sativa*) is completely sequenced.

Plants have also been engineered to be resistant to drought, excess soil salinity, excess soil acidity, and other soil stresses. The *Biotechnology* box on page 310 discusses some of the approaches that have been used to increase the salt tolerance of plants. In acidic soils, excess aluminum is brought into solution and is toxic to many plants. To address this problem, scientists have transferred genes coding for the enzyme citrate synthase from the bacterium *Pseudomonas aeruginosa* into *Arabidopsis* and papaya. Plants that have the genes overproduce citrate, which they pump out of their roots. In the soil, the citrate binds to aluminum ions, preventing the plants from taking up aluminum and allowing them to grow in aluminum-rich soil. *Arabidopsis* has also been genetically engineered to express a bacterial gene that enables the plant to convert methylmercury (an extremely toxic form of mercury that accumulates in the food chain) to less-toxic mercury compounds. Plants with this gene can grow in the presence of methylmercury concentrations that block the germination of normal plants. The transgenic plants might be useful in efforts to clean up soils contaminated with methylmercury.

Genetic engineers have increased the productivity of crop plants by making transgenic plants that produce more seeds or fruits, that have growth forms that allow more efficient cultivation, and that are resistant to herbicides used to kill weeds in the field. Seeds of many lines of herbicide-resistant plants are now on the market. For instance, some are resistant to glyphosate, a strong but biodegradable herbicide that inhibits the synthesis of aromatic amino acids and therefore kills most plants. Some glyphosate-resistant plants have many extra copies of the gene that codes for EPSP synthetase, the enzyme inhibited by glyphosate. The extra copies allow the plants to produce sufficient amounts of aromatic amino acids when glyphosate is applied. Other glyphosate-resistant plants have enzymes from bacteria that aren't inhibited by glyphosate. Both types of genetically engineered plants can survive while other plants are killed by the herbicide (Figure 14.9).

A line of transgenic rice developed in China contains an anti-senescence gene. Plants with this gene fill their grains with starch for a longer period than normal plants do. As a result, this line shows initial increases in crop yield of up to 40% over normal rice.

Figure 14.9 Herbicide-resistant corn. The corn plants on the left were given extra copies of the gene coding for EPSP synthetase, which is inhibited by the herbicide glyphosate. These plants survived when glyphosate was applied to rid the rows of weeds. Normal corn plants with two copies of the gene (right) must be weeded by hand.

Transgenic plants contribute to human health and nutrition

Genetically altered plants now produce a number of medically important proteins for human use. Alkaloids with anticancer properties are manufactured by cultured plant cells that have been given additional copies of genes coding for key enzymes in alkaloid production. Transgenic tobacco cells synthesize both the α and β polypeptide chains of human hemoglobin, the molecule that transports oxygen in the blood. In Chapter 1, you read about how genetically engineered potatoes (and fruits such as bananas) are being used to make edible vaccines for a strain of *E. coli* that causes severe diarrhea. With similar methods, scientists have been able to stimulate the production of rabies virus antibodies as well as resistance to rabies by feeding mice spinach leaves that contain viral genes. Edible vaccines could be of immense value in countries that lack the transportation, refrigeration, and money to pay for traditional vaccines. The World Health Organization of the United Nations estimates that 10 million children a year die in developing countries from diseases preventable by vaccination.

Plants can also be engineered to produce vitamins and to accumulate minerals lacking in human diets in particular regions of the world. Up to half the world's population has a diet deficient in particular vitamins or minerals, so crop plants with enhanced nutrient profiles are incredibly important in our future. An important breakthrough by plant genetic engineers in this area is the induction of provitamin A biosynthesis in rice endosperm, resulting in so-called golden rice (see Figure 1.6). Scientists have also used genetic engineering to markedly increase the content of carotenoids, which give rise to vitamin A, in plants that produce cooking oil. An estimated 154 million children worldwide are deficient in vitamin A. As you learned in Chapter 1, vitamin A deficiency is a major cause of blindness and death in areas where rice is a dietary staple.

Recently, genetic engineers succeeded in transferring the gene that codes for the iron-binding protein ferritin from soybean to rice. Plants that have the gene accumulate up to three times as much iron as do plants that lack the gene. A meal-sized portion of this iron-fortified rice could provide 30–50% of the daily iron requirement for adults. The need for iron-fortified rice is great, since rice provides lower levels of iron than are needed by people who rely on the plant as a primary food source and cannot afford mineral supplements. In the world, 2 billion people are at risk of serious iron deficiency, particularly those with largely vegetable-based diets. Those at risk include 400 million women of child-bearing age who suffer from anemia. Such women tend to give birth to stillborn or underweight babies and are more likely to die in childbirth. Iron deficiency affects more than 500 million children, many of whom die from it. As an added benefit, the genetically engineered rice has 20% more protein than normal rice and therefore may help prevent diseases caused by severe protein deficiency.

BIOTECHNOLOGY

Genetic Engineering of Salt-Tolerant Plants

Excess soil salinity reduces food production on almost one-third of the world's cultivated land. Highly productive irrigated farmland is particularly vulnerable to salinity because each cycle of irrigation leaves behind dissolved salts, which gradually become more concentrated in the soil. In ancient Mesopotamia, in the land between the Tigris and Euphrates rivers in what is now Iraq, salt buildup in the soil caused ancient civilizations to switch from salt-sensitive wheat to barley, a less useful grain that is salt tolerant. Ultimately, food production faltered, and the civilizations declined. Many irrigated areas in the United States have soil salt concentrations that reduce yields and dictate which crops can be grown. Rivers increase in salinity from their headwaters in the mountains to their mouths in the ocean. Up to half of this salt loading comes from soil, rocks, and hot springs. The other half comes from human activities. The water in rivers as they approach the ocean is frequently too salty for either plants or animals.

Most salinity problems involve excess sodium chloride, or table salt, a common component of Earth's crust. Animals need ionized salt—sodium and chloride ions—to survive, but the salt is required in the intercellular space. In the cytoplasm, chloride ions do no harm, but excess sodium ions are toxic. Animal cells spend up to a third of their metabolic energy pumping sodium ions out of the cytoplasm. Although plants don't require sodium chloride, their cells find sodium ions equally toxic, and that's why excess soil salinity interferes with plant growth and food production.

Scientists use traditional breeding, tissue culture, and genetic engineering to increase the salt tolerance of crop plants. Plant breeders cross salt-sensitive varieties of plants such as wheat with salt-tolerant varieties of the same species or, more frequently, of related wild grasses. Such crosses require several generations of field experiments for genetic stabilization to occur. In tissue culture experiments, researchers grow large numbers of cells and look for cells that tolerate high salt concentrations in the culture medium. Salt-tolerant cells are then regenerated into whole plants. Using these techniques, scientists have produced a number of salt-tolerant lines of wheat, rice, and sorghum.

Genetic engineering of increased salt tolerance has so far focused on producing plants with extra copies of a gene that codes for a sodium-ion pump. Adding extra copies of a particular gene to an organism is referred to as *gene overexpression*. The salt-tolerant *Arabidopsis* and tomato plants produced by this method tolerate high external salt levels by pumping excess sodium ions from the cytoplasm into either the intercellular space or the central vacuole. Cells compensate for the increased solute concentration in the intercellular space and central vacuole by producing more cytoplasmic solutes that are nontoxic, such as the amino acid hydroxyproline. One key to the market success of such genetically engineered plants will be the production of fruits that do not have such high sodium concentrations that they taste salty.

Scientists have also discussed transferring genes from highly salt-tolerant, or halophytic, bacteria to crop plants. Halophytic bacteria sometimes grow in conditions of unbelievable salinity, such as in ponds where seawater is evaporated to obtain salt. In fact, these bacteria may require such conditions to sustain their growth. The genes that enable halophytic bacteria to withstand extreme salinity will be interesting to investigate and may eventually lead to increased food production on salt-stressed soils.

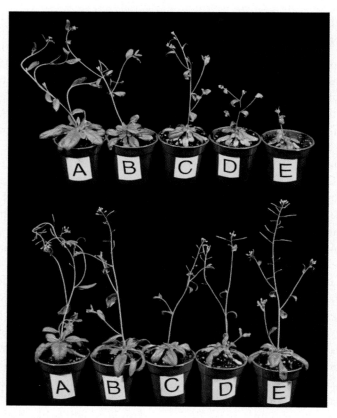

The upper five *Arabidopsis* plants have two copies of a gene for a sodium-ion pump. The lower five plants have extra copies of that gene and are salt tolerant. The concentration of sodium chloride was near zero in the pots labeled A, 50 mM in B, 100 mM in C, 150 mM in D, and 200 mM in E.

Genetically engineered crops require extensive field and market testing before they are released

Advances in biotechnology take time. Let's say a new genetically engineered plant uses a bacterial gene to produce a protein with high levels of an amino acid required in the human diet. Following such a research success, several years of field tests and nutritional studies follow before the plant can be deemed genetically stable, useful, and safe. More often than not, field tests will reveal some glitch that requires minor adjustments and changes, which take even more time. For example, the sensitivity of the new plant to a particular fungal disease may have increased, or the plant may not respond well to periods of drought. Altogether, it may take at least six years from the beginning of research for the modified plant to become available on the market.

One example of the practical challenges that confront plant biotechnology is provided by the genetic engineering of cotton plants (*Gossypium* sp.). As you may know, one problem with cotton clothing is that it wrinkles more easily than clothing made with cotton-polyester blends. Some genetic engineers are trying to solve this problem by producing cotton plants that make polyester as well as cotton. However, the transgenic plants they have created so far do not make significant amounts of polyester in the boll, or fruit, where the cotton fibers are found. Clearly, more research is needed into the mechanisms by which expression of the gene for polyester synthesis is controlled in various tissues.

The Flavr Savr® tomato (Figure 14.10) presents another example of the challenges of biotechnology. Tomatoes that taste good must ripen on the vine, but ripe tomatoes rot during shipping. On the other hand, tomatoes that ship without spoiling are picked green, but they tend to be tasteless. The Flavr Savr® was genetically engineered to have a gene that reduces the production of the enzyme polygalacturonase, which degrades cell walls and causes tomatoes to soften during ripening. Flavr Savr® tomatoes rotted more slowly than normal tomatoes, so they could be picked when they were ripe instead of green. In 1994, the Flavr Savr® became the first genetically engineered plant to be approved by the U.S. Food and Drug Administration (FDA). However, it was removed from the market after several years because mechanized harvesting and packing equipment damaged the vine-ripened fruit. The tomato also had less disease resistance and provided lower yields than normal tomatoes and was therefore more expensive. Finally, it did not grow well in the sandy soils of Florida, a prime tomato-producing region of the United States. Further research and consultations between tomato breeders and producers may eventually overcome these difficulties, allowing the original promise of this genetically engineered crop to be realized.

Genetically engineered crops must be safe for the environment and for consumers

As we noted earlier, plant breeding by farmers and later by scientists has introduced new alleles into plants for thousands of years. Even so, the introduction of foreign genes into plants through genetic engineering has raised concerns about the safety of the plants and of the food they produce. For example, suppose an edible vaccine produced by a genetically engineered plant caused a severe allergic reaction in some people who ate the plant. Or suppose genes for herbicide resistance moved from crop plants to weeds. Would "super weeds" be produced?

Figure 14.10 Spoilage-resistant tomatoes. Flavr Savr® tomatoes (left) were genetically engineered to resist spoilage during shipping, so they could be picked after they had ripened on the vine. Normal tomatoes (right) spoil soon after they become ripe.

THE INTRIGUING WORLD OF PLANTS

Wide Crosses Between Plants

In animals, crossing organisms of different species or genera is rarely successful because, even if fertilization occurs, the embryos generally fail to develop. In plants, however, such crosses frequently succeed. The success of such wide crosses may imply that the plants are more closely related than they appear.

A good example of a wide cross in plants is the production of triticale, a hybrid of wheat (*Triticum aestivum*) and rye (*Secale cereale*). This cross occurs occasionally in nature as well as in the laboratory. Plant breeders are interested in triticale because they would like to transfer certain desirable traits of rye—including its resistance to various diseases, drought, and other environmental stresses—to wheat.

Bread wheat has 42 chromosomes and rye has 14. When wheat gametes ($n = 21$) and rye gametes ($n = 7$) combine in a cross, the resulting hybrid plant has 28 chromosomes and is sterile. If the hybrid's chromosome number is doubled to 56, a fertile hybrid, called *octaploid triticale,* results. Octaploid

Indian mustard (left), stone cress (right), and a hybrid (middle)
growing in soil that has high levels of lead, zinc, and nickel.

triticale is unstable. Over a period of years, the chromosome number gradually decreases to 42 due to the loss of rye chromosomes. The hybrid with 42 chromosomes, known as *hexaploid triticale,* is genetically stable and contains some segments of rye chromosomes due to translocations.

Wide crosses can also be made by fusing protoplasts from different species and regenerating a whole plant from the resulting hybrid protoplast. If the protoplasts were derived from closely related plants, translocations between chromosomes will produce interesting and potentially useful hybrids. For example, protoplast fusion has been used to produce a hybrid of Indian mustard (*Brassica juncea*) and stone cress (*Thlaspi caerulescens*). Indian mustard tolerates lead but not zinc or nickel, whereas stone cress tolerates zinc and nickel but not lead. The hybrid plant tolerates all three metals. When grown in soil that contains high levels of these three metals, the parent plants grow poorly while the hybrid thrives.

Such gene movement could occur via natural viral or bacterial vectors or by **wide crosses,** breedings that occur occasionally in nature between relatively unrelated plants (see *The Intriguing World of Plants* box on this page).

Golden rice, the genetically engineered rice that contains increased levels of vitamin A, has been widely criticized on both environmental and nutritional grounds. In terms of environmental safety, critics worry that the antibiotic-resistance genes used in constructing golden rice could be transferred to bacteria that cause diseases in plants and animals. While such transfers are unlikely, they could occur, for example, in the guts of bees that consume pollen from transgenic plants.

With regard to the nutritional value of golden rice, some people have argued that golden rice might supply too much vitamin A—and therefore cause vitamin A toxicity—in the diet of people who eat mostly rice. Nutritional experts point out that malnourished individuals usually lack several nutrients in addition to vitamin A and that

vitamin uptake is reduced in such individuals. Other critics claim that golden rice doesn't have *enough* vitamin A. Estimates of how much golden rice a person would have to eat to obtain a full day's supply of vitamin A range from less than 0.5 kilogram to almost 7 kg (1.1–1.5 pounds). Critics also note that the popularity of polished rice, which has the vitamin A–rich aleurone layer removed, may be due to marketing efforts that persuaded many Asians to prefer polished rice over unpolished rice, which has an intact aleurone layer and therefore contains sufficient vitamin A. Also, vitamin A produced in green, leafy vegetables could easily add sufficient vitamin A to most diets. These critics say that golden rice is unnecessary if people return to eating unpolished rice and plant a variety of crops to obtain adequate nutrition.

Marketing and profit concerns surround the introduction of golden rice. Due to an agreement with the genetic engineering company that supported its development, golden rice seeds will be supplied free to farmers who have incomes less than $10,000 a year. However, at least

70 patents lie behind the techniques that led to golden rice, and not all patent holders have agreed to the free distribution of seeds. Also unsettled is whether farmers will be allowed to keep seeds for replanting and whether golden rice requires expensive fertilizers and pesticides for maximum production.

Clearly, if everyone lived on a farm, grew a variety of fruits and vegetables, and paid attention to their diet, golden rice would be unnecessary. In fact, most people live in cities and rely on agricultural production and distribution systems to supply sufficient quantities of nutritious food. In developed countries, many of our foods are vitamin enriched. While some of the concerns of golden rice critics appear legitimate, other arguments against golden rice and GM crops in general seem to originate from competition both in the domestic marketplace and between countries vying for a niche in the world market. Those trying to sell traditional crops don't want GM crops to gain a market share and so may try to raise undue public concern. Some of the complaints against GM plants are similar to ones that were made when tractors threatened to replace horses. The advantages of horses, which supply fertilizer and transportation in addition to pulling plows, were not replaced by tractors, which add air pollution and require fossil fuel. Nevertheless, a farmer who uses a tractor can feed more people than a farmer who uses a horse. Likewise, carefully tested GM crops offer the possibility of feeding many more people than traditional crops can.

Obviously, foods from genetically engineered plants must be thoroughly and carefully tested to ensure that they don't transfer harmful genes to other organisms and are safe for human consumption before they are approved for production. A balance must be achieved between people who see GM plants as providing only solutions and those who see only problems.

The future holds many opportunities for plant biotechnology

Many exciting efforts in plant biotechnology are in progress. In the next 25 years, another 50% increase in rice production will be required to feed Asia's growing population. Similar increases in wheat and corn production will also be needed. Plant breeders are therefore very interested in transferring genes responsible for increased productivity to crop plants that lack such genes but may have other useful traits, such as disease resistance or drought tolerance. For example, scientists intend to use both genetic engineering and traditional breeding to redesign the rice plant from the ground up. Their goal is to further increase yields while decreasing the amounts of water and fertilizer required by rice. This program illustrates the general strategy now envisioned for all major crops: Ongoing efforts of traditional breeding will interface with new efforts of plant biotechnology.

Other scientists are trying to increase crop yields by introducing the genes for C_4 photosynthesis into rice and other C_3 plants. Recall from Chapter 10 that C_4 plants are much more efficient at carbon fixation because they incorporate carbon dioxide into four-carbon compounds in mesophyll cells, thereby limiting photorespiration. Successful introduction of the C_4 photosynthetic pathway into rice should markedly increase rice production in areas with high temperatures and light intensities where rice is typically grown. One potential obstacle to this introduction is that rice lacks the distinctive Kranz anatomy frequently associated with C_4 plants. Recently, however, an aquatic angiosperm, *Hydrilla verticillata*, was discovered that carries out C_4 photosynthesis but does not have Kranz anatomy. By 2003, scientists had genetically engineered rice to express two of the three genes required for C_4 photosynthesis. Rice with these two genes fixed carbon as C_4 plants do. In the field, rice plants containing either one of these genes produced 10–35% more grain than unmodified rice. Efforts continue to introduce all three C_4 genes into rice.

A different team of researchers has identified two genes that control seed release in *Arabidopsis* fruits. These genes prevent a phenomenon called *catastrophic seed release*, or *shattering*, which happens when fruits suddenly break apart and release seeds. Transfer of these genes through genetic engineering would be very useful for certain crop plants. For example, farmers who grow canola for oil (which comes from the seeds) can lose half their harvest if shattering occurs. For these farmers, a canola variety with delayed seed release would be economically quite valuable.

Other researchers are working with *Arabidopsis* mutants that have an increased uptake of hydrogen ions into their root tips. Removal of hydrogen ions from the soil makes the soil less acidic, and the lowered acidity causes aluminum ions to precipitate, thereby reducing aluminum uptake by the plant. Once the mutated gene is identified, the researchers should be able to transfer it from *Arabidopsis* to crop plants, increasing the yields of those plants in soils that have toxic levels of aluminum ions.

Unaccomplished goals of plant biotechnology involve traits for which genes have not been isolated or traits that are controlled by multiple genes. One such trait is nitrogen fixation. As you learned in Chapter 10, bacteria that live in the soil or in the roots of certain plants convert atmospheric nitrogen into nitrogen compounds that plants can use. Because nitrogen fixation involves an association between plants and bacteria, several genes are involved. Moreover, plants that have this association make a considerable contribution of ATP to the process. Therefore, endowing plants with the ability to fix nitrogen will involve more than simply transferring a single gene from bacteria to plants. It may require the modification of ATP production and the cell-specific expression of transferred genes in the transgenic plants. More research on the mechanisms of gene activation in plants will be helpful in solving these sorts of problems.

Genomics and proteomics will provide information needed for future efforts in genetic engineering

The genes that control many plant processes have not yet been identified, so of course it is impossible to transfer those genes from one plant to another. Therefore, botanists are very interested in discovering the location and function of plant genes. The complete set of an organism's DNA is known as its **genome,** and the science of determining the nucleotide sequence of whole genomes is called **genomics.** Since the nucleotide sequence of a gene specifies the amino acid sequence of a protein, genomics is related to **proteomics,** the science of sequencing all of an organism's proteins and of understanding their function. Advances in one field lead to advances in the other. For example, if a few amino acids in a protein can be sequenced, then the corresponding nucleotide sequence in the gene coding for that protein can be linked to a specific chromosomal location.

Scientists determine the nucleotide sequence of a genome by using restriction enzymes to split chromosomes into many fragments (Figure 14.11). Each fragment is cloned and sequenced (see *Biotechnology* box on page 315), and overlapping sequences in the fragments are analyzed to establish the sequence of each chromosome. The sequence of amino acids in a protein is determined in a fundamentally similar way. The protein is broken into fragments by proteases, enzymes that break polypeptide chains between specific amino acids. Each fragment is sequenced, and analysis of overlapping sequences in different fragments reveals the overall protein sequence. Automation and computers simplify the sequencing of both DNA and proteins.

Sequencing of the *Arabidopsis* and rice (*Oryza sativa*) genomes was completed in 2000 and 2002, respectively. The *Arabidopsis* genome contains about 25,000 genes and the rice genome around 32,000 to 55,000; both genomes are much smaller than those of corn and wheat. *Arabidopsis* is related to a number of dicotyledonous crop plants,

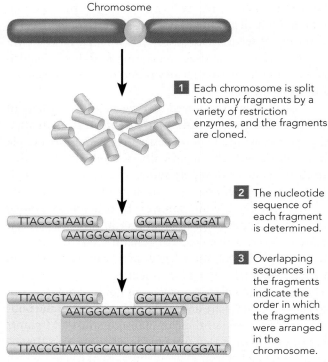

1 Each chromosome is split into many fragments by a variety of restriction enzymes, and the fragments are cloned.

2 The nucleotide sequence of each fragment is determined.

3 Overlapping sequences in the fragments indicate the order in which the fragments were arranged in the chromosome.

This procedure is repeated for each chromosome in the genome.

Figure 14.11 Sequencing a genome. There are three basic steps to establishing the sequence of nucleotides in an organism's genome.

including canola, cabbage, turnip, radish, and broccoli, but is not closely related to any of the world's three most important food plants—rice, wheat, and corn—all of which are monocots. However, all of these plants are flowering plants, so if the function of a gene is known in any one of them, a similar gene can probably be found in the others. Monocot cereals, such as rice, maize, and barley, have considerable **synteny,** meaning there are many regions of chromosomes in which genes occur in the same order. Dicots, including *Arabidopsis,* may turn out to have a good deal of synteny as well. One sequencing study showed that 81% of *Arabidopsis* genes were similar to genes in rice.

Genomic studies provide interesting information about the relatedness of all organisms. For example, humans and plants share a surprising number of genes—somewhere between 15–40%. Shared genes likely code for proteins that are essential for all organisms, although some shared homeotic genes (Chapter 13) are involved in quite different developmental events in plants and animals.

Scientists are very interested in the point mutations (single nucleotide polymorphisms) that make individual plants different (see Chapter 13). For example, we know

BIOTECHNOLOGY

DNA Sequencing

Our ability to rapidly determine the nucleotide sequence of genes and genomes has played a pivotal role in the advance of plant biotechnology and genetic engineering. DNA sequencing is mostly automated and relies on sophisticated machines, computers, and a method developed in the 1970s by British scientist Frederick Sanger, for which he received a Nobel Prize in 1980.

Recall from Chapter 13 that when DNA is replicated, DNA polymerase synthesizes two new strands using the existing strands as templates. In the Sanger method, the new strands are formed from the four normal nucleotides (dATP, dCTP, dGTP, and dTTP) as well as four modified nucleotides (ddATP, ddCTP, ddGTP, and ddTTP). DNA polymerase randomly chooses between the normal and modified nucleotides during DNA synthesis. However, as soon as it chooses a modified nucleotide, synthesis of that particular chain ceases because the modified nucleotides cannot continue a DNA chain. Each modified nucleotide is linked to a different fluorescent molecule, so DNA chains that end with different nucleotides appear as different colors.

The key to the Sanger method is to use enough of all eight nucleotides to ensure that the synthesis of some chains stops while that of others continues. As a result, DNA chains of various lengths are produced. The chains are then separated and ordered by length using gel electrophoresis. The color of each chain indicates the last nucleotide that was added. The sequence of nucleotides in the entire DNA molecule can be deduced from the nucleotide sequence in the newly synthesized chains.

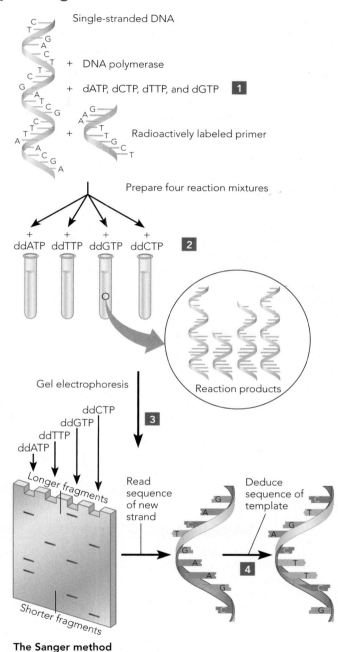

The Sanger method

1 Many copies of the segment of single-stranded DNA to be sequenced are divided four ways and incubated with labeled primer, four deoxynucleotide triphosphates, DNA polymerase, and one dideoxynucleoside triphosphate (A, T, C, or G).

2 The primer initiates the synthesis of the new DNA, which is terminated at various lengths when dideoxynucleotide triphosphate is incorporated.

3 The resulting strands of various lengths are separated by electrophoresis through polyacrylamide gel. Bands are identified by the radioactive label. Long DNA strands travel less rapidly than short strands.

4 The sequence of nucleotides in the single-stranded DNA is determined by ordering the bands with respect to length of travel and type of terminal nucleotide.

that Mendel's tall and short pea plants differed with respect to a single nucleotide in the *Le* gene, which codes for an enzyme that catalyzes the synthesis of a growth-promoting hormone. Scientists are now determining the effects of numerous point mutations and other genetic changes in a variety of organisms. The *Biotechnology* box

on page 316 details a method for distinguishing between fragments of DNA that differ in one or more nucleotides.

Sometimes, researchers can determine the function of a gene by using computer programs such as BLAST (Basic Local Alignment Search Tool) to search existing gene-sequence databases. The searches may locate genes of similar

BIOTECHNOLOGY

Analyzing DNA Fragments and Solving Crimes

In 1992, a man was convicted of the murder of a young woman in Phoenix, Arizona, based on DNA evidence obtained from a plant. Police were led to the man when they discovered his pager at the scene of the crime. He claimed that he had never been at the crime scene and that the woman had stolen his pager and left it there. However, seed pods found in the man's pickup truck would prove that he was lying. The pods were fruits of the palo verde tree (*Cercidium* sp.), which grew at the crime scene as well as throughout much of southern Arizona. By analyzing DNA samples from many palo verde trees, forensic scientists established that each tree is genetically unique and that the pods found in the truck came from one of the trees at the crime scene.

Restriction enzymes are essential tools in forensic DNA analysis. A single restriction enzyme will cut an organism's DNA into thousands of fragments with different lengths and nucleotide sequences. If two organisms had identical genomes, treatment of their DNA with the same enzyme would produce two identical sets of fragments. However, even closely related individuals in the same species have numerous genetic differences due to point mutations. Consequently, there will be many differences in the sets of fragments that are produced. These differences are known as restriction fragment length polymorphisms (RFLPs, pronounced "RIF-lips").

In a procedure called gel electrophoresis, DNA fragments are separated by size as they move through a polymeric gel in response to an electric current. Since the phosphate molecules of DNA carry negative charges, the fragments move toward the positive electrode. They move at a rate that depends on their size: Larger fragments move more slowly because they encounter more drag.

A gene can be associated with a particular fragment by a technique known as Southern blotting, developed by E. M. Southern in 1975. A radioactive probe consisting of a short sequence of nucleotides from the gene is synthesized. The fragments and the probe are separated into single strands by an alkaline solution. A single-stranded probe will bind to the fragment that has a complementary nucleotide sequence, thus locating the gene of interest.

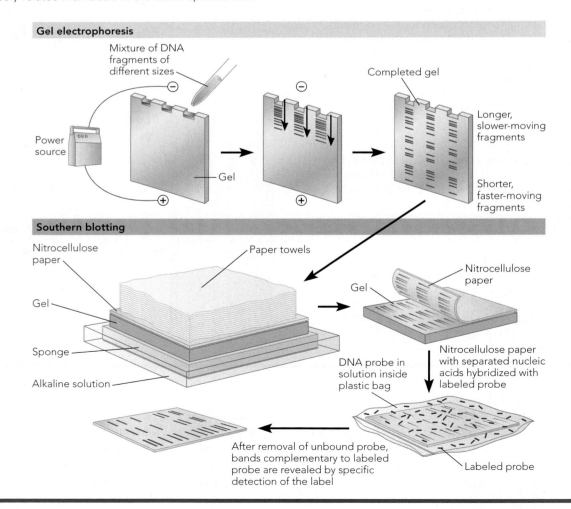

Gel electrophoresis

Mixture of DNA fragments of different sizes

Power source

Gel

Completed gel

Longer, slower-moving fragments

Shorter, faster-moving fragments

Southern blotting

Nitrocellulose paper

Paper towels

Gel

Sponge

Alkaline solution

Nitrocellulose paper

Gel

DNA probe in solution inside plastic bag

Nitrocellulose paper with separated nucleic acids hybridized with labeled probe

After removal of unbound probe, bands complementary to labeled probe are revealed by specific detection of the label

Labeled probe

structure whose function has been established in other organisms. Of course, this is not a foolproof method because the function of the gene may have changed as the result of mutation. Database searches can also be used to find out if a particular DNA sequence is part of any known genes.

Investigators studying *Arabidopsis* have another method for studying gene function. They can obtain *Arabidopsis* seeds in which every cell contains a transposon inserted in a particular nucleotide sequence. If that sequence is part of a gene, the gene will be silenced, or knocked out, as explained in Chapter 13. Looking for missing structures or functions in the plants that develop from those seeds will provide information about the normal function of the gene.

Genomics and proteomics will be increasingly important areas of plant biotechnology as more and more genomes are partially or totally sequenced. At present, the functions of most plant genes are not known, so much remains to be accomplished in genomics in the foreseeable future. As more is learned about the role of specific proteins in particular physiological or developmental processes, these proteins can be rapidly associated with the genes that code for them.

Ultimately, proteomics will fuse with genomics to make an integrated "proteogenomics" in which information about the genes and proteins of plants becomes more nearly complete. Exceedingly powerful computers will be required to synthesize the immense amount of information generated. The computers may allow three-dimensional, holographic modeling of organisms, both real and hypothetical, using the data supplied solely by gene sequences. Some day, scientists may be able to predict the effect of a single nucleotide change on the shape and catalytic action of a protein and on the structure and function of a plant.

Section Review

1. How have glyphosate-resistant plants been produced, and why are they useful?
2. Explain how genetic engineering has resulted in iron-fortified rice.
3. What are some of the criticisms that have been voiced against golden rice?
4. What are genomics and proteomics?

SUMMARY

The Methods of Plant Biotechnology

Genes can be transferred between species through genetic engineering (pp. 301–302)
Genetic engineering uses molecular techniques to move genes from one organism to another, thereby producing transgenic organisms. The introduction of new traits into plants is much faster through genetic engineering than through traditional plant breeding.

Plasmids often serve as vectors for gene transfer in plants (p. 302)
Genetic engineering in plants is often accomplished with plasmids, circular DNA molecules that are found in bacteria. Some plasmids, such as the Ti plasmid, can carry pieces of DNA from another organism into plant cells and incorporate that DNA into the plant's chromosomes.

Restriction enzymes and DNA ligase are used to make recombinant DNA (pp. 302–303)
Restriction enzymes cut DNA into fragments wherever specific sequences of nucleotides occur in the DNA. Fragments created by most restriction enzymes have single-stranded sticky ends that bind to complementary sequences on other fragments. Fragments from different sources of DNA can bind to each other at their sticky ends and become permanently linked by DNA ligase, forming recombinant DNA.

Cloning produces multiple copies of recombinant DNA (p. 303)
If plasmids containing recombinant DNA are reinserted into bacteria, the plasmids will be copied when the bacteria reproduce. Many copies of the recombinant DNA, called *clones*, can be produced in a few hours. Nucleic acid probes are used to determine which clones contain a specific gene of interest.

The polymerase chain reaction clones DNA without using cells (pp. 303–304)
The polymerase chain reaction is an automated procedure that quickly makes many copies of a specific DNA fragment in a tube using DNA polymerase and repeated cycles of heating and cooling.

Several methods can be used to insert cloned genes into plant cells (pp. 304–306)
Genetic engineers can use the Ti plasmid, gene guns, electroporation, microinjection, or liposomes to introduce cloned genes into plant cells.

In tissue culture, whole plants are grown from isolated cells or tissues (pp. 306–307)

Plant cells that contain foreign genes can be induced to develop into whole plants if they are placed in an artificial medium containing nutrients and hormones. The resulting plants will have the foreign gene in every cell.

The Accomplishments and Opportunities of Plant Biotechnology

Genetic engineering has made plants that are more resistant to pests and harsh soil conditions as well as more productive (pp. 307–309)

Plants have been engineered for resistance to insects, fungi, viruses, drought, salty soil, acidic soil, toxic metals, and herbicides. Other transgenic plants produce more seeds or fruits.

Transgenic plants contribute to human health and nutrition (pp. 309–310)

Transgenic plants produce medicines, the polypeptides in human hemoglobin, and edible vaccines. Rice has been genetically engineered to contain more vitamin A and more iron.

Genetically engineered crops require extensive field and market testing before they are released (p. 311)

Several years of testing are needed to determine whether transgenic crop plants are genetically stable and can be produced economically.

Genetically engineered crops must be safe for the environment and for consumers (pp. 311–313)

Many people are concerned that transgenic plants may transfer foreign genes to other organisms or have adverse health effects on people who eat the plants. Careful testing of transgenic plants must be performed to address such concerns.

The future holds many opportunities for plant biotechnology (pp. 313–314)

Future efforts in genetic engineering will focus on areas such as transferring genes for C_4 photosynthesis and nitrogen fixation to plants. Many of these efforts will have the overall goal of increasing the productivity of crop plants.

Genomics and proteomics will provide information needed for future efforts in genetic engineering (pp. 314–317)

Genomics, the science of determining the nucleotide sequence of genomes, and proteomics, the science of sequencing all of an organism's proteins, help genetic engineers identify genes that could be transferred to make useful transgenic plants.

Review Questions

1. What is a transgenic organism?
2. Why are plasmids used as vectors to transfer DNA into different organisms?
3. What are the basic components of artificial chromosomes?
4. What determines where a restriction enzyme will cut DNA into fragments?
5. What is the role of DNA ligase in making recombinant DNA?
6. What purpose do nucleic acid probes serve in the cloning process?
7. Describe how you might make a gene library of a wheat plant.
8. What does a PCR machine do, and how does it work?
9. How do you make a plant protoplast, and what can you use it for?
10. How can tissue culture be used to clone valuable plants?
11. How can genetic engineers produce virus-resistant plants?
12. How can you make an edible vaccine?
13. What possible environmental problems can occur with the use of crop plants that contain the *Bt* gene?
14. What fraction of our genes do we share with plants? What is the likely function of these shared genes?

Questions for Thought and Discussion

1. What do you think are the main advantages of biotechnology? In your opinion, what are the most serious risks?
2. What unforeseen dangers to the environment might be caused by GM crops?
3. How can we monitor the effects of GM foods on the environment and on human health while taking advantage of their benefits?
4. How can we ensure that improved crops and medicines are available to poor people in developing countries while providing sufficient profit to promote the continued development of those products?
5. What are some advantages and disadvantages of using plants to produce vaccines?
6. Defend or refute the following argument: We should not try to produce salt-tolerant and toxic-metal-tolerant plants because doing so simply allows soil contamination to continue. Instead, we should focus on changing practices that cause soil contamination in the first place.
7. Suppose you've been asked to invent a machine that can analyze the DNA of a plant and make accurate predictions about the appearance and physiology of the plant. How would such a machine work? How would it predict the phenotypic effect of a particular point mutation?
8. Make a flow diagram to show the steps necessary to identify and to subsequently clone the genes for nitrogen fixation in a soil bacterium.

Evolution Connection

How might the course of evolution of plants have been changed as a consequence of the establishment of pathogenic or mutualistic associations with viruses such as the tobacco mosaic virus or with bacteria such as *Agrobacterium tumefaciens*?

To Learn More

Visit The Botany Place Website at www.thebotanyplace.com for quizzes, exercises, and Web links to new and interesting information related to this chapter.

Lurquin, Paul F. *The Green Phoenix.* New York: Columbia University Press, 2001. The author traces the history, accomplishments, and problems of plant biotechnology.

Male, Carolyn J. et al. *100 Heirloom Tomatoes for the American Garden* (*Smith and Hawken*). New York: Workman Publishing Company, 1999. Heirloom tomatoes are the old-time, tasty varieties that often don't ship well but taste great.

Nicholl, S. T. Desmond. *An Introduction to Genetic Engineering* (*Studies in Biology*). New York: Avon, 2002. This book provides the basic concepts of genetic engineering, molecular biology, and genetics while also introducing applications of gene therapy and transgenic organisms.

Pinstrup-Anderson, Per, and Ebbe Schioler. *Seeds of Contention: World Hunger and the Global Controversy over GM (Genetically Modified) Crops.* Washington, DC: International Food Policy Research Institute, 2001.

Silver, M. Lee. *Remaking Eden: How Genetic Engineering and Cloning Will Transform the American Family.* New York: Avon, 1998. This Pulitzer Prize–winning author presents the pros and cons of genetic engineering with a specific emphasis on cloning.

Wiebe, Keith, Nicole Ballenger, and Per Pinstrup-Anderson. *Who Will Be Fed in the 21st Century?* Washington, DC: International Food Policy Research Institute, 2002. Dr. Anderson is director general of the International Food Policy Research Institute in Washington and winner of the World Food Prize in 2001.

The important and exciting discoveries of plant biotechnology are covered in several popular science magazines, including *Discover* and *Science Digest,* which are available at some newsstands and bookstores.

Evolution and Diversity

15
Evolution

A hummingbird feeding at and pollinating an orchid (*Sobralia amabilis*).

History of Evolution on Earth

Fossils and molecular dating provide evidence of evolution

Biogeography, anatomy, embryology, and physiology supply further evidence of evolution

Chemosynthesis may have been the first event in the origin of life on Earth

Prokaryotes were the predominant form of life for more than a billion years

Plate tectonics and celestial cycles have shaped evolution on Earth

Extinction is a fact of life on Earth

Mechanisms of Evolution

Evolution is a change in the frequency of alleles in a population over time

Most organisms have the potential to overproduce offspring

Individuals in a population have many phenotypic differences

Some traits confer an adaptive advantage

Natural selection favors individuals with the best-adapted phenotypes

Evolution can occur rapidly

In coevolution, two species evolve in response to each other

The Origin of Species

A biological species is a population of potentially interbreeding organisms

Both natural selection and geographical isolation drive speciation

Reproductive isolation can be prezygotic or postzygotic

Reproductive isolation in sympatric populations can occur because of polyploidy

The types of organisms that inhabit Earth have changed dramatically since life began more than 3.5 billion years ago. For example, if we could travel back in time 145 million years to the Jurassic period, we would find a very different collection of plants and animals than we see today. Gymnosperms (conifers and other nonflowering seed plants) and dinosaurs would be the dominant plant and animal groups on land. Flowering plants, mammals, and birds would not exist or would be quite rare.

As you learned in Chapter 1, change in living things over time is called *evolution*. Evolution is a constant feature of the universe that also applies to nonliving things. When we talk about the evolution of something, we often try to explain what causes the evolving. For instance, we might say that popular music evolves because musicians continually create new compositions and because people either like or do not like what they hear. Similarly, automobiles evolve because designers create new models, which appeal to different car buyers to varying degrees. The marketplace favors some styles and designs and not others. Competition for the limited time and money of consumers is thus a frequent component of evolutionary change in the marketplace.

The idea that organisms evolve over time occurred to more than one scientist in the mid-1800s. In fact, two British naturalists, Charles Darwin (1809–1882) and Alfred Wallace (1823–1913), independently developed similar ideas on the subject at about the same time. They both suggested that organisms evolve because some individuals are more successful than others at reproducing. As a result, the particular traits of the individuals that produce the most offspring become more common in future generations. The theory of evolution developed by Darwin and Wallace, known as *evolution by natural selection,* rapidly became a central organizing idea, or a paradigm, of biology. The *Plants & People* box on page 326 discusses the historical context of this idea.

In this chapter, we discuss evolution with a focus on plants and other photosynthetic organisms. We begin by looking at the history of major evolutionary changes on Earth. Then we consider the mechanisms of evolution, including natural selection. Finally, we investigate the concept of "species" and how evolution can produce new species.

An artist's conception of a Jurassic landscape.

History of Evolution on Earth

It is not easy to see evolution as it happens because noticeable changes in the phenotypes of populations often take hundreds, thousands, or even millions of years. Although it is possible to observe evolution in natural and laboratory settings, doing so requires scientific training and an overall understanding of the process. Nevertheless, comparisons of living organisms with each other and with fossilized forms reveal many similarities, which can be interpreted as indicating common evolutionary origins. Several lines of evidence clearly show that evolution has occurred in the past and is still occurring today.

Fossils and molecular dating provide evidence of evolution

Fossils give us information about life in the past. By comparing fossilized organisms from different periods in Earth's history with present-day organisms, we can infer how various forms of life have changed over time. Fossils can form in several ways. *Compressions* form when an organism or its parts, such as leaves, are covered by dirt or sediment (Figure 15.1a). These fossils are often found at the bottom of lakes or oceans or under volcanic ash, and they always contain residual organic material (carbon compounds). *Permineralizations* are made when minerals gradually replace the contents of dead cells, leaving the basic shape of a structure intact. Permineralizations are indeed rocks, although many also contain some organic material. Petrified wood is a good example of this type of fossil (Figure 15.1b).

If fossils are to provide useful information, their age must be accurately determined. That is usually accomplished by *radiometric dating,* a procedure that measures the amount of a radioactive isotope of an element in a fossil or the surrounding rock. One such isotope is carbon-14. While an organism is alive, it incorporates carbon-14 into its body, along with the more common isotope of carbon, carbon-12. When the organism dies, the ratio of carbon-14 to carbon-12 in its remains slowly decreases, as carbon-14 decays and becomes another element (nitrogen-14). The time required for half of a sample of radioactive isotope to decay, called the isotope's **half-life,** is constant for each isotope. Carbon-14 has a half-life of 5,730 years. Therefore, paleontologists can determine the age of fossils up to about 50,000 years old by measuring the ratio of carbon-14 to carbon-12 in the fossils.

Radioactive isotopes that decay more slowly than carbon-14 are used to date much older fossils. For example, potassium-40 decays into argon-40 (a gas) or calcium-40 with a half-life of 1.3 billion years. Thus, potassium-argon dating can be used to estimate the age of rocks that are as old as Earth itself. This method is especially useful for dating volcanic rocks because, when the rocks were molten, they released all their argon-40 into the atmosphere. Consequently, any argon-40 found in a volcanic rock must have formed since the rock solidified.

Paleontologists also make use of *index fossils,* which are fossils of organisms that lived in many places during the same, relatively brief (in geological terms) period of time. The wide distribution of index fossils allows paleontologists to match up layers of rock found at different locations. Rock layers that contain the same index fossils must have formed during the same geological period.

Figure 15.1 Fossils. (a) A compression of a *Dicroidium* leaf. **(b)** Petrified wood in Petrified Forest National Park, Arizona.

(a) (b)

PLANTS & PEOPLE

Germinating an Idea: Evolution by Means of Natural Selection

Both Charles Darwin and Alfred Wallace came up with the idea that evolution occurs by natural selection. In 1858, Wallace sent a paper detailing his idea to Darwin, who had put off the submission of his own manuscript on the topic for quite some time. Geologist Charles Lyell presented both men's papers at a meeting of the Linnaean Society of London on July 1, 1858. In 1859, Darwin published *On the Origin of Species*, in which he discussed in great detail his observations on evolution and his explanation for its cause.

Charles Darwin.

Alfred Wallace.

that organisms had evolved in the past. It took only a short intuitive leap to suggest that evolution was still occurring. If neither Darwin nor Wallace had proposed that evolution occurs via natural selection, someone else undoubtedly would have done so. The times were ripe for the idea.

Darwin's and Wallace's work was based largely on observations made by naturalists on expeditions to distant lands. During the European voyages of discovery from the 1400s through the 1800s, ships commonly carried naturalists whose job was to collect and catalogue new animals and plants. Darwin himself served as ship's naturalist aboard HMS *Beagle* on its around-the-world voyage from 1832–1836. Wallace spent considerable time in South America and southeast Asia beginning in 1850. He collected more than 125,000 specimens, including more than 1,000 new species.

Traveling European naturalists brought the specimens they collected home with them. The exhibition and sale of new species of plants and animals from the Americas, Asia, and Africa frequently helped finance the voyages themselves. Wallace, who, unlike Darwin, was not from a wealthy family, sold his collections to pay his bills.

In the United States, President Thomas Jefferson instructed Meriwether Lewis and William Clark to make similar collections as they searched for a route by water across North America between 1804 and 1806. Seeds sent back by Lewis and Clark were planted on the grounds of Monticello, Jefferson's home in Virginia. A live prairie dog even made the return trip successfully. Lewis and other expedition members wrote about many of the biological novelties they found, often accompanying their notes with hand-drawn illustrations.

The writings of other scientists also helped shape the thinking of Darwin and Wallace as they formulated their ideas. The scientific literature of the 1800s presented a variety of evolutionary interpretations of geology and biology. Darwin, in particular, was strongly influenced by geologists of his day who took the existence of fossils and the many layers of sedimentary rock as evidence that Earth is very old—much older than the 4,000 to 6,000 years proposed by many prominent theologians of the time. Increasing knowledge of the fossil record led many scientists to conclude

Darwin was not the first biologist to consider current geological theories while speculating on the evolution of organisms. Jean Baptiste Lamarck (1744–1829) was a French museum worker who noted that the traits of species seemed to change gradually over time. He incorporated his observations into a theory of evolution, published in 1809, which started with the spontaneous generation of simple microscopic organisms and ended with complex plants and animals.

Lamarck did not envision anything like natural selection as driving evolution. Rather, he thought that organisms had an innate drive to become both more complex and more nearly perfect. He believed that organisms changed over the generations by the inheritance of *acquired* traits and by use and disuse. Thus, Lamarck proposed, individual plants that grew particularly tall because of good exposure to sunlight would pass their increased height on to their offspring. Now, of course, we know that tall plants pass on this trait only if they have alleles for tallness and that these alleles are present whether a plant grows tall in the sun or short in the shade.

Despite the inaccuracy of his mechanism of evolution, Lamarck was right in proposing that evolution explains both the phenotypic changes in the fossil record and the existing species of today. He realized that Earth is quite old, that all species were not created at one time, and that species must be well adapted to their environment to survive.

The idea that natural selection can lead to the formation of new species was not entirely unique to Darwin and Wallace. In 1825, for example, Leopold von Buch wrote as follows:

The individuals of a genus strike out over the continents, move to far-distant places, form varieties (on account of the differences of the localities, of the food, and the soil), which owing to their segregation cannot interbreed with other varieties and thus be returned to the original main type. Finally, these varieties become constant and turn into separate species. Later they may again reach the range of other varieties which have changed in a like manner, and the two will now no longer cross and thus they behave as "two very different species."

Molecular dating methods compare the primary structure of DNA, RNA, and proteins in different organisms. The amount of similarity can be used to estimate the degree of relatedness between organisms. If the organisms are closely related by evolution—that is, if they evolved fairly recently from a common ancestor—they will have very similar sequences of nucleotides in their DNA or RNA and of amino acids in their proteins. If the organisms are distantly related, the sequences will be less similar. Molecular dating will be discussed in more detail in Chapter 16.

Biogeography, anatomy, embryology, and physiology supply further evidence of evolution

Fossils and molecular dating are not the only lines of evidence used by scientists studying evolution. Additional evidence comes from other areas of science that demonstrate similarity, and therefore probable relatedness, between various groups of organisms.

Biogeography is the study of where particular species of organisms occur and when they colonized a particular region. For example, newly formed volcanic islands provide living laboratories where scientists can study how plants that arrived from other locations have changed over time in response to the varied environments presented by newly formed land. The volcanic rocks on islands that formed at different times can be easily dated using the potassium-argon method. Thus, the evolution of plants on newer islands can be compared with that on older islands to determine how populations have changed over time and have adapted to unique climatic conditions on different islands.

Comparative anatomical and developmental studies provide more evidence of the relatedness of various groups of organisms. For example, all vascular plants have the same types of conducting cells in xylem and phloem. They also have a similar life cycle. In particular, embryology provides unique examples demonstrating the relatedness of plants, both within and between major groups. In flowering plants, for instance, the anatomical structure of the female gametophyte follows one of several patterns with respect to the number of nuclei and cells formed and the cells' role in the formation and development of the embryo. Flowering plants can be organized into groups of evolutionarily related species based on these embryological features.

Physiologically, all plants use the same basic biochemical mechanisms and molecules to carry out photosynthesis, respiration, DNA synthesis, transcription, translation, and many other cellular functions. In fact, similarities in biochemistry and physiology provide numerous examples and strong evidence for the relatedness of all organisms.

Chemosynthesis may have been the first event in the origin of life on Earth

Life appeared on Earth sometime between 4 billion years ago, when the crust solidified, and 3.5 billion years ago, when the earliest known fossils were formed. What series of events occurred during this 500-million-year period that resulted in life? The first event may have been the spontaneous formation of increasingly complex organic molecules from inorganic precursors, a process known as *chemosynthesis.*

In the 1920s, A. I. Oparin of Russia and J. B. S. Haldane of Great Britain independently hypothesized that the atmosphere of primitive Earth contained gases that could spontaneously react to form organic compounds. Such spontaneous reactions could not take place on Earth today, they reasoned, because the atmosphere now contains high levels of oxygen, which attacks chemical bonds. In contrast, the atmosphere of primitive Earth had very little free oxygen. In 1953, Stanley Miller and Harold Urey tested this hypothesis in a classic experiment. They were able to produce an "organic soup" containing amino acids and other simple organic compounds by adding energy in the form of electric sparks to a mixture of water vapor, methane, hydrogen, and ammonia (Figure 15.2). Although their mixture of gases was thought at the time to resemble the composition of Earth's early atmosphere, we now know that the early atmosphere probably also contained carbon monoxide, carbon dioxide, and nitrogen gas. Nevertheless, many other investigators have repeated the Miller–Urey experiment using different mixtures of gases and various forms of energy and have obtained basically the same result. These experiments have produced organic soups containing all 20 amino acids, sugars, lipids, the bases of DNA and RNA, nucleotides, and ATP.

The amino acids, sugars, and nucleotides found in simple organic soups are capable of spontaneously polymerizing to form peptides, polysaccharides, and nucleic acids. These polymerization reactions can be accelerated by minerals in particles of sand, clay, and rock. The polymers that form most readily will use the most precursors and will become the most numerous in the organic soup.

Under certain conditions, the polymers and other substances in organic soups can aggregate spontaneously to form cell-like structures called **protobionts** with various degrees of organization. Some protobionts have simple membranes and can carry out multistep chemical reactions. For example, protobionts that assemble from a mixture

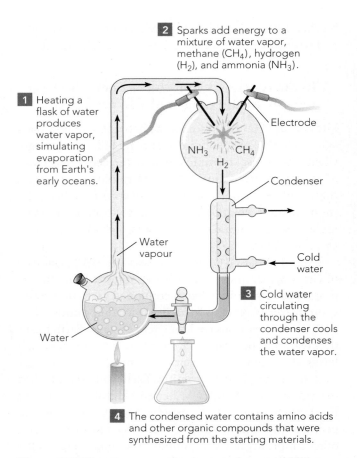

2 Sparks add energy to a mixture of water vapor, methane (CH_4), hydrogen (H_2), and ammonia (NH_3).

1 Heating a flask of water produces water vapor, simulating evaporation from Earth's early oceans.

Electrode

NH_3 CH_4 H_2

Condenser

Water vapour

Cold water

3 Cold water circulating through the condenser cools and condenses the water vapor.

Water

4 The condensed water contains amino acids and other organic compounds that were synthesized from the starting materials.

Figure 15.2 The chemosynthesis experiment of Miller and Urey.

Figure 15.3 Stromatolites. Stromatolites are fossils of domed structures produced over many years by layer upon layer of prokaryotes. The top layer may contain living cells.

of lipids and contain the enzymes phosphorylase or amylase can use the enzymes to break down starch into glucose phosphate or glucose. Protobionts with short sequences of RNA or DNA can also be produced. In some cases, these nucleic acids even make complementary copies of themselves. Other protobionts take up water and then spontaneously divide.

If the origin of life on Earth involved the evolution of protobionts into the first cells, at what point did these cell-like collections of molecules become alive? According to the chemosynthetic hypothesis, life is a product of continuity and of increasing complexity. If a protobiont/cell can divide, replicate its nucleic acids, and obtain the substances required to continue its existence, then it is alive.

Prokaryotes were the predominant form of life for more than a billion years

The first living cells may or may not have left fossil remains. Even if they did, their fossils would tell us little about the important biochemistry of these cells. Some sci-

entists have suggested that the first cells were autotrophs. Others hypothesize that life began with heterotrophic cells that broke down organic molecules from their environment to release energy by a primitive form of fermentation. These cells may have used RNA as both a genetic material and an enzyme and probably had relatively few genes. There likely were many different types of early cells, most of which did not survive for long in geological time.

The first organisms to leave fossils were primitive prokaryotes. As we noted earlier, these fossils were formed 3.5 billion years ago. Many occur in **stromatolites,** which are fossilized towers consisting of numerous layers of organic matter and sediment (Figure 15.3). The layers were formed by mats of prokaryotes, which trapped the sediment and lived in the uppermost layers.

Photosynthesis began in prokaryotes, probably as a process that used hydrogen sulfide (H_2S) as a source of electrons. Later, as H_2S became scarcer, most photosynthetic species began to use water as an electron source, generating oxygen as a by-product. More energy is required to remove electrons from water than from hydrogen sulfide. Geological evidence shows that oxygen began accumulating in the atmosphere at least 2.7 billion years ago, so the photosynthetic transition from H_2S to water must have begun by then.

Another major event in the evolution of photosynthetic organisms was the appearance of photosynthetic eukaryotes. The oldest fossils that are thought to be eukaryotic resemble those of simple unicellular algae and

Table 15.1 The Geologic Time Scale

Era	Period	Epoch	Beginning Date (mya)	Important Events
Cenozoic	Quaternary	Recent	0.01	Modern humans appear.
		Pleistocene	1.8	Humans appear.
	Tertiary	Pliocene	5	Apelike ancestors of humans appear.
		Miocene	23	Grazing mammals and apes appear.
		Oligocene	35	Browsing mammals and primates appear.
		Eocene	57	Grasslands appear.
		Paleocene	65	Mammals, birds, and pollinating insects diversify. Angiosperms become dominant land plants.
Mesozoic	Cretaceous		144	Angiosperms appear. Many groups of organisms (including dinosaurs) become extinct.
	Jurassic		206	Dinosaurs diversify. Birds appear.
	Triassic		245	Gymnosperms become dominant land plants. Dinosaurs and mammals appear.
Paleozoic	Permian		290	Many marine and land animals become extinct. Reptiles diversify.
	Carboniferous		363	Forests of seedless vascular plants are widespread. Seed plants and reptiles appear.
	Devonian		409	Bony fishes diversify. Amphibians and insects appear.
	Silurian		439	Early vascular plants diversify. Jawed fishes appear.
	Ordovician		510	Plants and animals colonize land.
	Cambrian		543	Most modern animal phyla appear.
Precambrian			4,600	Prokaryotes appear, followed by eukaryotic cells. Invertebrate animals and algae appear.

are about 2.1–2.2 billion years old. By the end of Earth's first era, the Precambrian, there were many forms of algae as well as a wide variety of marine invertebrate animals (Table 15.1). The second era, the Paleozoic (543–245 million years ago, mya), featured the colonization of land by plants and animals. The first vascular plants appeared during the Paleozoic; although seedless forms dominated the landscape, some seed-producing species were also present. The Mesozoic era (245–65 mya) saw gymnosperms (which produce seeds) and reptiles become the ruling plant and animal groups on land. The most recent era, the Cenozoic (65 mya to the present), has been dominated by flowering plants and mammals.

Plate tectonics and celestial cycles have shaped evolution on Earth

It would be impossible to make sense of the evolution of life on Earth without considering two major environmental influences. The first is **plate tectonics,** a unifying theory of modern geology that arose from the work of geologist Alfred Wegener in 1912. Plate tectonics is based on the finding that Earth's outer shell is composed of seafloor plates made of heavy rocks and continental plates made of lighter rocks. New ocean floor is created by the upwelling of molten rock from the mantle, which lies beneath Earth's crust. Upwelling pushes the seafloor plates apart, causing them to collide with the continental plates. At the collision points, called subduction zones, the seafloor plates dive under the continental plates, raising the continental plates and producing volcanoes, earthquakes, and eventually mountains.

The shifting of plates is responsible for **continental drift,** the movement of landmasses over the surface of the Earth. Although continental drift is slow, amounting typically to a few centimeters per year, over millions of years it has significantly changed the location and boundaries of continents (Figure 15.4). Continental drift accounts for a number of otherwise unexplainable observations about the distribution of plants and animals. For example, fossils of an extinct tropical seed fern, *Glossopteris,* are found in India, South America, southern Africa, Australia, and Antarctica, which were parts of the same land mass during the early Mesozoic. Also, Antarctica has fossils of

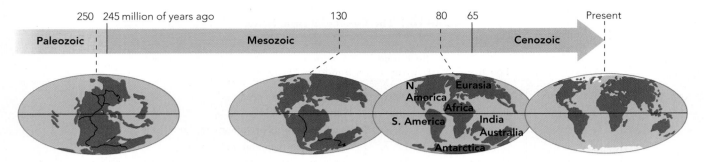

Figure 15.4 Continental drift. Because of plate movement, the relative positions of the continents have changed slowly over time. About 250 mya, all the continents were joined in a single landmass called Pangaea. It began to split apart around 180 mya into two supercontinents, called Laurasia (North America, Europe, and Asia) and Gondwanaland (Africa, South America, India, Australia, and Antarctica), separated by the Tethys Sea. The current arrangement of continents gradually came into being after the Mesozoic.

tropical plants because it was once located much closer to the equator. Australia has many unique species of plants and animals because it has been separated from other continents for the past 50 million years.

During his travels in Indonesia, Alfred Wallace noticed that the islands of Bali and Lombok, which are separated by only 30 kilometers (18.4 miles), had very different types of plants and animals: Bali had tropical rain forests inhabited by tigers, elephants, and monkeys, while Lombok had thorny, dry-land types of vegetation and animals such as kangaroos, wombats, and koalas. The abrupt biological division between these two islands, now known as *Wallace's Line,* represents a region where formerly separated continental plates have been colliding for the past 15 million years. Wallace's Line is much more distinct for animal types than for plant types because seeds and fruits cross the stretch of ocean between the two islands more readily than animals do.

The second major environmental influence on evolution is linked to cyclic changes in Earth's orientation and distance from the Sun. For example, the tilt angle of Earth's axis varies from 22 to 24.5 degrees every 41,000 years (Figure 15.5a). At present, the angle is 23.5 degrees and is decreasing. The direction of the tilt also varies, and that variation has a cycle of about 26,000 years (Figure 15.5b). Another cyclic variation is in the shape of Earth's orbit around the Sun: In a cycle that lasts about 93,000 years, the orbit changes from a near-circle to an ellipse and back again (Figure 15.5c). Currently, the Earth has a nearly circular orbit, and its distance from the Sun varies by only 6% over the course of a year. When the orbit is most elliptical, that distance varies by 30%. These cyclic changes in Earth's rotation and orbit are called *Milankovitch cycles,* after the Serbian astronomer, Milutin Milankovitch, who

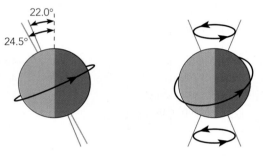

(a) Obliquity. The earth is aligned with the sun at an axial tilt known as obliquity. This axial tilt ranges from 22.0° to 24.5° over a cycle of 41,000 years.

(b) Precession. The earth also rotates around this axis forming different directions of tilt known as precession. This cycle occurs over 26,000 years.

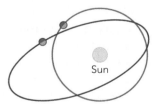

(c) Orbital variations. The earth also changes its orbital pattern from an eclipse to a circular pattern over 93,000 years.

Figure 15.5 Milankovitch cycles. The Milankovitch cycles affect the seasons and contribute to the alternation of warm and cold periods over geological time. **(a)** The tilt angle of Earth's axis varies between 22 degrees and 24.5 degrees in a cycle that lasts 41,000 years. **(b)** The direction of the tilt varies in a cycle that lasts about 26,000 years. **(c)** The shape of Earth's orbit around the Sun varies from nearly circular to elliptical in a cycle that lasts about 93,000 years.

developed the mathematical formulas that explain them in the early 1900s.

The tilt of Earth's axis is responsible for the seasons, and Milankovitch cycles can have major effects on the seasons. Variation in the tilt angle changes the intensity of the seasons. As the angle increases, the seasons become more extreme—summers are hotter and winters are colder. As it decreases, seasonal changes become less noticeable. Variation in the direction of tilt slowly reverses the seasons: In another 11,500 years, summer in the Northern Hemisphere will begin in December! As Earth's orbit becomes more elliptical, seasonal changes may become more extreme for one hemisphere (northern or southern) and less extreme for the other.

In temperate regions, seasons have played a major role in the evolution of anatomical features, developmental control mechanisms, and physiological systems of plants and other photosynthetic organisms. Many plants begin growth in spring and flower in response to specific day lengths in summer. As fall days become shorter and cooler, plants begin changes that lead to dormancy in winter. Humans and other animals that depend on plants for food synchronize their activities to match the rhythm of plant life. Because crops are tended and harvested in summer, for example, school has historically been scheduled for parts of the year with less agricultural work.

The Milankovitch cycles are primarily responsible for the alternation of warm and cold periods in geological history. During the warm periods, polar ice caps have melted, causing a rise in sea level and widespread flooding of coastal areas. During the cold periods, ice ages have occurred. At the peak of the last ice age around 18,000 years ago, ice covered about 29% of Earth's land area, almost three times as much it covers today (Figure 15.6). Static sheets of ice and slowly moving glaciers thousands of meters thick blanketed a large part of North America. Ice sheets and glaciers obliterate vegetation and cool nonglaciated regions of continents. They also markedly change the land they cover, so when they retreat, the areas that are exposed are vastly different than those that had existed there before.

Like continental drift, ice ages progress slowly, but their effects on plant and animal distributions are quite dramatic. Pollen recovered from lake-bottom sediments and plant material found in the middens (refuse piles) left by pack rats show that vegetation patterns have migrated as ice sheets have advanced and retreated. For example, pack rat middens reveal that the vegetation around the Grand Canyon was quite different 9,000 years ago, during the last ice age. Ponderosa pine (*Pinus ponderosa*), a common tree in present-day forests of the region, was

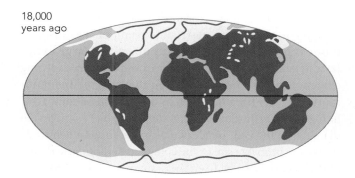

18,000 years ago

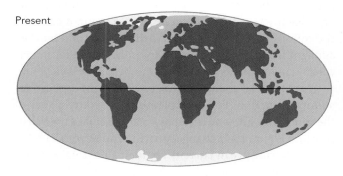

Present

Figure 15.6 The last ice age. A considerable portion of North America and Europe was covered by thick sheets of ice at the peak of the last ice age, 18,000 years ago.

practically absent; it was more abundant farther south. Limber pine (*Pinus flexilis*), which today grows farther north, was the predominant pine species. In the Northern Hemisphere, plant species generally grew 400–700 km (250–440 miles) farther south and 700–900 meters (2,300–3,000 feet) lower in elevation during the last ice age than they do now.

Extinction is a fact of life on Earth

Over the long course of evolutionary history, most species have become extinct. The fossil record indicates that at least five large mass extinctions have taken place since the beginning of the Paleozoic era. One mass extinction 250 million years ago eliminated 90% of all species on Earth. Another occurred 65 million years ago, at the end of the Mesozoic, and resulted in the disappearance of half of all marine species as well as many species of plants and land animals, including the dinosaurs.

In general, extinctions take place when the environment changes more rapidly than populations of organisms can adjust genetically. Such adjustments are based on new phenotypes that arise because of mutations. When the rate of environmental change is greater than the rate of mutation, some populations may not survive.

Extinctions can be caused by drastic environmental changes initiated by a catastrophic event, such as an enormous volcanic eruption or the impact of a large extraterrestrial object with Earth. For example, there is strong evidence that the mass extinction that claimed the dinosaurs may have been triggered by an asteroid or comet colliding with Earth. Chicxulub crater 185 km (11.3 miles) in diameter in the Gulf of Mexico near the Yucatán Peninsula marks the probable site of the collision. Radiometric dating of the material ejected from the crater revealed that the collision, like the mass extinction, occurred 65 million years ago. Scientists suspect that a huge dust cloud created by the collision may have cooled Earth's surface and blocked sunlight for years, reducing photosynthesis and ultimately causing many plants and animals to become extinct.

More gradual changes in the environment can also result in extinction. For example, the slow convergence of the continents into one landmass near the end of the Paleozoic (see Figure 15.4) changed marine and terrestrial habitats by altering ocean currents, reducing the total amount of shoreline, and increasing the area of arid land in the continental interior. These changes may have contributed to the mass extinction that happened around that time. More recently, a variety of other environmental changes brought about by human activities have increased the rate of extinction by a *thousandfold*. In many cases, these extinctions are the result of habitat destruction, as when temperate or tropical forests are cut for lumber or burned to clear ground for agriculture (Figure 15.7). Researchers estimate that habitat destruction by humans causes the extinction of thousands of species every year.

Figure 15.7 Habitat destruction in a tropical rain forest. Humans destroy 200,000 square kilometers (77,200 square miles) of tropical rain forest every year. As the forests disappear, thousands of species of plants and animals are becoming extinct.

Another significant factor in the extinction of species is competition between species for limited resources. We will explore this factor in a later section.

Section Review

1. **List two ways fossils can form.**
2. **How is radiometric dating used to establish the age of fossils?**
3. **What is an organic soup?**
4. **Explain plate tectonics.**
5. **How do ice ages change the geographical ranges of species?**

Mechanisms of Evolution

Look around you. We all look different, primarily because each of us has a different combination of alleles for the genes that control our visible features. A similar variation in alleles occurs among individuals in a population of plants or of any other organisms. While Mendel's laws of inheritance relate to genes in individuals (see Chapter 12), evolution applies to populations. The combined analysis of inheritance and evolution gave rise to **population genetics,** the study of the behavior of genes in populations.

Evolution is a change in the frequency of alleles in a population over time

Population genetics defines evolution as "a change in the frequency of alleles in a population over time." Consider, for example, a population of 1,000 garden pea plants. Each plant has two alleles for height (either *TT, Tt,* or *tt*). Therefore, there are 2,000 height alleles in the population. If 1,000 of the alleles are *T* and 1,000 are *t*, then the frequency of each allele in the population is 0.5. If, from generation to generation, the frequency of *T* increases while that of *t* falls, then according to the definition given above, evolution is occurring. The same would be true if the frequency of *T* decreases while that of *t* increases. On the other hand, if the frequency of both alleles remains the same over successive generations, then the population is not evolving.

In 1908, an English mathematician, G. H. Hardy, and a German physician, G. Weinberg, proposed that the frequency of alleles in a population will remain the same when the following five conditions are met:

1. The size of the population is large.
2. There are no mutations.
3. No migration occurs.
4. Mating is random.
5. Natural selection does not take place.

Under these conditions, the frequency of alleles is in a constant state, or equilibrium, known as *Hardy-Weinberg equilibrium*. On the other hand, if any one of the five conditions is not met, the frequency of alleles will change and evolution will occur. In any natural population, it is in fact extremely rare for all five conditions to be met. Therefore, it is more useful to think of Hardy-Weinberg equilibrium as a theoretical concept that allows us to analyze specific conditions that would *prevent* a population from evolving. Let us now examine each condition separately.

1. *The size of the population is large.* When you toss a coin multiple times, you are much more likely to get a 1:1 ratio of heads to tails if the number of coin tosses (the sample size) is very large. If the sample size is small, however, the ratio you get might deviate from a 1:1 ratio simply due to chance. Similarly, in small populations of organisms, the frequency of alleles can change over generations due to chance, a phenomenon known as **genetic drift.**

There are two situations in which population size can decrease enough for genetic drift to affect the frequency of alleles. The first, called the *bottleneck effect*, occurs when a drought, volcanic eruption, flood, or other disaster drastically and nonselectively reduces the size of a population (Figure 15.8a). If the frequency of alleles in the smaller population does not equal that in the original population, evolution has occurred.

The second situation that permits genetic drift, called the *founder effect*, occurs when a small number of individuals from a large population colonize a new area, such as an island (Figure 15.8b). If the population in the new area has a different frequency of alleles than the parent population, the population in the new area has evolved. The common cocklebur (*Xanthium strumarium*) provides a good example of the founder effect. Cocklebur seeds attach themselves to animal fur and feathers with tenacity. A single seed, carried to a new area by an animal, can establish a population with that seed's particular collection of alleles.

2. *There are no mutations.* As you learned in Chapter 13, mutations alter the nucleotide sequence of genes and can change one allele into another. Under most conditions, the rate of mutation is quite low: Often only one gamete in a million has a mutation in a particular gene. If a mutation does occur in an individual's gametes and is transmitted to its offspring, the frequency of alleles in the population will change. In a large population, however, the change in frequency will be small because the offspring

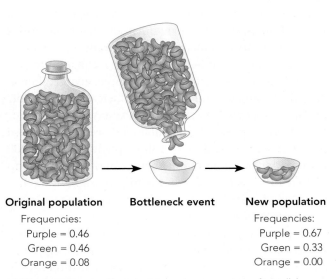

Original population   **Bottleneck event**   **New population**

Frequencies: Frequencies:
Purple = 0.46   Purple = 0.67
Green = 0.46   Green = 0.33
Orange = 0.08   Orange = 0.00

(a) The bottleneck effect. In this analogy, pouring a few jellybeans out of a bottle with a narrow neck is like sharply reducing the size of a population of organisms. The frequency of jellybeans (alleles) in the new population may differ from that in the original population.

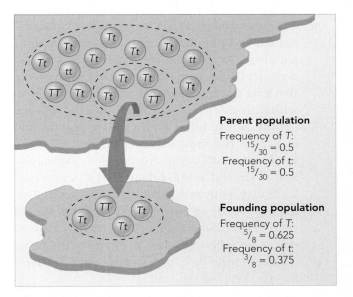

Parent population
Frequency of T:
$^{15}/_{30} = 0.5$
Frequency of t:
$^{15}/_{30} = 0.5$

Founding population
Frequency of T:
$^{5}/_{8} = 0.625$
Frequency of t:
$^{3}/_{8} = 0.375$

(b) The founder effect. When a few individuals from a large population move to a new area, the frequency of alleles in the founding population may differ from that in the parent population.

Figure 15.8 Situations that lead to genetic drift.

make up a tiny fraction of the total population. Therefore, mutations alone do not significantly disrupt a Hardy-Weinberg equilibrium. Nevertheless, it is important to remember that mutations are the original source of all genetic variation in populations.

As you will learn later, if a new allele that arises by mutation gives an individual a selective advantage in a particular environment, the frequency of that allele may increase rapidly due to natural selection. For example, a plant with a mutation that confers tolerance to a toxic metal may have a selective advantage in areas where the soil contains a high concentration of that metal. Mutations do not occur "on demand" because they are useful for a particular population. Instead, mutations occur at random in all populations, and natural selection may increase the frequency of the alleles that result from those mutations.

3. *No migration occurs.* Migration of individuals or gametes into or out of a population can change the frequency of alleles in the population. In plants, migration often occurs when pollen, seeds, or fruits from one population are carried into other populations by wind or animals. Alternatively, various forms of vegetative reproduction, such as the production of runners or horizontal stems, can bring together isolated populations with different allelic frequencies. The movement of alleles from one population to another is known as **gene flow.**

4. *Mating is random.* When mating is random, individuals in a population mate without regard to genotype. In contrast, when mating is nonrandom, individuals mate with other individuals that have a particular genotype, while individuals with different genotypes may be excluded from mating. Self-fertilization (in garden peas, for instance) is the most extreme example of nonrandom mating. Another example is *assortative mating,* in which individuals that share certain traits mate with each other. In plants, assortative mating may be an outcome of pollinator preference. Hawkmoths, for instance, tend to feed on nectar from flowers with long corolla tubes and thus are more likely to transfer pollen between flowers that have this trait.

5. *Natural selection does not take place.* The last condition required for Hardy-Weinberg equilibrium states that no particular combination of alleles is favored, or selected, over any other. This condition is almost never met in natural populations. Natural selection changes the frequency of alleles in a pop-

ulation because individuals with certain genotypes produce more offspring than individuals with other genotypes. Of the various mechanisms that can cause a population to evolve—genetic drift, mutation, gene flow, nonrandom mating, and natural selection—only natural selection can adapt a population to its environment. We will use the remainder of this section to explore natural selection in more detail.

Most organisms have the potential to overproduce offspring

If you have had a lawn overgrown by dandelions, observed an aquarium with a population explosion of guppies, or seen a lake turn bright green with algae in summer, then you have witnessed the ability of organisms to produce numerous offspring. Populations of sexually reproducing organisms are stable if each pair of individuals produce just two offspring, and those offspring survive and produce two offspring. However, most organisms have the ability to produce more offspring than can survive and reproduce. This allows for the occasional population explosion when the environment provides a suitable opportunity. More importantly, when the environment becomes even more challenging than usual, resulting in the death of many individuals, overproduction of offspring may increase the chances that at least a few individuals will survive to continue the population.

In his *Essay on the Principle of Population,* published in 1798, Thomas Malthus made the point that populations of organisms tend to increase geometrically. In a geometric progression, numbers increase by a constant factor. Multiplying by the factor 2, for example, gives 2, then 4, 8, 16, 32, 64, and so on. The numbers increase more rapidly for factors that are larger. For instance, a typical, medium-sized garden tomato might contain 300 seeds. If each seed produced a plant, and each of those plants produced 300 plants, after five generations there would be 2.43 trillion (2.43×10^{12}) plants! Clearly, the maximum reproductive potential of organisms is rarely realized. Other factors act to restrict the sizes of populations.

One reason population sizes are restricted is that resources are limited. Malthus argued that although populations tend to increase geometrically, their food supply does not. Therefore, populations have the potential to far outstrip their food supply. Of course, food is not the only resource needed by an organism. For plants and other photosynthetic organisms, resources include light, mineral nutrients, water, growing space, and, in the case of some flowering plants, the services of pollinators.

Individuals in a population have many phenotypic differences

With the exception of identical siblings, every person is unique. The same is true in most naturally occurring populations of any organism, including plants. Unless we are knowledgeable about the organism in question, we might not notice the differences as readily as we do for people, but they are present nevertheless. For example, the garden peas that Mendel studied varied with respect to height, flower color, seed shape, and other characters. Trees of a particular species often show considerable variation in height, longevity, shape and size of flower parts and seeds, and virtually every other measurable anatomical, physiological, or biochemical character. Darwin was well aware of individual variation within species, although he did not understand its source. Today, we know that many of the phenotypic differences between individuals are due to differences in genotype.

As we noted earlier, mutations are the original source of genetic variation in organisms. Recall from Chapter 13 that a point mutation, consisting of a single nucleotide change in DNA, may result in a single amino acid change in a protein. If that protein is an enzyme, this change in amino acid composition may be enough to change the catalytic activity of the enzyme, which could appear phenotypically as an altered structure or function.

Recombination and crossing over are additional sources of genetic variation in sexually reproducing organisms. *Recombination* refers to the shuffling of alleles that occurs during meiosis. Even if no new mutations have occurred, the independent segregation of homologous chromosomes in anaphase I of meiosis makes it very likely that each gamete carries a unique combination of alleles (see Figure 12.6). For example, in a pea plant with 7 pairs of chromosomes and different alleles for one gene on each chromosome, the chance that two identical gametes will be produced is 1 divided by 2^7, or 1/128. Crossing over during meiosis increases genetic variation still further by producing chromosomes with new combinations of alleles (see Figure 12.10).

Transposons are a fourth source of genetic variation. As discussed in Chapter 14, these sequences of DNA have the ability to move from place to place in chromosomes and can affect the phenotypic expression of the genes they enter.

Some traits confer an adaptive advantage

The Darwinian view of organisms is that they have particular traits because those traits provide superior adaptation in a particular environment. Thus, certain traits are more common among the individuals in a population, while traits that do not confer as great a selective advantage are less common. Let us examine this idea as it relates to leaf shape.

Large, undivided leaves with smooth edges (Figure 15.9a) are generally more common in plants that grow in regions with relatively high rainfall and high temperatures and in plants that live in the shade. Plants of tropical rain forests, for example, typically have such leaves. One survey found that 90% of trees in the Amazon rain forest had undivided leaves. In a rain forest, the lack of wind and low light intensity beneath the canopy may give plants with large, undivided leaves an adaptive advantage. Furthermore, the lower a plant's leaves are in a rain forest, the larger they tend to be and the more likely it is that they have an *acumen,* a long tip that drains the surface of the leaf of water dropping onto it from the vegetation above.

In contrast, highly divided, lobed leaves with less surface area (Figure 15.9b) are more often associated with plants that grow in temperate regions with less rainfall and lower temperatures and in plants that live in direct sunlight. In open deciduous forests in the United States, divided leaves more effectively dissipate wind without becoming damaged, and the greater light intensity makes a large surface area less important.

Plants that live under extreme conditions, such as those in deserts or tundra, tend to have small, thick, undivided leaves. Small leaves are less affected by wind, and thick leaves are better at retaining water.

Figure 15.9 Adaptive advantages of leaf shapes. (a) Undivided leaves with smooth edges are characteristic of plants in areas with high rainfall, such as tropical rain forests. Plants that grow lower in the forest, such as this specimen from a Malaysian rain forest, tend to have leaves that have an acumen, or drip tip. **(b)** Plants with highly divided, lobed leaves, such as this maple, are generally found in drier, temperate regions.

THE INTRIGUING WORLD OF PLANTS

Artificial Selection

For thousands of years, people have saved seeds from the best plants or made cuttings of remarkable plants in hopes of improving the quality and quantity of flowers, fruits, and vegetables. The process of selectively breeding plants or animals to favor the production of offspring with desirable traits is called *artificial selection*. On occasion, artificial selection has resulted in new types of plants. A good example is the diverse array of vegetables that have a common ancestor in wild mustard. Through artificial selection, plant breeders have exploited different aspects of the original plant to produce vegetables that look and taste very different. The same forces act in artificial selection as in natural selection, but the selective agent is the breeder rather than the constraints imposed by the environment.

Today, professional plant breeders practice artificial selection on a large and more sophisticated scale. They look for increased yields, disease resistance, cold tolerance, improved nutritional quality, and other useful traits. When they find such a trait, it may be in a cultivated crop or in a wild, noncultivated relative. Plant breeders then cross plants that have the useful trait with cultivated plants that lack the trait. The goal is to introduce the allele coding for the desired trait into the genome of the cultivated plants. For example, wheat breeders for many years tried to produce dwarf or semidwarf plants resistant to wind damage. They also wanted plants that put most of their photosynthetic products into grain and that produced no more leaves than were necessary to sustain high grain yields. Such plants were finally produced and have increased wheat production in wind-prone areas.

As you learned in Chapter 14, plant genetic engineers engage in the same sort of selection activities to obtain improved plants. Since genes can be moved from one species to another and still function, scientists can look for a desirable trait in virtually any other plant—or even organisms other than plants—and try to introduce it into a crop they want to improve. In nature, gene flow occurs between members of the same species and occasionally between related species. With genetic engineering, gene flow can occur between kingdoms.

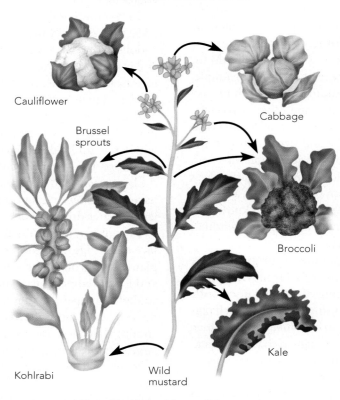

Vegetables derived from wild mustard.

Natural selection favors individuals with the best-adapted phenotypes

What accounts for the varied shapes of leaves? Certainly, leaf shape is genetically controlled, coded for by developmental genes that regulate the location and extent of cell division and growth. According to the Darwinian view, the shapes of leaves, like other traits, are subject to natural selection. Natural selection means that individuals that are best adapted to their environment have the greatest chance of surviving and successfully reproducing. Individuals that make a greater contribution to the next generation are said to have greater *fitness* (see *The Intriguing World of Plants* box on this page).

Natural selection occurs both within a species and between species. Within a species, the best-adapted individuals are more likely to reproduce, and their particular set of alleles will increase in frequency in the population. Between species, a similar phenomenon occurs, and the best-adapted species will increase its population size.

Some early evolutionists saw natural selection as a bloody battle between mighty carnivores over the carcasses of their prey. While this sort of thing does occur, competition can involve resources of many sorts and is more often indirect than direct. For instance, two plants growing in the same location may compete for light, minerals, and water.

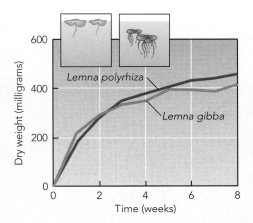

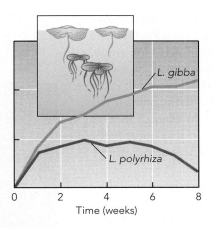

Figure 15.10 Competition between two species of duckweed. (a) When each species is grown separately, *Lemna polyrhiza* grows faster than *Lemna gibba*. (b) When the two species are grown together, *L. gibba* floats on top and receives more light, so it grows faster than *L. polyrhiza*.

As a specific example of natural selection and adaptation involving competition between species, consider an experiment that was carried out by Dr. J. Clatworthy and Dr. John Harper at the University of Oxford with two species of duckweed, *Lemna polyrhiza* and *L. gibba* (Figure 15.10). Both are tiny flowering plants that grow in quiet freshwater ponds and lakes. When grown separately, *L. polyrhiza* grows faster than *L. gibba*. However, when the two species are grown together, *L. gibba* grows faster than *L. polyrhiza*. That is because *L. gibba* has air-filled sacs that enable it to float at the surface where it maximizes its photosynthesis while shading *L. polyrhiza*. In this case, the plants are competing for a resource (light) but they are also competing for surface area where they can be exposed to the light.

As the example just described shows, competition can lead to selection of the best-adapted phenotype when re-sources are limiting. However, competition for resources is not necessary for natural selection to occur. Natural selection can take place in the absence of limiting resources if certain individuals or species simply produce more offspring than others. In this case, natural selection favors phenotypes that have the greatest reproductive ability.

Natural selection can change the frequency of phenotypes in a population in three ways. Figure 15.11 illustrates these ways using the example of plant height as a character for which multiple phenotypes are possible.

◆ **Stabilizing selection** reduces variation in a population by eliminating individuals that have extreme phenotypes—very tall and very short plants, in our example. As a result, the greatest frequency of individuals is at the middle of the phenotypic range.

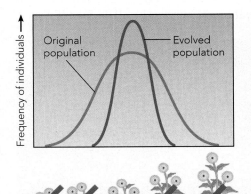

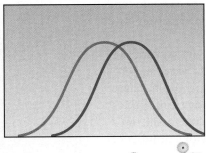

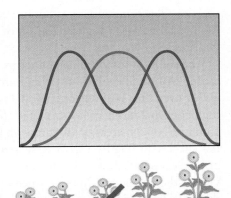

(a) In **stabilizing selection,** individual's with extreme phenotypes (tall plants and short plants) are eliminated.

(b) In **directional selection,** individual's with one extreme phenotype (in this case, short plants) are eliminated.

(c) In **diversifying selection,** individual's with intermediate phenotypes (plants of medium height) are eliminated.

Figure 15.11 Three modes of natural selection. Natural selection can change the frequency of phenotypes in a population by eliminating certain phenotypes and favoring others. These graphs show the effects of natural selection on an original population of plants of varying heights.

◆ **Directional selection** shifts the average or typical phenotype in one direction by favoring individuals that have one extreme phenotype. Directional selection would occur, for example, if tall plants were favored over short plants. If it continues, directional selection can eventually cause a population to evolve to the point where it becomes different enough from its original form to be considered a new species. The transformation of one species into another is known as **anagenesis,** or phyletic evolution.

◆ **Diversifying selection** splits a population into two parts by favoring individuals at both extremes of the phenotypic range—such as tall plants and short plants. It decreases the frequency of individuals with intermediate phenotypes. If diversifying selection continues, it can cause one species to evolve into two species, a process called **cladogenesis,** or branching evolution. We will delve further into the origin of species later in this chapter.

Evolution can occur rapidly

Darwin believed that noticeable evolutionary change occurs over long periods of geological time. As evolution continues, species continue to separate and eventually move into different genera or other taxonomic categories. In 1972, Niles Eldredge and Stephen Jay Gould proposed that, under some circumstances, evolution can be very rapid. In their model of **punctuated equilibrium,** long periods of little or no evolutionary change are punctuated with short periods of rapid change. The rapid sort of evolution proposed by Eldredge and Gould can take a few years or a few thousand years, but not the millions of years required for the gradual separation of a population into two species.

The colonization of mine tailings by plants provides an example of rapid evolution. Tailings are the waste materials left over after a mining operation. Frequently, they have high concentrations of heavy metals. Many plant species common to such areas do not grow on tailings at all, but a few species do (Figure 15.12). Research with a number of grass species has shown that the only species that can survive on the tailings are those in which mutations have produced individuals that are tolerant to heavy metals. The tolerant plants rapidly proliferate, whereas intolerant plants do not mature. Because the selection level is high, sizeable populations of tolerant plants develop in a few generations, and the frequency of alleles for tolerance increases rapidly. On normal soils, the tolerant plants survive but do not grow nearly as well as intolerant plants.

Figure 15.12 Evolution of grasses on mine tailings. The bent grass (*Agrostis tenuis*) in the foreground of this photo is growing on the tailings of a lead mine in Wales. Unlike many species, bent grass has alleles that give some individuals the ability to tolerate the heavy metals in the tailings.

Another example of rapid evolution can be seen when a pasture or lawn is subjected to mowing or grazing. *Prunella vulgaris* is a common herb in the mint family that can grow up to 30 centimeters tall in undisturbed pastures, but it also can produce a short, compact form that survives in lawns and well-grazed pastures (Figure 15.13). The compactness trait is sometimes passed on to offspring and sometimes not. Apparently, the allele for compactness is normally present in the population at low frequencies. This allele seems to have adaptive value only when the allele for tallness is

Figure 15.13 Evolution in response to grazing. The herb *Prunella vulgaris* exists in a tall form and a compact form. Only the compact form is found in grazed pastures and lawns.

EVOLUTION

Plants of the Galápagos Islands

The Galápagos finches, studied by Darwin, are usually cited as an example of how the microhabitats in an archipelago of small islands can cause the adaptive radiation and phenotypic diversification of a single species. These birds evolved many different beak structures depending on the diet they adopted in each microhabitat.

The plants of the Galápagos provide equally interesting but less well-known examples. Consider, for example, the flowering plants in the genus *Scalesia,* which is native to the Galápagos. These plants probably colonized the Galápagos soon after the islands began to form around 3 to 5 million years ago. The populations from which these colonists arose most likely were located on the mainland of western South America.

S. pedunculata

S. gordilloi

S. helleri

S. stewartii

The first species of *Scalesia* to reach the Galápagos Islands probably no longer exists but, it is thought to have lived on San Cristóbal, the oldest island. Plants in the genus *Scalesia* have evolved into a number of different species in the many microenvironments found on the Galápagos Islands. These photos, provided by Dr. Conley McMullen at James Madison University, show that *S. pedunculata* is a tall forest tree, with undivided leaves, that grows in moist uplands. *S. gordilloi* is a creeping, herbaceous form, with undivided leaves, that grows near the ocean but on dry soil. *S. helleri,* a small shrub found in arid lowlands away from the ocean, has finely divided leaves. *S. stewartii* is a shrub or small tree, with undivided leaves, that also grows in the arid lowlands.

Scalesia fruits do not have a hairy pappus (a structure that aids wind dispersal in some fruits), so it is likely that they were carried to the islands by a bird. In fact, *Scalesia* fruits are often coated with a sticky resin produced by the plant, which helps them adhere to the feathers of birds.

In the Galápagos, *Scalesia* evolved into a group of at least 15 species and numerous varieties, ranging from shrubs to trees, which occupy a number of distinct and diverse environments. The leaf shapes are indicative of the phenotypic diversity that has evolved. In general, *Scalesia* species with deeply lobed or highly divided leaves are restricted to dry, low areas, while species with oblong or elliptical leaves are found at higher elevations with more rainfall.

---

under intense selective pressure, as happens when a lawn mower or grazing animal cuts down tall plants but leaves short plants unscathed. Under such pressure, the population rapidly evolves into one consisting only of compact plants.

Evolution can also be rapid during **adaptive radiations,** which occur when a species moves into a previously unoccupied environment, such as an island, or an occupied environment that still has many opportunities for the species to succeed. Often, several new species will rapidly evolve from the original colonizing species. For example, when green algae began to produce forms that could survive out of water around 500 million years ago, they evolved into numerous types of primitive land plants. Evolution was rapid because the land had no existing plants, so a huge variety of environmental opportunities were available. The *Evolution* box on this page describes the adaptive radiation of one genus of plants in the Galápagos.

In coevolution, two species evolve in response to each other

It is possible to mow a lawn and not kill the grass because the meristem of a grass plant, at the bottom of the leaves, is not damaged by the mower's cutting blade. From an evolutionary standpoint, grazing animals allowed us to have lawns: Through the act of grazing, they have selected for plants that have short stems with meristems below the grazers' reach. Grasses and grazing animals have *coevolved.* Grass stems became shorter as browsing animals, which ate all sorts of vegetation, evolved into grazing animals that concentrated on grasses. The coevolution continues, as some grasses produce distasteful compounds that discourage grazing, while some grazing animals use their hooves to root out tasty grass meristems and roots.

Coevolution can also have a cooperative side. Many flowering plants came to depend on insects, birds, and bats to carry pollen from one plant to another for cross-pollination, which increases valuable phenotypic variation. Flowering plants evolved several types of innovations that attract effective pollinators. For example, many flowers produce a sugary nectar that serves as food for animals such as bees, which make it into honey. The shapes of many flowers were modified by evolution to fit particular pollinators, which had their shapes modified, in turn, to fit particular flowers. The photograph at the beginning of this chapter (page 323) shows one example of this sort of coevolution. Both the plant (the orchid) and the animal (the hummingbird) benefit. The plant gains an exclusive pollinator, which will search for the plant. The animal gains a food source it does not have to share with competitors of other species.

Flowering plants have also coevolved with animals with regard to seed dispersal. Animals that eat fruits and seeds serve as effective seed dispersal agents, provided the seeds survive. Fruits often taste quite sour until their seeds are mature, whereupon the fruit ripens and becomes sweet. Many fruits, such as grapes, have seeds that can pass through an animal's digestive system unharmed. Some, such as plums, contain compounds that accelerate passage through the digestive system. The seeds of a few plants actually require an acidic environment, such as occurs in an animal's stomach, to weaken the seed coat enough that germination can occur. Tomatoes and some other fruits have sticky seeds that cling to an animal's fur or feathers, aiding dispersal.

Section Review

1. What five conditions must be met for a population to be in Hardy-Weinberg equilibrium?
2. Define *natural selection*.
3. What are four sources of genetic variation?
4. Give some examples of resources that might be limiting and therefore cause competition between plants.

The Origin of Species

When a population of interbreeding organisms separates into two populations, each population may be subjected to different forces of natural selection and may evolve differently. Eventually, the two populations may become dissimilar enough to be considered different species.

A biological species is a population of potentially interbreeding organisms

Most botanists today define a *biological* species as a natural population of interbreeding or *potentially* interbreeding organisms that cannot breed with other populations. Such populations are said to be reproductively isolated, and gene flow between them does not occur. In short, by this definition a species is an independent evolutionary unit.

Everyone knows that a coconut palm is different from a giant sequoia. The two plants look quite different and do not interbreed, so they are indeed separate species according to the definition given above. With more closely related plants, the term *species* is more difficult to define. For example, what about different types of pines? Pines belong to the genus *Pinus,* which contains more than 90 species. The bristlecone pine (*Pinus longaeva*) and the longleaf pine (*P. palustris*) are separate *morphological* species because they differ in appearance. They are also separate *allopatric* or *geographical* species because they have different geographical ranges (Figure 15.14). They do not interbreed in nature due to their nonoverlapping ranges, but what if they were grown at the same location?

We really cannot know if two populations of plants are reproductively isolated without moving them to a common location for testing. If we do that, we may find that they do not interbreed. That result would indicate that we are dealing with two *biological* species instead of one. On the other hand, we may find the opposite result. For instance, *Platanus occidentalis,* native to North America, and *Platanus orientalis,* native to Asia, are two species of sycamore trees that do not interbreed in nature. However, when *P. occidentalis* and *P. orientalis* are cultivated together, the two plants interbreed quite readily, producing a fertile hybrid (*Platanus × hybrida*) called the London plane tree. In this case, their reproductive isolation is simply geographical and does not have a biochemical or genetic basis. According to the definition given above, *P. occidentalis* and *P. orientalis* would be considered one *biological* species if they grew in the same area.

Reproductive isolation is a matter of degree rather than an absolute state. Population geneticists express the degree of reproductive isolation in terms of gene flow. Gene flow does not occur at all between populations that are completely reproductively isolated—palms and sequoias, for example. In contrast, populations that interbreed only occasionally are partially reproductively isolated, and there is a small amount of gene flow between them. For instance, some interbreeding occurs between two sunflower species, *Helianthus annuus* and *H. petiolaris,* found in the western United States and can result in a genetically stable

Figure 15.14 Two geographically isolated species of North American pines.
(**a**) The bristlecone pine (*Pinus longaeva*) grows in the west, and (**b**) the longleaf pine (*Pinus palustris*) in the east. Separated by distance, the pines have evolved into different species.

hybrid, *Helianthus anomalus.* In some areas the three species are partially separated by elevation: In Utah, for example, *H. annuus* is found from 1,060 to 2,280 m (3,478–7,480 ft), *H. petiolaris* from 970 to 1,670 m (3,182–5,479 ft), and *H. anomalus* from 1,060 to 1,940 m (3,478–6,365 ft).

Even species that exist in the same area may be partially reproductively isolated because they occupy different microhabitats, or subsets of an environment. Variations in rainfall, temperature, soil conditions, and other environmental factors create a wide range of microhabitats. An example of microhabitat isolation can be seen in four

species of maples found in the eastern United States: red maple (*Acer rubrum*), silver maple (*A. saccharinum*), black maple (*A. nigrum*), and sugar maple (*A. saccharum*). These species have overlapping ranges (Figure 15.15), but each species is typically found in a different microhabitat. Red maple is adapted to either wet and swampy or dry and poorly developed soils. Silver maple thrives in the moist, well-drained soils of river basins. Black maple and sugar maple are frequently found together, but black maple tends to grow in drier, better-drained soils with high calcium levels, whereas sugar maple grows best in somewhat

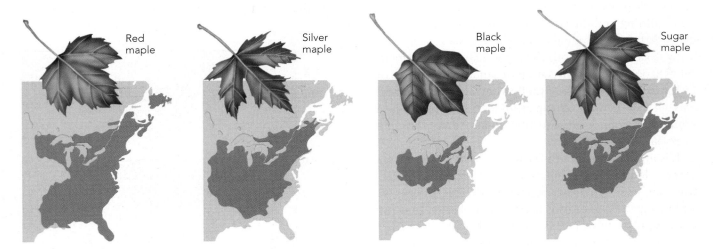

Figure 15.15 Microhabitat isolation in four species of maples. Red maple (*Acer rubrum*), silver maple (*A. saccharinum*), black maple (*A. nigrum*), and sugar maple (*A. saccharum*) have overlapping ranges. However, the species occupy different microhabitats within their ranges.

more acidic soils. Although the four species can interbreed, their microhabitat differences tend to keep each species reproductively isolated from the others.

Both natural selection and geographical isolation drive speciation

The Galápagos Archipelago consists of 13 major and many minor volcanic islands that lie about 950 km (590 miles) west of Ecuador in the Pacific Ocean. The islands originally had no life. Organisms migrated to each island from the mainland or from older islands. Darwin visited the Galápagos in 1835 and noticed that the species of animals and plants differed phenotypically from island to island, even though the islands were close together. He also noted, as Wallace would do later in Malaysia and Indonesia, that environmental conditions differed from island to island. For example, some islands were exposed to the wind and rain of arriving weather fronts, while others were not. Some islands were more exposed to wave action and salt spray than others. Some were dominated by high volcanic cones, while others lacked highlands.

Because different phenotypes are favored in different environments, Darwin believed that phenotypic variation acted upon natural selection to give rise to the differences he observed in populations of the same species on different islands. In short, Darwin proposed that environmental variation in an area provides the natural selection necessary to reproductively isolate populations, leading eventually to speciation, the formation of new species.

The dominant view of speciation in the twentieth century was articulated by Ernst Mayr of Harvard University in 1942. He proposed that geological events, such as mountain building and the widening of oceans as continents drift apart, separate populations of one species. Geographically isolated populations then gradually diverge genetically through the accumulation of random mutations or by natural selection. Eventually, they become different enough that they can remain reproductively isolated even if human actions or subsequent geological events brought them back together.

It is now clear that both Darwin's and Mayr's views are correct. Reproductive isolation and subsequent speciation can be driven either by environmental differences within a given area or by geographical isolation. In either case, populations or parts of a population must be separated in some sense. The separation may be distinct on a map, in which case the populations are said to be **allopatric.** Alternatively, the populations may be **sympatric** and have overlapping ranges. As we noted earlier, sympatric populations may, in fact, be effectively isolated if they have different microhabitat preferences.

Mountain ranges can serve as both a geographical and an environmental cause of speciation for two reasons. First, the opposite sides of mountain ranges often receive different amounts of precipitation. In Oregon and Washington, for example, air from the Pacific Ocean loses moisture as rain and snow that falls on the western slopes of the Cascades. The eastern slopes, in the so-called *rain shadow,* receive substantially less precipitation. Plants on the wetter western slopes often have quite different phenotypes from those on the drier eastern slopes. Second, at different elevations in the mountains there are differences in the length and severity of seasons, the amounts of snow and rain, and the frequency and severity of wind. Plants growing at different elevations often have different phenotypes. In the example illustrated in Figure 15.16, yarrow plants (*Achillea*) growing in the Sierra Nevada in California are generally smaller at higher altitudes. The change in the height of the plants with altitude is an example of a **cline,** a variation in phenotype that occurs along with some variation in the environment. When yarrow plants collected from a variety of elevations were grown in a common garden, they retained their size differences, suggesting that size is genetically controlled. Both of these examples show that mountains create different environments, which drive selection in different ways.

Reproductive isolation can be prezygotic or postzygotic

Different types and degrees of reproductive isolation can occur between two populations. *Prezygotic isolation* means that sperm from one population does not fertilize eggs from the other population, so no zygote is produced. In

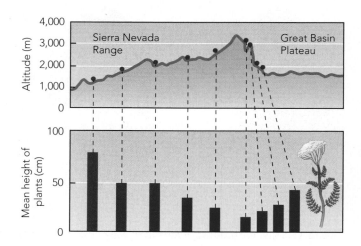

Figure 15.16 Phenotypic variation related to elevation. Yarrow (*Achillea*) grows at different elevations in the Sierra Nevada. Plants at lower elevations are generally taller than those at higher elevations.

plants, prezygotic isolation may be due to a failure to pollinate. Although pollen can be carried long distances by wind or animals, the farther apart two populations are, the less likely it is that living pollen will travel from one to the other. Even within a small area, pollination may fail to occur because one population opens its flowers in the morning while the other opens its in the late afternoon. Alternatively, one population may flower in June while the other waits until August, or one may be pollinated by insects while the other relies on birds. The four species of maple trees discussed above are prezygotically isolated because they occupy different habitats within their overlapping geographical ranges.

Even when pollination does take place, two populations of plants can be prezygotically isolated if the pollen does not grow successfully and accomplish fertilization. Pollen is deposited on the stigma of the flower. Here it must germinate, and the pollen tube must grow through the style to the ovary where the egg is located. Pollen grains germinate in response to specific concentrations of hormones, ions such as calcium, sugars, and other compounds. If the stigma does not provide the correct biochemical environment, pollen germination will fail. Various pollen-incompatibility genes control the biochemical environment of the stigma (see Chapter 12). Matching of the stigma's biochemical environment with the germination needs of the pollen makes it less likely that pollen of other species will germinate. That is important because pollen of many species may be deposited on a stigma, and the style would be filled with pollen tubes if every pollen grain germinated.

Postzygotic isolation means that fertilization occurs, resulting in the formation of a zygote, but the zygote or embryo does not survive or the adult plant is infertile. The embryo may die because the endosperm, a storage tissue that nourishes the growing seed, fails to develop. Postzygotic isolation in adult plants is most often due to problems with tetrad formation in meiosis (see Chapter 6). When homologous chromosomes do not pair correctly to form tetrads, then the daughter chromosomes are distributed unevenly, resulting in sperm and eggs without a full chromosomal complement. Hence, such adults are usually sterile. In Chapter 14, you read about an example of hybrid sterility in triticale, the product of a wide cross between wheat and rye.

Reproductive isolation in sympatric populations can occur because of polyploidy

Recall from Chapter 13 that *polyploidy* refers to cells that contain more than two complete sets of chromosomes. Polyploidy is common in the plant kingdom. More than 50% of flowering plants are polyploid, including 80% of species in the grass family. In nature, polyploidy often results from breeding between two different species, in which case it is known as **allopolyploidy.** As we noted in the preceding paragraph, interspecific hybrids are normally sterile because their chromosomes cannot pair during meiosis. In some hybrids, however, a spontaneous doubling of chromosomes occurs after fertilization. The resulting plants are now fertile because meiotic pairing can occur.

One example of an allopolyploidy is found in salt-marsh grasses of the genus *Spartina*. In the 1800s, interbreeding between *S. maritima* ($2n = 60$) and *S. alterniflora* ($2n = 62$) produced a sterile hybrid ($2n = 62$). By about 1890, spontaneous chromosome doubling had occurred in the hybrid, resulting in a new species, *S. anglica* ($2n = 122$; apparently, one pair of chromosomes was lost in the process). Also known as common cord-grass, *S. anglica* is fertile and has made itself a very successful pest in salt marshes along the coasts of England and northwestern France (Figure 15.17). Because it is polyploid, it is reproductively isolated from both of the parent species.

Figure 15.17 Common cord-grass (*Spartina anglica*), an allopolyploid.

Interspecific and even intergeneric hybrids can form large populations without chromosome doubling if they reproduce asexually. For example, Kentucky bluegrass (*Poa pratensis*) hybridizes freely with other grasses, which has resulted in a wide variety of hybrids. The hybrids reproduce by apomixis, in which nonsexually produced diploid cells give rise to seeds (see Chapter 6).

Section Review

1. Define *biological species.*
2. Distinguish between allopatric and sympatric populations.
3. How can a doubling of chromosome number result in speciation?
4. Distinguish between prezygotic and postzygotic isolation, and list the ways that each can occur.

SUMMARY

History of Evolution on Earth

Fossils and molecular dating provide evidence of evolution (pp. 325–327)

Fossils, such as compressions and permineralizations, provide information about how forms of life have changed over time. Radiometric dating, which measures the decay of isotopes in fossils or in surrounding rock, can accurately determine the age of fossils. Molecular dating compares the structure of DNA, RNA, and proteins in different organisms to estimate how long it has been since they had a common ancestor.

Biogeography, anatomy, embryology, and physiology supply further evidence of evolution (p. 327)

Biogeography, the study of where particular species of organisms occur, reveals how the distributions of organisms have changed over time. Comparative anatomy, embryology, and physiology provide evidence of the relatedness of different groups of organisms.

Chemosynthesis may have been the first event in the origin of life on Earth (pp. 327–328)

Organic compounds can form spontaneously from inorganic precursors similar to those thought to have been present on Earth before life appeared. Some of these compounds can react to form polymers, some of which can aggregate spontaneously to form cell-like protobionts.

Prokaryotes were the predominant form of life for more than a billion years (pp. 328–329)

The oldest fossils were formed 3.5 billion years ago by primitive prokaryotes. Photosynthesis began in prokaryotes and caused a rise in the concentration of oxygen in the atmosphere about 2.7 billion years ago. The first photosynthetic eukaryotes appeared at least 2.1–2.2 billion years ago.

Plate tectonics and celestial cycles have shaped evolution on Earth (pp. 329–331)

The shifting of tectonic plates is responsible for continental drift, which has influenced the distribution of plants and animals over time. Cyclic changes in Earth's orientation and distance from the Sun affect the seasons and cause warm periods to alternate with cold periods, including ice ages.

Extinction is a fact of life on Earth (pp. 331–332)

Most species that have existed on Earth have become extinct. Extinctions can be caused by catastrophic events, gradual changes in the environment, and competition between species.

Mechanisms of Evolution

Evolution is a change in the frequency of alleles in a population over time (pp. 332–334)

A population will be in Hardy-Weinberg equilibrium if five conditions are met: The size of the population is large, there are no mutations, individuals do not migrate into or out of the population, mating is random, and natural selection does not occur. If any of these conditions is not met, the frequency of alleles in the population will change, and the population will evolve.

Most organisms have the potential to overproduce offspring (p. 334)

Most organisms have the ability to produce more offspring than can survive and reproduce. However, population sizes are restricted, in part because resources are limited.

Individuals in a population have many phenotypic differences (p. 335)

The individuals in most naturally occurring populations are phenotypically different, largely due to differences in genotype. Genotypic differences can arise by mutation, recombination, crossing over, and the effects of transposons.

Some traits confer an adaptive advantage (p. 335)

Particular traits are more common in a population because they provide superior adaptation. For example, large, undivided leaves are adaptive for plants in some environments, while highly divided, lobed leaves are more adaptive in other environments.

Natural selection favors individuals with the best-adapted phenotypes (pp. 336–338)

Individuals that are best adapted to their environment have the greatest chance of surviving and reproducing. When resources are limiting, competition leads to selection of the best-adapted phenotype. Natural selection can be stabilizing, directional, or diversifying based on its effect on the frequency of phenotypes in a population.

Evolution can occur rapidly (pp. 338–339)

According to the model of punctuated equilibrium, short periods of rapid evolutionary change interrupt long periods with little or no change. Evolution can be rapid during the adaptive radiations of species in new environments.

In coevolution, two species evolve in response to each other (pp. 339–340)

The mutual evolution of grasses and grazing animals and of flowering plants and pollinators are examples of coevolution.

The Origin of Species

A biological species is a population of potentially interbreeding organisms (pp. 340–342)

Individuals in a biological species are potentially capable of breeding with others of the same species but are reproductively isolated from other species.

Both natural selection and geographical isolation drive speciation (p. 342)

Reproductive isolation and speciation can be driven by environmental differences within an area or by geographical isolation. Mountain ranges can serve as both causes of speciation.

Reproductive isolation can be prezygotic or postzygotic (pp. 342–343)

In prezygotic isolation, either pollination does not take place, or the pollen does not grow properly and lead to fertilization. In postzygotic isolation, fertilization occurs, but the zygote or embryo does not survive or the adult plant is infertile.

Reproductive isolation in sympatric populations can occur because of polyploidy (pp. 343–344)

Allopolyploidy results when a spontaneous doubling of chromosomes occurs in an interspecific hybrid. The resulting individuals are fertile.

Review Questions

1. Define *evolution* in general terms and in terms of population genetics.
2. What is the difference between compressions and permineralizations?

3. How did the atmosphere of primitive Earth compare with Earth's atmosphere today?
4. What are stromatolites, and why are they important?
5. Why do the continents drift?
6. What causes the seasons?
7. What will happen to a population if any of the conditions for Hardy-Weinberg equilibrium is *not* met?
8. Of the four sources of genetic variation, which is the ultimate source?
9. Who was Thomas Malthus?
10. How do leaves of plants in a tropical rain forest generally differ in shape from leaves of plants in temperate regions?
11. Define *anagenesis*.
12. What is punctuated equilibrium?
13. Define the term *adaptive radiation*.
14. How do plants and insects coevolve?
15. How could you tell if similar maple trees in California and Maine were really separate species?
16. According to Ernst Mayr, what causes speciation?
17. What is allopolyploidy?

Questions for Thought and Discussion

1. Suppose someone produces a protobiont that takes up molecules from its environment, makes them into membranes, and divides when it becomes large enough to collapse on itself. Would it be correct to say that this protobiont is alive?
2. How could you simulate an adaptive radiation of plants in a laboratory or field experiment?
3. Support or refute the statement that competition between plants for resources is "kinder" than competition between animals.
4. Set up an experiment to test Darwin's theory with plants in a lab. How would you impose natural selection on the plants?
5. Some philosophers have claimed that Darwin's theory is a definition and not a theory. For example, the statement, "A bachelor is an unmarried man," is not a theory. In the case of natural selection, the definition would be "The fittest species are those that survive." Is survival of the fittest a definition or a testable theory? Support your position.
6. In the desert, you observe that sagebrush plants are quite evenly spaced. What might cause this spacing, and how might it be an adaptive advantage to the plant?
7. You survey a population of plants and find the following flower–color frequencies: 10% pale yellow, 20% medium yellow, 40% dark yellow, 20% pale orange, and 10% orange. Draw a graph of these frequencies. Next, draw a graph depicting flower–color frequencies following the introduction into the population of a herbivore that feeds most heavily on the plants with dark

yellow flowers, less heavily on those with medium yellow flowers, and not at all on the other three classes.

Evolution Connection

The Mesozoic era was the age of dinosaurs and of plants called cycads (see Chapter 22), which had tough, sharp leaves, as do their descendants today. Cycads became gradually less abundant during the later Mesozoic era. During the early Mesozoic, cycads coexisted with softer-leaved ferns; the dinosaurs characteristic of this period had relatively soft, non-grinding teeth. What effect do you think these dinosaurs had on the relative abundance of cycads and ferns during the early Mesozoic? Could this have led to coevolution between the dinosaur and plant populations? Explain.

To Learn More

Visit The Botany Place Website at www.thebotanyplace.com for quizzes, exercises, and Web links to new and interesting information related to this chapter.

Darwin, Charles. *On the Origin of Species.* Indypublish.com, 2002. Darwin's book is detailed and interesting.

Gould, Stephen J. *The Structure of Evolutionary Theory.* Cambridge: Harvard University Press, 2002. While a difficult and complex read, this book is the most complete exposition of Darwin's theory since Darwin's book itself.

Gould, Stephen J. *The Hedgehog, the Fox, and the Magister's Pox: Ending the False War Between Science and the Humanities.* New York: Harmony Books, 2003. Gould's books are great reading and contain lots of information about science, viewed from an evolutionary perspective.

Graur, Dan, and Wen-Hsiung Li. *Fundamentals of Molecular Evolution.* Sunderland, MA: Sinauer Associates, 2000. This book describes all facets of molecular evolution using numerous examples and both mathematical and intuitive explanations.

Stone, Irving, and Jean Stone. *The Origin: A Biographical Novel of Charles Darwin.* New York: Doubleday, 1980. This excellent and exciting historical novel is out of print but is available in used bookstores and libraries.

Zimmer, Carl. *Evolution: The Triumph of an Idea.* New York: Harper Collins, 2001. This well-written and nicely illustrated book is the companion to an epic PBS series detailing all aspects of evolutionary theory.

16
Classification

Members of the genus *Capsicum,* or peppers.

Classification Before Darwin

Classification of organisms dates back to ancient times

Linnaeus laid the foundation for modern naming of species

Classification and Evolution

Systematists use a variety of characters to classify organisms

Molecular data play a key role in phylogenetic classification

Organisms are classified into a hierarchy

Systematists form hypotheses about evolutionary relationships

Cladograms are branching diagrams that show evolutionary relationships

Systematists often disagree about how to classify organisms

Major Groups of Organisms

Systematists have revised the number of kingdoms

Molecular data have led to identifying "super kingdoms" called domains

The domain Archaea and the domain Bacteria are two very different groups of prokaryotes

The domain Eukarya includes protists, animals, fungi, and plants

The Future of Classification

New species remain to be discovered

Systematists are studying speciation in action

Molecular data will continue to provide insights into evolution

The classification of organisms has practical benefits

Take a look at the four plants on this page. You would want to avoid the plant known by the common names jimsonweed, thorn apple, and devil's apple. It contains several very toxic alkaloid chemicals: atropine, scopolamine, and hyoscyamine. All of its parts are poisonous, particularly the seeds. Hundreds of cases of poisoning are reported each year in the United States alone, most involving attempted use as a recreational drug. Symptoms include hot skin, incoherent speech, lack of muscle coordination, and rapid heartbeat. High doses can cause seizures and heart attacks.

Every part of the common potato is also poisonous, except the swollen ends of its underground stems. The potato is the source of one of the world's most important foods. The plant contains a number of alkaloids, including a poisonous alkaloid called solanine, found in potato leaves and the skin of green potatoes. While not nearly as toxic as the alkaloids in jimsonweed, solanine can still make you sick.

Jimsonweed

The leaves and stems of tomatoes are also poisonous, containing the alkaloids tomatine and nicotine. However, the fruit is one we see almost daily.

Think twice, though, before biting into the fruit of the habanero, for it will bite back. The habanero is one of the world's hottest peppers—much hotter than the jalapeño and favored by fans of hot salsa. Peppers contain hot-tasting alkaloids called capsaicins.

Common potato

Clearly there are important differences between the jimsonweed, potato, tomato, and habanero pepper plants—at least in their uses and effects on humans. Therefore, you may be surprised at first to learn that botanists consider these four flowering plants—as well as tobacco, eggplant, and petunias, among others—to be related. These plants are therefore classified as members of the same family, known by the scientific name Solanaceae. Members of the Solanaceae family have similar leaf, flower, and fruit structures, and they contain particular toxic chemicals. In fact, when the tomato plant was first brought to Europe

from South America in the 1500s, many thought its fruit was poisonous. After all, the plant's flowers were similar in structure to those of the deadly nightshade, a poisonous plant familiar to Europeans. As you may have guessed, deadly nightshade is a common name for another member of the Solanaceae family.

Tomato

Rather than distinguishing plants and other organisms based on their human uses, scientists classify them into groups according to their characters, inherited traits that are observable or measurable. A scientific system of naming and classifying organisms is important for two reasons. Giving an organism one scientific name worldwide eliminates the confusion that results from the same organism having a different common name—or different organisms having the same common name—in different locations. For example, three different species of small trees—the North American *Cercis canadensis,* the Chinese *Cercis chinensis,* and the Mediterranean *Cercis siliquastrum*—are all known by the common name redbud. The Mediterranean species is also called the Judas tree, named after the betrayer of Christ, Judas Iscariot, who is said to have hanged himself from such a tree. In addition to avoiding such confusion over common names, a classification system reflects scientists' hypotheses about evolutionary relationships between organisms. The plants in the Solanaceae family, for instance, are all considered to be evolutionarily related to each other.

The modern scientific study of the evolutionary relationships between organisms is called systematics. This study includes taxonomy, the naming and classifying of species (from the Greek *taktos,* ordered, and *onoma,* name). The idea that species evolve, however, is relatively new. For more than two thousand years, people assumed that organisms did not change. As you explore the science of classification, notice how it has changed over the years and continues to be marked by new discoveries.

Habanero

Classification Before Darwin

Humans excel at observing and making use of their environment. Throughout history, people have often distinguished between various plants and animals based on practical concerns: Is this animal dangerous or not? Is this plant poisonous, or is it edible? What plants can be used as medicines for this illness? Questions like these have led to knowledge of the physical characteristics of various organisms, gradually contributing to the development of a scientific system of classification. First we will look at methods of classification before Darwin and the theory of evolution.

Classification of organisms dates back to ancient times

Early efforts to distinguish between organisms often related to the use of plants as medicine. Knowledge of medicinal plants has been passed down orally in many cultures for thousands of years. One of the first written guides on medicinal plants dates back to Sumeria around 2700 B.C. A notable guide compiled by Chinese scholar Li Shizhen (1518–1593) contained more than 12,000 prescriptions using 1,074 plant substances. He classified herbs into 16 categories and 62 subcategories.

In Europe, the Greeks began the written process of naming and classifying plants and animals. The philosopher Aristotle (384–322 B.C.) believed there were fixed types of plants and animals that could be defined based on form and function. Theophrastus (370–285 B.C.), a student of Aristotle, categorized more than five hundred plants on the basis of growth habit, identifying them as herbs, shrubs, or trees. He distinguished flowering from nonflowering plants and understood many aspects of basic plant structure. Around A.D. 70, the Greek pharmacologist Dioscorides (A.D. 40–90) wrote *De Materia Medica,* which was used as a reference on medicinal plants well into the 1600s. *The Canon of Medicine* by Muslim scholar Ibn Sina (A.D. 980–1037) also contained information about useful plants. Based on Greek and Muslim knowledge, it strongly influenced European medicine for centuries.

With the development of the printing press in the 1400s, descriptions of plants began to appear in printed books known as herbals. Most herbals focused on medicinal uses of plants, as well as on how to identify and collect appropriate plants. Many medieval and Renaissance herbalists believed in the doctrine of signatures, which said that every plant has a telltale feature, or "signature," hinting at its medicinal use. Since the walnut looked like a crinkled skull, for example, it was considered useful for ailments of the head. The mandrake root, with its humanlike shape, was thought to be good for the entire person and was often recommended as an aphrodisiac or as a cure for sterility (Figure 16.1). In truth, the mandrake contains alkaloids with highly narcotic properties.

Figure 16.1. The doctrine of signatures. According to the medieval doctrine of signatures, physical resemblances between plants and humans were considered signs of how plants could be used as medicines. To imaginatively inclined observers, dried roots from mandrake plants (*Mandragora officinarum*) resembled men or women and were thus considered useful in treating disorders—particularly sexual disorders—specific to one sex or the other.

Although the doctrine of signatures was distinctly unscientific, herbalists did contribute to scientific classification. They made accurate illustrations that were useful for identifying plants. Also, they typically described plants in Latin, the standard written language for scholars throughout Europe. Scientists could therefore easily share information about a particular plant, even though it might have different common names in different languages. The descriptions in herbals helped to further the process of naming plants and arranging them in groups according to their characters.

Linnaeus laid the foundation for modern naming of species

Despite the collection of information in herbals, the Middle Ages saw little progress in scientific classification. Then came the European voyages of exploration during the 1400s, 1500s, and 1600s, spurred by the desire for sea routes to Asia after the Turks had blocked overland trade. The new trade routes increased European access to Asian goods, including such prized plant products as spices and medicines. Meanwhile, the voyages introduced Europeans to many plants and animals from the Americas and Africa, sparking a growing interest in scientific classification. The most notable European taxonomist of the time was English naturalist John Ray (1627–1705), who established the species as the basis for classifying organisms. Ray used a variety of characters to describe each species. He also recognized the difference between monocots and dicots, the two main types of flowering plants.

Ray's work helped set the stage for the contributions of Carl von Linné (1707–1778), a Swedish professor of botany and medicine who is better known as Carolus Linnaeus (his latinized name). Linnaeus has been called the "father of modern taxonomy" because he popularized the system of scientific naming of species that we use today.

With the growing number of plants and animals being described, long lists of characters were needed to distinguish each species. The traditional descriptions listed the characters of a species but did not provide a concise name. For example, one description of a member of the genus *Physalis,* which consists of about 100 species of small herbs, translated as "Physalis, annual, much-branched, with strongly-angled, glabrous branches and leaves with saw-toothed edges." Also, different taxonomists might describe the same species in slightly different ways. To provide a brief, consistent way to refer to each species, Linnaeus used a two-part Latin name. The first part identified the genus, and the second part was usually an adjective. His name for the *Physalis* species just described was *Physalis angulata* ("angled Physalis"). Some common names for this species

are cape gooseberry and cutleaf ground-cherry. He named our own species *Homo sapiens* ("wise man"). The two-part species name served as a convenient shorthand, but each species was still defined by all its characters.

Linnaeus was not the first to use two-part Latin names, but he was the first to apply such names consistently and extensively to many species. This was his greatest contribution to the science of taxonomy: establishing a thorough system that was easy to use. In 1753 Linnaeus published the first edition of his *Species Plantarum* ("The Kinds of Plants"), which identified more than seven thousand species and influenced many scientists in his own time and long afterward.

The two-part name of a species, called a **binomial,** is always italicized or underlined. The part identifying the genus is capitalized. The second part, called the **specific epithet,** is not capitalized. The two parts must be used together to name a species. Consider, for example, the species name for the garden pea, *Pisum sativum.* The first part tells you that it belongs to the genus *Pisum.* However, *sativum* simply means "grown as a crop." In fact, a species in one genus often has the same specific epithet as a species in a different genus. For instance, garlic is *Allium sativum.* So you can see why the specific epithet alone cannot identify the species. In official listings of plant names, the binomial is often followed by the name or initial of the authority who named the plant. For example, a white rose named by Linnaeus would be *Rosa alba* L. However, the usual practice in textbook references is just to give the binomial.

Why did Linnaeus become the taxonomist cited in every botany textbook, to the exclusion of most, if not all, others? His fame is largely the result of being such a productive scholar who trained many students and produced a simple, usable system for naming species. At a conference in Paris in 1867, botanists decided to use Linnaeus's binomial system to classify plants. Zoologists soon decided to use binomials as well.

While modern botanists and zoologists have embraced Linnaeus' system of naming species, they consider his classification of organisms into more general groups to be arbitrary and oversimplified. For example, he defined major "classes" of plants by using just the number and position of stamens, the male pollen-producing part of a flower. In contrast, Bernard de Jussieu (1699–1777) of the Royal Botanical Garden in Paris used as many characters as possible when classifying, resulting in more natural groupings. In Linnaeus' time, his approach was also questioned on moral grounds, as some critics argued that the sexual life of plants was an inappropriate topic, even for scientific discussion.

In the next section, we will look at the present hierarchical system of classifying organisms. Although most of that system cannot be credited to Linnaeus, it is safe to say that he got the ball rolling.

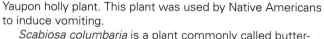

THE INTRIGUING WORLD OF PLANTS

What's in a Plant Name?

There is often an interesting history behind a scientific name or a common name for a plant. Here are just a few examples.

The first part of the scientific name for foxglove, *Digitalis purpurea,* is derived from the Latin word for "finger" because the little flowers resemble thimbles. The plant is a source of digitalis, a drug prescribed by doctors to strengthen the heart and regulate its beat. The common name foxglove is derived from the legend that fairies gave these moccasin-like flowers to the foxes so they could approach prey quietly.

The scientific name for the forget-me-not, *Myosotis pulvinaris,* is derived partly from the Greek *mys* ("mouse") and *otikos* ("ear"). The soft, short leaves suggest the ears of a mouse.

Hemerocallis fulva, the day lily, received its name from the Greek *hemera* ("day") and *kalos* ("beautiful"). That is because the flowers of the day lily close at night.

Atropa belladonna produces the alkaloid atropine, which can be used to dilate the pupils. The specific epithet *belladonna,* Italian for "beautiful lady," may refer to Italian women during the Renaissance who put drops of extract from the plant into their eyes to look more alluring. However, atropine can also be lethal, which explains the common name deadly nightshade and the fact that both the chemical and the genus are named after *Atropos,* the Greek Fate who held the shears to cut the thread of human life.

Ilex vomitoria is the scientific name for a plant commonly known as the

Foxglove (*Digitalis purpurea*).

Deadly nightshade (*Atropa belladonna*).

Yaupon holly plant. This plant was used by Native Americans to induce vomiting.

Scabiosa columbaria is a plant commonly called butterfly blue. It was considered a cure for scabies and various other skin diseases.

The scientific name for the buttercup genus, *Ranunculus,* means "little frog." Like frogs, buttercups prefer wet places.

The scientific name for the tickseed, *Coreopsis tinctoria,* is derived from the Greek *koreos* ("bug") and *opsis* ("like") because the seeds resemble a tick.

The common names elephant's head and elephanthead lousewort refer to the flower shape of *Pedicularis groenlandica.* The plant grows to be about a foot tall in high mountain bogs in the western United States.

The little elephants (*Pedicularis groenlandica*).

Section Review

1. Did classification become progressively more scientific? Explain.
2. What are the advantages of having a scientific name for a species?
3. Should Linnaeus be rightfully called the "father of modern taxonomy"? Explain.

Classification and Evolution

For most of his life, Linnaeus believed that all species remained unchanged after being created in the beginning by God (see the *Plants & People* box on page 352). His motto was *Nullae species novae*—Latin for "no new species." In this view, the formation of all species occurred at one time, much like an unchanging painting. Toward the end

PLANTS & PEOPLE

Linnaeus and the Lure of Plants

Linnaeus was fascinated by plants from a very early age. When he was 25, the Swedish Royal Society sent him to make the first scientific survey of Lapland, north of the Arctic Circle. On this trip he collected plants and studied the animals and geology. One plant he observed there is named for him: *Linnaea borealis,* the twinflower. By the time he returned, his lifelong interest in botany was assured. Eventually he became a professor of medicine and botany at the University of Uppsala. Many of his students later obtained prominent positions at universities throughout Europe, ensuring his future fame.

Linnaeus was particularly intrigued by plant sexuality. In fact, he classified plants into groups based on the number, union, and length of male and female parts. He lumped together the nonflowering plants simply because they did not have obvious sexual parts. He often compared plant and human sexuality, describing leaves as "bridal beds, which the Creator has so gloriously arranged" and stamens as "husbands in the same marriage." As can be imagined, this caused a considerable ruckus among

The young Linnaeus, holding a twinflower.

some of his critics. After one botanist referred to such sexual descriptions as "loathsome harlotry," Linnaeus named a genus of weeds, *Siegesbeckia,* after him.

As a professor, Linnaeus was very popular as a leader of field trips into the countryside. On these occasions as many as two hundred people went along, accompanied by drums and French horns and banners, if not an entire brass band. Linnaeus was in the lead, followed by a notetaker and a person in charge of order and discipline. Other people were in charge of collecting specimens or carrying food and drink. On arriving back at the campus, everyone shouted "*Vivat* [long live] Linnaeus."

Linnaeus' name has an interesting botanical history. It would have been Carl Nilson, as he was the son of Nils Ingemarson. In Sweden at that time, a son's last name was his father's first name with "son" added. However, two of his uncles changed their name to Liliander after a linden tree that grew on the family's land. When he was granted a noble title, Carl changed his name to Carl von Linné, which he then latinized to Carolus Linnaeus.

---

of his days, however, he noticed the great amount of variation that occurred within most species of plants. He also recognized that new varieties could be cultivated, and he noted examples of what we now call mutations. It seemed that new types of plants were appearing considerably after the first creation. Eventually Linnaeus abandoned his insistence that no new species could arise.

Prior to the work of Darwin, other scientists had also noticed mutations, so the idea that species were unchanging was already being questioned. After Darwin's *On the Origin of Species* was published in 1859, and the theory of evolution became increasingly accepted, taxonomists began to view classification quite differently. Instead of grouping organisms based only on physical appearance, they created family trees reflecting views of **phylogeny,** the evolutionary history of related species.

Classifications based on evolutionary relationships are known as phylogenetic classifications. They can be quite

different from classifications based solely on physical appearance. For evolutionary taxonomists, known as systematists, the basic question is not simply whether certain organisms are similar. Instead, they ask whether the similarities are the result of evolution from a common ancestor.

How can systematists answer that question, though? After all, there are no eyewitness accounts of how species actually evolved over many millions of years. If the static view of unchanging species is like a painting, then the evolution of species over time can be compared to a very long "movie"—one that is, of course, impossible to view. Some "scenes" have survived in the form of fossils— "snapshots" of organisms from the past—but the fossil record is incomplete and often difficult to interpret. Therefore, systematists must also use indirect methods to form hypotheses about how various species evolved. In this section, we will explore the usefulness and limitations of these methods of classification.

Systematists use a variety of characters to classify organisms

In classifying organisms, modern systematists analyze many of the same characters that have been observed for hundreds of years, as well as characters more recently revealed by electron microscopes. Some general categories of characters are structure, function, life cycles, and molecular data (DNA, RNA, and proteins).

Plants, for instance, are often classified by reproductive structures, such as seeds, cones, flowers, and fruits. Some major groupings of plants are based on whether or not they have seeds and what kinds of seeds they have. For example, flowering plants have seeds inside fruits, while conifers have seeds on cones. Stem shape, leaf shape, and pattern of leaf arrangement are also important structural characters. Suppose a particular tree has squarish, lobed leaves about six inches long that turn yellow in autumn; a tuliplike, orange-green flower; and a cone-shaped fruit. These characters describe the tulip poplar tree (*Liriodendron tulipifera*), common in deciduous forests of eastern North America (Figure 16.2). Systematists also compare microscopic structures, such as patterns of vascular tissue and the numbers of chromosomes within cells.

Characters relating to plant functions are also useful. Certain enzymes, for instance, may have similar functions in related plants, such as producing particular alkaloids. Types of photosynthesis can be another clue. Within some plant groups, C_4 photosynthesis has evolved in some members but not others and can therefore be used to distinguish subgroups.

Life cycles are particularly important clues to phylogeny. For example, some plants can be distinguished by methods of pollination. Some are wind pollinated, while others rely solely or mainly on insects, birds, or bats. Embryo development can be another clue. The major division of flowering plants into monocots and dicots is based on whether the embryo has one or two cotyledons, or seed leaves. The embryo shape, such as whether it is curved or straight, may also be used in classification. Furthermore, embryos of related species often have similar characteristics not apparent in the adult organisms.

Since the 1960s, new methods of analyzing DNA, RNA, and protein structures of organisms have provided molecular data that many systematists consider to be the ultimate

Figure 16.2 The tulip poplar. Characteristic leaves, flowers, and fruit aid in identification of *Liriodendron tulipifera*, known by the common names tulip poplar, yellow poplar, tulip tree, white poplar, and whitewood. It is the state tree of Indiana, Kentucky, and Tennessee. Edgar Allan Poe's story "The Gold Bug" describes this imposing tree, which can reach heights of up to two hundred feet: "[He] proceeded to clear for us a path to the foot of an enormously tall tulip tree, which stood, with some eight or ten oaks, upon the level, and far surpassed them all, and all other trees which I had then ever seen, in the beauty of its foliage and form, in the wide spread of its branches, and in the general majesty of its appearance."

taxonomic tool. However, as you will see, interpreting molecular data can be difficult.

Molecular data play a key role in phylogenetic classification

Molecular data can be a valuable tool for identifying evolutionary relationships because similarities in DNA, RNA, and proteins indicate that organisms are closely related. Molecular data can also provide strong evidence that physical similarities were inherited from a common ancestor. After all, the genetic information from DNA and RNA codes for the proteins actually give an organism its physical characteristics.

One well-established technique is to compare amino acid sequences of proteins from different organisms. The protein cytochrome *c* is often used because it is a relatively simple protein that is easily sequenced and is found in all organisms with aerobic respiration. The fewer the differences in the amino acid sequences, the more closely the organisms are related. For example, there are 38 differences between the cytochrome *c* amino acid sequences of wheat and humans, only 11 differences between horses and humans, and no differences between chimpanzees and humans. Roughly speaking, there are four or five changes in the cytochrome *c* sequences for every 100 million years since two species had a common ancestor. Therefore, cytochrome *c* can be used as a kind of **molecular clock** to estimate the extent of evolutionary separation—the length of time that two species have been reproductively isolated from each other.

In recent years, systematists have also been able to compare DNA and RNA nucleotide sequences of different organisms. Initially, they analyzed partial sequences by constructing a restriction map of DNA (see Chapter 14). Since restriction enzymes cut DNA at the locations of specified base sequences, a map of where these breaks occur provides a good sample of the DNA differences between two species. More recently, the approach has been to compare the complete sequences of genes or entire genomes. Scientists are gradually determining the complete DNA sequences of more and more species (Chapter 14).

Methods of analyzing molecular data, however, are sophisticated and can be difficult to apply uniformly to specimens from different species. Sources of variation in the structure of DNA, RNA, and proteins can cause problems. For example, DNA sequences can vary among individuals within a species. Also, some regions of chromosomes mutate more frequently than others. Furthermore, inversions, translocations, or duplications of chromosome regions may have occurred. These sources of error may be compounded when analyzing partial sequences instead of complete sequences. Deletions and insertions can also cause problems for comparing DNA or RNA sequences, but in these cases computers can help reconstruct the sequences.

Another problem is that some molecular data are not currently available. Information on DNA, RNA, and protein sequences has not been compiled, or is incomplete, for most living species. Furthermore, fossils do not typically provide useful molecular data, mainly because many types of fossils do not contain organic material. Even when they do contain organic matter, decay may have rapidly degraded the DNA, often leaving incomplete or distorted sequences. Despite the claims of movies such as *Jurassic Park,* accurate DNA sequences usually cannot be obtained from fossils.

Despite these limitations, molecular data are very useful for forming hypotheses about evolutionary relationships. Systematists compare classifications based on molecular data with those based on other characters. Often the classifications are remarkably similar. Analysis of any differences can result in refining the characters selected or the methods used, producing a more accurate phylogenetic classification.

Organisms are classified into a hierarchy

Based on the analysis of characters, an organism is classified into a hierarchy of categories, or levels. The most general category is a **domain,** a relatively new level that will be explained later in this chapter. The next category is **kingdom.** Each related group within a kingdom is called a **phylum** (plural, phyla). Although botanists have long used the term *division* instead of *phylum,* the two terms were declared equivalent in 1993 in the International Code of Botanical Nomenclature. We will use the term *phylum.* You will find a list of phyla in Appendix C. Each group within a phylum is called a **class.** Within a class, each group is called an **order.** Each related group within an order is called a **family.** Within a family, each group is called a **genus** (plural, genera). A genus is composed of one or more **species.** If you like mnemonic devices, try this for memorizing the levels within a domain: **K**ing **P**hilip **C**ame **O**ver **F**earing **G**reen **S**nakes (**K**ingdom, **P**hylum, **C**lass, **O**rder, **F**amily, **G**enus, **S**pecies).

Similarities in reproductive structures play a key role in the classification of plants because such similarities are frequently the result of genetic relatedness rather than environmental adaptation. Some plants are classified into phyla based on the presence or absence of seeds. Seed plants include four phyla of plants with naked seeds, called gymnosperms, and one phylum of plants with enclosed seeds, called angiosperms. Differences in reproductive structures and in stem and leaf structure help characterize gymnosperm phyla. Seedless plants include three phyla of bryophytes and four phyla of seedless vascular plants. Differences in organization and structure of sporangia help distinguish the various phyla of seedless plants.

Within a phylum, plants are grouped into classes based on characters that are widely shared. For example, since all angiosperms have cotyledons, or seed leaves, they can be divided into classes based on cotyledon number, with monocots as one class and dicots as another. In fact, monocots and dicots each have a number of other distinctive traits. Plants are differentiated into classes, orders, families, and genera based on more and more specific characters that are uniform and distinctive for the particular taxonomic level.

Even within a single species, individual variation in some characters occurs. As character differences accumulate between plant groups, they are eventually unable to

successfully interbreed. Though it is rare for animal species to successfully interbreed, plants in different species and even different genera can sometimes interbreed to produce fertile offspring.

As an example of classification, consider the common potato plant, *Solanum tuberosum.* It is, of course, a member of the kingdom Plantae. As a flowering plant, it belongs to the phylum Anthophyta. As a dicot, it is in the class Eudicotyledones. Among the dicots, it is in the order Solanales and then the family Solanaceae. Among about 90 genera in that family, it belongs to the genus *Solanum,* which includes about 2,300 species. Each named group at any level is called a **taxon** (plural, taxa). For example, Solanaceae is a taxon at the family level. As you can see from the descriptions of the taxa in Figure 16.3, the different taxa at each level have all

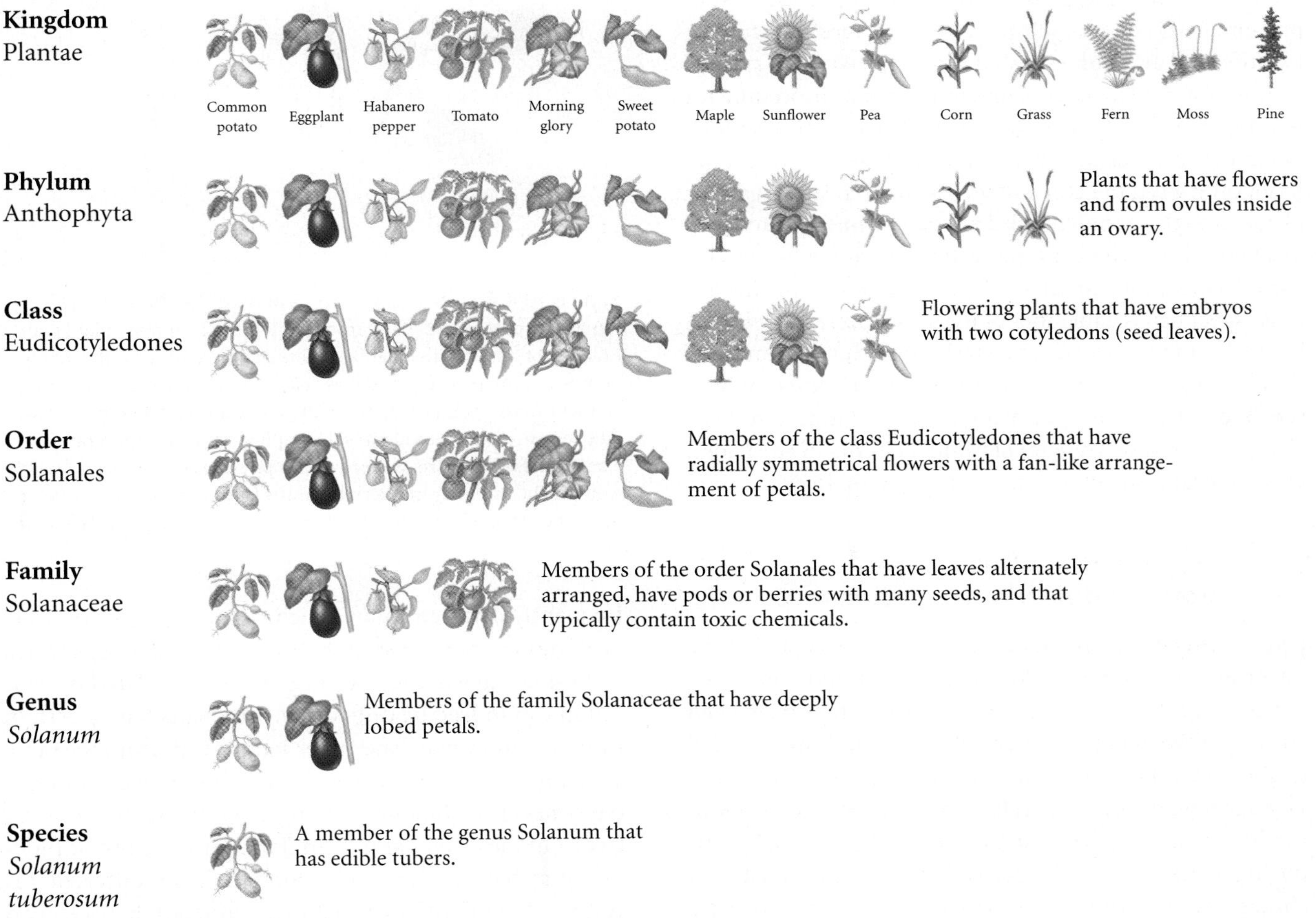

Kingdom
Plantae

Common potato   Eggplant   Habanero pepper   Tomato   Morning glory   Sweet potato   Maple   Sunflower   Pea   Corn   Grass   Fern   Moss   Pine

Phylum
Anthophyta

Plants that have flowers and form ovules inside an ovary.

Class
Eudicotyledones

Flowering plants that have embryos with two cotyledons (seed leaves).

Order
Solanales

Members of the class Eudicotyledones that have radially symmetrical flowers with a fan-like arrangement of petals.

Family
Solanaceae

Members of the order Solanales that have leaves alternately arranged, have pods or berries with many seeds, and that typically contain toxic chemicals.

Genus
Solanum

Members of the family Solanaceae that have deeply lobed petals.

Species
Solanum tuberosum

A member of the genus Solanum that has edible tubers.

Figure 16.3 Classification of the common potato plant, *Solanum tuberosum*. This hierarchy shows some examples for each level. Kingdom Plantae includes multicellular photosynthetic organisms that live on land and have an embryo protected within the mother plant. As you move down, each level becomes more limited because fewer plants share all those characteristics. For instance, the Solanaceae have traits not shared by all members of the order Solanales. At the species level, the common potato plant is unique because it is the only member of the genus *Solanum* with edible tubers.

the characters of the levels above *and* other similarities. As you move from the kingdom level to the species level, each taxon is more closely related. For example, members of the genus *Solanum* are more closely related to each other than they are to the other species within the Solanaceae family. You can recognize the level of some plant taxa by their names, as most plant family names end in *-aceae,* and most names of orders end in *-ales.* In common usage, only genus and species names are italicized. All taxa names, regardless of level, begin with a capital letter.

Given the great variety within groups of organisms, many systematists have expanded the number of levels in the hierarchy. Some additional levels have the prefixes *sub-* and *super-,* such as subphylum, superclass, subclass, superorder, suborder, superfamily, subfamily, subgenus, and subspecies. For example, some classes might be grouped into a superclass, and some superclasses might be grouped into a subphylum. Levels such as subspecies, variety, race, and cultivar identify wild and domesticated variations within a species.

Botanists and zoologists follow a standard procedure for naming and classifying new species. When a species is discovered, it is described in a major scientific journal. A description is given in Latin and several modern languages. For plants, a single specimen called a **type specimen** is preserved in a herbarium, where it is protected from moisture and insects. A type specimen can be used to determine whether another specimen is a member of the same species. The largest herbarium in the world, associated with the Museum National d'Histoire Naturelle in Paris, contains seven million specimens. Linnaeus' specimens are in the Linnean herbarium at the Swedish Museum of Natural History in Stockholm and at the Linnean Society in London (Figure 16.4).

Figure 16.4 A specimen from Linnaeus' herbarium. This tomato plant specimen dates from the 1700s. The tomato plant is an example of a species that has been named differently by different taxonomists. Linnaeus classified it as *Solanum lycopersicum.* The specific epithet means "wolf peach," reflecting the suspicions that the fruit was poisonous—dangerous like a wolf, despite its inviting appearance. The species was later reclassified as *Lycopersicon esculentum,* or "edible wolf peach," but many modern botanists favor Linnaeus' name because the tomato is closely related to members of the genus *Solanum.*

Systematists form hypotheses about evolutionary relationships

Ever since Darwin's theory became widely accepted in the scientific community, phylogenetic classification has taken center stage. Since the late 1800s, systematists have been creating **phylogenetic trees,** branching diagrams intended to show evolutionary relationships over time. The first phylogenetic trees mainly reflected what was known about the life cycles and basic structure of organisms. Today, phylogenetic trees can be derived from a wider variety of characters, including molecular data. However, they are not necessarily based on many characters; some are derived from as few as one character. The most reliable phylogenetic trees are typically based on more than one character, although trees based on a single molecular character can also be considered reasonable. Regardless of the number of characters used, all phylogenetic trees are hypotheses about how organisms are related. Systematists frequently disagree about which characters most indicate evolutionary relationships between particular organisms.

Evaluating whether two organisms are related by evolution can indeed be difficult. If two plants have a certain similarity, they may have inherited the trait from the same ancestor. Such a similarity is called a **homology.** For example, cones are a homology among conifers, the cone-bearing trees. On the other hand, a similarity in structure or function may have evolved independently from different ancestors. A similarity in structure or function between two species that are not closely related is called an **analogy.** An analogy is the result of **convergent evolution.** Just as different roads may meet, different evolutionary paths may converge, resulting in a similarity in a particular character. For example, even though different desert plants have thick, fleshy leaves that store water, this trait may have evolved independently in each type of plant. Spines have also evolved independently in different types of desert plants (Figure 16.5). In short, a similarity in structure or function does not necessarily mean that organisms are closely related. When comparing particular plants, systematists must look at each similarity in relation to other characters. A similarity may be outweighed by a greater number of differences or by more significant differences. Also, some similarities are superficial, while others are complex. A complex similarity is more likely to have evolved from a common ancestor than independently. Molecular data, if available, can be helpful because closely related plants will have more similar DNA, RNA, and proteins.

Systematists evaluate similarities in different ways. Some focus on comparing sheer quantities of traits without distinguishing between homologies and analogies. They believe that the more similarities two organisms have, the more closely related they are. Others compare only the numbers of homologies. However, the most common approach is to classify organisms according to the sequence in time at which evolutionary branches arose. This approach, known as **cladistics** (from the Greek *klados,* branch), focuses not on the number of homologies but rather on the order in time when homologies were inherited.

In comparing particular organisms, cladistics distinguishes whether a homology is a "primitive," or ancestral, character inherited from a more distant ancestor or a "derived" character inherited from a more recent ancestor. A **shared primitive character** is a homology that is not unique to the organisms being studied. In other words, organisms outside the group also inherited that character from the same distant ancestor. For example, seeds are a shared primitive character among angiosperms because angiosperms are not the only plants that have seeds. Gymnosperms also inherited this character from the ancestor of all seed plants. In contrast, a **shared derived character** is a homology that is unique to a particular group. For example, flowers are a shared derived character among angiosperms because only angiosperms and their common ancestor share that trait.

The following flowchart summarizes the types of similar characters:

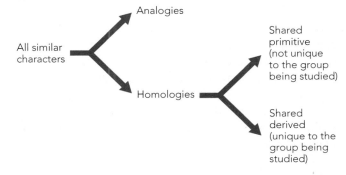

Figure 16.5 Convergent evolution. Members of at least three unrelated families of flowering plants have developed the spines and full, fleshy bodies of cactus plants. These include the cactus family (Cactaceae), the spurge family (Euphorbiaceae), and the milkweed family (Asclepiadaceae).

Cladograms are branching diagrams that show evolutionary relationships

Since the 1980s, with the growing popularity of cladistics, a common type of phylogenetic tree has been the **cladogram,** a tree-like branching diagram that shows evolutionary relationships. A **clade** is a branch of a cladogram that consists of an ancestor and all its descendents, all of whom share one or more characters that make them unique as an evolutionary branch. A clade is by definition **monophyletic,** meaning that it is a "single tribe" of organisms that evolved from a common ancestor. In defining a clade, systematists use shared derived characters because those reflect the closest evolutionary relationships.

A cladogram can be constructed in order to identify proposed evolutionary relationships between taxa at any level of the classification hierarchy, such as particular species, genera, families, orders, classes, phyla, or kingdoms. When creating a cladogram, the first step is to select a group to study. The group may be different species, or it may consist of different groups of species (such as different families or phyla). The types of organisms being studied are known as the **ingroup.** Depending on the plant material available and the interests of the investiga-

tor, the ingroup might be large or small and might be focused on a particular taxon or be more widespread. Suppose we decide to compare three phyla of vascular plants: ferns, conifers, and flowering plants. The next step is to choose an **outgroup,** a species or group of species closely related to the ingroup, but not as closely related as the ingroup members are to each other. The outgroup provides a basis for distinguishing shared primitive characters (those that are also present in the outgroup) from shared derived characters. As an outgroup, we choose a phylum of nonvascular plants consisting of mosses.

Next, we create a character table, with the taxa listed on one axis and the characters on the other axis. For simplicity, the table shown in Figure 16.6a uses only three characters, with 0 indicating the character is missing and 1 indicating the character is present. Character tables can include many characters and can show more than two states for each character, such as the number of sepals in a flower. Numerical entries make it easy to analyze the data by computer.

The data are then used to create a cladogram (Figure 16.6b). Each branch point, or node, represents the loss or gain of a particular character, indicating separate evolutionary paths. That gain or loss of a character defines a particular clade (Figure 16.6c). For example, the presence

Taxa

Characters	Mosses (outgroup)	Ferns	Conifers	Flowering plants
Flowers	0	0	0	1
Seeds	0	0	1	1
Vascular tissue	0	1	1	1

(a) Character table

Figure 16.6 A simple character table and cladogram. **(a)** A character table translates information into numbers that a computer can analyze. For instance, 1 is typically used to indicate a character is present. This table was used to prepare the cladogram in Figure 16.6b. The presence of a particular character, such as vascular tissue, is used to determine a branch point. **(b)** This cladogram shows a hypothesis about the evolutionary relationships between three phyla of vascular plants: ferns, conifers, and flowering plants. **(c)** As you move up the cladogram, members of each clade are more closely related. Ferns, conifers, and flowering plants all have vascular tissue, a character that mosses lack. However, conifers and flowering plants are even more closely related because both also have seeds. Finally, flowering plants are distinguished as a group from conifers, which lack flowers.

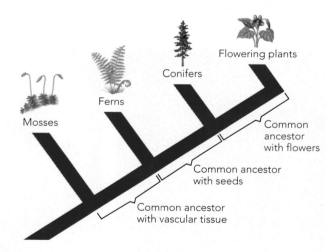

(b) Cladogram

Fern-conifer-flowering plant clade (shared derived character: vascular tissue)

Conifer-flowering plant clade (shared derived character: seeds)

Flowering plant clade (shared derived character: flowers)

(c) Examples of clades

of vascular tissue defines one clade consisting of an ancestor with vascular tissue and all of its descendants. Note also that a clade can be nestled within another clade. Within the clade of plants with vascular tissue is a clade of plants with vascular tissue *and* seeds. Another clade consists of plants with vascular tissue, seeds, *and* flowers.

A cladogram is most often constructed by a computer program that organizes the data from a character table into a graphical representation. Frequently, more than one cladogram can be constructed from a particular set of data. In such cases, systematists usually follow the rule of **parsimony,** which states that the simplest cladogram is probably correct. Recall that each branch point represents the loss or gain of a character, which may have been caused by one or more evolutionary events. The simplest cladogram represents the path with the fewest evolutionary changes. The rule of parsimony is one example of the use of Occam's Razor. William of Occam (1280–1349) was a philosopher who believed that the simplest explanation for an event is most likely to be correct.

Cladograms produced with different characters can be compared to see if they produce similar phylogenetic trees. Characters that do not produce similar cladograms may be analogies instead of homologies.

Systematists often disagree about how to classify organisms

Like any other type of phylogenetic tree, a cladogram is a hypothesis, as systematists continue to disagree about how to use characters to classify organisms. One conflict is between the "splitters" and the "lumpers." The "splitters" favor identifying a greater number of species or groups of species. The "lumpers" believe it makes more sense to have fewer species and groups of species. For example, lumpers have recently argued that dogs (*Canis familiaris*) and wolves (*Canis lupus*) should be considered one species, *Canis lupus,* because they can interbreed and produce fertile offspring. Splitters say it is problematic to combine a wild, endangered species with a domestic species in which artificial selection has produced many subspecies, or breeds. Similarly, systematists disagree about the classifications of the *Capsicum* genus, which includes wild and domesticated species of hot and mild chile peppers, as well as bell peppers. One reason for the disagreement is that fruit shape can vary widely, even among members of one species within the *Capsicum* genus. Some systematists have argued that different fruit shapes indicate different species of peppers (Figure 16.7).

Figure 16.7 One pepper species or many? These peppers reflect the extreme varieties within the species *Capsicum annuum,* due to artificial selection (clockwise from top left: bell pepper, Hungarian wax pepper, habanero pepper, jalapeño pepper). Some systematists believe that the different varieties should be classified as separate species, based on differences in shape and color, as well as other characters, including molecular data.

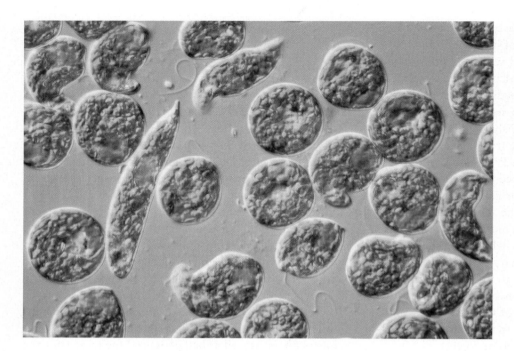

Figure 16.8 *Euglena*. This microorganism is one of many creatures that do not fit the traditional definitions of either animals or plants. It has chloroplasts, enabling it to photosynthesize like a plant. However, it is not solely an autrotroph. If placed in the dark, where it is unable to make food through photosynthesis, it can still get nutrients in a fungi-like manner by absorbing food. Meanwhile, flagella enable *Euglena* to move, a characteristic associated with animals.

The ability of plants to hybridize between species and genera also complicates cladistic analysis, as does the ability of plants to produce new species by forming polyploids (Chapter 15). Furthermore, keep in mind that two experts on a taxon, such as a certain genus of plants, might disagree on the exact classification of a particular species. Molecular taxonomy helps resolve some differences of opinion but does not eliminate them. Forming any hypothesis about evolutionary relationships is, after all, attempting to reconstruct the past from the inadequate vantage point of the present.

Section Review

1. Why might some types of characters be more important than others when classifying organisms?
2. Molecular data may either support or conflict with classifications based on other data. How would you explain this?
3. What is the hierarchy of taxonomic levels, from most general to most specific? How can a hierarchy reflect phylogenetic relationships?
4. Explain how types of similarities are evaluated in cladistics.

Major Groups of Organisms

You have seen how classification has changed over the years, and how systematists disagree about evolutionary relationships between species. It should come as no surprise, therefore, that even the most general groupings of organisms remain a subject of debate.

Systematists have revised the number of kingdoms

Until the mid-1800s, organisms were classified as either plants or animals. Photosynthetic organisms that did not move were classified as plants. Organisms that moved and ingested food were considered animals. This distinction seemed to work well for most organisms that are commonly seen. Some, however, did not fit into either category. For example, fungi were often regarded as plants simply because they do not move, even though they are not photosynthetic. Meanwhile, many organisms have a combination of plant-like, fungi-like, or animal-like characteristics. For example, the microscopic organism *Euglena* (Figure 16.8), commonly found in pond water, carries out photosynthesis like a plant but absorbs food like a fungus and moves like an animal.

As systematists saw the inaccuracy of classifying all organisms as either plants or animals, some proposed additional kingdoms. During the 1860s, biologists John Hogg and Ernst Haeckel suggested a new kingdom Protoctista that would include fungi, bacteria, single-celled algae, and many other single-celled organisms. Later, as the unique traits of fungi were recognized, most systematists placed fungi in their own kingdom. By the late 1930s, microscopic studies revealed that bacteria were single-celled prokaryotes, while all other organisms were eukaryotes. Most systematists therefore began classifying bacteria separately as the kingdom Monera (from the Greek *moneres,* single). With the removal of fungi and bacteria, the revised kingdom Protoctista contained algae, slime and water molds, and many non-photosynthetic single-celled organisms. The kingdom was usually referred to by the more easily pronounceable name Protista.

In 1969, Robert Whittaker of Cornell University proposed a five-kingdom system that became widely accepted. It consisted of the kingdoms Animalia, Fungi, Plantae, Protista, and Monera. The animal, fungi, and plant kingdoms were distinguished from each other primarily by their means of obtaining food. Animals ingest food; fungi absorb food; and plants make their own food through photosynthesis. All other eukaryotes were lumped together as the kingdom Protista. These organisms, many of which are microscopic, are known informally as protists. The other kingdom, Monera, consisted of all the prokaryotes, the microscopic organisms commonly referred to as bacteria. Viruses, which are even smaller and simpler in structure, were not classified because they cannot live and reproduce by themselves outside of a host organism.

Although Whittaker's scheme became popular, it was not the last word on classifying organisms into kingdoms. Issues remained regarding classification of protists and prokaryotes. In Whittaker's classification system, the kingdom Protista was still a catchall that grouped diverse eukaryotes together mainly because they did not fit the descriptions of animals, fungi, or plants. Many systematists are now exploring how to divide protists into several kingdoms. Furthermore, molecular data have shown that it is inaccurate to group all prokaryotes into a single kingdom.

Molecular data have led to identifying "super kingdoms" called domains

The five-kingdom system was based on comparing the visible structures and functions of organisms. In recent years, however, molecular data have yielded new insights into the differences between prokaryotes. Instead of the five-kingdom system, most systematists now consider the highest level of classification to be a "super kingdom" called a domain. According to this view, all

organisms can be classified into one of three domains: Archaea, Bacteria, or Eukarya. The organisms in the domain Archaea and the domain Bacteria are different types of prokaryotes, while the domain Eukarya consists of all eukaryotes.

The domain Archaea and the domain Bacteria are two very different groups of prokaryotes

It may seem strange that two of the three domains of life consist of microscopic creatures. After all, the average person probably thinks "a bacterium is a bacterium." However, beginning in the late 1970s, molecular studies of DNA and RNA sequences revealed that prokaryotes living in extreme conditions, such as very hot or very salty environments, differ significantly from other prokaryotes in their cell structures and particularly their RNA. At first these prokaryotes were called "archaebacteria," and the other prokaryotes were called "eubacteria" ("true bacteria"). However, most systematists now believe that archaea are so different that they should not even be called a type of bacteria. In fact, in some aspects of DNA and RNA structure, archaea are just as different from bacteria as we are. That is why the two "super kingdoms" of prokaryotes are called the domain Archaea and the domain Bacteria.

Systematists disagree about the evolution of archaea and bacteria, but it is clear that both archaea (from the Greek *archaios,* ancient) and bacteria have long evolutionary histories. Prokaryotes are the most ancient organisms on Earth, with fossils dating as far back as 3.5 billion years.

Classification of prokaryotes is still very much a work in progress, as systematists discover new species and gather molecular data. About four thousand prokaryotic species have been described so far, but as many as four million may exist. Scientists recognize, for example, that up to 95 percent of bacteria in a soil sample are unknown species. Most of the known prokaryotes are classified as bacteria, but debate continues over how to group them based on evolutionary relationships. Even less is known about the evolution of archaea, with species being tentatively grouped by where they live. For instance, those dwelling in extremely hot thermal vents are called extreme thermophiles ("heat lovers"), while those living in very salty environments such as the Great Salt Lake are called extreme halophiles ("salt lovers"). However, such classification does not necessarily reflect phylogenetic relationships.

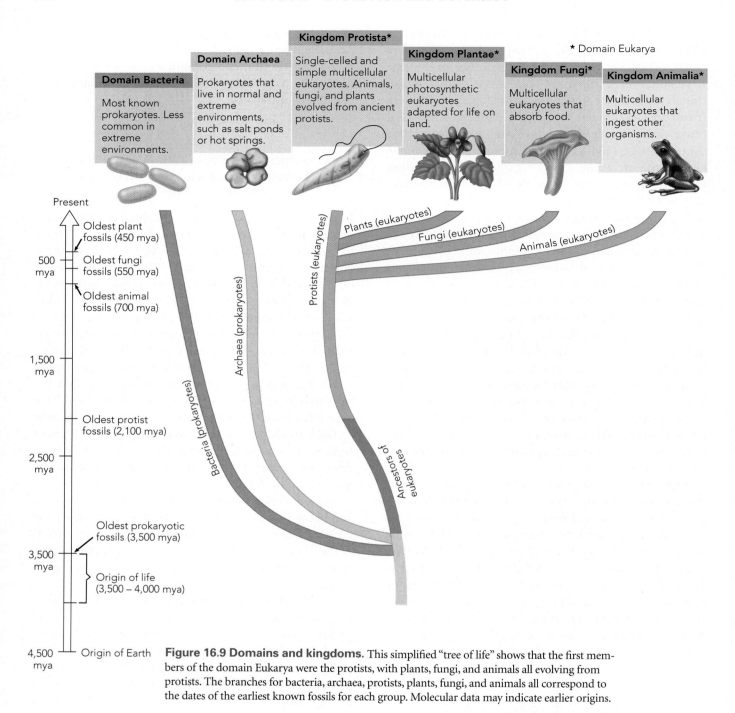

Domain Bacteria

Most known prokaryotes. Less common in extreme environments.

Domain Archaea

Prokaryotes that live in normal and extreme environments, such as salt ponds or hot springs.

Kingdom Protista*

Single-celled and simple multicellular eukaryotes. Animals, fungi, and plants evolved from ancient protists.

Kingdom Plantae*

Multicellular photosynthetic eukaryotes adapted for life on land.

Kingdom Fungi*

Multicellular eukaryotes that absorb food.

Kingdom Animalia*

Multicellular eukaryotes that ingest other organisms.

* Domain Eukarya

Present

500 mya

1,500 mya

2,500 mya

3,500 mya

4,500 mya

Oldest plant fossils (450 mya)

Oldest fungi fossils (550 mya)

Oldest animal fossils (700 mya)

Oldest protist fossils (2,100 mya)

Oldest prokaryotic fossils (3,500 mya)

Origin of life (3,500 – 4,000 mya)

Origin of Earth

Bacteria (prokaryotes)

Archaea (prokaryotes)

Protists (eukaryotes)

Plants (eukaryotes)

Fungi (eukaryotes)

Animals (eukaryotes)

Ancestors of eukaryotes

Figure 16.9 Domains and kingdoms. This simplified "tree of life" shows that the first members of the domain Eukarya were the protists, with plants, fungi, and animals all evolving from protists. The branches for bacteria, archaea, protists, plants, fungi, and animals all correspond to the dates of the earliest known fossils for each group. Molecular data may indicate earlier origins.

The domain Eukarya includes protists, animals, fungi, and plants

As you may recall from the discussion of cells in Chapter 2, eukaryotes evolved from prokaryotes. Unlike the two domains of prokaryotes, members of the domain Eukarya have cells with an enclosed nucleus. Also, unlike prokary-otes, which are single-celled, many eukaryotes are multi-cellular. Eukaryotes are the most-studied and best-under-stood organisms. Domain Eukarya includes the kingdoms Protista, Animalia, Fungi, and Plantae. The classification of animals, fungi, and plants into three kingdoms is fairly well established, but systematists are debating about how to divide the protists into several kingdoms. Figure 16.9

provides an overview of the evolution of the current domains and kingdoms.

In evolutionary history, the protists are the oldest eukaryotes, originating at least 2.1 billion years ago. They gave rise to the other eukaryotic organisms: animals, fungi, and plants. Kingdom Protista includes at least 50,000 named living species. Compared with the other three eukaryotic kingdoms, the protists are the most diverse in structure, ranging from single-celled organisms like *Euglena* (see Figure 16.8) to the giant seaweeds called kelp. They also vary in their methods of obtaining food. Some are photosynthetic like plants; some absorb food like fungi; and others ingest food like animals. Some, like *Euglena,* use more than one method of nutrition. Given this great variety, systematists are using molecular data and cladistic analysis to divide the protists into several "candidate kingdoms."

Kingdom Animalia consists of multicellular eukaryotes whose cells, unlike those of fungi and plants, do not have cell walls. Almost all animals ingest their food, and none are photosynthetic. The first animals appeared more than seven hundred million years ago. Today the kingdom contains well over a million species and has the most phyla. Since you are yourself an animal, you are perhaps most familiar with this kingdom. At first sight, however, you may not recognize some organisms as being animals. The green sea anemone below, for example, is sometimes thought to be a flower but is actually a carnivorous predator, using barbs in its tentacles to catch small swimming animals. The animal kingdom is extremely diverse, ranging from the phylum Porifera, commonly known as sponges, to the phylum Chordata, which includes vertebrates like us.

Kingdom Fungi consists of eukaryotes that absorb food, rather than ingesting it as animals do or making their own food as plants do. A fungus secretes enzymes that digest the food so it can be absorbed into the fungus. Most fungi absorb nutrients from dead organisms, but some are parasites on living organisms. Unlike plants, most fungi have cell walls made of chitin rather than cellulose. Most fungi are multicellular, but some are single-celled, like yeast. Fungi have existed for at least 550 million years, with more than 60,000 species surviving today. They were long thought to be related to plants, but molecular evidence now shows them as being much more closely related to animals. Many fungi cause plant diseases, which considerably decrease food production.

Kingdom Plantae consists of the multicellular eukaryotes that we know as land plants. With very few exceptions, plants can make their own food through photosynthesis. Also, they have cell walls mainly formed of cellulose. Reproductive characteristics include multicellular embryos retained in the female parent and the alternation of generations of haploid and diploid organisms. Based on the fossil record, plants have existed for at least 450 million years, and there are nearly 300,000 living species. The major kinds of plants are the bryophytes, such as mosses; seedless vascular plants, such as ferns; gymnosperms, which are vascular plants with exposed seeds, such as conifers; and angiosperms, which are flowering vascular plants with seeds enclosed in fruits. Later chapters in this unit will describe each of these kinds of plants.

For a list of the plant, fungi, and animal phyla, as well as the "candidate kingdoms" of protists and prokaryotes (see Appendix C). Although the list is set in type in this book, with the dynamic nature of classification, it is certainly not "set in stone."

Section Review

1. **Why might it be considered inevitable that the number of kingdoms would change?**
2. **How would you explain the difference between a domain and a kingdom?**
3. **What characteristics seem to be the most important in distinguishing the animal, fungi, and plant kingdoms?**

The Future of Classification

When the Lewis and Clark Expedition set out to explore the American West in 1804, part of their mission was to find new species of plants and animals. President Thomas Jefferson instructed them to take note of "everything worthy of notice . . . the soil and face of the country, its growth and vegetable productions . . . the dates at which particular plants put forth, or lose their flowers or leaf." That expedition alone collected specimens of 122 animal species and 155 plant species previously unknown to scientists (Figure 16.10). Two centuries later, the era of discovery is far from over. New species are being found in the United States and throughout the world. Meanwhile, modern techniques such as the use of molecular data provide new revelations about evolutionary relationships. The field of taxonomy is still full of surprises.

New species remain to be discovered

In recent years, systematists have embarked on several large-scale projects to list and map species in particular

Figure 16.10 *Mimulus lewisii.* Of Meriwether Lewis and William Clark's many botanical discoveries, several plants were named after them. *Mimulus lewisii*, also known as the monkey flower, is native to middle to high elevations of the Sierra Nevada and the higher elevations of the Rocky Mountains.

geographic areas. Some projects focus on cataloging all species native to or found in a particular area, such as North America. A more general effort known as the All Species Foundation is attempting to inventory all species of life on Earth by the year 2025. Currently, around 2 million species have been identified. Estimates of undiscovered species range as high as 100 million. Such estimates are based on surveys of previously unknown species found in particular regions.

Most undiscovered species live in tropical rain forests, which now cover less than 2 percent of Earth's surface but may contain 50 to 70 percent of all the species on the planet. However, rain forests are being destroyed at an alarming rate, with estimates of losses ranging between about 52,000 and 123,000 square miles per year. The National Academy of Sciences, for instance, estimates a yearly loss of 50 million acres or around 78,000 square miles—an area roughly the size of Nebraska. By the year 2020, 80 to 90 percent of tropical rain forests may be gone. As habitat is destroyed, many species may become extinct before scientists even know they exist.

Systematists are studying speciation in action

In addition to discovering species that already exist, systematists are observing the process of speciation in action. The world is a giant laboratory of species formation, as particular groups of organisms are becoming reproductively isolated. For example, horses and donkeys can still interbreed. However, the fact that their offspring—mules or hinnys—are usually infertile indicates that horses and donkeys are becoming reproductively isolated from each other. Accordingly, horses (*Equus caballus*) and donkeys (*Equus asinus*) are already classified as separate species.

Similarly, plants provide countless examples of the process of speciation. Consider the silversword alliance, 28 closely related Hawaiian species of flowering plants known as tarweeds (Figure 16.11). Found in three genera—*Argyroxiphium* (containing silverswords), *Dubautia*, and *Wilkesia*—these plants evolved from one species, probably from California, that colonized the islands. Over several million years, a relatively short span in evolutionary time, this colonizing species evolved differently on the various Hawaiian Islands, and even within each island because of differing environments such as rain forests, bogs, dry plains, and recently deposited lava. The 28 species of tarweeds range from vines to tall trees to several types of shrubs. It appears that most, if not all, of the species can interbreed, with varying degrees of fertility among offspring.

As you may recall from Chapter 15, speciation in plants can occur through so-called "wide crosses" between different species and even different genera. Such crosses occur in nature and were responsible for the origin of wheat. Wide-cross hybrids are sterile unless the chromosomes spontaneously double, in which case a new species is created. Speciation in plants also occurs by spontaneous chromosome doubling within one species. The coming years will see a rapid increase in information about how new species develop.

(a) Haleakala silversword (*Argyroxiphium sandwicense*).

(b) Dwarf liau (*Wilkesia hobdyi*).

(c) Na'ena'e (*Dubautia waialealae*).

(d) Na'ena'e (*Dubautia reticulata*).

Figure 16.11 The silversword alliance. These closely related plants, commonly known as tarweeds, vary remarkably in their structure as a result of evolving in different environments in the Hawaiian Islands. Although they can still interbreed when brought together in a garden, many of the populations are becoming separate species.

Molecular data will continue to provide insights into evolution

As you have read, our ability to sequence the amino acids in proteins and the nucleotides in DNA and RNA has greatly increased the amount and accuracy of the information we can obtain about living species. As molecular data become available on more species, systematists will gain an increased understanding of their evolution. For instance, they will know the complete nucleotide sequence of particular genes in many plants and calculate molecular clocks for more organisms.

Although often incomplete and open to interpretation, molecular data can provide considerable information about evolutionary relationships that is not available through other methods. For example, comparisons of DNA, RNA, and proteins show clearly and surprisingly that fungi are much more closely related to animals than to plants. Also, molecular data can help clarify relationships between the many phyla of protists, as their classification continues to undergo revision. Furthermore, some animals, plants, and fungi may be reclassified as a result of insights from molecular data. Comparing the genes and proteins of organisms can help systematists evaluate competing classifications derived by other methods, often putting proposed evolutionary relationships on a more secure footing.

The classification of organisms has practical benefits

The classification and preservation of species can lead to many medical benefits. For example, identifying closely related plants can help uncover sources of medicines. If a particular species contains a chemical used as a medicine, a related species might have the same chemical. Today about 25 percent of all prescriptions have a product obtained directly from plants, and undiscovered species hold the potential for new medicines. The fields of medicine and plant pathology can also profit greatly from the classification of prokaryotes, many of which cause diseases.

Plant systematics can also help increase agricultural productivity. Classification studies help locate wild relatives of cultivated plants. Genes from these wild relatives that provide resistance to disease or environmental stress can be transferred to crop plants by traditional breeding or genetic engineering. Also, the discovery of new plant species can yield new crops and add variety to the restricted diets of most of the world's people. In addition, the discovery of new pharmaceutical or food crops can provide a livelihood for farmers who can sell them as export crops.

The information that systematists gather about different types of organisms can also help protect species from extinction. For example, cataloging projects will reveal locations and population sizes of species near extinction that should be placed on endangered species or threatened species lists. This information can be used by urban planners and by managers of parks or nature preserves. The sense among many systematists is that we know human activity is causing extinctions of species, but that we do not have a firm grasp of "what is out there." Even in heavily populated regions, new species are occasionally uncovered.

As new species are discovered and new data are uncovered about known species, scientific classification will continue to be a dynamic field of study. In the remaining chapters of this unit on evolution and diversity, you will be exploring in more detail the rich variety of prokaryotes, protists, fungi, and plants.

Section Review

1. In what ways is scientific classification a dynamic field of study?
2. Why are molecular data important to the future of classification?
3. How would you explain the importance of classification of species?

SUMMARY

Scientists classify organisms using characters, or inherited traits. The modern study of evolutionary relationships is called systematics, which includes taxonomy, the naming and classifying of species.

Classification Before Darwin

Classification of organisms dates back to ancient times (pp. 349–350)
Early classifications were based primarily on practical needs, such as using plants as medicine. Descriptions of plants in herbals contributed to development of a scientific system of classification.

Linnaeus laid the foundation for modern naming of species (p. 350)

Linnaeus popularized the system of scientific names used today. The scientific name of a species, called a binomial, has two parts: the genus name and the specific epithet.

Classification and Evolution

Prior to Darwin, most scientists assumed that species did not change. As the theory of evolution became accepted, classification focused on phylogeny, the evolutionary history of related species. Since the fossil record is incomplete, systematists use indirect methods to form hypotheses about evolutionary relationships.

Systematists use a variety of characters to classify organisms (pp. 353–354)

Modern taxonomy uses characters relating to structure, function, life cycles, and molecular data, such as nucleotide sequences of DNA and RNA and amino acid sequences of proteins.

Molecular data play a key role in phylogenetic classification (p. 354)

Comparisons of DNA, RNA, and proteins provide strong evidence of whether organisms are closely related. Protein amino acid sequences can serve as molecular clocks to estimate when various species diverged in their evolution. Increasingly, systematists are also analyzing partial and complete DNA and RNA sequences. Mutations and incomplete data can complicate molecular analysis.

Organisms are classified into a hierarchy (pp. 354–356)

The main categories of classification are domains, kingdoms, phyla, classes, orders, families, genera, and species. A taxon is a named group at any level.

Systematists form hypotheses about evolutionary relationships (pp. 356–357)

All phylogenetic trees (diagrams of evolutionary relationships) are hypotheses. Systematists distinguish between homologies, or similarities inherited from the same ancestor, and analogies, or similarities resulting from convergent evolution. Cladistics distinguishes between shared primitive characters (homologies inherited from a distant ancestor) and shared derived characters (homologies unique to a particular group).

Cladograms are branching diagrams that show evolutionary relationships (pp. 358–359)

Cladistics uses cladograms, branching diagrams that show evolutionary relationships. A clade consists of an ancestor and all of its descendents. Cladistics follows the principle of parsimony, which assumes that the most likely evolutionary path involves the fewest changes.

Systematists often disagree about how to classify organisms (pp. 359–360)

Systematists can disagree whether organisms are separate species, with "splitters" favoring more species and "lumpers" arguing for fewer species. Disagreements also arise over which characters most reflect phylogenetic relationships.

Major Groups of Organisms

Systematists have revised the number of kingdoms (pp. 360–361)

Until the mid-1800s, organisms were classified as either plants or animals. To account for fungi, bacteria, and organisms called protists, eventually a five-kingdom system became widely accepted: Animalia, Fungi, Plantae, Protista, and Monera. Kingdom Protista included eukaryotes that were not animals, fungi, or plants. Kingdom Monera included all prokaryotes.

Molecular data have led to identifying "super kingdoms" called domains (p. 361)

Most systematists have now adopted a three-domain system, based on molecular differences. Domains Archaea and Bacteria are different types of prokaryotes, and Domain Eukarya consists of all eukaryotes.

The domain Archaea and the domain Bacteria are two very different groups of prokaryotes (p. 361)

Archaea are prokaryotes that live in extreme conditions, such as very hot or very salty environments. In their DNA and RNA, archaea differ from bacteria as much as eukaryotes do.

The domain Eukarya includes protists, animals, fungi, and plants (pp. 362–363)

Systematists are evaluating how to divide the kingdom Protista into several kingdoms. Kingdom Animalia consists of multicellular eukaryotes that ingest food. Kingdom Fungi consists of eukaryotes that absorb food and have cell walls made of chitin. Kingdom Plantae consists of multicellular photosynthetic eukaryotes with cell walls made of cellulose and multicellular embryos retained in the female parent.

The Future of Classification

New species remain to be discovered (pp. 363–364)

Systematists are cataloging known species and finding new species. Most undiscovered species live in tropical rain forests. Most prokaryotic species are yet to be discovered.

Systematists are studying speciation in action (p. 364)

Many groups of organisms are currently going through the process of speciation, or reproductive isolation.

Molecular data will continue to provide insights into evolution (p. 366)

Molecular data will help systematists understand evolution of species, clarify relationships between phyla, and lead to reclassifications of many organisms.

The classification of organisms has practical benefits (p. 366)

Classification can identify sources of medicines, increase crop productivity by identifying wild plants whose genes can improve related crops, and identify endangered species.

Review Questions

1. Explain how the terms *systematics* and *taxonomy* are not synonymous.
2. What constitutes a scientific approach to classification of organisms? How does it differ from a pragmatic approach, as in the classification of plants in medieval herbals?
3. In what ways was Linnaeus scientific in his method of classification? What was the historical significance of Linnaeus's work?
4. Explain the importance of having scientific names for species.
5. How did the theory of evolution revolutionize the classification of organisms?
6. How might the terms *static* and *dynamic* be applied in contrasting the view of Linnaeus with phylogenetic classification?
7. Why might molecular data be considered the most important taxonomic tool?
8. What are some characters used to classify organisms? Why do you think a large variety of characters must be analyzed?
9. Explain how a hierarchy is used to classify plants and other organisms.
10. Why do you think that much of Linnaeus' system of classification is still applicable, even though it was not based on phylogenetic relationships?
11. Explain the importance of collecting type specimens.
12. What might be some weaknesses in a solely quantitative approach to classification—one that focuses on comparing sheer numbers of similarities?
13. Explain why convergent evolution poses problems for systematists.
14. Why might it be said of cladistic analysis that "not all characters are considered equal"?
15. If you use a flower identification book that classifies flowers by color, is this method likely to be phylogenetic? Explain.
16. Explain why a cladogram cannot be described as being a "tree of life" diagram.
17. Give some examples to illustrate the difference between a shared primitive character and a shared derived character.
18. Why do you think disagreements arise in classification, such as those between "splitters" and "lumpers"?
19. Describe how and why the classification of major groups of organisms has become increasingly complex and remains somewhat fluid.
20. What future challenges do systematists face?

Questions for Thought and Discussion

1. Some characters in the flowers of two species indicate that the species are related, while other characters do not. List some possible examples of each and tell why both types of traits occur.
2. In the arctic tundra, what traits of plants would you expect to find that could be attributed to convergent evolution?
3. You have identified ten populations of cottonwood trees that live in separate regions. How would you determine if each population is a separate species?
4. Do you think molecular data will eventually make other types of analysis irrelevant? Why or why not?
5. Will any phylogenetic tree always remain a hypothesis? Explain.
6. Explain the following statement: A cladogram may look simple but is not easy to create.
7. Does the principle of parsimony seem reasonable with respect to evolution? Explain.
8. Why do you think there are so many disagreements about classification?
9. Will there ever be a final number of kingdoms? Explain.
10. Some say that taxonomy is just like stamp collecting. Do you agree? Explain.
11. Use the character table below to construct the most parsimonious cladogram that shows evolutionary relationships among the groups of plants listed.

	Conifers	Dicots	Monocots	Mosses	Gnetales	Ferns	Green algal ancestor
Vascular tissues	1	1	1	0	1	1	0
Heterospory	1	1	1	0	1	0	0
Xylem vessels	0	1	1	0	1	0	0
# Cotyledons	3	2	1	0	2	0	0
Flowers	0	1	1	0	0	0	0
Retention of zygote and embryo	1	1	1	1	1	1	0

Evolution Connection

During an examination of two flowering plants belonging to different orders (*Solanum tuberosum* (see Figure 16.3) and *Stellaria media,* Family Caryophyllaceae, Order Caryophyllales), you observe that both plants have deeply lobed petals. What evidence would you need in order to determine whether this represents a homology or an analogy?

To Learn More

Visit The Botany Place Website at *www.thebotanyplace.com* for quizzes, exercises, and Web links to new and interesting information related to this chapter.

Albert, Susan Wittig. *Chile Death.* New York: Berkley Publishing, 1998. One of a series of very readable murder mysteries relating to plants.

Ambrose, Stephen E. *Lewis and Clark: Voyage of Discovery.* Washington, DC: National Geographic, 2002. One of the best books about Lewis and Clark. Also, an excellent video by Ken Burns is available from many libraries.

Judd, Walter, Christopher Campbell, Elizabeth Kellogg, Peter Stevens, and Michael Donoghue. *Plant Systematics—A Phylogenetic Approach.* 2nd ed. Sunderland, MA: Sinauer Associates, 2002. A comprehensive view of the plant kingdom from the standpoint of a systematist.

Whitmore, Timothy. *An Introduction to Tropical Rain Forests.* 2nd ed. New York: Oxford University Press, 1998. A book that links rainforest biology with possible scenarios to prevent extinction of species.

17
Viruses and Prokaryotes

An *Agrobacterium tumefaciens* tumor on a *Kalanchoe* stem.

Viruses and the Botanical World

Viruses are complexes of nucleic acid and protein that reproduce inside cells

Viruses cause many important plant diseases

Several approaches are used to prevent viral diseases in plants

Viroids are infectious RNA molecules

Prokaryotes and the Botanical World

Prokaryotes are unicellular organisms with diverse characteristics

Some bacteria are photosynthetic, and some fix nitrogen

Bacteria cause a variety of diseases in plants

Prokaryotes have many uses in industry, medicine, and biotechnology

Until recently, viruses and prokaryotes were studied chiefly because they cause diseases in humans, livestock, and crop plants. We know, for example, that more than 400 types of viruses cause diseases in humans. While it may seem more important to understand and cure human diseases, fighting disease in agricultural animals and plants is equally vital because they supply our food. Diseases caused by viruses, prokaryotes, and fungi destroy 10–22% of the world's crops each year.

In the past half century, biologists have used viruses and prokaryotes extensively in basic research to understand development, physiology, genetics, and biochemistry in many organisms. Studies of the genes and enzymes of viruses and prokaryotes have yielded some of the essential tools of genetic engineering. For example, restriction enzymes, produced by bacteria to destroy the DNA of invading viruses, are key components in techniques to sequence genomes and to isolate particular genes. Also, understanding the process by which the prokaryote *Agrobacterium tumefaciens* infects plants to produce crown gall tumors on stems resulted in the discovery of the Ti plasmid, now commonly used as a vector to carry foreign genes into crop plants The bacterium infects members of the rose family in the crown region where the stem leaves the soil. The infection process involves insertion of DNA from the Ti (tumor-inducing) plasmid into the plant chromosome. Genetic engineers take advantage of this fact and make disabled plasmids that don't cause the disease but transfer valuable, useful genes.

Another example of the use of prokaryotes in genetic engineering comes from studies of prokaryotes that—along with algae—produce the beautiful colors of hot springs. While conducting basic research on *Thermus aquaticus*, a hot-springs prokaryote found in Yellowstone National Park in the 1960s, scientists learned that it has a form of DNA polymerase (called *TAQ polymerase*) that remains enzymatically active at high temperatures. TAQ polymerase is now used worldwide in the polymerase chain reaction (PCR), which replicates DNA through repeated cycles of heating and cooling. A Swiss company patented TAQ polymerase, the sales of which bring in $300 million a year for the company. Future development of biological resources from Yellowstone will involve "shared benefits contracts," in which the national park benefits from development and sales of products. It is possible that other endangered natural areas, such as rain forests, could be managed in a similar fashion, providing both conservation and economic benefits for companies.

The prokaryotes of hot springs are often quite colorful and therefore easily noticed, but most have not been well studied. Many other prokaryotes are unseen and unknown. Microbiologists estimate that more than 95% of the types of prokaryotes that inhabit soil have not been identified. We know nothing of these organisms' mode of nutrition, role in soil ecology, or potential usefulness in medicine, industry, or research.

In this chapter, we will investigate the diversity of viruses and prokaryotes. After exploring their form and function, we will examine the impact they have on the life of plants.

Morning Glory Pool in Yellowstone National Park.

PLANTS & PEOPLE

The Discovery of Viruses in Tobacco

The history of the discovery of viruses provides an interesting and typical example of the slow accumulation of scientific knowledge that results eventually in increased understanding. In 1883, Adolf Mayer, a German scientist, was looking for the cause of tobacco mosaic disease, which results in stunted plants with mottled, discolored leaves. Meyer found that he could transfer the disease to healthy plants by spraying them with sap collected from an infected plant. He concluded that the disease was caused by bacteria too small to detect with a microscope.

In 1884, Charles Chamberland, a colleague of Louis Pasteur, invented the porcelain bacterial filter. Dmitri Iwanowski, a Russian scientist, put sap from diseased plants through one of Chamberland's filters in 1892. The infectious agent passed through, and Iwanowski also concluded that he was working with tiny bacteria.

In 1898, Martinus Beijerinck, a Dutch scientist, found that the pathogen responsible for tobacco mosaic disease was not diluted when passed from plant to plant in sap. Therefore, the infectious agent must have reproduced in each plant and was not simply a chemical toxin. Beijerinck showed that the infectious agent reproduced only

TMV

A tobacco plant infected with tobacco mosaic disease.

in living cells, although it could survive for a long time when dried. He proposed that a filterable bacterium was involved. Around 1900, Friedrich Loeffler and Paul Frosch in Germany showed that a filterable infectious agent was the cause of hoof-and-mouth disease in cattle. Walter Reed came to the same conclusion about yellow fever.

After World War I, Felix d'Herelle proved the existence of bacteriophages, viruses that infect bacteria. He spread a liquid containing viruses on a layer of bacteria on agar and observed that clear areas soon formed. Each clear area contained broken bacterial cells due to the presence of one or more bacteriophages.

In 1935, Wendell Stanley, an American, crystallized particles of tobacco mosaic virus (TMV), the infectious agent in tobacco mosaic disease. He determined that the particles were composed of proteins. The nucleic acids in TMV were discovered later by Friederick Bawden and Norman Pirie. In 1955, Rosalind Franklin and others visualized TMV using X-ray diffraction and electron microscopy. In 1956, Heinz Frankel-Conrat and his co-workers discovered the infectivity of RNA from TMV.

Viruses and the Botanical World

When most people think of viruses, they think of viral diseases that affect human health. In fact, the word *virus* comes from a Latin root meaning "poison." However, viruses infect all organisms, including bacteria. Viruses that infect agricultural crops substantially lower yields and increase production costs. The key to controlling viral diseases is to understand the genetics, physiology, and biochemistry of viruses.

Viruses are complexes of nucleic acid and protein that reproduce inside cells

Early microbiologists considered viruses to be very small bacteria since they caused disease as bacteria do but could

not be seen with a light microscope and passed through filters that retained most bacteria. Scientists did not begin to discover the actual nature of viruses until 1935, when particles of tobacco mosaic virus were crystallized from an infected plant (see the *Plants & People* box on this page). We now know that viruses are distinct from bacteria and all other kinds of organisms. Most viruses are only 20–60 nanometers in diameter and thus are less than half the size of the tiniest cells. Although viruses have a variety of shapes (Figure 17.1), most consist of just two types of molecules: nucleic acid, either RNA or DNA; and protein, which forms an outer coat, or capsid. Some viruses also have a membrane called a *viral envelope* surrounding the capsid. The viral envelope is composed of lipids and proteins. The nucleic acid in a virus codes for between 3 and 200 proteins. Viruses have no organelles or other cellular structures and can reproduce only inside the cells of organisms, including plants. Because of their molecular sim-

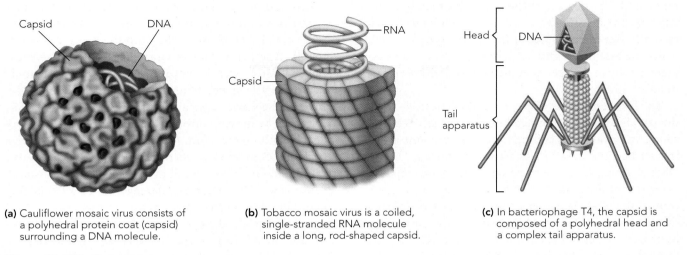

(a) Cauliflower mosaic virus consists of a polyhedral protein coat (capsid) surrounding a DNA molecule.

(b) Tobacco mosaic virus is a coiled, single-stranded RNA molecule inside a long, rod-shaped capsid.

(c) In bacteriophage T4, the capsid is composed of a polyhedral head and a complex tail apparatus.

Figure 17.1 Viral structure.

plicity and inability to perform any life functions outside living cells, viruses generally are not classified as living things. Some scientists speculate that viruses may have evolved from plasmids or transposons. Analyzing the sequences of nucleotides in viral nucleic acids should provide valuable clues about the evolution of viruses.

The best-studied viruses with respect to their reproductive cycles and genetics are those that infect bacteria. Such viruses are called *bacteriophages* (literally, "bacteria

eaters"), or *phages* for short. Bacteriophages can reproduce by either of two possible mechanisms: the lytic cycle and the lysogenic cycle (Figure 17.2).

In the typical **lytic cycle,** a virus attaches by its capsid to binding sites on the host cell's plasma membrane. The viral DNA then enters the cell and is transcribed, and the resulting RNA is translated. The DNA contains genes for one or more capsid proteins and, in some cases, for enzymes needed to make new viruses. The viral DNA is also

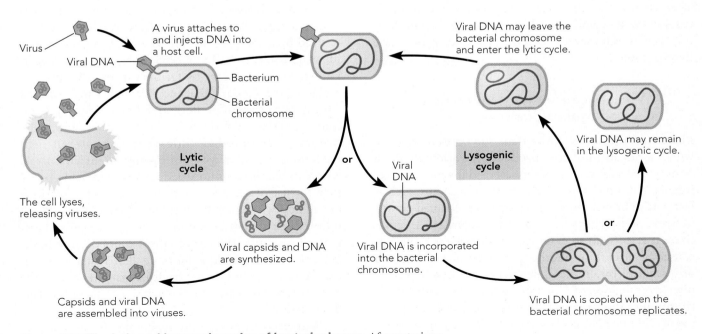

Figure 17.2 The lytic and lysogenic cycles of bacteriophages. After entering a bacterium, viral DNA may immediately enter the lytic cycle and begin directing the production of new viruses. Alternatively, the viral DNA may insert itself into the bacterial chromosome and enter the lysogenic cycle. Under certain conditions, viral DNA may switch from the lysogenic cycle to the lytic cycle.

replicated. The virus uses the host cell's metabolic machinery as well as the cell's amino acids and nucleic acids to synthesize many copies of the capsid and viral DNA, which are assembled into new viruses. The host cell eventually ruptures, or lyses, and the new viruses are released to infect other cells.

The **lysogenic cycle** differs from the lytic cycle in that the viral DNA is incorporated into the host cell's chromosome, but viral genes coding for capsid proteins are not transcribed and new viruses are not produced. The viral DNA is copied when the bacterial chromosome replicates prior to cell division, and each daughter cell receives a copy of the viral DNA. Thus, the virus's genetic information is maintained within the genome of the host cell and its progeny. Viruses that enter a lysogenic cycle can persist through periods when bacteria become dormant, such as when food is unavailable. Under these circumstances, bacteria often destroy their mRNA and protein, so viruses would have no raw materials with which to produce new viruses. When environmental conditions improve, bacteria return to a normal metabolism, and the viral DNA may leave the bacterial chromosome and enter the lytic cycle. Certain environmental influences, such as radiation or particular chemicals, may also cause a switch from the lysogenic cycle to the lytic cycle.

Eukaryotes have acute, lytic viral infections as well as persistent viral infections in which the virus may become dormant or increase in number very slowly. It is not yet clear whether the latter type of infection is caused by viruses in the lysogenic cycle, because such infections have not been as well characterized in eukaryotes as they have in prokaryotes.

Viruses cause many important plant diseases

Plants are infected by more than 500 types of viruses, which seriously limit crop production worldwide. Each virus typically causes diseases in several different crops. Infected plants often show characteristic symptoms, such as bleached or brown spots or rings on leaves, stunted growth of stems, or damaged flowers or roots (Figure 17.3). Sometimes it is difficult to determine if damage to a plant is caused by viruses, by another pathogen such as bacteria or insects, or by a mineral deficiency in the soil. A definite determination of viral infection can be made by purification of the virus and electron microscopy or by the use of DNA or RNA probes to viral-specific genes.

Various organisms carry viruses from one plant to another. Insects are the most common vectors, although birds, fungi, and even humans can also serve as agents of transmission. Viruses most often enter plants through cracks or wounds that breach plants' thick cell walls and waxy cuticle, which normally limit the access of viruses to cells. Another method of viral entry is through infected pollen grains. Viruses reproduce in plant cells via a modified lytic cycle in which infected cells usually do not lyse. Instead, the viruses typically spread from one cell to adjacent cells by moving through plasmodesmata. They may also be able to exit from infected cells by exocytosis or budding. Once viruses enter the phloem, they can spread rapidly to virtually any cell in the plant. A virus usually causes a localized region of damage, called a *lesion,* followed by widespread damage as the virus spreads.

Let us consider three important families of plant viruses in more detail. *Potyviruses* are a very large family of rod-shaped RNA viruses. They cause a number of mosaic diseases of crop plants, including potatoes, beans, soybeans, sugarcane, and peppers (Figure 17.4). The diseases are characterized by yellowish mottling on leaves, stems, and fruits and may spread to every plant in a field. Infected plants are usually stunted and produce deformed fruits. Potyviruses are transmitted from plant to plant by aphids, and some are carried into uninfected fields inside seeds from infected plants. No effective control procedures have been found. Destroying infected plants and using insecticides to kill aphids only delay the spread of the disease.

Waikaviruses and *badnaviruses* are responsible for several serious diseases in cereals, including tungro, the disease that currently causes the biggest crop losses in rice. Caused by a complex of both viruses, tungro was first identified scientifically in 1965, although it had been known for many years by farmers in the Philippines, Malaysia, Indonesia, India, and Pakistan. Tungro means "degenerated growth" in Ilocano (a Philippine language). Infected plants are stunted; have deformed, yellow, mottled leaves; and produce little or no rice. Overall yield reductions in Asia range from 10–60% in infected areas. No treatment is recommended for infected fields, other than the use of insecticides to cut down on populations of leaf hoppers, which transmit the disease. The best solution to persistent infections is to use rice germplasm that is resistant to the viruses, the leaf hoppers, or both. Since rice is the main food for more than 3 billion people, tungro has very important health consequences. In regions where it occurs, tungro is a major contributing factor in malnutrition.

Plant viral diseases are also highly significant economically. In 2000, the value of plant food crops worldwide was approximately $2 trillion. Approximately $800 billion of that value was lost to diseases (viral, bacterial, and

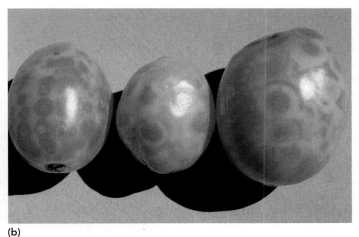

Figure 17.3 Effects of viral infections in plants. (a) Tomato spotted wilt virus in *Alstroemeria*. (b) Spotting on fruit infected with tomato spotted wilt virus.

(a)

(b)

fungal), insects, and weeds. The losses vary for specific crops and are higher in developing countries in which disease-resistant seed is expensive and difficult to obtain. Since many countries regularly straddle the line between just enough food and starvation, these crop losses are a major concern.

Several approaches are used to prevent viral diseases in plants

In general, it is much easier to prevent viral diseases than it is to cure them. In animals, vaccination has been effective at preventing many viral diseases since Edward Jenner used it to protect against smallpox in 1796. In plants, the development of resistant varieties is the most widely used method of avoiding viral infections. For some perennial crop plants, viral infections take several years to become

well established. The use of new, uninfected germplasm (such as seeds and cuttings) can avoid levels of infection that significantly reduce yields.

Virus-free plants are also sometimes obtained by removing the apical meristem at the very top of the plant and growing it into a new plant in tissue culture (see Chapter 14). The shoot apical meristem lacks vascular tissue, so the movement of viruses into meristem cells is radically reduced.

In 1986, Roger Beachy and other researchers showed that if viral genes are incorporated into the genome of plants, the transgenic plants are often subsequently resistant to viral infection. Typically, viral coat protein genes have been used, but other genes have worked as well. As you learned in Chapter 14, the transgenic plants block one or more early events in the process by which viruses reproduce inside plant cells. Resistance is usually specific for the same type of virus from which the genes were transferred, although resistance to related viruses may also occur.

Several plant genes that confer resistance to viral, bacterial, or fungal diseases have been isolated, and some of these genes have been found to induce disease resistance in transgenic plants. Disease resistance has also

(a)

(b)

Figure 17.4 Potyviruses. (a) Double-virus streak caused by a combination of potato virus X and tobacco mosaic virus leaving the fruit rough with irregular brown patches. (b) Alfalfa mosaic virus in potato.

been produced by giving plants genes that code for animal antibodies, another technique described in Chapter 14. For example, transgenic artichoke plants produced by this technique make antibodies to coat proteins of the artichoke mottle crinkle virus and can thus defend themselves against the virus to some extent.

Viroids are infectious RNA molecules

You might think that infectious agents could not get any simpler than a small virus, but that is not the case. Almost two dozen plant diseases are caused by **viroids,** which are circular strands of RNA containing 250–370 nucleotides. The RNA is not encased in a protein capsid as it is in an RNA virus, and it does not appear to be translated into protein. Viroids are found in the nucleolus of infected plants, but why they are there and how they cause disease is not known.

Viroids cause diseases in a number of plants, often discoloring leaves, cracking fruits or stems, stunting growth, and destroying flowers. Cadang-cadang is a viroid disease of coconut palms, which kills more than 1 million plants each year, particularly in the Philippines where coconuts are an important crop (Figure 17.5). The disease takes ten years or more to develop. Cadang-cadang is a devastating problem because it kills older coconut palms, in which growers have already invested considerable money and effort. Neither resistant cultivars nor a way to stop the spread of the disease has been found.

Section Review

1. Why aren't viruses generally classified as living things?
2. Explain the difference between the lytic cycle and the lysogenic cycle.
3. What is tungro?

Prokaryotes and the Botanical World

The seventeenth-century Dutch microscopist Anton van Leeuwenhoek realized that bacteria are both extremely numerous and quite diverse. Scientists since his time have confirmed and broadened that view. Understanding how bacteria and other prokaryotes survive and reproduce has led to methods for controlling many diseases of plants and animals.

Figure 17.5 Cadang-cadang. A viroid disease of coconut palms has killed all of the trees in this plantation.

Prokaryotes are unicellular organisms with diverse characteristics

You learned in Chapter 2 that prokaryotes are unicellular organisms that lack an enclosed nucleus and membrane-bounded organelles. Most of the DNA in a prokaryote occurs in the form of a small, circular chromosome that contains an average of several thousand genes. Prokaryotes also have even smaller loops of DNA called *plasmids,* each of which may contain about ten genes. Bacteria can exchange DNA by several methods, and plasmids can move between species. In Chapter 14 we saw how genetic engineers use the Ti plasmid of *Agrobacterium* to produce transgenic plants.

Prokaryotes typically have one of three characteristic shapes (Figure 17.6): a sphere called a *coccus* (plural, *cocci*), a cylinder called a *rod* or *bacillus* (plural, *bacilli*), and a curved or spiral rod called a *spirillum* (plural, *spirilla*). Most prokaryotic cells are considerably smaller than the cells of eukaryotes. The reason for this difference in size may be that in prokaryotes a number of important cellular functions, including transport, photosynthesis, and respiration, are carried out by the cell membrane and its extensions. In eukaryotes, photosynthesis and respiration occur in separate, membrane-bounded organelles within the cell, while the cell membrane is dedicated almost exclusively to controlling the molecular traffic into and out of the cell.

Most prokaryotes have a cell wall, and the structure of the cell wall is used in identifying and classifying prokaryotes. You know from your reading in Chapter 16 that prokaryotes are classified in two domains, Archaea and

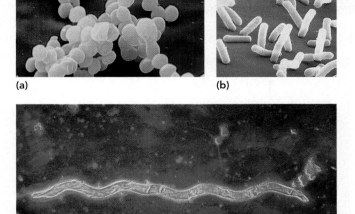

Figure 17.6 Characteristic shapes of prokaryotes.
(a) Cocci are spherical. (b) Bacilli are cylindrical or rod-shaped.
(c) Spirilla are spiral or curved.

Bacteria. The cell walls of archaea vary in structure but never contain muramic acid. Bacteria can be divided into three groups based on cell wall structure: (1) Gram-positive bacteria have a thick cell wall that traps the violet Gram stain and contains muramic acid and large amounts of peptidoglycan, a sugar polymer attached to protein. The protein cross-links between the long sugar molecules in peptidoglycan give this type of cell wall its strength. (2) Gram-negative bacteria have a thin cell wall that contains muramic acid but much less peptidoglycan. (3) Mycoplasmas have no cell wall at all. Some prokaryotes also have either a slime layer or a more substantial capsule outside the cell wall. Both are composed of polysaccharides, and the term *glycocalyx* is sometimes used to describe the polysaccharide complex that includes the cell wall and any external layers.

Recall from Chapter 15 that the oldest prokaryotic fossils are 3.5 billion years old, whereas the oldest fossils thought to be eukaryotic are about 2.1–2.2 billion years old. Thus, prokaryotes were probably the only life-form on Earth for more than 1 billion years. During this time, they underwent an adaptive radiation into many thousands of species suited to the range of environments that existed on ancient Earth. About 4,000 species of prokaryotes have been described and named, and microbiologists estimate that another 500,000 to 4 million remain to be discovered.

The prokaryotes that have been characterized are tremendously diverse in terms of where they live and how they obtain energy. For example, the *Conservation Biology* box on page 378 describes prokaryotes that live in dry soil and on the surfaces of rocks in North American deserts. A number of types of prokaryotes live in the digestive tracts of herbivores, including cows, giraffes, horses, rhinoceroses, rabbits, termites, caterpillars, and earthworms. These prokaryotes use the enzyme cellulase to break down celluose into glucose, much of which is absorbed by their hosts. Many prokaryotes, including *Escherichia coli,* also inhabit the human digestive tract, but they do not digest cellulose. Considering the prevalence of cellulose in plants, it is curious that animals do not produce cellulase themselves but instead rely on prokaryotes and certain other organisms to make cellulose usable as a food source.

Many archaea thrive under extreme conditions that are lethal to most other organisms. Methanogens, which produce methane gas and are killed by oxygen, live in anaerobic environments, such as marsh and swamp sediments. Thermophilic, or heat-loving, archaea are found in geothermal springs and undersea vents, where the water is hot enough to kill other organisms. Some deep-sea vent archaea can grow and reproduce under high pressures and at water temperatures of 113°C (235°F)! Halophilic, or salt-loving, archaea live in waters with very high salinity. They carry out a primitive form of photosynthesis that traps energy in ATP using a pigment known as *bacteriorhodopsin.*

Some bacteria are photosynthetic, and some fix nitrogen

All photosynthetic bacteria are Gram-negative. They include green sulfur bacteria, purple sulfur bacteria, prochlorophytes, and cyanobacteria. By definition, all photosynthetic bacteria use carbon dioxide (CO_2) as a source of carbon atoms, but they have variable sources of electrons. The sulfur bacteria are anaerobic and reduce CO_2 to carbohydrate with electrons from hydrogen sulfide (H_2S), sulfur (S), or hydrogen gas (H_2):

$$CO_2 + 2H_2S \rightarrow (CH_2O) + H_2O + 2S$$

For comparison and review, cyanobacteria, algae, and plants reduce CO_2 to carbohydrate with electrons from water:

$$CO_2 + 2H_2O \rightarrow (CH_2O) + H_2O + O_2$$

Sulfur bacteria use bacteriochlorophylls that differ in structure from chlorophyll *a* and that absorb light at longer wavelengths. These bacteria have one photosystem instead of two. The chlorophyll in purple sulfur bacteria is located in vesicles attached to the cell membrane, while that in green sulfur bacteria is in internal layered extensions of the cell membrane.

CONSERVATION BIOLOGY

Cryptobiotic Crust and Desert Varnish

Prokaryotes are vital members of inconspicuous communities known as *cryptobiotic crusts* and *desert varnish*. Cryptobiotic crusts are composites of lichens (see Chapter 19) and cyanobacteria that occur in the soil of desert areas in northern Mexico and the southwestern United States. These pioneer organisms secure the thin soil with sticky compounds produced by the cyanobacteria, which also enrich the soil by fixing nitrogen. Their effect on the soil paves the way for the eventual establishment of desert shrubs and other plants.

Cryptobiotic crust.

Desert varnish.

Cryptobiotic crusts are dry and gray most of the year but turn green soon after a rain. Full development of a mature crust requires decades and can involve more than one species of lichen and cyanobacteria. However, the crusts are so fragile that a single footprint can destroy years of growth.

Desert varnish is a thin black, brown, or reddish-brown coating that forms on rocks exposed to the hot desert sun. The varnish is produced by bacteria of several genera, including *Metallogenium* and *Pedomicrobium*, which oxidize trace amounts of iron and manganese, principally from rainwater and the atmosphere. The oxidation contributes electrons to an electron transport chain that provides the energy to synthesize ATP. The iron oxide and manganese oxide by-products are deposited over thousands of years to produce the varnish, which is seen most dramatically when a rock splits and reveals its unvarnished interior.

Scientists are very interested in studying the genes, enzymes, membranes, and physiological systems that enable prokaryotes to thrive in deserts and other marginal environments. Such studies may provide clues about the origins of prokaryotes on Earth.

Prochlorophytes are a group of photosynthetic bacteria that use chlorophyll *a* as their primary photosynthetic pigment and, like plants, chlorophyll *b* and carotenoids as accessory pigments. Therefore, they may be relatives of the prokaryotes that became chloroplasts in green algae and plants. Both free-living and symbiotic species have been found. Prochlorophytes occur commonly in ocean waters as part of phytoplankton (see *The Intriguing World of Plants* box on page 379).

Cyanobacteria carry out photosynthesis using chlorophyll *a* along with carotenoids and two unique accessory pigments called *phycobilins:* phycocyanin, which is blue, and phycoerythrin, which is red. Cyanobacteria have many parallel layers of membranes derived from the cell membrane that serve the same purpose as thylakoids in chloroplasts. In fact, chloroplasts probably originated as endosymbiotic cyanobacteria. The chloroplasts of red algae (phylum Rhodophyta) show a particularly close resemblance to cyanobacteria.

Most cyanobacteria form filaments of cells (Figure 17.7), a cellular association that led to their old, erroneous name, "blue-green algae." (Remember that algae are eukaryotes, whereas cyanobacteria are prokaryotes.) Filamentous cyanobacteria often have gas vesicles, which provide flotation, as well as resistant, thick-walled cells called *akinetes*,

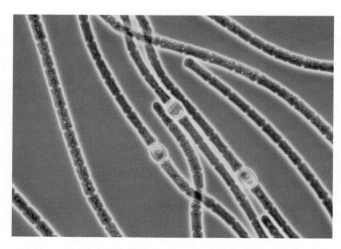

Figure 17.7 A filamentous cyanobacterium. *Anabaena* is a common cyanobacterium that forms filaments in freshwater habitats. The enlarged structures in each filament are heterocysts, specialized cells in which nitrogen fixation takes place.

THE INTRIGUING WORLD OF PLANTS

The Huge Importance of Tiny Photosynthesizers

Nearly half of all the photosynthesis on Earth occurs in the oceans. Until recently, single-celled algae (see Chapter 18) were thought to be the major contributors to marine photosynthesis, and prokaryotes were considered only minor contributors. This view changed in the 1970s and 1980s, when the marine bacteria *Synechococcus* and *Prochlorococcus* were discovered. Both belong to the group of photosynthetic prokaryotes known as *prochlorophytes*.

The concentration of prochlorophytes in the ocean can reach 1 billion cells per liter. Some investigators think these bacteria may be the single most important group of organisms on Earth in terms of their total photosynthetic output. In warm oceans, between 40° south and 40° north latitude, up to 40% of carbon fixation is carried out by *Synechococcus* and *Prochlorococcus* alone. The contribution of prochlorophytes to global photosynthesis may have been underestimated originally

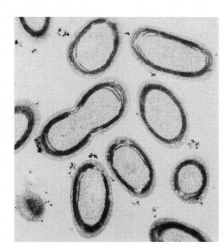

Prochlorococcus.

because they are extremely small—*Synechococcus* is 0.8–1.5 micrometers (μm) in diameter and *Prochlorococcus* only 0.5–0.7 μm—so they cannot be removed from seawater except with very fine filters.

Prochlorococcus live as deep as 200 meters (656 feet) below the ocean surface, where the intensity of light available for photosynthesis is only 0.1% of that at the surface. Its forms of chlorophylls *a* and *b* can absorb the wavelengths of light that penetrate to that depth. At least two types of *Prochlorococcus* occur, one adapted to high light intensity and one to low. The role of this important organism in ocean ecology is not completely understood. *Prochlorococcus* is believed to have evolved from a cyanobacterium with a reduction in cell and genome size. The organism's 1,700 genes have been completely sequenced, and scientists are actively studying the function of each gene product.

---

which can allow the bacteria to endure hostile conditions for several years.

More than 1,500 species of cyanobacteria live in a variety of environments—hot springs, farm ponds, tropical ocean waters, and even frozen lakes in Antarctica. Cyanobacteria also exist in symbiotic relationships with other organisms, including protozoa, sponges, algae that lack chlorophyll, bryophytes, and vascular plants. They are the photosynthetic partners of fungi in some lichens.

Many genera and species of bacteria, including cyanobacteria, are capable of fixing nitrogen. Recall from Chapters 4 and 10 that nitrogen-fixing bacteria have symbiotic associations with the roots of legumes, such as peas, soybeans, and alfalfa. Some filamentous cyanobacteria form large, thick-walled cells called *heterocysts* that fix nitrogen while keeping out oxygen, which inhibits the fixation process (Figure 17.7). Tropical cyanobacteria are major contributors to nitrogen fixation in the ocean. In rice paddies, the nitrogen-fixing cyanobacterium *Anabaena azollae* lives in cavities on the underside of leaves of *Azolla filiculonides*, a water fern. The fixed nitrogen provided by *Anabaena* is an important nutrient for *Azolla* and a natural fertilizer for the rice plants when the water ferns die and decompose.

Bacteria cause a variety of diseases in plants

Around 100 species of bacteria are known to cause plant diseases. Such diseases are commonly called *blights, blasts, wilts, soft rots, galls, cancers,* and *scabs.* The damage usually results from exotoxins, protein toxins secreted by the bacteria, although it is sometimes caused by endotoxins, which are toxins bound to bacterial cell walls or plasma membranes.

Bacterial soft rots typically affect juicy vegetables and fruits, such as carrots, potatoes, onions, tomatoes, and peppers (Figure 17.8). Altogether, they cause more damage

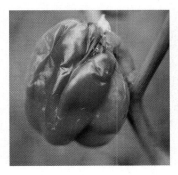

Figure 17.8 Bacterial soft rot. This pepper is infected with soft rot, caused by the bacterium *Erwinia carotovora.*

and loss to vegetable and fruit crops than any other bacterial disease. Various species of *Erwinia*, *Pseudomonas*, *Bacillus*, and *Clostridium* are associated with these diseases. The bacteria enter small wounds on the surface of the vegetable or fruit or are carried in by various insect larvae. Bacteria first feed on liquids released by wounded cells. They then release enzymes that break down the middle lamella between cells and, eventually, the cells themselves.

The first symptom of a soft rot is a watery area that begins to expand and deepen. The bacteria turn the plant tissue into a creamy, slimy mush, which often smells foul in later stages. Soft rots may develop in the field, during marketing, or after purchase. In the field, root crops are often infected, leading to death of the entire plant. If products even minimally infected are stored in plastic bags or containers, they often turn to mush within a few days.

Farmers use several approaches to control soft rots. They grow plants in well-drained soils and store harvested products in well-ventilated, dry areas. They also rotate vegetable crops with cereal crops to minimize disease carryover from year to year. The effectiveness of the latter approach is limited, however, because the bacteria that cause soft rots can overwinter in the soil or in insects and their larvae.

Lethal yellowing (Figure 17.9) is a fatal disease of palms that is caused by a mycoplasma and is spread by a leaf hopper. Around 75% of the coconut palms in Florida were destroyed by lethal yellowing by the 1960s. Currently, the disease is killing up to 90% of the coconut palms in Belize. Lethal yellowing can be prevented by injecting palms with antibiotics. That method sometimes works against other bacterial diseases in plants, but in general antibiotics are difficult to get into plants because of the cell walls, and they do not circulate well within most plants when they do get in.

Plants themselves have a surprising number of mechanisms for warding off bacterial infections and for containing or ending them once they occur. The first line of defense is the waxy cuticle, which discourages entry through the epidermis. Bacteria generally enter a plant through wounds or stomata. Some plants also release chemical inhibitors of bacteria into the soil or have such inhibitors in their cell walls. The presence of pathogenic bacteria inside a plant triggers other defensive mechanisms. Plant cells near the site of an infection produce antimicrobial chemicals called *phytoalexins*. The plant also recognizes specific pathogen molecules and responds by initiating a number of biochemical reactions that lead to cell death localized to the site of infection. This process, known as the *hypersensitive response* (see Chapter 11), limits the spread of the bacteria inside the plant. Invading bacteria may also be isolated by the formation of cork layers, while abscission layers may result in the shedding of infected leaves.

Usually the best way to control bacterial diseases in plants is to replant with varieties that are resistant to the pathogen. Resistance to a specific strain or race of a pathogen is called *vertical resistance*. A generalized resistance to a group of pathogens—for example, to all strains of a particular species of bacterium—is called *horizontal resistance*. Infection by one pathogen may induce systemic acquired resistance in which the entire plant gains resistance to a number of bacterial, viral, and fungal diseases. This process requires the activation of many genes and involves salicylic acid as a mediator (see Chapter 11).

Unfortunately, no plant can be made resistant to all diseases or even to all genetic variants of one disease. The tendency of modern food production systems to grow "monocultures," or large areas of a single variety of plant, increases the susceptibility of our food supply to diseases, including those that might be introduced by bioterrorists. Producing many varieties of crops with different spectra of disease resistance would be a major step toward lessening the impact of crop diseases and slowing their spread.

Prokaryotes have many uses in industry, medicine, and biotechnology

Humans have successfully harnessed the varied metabolic abilities of prokaryotes to make a number of useful products. Bacteria have been used for millennia to make cheese, buttermilk, and yogurt. Almost all varieties of these products are produced and flavored by bacteria of the genus *Lactobacillus*. Various sorts of meat and fish are also fermented by bacteria, as are some vegetables, such as the cabbage and the cucumber, which are converted into sauerkraut and pickles, respectively. These products probably came about originally by accident as various

Figure 17.9 Lethal yellowing. Most of the coconut palms in this grove in Ghana, Africa, were killed by lethal yellowing, a disease caused by a mycoplasma.

Figure 17.10 Making use of bacteria. These workers are spraying fertilizers on ground contaminated by an oil spill. The fertilizers promote the growth of soil bacteria, which help break down the oil.

foods were allowed to sit around until they spoiled in particular ways.

Bacteria are grown in large fermenters to produce vitamins and amino acids, which are sold individually and used as food additives. Some cyanobacteria, including *Aphanothece, Nostoc, Brachytrichia,* and *Spirulina,* serve as food in Japanese and Chinese cooking. Native populations in Africa and Mexico also used *Spirulina* for food. According to Spanish records, the Aztecs in central Mexico knew *Spirulina* as *tecuitlatl,* which was dried into cakes and used in a wide variety of dishes. Dried preparations of this cyanobacterium are now sold in health-food stores because of their high vitamin content.

Bacteria produce many of the antibiotics used in medicine, including streptomycin, tetracycline, and cycloheximide, which are synthesized by actinomycetes, soil bacteria in the genus *Streptomyces.* In nature, these compounds inhibit the growth of other bacteria that compete with *Streptomyces* for growing space and food. In the laboratory, colonies of *Streptomyces* can be grown on nutrient

medium coated with a layer of other bacteria. The colonies that produce antibiotics are readily recognized because each is surrounded by a clear area where the other bacteria have been killed by the antibiotic.

Many new uses for prokaryotes have been discovered in recent years. Biodegradable plastics can be both produced and broken down by bacteria such as *Alcaligenes eutrophas.* Use of these plastics may help reduce the need for new landfills, which are rapidly filling with plastics that do not degrade. The bacterium *Thiobacillus ferrooxidans* oxidizes metals such as copper, gold, and uranium, producing water-soluble compounds that can be readily extracted from low-grade ores. Prokaryotes are also used to clean up oil spills and to decontaminate soils containing pesticides or other toxic substances, a process known as **bioremediation** (Figure 17.10).

Certain strains of the soil bacterium *Pseudomonas syringae* can promote the formation of ice crystals and are being investigated for their potential usefulness in increasing snowfall. Other strains of the same bacterium retard ice formation and are sprayed on plants to save them from frost. Even grander weather modification schemes involving *P. syringae* are envisioned. Some biologists have suggested that this bacterium already influences the weather through its ability to affect ice formation.

Section Review

1. **Describe and name the three characteristic shapes of prokaryotes.**
2. **What substances do sulfur bacteria use as sources of electrons in photosynthesis?**
3. **What is the difference between vertical resistance and horizontal resistance?**
4. **List four ways that people use bacteria in industry and medicine.**

SUMMARY

Viruses and the Botanical World

Viruses are complexes of nucleic acid and protein that reproduce inside cells (pp. 372–374)
Viruses are nonliving complexes of DNA or RNA surrounded by a protein coat. Some viruses reproduce by means of the lytic

cycle, in which their DNA enters a host cell and leads to the production of new viruses that are released when the cell lyses. Other viruses incorporate their DNA into a cell's genome in the lysogenic cycle.

Viruses cause many important plant diseases (pp. 374–375)
Viruses are carried from plant to plant by vector organisms, usually enter plants through cracks or wounds, and move within plants via plasmodesmata and the phloem. Potyviruses cause

mosaic diseases of many crop plants, and waikaviruses and badnaviruses cause diseases in cereals, including the rice disease tungro.

Several approaches are used to prevent viral diseases in plants (pp. 375–376)
Botanists work to avoid viral infections in plants by developing resistant varieties and by using uninfected germplasm to grow virus-free plants. Genetic engineers have created plants that are resistant to some viruses by introducing genes for viral coat proteins or animal antibodies into the plants' genomes.

Viroids are infectious RNA molecules (p. 376)
Viroids are small, circular strands of RNA that cause a variety of plant diseases, including cadang-cadang, which affects coconut palms.

Prokaryotes and the Botanical World

Prokaryotes are unicellular organisms with diverse characteristics (pp. 376–377)
Prokaryotes are small cells that usually have a spherical, cylindrical, or spiral shape. Most prokaryotes have a cell wall, the structure of which is one basis for their identification and classification. Many prokaryotes in the domain Archaea live in habitats that are inhospitable to most other organisms, such as anaerobic environments, hot springs, and very salty waters.

Some bacteria are photosynthetic, and some fix nitrogen (pp. 377–379)
Photosynthetic sulfur bacteria reduce carbon dioxide with electrons from hydrogen sulfide, sulfur, or hydrogen gas. Cyanobacteria carry out photosynthesis using electrons from water. Many bacteria can fix nitrogen, and some exist in symbiotic relationships in which they provide fixed nitrogen or other nutrients to other organisms, including plants.

Bacteria cause a variety of diseases in plants (pp. 379–380)
Soft rots are bacterial diseases that typically affect juicy vegetables and fruits. Lethal yellowing is a bacterial disease of palms that is spread by a leaf hopper. Plants have a variety of mechanisms for preventing, containing, and ending bacterial infections.

Prokaryotes have many uses in industry, medicine, and biotechnology (pp. 380–381)
Prokaryotes are used to make various foods and food additives, antibiotics, and biodegradable plastics. They are also used to extract valuable metals from ores and to clean contaminated soil.

Review Questions

1. Of what types of molecules are viruses composed?
2. What is a bacteriophage?
3. Under what conditions is it advantageous for viruses to enter the lysogenic cycle?

4. What symptoms are characteristic of viral infections in plants?
5. In a viral infection, what is a vector?
6. What are potyviruses?
7. What is cadang-cadang?
8. Describe two approaches that have been used to produce transgenic plants that are resistant to viral diseases.
9. What distinguishes Gram-positive from Gram-negative bacteria?
10. Where do thermophilic archaea live?
11. How are the photosynthetic membranes of cyanobacteria related to the cell membrane in these prokaryotes?
12. How do farmers minimize crop losses due to soft rots?
13. What are phytoalexins?
14. From the standpoint of controlling plant diseases, why should we cultivate several types of a crop plant with diverse genetic backgrounds?

Questions for Thought and Discussion

1. How do you think viruses might have evolved from transposons?
2. Some biologists take the position that viruses should be classified as simple organisms. Using what you know about viruses, present arguments for and against this position.
3. Many viruses and bacteria kill their hosts. Does this outcome provide a selective advantage for the virus or bacterium? Defend your answer.
4. Scientists have discovered a huge, unicellular bacterium that is visible to the unaided eye. How do you suppose this bacterium has gotten around the usual limitations on bacterial size?
5. Suppose you are asked to isolate a previously unknown bacterium from the soil and to find out as much as you can about it. What sorts of experiments would you try?
6. Draw labeled diagrams to illustrate the similarities and differences between a bacterial cell and a plant cell.

Evolution Connection

Discuss possible coevolutionary relationships between bacteria and plants by comparing bacteria that cause serious plant diseases with those that trigger hypersensitive responses in plant hosts.

To Learn More

Visit The Botany Place Website at www.thebotanyplace.com for quizzes, exercises, and Web links to new and interesting information related to this chapter.

Hull, Roger. *Matthews' Plant Virology, Fourth Edition.* New York: Academic Press, 2001. This text is updated every ten years. The new edition contains classic material as well as newer molecular topics.

Oldstone, Michael. *Viruses, Plagues, and History.* London: Oxford University Press, 1998. How would the New World have been different if smallpox and other European diseases had not raged through Native American populations? This book asks intriguing "what if" questions and fills you in on how diseases have changed human history.

Sutic, Dragoljub D., Richard E. Ford, and Malisa T. Tosic, eds. *Handbook of Plant Virus Diseases.* Boca Raton, FL: CRC Press, 1999. This handbook presents basic information about viral-caused and viral-like diseases in many cultivated crops, including corn, rice, and potatoes.

Tortora, Gerard, Berdell R. Funke, and Christine L. Case. *Microbiology: An Introduction, Eighth Edition.* San Francisco: Benjamin/Cummings, 2001. This interesting text provides detailed information about the science of studying viruses and microorganisms.

Algae

Giant kelp (*Macrocystis*).

Characteristics and Evolution of Algae

Algae are distinguished by their photosynthetic pigments and other characteristics

Endosymbiosis played a key role in the evolution of algae

Unicellular and Colonial Algae

Euglenoids (phylum Euglenophyta) have a pellicle beneath the plasma membrane

Many dinoflagellates (phylum Dinophyta) have hard cellulose plates

Diatoms (phylum Bacillariophyta) form cell walls of silica

Yellow-green algae (phylum Xanthophyta) are important members of freshwater phytoplankton

Golden-brown algae (phylum Chrysophyta) form unique, dormant spores

Cryptomonads (phylum Cryptophyta) use ejectisomes for sudden escape

Haptophytes (phylum Prymnesiophyta) have a distinctive, moveable haptonema

Multicellular Algae

In many brown algae (phylum Phaeophyta), alternate generations are heteromorphic

Red algae (phylum Rhodophyta) have complex life cycles with three multicellular phases

Green algae (phylum Chlorophyta) share a common ancestor with plants

A dense forest grows along the west coast of North America from Mexico to Alaska. The photosynthetic giants in this forest reach heights of 50 meters (160 feet). Nearly 100 species of invertebrate and vertebrate animals live in the forest, and many other species visit the forest to feed or to escape predators. What is unusual about this forest is that it exists not on land but in the sea, and the giants in this forest are not trees but seaweeds called *kelps.*

Complex relationships link the organisms in and around the kelp forest. For example, sea urchins (members of the phylum of invertebrates known as *echinoderms*) voraciously consume kelps, and sea otters feed on sea urchins. When sea otters are absent from an area, the sea urchin population explodes, and, before long, the kelps in the area are severely impacted. When sea otters are reintroduced, the kelps recover.

The kelp forest also affects life on land. Ocean storms rip many kelps from their moorings and toss them in piles on the beach, where they gradually decompose. A considerable population of arthropods, mollusks, and worms inhabit these compost heaps and, in turn, become food for rodents, raccoons, and other animals. Bald eagles are attracted to the scene by the rodents.

Charles Darwin knew about kelp forests from his term as ship's naturalist on HMS *Beagle,* and he recognized their importance. In *The Voyage of the Beagle* he noted that "the number of living creatures of all Orders whose existence intimately depends on kelp is wonderful. . . . I can only compare these great aquatic forests with the terrestrial ones in the intertropical regions. . . . Yet if in any country a forest was destroyed, I do not believe nearly so many species of animals would perish as would here from the destruction of kelp."

Kelps are harvested on a large scale for use as fertilizer and as a source of a glucose polymer known as *algin.* Constituting up to 35% of the dry weight of a kelp, algin serves to thicken many products, including ice cream, toothpaste, cosmetics, soaps, and paints. It is particularly useful for ice cream because it prevents the formation of ice crystals. It helps make paint flow easily so that brush marks do not show. Each year, more than 140 million kilograms (150,000 tons) of kelp, worth more than $40 million, are harvested off the coast of California alone. Kelps are in general a renewable resource and can be harvested sustainably if appropriate precautions are taken. Harvesters take only the top meter or so (about 3 feet), and since these seaweeds can grow 15 centimeters (6 inches) a day, the harvesting does not destroy them.

Kelps and other seaweeds are algae (singular, *alga*), but not all algae are seaweeds. Algae include unicellular as well as multicellular species and can be found in fresh water and on land as well as in the ocean. Most species are free living, but a few live as parasites or symbionts inside other organisms. In this chapter, we will explore the diversity of algae, beginning with seven phyla, the members of which are primarily unicellular or colonial. We will then investigate three phyla of algae that include species with true multicellularity.

A sea otter resting in kelp blades.

A sea urchin at the base of a kelp forest.

Characteristics and Evolution of Algae

Algae are members of the kingdom Protista. Recall from Chapter 16 that Protista encompasses a diverse assortment of eukaryotes, which are organisms whose cells have nuclei. Many systematists regard Protista as an artificial kingdom because its members are not all closely related phylogenetically. Most algae are photosynthetic and, therefore, are sometimes described as plantlike protists. Other protists are described as funguslike or animal-like. All of these descriptions are artificial as well. For example, some funguslike protists have cell walls containing cellulose, as the cell walls of plants do, while some algae can absorb food, as fungi do, or ingest food, as animals do.

Algae are distinguished by their photosynthetic pigments and other characteristics

Photosynthetic algae have one or more chloroplasts per cell and, like plants, use the light reactions and the Calvin cycle to convert light energy into chemical energy. All photosynthetic algae use chlorophyll a as a primary photosynthetic pigment and another form of chlorophyll as an accessory pigment: green algae and euglenoids have chlorophyll b (which is also found in plants); some red algae have chlorophyll d; and all other algae have chlorophyll c, which is unique to algae. A variety of additional accessory pigments complement the green chlorophylls to give some algae their characteristic red, brown, yellow, or golden hues. Table 18.1 lists the photosynthetic pigments and other characteristics of ten algal phyla.

Table 18.1 Characteristics of Algal Phyla

Phylum	Approximate number of known species	Photosynthetic pigments	Characteristics
Phyla containing unicellular or colonial algae			
Euglenophyta (euglenoids)	800	Chlorophyll a, b Carotenoids	Mostly freshwater Some nonphotosynthetic Flagellated
Dinophyta (dinoflagellates)	3,000	Chlorophyll a, c Carotenoids	Marine and freshwater phytoplankton in warm waters Some nonphotosynthetic Some produce nerve toxins Flagellated
Bacillariophyta (diatoms)	5,600	Chlorophyll a, c Carotenoids	Marine and freshwater phytoplankton in cool waters; some terrestrial species Cell walls made of silica
Xanthophyta (yellow-green algae)	600	Chlorophyll a, c Carotenoids	Mostly freshwater phytoplankton Flagellated
Chrysophyta (golden-brown algae)	1,000	Chlorophyll a, c Carotenoids	Marine and freshwater phytoplankton Flagellated Some nonphotosynthetic
Cryptophyta (cryptomonads)	200	Chlorophyll a, c Phycobiliproteins	Marine and freshwater phytoplankton in cold waters Flagellated
Prymnesiophyta (haptophytes)	300	Chlorophyll a, c Carotenoids	Mostly marine phytoplankton in warm waters Flagellated
Phyla containing multicellular algae			
Phaeophyta (brown algae)	1,500	Chlorophyll a, c Carotenoid (fucoxanthin)	Mostly found in intertidal and shallow marine zones Include kelps Flagellated Produce spores and gametes
Rhodophyta (red algae)	5,000	Chlorophyll a, d Phycobiliproteins	Mostly marine No flagellated cells in life cycle
Chlorophyta (green algae)	7,500	Chlorophyll a, b	Mostly freshwater Some related to plants Sometimes flagellated

The chloroplasts of many algae contain a protein-rich structure called a **pyrenoid,** which contains rubisco, the enzyme that catalyzes the first step in the Calvin cycle (see Chapter 8). Although the presence of a pyrenoid is considered a primitive trait, at least some species in every phylum of algae have a pyrenoid. Various compounds resulting from carbon fixation may be stored in the pyrenoid or the cytoplasm. The type of photosynthetic product that is stored is one of the characteristics that systematists use to classify algae. Another characteristic is the number of membranes surrounding the chloroplast (see below).

Many algal cells have a light detector consisting of a complex of pigments, including 11-*cis*-retinol, a rhodopsin that also serves as a light receptor in the visual systems of animals. Another pigmented structure, the eyespot, or *stigma,* lies in the cytoplasm or inside the chloroplast near the light detector. The eyespot acts as a shade, preventing light that comes from certain directions from striking the light detector. The amount of light received by the light detector is translated into flagellar movement. Together, the light detector and eyespot enable some algal cells to orient themselves and to move in relation to a source of light.

Endosymbiosis played a key role in the evolution of algae

You learned in Chapter 2 that some organelles, including mitochondria and chloroplasts, are thought to have resulted from endosymbiosis, a process in which one cell ingested another cell. The number of membranes surrounding chloroplasts in different algal phyla suggests that these organelles went through anywhere from one to three different endosymbiotic events (Figure 18.1). In red algae and green algae, chloroplasts are surrounded by *two* membranes: an inner membrane, which originally surrounded a photosynthetic prokaryote, and an outer membrane, which is derived from a food vacuole of a heterotrophic cell that engulfed the prokaryote. This initial endosymbiotic event is referred to as *primary endosymbiosis.*

In two other phyla of algae, euglenoids and most dinoflagellates, chloroplasts are surrounded by *three* membranes. The most likely explanation for chloroplasts with three membranes is that they are products of *secondary endosymbiosis:* an alga containing a chloroplast surrounded by two membranes was ingested by a heterotrophic cell. According to this explanation, most parts of the alga were digested inside the heterotrophic cell's food vacuole, but the alga's chloroplast persisted and became an endosymbiont. The outermost membrane of the chloroplast is derived from the food vacuole. It is studded with ribosomes because the food vacuole was associated for a time with the endoplasmic reticulum (ER) of the heterotrophic cell. Therefore, the outermost chloroplast membrane is known as the *chloroplast ER.*

Many other algae, including brown algae, have chloroplasts that are surrounded by *four* membranes. These chloroplasts may have resulted from a third endosymbiotic event like the first two, in the process picking up another

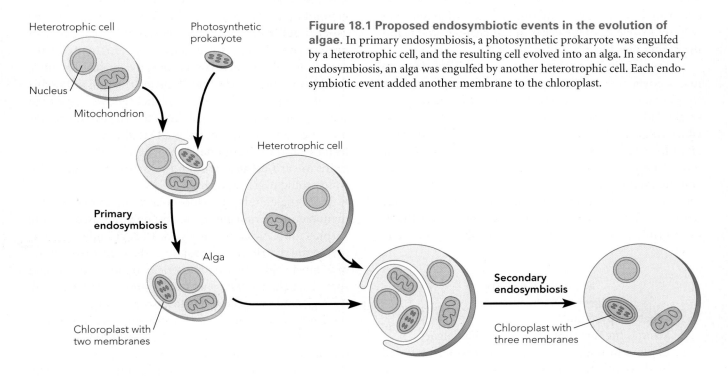

Figure 18.1 Proposed endosymbiotic events in the evolution of algae. In primary endosymbiosis, a photosynthetic prokaryote was engulfed by a heterotrophic cell, and the resulting cell evolved into an alga. In secondary endosymbiosis, an alga was engulfed by another heterotrophic cell. Each endosymbiotic event added another membrane to the chloroplast.

Heterotrophic cell

Photosynthetic prokaryote

Nucleus

Mitochondrion

Primary endosymbiosis

Alga

Heterotrophic cell

Chloroplast with two membranes

Secondary endosymbiosis

Chloroplast with three membranes

membrane derived from a food vacuole. Alternatively, their evolutionary history may include just two endosymbiotic events. In this scenario, the plasma membrane of the ingested alga remained intact inside the food vacuole and became the third chloroplast membrane, while the membrane of the food vacuole became the fourth chloroplast membrane.

At first glance, having extra chloroplast membranes might seem like a handicap because they impose additional barriers that substrates and products must cross in moving between the interior of a chloroplast and the cytoplasm. However, Robert Lee and Paul Kugrens of Colorado State University have found circumstantial evidence suggesting that the space between the second chloroplast membrane and the chloroplast ER is acidic and therefore high in dissolved carbon dioxide. If that is indeed the case, then the chloroplast ER would help promote Calvin cycle activity in the chloroplast, giving algal cells with three or four chloroplast membranes a higher rate of glucose production and thus a selective advantage.

Researchers are using molecular studies to determine the evolutionary relationships within algal phyla and between algae and plants. Such studies have revealed that plants and certain groups of algae are more similar than they were once thought to be. In light of these findings, some investigators have proposed various reclassifications of photosynthetic eukaryotes. One reclassification would move green algae and red algae from the protist kingdom to the plant kingdom. Another would create a new kingdom, called Viridiplantae, consisting of green algae and plants. Yet a third would put one class of green algae (Charophyceae, discussed later in this chapter) and plants in another new kingdom, Streptophyta.

In the remainder of this chapter, we will consider ten major phyla of algae. Four phyla—Bacillariophyta, Xanthophyta, Chrysophyta, and Phaeophyta—belong to a single clade called Stramenopila. (The name of this clade, which comes from Latin words meaning "straw" and "hair," refers to hairlike projections on the flagella of algae in these phyla.) Each of the other six phyla is in a separate clade, and the evolutionary relationships between them are uncertain at present. Given this uncertainty, we will divide our discussion of these ten phyla based on a theme that is more firmly established: their level of cellular organization.

Section Review

1. **What is a pyrenoid?**
2. **What is the function of the eyespot in algae?**
3. **Explain how algae with chloroplasts surrounded by three membranes are thought to have evolved.**

Unicellular and Colonial Algae

Most unicellular and small colonial algae belong to one of seven phyla: Euglenophyta, Dinophyta, Bacillariophyta, Xanthophyta, Chrysophyta, Cryptophyta, and Prymnesiophyta. While some of these algae live on land or attached to substrates in water, the majority exist as **phytoplankton,** the collection of microscopic, photosynthetic organisms that float freely near the surface of oceans and lakes. Phytoplankton also include a few species of cyanobacteria and prochlorophytes (see Chapter 17).

Phytoplankton carry out half of the world's photosynthesis and serve as the bases for all oceanic food chains, much as plants do for terrestrial food chains. The organisms in phytoplankton are extremely sensitive to temperature variations and pollution. A change of a few degrees in the water temperature or an increase in pollution has a major effect on phytoplankton survival, which ultimately affects all organisms higher in the food chain, including humans.

Euglenoids (phylum Euglenophyta) have a pellicle beneath the plasma membrane

Euglenoids are elaborately differentiated cells with one or two flagella that are used for locomotion. Most of the approximately 800 known species of euglenoids, such as *Euglena* (Figure 18.2), live in fresh water. Beneath a euglenoid's plasma membrane lies a supporting structure called a **pellicle,** which is composed of helical bands of protein connected to the endoplasmic reticulum by microtubules. The pellicle is rigid in some euglenoids but is flexible in others, including *Euglena.* The pull of the microtubules against the flexible pellicle allows *Euglena* to change shape as it swims, enabling it to maneuver in muddy pond bottoms full of debris. Euglenoids have a light detector near the base of one of their flagella. They typically swim toward diffuse light and away from bright light, which could overheat them.

Most euglenoids have chloroplasts and pyrenoids that manufacture *paramylon,* a glucose polymer used to store surplus food. As shown in Figure 18.2, granules of paramylon are distributed throughout the cytoplasm. However, most euglenoids are not strictly photosynthetic, and some lack chloroplasts and are not photosynthetic at all. All euglenoids have the ability to absorb organic molecules, such as acetate, from their environment. Organisms that can produce organic molecules through photosynthesis (autotrophy) *and* can absorb or ingest organic molecules (heterotrophy) are known as **mixotrophs.** Mixotrophy

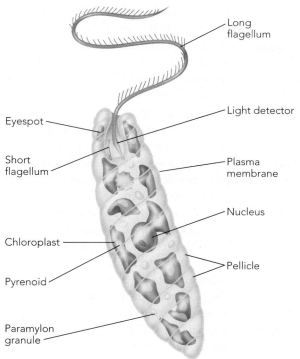

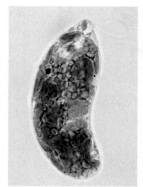

Figure 18.2 *Euglena*, a unicellular euglenoid. *Euglena* has two flagella, but only the long flagellum is used for locomotion. Helical bands that make up the pellicle provide support and flexibility. Pyrenoids in the chloroplasts produce paramylon, a food reserve that is stored in the cytoplasm.

gives many euglenoids, including *Euglena,* the ability to survive in environments where light is available but food is scarce, or vice versa.

Euglenoids do not reproduce sexually. Reproduction occurs by mitosis without the disappearance of the nuclear membrane, which simply pinches in half beginning in anaphase.

Many dinoflagellates (phylum Dinophyta) have hard cellulose plates

Dinoflagellates are important components of marine and freshwater phytoplankton. There are about 3,000 species, each with a characteristic shape. Often, the shape is determined by hard cellulose plates located in vesicles beneath the plasma membrane (Figure 18.3). Like many euglenoids, dinoflagellates have two flagella, but the flagella of dinoflagellates are unique in that they lie within two grooves in the plates. One groove wraps around the center of the cell like a girdle; the flagellum's beating in this groove causes the cell to spin. (*Dino-* comes from the Greek word *dinos,* meaning "rotation.") The other groove is oriented perpendicular to the first; the flagellum's beating in this groove is primarily responsible for forward motion.

About half of all dinoflagellate species are photosynthetic or mixotrophic. A few have green algae living endosymbiotically inside vacuoles in their cytoplasm. The rest are exclusively heterotrophic. Mixotrophic and heterotrophic dinoflagellates absorb dissolved organic molecules or ingest food particles. Some species feed through a tubular stalk, or peduncle, which they temporarily extend through a gap in the plates.

Some photosynthetic dinoflagellates, called *zooxanthellae,* live symbiotically in sponges, sea anemones, corals, mollusks, and other animals (Figure 18.4). The algae are protected from predators by the host and may secrete 50% or more of their photosynthetic product, mostly in the form of glycerol, into the host. The growth of coral reefs in tropical seas is made possible largely by zooxanthellae. Researchers have suggested that the complex and varied shapes of corals are adaptations that maximize the rate of photosynthesis by these algae.

Dinoflagellates synthesize a number of lethal compounds that interfere with the function of animal nervous systems. One of these compounds, saxitoxin, is produced by the dinoflagellate *Gonyaulax.* Saxitoxin blocks sodium ion channels in the plasma membrane of nerve cells, thereby preventing the cells from generating

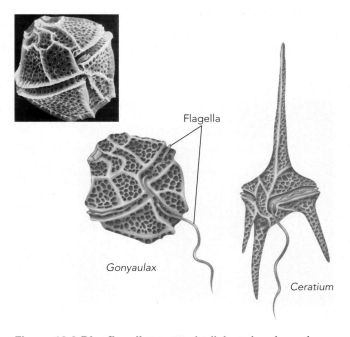

Figure 18.3 Dinoflagellates. Hard cellulose plates beneath the plasma membrane give many dinoflagellates their peculiar shapes. These unicellular algae spin as they move through the water, propelled by two flagella in perpendicular grooves.

Figure 18.4 Zooxanthellae in a giant clam. Giant clams, such as this *Tridacna,* serve as hosts for millions of dinoflagellates (*Symbiodinium microadriaticum*), which live symbiotically as zooxanthellae in the clams' colorful mantles. Photosynthesis by the zooxanthellae provides the clams with most of their food.

nerve impulses. Filter-feeding mollusks that ingest *Gonyaulax,* such as clams, oysters, scallops, and mussels, can absorb and concentrate the toxin without being injured themselves. However, people who eat these toxin-laden mollusks may suffer from paralytic shellfish poisoning, a condition that begins with tingling sensations in the mouth and face and is followed by paralysis that spreads throughout the body. Death occurs in 12 hours. No antidote for saxitoxin exists.

Dinoflagellate toxins become a problem when conditions are optimal for the asexual reproduction of these algae. Under such conditions, the population rapidly expands, a phenomenon referred to as an **algal bloom.**

Because dinoflagellates contain yellowish to reddish accessory pigments (xanthophylls), algal blooms involving marine dinoflagellates can change the color of seawater, producing what is commonly called a "red tide" (Figure 18.5a). Red tides are often triggered by the introduction of nutrients into surface water, either from the upwelling of deeper water or from agricultural runoff containing fertilizers and livestock wastes. Other factors that contribute to red tides include winds that move phytoplankton closer to the shore, high water temperatures near the surface, and bright, sunny days. As a result, poisonings of fish and other animals usually happen during summer. The *Bible* (Exodus 7:17) recounts the first of the plagues that struck Egypt by noting that the water in the river turned to blood, the fish died, the water stank, and the Egyptians could not drink it. Quite likely, this plague was an algal bloom of dinoflagellates. Some oil deposits, including those in the North Sea off the coast of England, are the product of repeated massive blooms of dinoflagellates. Oil shales associated with these deposits are often rich in dinoflagellate remains and in compounds derived from those typically found in dinoflagellates.

Gigantic fish kills in several North Carolina estuaries in the 1980s were linked to blooms of the dinoflagellate *Pfiesteria piscidia.* The blooms probably began when nutrient-rich wastes from hog farms escaped from impoundments and entered the estuaries. Determining whether *Pfiesteria* blooms caused the fish kills or simply accompanied them will require further research. *Pfiesteria* was unknown to science before the 1980s, and its life cycle still has not been completely described. Originally, the organism was thought to have a complex life cycle with at least two dozen stages,

(a)

(b)

Figure 18.5 Dinoflagellate displays. (**a**) A "red tide" is caused by the rapid growth of dinoflagellate populations in coastal waters. (**b**) Mosquito Bay in Puerto Rico is well known for its displays of bioluminescence, which are produced when the water is disturbed.

including toxin-producing amoeboid forms. However, recent research suggests that *Pfiesteria* may have only seven stages, more typical of other dinoflagellates, and that there are no amoeboid forms in its life cycle. Researchers have now produced fluorescent probes that bind to unique sequences of nucleotides in *Pfiesteria* DNA, thus providing a tool for positive identification of the dinoflagellate's life stages.

Pfiesteria is a mixotroph that engulfs other algae and uses their chloroplasts in photosynthesis for a few weeks. Scientists are unsure whether it can also manufacture its own chloroplasts. When encysted *Pfiesteria* cells detect substances secreted by living fish, the dinoflagellate may begin to produce toxins that trigger the development of predatory, flagellated cells. The toxins immobilize fish and other aquatic animals and may contribute to the production of open sores in which the flagellated cells feed. The sores may also attract other predatory, toxin-producing organisms. In humans, toxins associated with *Pfiesteria* blooms cause a variety of symptoms, including nausea, headaches, burning eyes, breathing difficulty, and speech impairment.

Some dinoflagellates are bioluminescent, emitting light by using the enzyme luciferase to catalyze the oxidation of luciferin. Each cell produces one flash a day or can glow dimly for longer periods when disturbed. While the adaptive value of dinoflagellate bioluminescence remains debatable, the combined light output of millions of these microscopic algae is enough to make ocean waters glow at night (Figure 18.5b). The nocturnal luminescence of tropical seas has been noted and pondered by many authors, including Samuel Taylor Coleridge in *The Rime of the Ancient Mariner*:

> About, about, in reel and rout
> The death-fires danced at night;
> The water, like a witch's oils
> Burnt green, and blue and white.

Diatoms (phylum Bacillariophyta) form cell walls of silica

Diatoms occur in fresh water and salt water and in moist vegetation on land. In the ocean they are found most commonly in cool or cold regions, including around and even in sea ice. Some species live attached to a substrate, but most are free swimming and, along with dinoflagellates, are major constituents of phytoplankton. Planktonic diatoms may be responsible for one-quarter of Earth's photosynthesis. Diatoms have existed for about 250 million years, and more than 5,600 living species have been identified. Some botanists estimate that the actual number of living species may be more than 100,000.

The unique structural feature of diatoms is their cell walls, or *frustules*, which have elaborate, ornamented patterns and numerous tiny pores (Figure 18.6). Some diatoms secrete a gelatinous substance called mucilage from the pores, which allows movement by gliding. Each frustule consists of two halves, one slightly larger than the other, which fit together like the top and bottom of a petri dish. Frustules are composed of silica (silicon dioxide, SiO_2), the main component of glass. As a result, the growth of diatoms is highly dependent on the presence of sufficient dissolved silica in the water. Attached diatoms in streams usually do best in a strong current, which insures a continuous supply of silica.

The accumulation of silica in their frustules gives diatoms a density approximately two-and-a-half times that of seawater. However, diatoms remain buoyant by storing oil, which is less dense than water. The oil also serves as a reservoir of food. During the day, free-swimming diatoms vary their density by producing or using oil and thus their vertical position in the water column. In some marine locations, a layer of diatoms is found at a depth of around 100 meters (330 feet). The light intensity is quite low at that depth, but the diatoms double their amount of chlorophyll and use the blue-green light that reaches them to increase the efficiency of carbon fixation.

Diatoms reproduce mainly asexually, by mitosis. Each daughter cell inherits one half of the parental cell's frustule and then manufactures the missing half. In both daughter cells, the newly manufactured half fits inside the half that came from the parental cell. Therefore, the

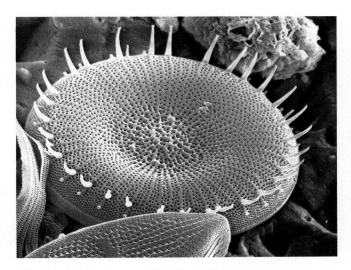

Figure 18.6 Diatoms. The varied and intricate shapes of diatoms are due to the cells' silica-containing cell walls, or frustules. This colored scanning electron micrograph of a diatom shows the numerous pores in its frustule.

daughter cell that inherits the larger half will be the same size as the parental cell, while the daughter cell that inherits the smaller half will be smaller than the parental cell. This process causes the size of cells in some lines to decrease with each generation. Sexual reproduction occurs when cells that reach a certain minimum size undergo meiosis and produce either eggs or sperm. Fertilization results in the formation of a zygote, which enlarges, manufactures a new frustule, and develops into a full-sized diatom.

The frustules of most dead diatoms dissolve, but those that do not dissolve fall to the bottom of oceans or lakes and fossilize. Because of their silica content, frustules make extremely good fossils. By noting changes in the frequency of specific diatom fossils in different ocean- and lake-bottom layers, paleontologists can track prehistoric climate changes. Accumulations of fossilized frustules are the main component of diatomaceous earth. Huge deposits of diatomaceous earth up to 200 meters (656 feet) thick exist near Lompoc, California, where extensive mining operations occur. Diatomaceous earth is used as an abrasive in polishes, as a filter for liquids in the manufacture of sugar, and as insulation for blast furnaces. It was also once used as an ingredient in toothpaste, until dentists discovered that it effectively wears away tooth enamel.

Yellow-green algae (phylum Xanthophyta) are important members of freshwater phytoplankton

The more than 600 species of yellow-green algae live mostly in fresh water, although some are found in the ocean or in damp soil. The free-living forms are an important part of phytoplankton, especially in fresh water and in some salt marshes. Although yellow-green algae typically are unicellular, some species exist as colonies or as long filaments of cells (Figure 18.7). Others are **coenocytic,** consisting of a single cytoplasmic mass that contains many nuclei, with no internal partitions separating the nuclei. Most yellow-green algae have two flagella that arise from opposite ends of the cell. One, called the *tinsel flagellum,* bears minute hairlike projections and pulls the cell forward. The other, called the *whiplash flagellum,* is smooth and moves the cell backward.

Some species of yellow-green algae serve as useful model systems for investigating chloroplast movement, which is common in other algae as well as plants. In the coenocytic yellow-green alga *Vaucheria,* for example, chloroplasts move to the center of the cell in dim light and to the edges of the cell in bright light. In darkness, the

Figure 18.7 Yellow-green algae. *Vaucheria,* a tubular coenocytic alga that grows on rocks in the intertidal zone.

chloroplasts are uniformly distributed. The cell acts as a lens to focus light on the cell's center, so the bright-light response may protect the chloroplasts from light-induced damage. The response is triggered by a blue-light receptor, which activates a network of actin fibers that moves the chloroplasts. In fact, the entire cytoplasm moves, not just the chloroplasts, as other cell structures are carried along.

Reproduction in yellow-green algae is predominantly asexual, involving fragmentation of filaments or spore formation, among other methods. Spores are formed inside the cell wall and are released when it ruptures. Only two genera, including *Vaucheria,* are known to use sexual reproduction.

Golden-brown algae (phylum Chrysophyta) form unique, dormant spores

Golden-brown algae (Figure 18.8) are about 1,000 species of mostly planktonic freshwater and marine algae. Their cells usually have one large chloroplast as well as two flagella of unequal length that emerge perpendicular to each other at one end of the cell. A light detector shaded by an eyespot is located at the base of the short flagellum, near the end of the chloroplast.

Some golden-brown algae are mixotrophic and feed on bacteria and nonliving organic matter, which they draw toward the flagellar end of the cell by waving the flagella. Food is ingested by *phagocytosis,* a form of endocytosis (see Chapter 10) in which large particles or whole cells are engulfed. Mixotrophic golden-brown algae often

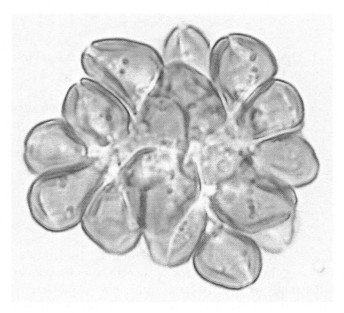

Figure 18.8 *Synura,* a colonial golden-brown alga.

Each cryptomonad cell has two flagella, one with long hairs on both sides, the other with short hairs on only one side (Figure 18.9). Just inside the plasma membrane and attached to it are protein plates, the shapes of which depend on the species. The membrane and plates together are called the *periplast*. In addition to chlorophylls *a* and *c*, cryptomonads have a phycobilin accessory pigment, either phycoerythrin or phycocyanin. Different combinations of pigments produce a range of colors from yellow-green to blue, red, or brown.

A characteristic feature of cryptomonads is the presence of structures called *ejectisomes,* which line the cell periphery and the depression from which the flagella arise. Ejectisomes are long, narrow strips of protein that are tightly coiled, like a spiral tape measure. They uncoil rapidly when released, propelling the cell in the opposite direction. Their release may enable cryptomonads to escape from predators or, in the case of heterotrophic or mixotrophic cryptomonads, to capture their own prey.

decrease their rate of photosynthesis and the size of their chloroplast when food is plentiful.

One unique feature of golden-brown algae is their formation of dormant spores, called *statospores,* which are encased in a wall made of silica. Statospores contain the nucleus, chloroplast, basal bodies, Golgi apparatus, and many mitochondria and ribosomes. Vacuoles and some ribosomes and mitochondria are lost, as are both flagella. Golden-brown algae typically form statospores in autumn which germinate in spring. For species living in ponds and shallow lakes that freeze completely during winter, statospores have distinct survival value. They fall to the bottom and settle in the mud, carrying the algal cells safely through the winter months.

Cryptomonads (phylum Cryptophyta) use ejectisomes for sudden escape

Cryptomonads (from the Greek *kryptos,* "hidden," and *monos,* "single") are so named because they are generally less than 50 micrometers in diameter and therefore easy to overlook. Some of the 200 species of cryptomonads are photosynthetic, some are heterotrophic, and many are probably mixotrophic. They are found mainly in cold water, both in the ocean and in lakes. Despite their small size, cryptomonads can constitute the bulk of phytoplankton during seasons when dinoflagellate and diatom populations decrease. Large blooms of cryptomonads commonly occur in ocean waters near the poles and in other locations, including Chesapeake Bay.

Haptophytes (phylum Prymnesiophyta) have a distinctive, moveable haptonema

The majority of the 300 or so known species of haptophytes live in the ocean, where they are significant members of phytoplankton, especially in the Tropics. In the

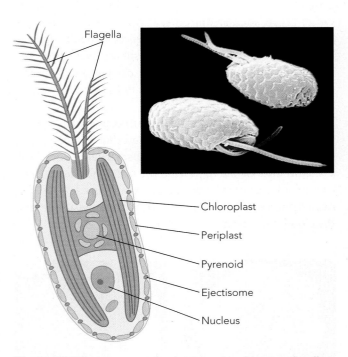

Figure 18.9 Cryptomonads. A cryptomonad has two flagella, a periplast composed of protein plates, and ejectisomes distributed around the periphery of the cell.

mid-Atlantic they account for nearly 50% of photosynthesis. There are also a few freshwater and terrestrial species. These algae are extremely small—generally only a few micrometers in diameter. Each cell has two disk-shaped chloroplasts and a number of golden-yellow plastids. In most species, the surface of the cell is covered with small, flat scales, which may be made of cellulose or calcium carbonate. Those containing calcium carbonate are called *coccoliths* (Figure 18.10). The white cliffs of Dover, in England, are composed largely of fossilized coccoliths.

The definitive characteristic of haptophytes is their *haptonema,* a moveable filament composed of three membranes surrounding seven microtubules. The haptonema is situated between two whiplash flagella but is not a flagellum itself, as it lacks the 9 + 2 pattern of microtubules typical of eukaryotic flagella. It does not beat like a flagellum and is not used for locomotion. Instead, the haptonema is used to attach the cell to surfaces and may help the cell avoid obstacles. It is also used to attract and gather food, supporting the mixotrophic nutrition of many haptophytes.

Cells of the haptophyte *Phaeocystis* clump together to form gelatinous colonies. The cells produce tightly coiled, chitinous filaments (apparently not haptonemata) that spring out of the cell, forming a web that may hold the colony together and attach it to structures such as fishnets. *Phaeocystis* contains high concentrations of compounds that absorb ultraviolet (UV) light, which apparently protect the cells from damage caused by UV irradiation. Other algae, such as diatoms, appear to be much more sensitive to UV light. As a consequence, while the appearance of the ozone hole in the atmosphere over Antarctica has caused a decline in the numbers of diatoms in Antarctic waters, *Phaeocystis* populations in the same area have increased dramatically.

Phaeocystis also releases into the atmosphere large quantities of dimethylsulfide, a compound that serves as a nucleus for cloud condensation. Clouds partially block both UV light and visible light used for photosynthesis. In an example of negative feedback, as *Phaeocystis* decreases its rate of photosynthesis, it releases less dimethylsulfide, and the cloud cover thins.

Section Review

1. Describe the structure and function of the pellicle in *Euglena*.
2. What causes a red tide?
3. Why do diatoms engage in sexual reproduction after a certain number of rounds of asexual reproduction by mitosis?
4. Where are ejectisomes found, and how are they used?

Multicellular Algae

Three phyla—Phaeophyta, Chlorophyta, and Rhodophyta—contain multicellular algae with complex cellular differentiation and tissue-level organization. While Chlorophyta also includes many unicellular species, Phaeophyta and Rhodophyta are almost exclusively multicellular. The marine multicellular forms are the algae known as *seaweeds.* Sexual reproduction is common in these three phyla, and many species have complex life cycles involving an alternation of generations. Recall from Chapter 6 that in such life cycles, two multicellular forms alternate with each other: a diploid, spore-producing form called a *sporophyte* and a haploid, gamete-producing form called a *gametophyte.*

In many brown algae (phylum Phaeophyta), alternate generations are heteromorphic

There are about 1,500 species of brown algae, most of which are marine. They include the giant kelps, such as *Macrocystis* and *Nereocystis,* as well as tiny species such as *Ralfsia expansa,* commonly known as *tar spot,* which resembles a dab of tar on a rock. Their plastids have large amounts of fucoxanthin, a carotenoid accessory pigment that gives these algae a brown or olive color.

All brown algae are multicellular and have a plantlike body called a **thallus** (from the Greek *thallos,* "sprout;"

Emiliania huxleyi

Florisphaera profunda

Umbellosphaera tenuis

Figure 18.10 Haptophytes. Many haptophytes are covered with plates made of calcium carbonate called *coccoliths.*

plural *thalli*). As shown in Figure 18.11, a kelp's thallus has three main parts: a rootlike **holdfast,** which anchors the thallus to the substrate; a stemlike, often hollow stalk called a **stipe;** and various numbers of flattened **blades,** which provide most of the surface area for photosynthesis. Some thalli also have gas-filled floats, or bladders, at the base of the blades. The floats help keep the blades near the surface, where sunlight is more intense. It is important to note, however, that despite their superficial similarity to plant structures, holdfasts, stipes, and blades lack the characteristic vascular tissue of true roots, stems, and leaves.

Many larger brown algae, including the kelp *Laminaria* (Figure 18.12), have a life cycle in which there is an alternation of **heteromorphic** generations. In such life cycles, the sporophyte and gametophyte are very different in appearance. In the case of kelps, the sporophyte is large and conspicuous, while the gametophyte is microscopic. Cells in the blades of the sporophyte develop into sporangia, which produce motile, haploid spores known as *zoospores.* All of the zoospores look alike, but some develop into male gametophytes and others develop into female gametophytes. Recent research indicates that the gametophytes of some brown algae are symbiotes in the cell walls

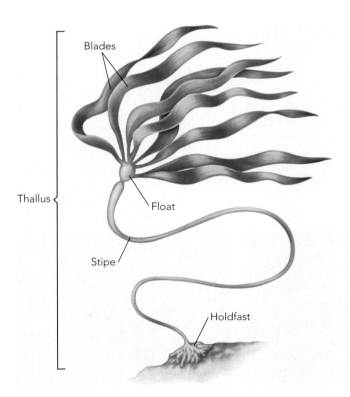

Figure 18.11 Structure of a kelp's thallus. The kelp *Nereocystis* has a body, or thallus, typical of many seaweeds. Photosynthetic blades are suspended by a gas-filled float. The stemlike stipe connects the float and blades to the anchoring holdfast.

of red algae. The gametophytes bear **gametangia,** which are structures (single cells, in *Laminaria* gametophytes) that produce gametes. Each male gametangium, called an **antheridium,** releases one motile sperm. Each female gametangium, called an **oogonium,** contains one egg. When fertilized, eggs remain attached to the female gametophyte and develop into new sporophytes.

In most brown algae with heteromorphic generations, the sporophyte is found in summer and the gametophyte in cooler seasons. The large sporophyte is adapted for maximizing photosynthesis, which is more productive during the long days of summer. During fall and winter, when the days are shorter and the weather often stormy, the tiny, more firmly attached form of the gametophyte may be more adaptive.

Red algae (phylum Rhodophyta) have complex life cycles with three multicellular phases

Red algae may have been the first eukaryotes formed by endosymbiosis involving photosynthetic prokaryotes. Most of the roughly 5,000 species are marine; fewer than 100 identified species live in fresh water. The vast majority of red algae are multicellular, with thalli measuring up to about 10 centimeters (3.9 inches) in length. Different species may be free living, epiphytic, or parasitic. Phycobilins and carotenoids give many red algae, such as *Rhodymenia pseudopalmata* (Figure 18.13a), their characteristic red or pink color.

The cell walls of red algae have cellulose as a framework but are mostly mucilages containing agars and carrageenans, both of which are polymers of galactose (a sugar similar to glucose) and are sold as food thickeners. Many red algae also form layers of calcium carbonate in their cell walls. These algae are commonly called *coralline algae* (Figure 18.13b). Not all red algae are red, however. Species that do not contain as much of those accessory pigments, such as *Halosaccion* (Figure 18.13c), are often blue-green or olive.

Red algae are perhaps best known for the complexity of their life cycles. Most have three multicellular phases: a haploid gametophyte and two diploid sporophytes. One of the sporophyte phases, known as the *tetrasporophyte,* produces spores, called *tetraspores,* by meiosis. The tetraspores germinate and grow into male or female gametophytes. Male gametophytes release nonflagellated gametes, called *spermatia,* which are carried by water currents to eggs on female gametophytes. (Red algae are the only algae that do not have flagellated cells during any stage in their life cycle.) Following fertilization, the zygote divides repeatedly by mitosis, producing the second sporophyte phase,

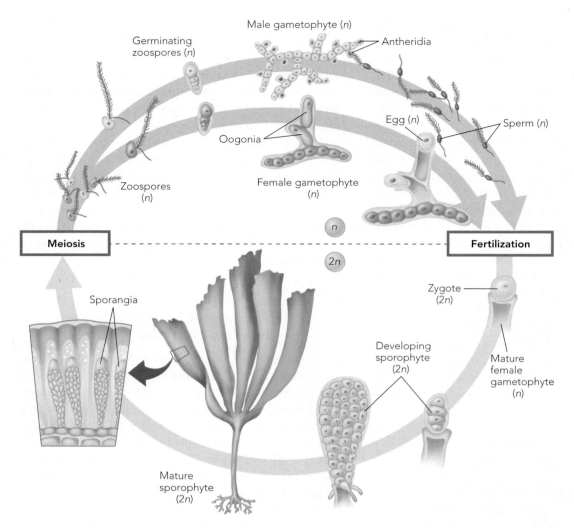

Figure 18.12 Life cycle of *Laminaria,* a brown alga. Sporangia in the blades of the large sporophyte produce zoospores, which develop into microscopic male and female gametophytes. Antheridia in male gametophytes release sperm, which fertilize eggs in the oogonia of female gametophytes. The resulting zygotes develop into new sporophytes.

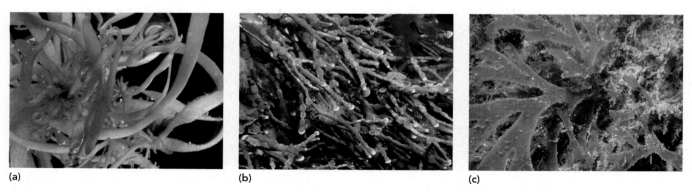

Figure 18.13 Red algae. (**a**) *Rhodymenia pseudopalmata.* (**b**) A coralline alga. (**c**) *Chandrus crispus.*

THE INTRIGUING WORLD OF PLANTS

Watermelon Snow

High in the mountains of western North America, hikers and campers frequently encounter snowfields that have a pink or red tint. The air above the colored snow smells like watermelons. Unfortunately, the snow does not have a watermelon's flavor.

Surprisingly, watermelon snow is caused by several dozen species of cyanobacteria and "snow algae," including the green alga *Chlamydomonas nivalis* (*nivalis* is Latin for "snowy"). In addition to chlorophyll, many snow algae contain high concentrations of orange or red carotenoids, which apparently

protect the cells from excess UV light that penetrates the thinner atmosphere at high elevations and is reflected by snowfields. From mid- to late summer, the algae bloom, producing huge areas of watermelon snow that contain millions of cells in a handful.

Snow algae obtain minerals from dirt and dust that blow onto the snow. The algae serve as food for a variety of other protists, invertebrates, birds, and mammals, which make up a food chain based on algal primary production. During the winter, the algae are dormant. They begin to grow again during the rapid snowmelts of alpine summers.

Watermelon snow.

---

the *carposporophyte,* which remains attached to the female gametophyte. The carposporophyte releases spores called *carpospores,* which develop into new tetrasporophytes.

No one understands for sure the evolutionary origin or adaptive advantage of such a complicated life cycle. One line of speculation is that, because the lack of flagellated gametes makes fertilization less likely, there is an advantage to having a life cycle that makes the most of fertilization when it does occur. According to this view, the carposporophyte represents an amplification stage, potentially producing many tetrasporophytes from each zygote.

Green algae (phylum Chlorophyta) share a common ancestor with plants

Most green algae live in fresh water, although many live in the ocean as seaweeds or as part of phytoplankton. Some others are terrestrial, growing in moist places favored by mosses and ferns or even in snow (see *The Intriguing World of Plants* on this page.) Green algae also exist in symbiotic relationships with other organisms. Some lichens (see Chapter 19), for example, are associations between fungi and green algae. The fungi provide protection and moisture, while the algae provide sugar produced by photosynthesis. Like plants, green algae have chlorophylls *a* and *b* and store starch inside plastids as a food reserve. These and a variety of other similarities strongly suggest that green algae and plants evolved from a common ancestor.

There are about 7,500 species of green algae in several classes, three of which we will explore in the rest of this chapter: Chlorophyceae, Ulvophyceae, and Charophyceae. The classes differ in the placement and anchoring of flagella, in when the spindles disappear in telophase, and in how cytokinesis occurs after nuclear division.

Class Chlorophyceae

Most chlorophyceans are unicellular or colonial. One of the best-studied is *Chlamydomonas,* a unicellular freshwater alga commonly found in ponds. Each cell has two flagella, a single chloroplast, a red eyespot, and a cell wall that lacks cellulose. One species of *Chlamydomonas* is being investigated for its potential as a source of fuel (see the *Biotechnology* box on page 399).

Chlamydomonas can reproduce either asexually or sexually (Figure 18.14). Both methods begin when a mature, haploid cell divides two or more times by mitosis, producing up to 16 daughter cells, which develop flagella before breaking out of the parental cell's wall. In asexual reproduction, the daughter cells are zoospores and develop directly into mature haploid cells. In sexual reproduction, the daughter cells are gametes. Each mature cell and all of the gametes it produces are of one *mating type,* designated either + or −. The terms *male* and *female* do not apply to *Chlamydomonas* because + and − gametes are identical in appearance. Such gametes are called **isogametes** instead of sperm or eggs. Fusion of a + gamete and a − gamete results in the formation of a zygote, which secretes a thick

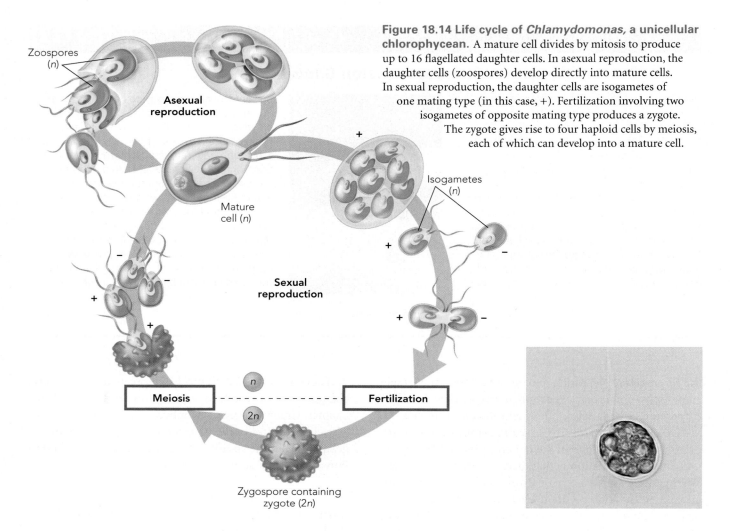

Zoospores (n)

Asexual reproduction

Mature cell (n)

Sexual reproduction

+

Isogametes (n)

+

–

+

–

–

+

+

Meiosis

n

2n

Fertilization

Zygospore containing zygote (2n)

Figure 18.14 Life cycle of *Chlamydomonas,* a unicellular chlorophycean. A mature cell divides by mitosis to produce up to 16 flagellated daughter cells. In asexual reproduction, the daughter cells (zoospores) develop directly into mature cells. In sexual reproduction, the daughter cells are isogametes of one mating type (in this case, +). Fertilization involving two isogametes of opposite mating type produces a zygote. The zygote gives rise to four haploid cells by meiosis, each of which can develop into a mature cell.

wall. The walled zygote is known as a *zygospore*. Inside the wall, the zygote produces four flagellated haploid cells (two of each mating type) by meiosis. The cells then break out of the wall and develop into mature cells.

Another unicellular chlorophycean, *Chlorella,* has been studied as a possible food source for humans. The cells of *Chlorella* produce large amounts of carotenoids but very little cellulose, so they are almost entirely digestible. The dry content of the cells is nearly 50% protein. *Chlorella* can grow rapidly using sewage or other wastes as a source of minerals.

The best-known colonial chlorophycean is *Volvox,* which consists of a few hundred to a few thousand photosynthetic cells arranged in a single layer at the surface of a hollow sphere (Figure 18.15). Each cell has two flagella on the outside of the sphere. Absorption of light by the cells' light detectors controls the beating of their flagella and directs the colony toward light. Nonflagellated reproductive cells are scattered throughout the surface of the sphere. These cells divide mitotically to produce a flat plate of cells, which breaks away from the interior surface to form a small daughter colony with its flagella facing inward. The daughter colony turns itself inside out and then escapes from the parental colony, using enzymes to digest a small hole in the gelatinous matrix that holds the parental colony together.

Numerous other chlorophyceans are colonial. For example, *Botrycoccus* forms floating colonies surrounded by a semirigid envelope. These algae produce oil as a storage product, and it has been suggested that they could be cultivated in large numbers to manufacture oil for fuel. Coal deposits and oil shales from the Tertiary period have remains of *Botrycoccus* and, therefore, may have been formed, in part, by this alga.

BIOTECHNOLOGY

Algae as a Source of Fuel

Amid dwindling supplies of fossil fuels like oil, coal, and natural gas, scientists are turning to alternate sources of energy. Fuel-cell technology is under investigation as a means of efficiently powering automobile engines. Fuel cells work by combining hydrogen and oxygen to produce water, with no air pollution. Energy is liberated in the process because electrons in the H–O bonds of water move closer to the oxygen nucleus. The energy released is harnessed to generate electricity, which can be used to run motors. Currently, hydrogen is purified from natural gas, of which it is a minor component. However, a sustainable and less expensive source of hydrogen is available from living cells.

Recent research on the green alga *Chlamydomonas reinhardtii* has

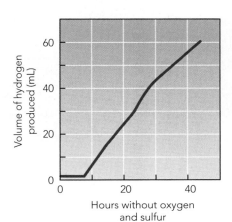

Hydrogen production by algal cells.

pointed to its potential as a major supplier of hydrogen for fuel. Researchers had known for some time that algae could produce small amounts of hydrogen when their oxygen supply was temporarily cut off. A team of scientists from the University of California at Berkeley and the National Renewable Energy Laboratory in Golden, Colorado, found that if algal cultures were deprived of both oxygen and sulfur, the algae could sustain hydrogen gas production at high levels for several days. Under these conditions, the cells apparently generate hydrogen as part of an alternate biochemical pathway to form needed ATP.

The economics of scaled-up production of hydrogen by algae appear to be attractive. The possibility also exists that genetic engineering (see Chapter 14) could produce algal strains with an increased ability to produce hydrogen.

It is likely that colonial algae gradually evolved from unicellular species. The appearance of colonial organization may have begun with a mutation that simply caused single cells to clump together. As larger colonies evolved, their increased size may have given them a selective advantage by making it easier for them to avoid predation.

Class Ulvophyceae

Ulva, or sea lettuce (Figure 18.16), is a common marine ulvophycean. It can be found attached to rocks in tide pools and in exposed areas when the tide is out. *Ulva*'s life cycle involves an alternation of **isomorphic** generations. That is, the gametophyte and sporophyte look nearly identical: both are bright green, flat thalli resembling a thin rubbery leaf of lettuce. The gametophytes of *Ulva*, like the mature cells of *Chlamydomonas,* are designated + or − because they produce similar-looking isogametes, each of which has two flagella. Zygotes formed by the fusion of + and − isogametes develop into sporophytes, which produce zoospores that have four flagella. The zoospores then germinate to form gametophytes. Isogametes and zoospores are released when the waters of an incoming tide first wet the thalli. Isogametes swim toward the light, while zoospores swim away from the light.

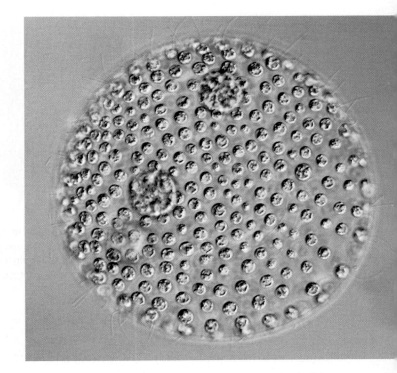

Figure 18.15 *Volvox*, a colonial chlorophycean. Each large sphere is a colony of a few hundred to a few thousand cells. The small spheres inside the large ones are daughter colonies.

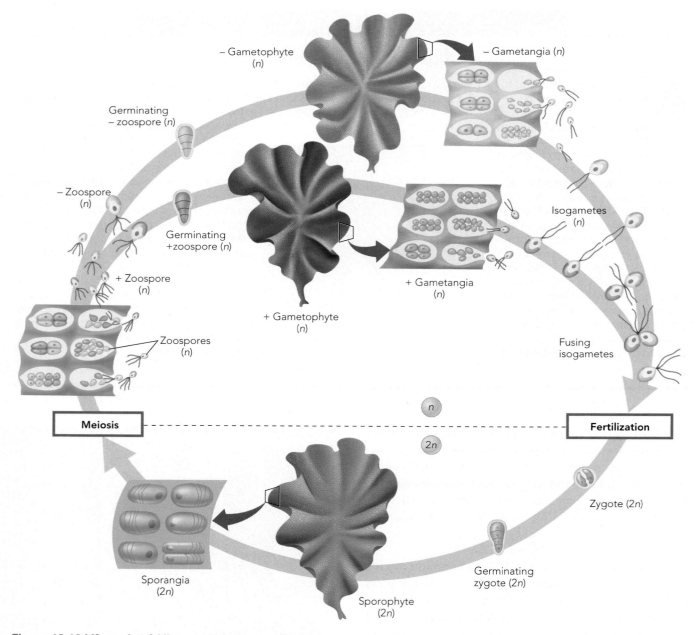

Figure 18.16 Life cycle of *Ulva*, an ulvophycean. Sporangia in the sporophyte produce zoospores, which germinate to form + and − gametophytes. The gametophytes and sporophytes are very similar in size and appearance. Gametangia in the gametophytes release + and − isogametes, which fuse. The resulting zygotes develop into new sporophytes.

Two other ulvophyceans worthy of note are *Acetabularia* and *Cephaleuros. Acetabularia* (see Figure 13.3) is found in warm, protected tropical waters and exists for most of its life as a single cell up to several centimeters long. As you learned in Chapter 13, *Acetabularia* was the subject of experiments demonstrating that the nucleus is the site of the hereditary material in eukary-

otes. *Cephaleuros* lives on the leaves of tea plants and is responsible for red rust, an important disease of tea and citrus trees in some regions of the world.

Class Charophyceae
Charophyceans include unicellular, colonial, and multicellular green algae. Two orders of charophyceans—Coleo-

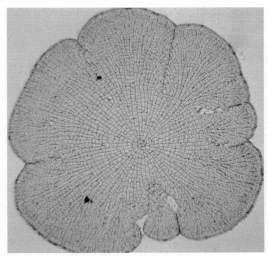

(a)

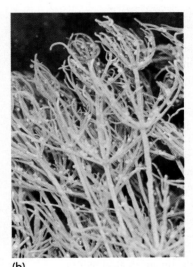

(b)

**Figure 18.17
Charophyceans.**
(**a**) *Coleochaete.*
(**b**) *Chara,* a stonewort.

chaetales and Charales—are the closest relatives of plants among all the algae. Coleochaetales consists of filamentous or disk-shaped algae that live in shallow regions of freshwater lakes, often attached to other organisms. An example is *Coleochaete* (Figure 18.17a), which can be found on leaves and other debris on lake bottoms. Charophyceans in the order Charales form mineralized cell walls containing calcium carbonate and magnesium carbonate. For this reason, they are commonly called stoneworts. Algae in this order, such as *Chara* (Figure 18.17b), have complex thalli with whorls of branches, nodes, and internodes. They resemble plants by having apical growth, tissues similar to those of vascular plants, and protective, nonreproductive cells covering oogonia and antheridia. However, the covering cells in algae and plants are probably of a different developmental origin and therefore not phylogenetically related.

Section Review

1. Name and describe the main parts of a kelp's thallus.
2. Distinguish between life cycles with heteromorphic generations and those with isomorphic generations.
3. How do the life cycles of most red algae differ from those of other multicellular algae?
4. Describe the cellular organization of *Volvox.*

SUMMARY

Characteristics and Evolution of Algae

Algae are distinguished by their photosynthetic pigments and other characteristics (pp. 386–387)
Photosynthetic algae have chlorophyll *a* and various accessory pigments. Many have a pyrenoid in their chloroplasts that contains rubisco and stores the products of carbon fixation. A light detector and a light-shielding eyespot enable some algal cells to orient themselves and to move in relation to a light source.

Endosymbiosis played a key role in the evolution of algae (pp. 387–388)
Algal chloroplasts are thought to have resulted from endosymbiosis. Chloroplasts that are surrounded by two membranes likely evolved from a single endosymbiotic event. Chloroplasts that are surrounded by three or four membranes may have evolved from two or three endosymbiotic events.

Unicellular and Colonial Algae

Euglenoids (phylum Euglenophyta) have a pellicle beneath the plasma membrane (pp. 388–389)
Euglenoids live mostly in fresh water and have helical bands of protein called a *pellicle,* which provides support. Euglenoids may be autotrophic, heterotrophic, or mixotrophic.

Many dinoflagellates (phylum Dinophyta) have hard cellulose plates (pp. 389–391)

Dinoflagellates have characteristic shapes determined, in many species, by hard cellulose plates beneath the plasma membrane. A pair of flagella that beat in perpendicular grooves cause dinoflagellates to spin as they move. Some dinoflagellates live as symbiotic zooxanthellae in a variety of animals. Others produce toxic compounds, which may reach dangerous concentrations in the water during algal blooms.

Diatoms (phylum Bacillariophyta) form cell walls of silica (pp. 391–392)

Diatoms produce two-part cell walls called *frustules* that contain silica. The transfer of frustule halves to daughter cells during asexual reproduction causes some lines of daughter cells to decrease in size, eventually triggering sexual reproduction. The accumulation of fossilized frustules at the bottom of oceans and lakes forms diatomaceous earth.

Yellow-green algae (phylum Xanthophyta) are important members of freshwater phytoplankton (p. 392)

Yellow-green algae may be unicellular, colonial, or coenocytic. Some species are used as model systems for studying chloroplast movement.

Golden-brown algae (phylum Chrysophyta) form unique, dormant spores (pp. 392–393)

Golden-brown algae are mostly planktonic and form spores called *statospores* that are encased in a wall of silica. Statospores allow some species to survive in bodies of water that freeze completely during winter.

Cryptomonads (phylum Cryptophyta) use ejectisomes for sudden escape (p. 393)

Cryptomonads have protein plates just inside the plasma membrane and long, coiled strips of protein called *ejectisomes* lining the cell periphery. Release of ejectisomes propels cryptomonads through the water.

Haptophytes (phylum Prymnesiophyta) have a distinctive, moveable haptonema (pp. 393–394)

Most haptophytes are marine and are covered with small, flat scales made of cellulose or calcium carbonate. They have a filament called a *haptonema* that may be used for anchoring, obstacle avoidance, or food gathering.

Multicellular Algae

In many brown algae (phylum Phaeophyta), alternate generations are heteromorphic (pp. 394–395)

All brown algae are multicellular and have a plantlike body called a *thallus*. Many, such as kelps, have a life cycle in which heteromorphic generations alternate: The sporophyte is large and the gametophyte is microscopic.

Red algae (phylum Rhodophyta) have complex life cycles with three multicellular phases (pp. 395–397)

Red algae are mostly marine, and almost all are multicellular. Many species have a life cycle that includes a haploid gametophyte and two diploid sporophytes. They are the only algae that do not have flagellated cells during any stage in their life cycle.

Green algae (phylum Chlorophyta) share a common ancestor with plants (pp. 397–401)

A variety of similarities between green algae and plants—including the presence of chlorophylls *a* and *b* and the storage of starch as a food reserve—suggest that both evolved from a common ancestor. Many green algae are unicellular (such as *Chlamydomonas*) or colonial (such as *Volvox*). Multicellular green algae include *Ulva*, which has a life cycle involving an alternation of isomorphic generations.

Review Questions

1. How is the term *plantlike protists* an accurate synonym for algae? How is it inaccurate?
2. What is the function of pyrenoids?
3. Explain the difference between the light detector and the eyespot of an algal cell.
4. How many membranes surround the chloroplasts of euglenoids and most dinoflagellates?
5. What might be the selective value of third and fourth chloroplast membranes?
6. What organisms are in the proposed kingdom Viridiplantae?
7. Why are phytoplankton important?
8. What is a mixotroph?
9. What are zooxanthellae?
10. Explain what happens during an algal bloom.
11. What is the ecological importance of *Pfiesteria piscidia*?
12. Why is it said that diatoms live in glass houses?
13. What does *coenocytic* mean?
14. Which group of algae forms statospores?
15. How do cryptomonads use ejectisomes?
16. What is the function of a haptonema?
17. Name the three parts of the thallus of a brown alga.
18. What do antheridia contain?
19. Briefly describe the life cycle of a typical red alga.
20. How does *Volvox* differ from *Chlamydomonas*?
21. Which green algae are most closely related to plants?

Questions for Thought and Discussion

1. Suppose you visit a lake in summer and find that the water contains small, green particles and long, green threads. How would you discover what the particles and threads are?

2. Algae can be found in salt water and in fresh water. With some exceptions, plants are not found in salt water. Why?

3. By living in the intertidal zone, kelps are at risk of being ripped from their mooring by waves and tossed onto the shore by ocean storms. Why don't kelps establish themselves in deeper water?

4. How would you determine which algae contribute the most to photosynthesis in a particular region of the ocean?

5. How do the chloroplast membranes of algae support the idea that "you are what you eat"?

6. Imagine an alga that lived at the mouth of a river. The alga would be bathed in fresh water at low tide and in salt water at high tide. How might it adapt to these different environments?

7. What selective advantage do large colonial algae like *Volvox* have over unicellular forms?

8. Draw a chloroplast of a brown algal cell. Add labels to your diagram to indicate the probable endosymbiotic origin of each of the membranes that surrounds the chloroplast you've drawn.

Evolution Connection

What evidence suggests that land plants shared a common ancestor with (a) green algae and, more specifically, (b) green algae in the class Charophyceae?

To Learn More

Visit The Botany Place Website at www.thebotanyplace.com for quizzes, exercises, and Web links to new and interesting information related to this chapter.

Barker, Rodney. *And the Waters Turned to Blood: The Ultimate Biological Threat.* Carmichael, CA: Touchstone Books, 1998. A blow-by-blow account of the polymorphic "killer" alga, *Pfiesteria.*

Lee, Robert E. *Phycology,* third edition. Cambridge: Cambridge University Press, 1999. A comprehensive, detailed text on algae with many specific examples to illustrate each algal group.

Lewallen, Eleanor, and John Lewallen. *Sea Vegetable Gourmet Cookbook and Wildcrafter's Guide.* Mendocino, CA: Mendocino Sea Vegetable Company, 1996. Algae are important components of many national cuisines.

Meinesz, Alexandre, and Daniel Simberloff, trans. *Killer Algae: The True Tale of a Biological Invasion.* Chicago: University of Chicago Press, 1999. This book tells how a hybrid, cold-resistant variety of the tropical green alga *Caulerpa taxifolia* was released from an aquarium and became a major problem in the Mediterranean Sea.

Thomas, David. *Seaweeds.* Washington, DC: Smithsonian Institution Press, 2002. Thomas provides a detailed look at seaweeds of all types and discusses the curious biology of algae as well as their economic importance.

19
Fungi

Antler fungi (*Clavariaceae*).

Characteristics and Evolutionary History of Fungi

A combination of morphological and developmental characteristics distinguish fungi from other organisms

Fungi probably evolved from flagellated protists

The Diversity of Fungi

Chytridiomycetes (phylum Chytridiomycota) produce flagellated reproductive cells

Zygomycetes (phylum Zygomycota) form resistant zygosporangia prior to meiosis

Ascomycetes (phylum Ascomycota) produce sexual spores in sacs called *asci*

Basidiomycetes (phylum Basidiomycota) produce sexual spores on club-shaped cells called *basidia*

Fungal Associations with Other Organisms

Lichens are associations of fungi and photosynthetic algae or bacteria

Some fungi form mutualistic associations with insects

Imagine a world without decomposition. When an organism died, it might dry out and mummify or remain hydrated, but it would never decay or rot. Countless dead plants and animals would litter the landscape or constitute a major storage problem. All soil would be sand or clay, since it would contain no organic matter.

Of course, dead organisms do decompose, thanks to the activities of soil bacteria, animals such as earthworms and nematodes (roundworms), and fungi. Decomposers sustain life by allowing the nutrients that are locked up in the tissues of organisms to recirculate in a continuous molecular reincarnation. Through the action of decomposers, carbon is released into the atmosphere as carbon dioxide (CO_2), nitrogen is released into the atmosphere as N_2 or N_2O (nitrous oxide), and minerals are released into the soil as ions.

In some cases, nutrients are recycled without returning to CO_2 and mineral ions. This occurs, for example, when a robin eats a worm, when a person eats a mushroom, and when a fly eats cow dung or feeds on a dead cow. In fact, a great amount of organic matter is recycled by being consumed. It is a way of life for all heterotrophic organisms, and that is exactly what decomposers do. They simply consume what many other organisms have not.

Fungi accomplish decomposition by secreting enzymes that break down complex organic compounds into simpler molecules, which the fungi absorb. They can decompose an amazing variety of substances. For example, more than 30 species of fungi can digest petroleum, while others can digest plastics. In the Tropics, humans engage in a perpetual struggle to limit fungal decomposition, which is promoted by the warm, humid air. During wars fought in the Tropics, fungi have frequently produced as much damage to supplies and soldiers as the human enemy has.

We have just begun to explore the diversity and potential usefulness of fungi. Fungi supplied the first antibiotic, penicillin, and are now a major source of many other medically useful compounds. Without fungi, bread would not rise, and bleu cheese would not be blue. Scientists are now investigating how to use petroleum-digesting fungi to clean up oil spills and other chemical messes. Yet undiscovered fungi represent a potential treasure trove of gifts.

In this chapter, we will consider the characteristics that distinguish fungi from other organisms and have led biologists to place fungi in their own kingdom. We will then investigate the four phyla of fungi, as well as a number of fungi that cannot be classified at present because our knowledge of their life cycles is incomplete. Finally, we will look at two important kinds of symbiotic relationships between fungi and other organisms.

Fungi decomposing a log.

Characteristics and Evolutionary History of Fungi

As you learned in Chapter 16, all organisms at one time were classified as either animals or plants. Because most animals can move from place to place but fungi and plants cannot, fungi were generally included in the plant kingdom. Later, however, taxonomists came to recognize that fungi are in fact very different from plants and deserve to be classified in a separate kingdom, Fungi. The study of fungi is called **mycology,** which is derived from the Greek word *mykes,* meaning "fungus."

A combination of morphological and developmental characteristics distinguish fungi from other organisms

Along with plants, animals, and protists, fungi are eukaryotes, organisms whose cells have a membrane-enclosed nucleus. However, fungi have a combination of characteristics that justify their placement in a separate eukaryotic kingdom.

Most fungi are multicellular and are composed of long filaments called **hyphae** (singular, *hypha*). Some hyphae, called *septate hyphae,* are divided into cells by internal walls known as **septa** (Figure 19.1a). The septa usually have a central pore large enough to allow small organelles and, in some cases, even nuclei to pass between cells. Other hyphae lack septa and are coenocytic, with multiple nuclei in a common cytoplasm (Figure 19.1b). All the hyphae of a particular type in a fungus form an interwoven mass called a **mycelium** (plural, *mycelia;* Figure 19.1c). As it passes through the phases of its life cycle, an individual fungus might consist of a single mycelium or several types of mycelia.

Fungi are heterotrophs, but, as you already know, they do not engulf food like animals. Instead, they absorb food after breaking it down into small molecules, which then pass through the plasma membrane by diffusion or with the help of transport proteins. Most fungi are **saprobes,** organisms that feed on dead organic matter. Other fungi are **parasites,** organisms that feed on living hosts, or predators, organisms that kill what they feed on. For example, *Arthrobotrys anchonia* uses hyphal loops to capture amoebae (animal-like protists) and small animals such as nematodes (Figure 19.2). When an organism enters a loop, the hyphae take up water and expand, tightening the loop and trapping the prey. The fungus then secretes enzymes that digest the prey. Other predatory fungi use sticky hyphae to snare prey. Finally, many fungi

(a) Septate hypha   (b) Coenocytic hypha

Figure 19.1 Fungal hyphae and mycelia. (**a**) Septate hypha. (**b**) Coenocytic (nonseptate) hypha. (**c**) White mycelia of fungi growing on a dead tarantula. Note the yellow fruiting body.

live in mutually beneficial relationships with algae, photosynthetic bacteria, or plants and receive organic compounds from them.

Fungi produce spores during sexual and asexual reproduction. The spores serve to disperse the fungus to new locations, and some help the fungus survive adverse conditions, such as dehydration or freezing. In all but one phylum, however, the spores lack flagella and are therefore nonmotile. The absence of flagellated cells in their

Figure 19.2 A predatory fungus. The soil fungus *Arthrobotrys anchonia* uses hyphal loops to snare prey.

life cycle distinguishes most fungi from the majority of protists and animals and from many plants.

In fungi that reproduce sexually, nuclear fusion, or **karyogamy,** often takes place long after cytoplasmic fusion, or **plasmogamy.** During the time before karyogamy occurs, the mycelium formed by plasmogamy contains two different haploid nuclei per cell. Such mycelia are said to be **dikaryotic** ("two nuclei") or **heterokaryotic** ("different nuclei"), and their ploidy is represented as $n + n$, rather than n (haploid) or $2n$ (diploid).

Some fungi have a curious type of sexual reproduction called *parasexuality.* In this process, hyphae of different mating types fuse, producing a dikaryotic cell. Nuclear fusion follows. Typically, the next step would be meiosis, but in parasexuality half the chromosomes are gradually lost, a process called *haploidization.* Fragments of homologous chromosomes can be exchanged in the diploid nucleus before haploidization. Consequently, the haploid nucleus may be genetically different from either of the original nuclei in the dikaryotic cell.

Two other characteristics help set fungi apart from other organisms. First, in most fungi the nuclear envelope remains intact during mitosis and meiosis. Such nuclear division is found in some protists but not in plants or animals. Second, the cell walls of fungi contain a substantial amount of chitin, a nitrogenous glucose polymer. The external skeletons of arthropods (invertebrate animals such as insects, spiders, and crabs) are also made of chitin. However, chitin likely originated separately in fungi and arthropods, and it is rare in other groups of organisms.

Fungi probably evolved from flagellated protists

The earliest fossils resembling fungi were formed some 540 million years ago, in the early Cambrian period. Because most fungi have quite soft bodies that do not fossilize well, it will probably be difficult to extend the fossil record back to the earliest days of the kingdom. However, other lines of evidence also provide information about the evolution of fungi. For example, comparing the amino acid sequences of more than 100 proteins common to fungi, plants, and animals suggests that fungi appeared as a kingdom around 1.5 billion years ago and that the phyla of fungi may have begun to separate into clades between 1.4 and 1.1 billion years ago. Since plants and animals did not move onto land until perhaps 700 million years ago, early fungi must have been aquatic. Mycologists are working to extend the fossil record and to narrow the gap of nearly a billion years between the molecular and fossil dates for the origin of funguslike organisms.

Molecular evidence also suggests that fungi are more closely related to animals than to plants. Both fungi and animals appear to have evolved from a flagellated protist that, like modern fungi, absorbed nutrients after secreting digestive enzymes onto food. Present-day protists called *choanoflagellates* closely resemble that ancestral protist (Figure 19.3). Choanoflagellates exist as single cells or colonies and are strikingly similar to the feeding cells of sponges, which are among the simplest of animals. The single phylum of fungi that has flagellated cells (Chytridiomycota) is most likely the direct link between protists and other fungi, which probably lost their flagellated stages early in their evolution.

Both the association between fungi and plants in mycorrhizae (Chapter 4) and the association between fungi and algae or cyanobacteria in lichens (discussed later in this chapter) probably evolved around 700 million years ago. Many mycologists now believe these associations were essential factors in the establishment of eukaryotic life on land. Before plants colonized land, the soil consisted principally of rocks and sand. Plants would not have been successful at obtaining sufficient quantities of minerals from this early soil without the huge increase in root surface area for absorption provided by fungal hyphae. Indeed, the association between fungi and roots could have begun with the first bryophytes and seedless vascular plants. These plants grew in marshy regions, where the rocky surface was probably covered by decaying organic matter provided by invertebrate animals and

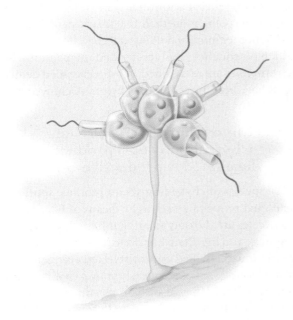

Figure 19.3 A colonial choanoflagellate. Choanoflagellates may be related to the ancestral protist from which fungi and animals are thought to have evolved.

algae. Fungi may have assisted early plants by decomposing organic material and releasing minerals, which the plants could absorb.

Section Review

1. How does the method of feeding used by fungi differ from that used by animals?
2. What is the relationship between a hypha and a mycelium?
3. Why is it thought that both fungi and animals evolved from a flagellated protist?

The Diversity of Fungi

Scientists have described well over 100,000 species of fungi, and many more remain to be discovered. Some mycologists estimate that more than a million species of fungi may exist! Fungi are classified primarily according to the details of their life cycle and morphology. Those species that have well-characterized life cycles are placed in one of four phyla: Chytridiomycota, Zygomycota, Ascomycota, and Basidiomycota.

Before the advent of DNA sequencing (Chapter 14), it was difficult or impossible to classify fungi that lacked a known sexual phase in their life cycle. Such fungi were collectively referred to as *deuteromycetes* ("second-class fungi," from the Greek word *deutero,* meaning "second") or *imperfect fungi* (because of their general lack of a sexual phase). At one point, more than 15,000 species of fungi were placed in this temporary holding category. Recently, analysis of DNA sequences has accelerated the pace at which deuteromycetes are reclassified. Most reclassified deuteromycetes have been placed in the phylum Ascomycota.

Chytridiomycetes (phylum Chytridiomycota) produce flagellated reproductive cells

The 700 species of chytridiomycetes produce spores and gametes that propel themselves by means of flagella. Chytridiomycetes are the only fungi that have flagellated cells at any stage during their life cycle, and for that reason they were once classified as protists. However, analysis of their nucleotide sequences clearly shows that chytridiomycetes are fungi. They also share several important enzymes and biochemical pathways with fungi and have the other fungal characteristics described earlier.

Most chytridiomycetes consist of spherical cells or coenocytic hyphae with only a few septa. In some chytri-

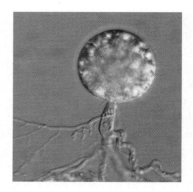

Figure 19.4 A chytrid.

diomycetes, the hyphae form slender, branched, rootlike structures called **rhizoids,** which grow into the food supply and hold the fungus in place (Figure 19.4). Chytridiomycetes most often live as water molds on dead leaves, branches, or animals in fresh water. Other species are marine, and some live in soil. Several species cause plant diseases such as black wart, a serious disease of potatoes.

Allomyces arbuscula, a well-studied water mold, has a sex hormone called *sirenin* that attracts male gametes to female gametes. Sirenin's name comes from the Sirens, mythological Greek figures who called to sailors in an attempt to lure their ships onto the rocks so they could capture the ships' goods. The life cycle of *Allomyces* features an alternation of isomorphic generations, which, as you remember from Chapter 18, means that the sporophyte and gametophyte are morphologically identical. In *Allomyces,* the diploid sporophyte can reproduce asexually by releasing diploid zoospores. It can also initiate the sexual cycle by releasing haploid zoospores, which germinate and develop into gametophytes. Alternation of generations is rare in fungi, but it appears in all plant phyla and in many algae.

Zygomycetes (phylum Zygomycota) form resistant zygosporangia prior to meiosis

More than 1,000 species of zygomycetes have been identified. Most form coenocytic hyphae and live on dead plants and animals or other organic matter, such as dung. Some live as endosymbionts in the digestive tract of arthropods, while others are the fungal components of endomycorrhizae (Chapter 4). Zygomycetes cause many types of soft rot in fruits and a few parasitic diseases in animals.

You are probably familiar with the zygomycete *Rhizopus stolonifer,* sometimes called *black bread mold* (Figure 19.5). *Rhizopus* is one of many fungi that commonly grow on bread, fruit, and other moist, carbohydrate-rich foods. Haploid mycelia of *Rhizopus* grow rapidly through the food, absorbing nutrients. Preservatives, such as calcium propionate and sodium benzoate, do a reasonable job of inhibiting the growth of *Rhizopus* and other fungi, at least for a while.

Like other zygomycetes, *Rhizopus* is capable of asexual and sexual reproduction, both involving spores. In a stable environment, asexual reproduction predominates (Figure 19.5). The mycelium extends specialized hyphae, called *stolons,* across the surface of the food. Wherever the stolons touch the surface, rhizoids grow into the food. The rhizoids anchor upright hyphae called **sporangiophores,** each of which forms a black sporangium at its tip. The sporangia are what you see when you spot the mold growing on a slice of bread. By the time they form, the lightly pigmented mycelium has grown for at least several days and has thoroughly penetrated the bread. (You may want to remember this the next time you throw away a moldy slice of bread and are about to eat the next slice in the loaf!) Nuclei and cytoplasm move through the sporangiophore into the

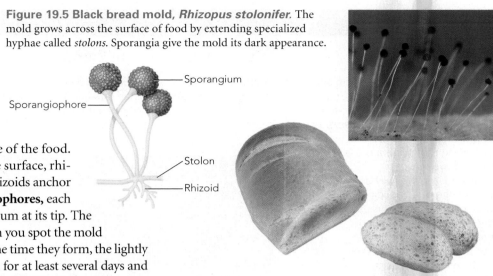

Figure 19.5 Black bread mold, *Rhizopus stolonifer*. The mold grows across the surface of food by extending specialized hyphae called *stolons.* Sporangia give the mold its dark appearance.

Sporangium

Sporangiophore

Stolon

Rhizoid

sporangium. Portions of cytoplasm, including one or more nuclei, are eventually partitioned into haploid spores. When the sporangium wall breaks, the spores are released, and if they land on a suitable source of food, they will germinate and begin the asexual cycle again.

A variety of conditions, including a dry environment, a shortage of food, or even the mere presence of opposite mating types, can trigger the sexual life cycle of *Rhizopus* (Figure 19.6). Mycelia of + and – mating types release chemicals that cause hyphae of opposite mating types to

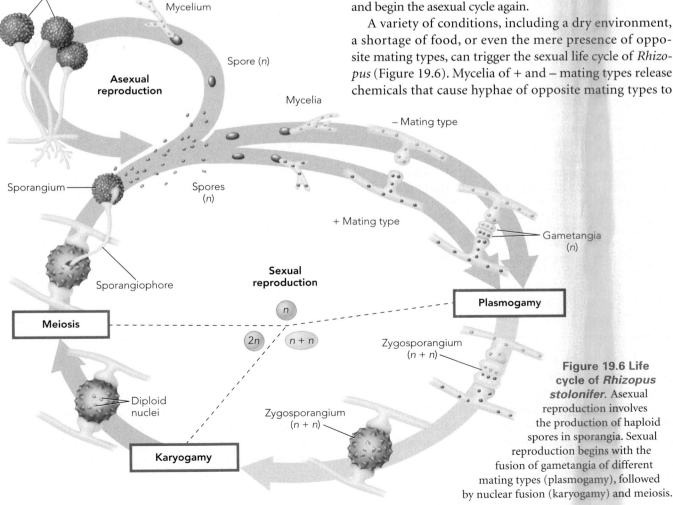

Sporangia

Mycelium

Spore (*n*)

Asexual reproduction

Mycelia

– Mating type

Sporangium

Spores (*n*)

+ Mating type

Gametangia (*n*)

Sporangiophore

Sexual reproduction

n

Plasmogamy

Meiosis

2*n*   *n* + *n*

Zygosporangium (*n* + *n*)

Diploid nuclei

Zygosporangium (*n* + *n*)

Karyogamy

Figure 19.6 Life cycle of *Rhizopus stolonifer*. Asexual reproduction involves the production of haploid spores in sporangia. Sexual reproduction begins with the fusion of gametangia of different mating types (plasmogamy), followed by nuclear fusion (karyogamy) and meiosis.

THE INTRIGUING WORLD OF PLANTS

Fungi That Live on Dung

Some fungi perform a valuable service by returning the nutrients in dung to the soil. Such fungi, called *coprophilous* (literally, "dung-loving") fungi, are found in the phyla Zygomycota, Ascomycota, and Basidiomycota. Usually zygomycetes sprout first from dung since they have a more rapid life cycle.

Each fungus typically grows on the dung of a specific type of animal. The fungus must have a mechanism that ensures that its spores travel with the animal, so that they are deposited along with their new food supply. Many coprophilous fungi produce spores that travel inside an animal's digestive tract. Often, the spores will not germinate unless they have been partially digested.

Pilobolus sporangiophores.

The mechanism of spore dispersal used by fungi that live on herbivore dung generally differs from that used by fungi that live on the dung of omnivores or carnivores. For example, consider the coprophilous zygomycete *Pilobolus*, which lives on the dung of herbivores such as cows. Near the tip of each *Pilobolus* sporangiophore, a swollen area focuses light, causing the sporangiophore to bend toward the Sun. High solute concentrations in the swollen area drive water uptake, which makes the area swell further and finally explode. When that happens, the sporangium at the tip of the sporangiophore is ejected a distance of up to 2 meters (6.5 feet) in the direction of the brightest light, where the grass is more likely to be growing and the spores are more likely to be eaten by cows.

Phycomyces, another coprophilous zygomycete, produces two types of asexual sporangiophores, called *macrophores* and *microphores*. Like *Pilobolus* sporangiophores, *Phycomyces* macrophores bend toward light. In contrast, the growth of *Phycomyces* microphores is inhibited by light. Thus, *Phycomyces* can reproduce asexually whether it is in a dark barn, scattering spores on nearby hay, or out in the field, scattering spores on grass!

Spirodactylon is a zygomycete that grows on rat dung. Since rats are omnivores and eat in many places, *Spirodactylon* has no guaranteed way to direct its spores toward the areas where rats eat. Therefore, bending its sporangiophores toward or away from light would not increase the chances of spore germination. Instead, *Spirodactylon* produces very long sporangiophores that have sticky, tightly coiled sections. The sporangiophores become entangled in the rats' fur upon contact, and such contact is nearly certain because the rats deposit their dung along the same trails they travel every day. When a rat grooms itself, it ingests the sporangiophores and the spores they carry.

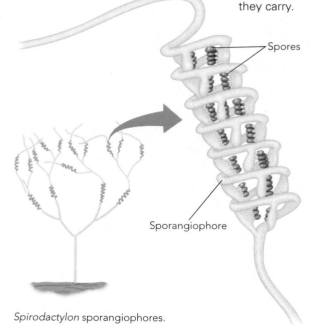

Spores

Sporangiophore

Spirodactylon sporangiophores.

grow toward each other. On contact, each hypha forms a gametangium consisting of a single cell with multiple nuclei. Fusion of the two gametangia (plasmogamy) produces a zygosporangium that contains nuclei of both mating types. The zygosporangium develops a thick, resistant wall and contains a single **zygospore.** Karyogamy occurs inside the zygospore, and the diploid nuclei undergo meiosis, forming haploid nuclei. Germination of the zygospore produces a sporangiophore with a sporangium at its tip. The haploid nuclei then undergo mitosis and move into the sporangium, where they become spores. Rupture of the sporangium releases the spores, which may germinate and produce new mycelia (see *The Intriguing World of Plants* box on this page).

(a) A cup fungus, *Sarcoscyha austriaca.*

(b) Powdery mildew on apple.

Figure 19.7 Ascomycetes.

Ascomycetes (phylum Ascomycota) produce sexual spores in sacs called *asci*

The phylum Ascomycota contains more than 30,000 species of fungi that live independently and close to 60,000 species if those involved in lichens (discussed later in this chapter) are considered. Most live on dry land and have hyphae with perforated septa. Ascomycetes include numerous cup fungi (Figure 19.7a), most yeasts, and various blue, green, pink, and brown molds commonly found on improperly preserved food. Several important plant diseases, including powdery mildews, are caused by ascomycetes (Figure 19.7b).

Like zygomycetes, ascomycetes can reproduce asexually or sexually, but asexual reproduction is more common (Figure 19.8). The asexual spores of ascomycetes, called **conidia** (singular, *conidium*), do not form inside sporangia. Rather, they are produced at the tips of modified hyphae called *conidiophores*. Often, conidia contain more than one nucleus.

A range of environmental variables triggers sexual reproduction in ascomycetes, which typically begins with the chemical attraction of haploid mycelia of different mating types. Each mycelium produces a large, multinucleated cell that functions as a gametangium. The two gametangia, one called an antheridium and the other called an **ascogonium**, form next to each other. Plasmogamy occurs when a thin outgrowth known as a *trichogyne* (literally, "female hair") extends from the ascogonium to the antheridium. Nuclei from the antheridium move through the trichogyne into the ascogonium, and nuclei of opposite mating types pair up. The ascogonium then begins to produce septate, dikaryotic hyphae, which are incorpo-

rated into a fruiting body called an **ascocarp,** or **ascoma.** The ascocarp also contains many haploid hyphae derived from the parent mycelium. Some ascocarps are microscopic, while others, like the ones shown in Figure 19.7a, may be several centimeters across. Cells at the tips of the dikaryotic hyphae expand and form saclike **asci** (singular, *ascus*) within the ascocarp. Karyogamy occurs in the ascus, and the diploid nucleus undergoes meiosis. The haploid daughter nuclei then undergo mitosis, producing eight nuclei that are incorporated into the ascospores, which are often arranged in a line. When the ascospores germinate, they produce new haploid mycelia.

Two types of edible ascomycetes are truffles and morels. Truffles, such as *Tuber melanosporum,* grow underground, often under oak trees (Figure 19.9a). They are highly prized in French cuisine and, depending on their type and quality, may sell for more than $600 per kilogram (2.2 pounds). Despite many attempts, however, no one has succeeded in growing truffles as an agricultural crop. Therefore, truffles are still collected from the wild, usually with the help of pigs or trained dogs. Pigs are very sensitive to the scent of truffles, which is produced by molecules that resemble pig sex pheromones. Morels (*Morchella* spp.) are another favorite gourmet ascomycete and are quite tasty when fully cooked (Figure 19.9b). Forest fires can create soil conditions that favor the fruiting of morels.

Yeasts are unicellular fungi, most of which are ascomycetes. A typical yeast is the ascomycete *Saccharomyces cerevisiae,* known as baker's yeast and brewer's yeast, which is used to cause fermentation during baking and brewing (Chapter 9). Fermentation by yeast was exploited 6,000 years ago in ancient Sumer, now located in Iraq, but the identity of the organism that causes it was not discovered until

Figure 19.8 Life cycle of an ascomycete. Asexual reproduction occurs by means of haploid spores called *conidia*. Sexual reproduction begins with the fusion of an ascogonium and an antheridium (plasmogamy), followed by karyogamy, meiosis, and mitosis inside a saclike ascus.

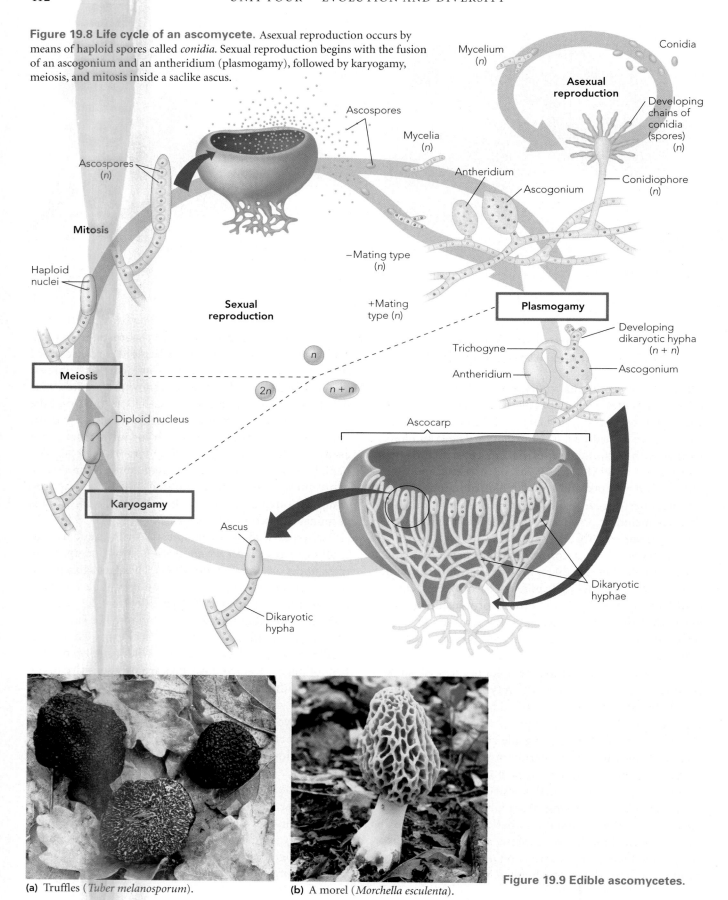

(a) Truffles (*Tuber melanosporum*).

(b) A morel (*Morchella esculenta*).

Figure 19.9 Edible ascomycetes.

the 1900s. In medieval Europe, fermentation was regarded as a miracle, and yeast was simply called "godisgood" in many manuscripts and books. Yeasts can be either diploid or haploid and typically reproduce asexually by budding, in which daughter cells are produced from a small pore in the side of the parent cell. A sexual cycle, resulting in the production of ascospores, can also occur. In recent years, *S. cerevisiae* has served as a model organism for genetics and the study of gene action because, like bacteria, it can be grown easily in the laboratory.

Ascomycetes include several important species in the genus *Aspergillus*. For example, large industrial cultures of *Aspergillus niger* produce most of the citric acid in sodas, while soybeans fermented with *A. oryzae* are made into soy sauce and soy paste, or miso. Not all species of *Aspergillus* are beneficial, however. *Aspergillus flavus* and *A. parasiticus* produce a metabolite, aflatoxin, which markedly increases the chances of getting liver cancer if it is ingested. Aflatoxin is sometimes found in products containing corn and wheat and in peanut meal, which is used to make peanut butter and to feed poultry. Poultry food is now monitored for aflatoxin contamination. Old jars of peanut butter still cause occasional problems, as does milk from cows that have consumed the toxin. Other species of *Aspergillus* cause the severe lung disease aspergillosis when inhaled.

Another lung disease, called *valley fever* or *coccidioidomycosis*, is caused by inhaling conidia of the ascomycete *Coccidioides immitis*. The disease usually produces mild, flu-like symptoms but can be fatal if many spores are inhaled, particularly by a person with a weakened immune system. In the skin, ascomycetes known as *dermatophytes* (from Greek words meaning "skin plants") cause athlete's foot, ringworm, and similar diseases.

The ascomycete *Claviceps purpurea* causes **ergot,** a disease of grains such as wheat and barley. Ergotized grain contains a number of toxic chemicals produced by the fungus, including lysergic acid amide, a precursor of lysergic acid diethylamide (LSD). Consuming ergotized grain can cause ergotism, a human disease also known as St. Anthony's fire, a toxic condition whose symptoms include hallucinations, disorientation, cramps, and convulsions and may result in death. Ergotism killed 40,000 people in a European epidemic in A.D. 994. The disease can also produce crazed behavior and delusions. The Salem witch trials, which were held in colonial Massachusetts in 1692, began after a group of young girls became hysterical while practicing magic. They were thought to be under the spell of witchcraft, but some people now suspect that they were showing symptoms of ergotism. Today, most cultivars of grain resist infection by *C. purpurea*, and granaries, where grain is stored, are well ventilated to retard fungal growth.

The chestnut blight fungus (*Cryphonectria parasitica*) is an ascomycete that has killed more than 3.5 billion American chestnut trees from Maine to Georgia. Once one of the largest and most common trees in eastern deciduous forests, American chestnuts (*Castanea dentata*) still grow from old rootstock but are killed by the fungus before they become large enough to reproduce. *C. parasitica* was introduced accidentally from Asia around 1900. See the *Conservation Biology* box on page 414 for a discussion of Dutch elm disease, another important plant disease caused by an ascomycete.

The first antibiotic, penicillin, was discovered in 1928 by Alexander Fleming, who noticed that an ascomycete, the mold *Penicillium*, prevented the growth of bacteria (Figure 19.10). Unfortunately, Fleming may not have fully appreciated the medical implications of his finding. At the beginning of World War II, British and American scientists continued Fleming's work, collecting *Penicillium* from many sources. They eventually obtained useful strains of the fungus that produced several hundred times as much penicillin as Fleming's original strain. Penicillin replaces

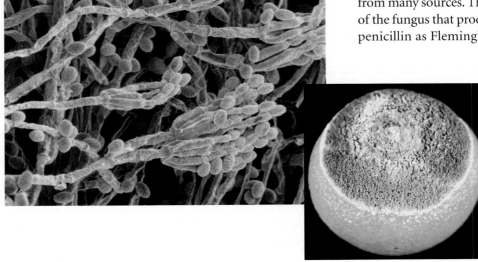

Figure 19.10 *Penicillium*. The mold *Penicillium* produces penicillin, which inhibits the growth of Gram-positive bacteria.

CONSERVATION BIOLOGY

Dutch Elm Disease

At one time, the American elm (*Ulmus americana*) was planted throughout the United States as a shade tree. It was favored over other trees because of its fast growth and high tolerance to a variety of suboptimal growing conditions. In the 1930s, however, Dutch elm disease (DED) was introduced to North America from Europe. Since then, DED has destroyed more than half the American elm trees in the United States. Avenues once shaded by rows of American elms have been left without trees or have been replanted with other, less desirable species.

The ascomycete *Ophiostoma ulmi* is the cause of DED. The disease is transmitted in two ways: by insects and by root grafts. The main insect vectors are two species of elm bark beetles, which, as their name suggests, feed on the bark of the tree. Although beetles that are fungus-free do not cause significant damage to the tree, beetles that have been exposed to the fungus infect the tree through the wounds they produce. Once infected, the tree produces cytoplasmic outgrowths from parenchyma cells that plug the water-conducting cells of xylem. This response by the tree stops the fungus from spreading, but it also cuts off the water supply to parts of the tree above the wound. The leaves wither and die, starting at the top of the tree and moving down. The entire tree dies within about a month.

The fungus then takes over the dead tree, producing spores throughout the xylem and under the bark. The spores that form under the bark are sticky. Adult beetles are attracted to the dead elm trees and breed there, laying eggs in the tree. The larvae that hatch from the eggs feed on the tree and develop into adults. When their adults bore their way out through the bark, they become covered with the

A diseased American elm tree and Conidiophores of *Ophiostoma ulmi.*

sticky spores and carry them to other trees.

DED can spread by root grafts if the trees are grown close to each other. Under these conditions, their roots grow together and share xylem, which is usually to the trees' mutual advantage. However, since the fungus spreads in this very part of the tree, it is shared by the trees along with water. The fungus can travel down an entire row of elms within a few days by way of grafted roots. One way to stop the transmission of the disease in the root systems is to disrupt the root grafts, usually by digging deep trenches between the trees. This solution is often difficult to apply in urban settings without damaging buried pipes and cables.

Dedicated and frequent pruning of dead wood where the beetles feed is another way to control the spread of DED. Such sanitation programs have saved 75% of the elms in some cities for 25 years. Other methods include using fungicides, insecticides, and traps that attract beetles to sticky paper infused with sex pheromones. Planting American elms with other species of trees can also reduce the likelihood of infection. The ultimate solution will be the production of DED-resistant American elms. Several laboratories have obtained resistant trees by selection of existing varieties, and some resistant lines are now available for planting. Attempts to introduce genes for resistance to the fungus into the genome of American elm trees are underway.

---

a crucial building block in the cell wall of Gram-positive bacteria blocking cell wall synthesis. The resulting wall is weakened, and the bacterium explodes due to unrestricted water uptake. Another group of medically important compounds, cyclosporines, come from the soil-dwelling ascomycete *Cordyceps subsessilis.* Cyclosporines are used to suppress organ rejection in organ-transplant patients.

Ascomycetes in the genus *Trichoderma* are attracted to the hyphae of other fungi, which they digest. Various species of *Trichoderma* are under intense investigation as a way to control harmful fungi of plants. Because *Trichoderma* produces enzymes that degrade wood, scientists are studying them as a possible means for producing ethanol from wood. *Trichoderma* is also being investigated as a source of enzymes that can be added to laundry detergents to soften fabrics.

Basidiomycetes (phylum Basidiomycota) produce sexual spores on club-shaped cells called *basidia*

We are all familiar with mushrooms, in the grocery store and on the lawn. Mushrooms are one of several types of reproductive structures of fungi in the phylum Basidiomycota, which is divided into three classes: Basidiomycetes, Teliomycetes, and Ustomycetes.

Class Basidiomycetes

The class Basidiomycetes contains more than 14,000 species of mushrooms, toadstools, stinkhorns, puffballs, shelf fungi, jelly fungi, and bird's-nest fungi (Figure 19.11). In common speech, *mushroom* refers to edible species and *toadstool* to inedible species, but these terms have no scientific meaning. Just because a fungus is called a mushroom does not mean it is edible!

(c)

(d)

(a)

(e)

(b)

(f)

Figure 19.11 Basidiomycetes. (a) Dryad's saddle shelf fungi (*Polyporus squamosus*). (b) A puffball (*Calostoma innabarina*). (c) A Witches' Butter jelly fungus (*Tremella mesenterica*). (d) A bird's nest fungus (*Crucibulum leave*). (e) A veiled stinkhorn. (f) Shaggy parasol mushrooms (*Macrolepiota rhacodes*).

A mushroom is actually an above-ground fruiting body produced during part of the fungus's life cycle. More than 90% of the volume and mass of the fungus may lie underground in the form of haploid mycelia of different mating types. In many species, each mating type is determined by a unique combination of alleles of two genes, *A* and *B*. For example, a species that has two *A* alleles (A_1 and A_2) and two *B* alleles (B_1 and B_2) has four mating types: A_1B_1, A_2B_1, A_1B_2, and A_2B_2. One species, the split gill fungus (*Schizophyllum commune*) has at least 300 *A* alleles and 90 *B* alleles, resulting in at least 27,000 mating types!

Many fungi in the class Basidiomycetes do not reproduce asexually, although some species produce asexual spores (conidia). In the life cycle of a typical mushroom, mycelia of different mating types are attracted to each other and fuse, producing dikaryotic hyphae (Figure 19.12). The dikaryotic hyphae elongate and branch,

forming a dikaryotic mycelium that eventually grows out of the ground and produces a mushroom, also known as a **basidiocarp** or **basidioma.** Inside the basidiocarp, large, club-shaped cells called **basidia** (singular, *basidium*) form at the ends of the dikaryotic hyphae. The nuclei in each basidium undergo karyogamy, producing a diploid nucleus that then goes through meiosis, yielding four haploid nuclei per basidium. Four small swellings form at the end of each basidium, and one haploid nucleus moves into each swelling. These swellings become haploid **basidiospores,** which produce new haploid mycelia when they germinate.

Mushrooms usually consist of a cap at the end of a stalk. Ones that are just forming, called *buttons*, are sometimes covered by a thin membrane that ruptures as the mushroom increases in size. Pieces of this membrane sometimes may remain on the top of the cap and around the lower

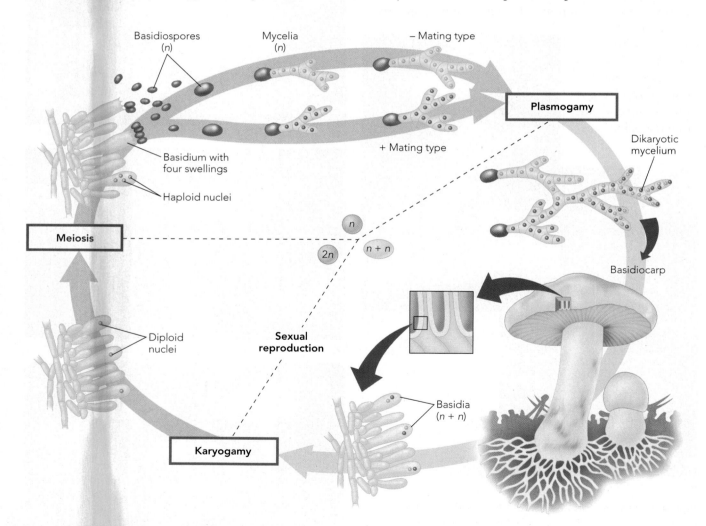

Figure 19.12 Life cycle of a mushroom. Haploid mycelia of opposite mating types fuse (plasmogamy) and produce dikaryotic mycelia, which form a basidiocarp. Karyogamy and meiosis occur inside basidia.

PLANTS & PEOPLE

Growing Mushrooms

Agaricus brunnescens, the common grocery store or pizza mushroom, accounts for 90% of commercial mushroom production in the United States and 40% worldwide. Portabella mushrooms are simply a larger, brown strain of the same species. A. brunnescens has also been named A. bisporus (meaning "two spores") because each basidium produces only two basidiospores, each of which contains two nuclei. Thus, when the basidiospores germinate, the mycelium they form is already dikaryotic.

The cultivation of these mushrooms involves several stages. First, horse manure is broken down, or composted, in huge piles by bacteria and other fungi. Then it is transferred to wooden trays, where it is sterilized to kill the microorganisms responsible for the original composting. Next, the manure is inoculated with A. brunnescens mycelia that have been growing on sterilized

Commercial mushroom production.

cereal grains. Once the compost is fully colonized by the mycelium, several centimeters of soil, known as casing, is placed atop the manure to encourage the formation of basidiocarps (mushrooms). Mushrooms suitable for sale can be collected after two to three weeks.

Lentinus edodes, the shiitake mushroom, is the second most commonly eaten species. Shiitake protein has all of the essential amino acids and so can be important in a vegetarian diet. Shiitake mushrooms also contain a polysaccharide known as lentinan. Some studies have suggested that lentinan may be useful in lowering cholesterol levels and in accentuating the cancer-killing effects of chemotherapy.

Pennsylvania and California are the states with the highest production of edible mushrooms. Yearly production in the United States is now more than 2.2 billion kilograms (1 billion pounds), with an estimated value of almost $1 billion.

---

part of the stalk. Some mushrooms have thin sheets of tissue, called *gills*, on the underside of the cap. Each gill is composed of many hyphae.

Frequently, a dikaryotic mycelium grows outward in all directions from the site where plasmogamy took place. A circle of mushrooms, commonly known as a *fairy ring*, may form periodically at the outer margin of growth (Figure 19.13). A dikaryotic mycelium may continue to grow

Figure 19.13 A fairy ring. The mushrooms in this ring are part of the same mycelium.

for centuries, using the food supply in the soil as it grows. In 1992, scientists discovered a giant mycelium on the Upper Peninsula of Michigan that covered 15 hectares (37 acres) and weighed an estimated 9,700 kilograms (21,000 pounds). At the time, it was thought to be the world's largest organism, but it soon lost that title when a mycelium in Oregon that covered 890 hectares (2,200 acres, or 3.4 square miles) was found! Both were mycelia of a species in the genus *Armillaria*.

Many mushrooms are edible, including *Agaricus brunnescens* (also known as *A. bisporus*), the most common grocery store mushroom (see the *Plants & People* box on this page). However, some mushrooms—less than 1%—are poisonous (Figure 19.14). At the button stage, edible and poisonous mushrooms often look quite similar. The toxins in poisonous mushrooms are usually alkaloids, and different species vary widely in the amount of toxins they contain. Ingestion of nonlethal amounts of the toxins in some mushrooms causes vivid hallucinations, accounting for the use of those mushrooms in certain religious ceremonies.

(a)

Figure 19.14 Poisonous mushrooms.
(a) Death cap (*Amanita phalloides*).
(b) Destroying angel (*Amanita virosa*).

(b)

Class Teliomycetes

The more than 7,000 species of rusts in class Teliomycetes do not form mushrooms, but they do form septate basidia that develop in regions called **sori** (singular, *sorus*) on the leaves or stems of infected plants. A number of rusts cause diseases in crop plants and trees, which have serious negative effects on food availability and cost. Farmers experience lower yields and higher costs from fighting the diseases. Consumers pay more because production decreases. Countries may have to import food or reduce cash-producing food exports.

Many rusts parasitize two plant species in their life cycle. For example, white pine blister rust (*Cronartium ribicola*) infects five-needled pines and currants or gooseberries in the genus *Ribes* (Figure 19.15a). Black

stem rust of wheat (*Puccinia graminis*) infects wheat and barberry (*Berberis*), causing yearly losses of more than a billion dollars in the United States and Canada alone (Figure 19.15b). Fungicides generally are not effective against rusts. To fight black stem rust, breeders strive for genetically resistant cultivars of wheat, but long-lasting resistance is difficult to obtain because the rust mutates frequently. More than 350 genetic races of the rust occur. Another approach is to interrupt the life cycle of rusts by eliminating the economically less-important host from an area. However, some rusts, such as black stem rust, can survive in the remaining host, especially where the climate is warm enough to allow that host to grow throughout the year.

Rusts have complex life cycles involving several different types of spores. Consider black stem rust as an example. In spring, haploid *basidiospores* produced on wheat germinate in barberry. The germinating spores form hyphae that enter the leaves through stomata and develop into mycelia, which produce haploid gametes. The mycelia release nectarlike secretions that attract insects, which carry the gametes from leaf to leaf and plant to plant. Plasmogamy between gametes and haploid

(a)

(b)

Figure 19.15 Rusts. (a) White pine blister rust (*Cronartium ribicola*). **(b)** Black stem rust (*Puccinia graminis*) on wheat.

hyphae of opposite mating types results in dikaryotic hyphae, which produce *aeciospores*. The aeciospores infect wheat, where they germinate into dikaryotic hyphae that produce *urediniospores* throughout summer. In late summer and early fall, the hyphae gradually stop producing urediniospores and begin producing dikaryotic *teliospores,* which go through karyogamy and then overwinter. In spring, meiosis gives rise to basidiospores, which reinitiate the life cycle.

Class Ustomycetes

Like rusts, smuts (class Ustomycetes) cause considerable economic damage to crops. More than 1,000 species of smuts parasitize four times that number of flowering plant species, including all commercially important cereals and grasses. A typical smut is *Ustilago maydis,* which infects corn (Figure 19.16). Although *U. maydis* is multicellular in nature, it can be grown as a unicellular fungus in the laboratory, where biologists can study and manipulate its genetics.

A corn smut infection begins with the germination of a single basidiospore, which produces numerous hyphae. Plasmogamy between hyphae of different mating types results in dikaryotic hyphae, which form a large sorus.

Figure 19.16 Corn smut (*Ustilago maydis*).

Nearby plant tissue swells because the fungal mycelium stimulates both enlargement and division of plant cells. Often, the fungus then spreads from the swollen region, called a tumor or **gall,** to other parts of the plant. *Ustilago maydis* commonly infects the ears, tassels, and stem, where it disrupts translocation in the xylem and phloem. The dikaryotic hyphae produce millions of teliospores. Smuts (from the German word *smutzen,* meaning "stain") get their name from the gray-green, sooty appearance of the teliospores. Karyogamy and meiosis occur in the teliospores, which overwinter. Like the teliospores of most smuts, those of *U. maydis* are spread by wind. When they germinate in spring, they produce septate basidia that form the customary four basidiospores. These spores infect other plants directly and also produce haploid infective cells known as sporidia.

Whereas most smuts are considered pests because of their destructive effects on plants, corn smut is regarded as a delicacy by some people. In Mexico, it is called *huitlacoche* and is eaten before the teliospores mature. Some farmers actually infect their corn with the fungus, knowing that huitlacoche commands a high price among its devotees. Smut-infected corn is used to make soup and creamed corn, among other dishes, which are pronounced delicious by most who have the courage to try them.

Section Review

1. **How do chytridiomycetes differ from other fungi?**
2. **Explain the relationship between stolons, rhizoids, sporangiophores, and sporangia in zygomycetes.**
3. **What are an ascocarp and an ascus?**
4. **What role do basidia play in the reproduction of mushrooms?**

Fungal Associations with Other Organisms

At least a quarter of all species of fungi are involved in a symbiotic relationship with an organism of another species. These close, long-term interactions can be parasitic, in which one species benefits while the other is harmed, or mutualistic, in which both species benefit. You read about several parasitic fungi earlier in this chapter, including the chestnut blight fungus, black stem rust, and corn smut. In Chapter 4, you learned about mycorrhizae, a mutualistic interaction between plant roots and fungi in which the plants supply organic molecules while the fungus increases the surface area of the root, accelerating the absorption of water and minerals from the soil. In this section, we will examine two additional symbiotic relationships that involve fungi.

Lichens are associations of fungi and photosynthetic algae or bacteria

Lichens are living associations between a fungus and a photosynthetic partner, either an alga or a cyanobacterium. The fungus in a lichen is called the *mycobiont,* and the alga or cyanobacterium is called the *photobiont.* At least 23 genera of algae and 15 genera of cyanobacteria occur in lichens. Although each lichen is composed of two species, lichens are given scientific names as though they were single species; the name given is that of the fungus. Estimates of the number of lichen species range from 13,500 to 30,000. Typically, lichens are classified according to the type of fungus they contain. Around 98% of fungi in lichens are ascomycetes, and the remaining 2% are basidiomycetes. Molecular evidence suggests that lichens evolved in several separate events and that some major groups of ascomycetes that do not form lichens evolved from lichen-forming fungi.

The body of a lichen, called a *thallus,* consists mostly of fungal hyphae. In some thalli, cells of the photobiont appear throughout the lichen. Usually, however, the photo-synthetic cells occur in a layer near the top of the thallus. As Figure 19.17 shows, these so-called stratified lichens can appear crustose (crusty), foliose (leafy), or fruticose (bushy).

Scientists who study lichens are unsure whether all or even most lichens represent mutualistic relationships. For the mycobiont, the benefit is clear: It receives carbon compounds and, in lichens that contain cyanobacteria, fixed nitrogen from the photobiont. For the photobiont, however, the benefit is less clear. Some algae and cyanobacteria exist as free-living species and as photobionts in the same habitat, and in these instances there is no obvious advantage to living with a fungus as part of a lichen. In other cases, the mycobiont may promote the survival of the photobiont by securing the thallus to rocks and other hard substrates and by providing a thick surface layer that prevents desiccation. Lichens can be dried until they consist of a very small percentage of water and will recover rapidly when rehydrated. In addition, fungal hyphae near the surface of many lichens contain compounds that protect the photobiont from damage by ultraviolet (UV) light. The concentration of these compounds in lichens correlates well with the intensity of UV light in the regions where they grow.

These unique features account for the ability of lichens to survive in many terrestrial environments, including some that are inhospitable to other life-forms. For example, lichens commonly grow on exposed rocks, often in

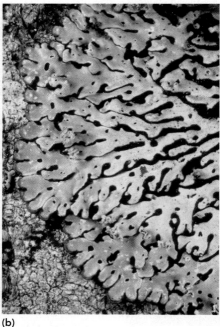

(a)   (b)   (c)

Figure 19.17 Lichens. (**a**) A crustose lichen (*Caloplaca ignea*). (**b**) A foliose lichen (*Menegazzia terebrata*). (**c**) A fruticose lichen (*Heterdermia echinata*).

windswept locations with extreme, widely varying temperatures. Between 200 and 300 species of lichens exist in Antarctica, where only a handful of plant species can survive. In particularly harsh regions of Antarctica, lichens grow between particles of sand and among the crystals in rocks as cryptoendoliths (from the Greek words *kryptos*, meaning "hidden," *endon*, meaning "within," and *lithos*, meaning "stone"). Lichens grow on mountains at elevations up to 7,300 meters (24,000 feet), in hot deserts, and on ocean shores subject to intermittent salt spray. In cities, lichens can be found on buildings, pavements, and even stained-glass windows. Lichens are also abundant in less extreme environments, such as rain forests, where they coat the trunks of trees as well as the forest floor. A single tree trunk in a forest may house several dozen lichen species.

Lichens are quite sensitive to alterations in their environment and to environmental pollution. For example, changes in the average humidity of a region due to logging or damming of rivers can cause some species to die out and allow others to appear. Lichens react to acid rain (produced when oxides of sulfur and nitrogen dissolve in rainwater), and they readily absorb particulate pollution from motor vehicles, power plants, and factories in which the particles contain heavy metals and other toxic compounds. The types of metals accumulated by lichens can frequently be used to identify the source of the pollution, creating the potential for corrective action. Measuring the growth rate of lichens often provides an accurate assessment of the identity of particular types of pollutants and their concentrations.

The importance of lichens to humans goes beyond their usefulness in monitoring air pollution. Lichens containing cyanobacteria as photobionts increase soil fertility by fixing nitrogen. Some species of lichens indicate the presence of particular metals in rocks and soil where they grow. Knowledge of this fact has helped prospectors since the time of ancient Rome. For thousands of years, people have also used lichens to make dyes for fabrics and paints. Before the production of synthetic dyes in the 1900s, collection of lichens was a common activity, and large factories for processing lichens into dyes existed throughout the world.

Lichens that grow on rocks frequently begin the process in which rocks break down to soil that can support the growth of plants (see Chapter 25). Lichens release acidic metabolites that decompose rocks much more rapidly than weathering caused by wind, rain, or freezing and thawing. These metabolites dissolve minerals in the rock, making them available to the mycobiont and photobiont in the lichen.

Lichens are important members of many biological communities. For example, lichens known as *reindeer moss* (*Cladonia*) provide around half the food consumed by reindeer, or caribou, living on the arctic tundra. In winter, when plants are scarce, caribou move the snow aside with their hooves to get at the lichen.

Some fungi form mutualistic associations with insects

Fungi participate in a variety of symbiotic associations with insects. Many of them are parasitic and may result in the death of the insect. Some, however, are mutualistic. One of the most interesting is between several species of basidiomycetes and Central American leaf-cutting ants in the genus *Atta* (Figure 19.18). The ants live in colonies of up to 8 million individuals. Each colony constructs a nest underground that can have as many as 1,000 chambers,

(a)

(b)

Figure 19.18 A mutualistic association between fungi and ants. (a) A Central American leaf-cutting ant (*Atta*) carries a piece of leaf back to the nest. (b) One of the fungal gardens tended by the ants.

each approximately 30 centimeters (1 foot) in diameter. In many of the chambers, the ants cultivate fungal gardens, where they tend mycelia that feed on pieces of leaves the ants cut from nearby vegetation. The fungi feed on the leaves, using the enzyme cellulase to break down the cellulose they contain. The ants, like most animals, lack cellulase. They harvest the swollen hyphal tips, called *bromatia*, as their food. The ants also patrol the gardens and assiduously remove other fungi, particularly those that consume the principal species that forms the garden. New colonies are established by queen ants, which carry fungal hyphae in a pouch in their mouth.

African and Asian termites in the subfamily Macrotermitinae also cultivate fungal gardens. Whereas many termites contain animal-like protists in their gut that can digest cellulose, these termites do not. Their relationship with fungal mycelia is similar to that of leaf-cutting ants, except that the termites usually bring the fungi pieces of wood or dead vegetation rich in cellulose. These termites live in aboveground mounds that can be up to 6 meters tall (19.7 feet) and 3 meters (9.8 feet) in diameter.

Section Review

1. In those lichens that appear to represent mutualistic associations, how does each partner in the association benefit?
2. What are cryptoendoliths?
3. Why do some ants and termites cultivate fungal gardens?

SUMMARY

Characteristics and Evolutionary History of Fungi

A combination of morphological and developmental characteristics distinguish fungi from other organisms (pp. 406–407)
Fungi are heterotrophs that absorb food after breaking it down into small molecules. Most fungi are multicellular and are composed of filaments of cells, hyphae, interwoven into mycelia. Fungi that reproduce sexually often form dikaryotic mycelia during the period between nuclear fusion (karyogamy) and cytoplasmic fusion (plasmogamy). The cell walls of fungi contain chitin.

Fungi probably evolved from flagellated protists (pp. 407–408)
Molecular analyses suggest that fungi evolved about 1.5 billion years ago, probably from a flagellated protist that also was ancestral to animals. Very early associations between fungi and other organisms in mycorrhizae and lichens may have been important in the establishment of eukaryotic life on land.

The Diversity of Fungi

Chytridiomycetes (phylum Chytridiomycota) produce flagellated reproductive cells (p. 408)
Chytridiomycetes are the only fungi that produce flagellated spores and gametes. They most often live as freshwater molds, and some have a life cycle featuring an alternation of generations, which is rare in fungi.

Zygomycetes (phylum Zygomycota) form resistant zygosporangia prior to meiosis (pp. 408–410)
Most zygomycetes, including black bread mold (*Rhizopus*), form coenocytic hyphae and have both asexual and sexual life cycles. Asexual reproduction predominates in stable environments. Unfavorable conditions trigger sexual reproduction, which involves the formation of thick-walled, resistant zygosporangia, each of which contains a zygospore.

Ascomycetes (phylum Ascomycota) produce sexual spores in sacs called *asci* (pp. 411–415)
Ascomycetes include cup fungi, truffles, morels, most yeasts, and a variety of molds and mildews. Most have septate hyphae. Ascomycetes form asexual spores called *conidia* at the tips of modified hyphae, and they form sexual spores in saclike asci, which are contained within a fruiting body called an *ascocarp*. Some ascomycetes are used to manufacture food products or medicines, such as penicillin. Others produce toxins that contaminate foods or cause diseases, including valley fever and athlete's foot.

Basidiomycetes (phylum Basidiomycota) produce sexual spores on club-shaped cells called *basidia* (pp. 415–419)
Basidiomycetes include mushrooms, puffballs, shelf fungi, rusts, and smuts, among other forms. Many lack an asexual life cycle. Basidiomycetes produce sexual spores that arise as swellings on large, club-shaped cells (basidia). In fungi that belong to the class Basidiomycetes, the basidia are contained in a basidiocarp. In rusts and smuts, the basidia develop in sori. Many rusts and smuts cause diseases in crop plants and trees.

Fungal Associations with Other Organisms

Lichens are associations of fungi and photosynthetic algae or bacteria (pp. 420–421)

The body of a lichen consists mostly of fungal hyphae, which secure the lichen to the substrate and prevent desiccation. Located among the hyphae are algal or cyanobacterial cells that provide the hyphae with carbon compounds and sometimes fixed nitrogen. Lichens can survive in many terrestrial environments but are sensitive to environmental alterations and pollution.

Some fungi form mutualistic associations with insects (pp. 421–422)

Leaf-cutting ants and certain termites cultivate fungal gardens. The insects grow the fungi in underground chambers or aboveground mounds and feed them vegetation or pieces of wood. The fungi produce swollen hyphal tips, which the insects eat.

Review Questions

1. Why is decomposition important? What would happen if it did not occur?
2. What is the difference between septate and coenocytic hyphae?
3. Explain the difference between plasmogamy and karyogamy. What is a dikaryotic mycelium?
4. How old are the oldest known fungal fossils?
5. Based on molecular sequencing, when are fungi thought to have appeared as a kingdom?
6. What conditions can cause black bread mold to begin sexual reproduction?
7. Name several examples of ascomycetes.
8. How do ascomycetes reproduce asexually?
9. Explain the steps involved in the formation of an ascus.
10. Why is the mold *Penicillium* important?
11. What is the difference between mushrooms, rusts, and smuts?
12. What is an ascocarp?
13. What do wheat and barberry have in common?
14. When is smut a delicacy?
15. What is the difference between a photobiont and a mycobiont?
16. How is the association of *Atta* ants and certain fungi mutualistic?

Questions for Thought and Discussion

1. You observe what appears to be a disease on a tree trunk. How would you determine if it is caused by an insect, a virus, a bacterium, or a fungus?
2. A friend collects wild mushrooms and tests their safety for human consumption by feeding a few to his dog first. What do you think of his method of testing?
3. What happens to the millions of spores produced by a typical mushroom?
4. Why do you suppose shelf fungi usually grow on dead trees?
5. How would you test the hypothesis that ergotism was involved in a historical event, such as the Salem witch trials?
6. How would the world be different if fungi did not exist?
7. 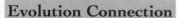 Hyphae are very versatile structures used by fungi for growth, food acquisition, reproductive processes (fertilization and meiosis), and the formation of complex reproductive structures. Draw labeled diagrams to illustrate this versatility in specific groups or species of fungi.

Evolution Connection

What characteristics of fungi indicate that these organisms belong in a different evolutionary line from that of the plants?

To Learn More

Visit The Botany Place Website at www.thebotanyplace.com for quizzes, exercises, and Web links to new and interesting information related to this chapter.

Hudler, George W. *Magical Mushrooms, Mischievous Molds.* Princeton, NJ: Princeton University Press, 1998. This Cornell professor makes the study of fungi seem like the most fascinating of topics. He relates fungi to history, health, and human interests of all sorts.

Lincoff, Gary A. *National Audubon Society Field Guide to North American Mushrooms.* London: Knopf, 1981. An excellent field guide with color photographs of more than 700 mushrooms.

Purvis, William. *Lichens.* Washington, DC: Smithsonian Institution Press, 2000. This comprehensive book is well illustrated and full of interesting details.

Schaechter, Elio. *In the Company of Mushrooms: A Biologist's Tale.* Cambridge, MA: Harvard University Press, 1998. This guide to classifying mushrooms has many interesting details about the fungal world.

20

Bryophytes

Mosses in Olympic National Park, WA.

An Overview of Bryophytes

Bryophytes were among the first land plants

Bryophytes have many similarities to green algae in the class Charophyceae and to vascular plants

In bryophytes, alternation of generations involves a dominant gametophyte and attached sporophyte

Bryophytes play important ecological roles

Many bryophyte species tolerate drought conditions

Liverworts: Phylum Hepatophyta

Liverwort gametophytes can be either thalloid or leafy

A liverwort life cycle demonstrates dominance of the gametophyte

Hornworts: Phylum Anthocerophyta

The hornwort life cycle features a hornlike sporophyte

The evolutionary history of hornworts, as with other bryophytes, is being debated

Mosses: Phylum Bryophyta

There are three main classes of mosses

The life cycle of *Polytrichum* demonstrates characteristic features of mosses

In the three previous chapters, we looked at viruses and at organisms in the domain Archaea, in the domain Bacteria, and in the kingdom Protista and kingdom Fungi of the domain Eukarya. We focused on photosynthetic life-forms and on organisms that cause plant diseases or that decompose complex organic molecules. Now we turn our attention to the kingdom Plantae, beginning with the simplest types of plants, the bryophytes (from the Greek *bryon*, "moss," and *phyton*, "plant"), which include liverworts, hornworts, and mosses. The most familiar bryophytes are mosses, which often grow in moist environments such as forests and wetlands but are also found in dry regions such as tundra. Being among the first land plants, bryophytes have existed, according to fossil evidence, for more than 400 million years—and perhaps, according to molecular evidence, as long as 700 million years.

Peat farmer, digging turf in Ireland.

Some of the most common bryophytes are the members of the moss genus *Sphagnum*, which typically live in bogs and cover between one and three percent of Earth's landmass. As new *Sphagnum* moss tissue grows on top of older tissue, only the top few centimeters are photosynthetic. The rest of the plant remains attached below, dying and decaying along with other bog plants to form an organic soil known as peat, which is why *Sphagnum* is commonly called *peat moss*. *Sphagnum* moss absorbs 10 to 20 times its dry weight in moisture, making it useful as a soil amendment to increase organic material and water retention of soils.

Tollund Man, found in Denmark, was preserved by the tannic acids of a peat moss bog.

Large areas of *Spaghnum* moss, known as *peatlands*, occupy about 400 million hectares (988 million acres) worldwide, equivalent to almost half of the area of the continental United States. *Sphagnum* is especially common in moist cool areas such as Ireland and parts of the northeastern United States and Canada. Peatlands store an estimated 400 billion tons of organic carbon, which can serve as a source of fuel. Peat produces 3.3 kilocalories per gram, greater than wood but considerably less than coal. Peat reserves in the United States contain as much

energy as 240 billion barrels of oil (a 38-year supply for the United States). As fossil fuels become scarce and expensive, peat will undoubtedly become more popular. Ireland already obtains 20% of its fuel from peat. *Sphagnum* is a renewable energy source that accumulates biomass at a rate twice that of corn, but overharvesting can damage fragile wetlands.

The water-absorbing capacity of *Sphagnum* has made it useful over the centuries in draining wounds and serving as a diaper material. Before cotton gauze was available, large quantities of *Sphagnum* were sold for those purposes. When gauze supplies ran short during the Civil War and World War I, doctors and nurses turned to tons of *Sphagnum* moss as a product that was readily available and could be sterilized. *Sphagnum* not only kept wounds clean and well drained but also seemed to prevent and help cure infections. The antibiotic effect of *Sphagnum* moss may be due simply to its acidic pH, or the moss may have antibacterial compounds.

Sphagnum moss does prevent decay both of the plant and of animals that die in it. Commercial peat mining operations have unearthed well-preserved human bodies as old as 3,000 years. Peat also preserves plant pollen, giving scientists an excellent idea of past climates and vegetation. Since peat resists decay, it removes a considerable amount of CO_2 that could otherwise contribute to global warming.

Sphagnum also accumulates heavy metals, such as lead, copper, and zinc, which are associated with human activities such as mining and manufacturing. When core drills are used in peat bogs, the layers can be dated using radioactive carbon and analyzed for the occurrence of heavy metals.

Although most bryophytes are not as important to humans as *Sphagnum*, all have intriguing natural histories and are vital members of their plant communities. We will look at the general evolutionary history and characteristics of bryophytes before exploring the distinguishing features of liverworts, hornworts, and mosses.

An Overview of Bryophytes

When the oxygen concentration in the atmosphere reached approximately 2%, about a tenth of today's level, multicellular eukaryotes could survive on land if they retained enough water. When land plants evolved from green algae, bryophytes were among the first plants to colonize dry land. They were typically found, as today, in moist environments where fresh water was readily available. The term *bryophytes* does not refer to a scientific classification but instead is an informal reference to all nonvascular plants. **Bryologists,** scientists who study bryophytes, formerly classified all nonvascular plants together as phylum Bryophyta, with three classes. However, each of these classes has now been elevated to a separate phylum: the phylum Hepatophyta, consisting of about 6,000 species of liverworts; the phylum Anthocerophyta, consisting of about 100 species of hornworts; and the phylum Bryophyta, consisting of about 9,250 species of mosses (Figure 20.1). The classification as three separate phyla reflects the view that liverworts, hornworts, and mosses evolved independently along separate paths from the same group of green algal ancestors.

Bryophytes were among the first land plants

Land plants first occur in the fossil record around 450 million years ago, in fossil fragments that appear to be of liverwort origin. Land offered more abundant sunlight and a substrate of rocks that was rich in minerals. Of course, green algae did not leap out of the water to take advantage of these conditions. Instead, the earliest land plants evolved after certain species of green algae became stranded during periods of seasonal drought without surrounding water. Adaptations that helped them survive drought had survival value and became common. These adaptations probably included vertical stems above ground and underground stems specialized for water and nutrient uptake.

Bryophytes evolved at about the same time as the amphibious animals. Like frogs, toads, and salamanders, their sperm requires free water for swimming to the egg. For this reason, bryophytes most commonly live in moist areas such as bogs (see the *Evolution* box on page 427), or forests that are frequently shrouded in clouds or fog. However, some bryophytes can survive in generally dry environments such as deserts and tundra.

Since bryophytes have soft bodies, they decompose before they can fossilize well, so there is little evidence of their shape, making it difficult to determine their early phylogeny. The earliest complete fossils of bryophytes date to near the end of the Devonian period, around 360 million years ago. Recent DNA, RNA, and protein sequencing data indicates that bryophytes may have arisen as early as 700 million years ago from the same group of green algae that gave rise to vascular plants. (This topic will be explored in more detail in Chapter 21.) If we could travel back in time to when the first vascular and nonvascular land plants evolved, we would undoubtedly see many intermediate types of "algae-plants" along with vascular and nonvascular plants. Evidence of such intermediate types of organisms would help in tracing plant evolution, especially because these organisms are all extinct.

Bryophytes are considered to be nonvascular plants because they do not have an extensive transport system with xylem and phloem, an absence that restricts their size and limits their distribution on dry land. It is often said that they do not have "true" roots, stems, and leaves because these terms traditionally apply to sporophytic organs in vascular plants (see Chapters 3 and 4), whereas in bryophytes the structures that serve as stems and leaves are found on the gametophytes. Given the similarities to vascular stems and leaves in function, and often in appearance, bryologists typically use the terms *stems* and *leaves,* a practice followed in this text. Meanwhile, since bryophytes also have similarities to green algae, the body of a bryophyte is often called a *thallus* (plural, *thalli*), the term also used to describe the bodies of algae as being less differentiated than those of vascular plants.

(a)

(b)

(c)

Figure 20.1 Bryophyte diversity. Bryophytes include **(a)** liverworts, **(b)** hornworts, and **(c)** mosses.

EVOLUTION

Bogs

By studying fossils, we can gather information to help envision the world when the first plants were evolving. While adapting to their environment, they were also altering the environment as they grew and reproduced. In observing how living bryophytes change the environment, we can imagine what Earth must have been like when land plants first evolved.

In freshwater environments, bryophytes are common around the edges of ponds and can eventually change the environment by accumulating biomass. For example, *Sphagnum* moss and other associated plants often grow slowly into a floating mat that gradually covers the top of a pond. Eventually, the mat becomes thick enough to walk on, converting the pond into a quaking bog. Walking on a quaking bog feels like walking on a waterbed covered with layers of thick blankets. With time, the entire pond will fill with dead *Sphagnum,* with a living layer of it on top.

Sphagnum grows out onto the surface of a pond and gradually turns it into a quaking bog, which really does quake when you walk or jump on it. Eventually it becomes a bog with no free water.

As you approach a typical floating bog, you can observe a succession of vegetation types, with trees giving way to shrubs and finally to herbs. Underlying all these vascular plants is *Sphagnum*. Near the bog's center, the *Sphagnum* is alive, while around the edge and extending above the water in some places is a layer of partially decayed *Sphagnum* known as *peat*. The first trees grow into older peat known as *woody peat*. Finally, and most distant from the bog, mature trees grow in black humus, which is the final breakdown product of *Sphagnum*. The *Sphagnum* gradually grows to cover the bog.

Bogs in which *Sphagnum* is the predominant plant are common in temperate zones, occupying at least 1% of Earth's surface. Peat bogs make wonderful evolutionary timelines. Core samples can be removed and the various layers can be carbon dated and observed to find different types of pollen and other evidence of past organisms.

Nevertheless, bryophytes clearly have differentiated structures. Instead of roots, they have **rhizoids** that serve primarily for anchorage rather than absorption, which occurs through whatever portion of the plant is in direct contact with water and nutrients. Many mosses have a transport system consisting of water-conducting cells called **hydroids** and food-conducting cells called **leptoids.** In their transport functions, these cells are somewhat similar to tracheids and sieve-tube members, but provide little structural support because they have thin cell walls. Hydroids are collectively called the *hadrom*, and leptoids are collectively called the *leptom*. Some bryophytes have stomata and are therefore able to transpire. As you will see later, bryophytes also have some remarkable reproductive mechanisms.

Bryophytes have many similarities to green algae in the class Charophyceae and to vascular plants

This text retains the traditional classification that places algae and plants in different kingdoms. Some botanists have proposed that algae in the class Charophyceae (Figure 20.2) should be included with plants in a new kingdom called Streptophyta. Others have suggested a new kingdom called Viridiplantae, which would include all plants and all green algae. This proposed kingdom is

Figure 20.2 Green algae gave rise to bryophytes.
Chara is a green alga that has features similar to the alga that may have been ancestral to the bryophytes. Its growth habit is plantlike, although this may simply result from convergent evolution.

probably too broad, but the proposed kingdom Streptophyta may have more merit. We do not have trouble today deciding if a particular living organism is a green alga or a plant. However, a view of organisms living when the evolutionary transition was occurring might have provided examples of organisms that would be truly difficult to distinguish as either green algae or plants.

It is not known whether bryophytes and vascular plants arose from the same species of green algae or from different species, and botanists have not found fossil representatives of species that bridge the evolutionary gap between green algae and plants. They are well aware, however, of the unique biochemical and structural similarities between green algae in the class Charophyceae and plants in general, such as the following:

- Cell walls are composed mainly of cellulose.
- Mitotic spindles remain during cytokinesis, which occurs through a phragmoplast.
- The pigment phytochrome is present.
- Chloroplasts contain chlorophyll *a* and *b,* as well as carotenoids.
- Thylakoids are stacked into grana.

In addition to traits shared with some green algae, bryophytes and vascular plants have further similarities as members of the plant kingdom. Many of these characteristics enhance survival on land by protecting gametes and spores from drying out. Here are a few examples:

- A layer of sterile cells protects structures that produce male and female gametes.
- A multicellular embryo is protected within the female parent.
- A multicellular, diploid sporophyte produces spores by meiosis.
- A layer of sterile cells protects multicellular sporangia.

In summary, bryophytes and other plants share some cellular features with green algae in the class Charophyceae but not with other types of algae. Furthermore, bryophytes and vascular plants have additional similarities relating to survival on land. However, differences between bryophytes and vascular plants indicate that, even though they may have common ancestors among green algae, natural selection has guided their evolution along different pathways.

In bryophytes, alternation of generations involves a dominant gametophyte and attached sporophyte

As with all plants, the sexual life cycle of bryophytes involves alternation of generations between a diploid sporophyte and a haploid gametophyte, with one form typically depending on the other for nutrition (see Chapter 6). However, bryophytes differ from vascular plants in the relative sizes of sporophytes and gametophytes. In vascular plants, the sporophyte is dominant, with the gametophyte being independent in some species and dependent on the sporophyte in others. In contrast, it is the gametophyte that is dominant in all three groups of bryophytes, with the sporophyte attached to and dependent on the gametophyte for most of its water and nutrition. As an example, Figure 20.3 shows a simplified version of a moss life cycle. Later in the chapter we will look at liverwort and moss life cycles in detail.

In most bryophyte species, gametophytes are about a centimeter or less (0.4 inch) in height. Bryophyte gametophytes have gamete-bearing structures known as **gametangia** (singular, *gametangium*). Male gametangia, called **antheridia** (singular, *antheridium*), contain sperm

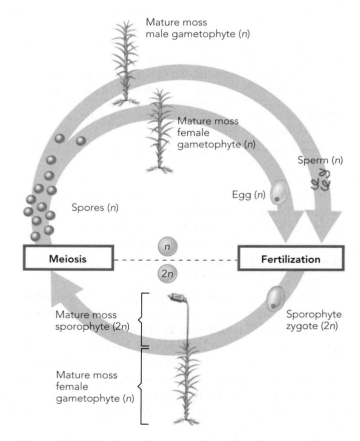

Figure 20.3 All plants have an alteration of generations. These generations are a diploid (2*n*) sporophyte and a haploid (*n*) gametophyte. In bryophytes, the gametophyte is dominant, and the sporophyte is attached to the gametophyte. The example shown in the diagram is a moss. In vascular plants, the sporophyte and gametophyte are separate plants, or the sporophyte is dominant and the gametophyte is attached to the sporophyte.

produced by mitosis. Female gametangia are called **archegonia** (singular, *archegonium*), with each archegonium containing one egg, also produced by mitosis. The archegonium is flask-shaped, with the egg inside its base. The sperm in each antheridium and the egg in each archegonium are surrounded by a protective layer of sterile cells that do not participate directly in reproduction. Many bryophyte species have bisexual gametophytes, having both antheridia and archegonia, while many others have separate female and male gametophytes. In some bryophyte species, sex is determined by sex chromosomes, as is the case in many animals. In plants, sex chromosomes were first discovered in bryophytes.

As in other types of plants, gametophytes and sporophytes alternate in producing each other, with gametophytes giving rise to a sporophyte through the process of fertilization, and a sporophyte giving rise to gametophytes through meiosis (Figure 20.3). In bryophytes, the gametophytes have anatomical features that facilitate fertilization by causing sperm-filled drops of water to splash about and sometimes attach to archegonia. Fertilization occurs when a sperm released from an antheridium unites with the egg in an archegonium, producing the zygote of the sporophyte. The sporophyte develops within the archegonium, receiving water and most nutrients from the gametophyte and remaining attached through maturity. At the tip of the mature sporophyte is a sporangium, which produces haploid spores by meiosis. After being released by the sporangium, a spore falls to the ground and may germinate to produce a **protonema** (plural, *protonemata*, from the Latin *proto*, "first," and Greek *nema*, "thread"), a typically threadlike structure that is most noticeable in mosses. A protonema forms one or more buds, each of which grows into a **gametophore,** a structure that bears gametangia. Gametophores can be either leafy or flattened (thalloid), and a gametophyte can have more than one gametophore. Later we will look at detailed versions of liverwort and moss life cycles.

Like all plants, bryophytes can also reproduce asexually. Liverworts, hornworts, and mosses can all reproduce through simple fragmentation, in which pieces of the plant—usually from the gametophyte—break off and establish new plants. Many liverwort and moss gametophytes can also reproduce asexually through various specialized structures called *brood bodies*. For example, **gemmae** (singular, *gemma;* from the Latin *gemma*, "bud") are small multicellular bodies that grow into new gametophytes when detached from the parent plant. Gemmae occur most frequently along the margins of gametophyte leaves and stems, but in some liverworts, such as those of the genus *Marchantia*, gemmae form inside a cuplike structure (Figure 20.4). Liverworts and mosses also have brood bodies called bulbils, which are asexual buds that

Figure 20.4 Gemmae are examples of asexual reproduction in bryophytes. The gemmae of *Marchantia thallus* are contained in gemmae cups around 1.0 millimeter (0.04 inch) in diameter. Drops of water splash gemmae onto adjacent soil, where they begin to grow into a new gametophyte.

detach from gametophytes and establish themselves as independent plants.

Bryophytes play important ecological roles

Bryophytes play an important role in plant succession. Mosses, for example, are usually the first plants to colonize rock surfaces and crevices, beginning the breakdown process that can eventually produce soil. Their rhizoids secrete acid that gradually dissolves the rock, leaving small pockets of soil to which successive generations of moss plants add organic matter. The seeds of other plants germinate in these pockets and establish more complex plant communities. Tree seedlings begin growing in crevices and may eventually split the rocks as roots expand.

Bryophytes are also important members of epiphytic plant communities that grow in trees of tropical and temperate rain forests. These communities also include flowering plants and seedless vascular plants such as ferns. In all rain forests, but particularly in temperate rain forests such as those in western Washington, bryophytes are major photosynthetic contributors to epiphytic communities. One reason that rain forests have such a high diversity of plant and animal life is that the high rainfall supports more photosynthesis, with the forest providing food not only in the form of tree leaves but also in the form of epiphytes.

Bryophytes are also major members of tundra ecosystems, where they serve as food for herbivores, along with lichens. However, reindeer moss (*Cladonia rangiferina*), the principal food of caribou, is actually a lichen.

Many bryophyte species tolerate drought conditions

Although the greatest variety of bryophyte species live in moist, warmer climates, some can survive in dry, seemingly hostile environments because of mechanisms enabling them to tolerate drought. For example, when conditions become dry, some liverworts simply roll up into a tubular form, which protects the upper surface of the plant from the Sun. Many mosses produce hairlike projections called *hair-points* at the tips of their leaves, creating a boundary layer that prevents excessive water loss. When hair-points are removed with scissors, water loss increases by as much as one third.

Mosses of the genus *Tortula* of Europe and southern North America are well known for being able to survive for years in a dry state. Within a few hours of receiving just a little water, the rehydrated *Tortula* mosses become golden yellow or green and begin to photosynthesize again (Figure 20.5). Among *Tortula* mosses, another secret to surviving drought may be that as the moss dries, it produces a type of mRNA that codes for proteins that will later help to repair the extensive cell damage caused by desiccation. The mRNA might be protected during desiccation by being bound to a protein. Scientists are considering inserting the genes that code for these proteins into crop plants to try to improve their drought tolerance.

Figure 20.5 A moss that survives drought. *Tortula rurali* is a genus of moss that lives in regions with only occasional rainfall. During dry periods, the plant looks completely dried out and dead. However, a few minutes after rain, the plants are rehydrated and fully functional.

Section Review

1. List several features shared by plants and green algae.
2. What are the principal differences between bryophytes and vascular plants?
3. Describe the life cycle of a typical bryophyte.
4. What are some ways in which bryophytes are ecologically important?

Liverworts: Phylum Hepatophyta

According to RNA sequence data, liverworts were possibly the earliest land plants and are the living plants most closely related to green algae. Further support for this hypothesis comes from experiments showing that liverworts and green algae lack particular pieces of DNA that are present in hornworts, mosses, and vascular plants.

In general, liverwort gametophytes are distinguishable as being more horizontal and flattened in appearance than those of most mosses. Most mosses have needlelike leaves, whereas leaves of liverworts, when present, are usually thin and flat. However, some liverworts and mosses are difficult to tell apart except by bryologists.

The name *liverwort* (from the Old English *wyrt,* "herb") reflects the medieval belief in the medicinal use of some of these small, herblike plants. According to the doctrine of signatures (see Chapter 16), the liver-shaped thallus of *Marchantia* liverworts was a "signature," or characteristic sign, that the plant was useful in treating liver ailments. The scientific name for the phylum, Hepatophyta (from the Latin *hepaticus,* "liver"), also alludes to this association.

Liverwort gametophytes can be either thalloid or leafy

Liverworts can be divided into two main categories: thalloid and leafy (Figure 20.6). In thalloid liverworts, the gametophyte is a flat, green structure that is sheetlike or algae-like in appearance, up to several centimeters across, and typically between one and ten cell layers thick. The thallus grows horizontally as a result of the division and elongation of meristematic cells at the tip of each branch. The branches divide to form two equal parts that grow away from each other at an angle. In leafy liverworts, the gametophyte looks more plantlike, usually with three ranks of flattened leaves that are one cell layer thick, on a branching structure that forms a mat. One rank of leaves is typically on the bottom of the stem. By contrast, the

(a) A leafy liverwort (*Plagiochila deltoidea*).

(b) A thalloid liverwort (*Aneura orbiculata*).

Figure 20.6 Liverwort diversity.

leaves of most mosses form a spiral and occur on all sides of the stem. Leafy liverworts, which constitute more than 80% of the known liverwort species, reach their greatest diversity in foggy, tropical regions with high rainfall. Some liverwort species are aquatic (Figure 20.7).

A liverwort life cycle demonstrates dominance of the gametophyte

In liverworts, as in all bryophytes, the gametophyte is dominant—the form most readily visible. Figure 20.8 shows the life cycle of the thalloid liverwort genus *Marchantia*, which is common in the Northern Hemisphere. Since the

vast majority of liverworts are leafy and tropical, *Marchantia* cannot be called a typical liverwort. However, textbooks often use it as an example of a liverwort life cycle because bryologists know more about its structures, in part because *Marchantia* species are relatively larger than most other liverwort species. The lobed gametophyte thallus can cover about a tenth of a square meter (about a square foot), obtaining mineral nutrients from single-celled rhizoids that penetrate the soil. *Marchantia* grow best in cool, moist locations that have diffuse light, with colonies of plants often carpeting a forest floor.

The gametophytes of *Marchantia* are more striking than those of most liverworts, with the gametophores looking

Figure 20.7 Aquatic liverworts have become popular as aquarium plants. Aquatic liverworts, such as *Riccia*, support aquarium life by releasing oxygen into the water and providing protective places for small fish. The plants will float or can be anchored to the bottom.

Figure 20.8 Life cycle of *Marchantia,* a thalloid liverwort.

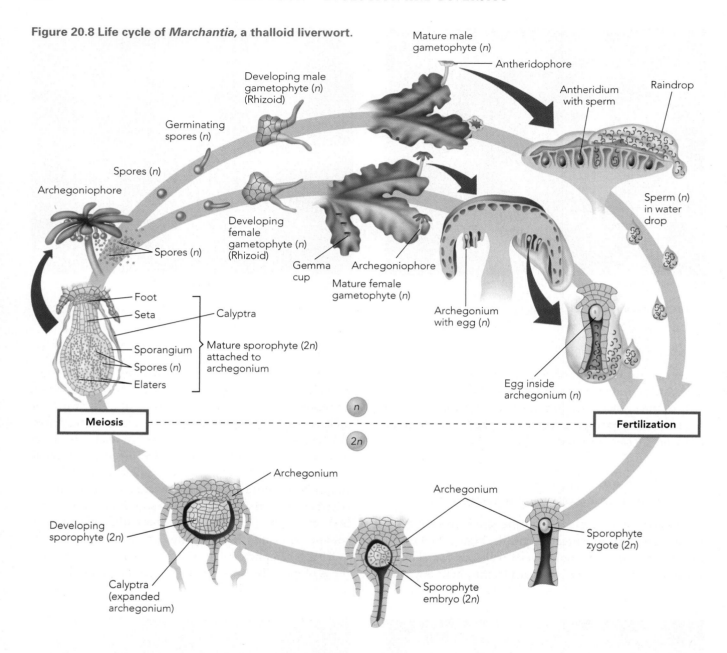

like tiny trees that elevate the antheridia and archegonia about a centimeter above the rest of the gametophyte thallus (Figure 20.8). In other liverworts, the gametophores do not stick up as noticeably. While many liverworts have bisexual gametophytes, *Marchantia* has separate male and female gametophytes. As in all bryophytes, fertilization requires free water so that the sperm can swim to the egg, a process facilitated by the gametophore structures. The **antheridiophores**—the gametophores of male gametophytes—have flattened, disk-shaped tops imbedded with antheridia. The top serves as a splash cup for raindrops or water dripping from other plants. As drops splash onto each antheridiophore, special cells between the antheridia absorb water and expand, putting pressure on the antheridia.

This pressure causes the antheridia to break open and release sperm, which are then carried in drops onto the surface of the gametophyte thallus. Meanwhile, the archegonia hang down as lobes underneath the umbrella-shaped tops of the female gametophores, which are known as **archegoniophores.** If a sperm-filled drop that is splashed up lands on an open, hanging archegonium, some of the water remains attached, enabling the sperm to swim to the egg.

While the gametophytes of *Marchantia* have gametophores that are more readily visible than in other liverworts, the *Marchantia* sporophyte is relatively inconspicuous, like those in most other liverworts. The sporophyte zygote develops within the archegonium, with the archegonium enlarging to form a protective **calyptra,** a thin, veil-like structure that at first completely covers the sporophyte (Figure 20.8). The typical liverwort sporophyte consists of three parts: a foot, a stalk, and a sporangium. The foot attaches to the archegonium, allowing the sporophyte to absorb water and nutrients from the gametophyte. The sporophyte's stalk, called a **seta** (plural, *setae*), connects the foot with the sporangium and is typically short in liverworts. The sporangium contains hundreds to thousands of spores produced by meiosis. When the sporangium opens, elongated cells called **elaters** twist and turn in the process of drying out, which has the effect of dispersing the spores.

Section Review

1. Describe the two types of liverwort gametophytes.
2. Describe how liverwort sporophytes and gametophytes produce each other.

Hornworts: Phylum Anthocerophyta

Conspicuous horn-shaped sporophytes are what mainly distinguish hornworts from other bryophytes, giving hornworts their common name and the phylum name Anthocerophyta (from the Greek *keras,* "horn") (Figure 20.1b). The sporophyte of hornworts, as well as that of mosses, has stomata with guard cells, structures that are absent in liverworts.

Hornwort gametophytes are somewhat similar to liverwort gametophytes in growing more horizontally than vertically, in contrast with typical moss gametophytes. The hornwort gametophyte is often shaped like a rumpled, round, green sheet extending a few centimeters across, with the edge turned up and ruffled. In some species, nitrogen-fixing bacteria live in internal cavities of the gametophytes. In this mutualistic association, the bacteria supply nitrogen fertilizer to the hornwort, which provides a home for the bacteria.

The hornwort life cycle features a hornlike sporophyte

In the life cycle of a typical hornwort, the gametophyte forms antheridia, initiating from cells under a surface layer of sterile cells. At maturity, the antheridia become visible when the sterile cells dry out and break open. Archegonia arise from surface cells, but normal thallus cells surround the egg itself, and the archegonium has a reduced, indistinct neck.

Fertilization results in a zygote that develops into a sporophyte with a characteristic spike or "horn" that can stick up several centimeters from the surface of the gametophyte. At the base of the horn is the sporophyte's foot, which anchors the sporophyte to the gametophyte. All water and minerals, and some food, for the sporophyte are absorbed from the gametophyte through the foot. Just above the foot is the meristem that lengthens the sporophyte. In this way, the hornwort sporophyte grows from its base rather than from its apex, which is unusual among plants. The sporophyte remains photosynthetic, which is also the case with most liverworts but not with most mosses.

The sporangium begins above the meristem and extends to the tip of the horn. A cross section reveals a central cylinder of nonreproductive tissue that is surrounded by a layer in which meiosis occurs and then by another cylinder of nonreproductive tissue. The sporangium produces spores along its entire length. Spores in the tip of the horn mature and are released first while tissue toward the base is still undergoing meiosis. The release of spores resembles the peeling of a banana, as the sporangium splits open, beginning at the top.

The evolutionary history of hornworts, as with other bryophytes, is being debated

Hornworts evolved from the same group of algal ancestors as did other plants, although whether they diverged into a distinct group before or after liverworts is a subject of debate. According to the molecular evidence, the evolution of hornworts from green algae occurred around 700 million years ago, although fossils older than 400 million years have yet to be found. The problems in deciphering hornwort evolution are similar to those with all the groups of plants, illustrating the difficulties in interpreting the evidence presented by molecular systematists and by **paleobotanists,** scientists who study plant fossils.

One problem is that the organisms that bridged the algal phyla and that gave rise to the bryophyte phyla and the phyla of vascular plants are all extinct and were probably of several different types—in species, genus, and family, at least. In some evolutionary scenarios for animals, the molecular evidence and fossil evidence agree on rough dates, but that is not yet the case with plant evolution. The 300 million years separating the fossil and molecular dates must be filled in with either discoveries of older fossils or with a calculation of increasingly younger molecular dates.

Another difficulty is that all plants have traits that link them closely to green algae, but some traits may be unique to particular plant groups. For example, most hornworts have cells with one large chloroplast containing a **pyrenoid,** a region with starch deposits resulting from photosynthesis. Hornworts are the only plants sharing this characteristic with algae—in particular, green algae in the Class Coleochaetes in the phylum Chlorophyta. Accordingly, some paleobotanists hypothesize that most hornworts evolved from a different group of green algae than other bryophytes or that they have retained primitive features linking them to green algae. Various groups of primitive plants may have arisen in more than one geographical location and from related but not identical groups of algae.

Debate has also arisen over the evolutionary relationship between hornworts and an extinct vascular plant called *Horneophyton lignieri,* known from fossils dating to around 400 million years ago (Figure 20.9). The fossils indicate a vascular plant that was up to 20 cm (7.9 inches) tall and had a branching, stemlike sporophyte with terminal sporangia that resembled hornwort sporophytes. Whether *Horneophyton* gave rise to modern hornworts or is related to a common ancestor is not known. The apparent gametophyte of *Horneophyton* is a separate fossil identified as *Langiophyton mackiei.* Since the sporophyte and gametophyte of *Horneophyton* are attached, deriving living hornworts from the fossil plants involves a theoretical assumption—for example, that the gametophyte of *Horneophyton* was originally attached to it. The classification dilemma of *Horneophyton* is a good example of the difficulties in tracing the evolutionary lineage of a fossil plant to its living relatives, if any. The fossil record reveals evidence of structure and shape, perhaps of just a part of the plant. Evolutionary linkages are therefore the result of interpretations by scientists.

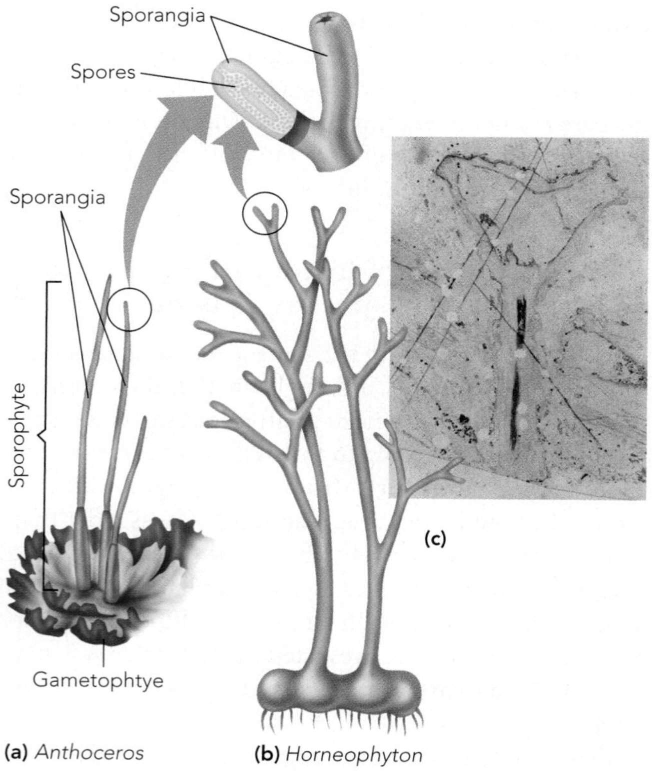

Section Review

1. Describe the hornwort sporophyte and gametophyte.
2. Compare hornworts with liverworts.
3. Why is there debate over the evolution of hornworts and other plants?

Figure 20.9 Hornworts and *Horneophyton:* An evolutionary relationship? Hornworts are similar to an early vascular plant, the extinct *Horneophyton,* which arose around 400 million years ago. Both types of plants have stemlike sporangia, prompting hypotheses that either *Horneophyton* had hornwort origins or hornworts evolved from *Horneophyton.* The current fossil evidence is inconclusive. (a) A living hornwort of the genus *Anthoceros.* (b) A reconstruction of *Horneophyton* from fossil evidence. (c) A cross section of a fossilized *Horneophyton* sporangium, showing the general similarity to hornwort sporophytes.

Mosses: Phylum Bryophyta

As you have seen, the term *bryophytes* does not apply only to the phylum Bryophyta but instead is a broader reference to *all* nonvascular plants: liverworts, hornworts, and mosses. In contrast, the phylum Bryophyta consists only of those bryophytes scientifically classified as mosses. Common use of the word *moss* can be confusing, often referring to plants and algae that look like moss but are not members of the phylum Bryophyta or any other group of bryophytes. For example, Irish moss is actually an alga; Spanish moss is a flowering plant; and club moss is a seedless vascular plant. In addition, lichens are often called *mosses,* even though they are a symbiotic association of a fungus and an alga. Compared with other bryophytes, mosses are distinctly "leafy" and less thalloid in appearance, and the gametophytes are frequently vertical.

Like other bryophytes, mosses tend to grow most abundantly and with the greatest variety of species in moist, forested regions and in wetlands. However, some species occur in deserts and on relatively dry rock outcroppings, where they typically live on north-facing slopes that receive less sun and are watered by occasional rainstorms. A few

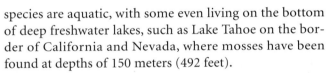

(a)

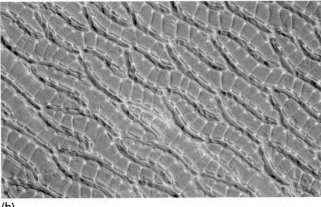

(b)

Figure 20.10 *Sphagnum* moss holds so much water because its leaves consist mostly of dead cells that can absorb water. (a) Clump of *Sphagnum* moss. (b) Close-up photomicrograph of leaf showing small photosynthetic cells and large, nonphotosynthetic cells.

species are aquatic, with some even living on the bottom of deep freshwater lakes, such as Lake Tahoe on the border of California and Nevada, where mosses have been found at depths of 150 meters (492 feet).

There are three main classes of mosses

Mosses consist of three main classes. The class Spagnopsida (also known as Sphagnidae) contains 150 species of peat mosses, which are particularly important for human use, as you saw in the chapter introduction. The class Andreaeopsida (also called Andreaeidae) comprises 100 species of granite mosses. The class Bryopsida (also known as Bryidae) features more than 9,000 species and includes the most familiar types of mosses, with many playing an increasing role as landscape plants. Since Bryopsida mosses are more commonly known, they are often informally called "true mosses."

The class Spagnopsida consists only of the genus *Sphagnum*, known as *peat moss*, one of the most widespread genera of mosses (Figure 20.10). *Sphagnum* has a sheet-like protonema, rather than the typical threadlike shape found in most bryophytes. The sheet is one cell layer thick and grows by cell divisions at its outer edge. Eventually, buds that have apical meristems develop from cells along the margin of the sheet. The leaves of *Sphagnum* consist of groups of large dead cells surrounded by thin living cells. The dead cells have many holes, accounting for the moss's remarkable ability to absorb water. *Sphagnum* sporophytes do not have setae. Instead, spherical sporangia are attached to stalks that are actually part of the gametophyte.

The members of the class Andreaeopsida are called *granite mosses* or *rock mosses* because some species live on rocks, often at high elevations (Figure 20.11). A number of species also live in soils of cold, temperate regions. Typically blackish green in color, granite mosses are frequently the only plant life encountered in dry, windy, cold mountainous microenvironments, where they live not only on rocks but also on the snow and ice itself. On glaciers in Kenya, Iceland, and Norway, granite mosses form what are called "moss balls" or "glacier mice"—rounded cushions that grow radially on all sides and roll around on the glacier as the wind blows. A unique feature of granite mosses is the formation of four vertical slits on the sporangium. Under dry conditions, the sporangium shrinks enough to open these slits, allowing spores to disperse in the wind.

Figure 20.11 Granite mosses, or rock mosses. (*Grimmia laerigata*)

THE INTRIGUING WORLD OF PLANTS

Unusual Mosses

Bryologists continue to discover new species of bryophytes, including unusual mosses, in a variety of habitats. *Scopelophila cataractae* and several species in the genus *Mielichhoferia* are known as copper mosses because they grow in soils with high levels of copper. Some of these species require high concentrations of copper ions, while others exclude copper ions. These mosses are found in nature in copper-containing soils, in mine tailings, and also under the drip lines of copper-roofed Buddhist temples. Some species absorb other heavy metals in addition to copper. Copper

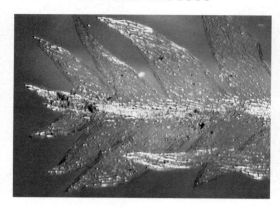

The *Schistostega pennata* gametophyte becomes luminous and gold-colored when light shines on it.

mosses may play a direct role in reclamation of contaminated soils, and scientists are studying their genes in an effort to produce other types of metal-resistant plants.

The moss *Schistostega pennata*, known as *goblin's gold*, lives in dark overhangs or in cave mouths. The protonemata and thalli of this moss produce a golden-green reflective glow, frequently noted by cave explorers. The spherical shape of its cells causes the reflection, which may be an adaptation to reflect light—from the Sun or flashlights—from one cell to the other for efficient use.

Growth patterns of mosses of the class Bryopsida vary greatly with species and environment (Figure 20.12). In areas where desiccation is a problem, species tend to be short and to grow in compact clumps, with gametophytes pressed tightly together, thereby reducing exposed surface area and limiting water loss. In areas where moisture is adequate, mosses may still grow in clumps but with the gametophytes more distinct and separated. In rain forests and other very moist areas, gametophytes are frequently larger and pendulous, hanging down from rocks or from tree branches, occasionally resembling ferns more than mosses. Many mosses live on rocks and on soil, but most tropical species are epiphytes living in trees. Others remain as protonemata and look like filamentous algae. Some grow year-round, but others are seasonal, turning brown and looking dead during dry periods but being re-

hydrated by rain to a rich green. Some are sensitive indicators of pollution, whereas others can thrive in polluted cities. A few species grow only on soils enriched with certain minerals, with some accumulating heavy metal ions or radioactive ions. Dung moss, *Splachnum*, produces a fly-attracting odor, with spores using flies as vehicles to find new dung on which to grow (see *The Intriguing World of Plants* box on this page).

The life cycle of *Polytrichum* demonstrates characteristic features of mosses

An example of the life cycle of a typical moss is found in the genus *Polytrichum* of the class Bryopsida (Figure 20.13). When you look at a carpet of moss, you are mainly

Figure 20.12 Growth habits of mosses.

seeing the leafy gametophytes, which typically live for many years and can often survive periods of drought. Each gametophyte originates from a spore that germinates into a protonema, which soon develops one or more buds. Each bud then gives rise to the leafy part of the gametophyte. Depending on the moss species, the top of a gametophyte can produce antheridia, archegonia, or both. Each antheridium contains many sperm surrounded by sterile cells, and each archegonium produces one egg, located at its base.

When a film of water is present, fertilization can take place. The zygote develops in the archegonium, becoming an attached sporophyte consisting of a foot, seta, and sporangium. If you look closely at moss, you may see the sporophytic setae—the tiny stalks protruding above the gametophyte "carpet." If moisture is insufficient, however, the moss will not produce sporophytes. In extreme cases, many years may pass without sporophyte formation. A rare moss found on old limestone walls in northwest England recently produced spores after a lapse of 130 years.

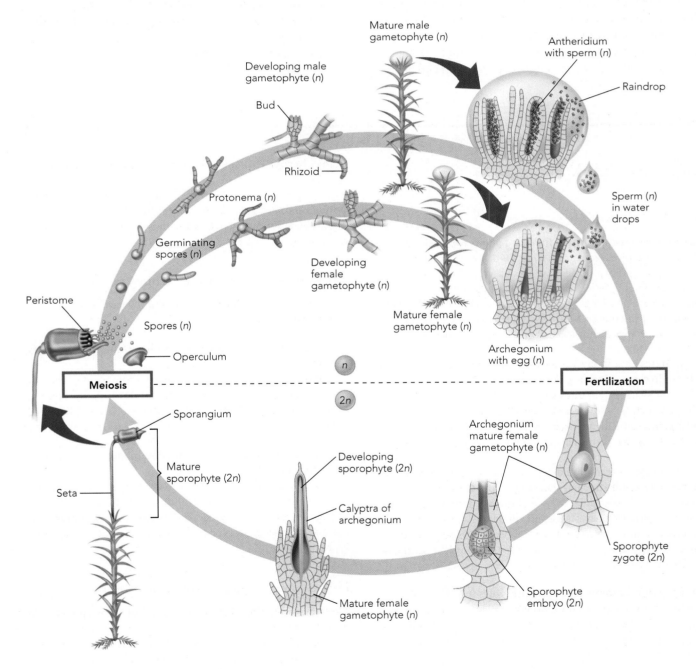

Figure 20.13 Life cycle of *Polytrichum,* a moss.

Typically, a moss sporophyte is photosynthetic early in its development but then turns brown and depends on the gametophyte for nutrition during the final phases of its development. The foot at the base of the seta protrudes into the gametophyte and absorbs food. Meiosis occurs in the sporangium capsule, and the spores are released when the **operculum,** the sporangium lid, falls off after a layer of cells at its base dries out. In mosses of the class Bryopsida, one or two rings of "teeth" surround the exposed opening of sporangium. These teeth regulate the passage of spores and are known collectively as the **peristome** (from the Greek *peran,* "to pass through," and *stoma,* "mouth"). In moist conditions, the teeth curl inward, closing the sporangium. Under dry conditions, in which spores are carried more easily by the wind, the teeth uncurl, opening the sporangium and gradu-ally dispersing hundreds of spores. In *Sphagnum* , the operculum bursts open in an explosive discharge of spores, caused by shrinkage of the sporangium to about a quarter of its original size without any release of air pressure.

Section Review

1. How does the structure of *Sphagnum* allow it to retain water?
2. Where do granite mosses typically occur?
3. Describe different growth patterns in Bryopsida mosses.
4. Describe a typical moss life cycle.

SUMMARY

An Overview of Bryophytes

Bryophytes were among the first land plants (pp. 426–427)
Bryophytes evolved from green algae about 450 million years ago. The earliest fossil bryophytes have been dated at about 360 million years ago. Bryophytes are considered nonvascular because vascular plants have sporophytes with roots, stems, and leaves. However, bryophyte gametophytes have structures that function as stems and leaves, as well as rhizoids that serve mainly for anchorage.

Bryophytes have many similarities to green algae in the class Charophyceae and to vascular plants (pp. 427–428)
All plants share certain traits with green algae of the class Charophyceae, such as cellulose and chloroplasts. Bryophytes and vascular plants share further traits that foster survival on land, such as structures that protect spores and gametes from drying out.

In bryophytes, alternation of generations involves a dominant gametophyte and attached sporophyte (pp. 428–429)
In contrast, vascular plants have a dominant sporophyte. Bryophyte gametophytes have antheridia containing sperm and archegonia containing eggs. The structures bearing these gametangia are called *gametophores.* Many bryophytes reproduce asexually by fragmenting or by producing brood bodies such as gemmae.

Bryophytes play important ecological roles (pp. 429–430)
Mosses are examples of how bryophytes play a key role in plant succession by helping to break down rocks into soil. Bryophytes contribute to the photosynthesis in epiphytic communities in rain forests.

Many bryophyte species tolerate drought conditions (p. 430)
Some liverworts can roll up to minimize sun exposure. Hairpoints on many mosses limit water loss. Production of special types of RNA may help some mosses repair damage from desiccation.

Liverworts: Phylum Hepatophyta

RNA and DNA evidence indicates that liverworts may be the earliest land plants.

Liverwort gametophytes can be either thalloid or leafy (pp. 430–431)
Thalloid liverwort gametophytes are flat, green plants, generally sheetlike or algae-like in appearance. About eighty percent of liverwort species are leafy, usually with three ranks of flattened leaves.

A liverwort life cycle demonstrates dominance of the gametophyte (pp. 431–433)
The life cycle of the thalloid liverwort *Marchantia* features a lobed gametophyte, which produces antheridiophores and archegoniophores. The zygote develops in the archegonium and becomes a small sporophyte, consisting of a foot, a seta, and a sporangium.

Hornworts: Phylum Anthocerophyta

The hornwort life cycle features a hornlike sporophyte (p. 433)
The sporophyte grows from its base, through a meristem above its foot, and splits from the top down to release spores.

The evolutionary history of hornworts, as with other bryophytes, is being debated (pp. 433–434)

Molecular evidence places hornwort origins around 700 mya, but the current fossil record extends only to 400 mya. The incomplete fossil record makes it difficult to determine evolutionary relationships between hornworts and other plants.

Mosses: Phylum Bryophyta

Whereas the term *bryophytes* refers informally to all nonvascular plants, the phylum Bryophyta is the scientific classification for mosses. Like other bryophytes, mosses are more diverse in moist environments but some species survive in arid environments.

There are three main classes of mosses (pp. 435–436)

The class Spagnopsida consists solely of mosses of the genus *Sphagnum*, which have sheetlike protonemata. The class Andreaeopsida consists of granite mosses, which are blackish green and live in cold, temperate regions, including rocks at high elevations. Most mosses are in the class Bryopsida, a diverse group consisting of the most familiar species, often called "true mosses." Bryopsida growth patterns are compact, loosely clumped, and pendulous (hanging from rocks or branches).

The life cycle of *Polytrichum* demonstrates characteristic features of mosses (pp. 436–438)

The sporangium can protrude several centimeters above the gametophyte. Spores germinate into protonemata that produce gametophytes. In the *Polytrichum* sporophyte, the operculum (sporangium lid) falls off, exposing the peristome that gradually releases spores.

Review Questions

1. Compare and contrast bryophytes and vascular plants.
2. How are bryophytes similar to green algae? How are they different?
3. Why is it difficult to trace the origins of plants?
4. Distinguish between a sporangium, antheridium, and archegonium.
5. Describe how bryophytes can reproduce asexually.
6. How can you tell if a particular bryophyte is a liverwort, hornwort, or moss?
7. What are some ways in which bryophytes are economically important?
8. Compare and contrast the sexual life cycles of a liverwort and a typical moss.
9. Explain the difference between the term *bryophyte* and the phylum Bryophyta.
10. Distinguish between the three classes of mosses.
11. Describe the ecological significance of bryophytes.
12. Give some examples challenging the notion that bryophytes are "simple" plants.

Questions for Thought and Discussion

1. Devise an experiment to test whether mosses produce compounds that prevent decay of organic materials.
2. What evidence do we have that the conducting system in mosses is less efficient than that of vascular plants? Why do you think it is less efficient?
3. If the Earth's gravity were only 10% of its actual value, how do you think plant evolution might have proceeded differently?
4. Why do you think bryophytes have not formed a lichen-like association with fungal cells?
5. By means of labeled drawings, compare and contrast the sporophyte stages of mosses, liverworts, and hornworts.

Evolution Connection

What adaptive features of bryophytes were most likely important in enabling these plants to be successful as epiphytes in moist habitats such as tropical rainforests?

To Learn More

Visit The Botany Place Website at www.thebotanyplace.com for quizzes, exercises, and Web links to new and interesting information related to this chapter.

Conard, Henry S. *How to Know the Mosses and Liverworts*. New York: McGraw-Hill, 1979. A useful identification guide.

Malcolm, Bill, Nancy Malcolm, and W. M. Malcolm. *Mosses and Other Bryophytes: An Illustrated Glossary*. Portland: Timber Press, 2000. An illustrated dictionary on mosses.

Schenk, George. *Moss Gardening: Including Lichens, Liverworts, and Other Miniatures*. Portland: Timber Press, 1997. An overview of gardening with bryophytes in Japan, North America, and Europe.

Shaw, A. Jonathan, and Bernard Goffinet, eds. *Bryophyte Biology*. Cambridge: Cambridge University Press, 2001. A source of recent classifications and information on advances, including molecular dating.

21

Seedless Vascular Plants

Tree ferns (*Cyathea smithii*), New Zealand.

The Evolution of Seedless Vascular Plants

Seedless vascular plants dominated the landscape around 350 million years ago

Land plants arose from green algae in the class Charophyceae

Three phyla of extinct vascular plants appear in the fossil record beginning 430 million years ago

In living seedless vascular plants, alternation of generations involves independent gametophytes and sporophytes

Types of Living Seedless Vascular Plants

Whisk ferns comprise most of the living members of phylum Psilotophyta

Living members of phylum Lycophyta include club mosses, spike mosses, and quillworts

Horsetails are the living members of phylum Sphenophyta

Phylum Pterophyta consists of ferns, the largest group of seedless vascular plants

During his botanical studies near Concord, Massachusetts, in 1851, Henry David Thoreau, author of *Walden*, discovered a rare, native climbing fern, *Lygodium palmatum*. "It is a most beautiful slender and delicate fern," he wrote, "twining like [a] vine about the meadow-sweet, panicled andromeda, goldenrods, etc., to the height of three feet or more and difficult to detach. . . . Our most beautiful fern, and most suitable for wreaths or garlands. It is rare."

In recent years, two exotic relatives of Thoreau's fern have posed a serious ecological challenge in regions of the southern United States, particularly Florida. The climbing fern *Lygodium microphyllum*, native to parts of Southeast Asia and Australia, and the Japanese climbing fern *Lygodium japonicum* both entered the United States as horticultural plants for hanging baskets. They then escaped cultivation and became exotic pests. Like many introduced species, these Asian ferns thrive because they encounter few growth restrictions in their new geographical range. They grow rapidly and spread by wind-borne spores that may be carried 40 miles or more. These hardy ferns currently cover 40,000 acres in south Florida and have increased their range 100-fold over six years, surviving floods and droughts.

Lygodium paponicum.

Although beautiful, the ferns can be deadly to other plants, covering other vegetation in masses up to 0.6 meters (2 feet) thick. They kill other plants by cutting off the light or by sheer mass, even causing some trees to collapse under their weight. The vines, actually climbing leaves, may be up to 30.5 meters (100 feet) long, sometimes acting as fire ladders that rapidly carry flames into dry, dead trees. Masses of ferns readily break off during fires, carrying flames to new locations and resulting in the destruction of valuable forests.

To control the spread of these ferns, Australian scientists are investigating insects as possible biological control agents. These include a leaf-mining beetle, which tunnels through leaves, eventually killing them. Introducing this beetle might help control *Lygodium,* but it might also have other unintended consequences, such as the beetle attacking native plants.

Lygodium overtaking cypress in Florida.

Lygodium is an example of a worldwide problem involving hundreds of species. When humans accidentally or intentionally introduce exotic plants to a new location, the plants frequently encounter new environments that are unsuitable for growth and reproduction. However, serious ecological problems can result if the new location proves ideal for the exotic species in terms of growing conditions and a lack of competition, disease, and herbivores.

While *Lygodium* can be a pesky fern in some environments, another fern plays a productive role in agriculture. The water fern *Azolla* forms a mutualistic association with a cyanobacterium, *Anabaena azollae*, which lives in cavities found on the lower sides of leaves. *Anabaena* fixes nitrogen; that is, it can convert nitrogen from the air into a form usable by plants. In rice paddies, *Azolla* is frequently introduced for the nitrogen it supplies and because it effectively shades weeds that otherwise would compete with rice for nutrients and sunlight.

Ferns and other seedless vascular plants were the first vascular plants and, along with bryophytes, were the principal plants for perhaps 100 million years. Now, they constitute only around 5% of living plant species, but they continue to thrive in some environments, sometimes outcompeting seed plants, as the climbing ferns bear witness. Seedless vascular plants have played a very important role in plant evolution, giving rise to the seed plants that dominate the modern plant world. In this chapter, we will first step back into the distant past to look at the origins of seedless vascular plants before exploring the characteristics of living species.

A bullfrog in *Azolla.*

The Evolution of Seedless Vascular Plants

All plants produce multicellular embryos, which are retained for a time within the tissues of the female gametophyte. For this reason, both bryophytes and vascular plants are known as *embryophytes*. Vascular plants themselves are known as *tracheophytes* because they have tracheids.

Despite the clarity of this classification, we do not know most of the details about the colonization of land by plants. An atmospheric oxygen level of at least 2%, far less than today's level of 20%, was necessary for life on land. Photosynthesis by algae and cyanobacteria in the ocean and in fresh water added enough oxygen to the atmosphere to support land-based aerobic respiration by the Silurian period around 430 million years ago, when the first plant fossils appear in the fossil record, and perhaps as early as 700 million years ago, when the molecular evidence indicates that plants diverged from green algae.

We also do not know much about the organisms that bridged the gap between algae and plants. While it is tempting to hypothesize that plants colonized the land and prepared the way for animals and other organisms by establishing the base of terrestrial food chains, the evidence argues otherwise. The fossil record and molecular data both indicate that mutualistic associations between plants and fungi (mycorrhizae) are ancient. Although early plants did not have roots as we know them today, fungi appear to have been associated with underground stems and probably supported the uptake of nutrients from mud or rocky soil. Animals—particularly arthropods, which had a chitinous exoskeleton preventing desiccation—may have colonized land at about the same time as plants. However, bryophytes and the first seedless vascular plants probably preceded most animals and served as the basis of the first terrestrial food chains. The first land plants and animals both required water for the sperm to swim to the egg and so were restricted to moist regions. Vast regions of the Earth were still uninhabited.

Seedless vascular plants dominated the landscape around 350 million years ago

Suppose that a time machine transported you back 350 million years. In most of what is now North America, you feel a wet, warm blanket of humidity as you stand in a steamy, tropical forest with huge swamps (Figure 21.1). This was the characteristic environment near the equator during the Carboniferous period (363–290 million years ago). At that time, all of the major continents were pushed together into a huge landmass called *Pangaea*. Large parts of what are now Eurasia, North America, northern South America, and northern Africa were quite near the equator—much farther south than today. What are now southern South America, southern Africa, Australia, Antarctica, and northern Asia were heavily glaciated because they were close to the poles.

As you compare the foliage of the carboniferous forest with information on modern plants, you find ancient bryophytes—the ancestors of modern liverworts, hornworts, and mosses. Gymnosperms have begun to evolve and are becoming common in drier, upland regions, but flowering plants will not arise for another 200 million years. Instead, it is seedless vascular plants that dominate the landscape. Some look similar to modern ferns, but others are unfamiliar, including tall trees with leaves more than a meter in length. You are able to classify some, but others are not mentioned in the information you brought from the present. Meanwhile, there are no birds, mam-

Figure 21.1 Artist's conception of a tropical Carboniferous forest. During the Carboniferous period, North America and Europe were considerably closer to the equator than they are today and had tropical swamp forests. Carboniferous forests in drier regions farther away from the equator were sparser and more open than the one depicted here.

mals, or reptiles, but amphibians are in abundance, as are insects and other invertebrates.

Seedless vascular plants were dominant during the Carboniferous period because they faced little competition from other plants and because they thrived in moist, tropical regions with high rainfall. Along with bryophytes (see Chapter 20), they were the first land plants. They differed from bryophytes in having a vascular system of xylem and phloem, but they were unlike most modern plants because they had neither seeds nor flowers. The large amounts of biomass they produced formed a major part of the world's extensive coal deposits—hence the name *Carboniferous period*. Modern seedless vascular plants, primarily consisting of ferns, still thrive in their environments but display much less species diversity than their ancient ancestors.

Land plants arose from green algae in the class Charophyceae

Now imagine traveling further back in time to explore the ancient origins of seedless vascular plants. You find yourself in a landscape of 450 million years ago, during the Silurian period. The land is bare of either plants or animals, but in and near shallow freshwater lakes live organisms that seem to be both plantlike and algae-like. It is difficult to classify these branched, leafless organisms, which are typically only a few centimeters tall and never more than about a meter tall. You are observing some of the earliest stages in the evolution of land plants.

Paleobotanists are not certain how seedless vascular plants evolved because the fossil record is incomplete both geographically and over time. However, the fossil record, molecular studies, and evidence from living plants can be combined to supply a reasonably coherent picture. Based on this evidence, paleobotanists think that vascular tissue, a waxy epidermis, and other characteristics evolved over millions of years in freshwater green algae, allowing them to survive periods when the water dried up. Eventually, they were able to complete their entire life cycle on land, as long as water was available to enable sperm to swim to eggs. Some fascinating transitional species must have existed, with characteristics of both algae and seedless vascular plants. Eventually, seedless vascular plants themselves evolved, sharing the land with bryophytes.

Paleobotanists hypothesize that the evolution of land plants from green algae of the class Charophyceae in the phylum Chlorophyta began between 700 and 450 million years ago (see Chapter 20). The current fossil record supports the later date, based on the first partial bryophyte fossils, whereas data on DNA and RNA sequences of living plants and algae support the earlier date. Paleobotanists

actively seek new fossils that may fill in our knowledge about the origin of vascular plants and narrow the gap between the two dates. Certainly by the time of the first complete bryophyte fossils (360 million years ago) and the first fossils of vascular plants (430 million years ago), considerable evolution from algal ancestors had already occurred. This is particularly true if the molecular date of 700 million years ago, based on a sequencing study of 119 proteins, is correct.

A number of living species of green algae resemble one land plant or another, although the direct ancestors are extinct. In particular, the green alga *Chara* most clearly resembles land plants, although it was definitely not the direct ancestor, as almost half a billion years have passed since algae gave rise to land plants.

As seedless vascular plants evolved, they underwent an adaptive radiation (see Chapter 15) into many different species that occupied a variety of land environments. Their success was phenomenal, as they quickly became major photosynthetic components of the biosphere and provided food for amphibians and other animals, remaining the dominant plants for about 100 million years.

Despite their success, seedless vascular plants did not remain dominant in all environments because their method of fertilization depended on water, thereby limiting their habitat and making them vulnerable to drought. Also, the developing embryos needed moisture and were unprotected from animals. By contrast, the first seed plants—ancient gymnosperms—could reproduce without fresh water because a pollen tube delivered the sperm to the egg. At first, the sperm of seed plants were probably all flagellated, but eventually nonflagellated sperm evolved in many species. Following fertilization, a seed protected the embryo from desiccation and provided a food supply to support germination. Because they did not require moist environments, gymnosperms could grow in places seedless vascular plants had not colonized, and by the Mesozoic era (245–66 million years ago), these had replaced seedless vascular plants as the dominant plants.

In several important respects, the evolution of early land animals resembled that of plants. When seedless vascular plants were thriving during the Carboniferous period, amphibians—the first tetrapods, or four-legged land animals—were the dominant form of animal life and were the ancestors of today's frogs, toads, and salamanders. As with seedless vascular plants, fertilization in amphibians requires water to enable sperm to swim to eggs. Also, the developing embryos need moisture and are generally unprotected from predation. Amphibians were eventually replaced in most environments by reptiles, the first amniotes, which had a complex egg with a protective shell and an internal sac, or amnion, containing supplies of moisture that protected the developing embryo from desiccation. Although plant seeds and reptilian eggs are not evolutionarily related, they do serve similar functions. Seeds protected plant embryos, enabling seed plants to survive in a greater variety of environments than seedless vascular plants.

Three phyla of extinct vascular plants appear in the fossil record beginning 430 million years ago

Figure 21.2 reflects one hypothesis of the evolutionary relationships between plants and green algae of the class Charophyceae, indicating a possible sequence in which each plant group diverged from common ancestors in the line of plant evolution. For instance, the diagram shows that mosses and vascular plants have common ancestors not shared with liverworts and hornworts, indicating that mosses are the bryophytes most closely related to vascular plants. Fossils of the plants that bridged the gap from algae to bryophytes and vascular plants have not been found. Conditions may not have been optimal for fossilization, or the fossils may exist in unexplored geological formations.

Existing fossil evidence traces the ancestry of modern vascular plants to plants that first arose during the Silurian period (439–409 million years ago). Fossil plants have given paleobotanists clues to the appearance and structure of these earliest vascular plants. In 1859, a Canadian scientist named John Dawson discovered the first fossils of seedless vascular plants in Quebec, but because of the fragmentary nature of the fossil remains, paleobotanists could not agree on what the plant looked like. Dawson's fossils, which showed plants that were very unlike any known living or extinct plants, turned out to be of Devonian origin. In 1917, the existence of more ancient seedless vascular plants became well established after the discovery near the town of Rhynie, Scotland, of fossils that showed internal structure. These fossils, imbedded in a type of flinty rock called *chert*, revealed seedless vascular plants that were all less than a meter in height, with most being considerably smaller. These extinct plants are currently classified as the phylum Rhyniophyta, commonly called *rhyniophytes*. They grew in swamps or bogs and consisted of photosynthetic, branching stem systems, often with sporangia on the branch tips. An example is *Cooksonia*, shown in Figure 21.3. Rhyniophytes appear in the fossil record in the early Silurian period, around 430 million years ago, and evolved from green algae through transitional groups not yet discovered as fossils. The rhyniophytes might be reclassified as several phyla if more fossil evidence becomes available.

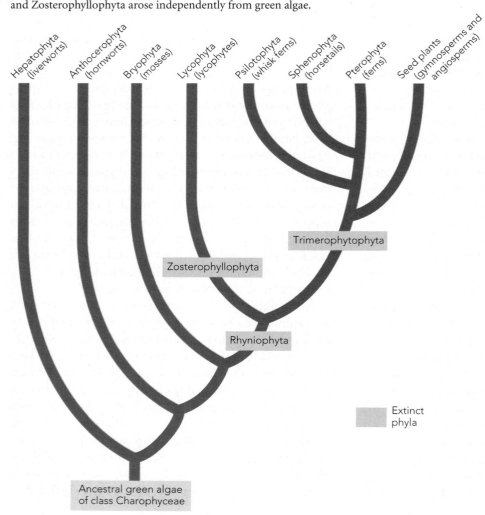

Figure 21.2 A hypothesis of the origins of seedless vascular plants. The earliest vascular plants strongly resembled one or more extinct species of green algae of the class Charophyceae. This diagram reflects one hypothesis of the origins of seedless vascular plants, according to which the earliest plants gave rise to the phylum Rhyniophyta, and the rhyniophytes gave rise to the phylum Zosterophyllophyta and phylum Trimerophytophyta. These two phyla are extinct but gave rise to the four living phyla of seedless vascular plants. Some paleobotanists believe that Rhyniophyta and Zosterophyllophyta arose independently from green algae.

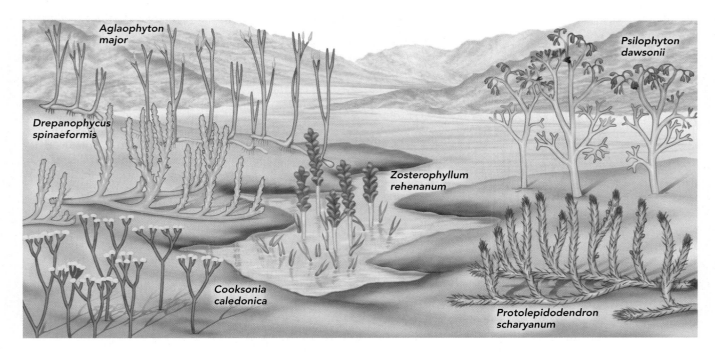

Figure 21.3 A Devonian landscape. By the beginning of the Devonian period (409–363 million years ago), a number of species of plants in three different phyla, Rhyniophyta, Zosterophyllophyta, and Trimerophytophyta, had colonized marshy landscapes. The drier landscapes away from water were devoid of life. By the end of the Devonian period, most species of these marsh plants were extinct. We do not know where the transition from green algae to these plants took place first or whether it occurred more than once.

By the end of the Devonian period, 363 million years ago, most rhyniophyte species were extinct. The rhyniophytes probably gave rise, either directly or indirectly, to two other groups of extinct vascular plants, classified as the phylum Zosterophyllophyta (called *zosterophylls*) and the phylum Trimerophytophyta (called *trimerophytes*) (see Figures 21.2 and 21.3). Both of these groups had also become extinct by the end of the Devonian period. However, their descendents were seedless vascular plants that dominated throughout the rest of the Paleozoic era, which ended 246 million years ago. Their distant relatives are still alive today.

As noted previously, Figure 21.2 presents one hypothesis. Some paleobotanists believe that trimerophytes, whose members have new features not found in the rhyniophytes, may have directly evolved from earlier transitional plants instead of from rhyniophytes. Also, some believe that zosterophylls evolved independently of rhyniophytes, either directly from green algae or from transitional plants that preceded the rhyniophytes. However, most paleobotanists agree that plants evolved ultimately from green algae. The disagreement arises over the details of the evolutionary path.

Rhyniophytes, zosterophylls, and trimerophytes all had branching photosynthetic stem systems without roots or leaves (Figure 21.4). In addition to aboveground stems, they had rhizomes (horizontal underground stems). They lived in marshy areas and probably arose from algae that could survive either in or out of fresh water. Their sporangia were surrounded by protective sterile cells and consisted simply of cells that produced spores.

Although similar in general structure, with dichotomous branching of stems, the rhyniophytes differed from the zosterophylls in their vascular system and in the arrangement of their sporangia:

◆ In the rhyniophytes, sporangia occurred at the end of the principal stems, which were sometimes shortened; in the zosterophylls, they were found at the end of very short lateral branches that were grouped together at the top of major stems.

◆ In the rhyniophytes, sporangia opened along the sides; in the zosterophylls, the sporangia opened at the tops.

◆ In rhyniophytes, xylem cells matured first in the center of stems, then outward around the periphery of the xylem strand. This is the pattern followed in each vascular bundle in most living plants, except lycophytes. In zosterophylls and lycophytes, the xylem matured in the opposite direction, first around the periphery and then inward toward the center of stems.

The rhyniophytes and trimerophytes had distinctive branching patterns. In the most primitive rhyniophytes,

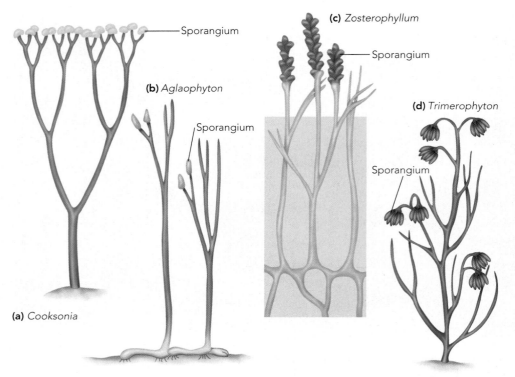

(c) *Zosterophyllum*

Sporangium

Sporangium

(b) *Aglaophyton*

Sporangium

(d) *Trimerophyton*

Sporangium

(a) *Cooksonia*

Figure 21.4 Extinct seedless vascular plants. (a) Rhyniophytes, such as *Cooksonia*, consisted solely of stems, which produced branches of more or less equal length, with sporangia at the tips. (b) *Aglaophyton* may be an intermediate between the rhyniophytes and the bryophytes. The plant lacked tracheids but had a primitive conducting system similar to mosses. (c) In the zosterophylls, such as *Zosterophyllum,* branches had many short lateral sporangia. (d) The trimerophytophytes, such as *Trimerophyton,* probably evolved from the rhyniophytes around 360 million years ago. They produced more branches than rhyniophytes, often bearing clusters of sporangia.

such as *Cooksonia,* dichotomous branches were of equal length and the sporangia were terminal. More advanced rhyniophytes showed a pattern of overtopping in which one branch grew longer than the other, with sporangia forming on the shorter branch, a pattern that also appears in zosterophylls. The sporangia-bearing branches are quite short and sometimes grouped near the end of longer branches. Trimerophytes had more complex branching than the rhyniophytes. Overtopping was extensive, until the plants developed a central axis with many side branches bearing terminal sporangia (Figure 21.5).

Fossils and molecular evidence indicate that certain trimerophytes gave rise to three of the four living phyla of seedless vascular plants: phylum Psilotophyta, phylum Sphenophyta, and phylum Pterophyta. Molecular evidence suggests that these three phyla are monophyletic and closely related. Trimerophytes also gave rise to seed plants. Some zosterophylls are the ancestors of the other living phylum of seedless vascular plants, Lycophyta.

In living seedless vascular plants, alternation of generations involves independent gametophytes and sporophytes

Some characteristics of the life cycles of seedless vascular plants are distinctive, while others are shared with bryophytes, seed plants, or both. As in life cycles of other plants,

alternation of generations involves gametophytes and sporophytes. The gametophytes produce sperm and eggs by mitosis, and the sporophytes produce spores by meiosis. In the production of sperm and eggs, seedless vascular plants are similar to bryophytes in having both antheridia and archegonia. However, the sporophyte–gametophyte relationship in seedless vascular plants differs from both bryophytes and seed plants. In seedless vascular plants, both the sporophytes and the gametophytes are independent at maturity, although the gametophytes are usually short-lived. In contrast, only the gametophytes are independent in bryophytes, and in seed plants only the sporophytes are independent. The sporophytes of seedless vascular plants, like those of seed plants, are much larger than the gametophytes.

Seedless vascular plants vary in their production of spores. Most species are **homosporous,** which means they produce only one type of spore, as is the case with bryophytes. This condition, known as **homospory,** can result in separate male and female gametophytes or in bisexual gametophytes, depending on the species. In contrast, some species of seedless vascular plants are **heterosporous**—producing two types of spores, megaspores and microspores, so named because megaspores are typically larger than microspores, although both are barely visible to the naked eye. Sporangia that produce megaspores are called **megasporangia,** while those that produce microspores are called **microsporangia.** This production of two types of spores on two different types of sporangia is known as

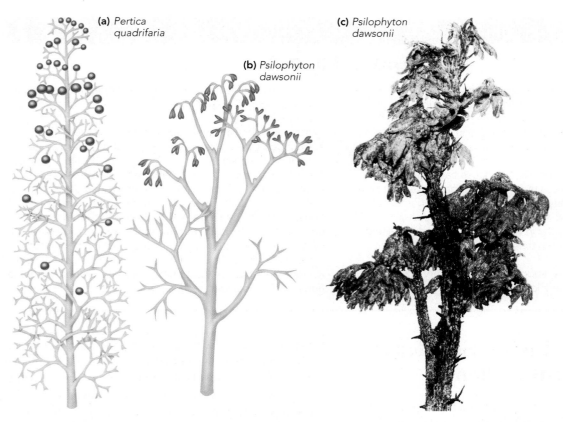

Figure 21.5 Phylum Trimerophytophyta. Trimerophytes probably derived from the rhyniophytes but had more extensive branching and extreme overtopping, resulting in formation of a central stem. **(a)** *Pertica quadrifaria* shows considerable branching, with globular clusters of sporangia. **(b)** This rendering shows *Psilophyton dawsonii* with fertile branches. **(c)** This is a fossilized fertile shoot of *Psilophyton dawsonii* obtained by gradually dissolving the rock matrix.

heterospory. Megaspores produce female gametophytes, known as **megagametophytes,** and microspores produce male gametophytes, called **microgametophytes.** Although heterospory is relatively rare in seedless vascular plants, it is the rule among all seed plants.

Gametophytes of living seedless vascular plants are usually only a few millimeters in diameter. In some species, the gametophytes live independently by being photosynthetic, while in others they absorb organic molecules from associated fungi. Like bryophytes, seedless vascular plants rely on water for fertilization, with the sperm swimming through water, such as raindrops or dew, to the egg in an archegonium. Later in the chapter, we will look at some specific life cycles of seedless vascular plants (see *The Intriguing World of Plants* box on page 448).

Although botanists fully understand the life cycles of living seedless vascular plants, the reproductive habits and life cycles of their fossil ancestors are largely unknown. For instance, fossil evidence is inconclusive regarding the type of alternation of generations in extinct seedless vascular plants. The lack of fossils is not surprising because the gametophytes of seedless vascular plants are typically small and fragile. Still, paleobotanists are trying to decipher clues from existing fossils, which are open to differing interpretations. Observing separate fossils of sporangia and stem fragments, for instance, may lead to conflicting conclusions—either that the sporangia and stem were from the same species or that they were from different species. Paleobotanists are also trying to determine whether extinct species had sporophytes and gametophytes that differed from each other in appearance, as is the case with living species. The exact nature of the gametophytes of the earliest fossil vascular plants awaits the discovery and study of more fossils. Fossil gametophytes have been discovered for some of the bryophyte-like plants from Rhynie.

Section Review

1. **When did vascular land plants first appear?**
2. **Describe the first vascular plants.**
3. **Describe the relationship of bryophytes to seedless vascular plants.**

THE INTRIGUING WORLD OF PLANTS

Alternative Life Cycles

In the normal life cycle of seedless vascular plants, gameto-phytes are haploid (*n*) while sporophytes are diploid (2*n*). Why should having one or two copies of the complete genome so markedly influence the size and shape of the adult as well as its role in the cycle of sexual reproduction? It turns out that, both in nature and in the lab, these rules are not absolute. Occasionally, a process called *apogamy* ("without gametes") results in haploid sporophytes, formed from cells of gametophytes but without the production of gametes or fertilization. Also, a process called *apospory* ("without spores") results in diploid gametophytes, arising from leaf cells of sporophytes but without meiosis and spore formation.

In laboratory studies, apogamy and apospory can be induced by hormone and light treatments. In general, rich nutrient media with added sucrose will encourage produc-tion of apogamous sporophytes. These lab results are con-sistent with conditions in nature, where the zygote of the sporophyte normally develops within the archegonium, surrounded by photosynthetic cells of the gametophyte, a nutrient-rich environment. Meanwhile, laboratory studies showed that poor nutrient media without sucrose will encourage the production of aposporous gametophytes. Again, these results seem consistent with natural condi-tions, in which the spore is shed from the plant and devel-ops on moist ground, an environment that would seem to be rather nutrient-poor. Nutritional status apparently acti-vates two quite different sets of genes, resulting in either apogamy or apospory, but neither process's signal transduc-tion pathway is understood.

Types of Living Seedless Vascular Plants

So far in this chapter, we have explored the origin of land plants from green algae and the three phyla of seedless vascular plants that existed during the Silurian and De-vonian periods. Now we will examine life cycles and char-acteristics of living seedless vascular plants. These "living fossils" also provide tantalizing glimpses into the evolu-tion of land plants.

Four phyla contain living plants evolutionarily related to the vascular plants from Carboniferous forests (Fig-ure 21.6). The surviving members of phylum Psilotophyta, also known as psilotophytes, consist of 142 species and are the simplest living vascular plants, having no roots. Most of these species are commonly known as *whisk ferns*. The members of phylum Lycophyta, called *lycophytes*, consist of more than 1,000 species of club mosses and re-lated plants. Lycophytes produce large numbers of simple leaves and sometimes superficially resemble large mosses. The members of phylum Sphenophyta, called *spheno-phytes*, include 15 species commonly known as *horsetails* because of their characteristic jointed structure with whorls of needlelike leaves at each joint. The vast major-ity of living species of seedless vascular plants belong to the phylum Pterophyta, which consists of more than 11,000 species of ferns. Ferns, also known as pterophytes, have larger, more complex leaves than other seedless vas-cular plants.

Recent molecular evidence suggests that psilotophytes, sphenophytes, and pterophytes are the seedless vascular plants most closely related to seed plants. Some botanists propose a reclassification that would recognize only two major groups of seedless vascular plants: the phylum Lycophyta, which evolved from zosterophylls, and a new phylum (or possibly a subkingdom) consisting of all other seedless vascular plants, which evolved from trimero-phytes, as did seed plants (see Figure 21.2). In this text, we take note of the proposed change but will still describe living seedless vascular plants as four phyla, pending for-mal reclassification. Also, they are still described as four phyla in much of the botanical literature.

Whisk ferns comprise most of the living members of phylum Psilotophyta

The living members of the phylum Psilotophyta consist of two genera: *Psilotum* and *Tmesipteris* (the *T* is silent). The vast majority of psilotophytes belong to the genus *Psilotum*, which contains 129 species of whisk ferns. They are found in tropical and subtropical regions of Asia and the Americas and also frequently grow as weeds in green-houses. In many respects, whisk ferns resemble rhynio-phyte fossils in having branching stem systems with protosteles. Unlike other living vascular plants, they have no true roots or leaves. Instead of leaves, the stem has small, scalelike, nonvascularized flaps of green tissue called **enations** (from the Latin *enatus*, "to rise out of"). The photosynthetic, whisklike stems bear tri-lobed, yellow spo-rangia, usually along the sides of the stem (Figure 21.6). Nutrients are absorbed by rhizomes that have rhizoids, or root hairlike structures. The other psilotophyte genus, *Tmesipteris*, consists of 13 species that are common in the

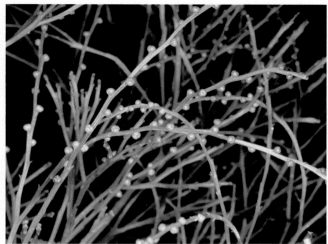

(a) Phylum Psilotophyta includes whisk ferns, such as *Psilotum nudum.*

(b) Phylum Lycophyta includes club mosses, as well as spike mosses and quillworts.

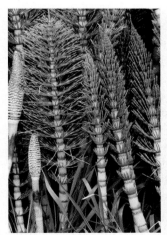

(c) Phylum Sphenophyta consists of horsetails, such as *Equisetum arvense.*

(d) Phylum Pterophyta consists of ferns, such as *Dryopteris affinis.*

Figure 21.6 Living seedless vascular plants.

South Pacific, usually as epiphytes hanging from rocks or attached to other plants, such as tree ferns (Figure 21.7). Like the whisk ferns, *Tmesipteris* lacks roots. However, instead of enations, the genus has single-veined leaves. Both genera of psilotophytes are closely related to ferns.

Like most seedless vascular plants, including its close relative *Tmesipteris*, *Psilotum* is homosporous and has bisexual gametophytes. The *Psilotum* gametophyte is a small underground structure less than a centimeter long (Figure 21.8). It sometimes contains vascular tissue but is not photosynthetic, relying instead on mutualistic fungi to supply nutrition. After fertilization occurs, the young sporophyte grows within the base of an archegonium, developing a foot that attaches it temporarily to the gametophyte. As in all seedless vascular plants, eventually the new sporophyte detaches from the gametophyte, becoming an independent plant. In the mature sporophyte, a photosynthetic branching stem system forms from buds on rhizomes. Sporangia contain diploid spore mother cells that produce haploid spores by meiosis, giving rise to the gametophytes and completing the life cycle.

Living members of phylum Lycophyta include club mosses, spike mosses, and quillworts

Whereas psilotophytes are most closely related to ferns, and therefore to extinct trimerophytes, lycophytes arose from extinct zosterophylls. Phylum Lycophyta contains about 1,000 living species, classified into three orders: Lycopodiales (club mosses), Selaginellales (spike mosses), and Isoetales (quillworts) (Figure 21.9).

Figure 21.7 *Tmesipteris*. Native to tropical regions of Australia and the South Pacific, *Tmesipteris* frequently grows as an epiphyte. The leaves have a single vein.

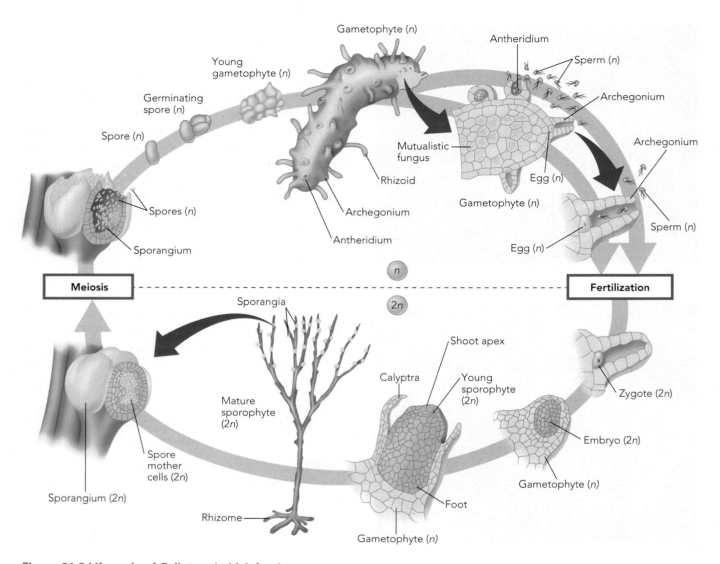

Figure 21.8 Life cycle of *Psilotum* (whisk fern).

Modern lycophytes are small herbaceous plants, but some of their ancient ancestors were trees that dominated moist tropical and semitropical forests about 325 to 280 million years ago, during the Carboniferous period. Ancient lycophytes were the most diverse and prevalent species of the time, ranging from tiny herbs to trees with trunks up to a third of a meter (about 1 foot) or more in diameter. *Lepidodendron* and *Sigillaria* are typical examples of extinct lycophyte trees, reaching 10 to 54 meters (33 to 177 feet) at maturity, with leaves up to about a meter in length (Figure 21.10). The trees could not grow taller for two reasons. First, with each dichotomous branching the stems became thinner, which eventually limited height. Also, the cambial cells produced only small amounts of secondary xylem and no secondary phloem, so the small conducting system also restricted height. The remains of lycophyte trees contributed greatly to the extensive coal

formation that gives the Carboniferous period its name. When the trees died and fell into the anaerobic swamps, only limited decomposition occurred, and the accumulated weight eventually changed the vegetative debris into coal.

All lycophytes have **microphylls**—leaves that have a single vascular trace, or leaf vein. Although the microphylls of modern lycophytes are generally small (hence the name, meaning "small leaf"), the microphylls of some extinct lycophyte trees reached up to about a meter (3 feet) in length. Microphylls are typically elongated and spirally arranged, and there is no leaf gap; that is, there is no break in the stem's vascular cylinder where the leaf branches from the main vascular system (Figure 21.11a). Sphenophytes (horsetails) also have microphylls, and some botanists consider *Tmesipteris* leaves, which have single veins, to be microphylls as well (see Figure 21.7). In lycophytes, fertile microphylls with sporangia often form small strobili

(a) The order Lycopodiales includes club mosses, which are mostly tropical but also occur in deciduous forests of the eastern United States. A rhizome gives rise to roots and shoots. The shoots produce numerous microphylls and also sporophylls, which are aggregated into club-shaped cones. *Lycopodium obscurum* is shown here.

(b) The order Selaginellales contains various species of *Selaginella*, known as *spike mosses*. These plants are smaller than club mosses and usually grow horizontally along the ground. Their leaves are small and delicate, as are their strobili.

(c) The order Isoetales consists of the genus *Isoetes*, known as quillworts and found in marshy regions. *Isoetes gunnii* is shown here.

Figure 21.9 Phylum Lycophyta.

(a)   (b)

Figure 21.10 Extinct lycophyte trees. Many species of giant trees, such as **(a)** *Lepidodendron* and **(b)** *Sigillaria*, dominated swamps and marshes of the Carboniferous period. Some species reached about 45 m (about 150 feet) in height.

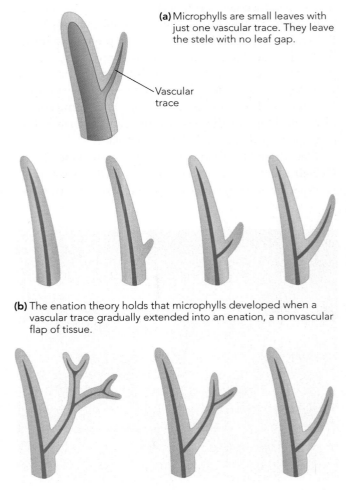

(a) Microphylls are small leaves with just one vascular trace. They leave the stele with no leaf gap.

Vascular trace

(b) The enation theory holds that microphylls developed when a vascular trace gradually extended into an enation, a nonvascular flap of tissue.

(c) According to the telome theory, microphylls evolved as a result of the reduction of existing telomes.

Figure 21.11 The structure and possible origin of microphylls.

(cones), which should not be confused with the seed-bearing cones of gymnosperms.

There are two main theories of the origin of microphylls. One view is that they evolved when vascular tissue extended into existing enations (Figure 21.11b). A competing theory is that microphylls are short branches that result from differing growth rates in the top two twigs, or **telomes** (from the Greek *telos,* "end"), of a dichotomous branch (Figure 21.11c).

Most living lycophytes belong to the order Lycopodiales, commonly known as *club mosses* because of the club-shaped strobili. Some are found in the United States, but most of the 200 species are tropical, and many of them are epiphytic. The growth habit is typically that of a branching rhizome that produces both underground rootlike structures and photosynthetic branches resembling giant mosses. The life cycle is similar to that of *Psilotum* (see Figure 21.8). Sporangia are usually found on the upper surface of sporophylls (fertile leaves), which may be grouped together, forming strobili. The spores are homosporous and germinate into bisexual gametophytes. Depending on the species, the gametophytes are sometimes photosynthetic and occasionally are found underground, where they rely on mutualistic fungi for nutrition. Gametophytes can take years to mature and may produce sporophytes for more than one year. The young sporophyte develops in the base of an archegonium before eventually becoming an independent plant.

The order Selaginellales contains only one family (Selaginellaceae) and one genus (*Selaginella*). Most of the 700 species of *Selaginella,* or spike mosses, live in moist

tropical environments (see Figure 21.9b), but some live in arid regions, such as the desert-dwelling *Selaginella lepidophylla* (Figure 21.12). Unlike club mosses and most other seedless vascular plants, *Selaginella* species are heterosporous, producing microspores and megaspores (Figure 21.13). Within each strobilus, sporangia appear on the surface of sporophylls. Sporophylls with microsporangia are called **microsporophylls,** while those with megasporangia are called **megasporophylls.** *Selaginella* species also differ from most other seedless vascular plants in gametophyte development, which is **endosporic,** taking place mostly *inside* the spore wall. In other seedless vascular plants, as well as in bryophytes, gametophyte growth is **exosporic,** taking place *outside* the spore wall. In *Selaginella,* each microgametophyte, consisting of little more than sperm cells, grows within a microspore and, after reaching maturity, releases the sperm. Meanwhile, each mature megagametophyte ruptures the megaspore wall, exposing the archegonia, which require water for fertilization. Following fertilization, the young sporophyte is at first attached to the megagametophyte but eventually becomes an independent plant.

The order Isoetales, commonly known as *quillworts,* contains only one family (Isoetaceae) and one genus (*Isoetes*) (see Figure 21.9c). Closely related to *Lepidodendron* and other lycophyte trees of the Carboniferous period, the 60 species of quillworts are the only living lycophytes that have a vascular cambium. Unlike their extinct relatives, quillworts are not large and consist of an expanded corm (underground stem) producing roots and quill-like microphylls that can all become sporophylls. Like *Sela-*

Figure 21.12 *Selaginella lepidophylla,* the resurrection plant. While most *Selaginella* species live in moist tropical regions, *Selaginella lepidophylla* lives in the arid deserts of the southwestern United States. Its common name alludes to its ability to revive and become photosynthetic again shortly after a rainstorm. The photos show the same plant in the dry state and after rehydration.

Figure 21.13 Life cycle of *Selaginella* (spike moss).

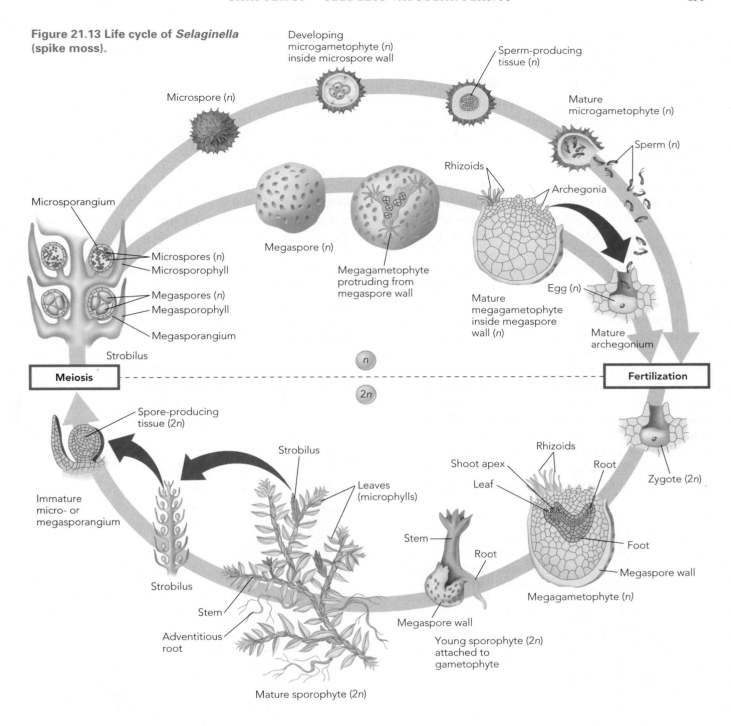

ginella, quillworts are heterosporous, with microsporangia and megasporangia forming on the surface of leaves, near their attachment to the plant. Quillworts live in regions that are underwater part or all of the year, and they are occasionally used as aquarium plants. Some species have no stomata, instead obtaining carbon for photosynthesis from the organic mud in which they live. During the day, photosynthetic bacteria and algae keep CO_2 concentrations in the water low. At night, respiration of bacteria and other organisms markedly increases CO_2 levels.

For this reason, such species have CAM photosynthesis (Chapter 8).

Horsetails are the living members of phylum Sphenophyta

Members of the phylum Sphenophyta probably evolved from extinct trimerophytes. Extinct treelike forms occur in the family Calamitaceae, reaching heights of about 20 meters (66 feet) and having trunks up to 30 centimeters

(1 foot) in diameter (Figure 21.14). Like the extinct lyco-phyte trees, they had vascular cambia that produced no secondary phloem. The only living sphenophyte genus is *Equisetum*, consisting of 15 species of plants commonly known as horsetails. As noted earlier, recent molecular evidence indicates that sphenophytes and psilotophytes are closely related to ferns, suggesting that all three groups should be placed in the same phylum.

Horsetails are among the world's most unusual plants. The sporophyte has a hollow, jointed stem with whorled microphylls at the nodes. The microphylls feel somewhat rough because their epidermal cells contain silica, which is why historically horsetails had been used to clean pots and have been commonly known as *scouring rushes*. Horse-tails have been called "living fossils" because today's plants are practically indistinguishable from fossils 400 million years old.

Like most seedless vascular plants, *Equisetum* is homo-sporous. Sporangia are clustered into umbrellalike spo-rangiophores that are grouped together geometrically into a strobilus (Figure 21.15). Some species have separate sterile and fertile shoots, whereas in others every shoot becomes fertile at maturity. Within each sporangium, spores are wrapped with elongated structures called *elaters,* which uncoil as the strobilus matures and dries, dispers-ing the spores. Each germinating spore develops within a few weeks into an independent, photosynthetic gameto-phyte that is typically bisexual. As in all seedless vascular plants, the sporophyte eventually detaches from the gametophyte and becomes an independent plant.

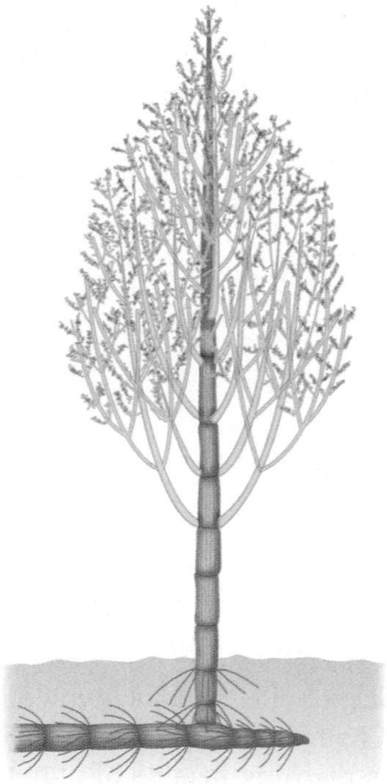

Phylum Pterophyta consists of ferns, the largest group of seedless vascular plants

Ferns evolved from extinct trimerophytes and first appeared during the Carboniferous period. Today they are the most successful and widespread group of seedless vascular plants. They usually occur in moist terrestrial environments and are less frequently found in fresh water, on mountains, and in deserts. Most of the 11,000 species are tropical, being adapted to moist, warm conditions. The phylum contains vines, epiphytes, and trees (Figure 21.16), but even the largest living tree ferns do not have secondary growth.

Ferns are the earliest group of plants to have **megaphylls,** leaves with a highly branched vascular system, in contrast with the single vascular trace in microphylls. Megaphylls are generally larger than microphylls and, unlike micro-phylls, have leaf gaps or similar areas of parenchyma where the vascular tissue leaves the stele. Megaphylls are also characteristic of all seed plants, making them the most common type of leaf in modern plants. Since early

Figure 21.14 An extinct giant horsetail. During the Carboniferous period, large treelike members of the Sphenophyta flourished. This drawing is a reconstruc-tion of *Calamites*, a giant horsetail that could reach 20 meters (66 feet) tall.

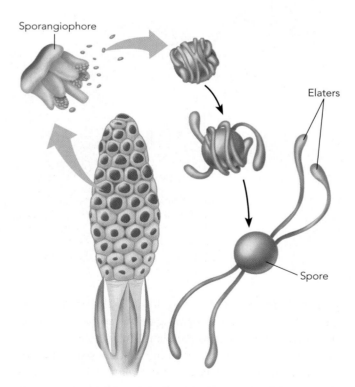

Figure 21.15 The *Equisetum* strobilus. In the strobilus, sporangiophores are displayed in a geometric array. On the outside of each spore, elaters unwind and catch the wind, aiding in dispersal.

Figure 21.16 Tree ferns. In the Tropics, many species of tree ferns, reaching as much as 6 meters (20 feet) in height, grow as remnants of tree-fern forests common in the Mesozoic era.

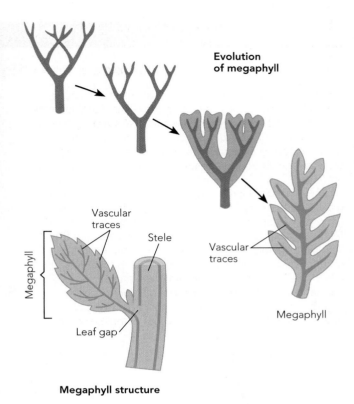

Megaphyll structure

Figure 21.17 The structure and possible origin of megaphylls. Ferns are the only seedless vascular plants with megaphylls, leaves with more than one vascular trace, which also occur in all seed plants. Megaphylls always produce a leaf gap or an analogous structure. They probably formed from flattened, dichotomous branches in which the spaces were gradually filled with tissue.

plants had branching stem systems, paleobotanists theorize that many structural features of modern plants were derived by changes in telome growth rates, with megaphylls arising by formation of leaf tissue connecting branch systems (Figure 21.17). Botanists have also attributed the development of sporangia on leaves to changes in telome growth (see the *Evolution* box on page 456).

With their generally larger surfaces and extensive veins, megaphylls may have given ferns a photosynthetic advantage over seedless vascular plants with microphylls. Fern leaves occur in a wide variety of shapes and sizes (Figure 21.18). In a few species, they can even become meristematic and develop new plants at their tips. Most fern sporangia have short stalks, with the sporangium itself surrounded by an **annulus,** a backbonelike line of cells with

thickened walls. When the spores are mature, the annulus dries and contracts, rapidly opening the sporangium and hurling the spores away from the plant.

(a) A floating fern, *Salvinia natans.*

(b) A lip fern, *Cheilanthes argentea.*

(c) An evergreen perennial fern, *Asplenium scolopendrium.*

(d) A staghorn fern, *Platycerium hillii.*

Figure 21.18 Fern megaphylls.

EVOLUTION

Telomes and Origins of Sporangia

The earliest vascular plants were simple branching stem systems that had no roots or leaves. Sporangia occurred on the tips of the stems. *Cooksonia* (see Figure 21.4a) is a good example of a fossil showing these primitive traits. Paleobotanists believe that changes in the growth rates of telomes resulted in the evolutionary development of many of the features we find today in plants.

Early vascular plants were branching stem systems in which the telomes were of equal length **(a).** Two basic growth processes could account for several anatomical features of vascular plants. The first is overtopping **(b),** in which one telome grows abnormally long while the other has normal growth. The second is reduction **(c),** in which one telome grows very little while the other has normal growth. Overtopping and reduction can occur together in a pair of telomes **(d).** The tri-lobed sporangia of *Psilotum* **(e)** could have arisen by a combination of

overtopping and reduction, resulting in the loss of one of the four original reduced telomes. In the lycophytes, a sporangium comes to lie atop a microphyll **(f),** forming what is known as a sporophyll, which again could be caused by overtopping and reduction. In *Equisetum* **(g),** multiple recurved sporangia result from overtopping and reduction. In ferns **(h),** sporangia occur frequently on the underside of leaf margins (Figure 21.20). Megaphylls arose from tissue filling in between closely spaced telomes (Figure 21.17).

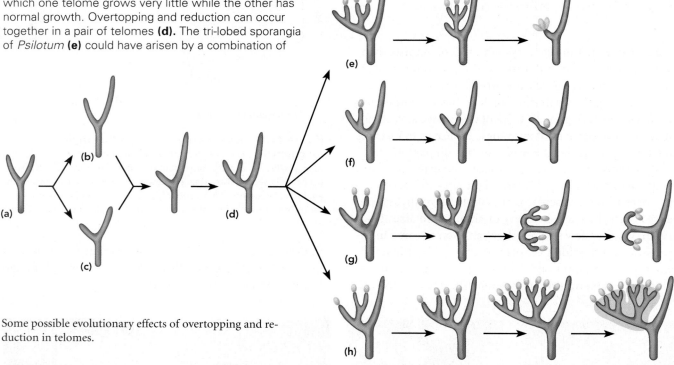

Some possible evolutionary effects of overtopping and reduction in telomes.

Figure 21.19 shows the sexual life cycle of a typical fern, which is homosporous and produces bisexual gametophytes. Usually photosynthetic, the gametophytes are one cell layer thick, less than half a centimeter across, and frequently heart-shaped. Sperm are coiled and multiflagellate, and after fertilization the embryo grows within the archegonium. At first, the young sporophyte is dependent, absorbing nutrients while attached to the gametophyte. However, it eventually sustains itself through photosynthesis and becomes a separate plant while the gametophyte shrivels and dies.

The production of spores occurs on the sporophyte megaphylls, known as **fronds.** Fronds are often compound, divided into leaflets called **pinnae,** which attach to the **rachis,** an extension of the petiole. Immature fronds are coiled, forming what are known as **fiddleheads,** or *crosiers,* that are edible in some species, although recent studies have shown that some are carcinogenic. Most fern species have one type of frond, which is both fertile and photosynthetic. Some species have separate sterile and fertile fronds, with the sterile fronds being mainly nonphotosynthetic. The sporangia on fertile fronds typically occur in

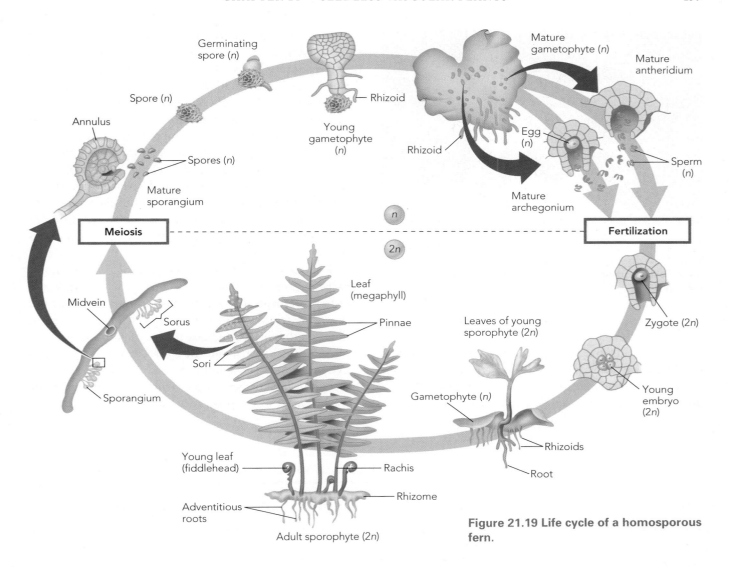

Figure 21.19 Life cycle of a homosporous fern.

groups called **sori** (singular, *sorus*), which are usually on the underside of the frond. Arrangements vary greatly among species, but sori typically appear as random dotlike structures on the frond surface or at the edge of fronds (Figure 21.20). Depending on the species, each sorus may be either "naked" or covered by a part of the frond. The covering may be an umbrellalike structure called an **indusium** (Latin for "tunic;" plural, *indusia*), or it may be simply the curled edge of the frond, often called a *false indusium* (Figure 21.21). Fern sporangia are formed by one of two possible developmental patterns: eusporangiate

Figure 21.20 Some patterns of fern sporangia.
A cluster of fern sporangia, called a *sorus,* may be round or may extend for some distance along the edge of a leaf. Although arrangements may vary, sori will be found near the leaf margin on the underside of the leaf.

Figure 21.21 Clusters of fern sporangia are often protected. Sori, or clusters of sporangia, may be shielded by umbrellalike structures called *indusia* (the brown circular structures shown here).

Most ferns are homosporous, with two orders of water ferns (Marsileales and Salviniales) being the only living heterosporous ferns. Many of the distinctive features of water ferns are adaptations to their aquatic environment. Heterospory itself might be easier to sustain in water because the sperm can swim distances of a centimeter or more to the egg without having to rely on dewdrops or raindrops to supply a suitable medium. The genus *Marsilia*, which has cloverlike floating leaves and usually lives in shallow lakes, is typical of the order Marsileales. Microspores and megaspores are born in nutlike sporocarps at the end of short stalks. Sporocarps are quite drought-resistant, able to withstand periods when the shallow lakes dry up. Each sporocarp forms from a modified fern leaf, which folds inward and fuses along the edges. When sporocarps germinate, a chain of sori emerges, with indusia and either microsporangia or megasporangia. *Salvinia* and *Azolla* are two small water ferns in which the entire plant floats on the surface. In these plants, each sporocarp contains a single sorus, and the sporocarp wall is a modified indusium. The sporangia produce a mass of mucus through which the sperm can swim without risk of being washed away as the plants are carried along by currents and wind.

or leptosporangiate. In eusporangiate ferns (the orders Ophioglossales and Marattiales), sporangia develop from a group of meristematic initial cells on the leaf. All other orders have leptosporangiate ferns, in which sporangia grow from a single initial cell.

Table 21.1 Summary of Seedless Vascular Plants

Phylum	Common name	Species	Status	Leaf type	Spore type	Comment
Rhyniophyta		A few dozen known	Extinct	No leaves	Probably homosporous	Probably will be reclassified as more than one phylum
Zosterophyllophyta		A few dozen known	Extinct	Microphylls	Probably homosporous	Gave rise to lycophytes
Trimerophytophyta		A few dozen known	Extinct	Microphylls	Probably homosporous	Gave rise to psilotophytes, sphenophytes, and pterophytes
Psilotophyta	Whisk ferns (common name for *Psilotum*)	142	Living	Enations (in whisk ferns); microphylls in *Tmesipteris*	Homosporous	Simplest vascular plants. Molecular evidence indicates they are ferns.
Lycophyta	Club mosses, spike mosses, quillworts	about 1,000	Living	Microphylls	Club mosses are homosporous; spike mosses and quillworts are heterosporous	Endosporic gametophyte in spike mosses and quillworts
Sphenophyta	Horsetails	15	Living	Microphylls	Homosporous	Jointed, hollow stems. Molecular evidence indicates they are ferns.
Pterophyta	Ferns	about 11,000	Living	Megaphylls	Mostly homosporous; water ferns are heterosporous	Most common seedless vascular plants

Asexual reproduction is quite common in many fern species. Typically, it occurs through the horizontal underground stems known as *rhizomes,* as in bracken ferns. A few fern species, such as the European *Trichomanes speciosum* and several species of *Hymenophyllum, Vittaria,* and *Trichomanes* found in Great Smoky Mountains National Park, lack sporophytes and can only reproduce asexually. They do so through special filaments that detach from gametophytes and develop into new plants. Such species form colonies that can be more than 1,000 years old, living in habitats also favored by mosses.

Considered as a group, the seedless vascular plants encompass a wide variety of plants and developmental patterns (Table 21.1). They continue to have a selective advantage in some warm, moist environments and even in some cold or dry environments. In their heyday, seedless vascular plants were the crown of botanical evolution. They trace their ancestry to the most ancient days of the plant kingdom. Although seedless vascular plants are not generally important agriculturally, some water ferns have an important role in the overall scheme of rice cultivation, as noted in the chapter introduction.

Section Review

1. **How does a homosporous species differ from a heterosporous species?**
2. **Compare the four living phyla of seedless vascular plants.**
3. **Compare and contrast the *Selaginella* and fern life cycles.**

SUMMARY

The Evolution of Seedless Vascular Plants

Seedless vascular plants dominated the landscape around 350 million years ago (pp. 442–443)
Seedless vascular plants, along with bryophytes, were the first plants. Unlike bryophytes, they had both xylem and phloem. During the Carboniferous period (363–290 million years ago), they were the most widespread types of plants.

Land plants arose from green algae in the class Charophyceae (pp. 443–444)
The first plant fossils, from seedless vascular plants, date to around 450 million years ago. Molecular data suggests that the first plants may have evolved as long as 700 million years ago. Like bryophytes, the first vascular plants did not produce seeds and required external water so that the sperm could swim to the egg for fertilization.

Three phyla of extinct vascular plants appear in the fossil record beginning 430 million years ago (pp. 444–446)
The earliest fossil evidence of vascular plants is of extinct plants called *rhyniophytes,* which were probably ancestors of two other extinct groups: zosterophylls and trimerophytes. Consisting of branching photosynthetic stems and living in marshy areas, all three became extinct by around 363 million years ago. Zosterophylls are ancestors of the phylum Lycophyta, while other existing phyla of seedless vascular plants, as well as seed plants, evolved from trimerophytes.

In living seedless vascular plants, alternation of generations involves independent gametophytes and sporophytes (pp. 446–447)
In seedless vascular plants, as in seed plants, the sporophyte is dominant. Most species are homosporous. Heterospory, which is characteristic of seed plants, occurs in some species.

Types of Living Seedless Vascular Plants

The four living phyla are Psilotophyta (mainly whisk ferns), Lycophyta (club mosses and related plants), Sphenophyta (horsetails), and Pterophyta (ferns), with ferns constituting the largest group.

Whisk ferns comprise most of the living members of phylum Psilotophyta (pp. 448–449)
The principal genus, *Psilotum,* contains whisk ferns, plants that consist of branching stem systems. They lack roots and have enations instead of leaves. Tri-lobed, yellowish sporangia occur at the end of stems. The genus *Tmesipteris* has leaves with single veins. Psilotophytes are homosporous and have bisexual gametophytes. As in other seedless vascular plants, the sporophyte becomes independent.

Living members of phylum Lycophyta include club mosses, spike mosses, and quillworts (pp. 449–453)
Although the phylum includes extinct trees, all modern lycophytes are herbaceous. All have microphylls, which in some species have sporangia that are arranged into conelike strobili. Club mosses are homosporous and have exosporic gametophytes, but spike mosses and quillworts are heterosporous and have endosporic gametophytes.

Horsetails are the living members of phylum Sphenophyta (pp. 453–454)

Horsetails (*Equistum*) are the only living genus of spheno-phytes, evolving from extinct trimerophytes. They are homo-sporous. The sporophyte has hollow, jointed stems with whorled microphylls. The independent, photosynthetic gametophytes are typically bisexual. Treelike species existed in the Carbonifer-ous period.

Phylum Pterophyta consists of ferns, the largest group of seed-less vascular plants (pp. 454–459)

Ferns evolved from extinct trimerophytes and today consist of herbaceous plants, vines, epiphytes, and tree ferns. Ferns are the first group of plants to have megaphylls. Most species are homo-sporous, and gametophytes are photosynthetic and usually bisex-ual. In most species, sporophytes have only fertile, photosynthetic fronds, but some species have separate sterile and fertile fronds.

Review Questions

1. How did the forests of the Carboniferous period differ from those of today and why?
2. Describe the evolutionary relationship between green algae, bryophytes, and seedless vascular plants.
3. Compare and contrast the three phyla of earliest vascu-lar plants.
4. How does the sporophyte–gametophyte relationship in seedless vascular plants differ from both bryophytes and seed plants?
5. In what sense can seedless vascular plants be called "liv-ing fossils"?
6. How do whisk ferns differ from other vascular plants?
7. Distinguish the three orders of lycophytes.
8. Compare and contrast enations, microphylls, and megaphylls.
9. Describe how microphylls may have evolved.
10. Describe the possible origin of megaphylls.
11. What is endosporic development of a gametophyte?
12. What are the characteristic features of horsetails?
13. How do ferns differ from other seedless vascular plants?
14. Describe the fern gametophyte and sporophyte.

Questions for Thought and Discussion

1. What do you think are the advantages and disadvantages of having independent gametophytes and sporophytes?
2. Why are there no longer giant forest trees in the phyla of seedless vascular plants?

3. Why do you think megaphyll shapes vary so much among fern species?
4. Why do you think seedless vascular plants no longer dominate the landscape?
5. Using labeled diagrams, illustrate the similarities and differences between a living horsetail and the extinct seedless vascular plant *Cooksonia*.

Evolution Connection

If we assume as a working hypothesis that bryophytes evolved from green algae and that the seedless vascular plants evolved from the bryophyte line of evolution, what evolutionary steps must have occurred to enable a divergence from the bryophytes and lead to the seedless vascular plants of today?

To Learn More

Visit The Botany Place Website at www.thebotanyplace.com for quizzes, exercises, and Web links to new and interesting infor-mation related to this chapter.

Cobb, Boughton. *A Field Guide to Ferns and Their Related Fam-ilies: Northeastern and Central North America.* Boston: Houghton Mifflin, 1999. The identification system tells exactly how to differentiate between different species.

Keator, Glenn, and Ruth M. Heady. *Pacific Coast Fern Finder.* Rochester, NY: Nature Study Guild Publishers, 1995. The system tells how to differentiate between different fern species.

Kenrick, Paul, and Peter R. Crane. *The Origin and Early Diversi-fication of Land Plants: A Cladistic Study.* Washington, DC: Smithsonian Press, 1997. A well-written, detailed book about evolutionary relationships among early vascular plants.

Sacks, Oliver. *Oaxaca Journal (National Geographic Directions).* Washington, DC: National Geographic, 2002. This book takes the reader on an expedition to Mexico to collect rare ferns.

22
Gymnosperms

Ponderosa pines (*Pinus ponderosa*), Yosemite National Park.

An Overview of Gymnosperms

Seed plants have significant selective advantages

Living gymnosperms are related to extinct plants from the Paleozoic and Mesozoic eras

In gymnosperms and other seed plants, dependent gametophytes develop within the parent sporophyte

The pine life cycle illustrates basic features of gymnosperm reproduction

Types of Living Gymnosperms

Phylum Coniferophyta contains conifers, which are the dominant forest trees in cooler climates

Phylum Cycadophyta contains cycads, which resemble tree ferns or palms

Phylum Ginkgophyta contains one living species

Phylum Gnetophyta contains three diverse genera found in tropical forests or in deserts

Conifers such as pine, spruce, and fir trees are the most familiar gymnosperms. However, other plants seemingly unrelated to conifers turn out to be gymnosperms as well. The beautiful maidenhair tree (*Ginkgo biloba*) resembles a broadleaf tree, which is a flowering plant, yet it is also a gymnosperm. Hikers in Utah and other southwestern states pass by scrubby bushes of *Ephedra* (Mormon tea), which look like a collection of half-dead twigs, little realizing that these are quite healthy relatives of pine trees. In the deserts of southern Africa grows an unusual gymnosperm known as *Welwitschia mirabilis*, which looks, at best, like a bedraggled lily with shredded leaves.

Ephedra.

What gymnosperms have in common is clearly not their outer appearance. Rather, they are similar because they produce so-called naked seeds that are exposed on modified leaves, rather than being enclosed within fruits. *Gymnos* in Greek means "naked," while *sperm* means "seed." Gymnosperms are also characterized by having strobili, or cones. Most species display recognizable pollen cones producing many pollen grains and ovulate cones producing multiple seeds. As you will see later, *Ginkgo biloba* and a few other species have unique modified leaves and branches that produce pollen and single-seeded ovulate cones protected by fleshy coverings, which make them look somewhat like fruits.

Gymnosperms have many human uses. Conifers are commonly seen in landscaping, but they are used as well in construction, in papermaking, as Christmas trees, and as firewood. *Ephedra* has become a popular shrub for landscaping in very dry areas, while *Ginkgo* trees are often planted in cities because they have beautiful yellow autumn foliage and are quite tolerant of air pollution. Both *Ginkgo* and *Ephedra* have medicinal uses. Ginkgo seeds have been roasted and used for cen-

Ginkgo tree.

turies in China as a digestive aid. Many people believe that an extract of ginkgo leaves improves blood circulation to the brain and thus boosts memory. However, recent studies have cast doubt on these claims. *Ephedra* has been used for millennia in China to cure coughs. It has long been a Native American remedy for digestive disorders, headaches, and burns. Western settlers used it to make an energy-boosting drink called *Mormon tea*. Its human uses relate to the plant's production of a secondary metabolite called *ephedrine,* a bitter-tasting, insect-repelling compound that is used pharmaceutically as a decongestant. Like most secondary metabolites of plants, ephedrine is potentially dangerous. In moderate doses it suppresses appetite, and it was used as an ingredient in diet pills until such side effects as high blood pressure, heart rate irregularities, seizures, and strokes were reported. Ephedrine can also be used to make methamphetamine, a dangerous street drug.

Welwitschia mirabilis.

In this chapter, we will first explore the evolution and general characteristics of gymnosperms. Then we will look at the distinctive characteristics of the four living phyla: Coniferophyta, Cycadophyta, Ginkgophyta, and Gnetophyta. The phylum Coniferophyta consists of the conifers. The phylum Cycadophyta includes palmlike or fernlike plants known commonly as *cycads.* The phylum Ginkgophyta has only one living species, *Ginkgo biloba*. The plants in the phylum Gnetophyta, which include the genera *Ephedra, Welwitschia,* and *Gnetum,* are called *gnetophytes* and are the gymnosperms most similar to angiosperms. Some systematists place all gymnosperms in the phylum Pinophyta, making the four phyla presented here into orders. The increasing use of molecular methods in plant systematics will undoubtedly introduce changes in the taxonomic view of the living and extinct gymnosperms.

An Overview of Gymnosperms

The vast majority of modern plants produce seeds and include about 760 species of gymnosperms and almost 250,000 species of angiosperms, so it is no wonder that seeds are so familiar—particularly those of flowering plants. To many people, seeds are a dried-up reproductive structure of a vegetable or flower planted in the garden. After watering, the seed germinates and a seedling emerges after some days or weeks. In other cases, seeds are something eaten for a snack or added to a salad. More importantly, most human food comes directly from the consumption of seeds and fruits of such flowering plants as rice, corn, and wheat. We will look at the general advantages of seeds before exploring gymnosperm evolution and characteristics.

Seed plants have significant selective advantages

Figure 22.1 reviews the locations of seeds in gymnosperms and angiosperms. Gymnosperm seeds form on the surface of leaves or branches of cones and are "exposed" in the sense that they are not completely enclosed within a fruit.

Gymnosperm seeds result from a single fertilization of sperm and egg. Angiosperm seeds result from a double fertilization of sperm with egg and with two nuclei of the megagametophyte (female gametophyte). In gymnosperm seeds, megagametophyte tissue nourishes the developing embryo. In angiosperm seeds, the nourishment is provided by endosperm, formed by the union of a sperm with two nuclei of the megagametophyte.

Seeds evolved relatively late in the history of plant life. For about 100 million years after the origin of plants from green algae, only seedless plants existed. The evolution of the seed—an embryonic plant combined with a food supply and surrounded by a protective seed coat—allowed plants to be much more successful on land (see Chapter 6). For the plants that produce them, seeds provide the sexual biological link between generations—the future and the past. To be successful in an evolutionary sense, a plant must transfer its genes to the next generation. Seeds allow plants to do this in an efficient manner with significant selective advantages over seedless plants:

◆ The dormant state of seeds enables seed plants to survive extended periods of cold winter or drought.
◆ The seed coat serves as a barrier against bacterial or fungal decay.

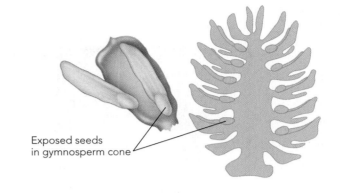

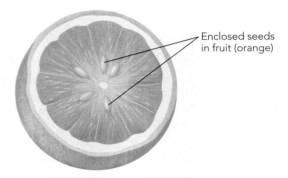

Exposed seeds in gymnosperm cone

Enclosed seeds in fruit (orange)

Figure 22.1 Gymnosperm "naked" seeds and angiosperm seeds in fruit. Gymnosperm seeds are exposed on the surface of modified branches or leaves. The enclosed seeds of flowering plants, or angiosperms, form inside fruits.

◆ Seeds attract seed-eating animals, which destroy some seeds but distribute others.
◆ Seeds include food for the developing embryo and germinating seedling.

In addition to the advantages of seeds, seed plants have other significant adaptations that facilitate survival on land:

◆ A hollow pollen tube produced by the microgametophyte (male gametophyte) delivers sperm to the egg, making fertilization possible without the presence of water. Accordingly, the vast majority of seed plants have nonflagellated sperm. The exceptions are a few gymnosperms: the cycads and *Ginkgo biloba*.
◆ The gametophytes are reduced in size and are protected and nourished within the sporophyte.

Overall, the seed, the pollen tube, and the reduced but protected gametophytes made possible an adaptive radiation in which seed plants thrived in many locations where seedless plants could not. Next, we will look at some possible paths by which gymnosperms evolved from seedless

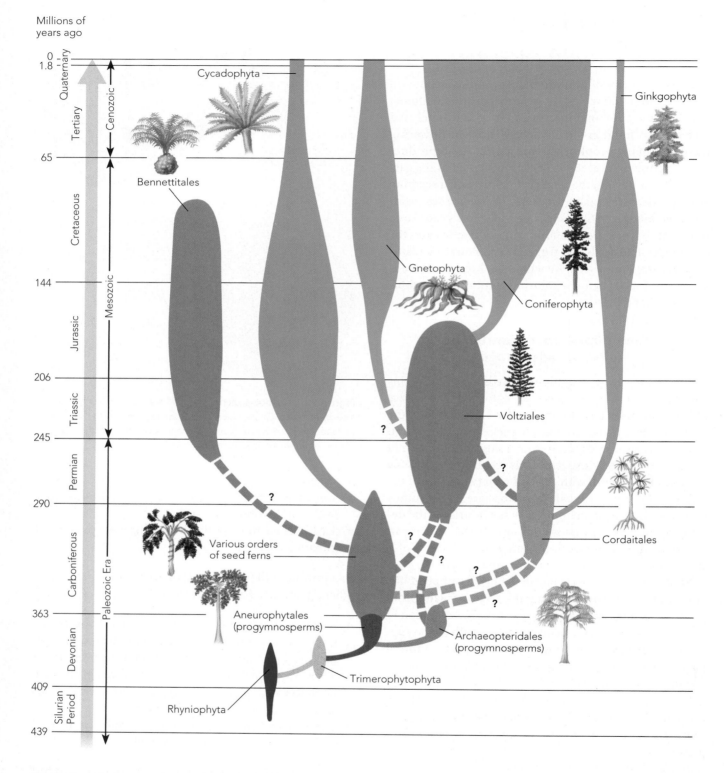

Millions of
years ago

Figure 22.2 A hypothesis for the evolution of gymnosperms. The evolution of
gymnosperms involved at least four extinct groups: progymnosperms (the Aneurophytales
and the Archaeopteridales); seed ferns; and two groups of primitive gymnosperms, the
Cordaitales and Voltziales. Many of the evolutionary links are uncertain.

vascular plants. Chapter 23 will examine the evolution of angiosperms.

Living gymnosperms are related to extinct plants from the Paleozoic and Mesozoic eras

Four groups of extinct plants played important roles in the evolution of modern gymnosperms: progymnosperms; seed ferns; and two groups of primitive gymnosperms, the Cordaitales and the Voltziales. Figure 22.2 reflects one hypothesis of gymnosperm evolution during the Paleozoic and Mesozoic eras. As the figure indicates, a number of the evolutionary relationships are uncertain. The fossil evidence is inconclusive regarding whether seeds evolved only once or multiple times in separate evolutionary lines.

The plants known as *progymnosperms* probably arose from the phylum Trimerophytophyta (trimerophytes) (Chapter 21) in the middle Devonian period and persisted until the Lower Carboniferous period. Progymnosperms did not produce seeds, but the presence of wood distinguished them from trimerophytes and seedless vascular plants. They had secondary xylem similar to that of living conifers, and their vascular cambium produced both secondary xylem and secondary phloem. There were two groups of progymnosperms: the Aneurophytales and the Archaeopteridales, neither of which looked like a gymnosperm. The Aneurophytales were homosporous and had complicated three-dimensional branching, resembling overgrown trimerophytes (Figure 21.5). The Aneurophytales gave rise to the Archaeopteridales, which had flattened, fernlike leaves with heterosporous sporangia.

The Aneurophytales may have given rise to a diverse group of plants known as *seed ferns,* or *pteridosperms,* which were the precursors of cycads and perhaps of the Cordaitales and the Voltziales. Seed ferns consisted of a number of unrelated plant groups that are currently classified together because they resembled tree ferns in shape but, unlike the progymnosperms, they produced seeds. Appearing in the late Devonian period (around 365 million years ago), they were the first seed plants, with integuments protecting ovules to various degrees (Figure 22.3). At least one group of seed ferns gave rise to the cycads. Another may have given rise to an extinct group of gymnosperms called the *Bennettitales,* which resembled cycads in their overall palmlike or fernlike shape. However, the Bennettitales sometimes had microsporophylls and ovules organized into structures that looked somewhat like flowers. Many paleobotanists believe that flowering plants and

Bennettitales have a common ancestor, although the evolutionary links between gymnosperms and angiosperms are uncertain (see Chapter 23).

The Archaeopteridales may have given rise to the Cordaitales and the Voltziales. The Cordaitales were shrubs and trees common during the Carboniferous and Permian periods in swamps and on dry land. Their slender leaves, often found at the end of short branches, were up to a meter (3 feet) in length, with many veins. The Cordaitales also had a vascular cambium, and separate pollen and ovulate cones. The Cordaitales apparently gave rise to the phylum Ginkgophyta, which persists to the present day. The Voltziales, which may have arisen from the Archaeopteridales, seed ferns, or the Cordaitales, lived during the Carboniferous, Permian, Triassic, and Jurassic periods. They resembled a living conifer called the *Norfolk Island*

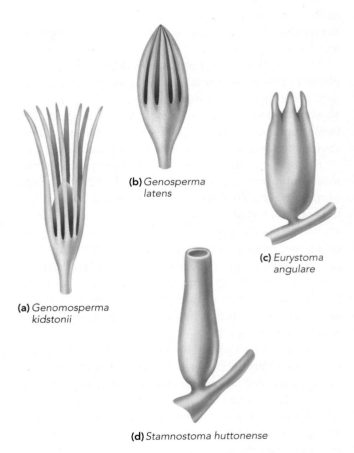

(b) *Genosperma latens*

(c) *Eurystoma angulare*

(a) *Genomosperma kidstonii*

(d) *Stamnostoma huttonense*

Figure 22.3 Early seeds. Many fossil plants, particularly seed ferns, produced early forms of seeds. The illustrations here are based on fossil evidence. Note that the ovules are surrounded to various degrees by integuments, with separated integuments in *Genomosperma kidstonii* and complete fusion of integuments in *Stamnostoma huttonense.*

pine, discussed later in the chapter. Short needles covered the entire length and radius of spirally arranged branches. The Voltziales probably gave rise to the Coniferophyta and may have given rise to the Gnetophyta.

In gymnosperms and other seed plants, dependent gametophytes develop within the parent sporophyte

The evolution of seeds is closely tied to the evolution of sporophylls and sporangia on sporophytes. Before the development of the large vascularized leaves known as *megaphylls,* sporophytes bore sporangia at the end of leafless branches known as *telomes* (see Chapter 21). Megaphylls evolved from a group of shortened branches with tissue developing between them. The association of sporangia with leaves began during the Devonian period (409 to 363 million years ago) and continued through the evolution of the sporophytes of seedless vascular plants, gymnosperms, and finally angiosperms.

The evolution of seed plants with dependent gametophytes from seedless plants that had separate and independent gametophytes involved a number of developments. Most of the intermediate forms of plants are extinct, but two key precursors to the emergence of seeds were heterospory and endosporic development (see Chapter 21).

The original vascular land plants were probably homosporous, with one type of spore produced by one type of sporangium. This characteristic is shared by most living seedless vascular plants. In contrast, all seed plants are heterosporous, producing two types of spores, microspores and megaspores, in two different types of sporangia, microsporangia and megasporangia. Heterospory occurred first among some species of seedless vascular plants and is evident in some living species, such as the genus *Selaginella* and a few ferns, which are not closely related to gymnosperms.

In most seedless plants, both extinct and living, spores germinate to produce gametophytes *outside* the confines of the spore wall, known as *exosporic development* (see Chapter 21). Many seedless vascular plants retain this type of gametophyte development. In contrast, the gametophytes of seed plants develop endosporically—growing *inside* the spore, a process also evident in a few seedless vascular plants, including *Selaginella.* The selective advantage of endosporic development may be that it protects the developing gametophyte from desiccation while supplying nutrition and water.

Although sharing with some seedless plants the characteristics of heterospory and endosporic development,

seed plants are unique in that the growth of spores into gametophytes, fertilization, and the initial development of the sporophyte embryo all take place within the parent sporophyte. As noted previously, this arrangement offers greater protection and nutrition to the developing sporophyte embryo. In contrast, seedless plants release spores into the environment, and all of these processes occur outside the parent sporophyte.

Figure 22.4 provides an overview of how alternation of generations in seed plants differs from that in seedless plants. In bryophytes (see Chapter 20) and seedless vascular plants (see Chapter 21), each dependent sporophyte embryo grows out of an independent gametophyte plant. The sporophyte remains dependent on the gametophyte for food, water, and support throughout its life. In contrast, gametophytes in seed plants are dependent on the sporophyte and remain attached to it. Within the megagametophyte, a sporophyte embryo develops from an egg fertilized by a sperm. The embryo becomes part of a seed, consisting of an embryo, seed coat, and a food supply. The seed is not usually released until development of the embryo is complete. After release, the seed can germinate and grow into an independent sporophyte.

The pine life cycle illustrates basic features of gymnosperm reproduction

Although asexual reproduction is found in a few species of gymnosperms, notably in redwoods, sexual reproduction is the rule in most gymnosperms. Since the vast majority of living species are conifers, we will look at a pine life cycle (Figure 22.5). Like most other gymnosperms, conifers are wind-pollinated. In some gymnosperm species, pollen and ovulate cones are on separate plants. Pines and many other conifer species have both types of strobili on each plant. Ovulate cones, also known as *female cones* or *seed cones,* usually occur on higher branches. Pollen cones, also known as *male cones,* typically occur on lower branches. This arrangement promotes cross-pollination, the transfer of pollen from one plant to another, because the windblown pollen is not usually distributed from the bottom to the top of the same tree.

Ovulate cones in conifers are typically more complex than pollen cones. Pollen cones in conifers are sometimes called *simple cones* because each cone consists of spirally arranged microsporophylls attached directly to a central axis. Each microsporophyll, known more commonly as a *scale,* has two pollen-containing microsporangia on its lower surface. The complex ovulate cones that are characteristic of pine and most other conifers are sometimes

Bryophytes and Seedless Vascular Plants

Moss (a bryophyte)

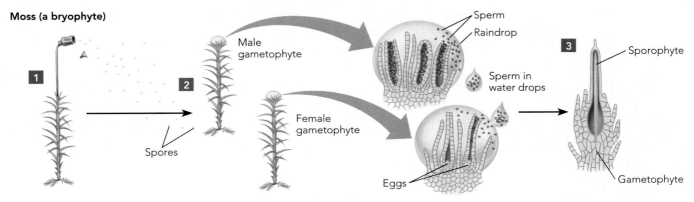

Sperm
Raindrop
Sperm in
water drops
Sporophyte
Male
gametophyte
Spores
Female
gametophyte
Eggs
Gametophyte

Fern (a seedless vascular plant)

Young
sporophyte
Spores
Egg
Gametophyte
Sperm
Gametophyte

1 Sporophyte releases spores, which develop into independent gametophytes.

2 Fertilization takes place inside an independent gametophyte, separate from the mature sporophyte.

3 Young sporophyte develops from an embryo within an independent gametophyte.

Seed Plants (Gymnosperms and Angiosperms)

Pine (a gymnosperm)

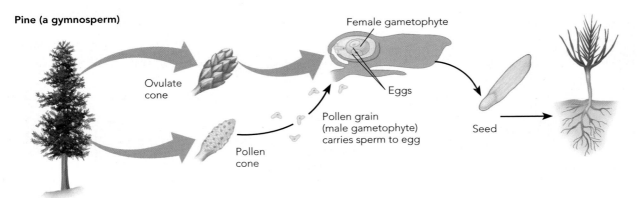

Female gametophyte
Ovulate
cone
Eggs
Pollen grain
(male gametophyte)
carries sperm to egg
Seed
Pollen
cone

1 Spores remain on sporophyte while developing into dependent gametophytes.

2 Fertilization occurs while female gametophyte is still attached to sporophyte.

3 Young sporophyte grows from seed released by mature sporophyte.

Figure 22.4 Comparing the sporophyte-gametophyte relationship in seedless plants and seed plants. In contrast with the two groups of seedless plants—bryophytes and seedless vascular plants—seed plant gametophytes, such as those in gymnosperms, are dependent on the mature sporophyte.

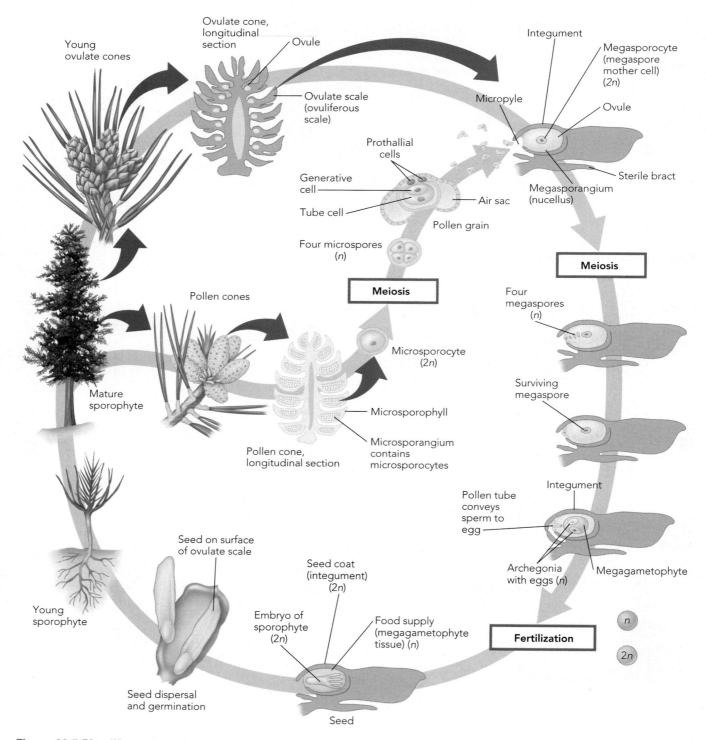

Figure 22.5 Pine life cycle.

called *compound cones* because they consist of a central axis and spirally arranged modified branches called *seed-scale complexes,* also known as *bract-scale complexes.* Each seed-scale complex includes a sterile bract and an ovulate scale, also called an *ovuliferous scale,* which consists of

fused megasporophylls. Each ovulate scale bears two ovules on its upper surface. Each ovule contains a megasporangium, which in seed plants is also called a **nucellus** (from the Latin *nucella,* "small nut"), surrounded by a large integument produced by the sporophyte. Pollen grains can enter through a small opening in the integument called a **micropyle** (from the Greek *pyle,* "gate").

In tracing the pine life cycle, we will start with events in the pollen cone, as shown in the middle of Figure 22.5. As with all seed plants, no antheridia develop. Unlike an antheridium, which contains many single-celled sperm, each microsporangium contains several hundred **microsporocytes,** also known as *microspore mother cells.* Each microsporocyte undergoes meiosis to produce four haploid microspores. Each microspore then gives rise to a microgametophyte—a pollen grain with four cells—that develops endosporically. Two of the cells, known as *prothallial cells,* have no known function. A third cell is called the *generative cell* because it will give rise to a sterile stalk cell as well as to what is known as a *body cell,* which eventually produces two sperm. The fourth cell is called the *tube cell* because it will produce the pollen tube, which is a device that facilitates delivery of sperm to the egg without water. Thus, the production of pollen is a significant evolutionary change. Each pollen grain has two air sacs that function as "wings." The yellowish pine pollen is produced and dispersed in the spring, riding the wind and coating everything it touches.

Pollen lands on drops of fluid produced by each megasporangium (nucellus) on each ovulate scale. As the pollination droplet evaporates, the pollen is brought through the micropyle into contact with the megasporangium, where it germinates. Often, more than one pollen grain germinates in each megasporangium. Germination stimulates the development of the megagametophyte, as shown along the right side of Figure 22.5. About a month after pollination, the **megasporocyte,** also known as a *megaspore mother cell,* divides by meiosis to produce four megaspores. Usually, only the megaspore farthest from the micropyle develops into the megagametophyte, while the other three megaspores abort. When the megagametophyte nears maturity, which takes about a year in pines, typically two to five archegonia form near the micropyle. The production of archegonia, each of which contains a single egg, is characteristic of most gymnosperms, including conifers, cycads, *Ginkgo biloba,* and *Ephedra.* Archegonia are a more primitive feature, which gymnosperms share with seedless plants. In angiosperms and a few gymnosperms, *Welwitschia* and *Gnetum,* the megagametophytes do not produce archegonia.

While the megagametophyte develops within the megasporangium, each pollen grain forms a pollen tube that grows through the megasporangium. The full development of a pollen tube, containing two sperm, takes about a year. As in all gymnosperms except cycads and *Ginkgo biloba,* the sperm are nonflagellated. About 15 months after pollination, fertilization takes place, when a pollen tube conveys two sperm to an archegonium. After one sperm fertilizes the egg, the other sperm degenerates. Since there are often multiple pollen tubes, multiple archegonia are fertilized and more than one embryo initially develop, a phenomenon known as **polyembryony.** Typically only one embryo survives, a result of that one being more vigorous or better positioned in obtaining nutrition. The pine embryo is uncurved and has many cotyledons. Each seed typically consists of only one embryo, a food supply of megagametophyte tissue, and a seed coat (formerly the integument). As a result of polyembryony, however, a small percentage of pine seeds have more than one embryo and therefore produce more than one seedling when the seed germinates.

An ovulate pinecone becomes quite woody as it matures. The scales grow closed after pollination and remain closed while the cone matures. Seeds are shed the second autumn after pollination. Pine seeds are winged, which aids in their transport by the wind. The seeds are frequently released by animals trying to eat the seeds or by old age as the cones begin to rot. Some pines, such as the lodgepole pine (*Pinus contorta*), require intense heat for the seeds to be released. Therefore, forest fires both create the sunny environments required by lodgepole seedlings and release the seeds from cones so they can germinate in an appropriate location. In national parks and national forests, controlled burns eliminate buildup of shrubs, dead trees, and other burnable materials, helping to ensure healthy forests that are more like those found before the arrival of humans.

Section Review

1. **What are the significant selective advantages of seeds?**
2. **What are progymnosperms, and what may be their evolutionary relationship to modern gymnosperms?**
3. **Describe in general how reproduction in seed plants differs from that in seedless plants.**
4. **Describe how alternation of generations is evident in the pine life cycle.**

Types of Living Gymnosperms

Gymnosperms were the dominant plants in the Mesozoic era, but by the Cenozoic era they had been replaced in many habitats by flowering plants. The four surviving phyla of gymnosperms vary greatly in appearance and in habitat.

Phylum Coniferophyta contains conifers, which are the dominant forest trees in cooler climates

Although the word *conifer* means "cone-bearing," keep in mind that all gymnosperms have cones (strobili). Members of the phylum Coniferophyta, the most common gymnosperms, consist of about 50 genera of trees with approximately 550 species worldwide, mostly in the Northern Hemisphere. Table 22.1 provides a sampling of conifer diversity.

Conifers include the world's tallest and largest plants. The tallest are coast redwoods (*Sequoia sempervirens*), which grow primarily in California and Oregon (see Chapter 3). The record holder is known as the Stratosphere Giant, soaring 112.34 meters (368.6 feet). Meanwhile, the most massive tree is a giant sequoia (*Sequoiadendron giganteum*) in Sequoia National Park. Called the General Sherman Tree, it measures 31 meters (101.7 feet) in maximum circumference and is estimated to weigh 6,028 metric tons (6,167 short tons). The General Sherman Tree contains enough wood to build more than 100 three-bedroom houses, and indeed most giant sequoias were cut down for lumber in the late 1800s. The few remaining groves are protected within national parks.

Conifer wood is called *softwood* because it cuts and nails easily. Many conifer species, such as Douglas fir (*Pseudotsuga menziesii*), are major sources of construction wood. Anatomically, softwood has no fibers and has thinner cell walls than the wood of most angiosperm trees, which is typically described as hardwood. The conducting cells in the xylem of conifers, like those of almost all gymnosperms, consist solely of tracheids. Conifers produce resin, which moves through the plant in resin ducts and helps protect the trees from attack, but many conifers remain vulnerable to disease-causing organisms and herbivores (see the *Biotechnology* box on page 472).

Conifers are often the dominant species at high elevations, in regions where winters are long and cold and often characterized by dry winds, as in the northern United States and countries at higher latitudes, such as Canada and Russia. In these areas, conifers have several selective advantages over flowering plants in tolerance of cold weather and dry winds. Since they lack vessel elements, for example, they are not prone to permanent disruption of water flow by freezing. Also, conifer leaves are narrow and typically needlelike, in contrast with the broader blades common in flowering plants. Narrower leaves expose less surface to the air and are therefore less susceptible to damage by freezing or by dry winds. The stomata are recessed and therefore lose water less readily (Figure 22.6). Beneath the epidermis is an area known as the *hypodermis,* which has thick-walled cells that prevent water loss. The epi-

Table 22.1 A Sampling of Conifer Diversity

Genus	Common Name	Comments
Abies	Fir	Soft foliage, upright cones
Araucaria	Norfolk Island pine, hoop pine	Found in Southern Hemisphere
Cedrus	Cedar	Native to Middle East and Far East; leaves are needles in dense clusters
Cupressus	Cypress	Mostly shrubs or small trees, fleshy cones
Juniperus	Juniper	Source of flavoring for gin
Larix	Larch	Deciduous
Metasequoia	Dawn redwood	Living fossil native to China
Picea	Spruce	Prickly foliage, pendulous cones
Pinus	Pine	Needlelike leaves unique among conifers
Podocarpus	Podocarp	House plant native to Southern Hemisphere
Pseudotsuga	Douglas fir	Most common wood for construction
Sequoia	Coast redwood	Tallest tree
Sequoiadendron	Sequoia, big tree	Most massive tree
Taxodium	Bald cypress	Deciduous, lives in swamps in the Southeastern United States
Taxus	Yew	Source of taxol, a cancer drug
Tsuga	Hemlock	Not the poisonous herb that killed Socrates

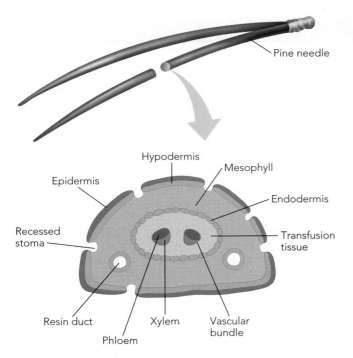

Figure 22.6 Drought adaptations of a pine needle.
The pine needle's recessed stomata, the thick epidermis, the hypodermis, the endodermis surrounding the vascular bundle, and the transfusion tissue are all adaptations that prevent water loss.

dermis itself has a waxy outer layer, as in most flowering plants.

Several anatomical features of conifer leaves efficiently conduct water to the photosynthetic mesophyll. Their vascular bundles are surrounded by an endodermis, which prevents water loss by directing water and mineral transport through cell membranes. Between the vascular bundles and the endodermis, a region of transfusion tissue moves liquid efficiently from the xylem into the mesophyll (Figure 22.6).

Conifer leaves are simple rather than compound and are borne singly or in clusters called *fascicles*. Leaves at the tops of tall conifers are often shorter and more rounded in cross section than the counterparts below. Conifer leaves can remain photosynthetically active for 2 to nearly 50 years before they are dropped to the forest floor. Pine trees and most other conifers retain individual leaves for a minimum of 2 to 5 years, leaving the visual impression that the leaves are evergreen. However, old leaves are shed gradually, and new leaves appear each year on branch tips. Some conifers are deciduous, such as larch (*Larix*), bald cypress (*Taxodium distichum*), and dawn redwood (*Metasequoia glyptostroboides*) (Figure 22.7).

The sporophylls of conifers occur in cones, which are spirally arranged shoots composed of modified leaves and branches. Pollen cones are usually a few centimeters in length, with papery sporophylls. Ovulate cones are typically woody and can be up to 60 centimeters (2 feet) in length. However, some conifers produce seed-bearing structures that resemble fleshy berries more than cones. For example, in junipers and cypresses, ovulate cone scales cover the seeds (Figure 22.8). The fused juniper scales turn various colors, depending on the species. In yew trees a fleshy covering called an *aril*

Figure 22.7 *Metasequoia*, a deciduous conifer. *Metasequoia glyptostroboides* (dawn redwood) was thought to be extinct until a live specimen was discovered in 1944 in south-central China. In 1948, an expedition discovered around 10,000 living trees in an isolated Chinese forest. The trees, which can become at least 45.7 meters (150 feet) tall, are remnants of huge *Metasequoia* forests that lived from 15 to 100 million years ago. *Metasequoia* was once the most common gymnosperm in North American forests.

BIOTECHNOLOGY

Improving and Protecting Trees

As a result of the decreasing size of Earth's forests and the ever-increasing need for wood, trees with higher quality wood are economically worth developing. Scientists are exploring a variety of techniques, including traditional breeding and genetic engineering.

One major focus has been to improve resistance to disease and insects. Trees with resistance to a particular disease can be selected and used in breeding experiments to transfer useful genes to a larger population. Since trees take so long to mature, most early efforts in biotechnology to improve trees focused on cloning methods. If a particularly valuable tree is located, the most efficient means of using its unique traits is to make many genetically identical copies of the tree and use them to plant new forests. Genetic engineering can also play a role, as when genes that might confer resistance to the disease are added to the chromosomes of susceptible species. Yet another approach has been used to protect hemlocks (*Tsuga* species), beautiful and important forest conifers in the eastern United States and Canada. In parts of New England, hemlock is the dominant forest species, with virgin stands more than 400 years old. However, the trees are threatened by the hemlock wooly adelgid, an aphid-like insect introduced from Japan in the 1920s. The insects suck sap from young twigs, eventually killing the tree. About half of the range of the trees in the United States is currently infected, with the trees having been completely eliminated in many regions. Scientists are currently experimenting with the use of a beetle from Japan, *Pseudoscymnus tsugae,* as a biological control agent, feeding on the adelgid.

Meanwhile, both traditional breeding and genetic engineering are being used to improve the quality of wood products. One example of genetic engineering is the effort to reduce lignin content to improve paper production. Lignin, which strengthens and hardens wood, is undesirable for making paper because it must be separated from cellulose fibers by water-polluting chemical treatments. The addition of two genes that alter lignin biosynthesis have genetically engineered aspen trees to contain half the normal lignin content and 30% more cellulose. Hemlock and other conifers, such as white spruce, are also important sources of wood pulp, so varieties with reduced lignin would be desirable. However, environmentalists worry that the genetic combination might escape into nature, decreasing the strength of wood and having dire effects on native trees. In addition, they point out that alternative sources of wood pulp, such as kenaf (see Chapter 5), do not contain lignin. In short, forest biologists must weigh the possible ecological effects of genetic engineering.

Hemlock wooly adelgid and the beetle that eats it.

(from the Latin *arillus,* "grape seed") partially surrounds each ovule (Figure 22.9a). Seeds of podocarps, a group of Southern Hemisphere conifers, have a completely covered ovule on top of a large fruitlike structure (Figure 22.9b). However, any gymnosperm structures that look like fruits are actually seeds with coverings arising from integuments. In contrast, fruits have ovaries, which are found only in flowering plants.

Although most conifer species are found in cooler regions of the Northern Hemisphere, members of a family known as the Araucariaceae are native to the Southern Hemisphere. The genus *Araucaria* is common in warmer regions of South America, southern Asia, and Australia, serving as an important source of wood for construction and fuel. Most *Araucaria* species have separate pollen and seed-producing plants. Some, notably the Norfolk Island "pine" (*Araucaria heterophylla*), have been widely cultivated elsewhere. Named after Norfolk Island, near New Zealand, the tree is a familiar house and greenhouse plant throughout the United States but grows to its full height outdoors only in parks and planted groves in semitropical regions (Figure 22.10). While cultivated plants rarely grow more than a meter (3.3 feet) tall, trees in the wild can reach more than 50 meters (164 feet) in height. Another commonly cultivated species is

(a) Ovulate cones of common juniper.

Figure 22.8 Ovulate cones in juniper and cypress. In the ovulate cones of some conifers, such as juniper and cypress, fleshy scales completely cover the seed, causing the cone to look very much like a berry.

(b) Ovulate cones of a cypress.

Figure 22.9 Fleshy seed coverings in yew and podocarps. In conifers such as yew and *Podocarpus,* the ovule is covered by a fleshy cuplike structure rather than by cone scales. (**a**) In the yew, the covering is red and is known as an *aril.* (**b**) The white *Podocarpus* ovule sits atop a fleshy purple structure favored by birds that disperse the seeds.

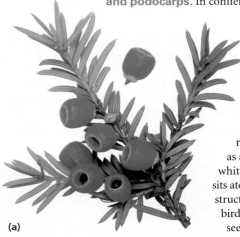

(a)

(b)

the monkey-puzzle tree (*Araucaria araucana*), which has short, sharply pointed leaves that surround the branches and in young trees cover the trunk (Figure 22.11). The origin of the common name is not certain. It may refer to a nineteenth-century Englishman's remark that even a monkey could not climb the tree, although monkeys are not found in the tree's native region of Chile. Another species, *Araucaria angustifolia,* has various common names, including the candelabra tree and the pinheiro tree. A bird known as the *blue gralha* is well known for its habit of removing the seeds from cones of the pinheiro tree, eating some, and planting the remainder in the ground.

Figure 22.10 Norfolk Island pine (*Araucaria heterophylla*). Grown in containers as a common houseplant in the United States, Norfolk Island pine is an important forest tree in the Southern Hemisphere.

THE INTRIGUING WORLD OF PLANTS

The Wollemi Pine: A Living Fossil

The *Wollemi* pine (*Wollemia nobilis*) is a large conifer with a trunk diameter at maturity of more than 1 meter (3 feet). Known previously from 150-million-year-old fossils, the pine was only recently discovered alive in Australia. In 1994, David Noble, a New South Wales National Parks and Wildlife Service officer, was bush-walking in Wollemi National Park, northwest of Sydney, when he saw, in a sheltered gorge, a grove of 40 trees that he did not recognize. Ultimately three small groves of seedlings were found growing on rain-fed ledges.

Wollemi pine trees in a nursery.

Indeed, *Wollemi* is an aboriginal word meaning "look around you." The *Wollemi* pine is related to the Norfolk Island pine and represents a third genus in the family Araucariaceae. This pine has now been grown from seed in nurseries, and an effort is underway to preserve existing small populations. Some of the trees may be 1,000 years old. Finding this population of living trees underlies the importance of national parks and wilderness areas throughout the world that preserve biological diversity.

During the Mesozoic era, vast forests of trees in the Araucariaceae were common. The members of this family have since been displaced from many of their ancient environments by flowering plants, and human activity now threatens some species. However, some good news is that several groves of a species now called the *Wollemi pine* (*Wollemia nobilis*), which was previously thought to be extinct, were recently discovered in Australia (see *The Intriguing World of Plants* box on this page).

Phylum Cycadophyta contains cycads, which resemble tree ferns or palms

The cycads are the second largest group of gymnosperms, with living cycads consisting of 11 genera and 140 species. With their frondlike or palmlike leaves, cycads are frequently mistaken for ferns or flowering palms, rather than

Figure 22.11 Monkey-puzzle tree (*Araucaria araucana*). Sharp, pointed leaves surround the stem. This tree is in Conguillo National Park, Chile.

being recognized as relatives of conifers. Like conifers and other gymnosperms, cycads have cones, but their cones are generally larger than those of conifers, sometimes several meters in length (Figure 22.12). Unlike most species of conifers, which have pollen and ovulate cones on the same tree, all cycad species have separate pollen and seed-producing plants. Pollen cones and ovulate cones are both large, with the ovulate cones of many species often being pollinated by beetles, an example of insect pollination, a common feature of flowering plants. Unlike the nonflagellated sperm of most gymnosperms, cycad sperm are flagellated, and swim the short distance to the egg, rupturing the pollen tube to bring about fertilization. The tallest cycads are 15 meters (50 feet), but many have short trunks. Trunks are covered with scaly leaf bases arranged spirally. The living species are a remnant of a much wider diversity found during the Mesozoic era (245 to 65 million years ago), which is sometimes referred to not only as the Age of Dinosaurs but also as the Age of Cycads.

Phylum Ginkgophyta contains one living species

The only living species of the phylum Ginkgophyta is the maidenhair tree (*Ginkgo biloba*), which you saw in the chapter introduction. It can grow to about 30 meters (about 100 feet) and is called the maidenhair tree (Figure 22.13) because its fan-shaped leaves, with two lobes, are similar to those of the maidenhair fern. Leaves on seedlings or long shoots are deeply lobed, while the majority of leaves on short side shoots are scarcely lobed. Living ginkgos seem unchanged from ginkgo fossils that are 150 million years old. Ginkgos would probably have become extinct if they had not been

(a)

Figure 22.12 Cycads produce prominent pollen and ovulate cones. (a) Cycad trees (*Lepidozamia hopei*), North Queensland, Australia. **(b)** A cycad ovulate cone.

(b)

(a) *Gingko* leaves.   **(b)** *Ginkgo* cones on a pollen-producing tree.   **(c)** Fleshy *Ginkgo* seeds.

Figure 22.13 *Ginkgo biloba*.

grown in Chinese monasteries for hundreds if not thousands of years. As in cycads, the pollen grain germinates and multiflagellated sperm swim to the egg. Megagametophytes give rise to fleshy seeds that look like small plums. The so-called fruits are anatomically quite different from true fruits in angiosperms because the flesh is simply a seed coat rather than an ovary surrounding a seed.

Like cycads and gnetophytes, *Ginkgo* has separate pollen and seed-producing plants. In the United States, the trees are commonly used for ornamental landscaping—typically only the pollen-producing trees are used because *Ginkgo* seeds contain an acid that smells like rancid butter. In Asia, however, more seed-producing trees are grown because the seeds are popular in some cuisines. To prepare edible seeds, the outer two layers of the integument are peeled away to reveal a stony inner integument or kernel, which is roasted and cracked open. The embryo and megagametophyte are eaten.

Phylum Gnetophyta contains three diverse genera found in tropical forests or in deserts

Members of the phylum Gnetophyta, known as gnetophytes, include 70 species, divided into three genera: *Ephedra, Gnetum,* and *Welwitschia.* With the exception of a few species of *Ephedra,* each gnetophyte species has separate pollen and seed-producing plants. In terms of their external appearance, the three genera differ distinctly but are categorized together because of molecular evidence and because of the fact that they have more angiosperm-like features than other gymnosperms. One such trait is the presence of vessel elements in addition to tracheids. All other gymnosperms have only tracheids. Also, *Welwitschia* and *Gnetum* are similar to flowering plants in not having archegonia, and some species of *Ephedra* and *Gnetum* are the only plants aside from angiosperms to go through a form of double fertilization. However, the process yields extra embryos, rather than the endosperm produced in angiosperms. This difference may indicate that double fertilization in angiosperms evolved along a separate evolutionary pathway.

The more than 30 species of *Ephedra,* also known as *Mormon tea* and *joint firs,* occur in deserts and other arid regions (Figure 22.14), including many parts of the western United States. *Ephedra* shrubs may appear to be composed solely of short green sticks but actually have tiny leaves that form at the nodes and soon turn brown.

The genus *Gnetum* contains more than 30 species of primarily African and Asian tropical plants, which can be vines, shrubs, or trees (Figure 22.15). Their broad, leathery leaves resemble those of some flowering plants.

The genus *Welwitschia* consists of only one species, *Welwitschia mirabilis,* which is native to the arid coastal deserts of southwestern Africa (Figure 22.16). Since these

(a)   (b)

Figure 22.14 *Ephedra* **is one of three living genera in phylum Gnetophyta.** (a) A pollen-producing *Ephedra* plant. (b) *Ephedra* ovulate cones.

Figure 22.15 *Gnetum* seeds and leaves. Gnetum seed coats are fleshy, giving them a fruitlike appearance, and the leaves are like those of some flowering plants.

(a) *Welwitschia* pollen cones.

(b) *Welwitschia* ovulate cones.

Figure 22.16 *Welwitschia* is one of three living genera in phylum Gnetophyta.

regions receive less than 2.4 centimeters (1 inch) of rain a year, *Welwitschia* survives by being extremely drought-tolerant and taking up water from the frequent coastal fogs. *Welwitschia* has a very unusual appearance, with a carrot-shaped stem up to a meter in diameter and growing up to 3 meters (9.8 feet) into the soil. Above ground, the stem produces two straplike leaves that can measure up to about 6 meters (about 20 feet) long. The leaves have a meristem at their base and continue growing throughout the plant's life but gradually split and break, giving the plant a ragtag appearance even if it is quite healthy.

Section Review

1. Describe the general characteristics of conifers.
2. What are some distinguishing features of cycads?
3. Describe the distinguishing features of *Ginkgo biloba*.
4. Why are gnetophytes classified as a phylum even though they are so diverse?

SUMMARY

An Overview of Gymnosperms

Seed plants have significant selective advantages (pp. 463–465)
Seeds provide gymnosperms with significant selective advantages for life on dry land. Seeds also enable the embryo to survive dry or cold periods of the year. The integument protects the embryo from desiccation. The pollen tube delivers sperm directly to eggs and eliminates the need for fresh water for sexual reproduction.

Living gymnosperms are related to extinct plants from the Paleozoic and Mesozoic eras (pp. 465–466)
Progymnosperms arose in the Middle Devonian from phylum Trimerophytophyta. The two main groups of progymnosperms produced coniferlike wood but had spores instead of seeds. In

the Carboniferous, progymnosperms gave rise to seed ferns, a group of taxonomically diverse plants that gave rise to cycads. Progymnosperms also may have given rise to primitive, independent lines of gymnosperms known as the Cordaitales and Voltziales. These plants lived during the Carboniferous and Permian periods and possibly gave rise to conifers, gnetophytes, and ginkgos.

In gymnosperms and other seed plants, dependent gametophytes develop within the parent sporophyte (p. 466)
Seed plants are heterosporous, with gametophytes developing endosporically. The gametophytes are dependent on the sporophyte and are quite reduced in size from those in most seedless plants.

The pine life cycle illustrates basic features of gymnosperm reproduction (pp. 466–469)
Microsporangia and megasporangia occur on separate pollen and ovulate cones. Two microsporangia occur on the lower surface of each microsporophyll. Two ovules occur on the upper surface of each ovulate scale. Each ovule contains a megasporangium (nucellus), in which the megagametophyte develops. After fertilization of the egg, the ovule becomes a seed, consisting of the embryo, food supply (former megagametophyte), and seed coat. The germinating seed develops into the independent sporophyte.

Types of Living Gymnosperms

Phylum Coniferophyta contains conifers, which are the dominant forest trees in cooler climates (pp. 470–474)
Living conifers consist of 550 species, most commonly found in cooler climates in the Northern Hemisphere. They contain the tallest and largest living trees. Conifer needles may be deciduous or evergreen and can remain active for up to 50 years. They have a number of adaptations facilitating survival in cold, windy environments.

Phylum Cycadophyta contains cycads, which resemble tree ferns or palms (pp. 474–475)
The 140 species of cycads are remnants of many more species that lived in the Mesozoic era. Their cones are usually larger than those of conifers, and their trunks are covered with scaly leaves.

Phylum Ginkgophyta contains one living species (pp. 475–476)
Ginkgo is a deciduous tree with characteristic fan-shaped leaves that survived in Chinese monasteries for centuries after it became extinct in the wild. The trees grow well in polluted city environments.

Phylum Gnetophyta contains three diverse genera found in tropical forests or in deserts (pp. 476–477)
The 70 species of gnetophytes include three genera. The genus *Ephedra* resembles shrubs composed of short green sticks. They

are common in many desert regions of the western United States. The genus *Gnetum* includes tropical plants found mostly in Asia and Africa. Their broad, leathery leaves resemble those of some flowering plants. *Welwitschia,* native to arid coastal deserts of southwest Africa, produces a pair of long, straplike leaves.

Review Questions

1. Explain how seeds and pollen tubes provided advantages in adaptation to life on land.
2. Why do most seed plants have nonflagellated sperm?
3. Explain the difference between exosporic and endosporic development.
4. Describe in general what is known about the evolution of gymnosperms, explaining how that knowledge is incomplete.
5. Draw a very simple, general life cycle showing alternation of generations in a gymnosperm.
6. Describe the production of sperm and eggs by a pine tree. How does fertilization take place?
7. In what ways is the pine life cycle representative of most gymnosperms? How do some gymnosperms differ from pines in their reproduction?
8. Describe some variations in cone structure among conifers.
9. How would you describe a cycad to someone who has not seen one?
10. What are some human uses of the *Ginkgo biloba* tree?
11. In what ways are the gnetophytes considered to be the gymnosperms that are most similar to angiosperms?
12. Describe *Ephedra* and *Welwitschia* to someone who has not seen either plant.

Questions for Thought and Discussion

1. From what you know of plants, why do you think that gymnosperms still have the selective advantage over most other plants in cool, windy, mountainous regions?
2. Compare the advantages and disadvantages of exosporic and endosporic development.
3. Producing leaves each year requires a large expenditure of energy. Explain why all plants do not retain their leaves for at least several seasons as conifers do.
4. People living in cold, windy areas are advised to water their trees in the winter when they are not growing and may have no leaves. Why?
5. In many spruce trees, the ovulate cones hang down, but in many firs they point up. What might be the selective advantage of each positioning?
6. Draw a labeled diagram of an ovule as seen in longitudinal section. What is the developmental origin of each component of the ovule you've drawn?

Evolution Connection

The production of separate male and female gametophytes, which is inevitable in heterosporous plants, can occur also in homosporous plants. What evidence suggests that the seed habit evolved from heterosporous rather than homosporous ancestors? Can you think of a reason why heterospory might be advantageous over homospory in terms of both ovule and pollen production?

To Learn More

Visit The Botany Place Website at www.thebotanyplace.com for quizzes, exercises, and Web links to new information related to this chapter.

Arno, Stephen F., and Steven Allison-Bunnell. *Flames in Our Forest: Disaster or Renewal?* Washington, DC: Island Press, 2002. This book tells about why our western forests are burning and what, if anything, we should do about it.

Lanner, Ronald M. *Made for Each Other: A Symbiosis of Birds and Pines.* Oxford: Oxford University Press, 1996. Lanner is a professor of forest resources at Utah State University who became interested in how the whitebark pine reproduces. The large pinecones are closed and do not release the seeds and the seeds are wingless, yet the pine is distributed over a wide area. An entertaining and well-written account.

Taylor, Murry A. *Jumping Fire: A Smokejumper's Memoir of Fighting Wildfire.* Orlando: Harcourt Paperbacks, 2000. A fascinating account of life as a smokejumper, reflecting a perspective on forest fires, which are particularly common in the coniferous forests of the western United States.

23
Angiosperms: Flowering Plants

The monster flower (*Rafflesia keithii*), Borneo.

Sexual Reproduction in Flowering Plants

Angiosperms, like gymnosperms, have a dominant sporophyte and dependent gametophyte

Self-pollination and cross-pollination are both common in angiosperms

The Evolution of Flowers and Fruits

The selective advantages of flowering plants account for their success

Flowers have evolved as collections of highly modified leaves

The evolution of angiosperms began during the Mesozoic era

During the Cretaceous period, angiosperms spread rapidly throughout the world

A Sampling of Angiosperm Diversity

Phylum Anthophyta contains more than 450 families, classified mainly by flower structure

Several families illustrate the diversity of floral and fruit structure

More than 90% of existing plant species are flowering plants, and they represent an astounding variety. Consider four very different genera of angiosperms: petunias, sunflowers, corn, and an unusual genus of succulents.

Petunias, such as *Petunia axillaris,* are a familiar group, making a prominent appearance, with many colorful variations, each summer in gardens and pots. The blossoms attract insects that eat the nectar and inadvertently carry pollen from one flower to the next.

A sectioned sunflower.

Sunflowers, such as *Helianthus,* are also readily recognizable. Few people, however, realize that a sunflower is not one flower but many. The center of the complex "head" consists of small components called disk flowers, while parts known as *ray flowers* form the outer ring. In large sunflower species, each head supplies many fruits known as *sunflower seeds.* Although much more complex than a petunia, a sunflower still functions as a single flower in attracting pollinators.

A third example of floral diversity is corn (*Zea mays*). Many people do not think of corn as a flowering plant, but it is. The tassel at the plant's top contains many small male flowers. The strands of corn silk are the long stigmas and styles of the female flowers. The female flowers mature to form the fruits known as corn kernels. Wind carries pollen from anthers at the top of the plant to the stigma on each ear.

Corn tassels (male flowers) and corn silk (stigmas and styles of female flowers).

Ceropegia haygarthii.

A fourth example of floral diversity is *Ceropegia haygarthii,* a member of a genus of succulents native mostly to Africa. These species have elaborate containers that temporarily trap insects feeding on the nectar. Many species have specially shaped stigmas that remove pollen from an insect's mouth. Meanwhile, the stigmas produce sticky substances that attach the flower's own pollen to the insect. These processes facilitate transfer of pollen between flowers and between plants.

Why are there many different types of flowering plants and uniquely shaped flowers? In their evolution, flowering plants have successfully adapted to many different types of environments by accumulating, through random mutation and natural selection, alleles that increased the likelihood of survival. Flowering plants continue to be more successful at producing offspring in most environments than other types of plants. Every generation represents an evolutionary testing ground. Meanwhile, breeders of new varieties use artificial selection to yield improvements for human use, such as flowers

An *Arabidopsis* flower and a mutant specimen with more petals.

that are more attractive and longer lasting or food crops that are better tasting, more resistant to disease, and more productive. Genetic engineering can also play a role. For instance, recent experiments in modifying thale cress plants (*Arabidopsis thaliana*) to produce more petals are providing insights into how flowers grow and may lead to applications in the gardening and landscaping industries.

Chapter 6 discussed the general structures and varieties of flowers and fruits, as well as pollination and methods of seed dispersal. Here we will look at the life cycle and pollination of flowering plants in more detail. Then we will examine angiosperm evolution and classification before looking at some of the more than 450 families as examples of diversity in structure and environmental adaptation.

Sexual Reproduction in Flowering Plants

As in all plants, the sexual life cycle of flowering plants involves alternation of generations between sporophytes and gametophytes. In bryophytes, the gametophyte is dominant, with an attached, dependent sporophyte. In seedless vascular plants, the sporophyte and gametophyte are typically separate plants, with the sporophyte being larger and always photosynthetic. In gymnosperms and angiosperms, sporophytes are larger, photosynthetic, and dominant. Specialized reproductive structures on the sporophyte are grouped together and develop from modified leaves. Gametophytes develop endosporically within microspores and megaspores, and the megagametophyte eventually becomes part of a seed. Unlike gymnosperms, however, the ovules of flowering plants are inside ovaries that develop into fruits.

Angiosperms, like gymnosperms, have a dominant sporophyte and dependent gametophyte

In angiosperms, the alternation of generations found in gymnosperms is further modified, and the gametophytes are further reduced in size and in cell number. In gymnosperms, the immature microgametophyte, or pollen grain, has four cells when it is shed from the microsporangium. In flowering plants it has only two or three cells when released from the anther. In gymnosperms, the mature microgametophyte has six cells, including two sperm. In flowering plants, it has only three cells, including two sperm. The megagametophyte is reduced to a much greater extent. In gymnosperms it consists of several hundred to several thousand cells, often including archegonia, as in pine. In flowering plants, it most often has eight nuclei and seven cells, and no archegonia are present. In some species, it consists of as few as four cells. Because angiosperm gametophytes, particularly the megagametophyte, are smaller than those in gymnosperms, less energy is expended to produce them, which could be a selective advantage. That is, it may increase the likelihood of a plant surviving through natural selection.

In most angiosperm species, male and female gametophytes occur not only on the same plant but also on the same structure: a bisexual (perfect) flower. Bisexual flowers have stamens (producing microgametophytes) and carpels (producing megagametophytes). Species that have unisexual (imperfect) flowers, lacking either stamens or carpels, are usually monoecious, with each plant having both staminate (male) and carpellate (female) flowers. However, some angiosperm species are dioecious, having staminate and carpellate flowers on separate plants. By comparison, the male and female gametophytes in gymnosperms are always on separate structures: pollen cones (male cones) and ovulate cones (female cones), with most species having both types of cones on the same plant.

Figure 23.1 shows the life cycle of a typical flowering plant species that has bisexual flowers. As with all plant life cycles, the key stages of reproduction are meiosis and fertilization, with growth occurring by mitosis throughout the life cycle. Beginning at the upper right, you can see how a stamen produces a pollen grain, or microgametophyte. Each stamen has one anther, each typically containing four microsporangia, also known as *pollen sacs*. Within each microsporangium, each microsporocyte (microspore mother cell) undergoes meiosis and thereby produces four haploid microspores. Each microspore then grows into an immature pollen grain consisting of a tube cell that will produce the pollen tube and a generative cell that will produce two sperm. Before being shed from the anther, each pollen grain develops a protective outer wall called the *exine* and an inner wall called the *intine*. The outer wall is made of a resistant polymer known as *sporopollenin,* found in all plant spores. In each angiosperm species, the outer wall of the pollen grain has a characteristic pattern. That is why scientists can analyze pollen grains to identify a plant species causing a particular allergy or determine that a certain species is associated with archaeological remains. Final development of the mature pollen grain includes growth of the pollen tube, which occurs after the pollen grain is deposited on a receptive stigma of a flower.

Now we will look at the development of the megagametophyte, tracing the stages shown in the center of Figure 23.1. In a flowering plant, the megasporangia are found inside the ovary, which contains one ovule or multiple ovules. Each ovule consists of a megasporangium, also called a *nucellus,* which is surrounded by one or two integuments. The integuments meet at the micropyle, the opening where the pollen tube will penetrate. Within each megasporangium, a megasporocyte (megaspore mother cell) undergoes meiosis and forms four haploid megaspores. Three of the megaspores usually disintegrate, with the one farthest from the micropyle surviving to grow into the megagametophyte. In about two-thirds of angiosperm species, the developing megagametophyte produces eight nuclei. Four nuclei congregate at each end, and then one nucleus from each end migrates to the cen-

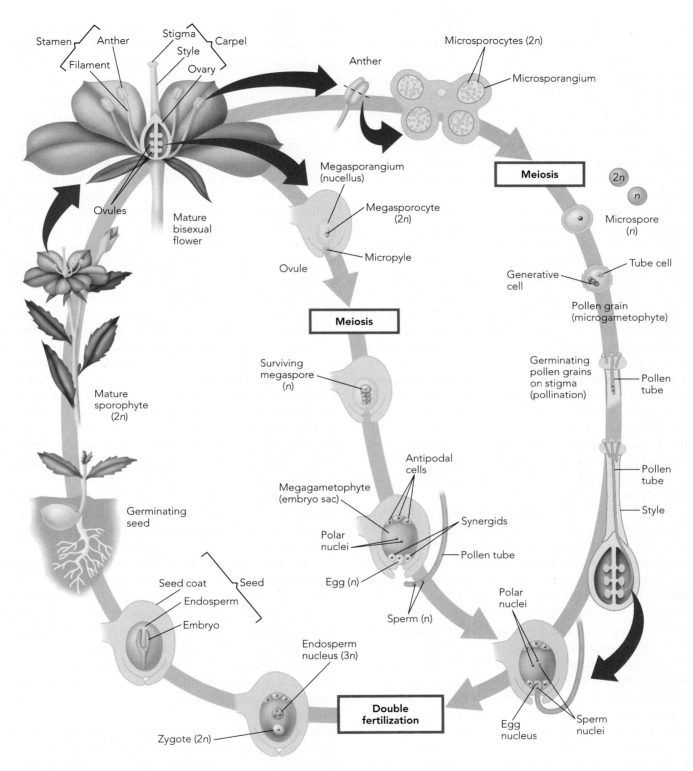

Figure 23.1 Life cycle of a typical flowering plant. Since the diagram focuses on generic flower and seed structure, it does not show the fruit structure surrounding the seed.

ter. The two nuclei that move to the center are known as *polar nuclei.* The three remaining nuclei at each end form cells, while a large central cell contains the polar nuclei. At the end near the micropyle, the middle cell is the egg, flanked by two short-lived cells called *synergids,* which will function in the process of fertilization. The three cells at the opposite end, known as *antipodal cells,* have no known function. The mature megagametophyte, which consists of seven cells and eight nuclei, is also known as an *embryo sac* because the embryo will develop there after fertilization.

In most angiosperm species, the megagametophyte is fully developed before pollination. The stigma of a typical flower may receive pollen from a variety of plants of the same species and even from a variety of species. However, various proteins and other molecules ensure that only pollen from "acceptable" plants can germinate. The surface of the stigma contains calcium ions necessary for pollen germination, as well as hormones that stimulate growth of the pollen tube. After the pollen grain comes in contact with the stigma, the tube cell elongates and produces a pollen tube, which typically grows over a period of hours to a few days. In contrast, the process takes more than a year in most gymnosperms, such as pines. The tube grows through so-called *transmitting tissue* in the style and into one of the synergids, which may produce substances that attract the growing pollen tube. The two nonflagellated sperm pass through the pollen tube and into the synergid.

Next, a unique form of double fertilization takes place. One sperm moves from the synergid and fertilizes the adjoining egg, resulting in a zygote that will develop into the embryo. The second sperm combines with the two polar nuclei in the center of the embryo sac, forming a triploid endosperm nucleus. This triploid nucleus divides by mitosis, giving rise to the endosperm that nourishes the developing embryo. In endosperm development of some species, a period of nuclear division precedes cytokinesis. In other species, the nuclear and cell divisions occur simultaneously. In a few angiosperms, megasporangium cells divide and produce a diploid nutritive tissue known as *perisperm.* In any case, in dicots the resulting tissue is digested and absorbed by the embryo as it expands in size within the embryo sac. In monocots, endosperm is not used during formation of the embryo but serves instead as a food supply for the germinating seedling. Cereal grains are important as food because of their abundant starchy endosperm.

Only angiosperms have a form of double fertilization that produces both an embryo and endosperm. Another type of double fertilization occurs in the two genera of gymnosperms that are most similar to angiosperms: *Ephedra* and *Gnetum.* In these gymnosperms, the second fertilization produces additional embryos rather than endosperm.

Self-pollination and cross-pollination are both common in angiosperms

Some angiosperm species rely either mainly or exclusively on self-pollination, often referred to as *selfing.* In self-pollination, a flower is pollinated by its own pollen or by pollen from another flower on the same plant. Since self-pollinating plants do not require pollinators or other plants for successful reproduction, they can establish themselves in satellite areas isolated from the general population, which can sometimes result in a more rapid expansion of the range of a species. However, some disadvantages of self-pollination are reduced genetic variability and the possibility that some seeds may not be viable because of the pairing of harmful recessive alleles.

Many angiosperm species rely either mainly or exclusively on cross-pollination, the transfer of pollen from one plant to another. Cross-pollination reduces the chance that harmful recessive alleles will end up paired in the same organism. By far the most important advantage of cross-pollination is the genetic diversity that results from mixing genotypes with different alleles at fertilization. Cross-pollination increases the number of genetic combinations in a population and therefore the likelihood that plants can successfully respond to environmental change. Also, it gives rise to hybrid vigor. For example, crossing purebred lines of corn produces hybrid offspring with larger, stronger plants and higher yields.

Cross-pollination is enforced by various mechanisms, depending on whether the species has bisexual or unisexual flowers. Among species that have unisexual flowers, dioecious species such as willows and date palms require cross-pollination because the male and female flowers are on separate plants. Even a species that is monoecious may rely on cross-pollination if the male and female flowers on the plant develop at different times of the growing season. Some examples are cucumbers, corn, maples, and oaks. Even many species that have bisexual flowers, such as apples and most sweet cherries, require cross-pollination, either because the male and female gametophytes develop at different times or because of self-incompatibility, a plant's ability to reject its own pollen. Self-incompatibility is the most common mechanism enforcing cross-pollination.

Some angiosperm species rely solely on wind or water to transfer pollen, whether by self-pollination or cross-pollination. Wind-pollinated species typically have small

BIOTECHNOLOGY

Superweeds

Pollen from one plant may be transferred to the stigmas of many different plants. Scientists worry that pollen of transgenic crops containing herbicide-resistant genes might transfer that resistance to wild relatives that commonly grow near cultivated fields. Such worries are valid because some plants can readily hybridize with related wild species. Scientists have already demonstrated that many types of genes are gradually transferred from crop plants to wild relatives, which could become herbicide-resistant "superweeds."

A related problem is that crop plants can become weeds. For example, canola is a crop grown in the western United States and Canada to produce cooking oil. However, herbicide-resistant canola becomes a weed for a farmer cultivating a different crop.

Canola is an example of a crop that can infiltrate other crop fields.

Superweeds resistant to herbicides are just one example of problem weeds that could result from cultivating useful transgenic plants. Drought tolerance, disease and pest resistance, frost tolerance, and even increased yields might be transferred, over time, to weedy plants.

Superweeds can be avoided by using cultivation, instead of herbicides, to remove weeds both in and around fields of crops. The likelihood of superweeds taking over a field can also be minimized by switching the type of herbicide used every year or two so that herbicide-resistant weeds do not become established. Crop rotation also reduces the possibility of growth of herbicide-resistant weeds.

---

flowers that are not colorful and that have no nectar or odor, with stamens thrust out, allowing pollen to be easily caught by the wind. Many grasses, such as corn, and some oak trees are examples of reliance on wind pollination. One concern about growing genetically modified crop plants, especially those that are wind-pollinated, is that pollen-carrying modified genes can easily escape from cultivation, as discussed in the Biotechnology box on this page. In most aquatic plants, pollination occurs above water and is facilitated by the wind or by insects. However, in some species the pollen floats on the surface, and other species, such as eel grass (species of *Zostera*), have adaptations allowing underwater pollination. Eel grass pollen grains are long and threadlike, which increases the likelihood of contacting a receptive stigma.

Most angiosperm species rely on pollinators such as insects, birds, and bats. Interaction with pollinators developed early in angiosperm evolution. You can imagine how it might have begun. As a result of chance mutations, plants produced brightly colored leaves near microsporangia. The bright colors attracted insects, which occasionally picked up pollen grains that were transmitted to other plants. Eventually, random mutations produced colorful flowers that sometimes had sweet or nutrient-rich nectars, which attracted animals such as insects. When feeding on nectar, they carried pollen from one plant to another, thereby becoming pollinators. Plants that were more brightly colored or produced sweeter,

richer nectar might be visited more frequently by insects and other pollinators. Accordingly, pollen containing the alleles that produced such characteristics might participate in more pollinations, thereby increasing the frequency of those alleles in the population.

We will probably never know exactly how pollinator–plant interactions began. Once initiated, however, any mutation that increased the frequency of pollinator visits to a flower would have had a selective advantage. Some angiosperm species are specialized for pollination by a particular type of pollinator, such as bees, or by a single species of pollinator. Others can be pollinated by more than one species or even by different types of pollinators. Meanwhile, pollinators rely on these plants for food. Over time, natural selection strengthened such mutually beneficial relationships, as the reproductive success of both plant and animal depended on their interaction. Such developments are examples of **coevolution** of different species, in which adaptations in one species have a selective effect on adaptations in another species.

Insects, birds, and bats are frequent participants in coevolution with plants. Flowers pollinated by birds produce a lot of nectar and are usually large, odorless, and frequently red, a color that most insects cannot see. Some examples are cacti, banana, many orchids, and poinsettia. Bats favor large, sturdy, nectar-rich flowers that bloom at night and often have wide corolla tubes. Examples are tropical plants such as mangos and bananas, and desert plants such as the *Agave* genus (Figure 23.2a). Bees and butterflies tend to pollinate brightly colored flowers, often with blue or yellow petals and distinct floral markings. Examples are foxglove, alfalfa, and clover. Flies tend to

(a) A bat pollinating a saguaro cactus.

(b) A syrphid fly pollinating a daisy.

(c) An ant pollinating a forget-me-not.

Figure 23.2 Pollinators carry pollen from plant to plant. In the coevolution of flowers and pollinators (mostly insects), flower shape, smell, color, and food value have evolved in conjunction with specific pollinators.

pollinate flowers that have strong, putrid odors, such as the Indonesian corpse flower you saw at the beginning of Chapter 9. Such flowers are sometimes called "carrion flowers" and include many members of the milkweed family (Asclepiadaceae), as well as some orchids, daisies, and lilies (Figure 23.2b). Beetles typically pollinate large flowers with strong odors that are often spicy, yeasty, or foul. Examples are magnolias, carrion flowers, and some poppies. Ants often pollinate plants with sugary secretions like cacti or even trees (Figure 23.2c). Moths are usually attracted to flowers that bloom late in the day or at night and have elongated corolla tubes with a heavy, sweet fragrance. Examples are tobacco, evening primrose, and many desert plants.

Pollination by animals is a more efficient method of delivering pollen to the stigma than wind pollination because pollinators carry pollen directly from one plant to another. In contrast, wind dispersal is typically random. Wind is unpredictable in intensity and can be highly directional, restricting both the extent and direction of pollination, which is why it works best in dense populations. Pollinators often travel a considerable distance during their daily activities, thereby facilitating the spread of pollen into areas not reached by the wind. Also, in plants with bisexual flowers, a pollinator can frequently deliver and acquire pollen at the same time.

Section Review

1. In flowering plants, how are sperm and eggs produced and then brought together? Compare this process with the process in gymnosperms.
2. Is cross-pollination more effective than self-pollination? Explain.
3. Explain how coevolution plays a role in angiosperm reproduction. What effect do you think it has on angiosperm diversity?

The Evolution of Flowers and Fruits

The phylum Anthophyta contains approximately 250,000 named living species, compared with only about 760 gymnosperm species, about 12,000 species of seedless vascular plants, and about 15,000 bryophyte species. However, if you could return to the world of 200 million years ago, you would find well-developed vegetation containing no flowering plants. Angiosperms do not appear in the fossil record until around 130 to145 million years ago, but they spread quickly because their distinctive features gave

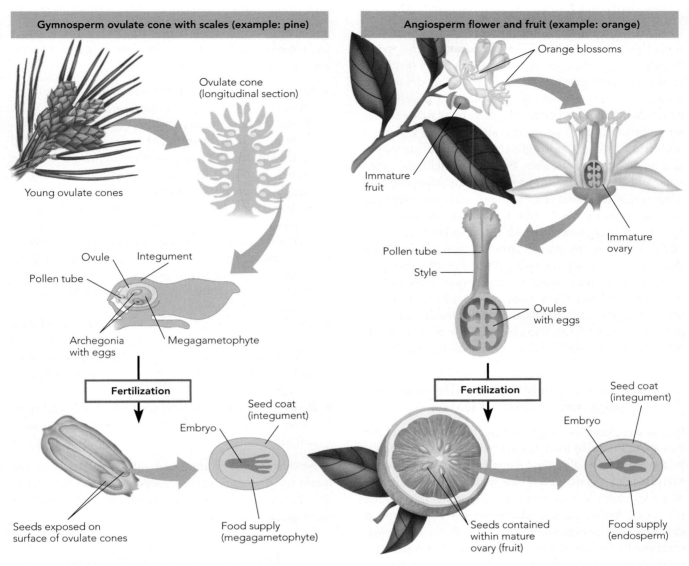

Figure 23.3 Comparison of seed origin and location in gymnosperms and angiosperms.

them significant selective advantages over other plants in many environments.

The selective advantages of flowering plants account for their success

Like gymnosperms, angiosperms have seeds and pollen tubes that facilitate survival and reproduction on land. Seed coats protect the developing embryo from desiccation, while pollen tubes allow fertilization without the fresh water needed by flagellated sperm. However, in angiosperms the eggs, and later seeds, have additional protection. As you may recall, the term *angiosperm* comes from the Greek words *angion* ("container") and *sperma* ("seed"). The name describes one of the new features of

flowering plants. Whereas gymnosperm seeds are exposed on the surface of modified branches, seeds of flowering plants are enclosed in a container called the *ovary*, which is the base of a modified leaf called a *carpel* (Figure 23.3). A fruit is typically a mature ovary. Thus, in angiosperms the developing embryo is protected from desiccation, disease, and herbivores not only by the seed coat but also by the surrounding tissues of the fruit. However, not all scientists agree that embryo protection is the chief selective advantage of having ovaries. For example, paleobotanist David Dilcher of the University of Florida suggests that the protection of eggs within immature ovaries may have arisen to prevent self-pollination in bisexual flowers.

In addition to the protective role of ovaries, angiosperms have other selective advantages. Some traits, such as vessel elements and deciduous leaves, relate to efficient use of water. Vessel elements allow more efficient conduction of water than tracheids. The deciduous habit, which appears in only a few gymnosperms, is present in many species of flowering plants, providing a way of surviving dry or cold seasons. Flowers provide an advantage in distributing pollen. While gymnosperms rely solely on wind or beetles for pollination, angiosperms also attract a variety of other pollinators through the bright colors and alluring shapes and scents of many flowers. Pollinators facilitate cross-pollination, which prevents or reduces inbreeding and may allow widely separated plant populations to exchange and recombine their genes. Double fertilization leads to development of the endosperm, which nourishes embryos. Fruits often aid in seed dispersal by attracting animals. Seeds are then scattered during the eating process or after they pass through the animal's digestive system. Some fruits, such as those of cocklebur (*Xanthium* species), have spines or hooks that attach to animals, thereby facilitating seed dispersal (see Figure 6.13). The very existence of seeds allows plants to survive periods that are unsuitable for growth.

Flowers have evolved as collections of highly modified leaves

In vascular plants, most spore-producing structures have evolved as modified leaves called *sporophylls.* Sporophylls first evolved in seedless vascular plants. In living seedless vascular plants, sporangia typically appear on the surface of the sporophyll. In gymnosperms, sporangia developed within sporophylls and modified branches that evolved as scales, usually organized into cones. In angiosperms, sporophylls evolved differently from those of other vascular plants, in both structure and organization. Sporophylls containing microsporangia evolved into stamens, while sporophylls with megasporangia evolved into carpels. Sepals and petals evolved as nonfertile modified leaves associated with sporophylls.

Depending on the angiosperm species, a flower may lack one or more of the four types of modified leaves: stamens, carpels, sepals, and petals. However, all flowers have at least one type of sporophyll—stamens or carpels. In contrast with indeterminate leafy shoots, which are long-lived and continue to produce leaves, all flowers are determinate shoots modified for reproduction. That is, when a flower reaches maturity, it stops growing and produces, or "sets," seeds. While each ovule within the ovary of a carpel develops into a seed, the stamens, sepals, and petals are typically shed. Meanwhile, each ovary—often fused with other ovaries and sometimes other flower parts—develops into a fruit that encloses the seed or seeds.

By studying fossils, paleobotanists have formed hypotheses about the evolution of stamens, carpels, sepals, and petals as modified leaves. The following major trends are evident as angiosperms evolved:

◆ Stamens and carpels become less leaflike, with carpels often fusing to form compound carpels. Figure 23.4a illustrates a hypothesis for the evolutionary development of stamens, from microsporangia being located on the surface of a flat sporophyll to being part of an anther. Figure 23.4b shows a possible evolution of carpels, from megasporangia being located on a sporophyll surface to being enclosed in an ovary of a carpel.

◆ Sepals and petals evolve from being very similar, or even identical, to being more distinct in appearance.

◆ The number of floral parts becomes fixed and often reduced. In many major groups of angiosperms, for example, stamens are reduced from many to four, five, or multiples of three.

◆ The arrangement of floral parts evolves from being spiral, similar to cone scales, to being whorled. A whorl consists of three or more parts attached to the same node. The number of whorls is reduced from four to three, two, or one.

◆ Radial symmetry of flowers gives way to bilateral symmetry in a number of independent evolutionary lines (Figure 6.8).

Next, we will look at the general characteristics of the major types of angiosperms. As you will see, characteristics that are primitive, or ancestral, are still present in some species.

The evolution of angiosperms began during the Mesozoic era

In the previous chapter, you learned that the earliest seed plant fossils date to approximately 365 million years ago, during the late Devonian period. Angiosperms first appear in the fossil record of the early Cretaceous period, around 142 million years ago (Figure 23.5). However, some angiosperm traits appear in fossils that are up to 200 million years old, and RNA and DNA sequence data suggest that the lineage giving rise to flowering plants may have been separate from other seed plants for at least 280 million years. This molecular evidence receives support from chemical analyses of rock strata, revealing the presence of the organic compound oleanane in rock deposits dating from 290 to 235 million years ago. Oleanane, a compound that acts as defense against insects, is produced by flowering plants but not by gymnosperms.

(a) Evolution of stamens

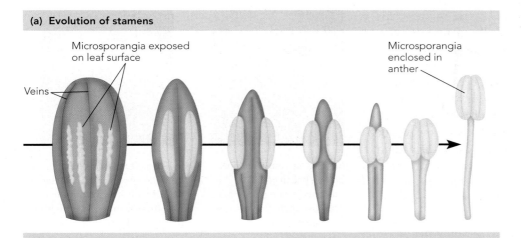

Veins

Microsporangia exposed on leaf surface

Microsporangia enclosed in anther

(b) Evolution of carpels

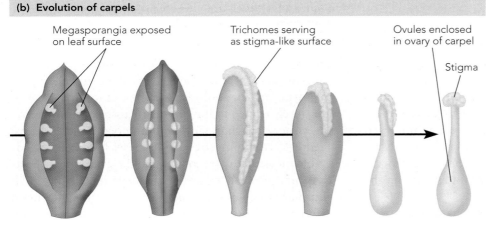

Megasporangia exposed on leaf surface

Trichomes serving as stigma-like surface

Ovules enclosed in ovary of carpel

Stigma

Figure 23.4 A hypothesis for the evolution of stamens and carpels. (**a**) Stamens may have evolved as exposed microsporangia became enclosed and the lower portion of the leaf became narrower, evolving into a filament. (**b**) Carpels may have evolved as leaves folded in, enclosing the megasporangia. The stigma of the carpel may be the result of a reduction of a larger stigmatic surface.

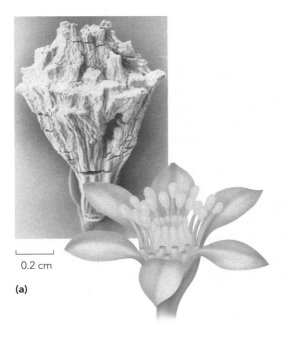

0.2 cm

(a)

(b)

1 cm

Figure 23.5 Fossilized flowers.

(**a**) The SEM micrograph shows a water lily from the early Cretaceous period of Portugal, about 130 million years ago. The SEM micrograph is accompanied by an artist's reconstruction of the flower.

(**b**) At 142 million years old, this fossil flowering plant, found in China, is the oldest on record. Note the leaf-like seed pods.

The evolutionary relationships between angiosperms and gymnosperms remain unclear. The groups of gymnosperms that are most similar to flowering plants are the Bennettitales and the gnetophytes (see Figure 22.2). The Bennettitales, an extinct group of cycad-like plants, have reproductive structures that are somewhat similar to flowers. Their external bracts, reminiscent of sepals or petals, are folded up beside sporophylls or on top of sporophylls. The gnetophytes, which include the genera *Ephedra, Gnetum,* and *Welwitschia,* have several traits shared by flowering plants. These include the presence of vessel-like structures in the xylem, the lack of archegonia (in *Gnetum* and *Welwitschia*), and similarities between their strobili and the inflorescences of some primitive flowering plants.

The fossil record for flowers is not particularly instructive as to the course of angiosperm evolution. Critical stages in the evolution of angiosperms are not yet adequately represented in the known fossil record. The species of living flowering plants are so numerous and so diverse that sorting them into a phylogenetic classification has been challenging. Systematists have found that the classifications of earlier taxonomists are generally accurate but that convergent evolution has caused occasional significant errors, particularly in classifying the early evolution of the phylum.

Pollen grain structure is one of the characters (traits) used in tracing angiosperm evolution. Like some gymnosperms and all monocots, the pollen grains of the most primitive angiosperms have a single germination opening, or aperture, and are therefore called *monoapertuate* (Figure 23.6a). In contrast, all eudicots—the vast majority of angiosperms—have triapertuate pollen grains, which facilitate pollination by providing three paths for the pollen tube to emerge (Figure 23.6b). If the aperture is shaped like a slit, the pollen is known as *colpate,* as in *tricolpate.*

The real breakthrough in classifying angiosperms has come with the use of molecular data (Chapter 16). The data from DNA and RNA sequencing support the hypothesis that the Bennettitales, the gnetophytes, and the angiosperms all had closely related ancestors. Still, the nature of these ancestors remains a subject of debate.

Traditionally, the phylum Anthophyta has been divided into two main classes: monocots and dicots. Species in which the embryo has one cotyledon have been classified as *monocots,* and those having embryos with two cotyledons have been called *dicots.* This distinction remains useful in broadly distinguishing types of angiosperms. However, recent molecular studies have revealed that dicots are not monophyletic, representing instead several lines of evolution. The vast majority of

 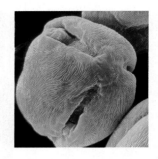

(a) Single germination openings on pollen grains are characteristic of gymnosperms, primitive angiosperms, and monocots.

(b) Three germination openings on pollen grains are characteristic of eudicots, a group that contains the majority of living flowering plants.

Figure 23.6 Pollen grain germination openings are a trait used in tracing evolution.

dicots are now known as *eudicots.* Molecular comparisons indicate that the other groups of dicots–called basal angiosperms, magnoliids, and the Ceratophyllaceae— are closely related to the earliest angiosperms, while having some similarities with both monocots and eudicots. Figure 23.7a summarizes a hypothesis, based on considerable molecular data, concerning the evolutionary relationships between groups of angiosperms. Figure 23.7b summarizes major characteristics of the four main groups: basal angiosperms, magnoliids, monocots, and eudicots.

Basal angiosperms consist of several families of herbs and woody shrubs. Although not monophyletic, they are grouped together as the most primitive flowering plants. Most are extinct, with their surviving relatives constituting about 0.5% of living angiosperm species. The extinct ancestors of basal angiosperms are thought to include the first angiosperms, which gave rise to all others. Living basal angiosperms, which occur mostly in the Tropics, have flowers that are insect-pollinated, bisexual, and radially symmetrical, typically with a spiral arrangement of flower parts, including sepals and petals that look alike. They share three important traits regarded as primitive in angiosperms. First, their pollen grains have a single opening. Second, their carpels form a tube with the edges sealed by secretions, whereas in most other angiosperms the carpel is folded lengthwise in the middle like a fan, with the edges fused by a continuous layer of epidermal cells. Third, the stigma extends down the side of the carpel where the edges meet, rather than being restricted to the top of the carpel, as in most angiosperms. In addition to pollen and carpel structure, several groups of basal angiosperms differ from most angiosperms in lacking vessels or having vessels that resemble tracheids.

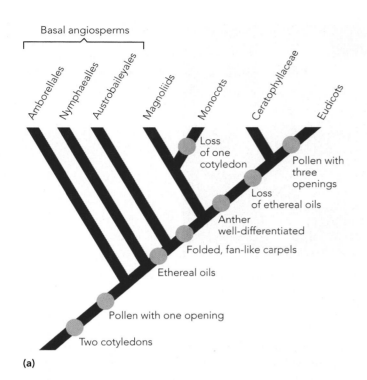

Basal angiosperms

Amborellales
Nymphaeales
Austrobaileyales
Magnoliids
Monocots
Ceratophyllaceae
Eudicots

Loss of one cotyledon
Pollen with three openings
Loss of ethereal oils
Anther well-differentiated
Folded, fan-like carpels
Ethereal oils
Pollen with one opening
Two cotyledons

(a)

Figure 23.7 An overview of angiosperms. (a) This diagram reflects possible evolutionary relationships of major groups of angiosperms, based on analysis of nucleotide sequencing. About 97% of angiosperm species are either monocots or eudicots. **(b)** A comparison of major groups of angiosperms.

	Basal angiosperms (about 0.5% of angiosperm species) Examples: *Amborella,* water lily family	**Magnoliids** (about 2.5% of angiosperm species) Examples: laurel family, magnolia family, black pepper	**Monocots** (about 28% of angiosperm species) Examples: grasses, palms, orchids, lilies, irises	**Eudicots** (about 69% of angiosperm species) Examples: roses, grapes, oaks, apples, peas
Flowers	Usually spiral arrangement	Usually spiral arrangement	Usually whorled arrangement	Usually whorled arrangement
	Often numerous floral parts	Few to numerous floral parts	Floral parts usually in multiples of three	Floral parts usually in multiples of four or five
	Carpels sealed by secretions	Carpels sealed by cells	Carpels sealed by cells	Carpels sealed by cells
	Stigma often extends down the carpel	Stigma sometimes extends down the carpel	Stigma reduced to tip of carpel	Stigma reduced to tip of carpel
	Carpels tubelike	Folded carpels	Folded carpels	Folded carpels
	Anthers and filaments poorly differentiated	Anthers and filaments poorly differentiated	Anthers and filaments frequently well differentiated	Anthers and filaments well differentiated
Pollen (see Figure 23.6)	One opening	One opening	One opening	Three openings
Seeds (see Figure 3.11)	Two cotyledons	Two cotyledons	One cotyledon	Two cotyledons
Leaf vein pattern (see Figure 4.19)	Usually netlike	Usually netlike	Usually parallel	Usually netlike
Vascular bundles in stem (see Figure 4.10)	Usually in a ring	Usually in a ring	Scattered arrangement	Usually in a ring
Root system (see Figure 4.1)	Usually taproot	Usually taproot	Usually fibrous	Usually taproot

(b)

Figure 23.8 Basal angiosperms. Six families of living basal angiosperms are closely related to the earliest flowering plants.

(a) *Amborella trichopoda* is the living plant most closely related to the earliest angiosperms.

(b) *Nymphaea,* a member of the water lily family, has numerous stamens, a primitive basal angiosperm characteristic.

Living basal angiosperms include members of the orders Amborellales, Nymphaeales, and Austrobaileyales. The only living member of the order Amborellales is *Amborella trichopoda,* a shrub found only on the South Pacific island of New Caledonia (Figure 23.8a). Molecular analysis has shown that *Amborella* belongs to an evolutionary line that emerged before the other basal angiosperms, making it a descendent of the most primitive angiosperms. Among other primitive traits, *Amborella* has no vessels and has pollen with a single, poorly defined opening. The order Nymphaeales includes the water lily family (Nymphaeaceae). The leaf structure of water lilies has been modified by convergent evolution for watery environments, but the floral structure is similar to most other basal angiosperms in having many stamens and also tubular carpels with edges sealed by secretions (Figure 23.8b). The order Austrobaileyales includes some families of tropical shrubs and herbs. A group of investigators in China and the United States have recently discovered fossils of a family of extinct basal angiosperms, Archaefructaceae, which may represent a fourth order.

Magnoliids are a monophyletic group of about 20 families. Unlike basal angiosperms, they have carpels sealed by cells, rather than the more primitive tubelike carpels sealed by secretions. However, magnoliids also have primitive traits, such as spirally arranged flower parts, often numerous stamens with poorly differentiated filaments and anthers, numerous carpels, and pollen grains with only one opening. Magnoliids make ethereal oils, which are responsible for the scents of nutmegs and laurel leaves. Ethereal oils also occur in monocots but not in most di-

cots. The magnoliids probably arose from a common ancestor with basal angiosperms around 130 million years ago, and living groups include both woody and herbaceous plants. Some examples of woody magnoliids are shrubs and small trees that constitute the laurel family (Lauraceae) and the magnolia family (Magnoliaceae) (Figure 23.9a). Some examples of herbaceous magnoliids are the pepper family (Piperaceae) (Figure 23.9b). Herbaceous magnoliids and basal angiosperms are sometimes loosely referred to as paleoherbs ("ancient herbs"). Overall, the magnoliids constitute about 2.5% of living angiosperm species.

Some scientists have suggested that flowering plants should be called phylum Magnoliophyta instead of phylum Anthophyta. However, this suggestion stems partly from the view that magnoliids are basal to other angiosperms. While magnoliids retain primitive pollen and vessel structure, recent molecular evidence clearly indicates that they are not basal in angiosperm evolution. As shown in Figure 23.7a, magnoliids share a common ancestor with monocots, the Ceratophyllaceae, and eudicots. However, this ancestor is not shared with basal angiosperms.

Like basal angiosperms and magnoliids, monocots have pollen with one opening. Unlike all other angiosperms, however, monocots have embryos with one cotyledon and typically have leaves with parallel rather than netted venation, stems with scattered vascular bundles, a fibrous root system, and floral parts in multiples of three. Monocots, along with the magnoliids, arose around 125 to130 million years ago from the common ancestors of eudicots, which were also probably the ancestors of the

(a) *Michelia figo*, a member of the magnolia family, is an example of a woody magnoliid.

(b) *Piper nigrum*, a member of the *Piper* (pepper) genus, is an example of a herbaceous magnoliid.

Figure 23.9 Magnoliids, consisting of 20 families of dicots, represent about 2.5% of living species of angiosperms. They are a monophyletic group that arose after basal angiosperms.

Ceratophyllaceae. About 28% of the living angiosperm species are monocots.

The only living members of the Ceratophyllaceae are the genus *Ceratophyllum*. These aquatic plants are reduced and simplified for life submerged in the water. They have no roots, and the small leaves lack stomata, relying instead on diffusion for passage of gases. Molecular sequencing studies do not agree on the placement of these plants in relation to the other groups of angiosperms.

The vast majority of angiosperms have embryos with two cotyledons, with the largest group among them now being known as the *eudicots* ("true dicots"), constituting around 69% of living angiosperm species. They are unique among angiosperms in having pollen with three openings. Unlike all other angiosperms except the Ceratophyllaceae, eudicots consistently have stamens with well-differentiated filaments and anthers, and most eudicots lack ethereal oils. Their floral parts are often in multiples of four or five. Eudicots arose from a common ancestor of monocots, magnoliids, and Ceratophyllaceae. This ancestor lived around 125 to130 million years ago.

On the basis of anatomical, developmental, and molecular evidence, systematists now divide the more than 150,000 species of eudicots into two principal groups: basal eudicots and core dicots. Basal eudicots, so called because they were the first eudicots, contain about a dozen main families in the orders Ranunculales and Proteales. Core eudicots contain most eudicot families and are currently divided into three groups: the carophyllids, rosids, and asterids. The carophyllids, with two orders, are a distinct clade (monophyletic group) based on characteristics of seed coats, vessel elements, pollen, and other structural traits. Some examples are cacti and carnations. The rosids include many groups that form root nodules containing nitrogen-fixing bacteria. Examples are grapes, geraniums, and violets. Molecular evidence suggests that the rosids are not monophyletic and should probably be divided into several groups. The asterids contain plants with single integuments, a thin-walled megasporangium, and frequent occurrence of chemical compounds called *iridoids*. Examples are tea, morning glorys, and sunflowers. Molecular evidence suggests that the asterids are monophyletic. Categorizing the large number of core eudicots into these groups serves as a convenient framework to divide the more than

450 families. In future years, the number of groups may increase as the taxonomy of the rosids is sorted out.

During the Cretaceous period, angiosperms spread rapidly throughout the world

The Cretaceous period began 144 million years ago and ended 65 million years ago. Angiosperms do not appear in large numbers in the fossil record until the last 30 million years of that period, a relatively short evolutionary time span. In evolutionary terms, therefore, the origin and spread of angiosperms can be described as being rather sudden, a development that Darwin himself called "an abominable mystery." We do not know exactly where the first angiosperms originated, or exactly why early angiosperms were better adapted than gymnosperms and seedless plants in many environments. The number of gymnosperm species began to decline, however, until only about 760 species remain today. The so-called primary adaptive radiation of flowering plants is similar in scope to the rapid spread of gymnosperms when they first evolved as successful seed plants.

When angiosperms first appeared around 130 million years ago, the major southern continents—Africa, South America, India, Antarctica, and Australia—were connected into the supercontinent of Gondwanaland (Figure 15.28b). A northern supercontinent called Laurasia—North America, Europe, and Asia—was connected to Gondwanaland at the northern end of Africa and through Central America. Climates ranged from extremely tropical along the equator to quite arid and cool in regions north and south of the equator. The continental linkages enabled the major families of angiosperms to spread onto a number of continents. As the supercontinents broke apart during the Cretaceous period, the variation in climates in different continents influenced the evolution of wide-ranging diversity in the adaptation of angiosperms.

The results of the primary radiation of angiosperms can be seen in the original geographic distribution of particular crop plants. Wheat, potatoes, strawberries, and other crop plants are now cultivated throughout the world. However, each crop plant arose in a particular geographic location during the primary adaptive radiation of angiosperms. Agriculture arose anywhere from 5,000 to 12,000 years ago, when people in various locations began to cultivate plants in addition to gathering from the wild. For example, wheat and barley cultivation began about 11,000 years ago in the Fertile Crescent—an area that today comprises part of Turkey, Iraq, part of Iran, Syria, Jordan, Israel, and part of Egypt—and gradually spread west and north into Europe.

Agriculture began more recently in China, India, South America, and Africa. In cultivating food crops, humans began a process of artificial selection in which seeds, seedlings, or fragments from the best plants were saved for the next planting. This process initiated a human-mediated migration of selected angiosperms, as seeds were carried long distances by trade caravans, explorers, and immigrants.

In 1916, Russian botanist N. I. Vavilov expanded the work of earlier botanists to try to identify geographic locations where crop plants were first domesticated. He looked for regions where wild relatives of cultivated plants are found and initially proposed eight major centers of crop diversity (Figure 23.10). Vavilov later expanded that number and proposed that a region with greatest genetic diversity of a species was also its center of origin. Some recent researchers have questioned that conclusion, saying that "centers of origin" may simply be where the most genetic recombination occurred. Also, the spread of crop cultivation throughout the world has made it more difficult to trace the origins of crop species. Although debate continues over the locations and numbers of centers of crop diversity, Vavilov's research and similar studies remain useful in efforts to preserve food-crop diversity. Worldwide, the three most important agricultural crops are rice, wheat, and corn. The *Evolution* box on page 496 provides an overview of their origins.

Section Review

1. **Explain how angiosperm characteristics might facilitate survival.**
2. **Describe the major evolutionary trends in flower structure.**
3. **Compare and contrast the four main groups of angiosperms: basal angiosperms, magnoliids, monocots, and eudicots.**
4. **Does the origin and spread of angiosperms remain a mystery? Explain.**

A Sampling of Angiosperm Diversity

Angiosperms are a very successful monophyletic group of plants. The fact that such a large group is monophyletic is an indication of the success of the structural and reproductive adaptations found in the phylum Anthophyta. The two largest groups of flowering plants, typically and historically designated as subphyla or classes, are monocots and eudicots. As described earlier, however, recent molecular data in-

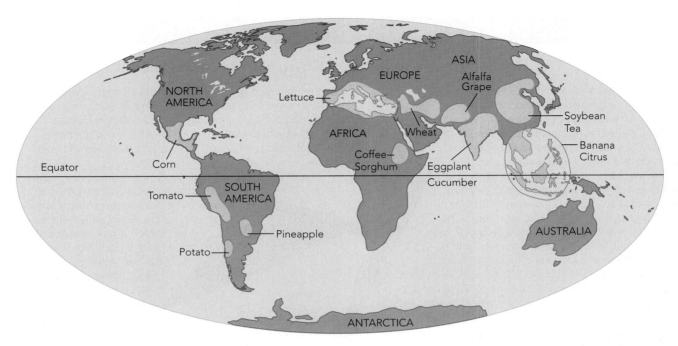

Figure 23.10 Vavilov's centers of crop diversity. This map indicates possible centers of diversity and origin of various crops, as described by Russian plant geneticist N. I. Vavilov. Selected crops listed for each center are examples of how distribution of plant species was not uniform throughout the world. Vavilov's research formed the foundation for ongoing efforts to maintain food-crop diversity.

dicate that the phylum consists of at least four main groups: basal angiosperms, magnoliids, monocots, and eudicots. As the largest group, eudicots are further divided into basal eudicots and core eudicots. Above the level of family and order, there is no agreement on the terms that should be applied to the various groups. Most plant systematists simply try to identify clades without determining whether groups are subphyla, superclasses, classes, or subclasses.

Angiosperm species are grouped into many families, each of which has characteristic traits and may be found in a number of specific habitats. Angiosperms are particularly diverse in floral and fruit anatomy, size and shape of the sporophyte, and degree of convergent adaptation to different environments. As discussed in Chapter 16, similarities may result from convergent evolution, rather than from having the same ancestor. Molecular sequencing has been a considerable aid in determining phylogenetic relationships.

Phylum Anthophyta contains more than 450 families, classified mainly by flower structure

The more than 250,000 species of flowering plants occur in more than 13,000 genera, which are grouped into more than 450 families. Families are usually classified by the structure of flowers, fruits, leaves, and stems. Biochemical characteristics, such as the presence or absence of particular alkaloids, are also important in classification. In practice, flower structure frequently plays a major role in the assignment of species to genera, genera to families, and families to orders. Members of a particular family usually share some traits related to these structures, even if convergent evolution has modified other traits.

Systematists who are experts on particular families learn to recognize many family members, whether or not they have traits easily observable to the naked eye. Systematists use three types of data to classify plants: (1) observable structural and biochemical traits; (2) traits that require microscopes or analytical equipment to determine; and (3) molecular data, which require DNA or RNA sequencing machines with accompanying computers to determine and analyze.

Each family is distinguished by a combination of characters, rather than by any single trait. However, specific characters can often be diagnostic for a family, as in the case of the square stem shape and aromatic leaves of many plants in the mint family, or Laminaceae. Saying that characters are diagnostic for a family means that they are strong indicators that a species is a family member but are not necessarily proof. For example, a square-stemmed,

EVOLUTION

The Origins of Domesticated Corn, Wheat, and Rice

Corn, or maize (*Zea mays,* subspecies *mays*), originated in Mexico or Central America. The domestication of corn took place by selective breeding of teosinte (*Zea mays,* subspecies *parviglumis*), a closely related wild grass in southern Mexico. Teosinte produces two rows of fruits or grains, with each grain surrounded by a woody fruit coat, which makes it difficult to grind the grains into a meal like corn meal. The key events in domesticating maize are not known for certain, but finding teosinte mutants with reduced or absent fruit coats and more grains were important events.

Wheat originated in the mountains of Syria, Jordan, Turkey, and southern Russia. Durum wheat is a naturally

Teosinte.

occurring hybrid of two grass species. Bread wheat adds genes from a third grass. By 5,000 B.C. bread wheat was cultivated in Egypt, India, China, and northern Europe. Bread wheat was apparently domesticated in the same general regions as durum wheat, which is used to make pasta.

The rice species *Oryza sativa* has three varieties, indicating that the species was domesticated on at least three occasions, probably somewhere in Asia. *Oryza sativa* variety *indica*, which is grown in southern China, India, and Southeast Asia, may have given rise to *Oryza sativa* variety *japonica*, which is grown in northern China, Japan, and Korea. It is also possible that the two varieties arose separately from similar primitive rice ancestors. A third variety, *Oryza sativa* variety *javonica*, is grown in Indonesia. *Oryza glaberrima*, another rice species, is cultivated in some regions of Africa. Since rice has been subjected to directed breeding for millennia in many locations, even molecular data may not be able to solve the mystery of its origins. Many Americans only know about one or two types of rice, but a typical Asian rice market may have as many as 50 varieties, with unique tastes and consistencies.

Rice.

Durum wheat.

aromatic plant is probably a mint, but some mints do not have square stems or aromatic leaves, and plants in other families can have either or both of these characters. Sometimes a combination of less distinctive characters will indicate that a species belongs to a particular family. In recent years, RNA and DNA sequencing has provided useful information about the relationships within and between families.

The Solanaceae (nightshade or potato family) is an example of how family members are identified by a col-

Figure 23.11 *Physalis,* **a member of the Solanaceae.** The fused, folded corolla of this *Physalis angulata,* which is often shaped like a pentagon, is typical of the family.

Figure 23.12 *Salvia,* **a member of the Lamiaceae.** The flowers of this sage, *Salvia guaranitica,* are typical of the Lamiaceae, or mint family. Flowers in the mint family are usually bilaterally symmetrical and located at the top of commonly square stems with opposite leaves.

lection of characters rather than any particular trait. The Solanaceae typically have flowers with parts in multiples of five, petals fused into a ruffled corolla, fruits that are berries, individual flowers with superior ovaries, round stems, and alternate or spiral leaves (Figure 23.11). Members of the Solanaceae, such as the tomato plant, frequently have distinctly unpleasant odors when the leaves are crushed. Despite the fact that most members of the family do not share a readily observable trait, they are recognizable to anyone familiar with the group.

In some families, diagnostic traits are more unusual and therefore more readily observable, such as the square stems and aromatic leaves in many members of the Lamiaceae. Species in the mint family usually have bilaterally symmetrical (irregular) flowers with parts in multiples of two or four, fused petals organized into two lips, superior ovaries, fruits consisting of four nutlets, square stems, and opposite leaves. Many mint species produce characteristic "minty" odors when the leaves are crushed. The combination of square stems, opposite leaves, and bilaterally symmetrical flowers is easily observed, used in guidebooks, and often diagnostic for the family (Figure 23.12).

Several families illustrate the diversity of floral and fruit structure

Families of flowering plants are distinguished by a collection of traits. A particular species in a family will have many, but not necessarily all, of these traits. We will focus mainly on flower structure as the most commonly used diagnostic trait for identifying families. You may wish to

review Chapter 6 for descriptions and images of different floral and fruit structures. The following examples include some of the largest angiosperm families and also reflect the extensive diversity within the phylum.

The grass family (Poaceae or Graminae) contains about 10,000 monocot species and includes nearly all cereals, the grain-producing plants, making it the most important family providing food for humans. Like many primitive angiosperms, most grasses are wind-pollinated. Most species have thin, relatively short stems, but in species such as tropical bamboos the stems are quite thick and tall. Grass flowers are not showy and may be bisexual or unisexual, depending on the species. As you saw in the chapter introduction, corn is a grass with unisexual flowers. Most grass species, however, have bisexual flowers (Figure 23.13). The anthers dangle on long, thin filaments, allowing the pollen to sift out into the breeze. The pollen grains themselves are thin-walled and dry, which allows the wind to carry them farther. Fruit is a single-seeded caryopsis—a grain with a single, prominent

THE INTRIGUING WORLD OF PLANTS

A Recently Discovered Orchid

In May 2002, a Virginia nursery owner traveling in Peru bought an orchid he had never seen before for $6.50 at a roadside stand. The magenta and purple blossom measured 15.2 cm (6 inches) across on a stem 30.5 cm (1 foot) high. It turned out to be an orchid species that was unknown to science. Realizing the importance of his find, he returned three days later to buy more but found that the entire hillside of plants—the only known location of the species—had already been stripped clean by collectors seeking to market the plants. The previously unclassified orchid was subsequently identified as a member of

the genus *Phragmipedium* and was named *Phragmipedium kovachii,* after the nursery owner's last name. The orchid's size and unique color make it potentially worth millions of dollars if cultivated successfully. However, the nursery owner was unable to capitalize on his find. By moving the plant from Peru to Florida, he had unintentionally violated the Convention on International Trade in Endangered Species.

A recently discovered orchid, *Phragmipedium kovachii.*

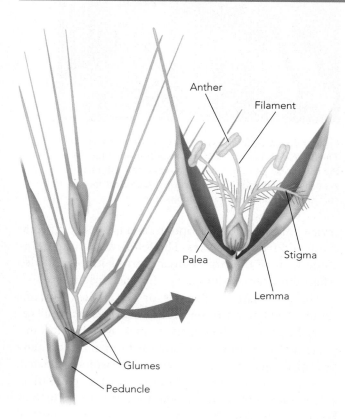

Figure 23.13 The Poaceae, or grass family, includes almost all of the cereals, or grain-producing plants. A typical grass flower is bisexual and is contained within two bracts called *glumes,* which separate as the inflorescence develops. Each flower is covered by two additional bracts, the palea and lemma, which open to reveal three stamens and usually two feathery stigmas.

38,000 because experts disagree on what is a variety and what is a species and also because rain forests contain many unknown species (see *The Intriguing World of Plants* box on this page). Most orchid flowers are bilaterally symmetrical and have three or fewer of each type of modified leaf. Petals, for instance, are typically in groups of three, whereas there are typically one or two stamens. The petals are large and showy, usually with two lateral petals flanking a central, flask-shaped petal with a large lip that accommodates pollinators (Figure 23.14a). Paired anthers frequently detach from the plant as free structures called *pollinia.* The fruit is a capsule and contains seeds that have no endosperm. Embryos are quite small compared with those of other plants, and seeds are shed when embryos are immature. Orchid seeds typically need a fungal partner in order to germinate and initiate seedling development.

The sunflower family (Asteraceae or Compositae) is the largest family of eudicots and the second largest family in the plant kingdom, with more than 23,000 species of herbs, shrubs, and trees. Species are found worldwide but especially in temperate regions. The flowers have complex heads that are radially symmetrical. They appear from a distance to be a single flower but are actually inflorescences, composites of many individual flowers, as you saw in the sun-

cotyledon called a scutellum (from the Latin *scutella,* "small shield").

The Orchidaceae, or orchid family, consist of monocots that form the largest family of plants. Usually found in the Tropics, orchids are herbaceous plants, many of which are epiphytes. Species estimates range between 20,000 and

(a) A typical orchid, *Paphiopedilum fairrieanum.*

(b) A daisy fleabane has a complex head composed of small yellow disk flowers surrounded by larger ray flowers.

(c) The sweet pea, *Lathyrus odoratus,* is a legume with bilaterally symmetrical flowers.

(d) In these zucchini flowers (*Cucurbita pepo*), the female flower is larger than the male flower, typical for the Cucurbitaceae.

(f) The world's smallest flower is the duckweed *Wolffia microscopica,* commonly known as *water meal* because it looks like mealy particles in the water. Each leaf is less than 1 millimeter (0.04 inch) across, with stamens protruding only about half a millimeter.

(e) Male and female catkins of *salix discolor.*

Figure 23.14 Examples of diversity in floral structure.

flower example in the chapter introduction. Daisies and fleabanes are another familiar family member (Figure 23.14b). Flies, butterflies, and bees are all attracted to the inflorescences of the various species in the family. The sepals of many species are considerably modified to form bristles or scales that can be barbed or fluffy and assist in seed distribution, as in the case of dandelions. The fruit is an achene, which is a thin, dry fruit containing a single seed that is attached to the fruit wall at only one point. The hairs associated with the dandelion fruit are modified sepals. What we call a sunflower "seed" is actually an achene.

The legume family (Fabaceae or Leguminosae) is the third largest in the plant kingdom, containing more than 18,860 species of eudicots, including herbs, shrubs, trees, and vines found worldwide. In most species, the bilaterally symmetrical flowers have one carpel, with other parts occurring in multiples of five (Figure 23.14c). Fruits are frequently legumes, such as peas or beans. Seeds often have little or no endosperm, but embryos frequently have fleshy cotyledons that store considerable amounts of food. Leaves are alternate, compound, and pinnate. Some legumes display "sleep movements," elevating leaves in the morning and lowering them at night. Many species form mutualistic associations with nitrogen-fixing bacteria, which live in nodules on the roots. Thus, these legumes increase soil fertility and are extremely important to agriculture.

The Cucurbitaceae, or gourd family, is a group of eudicots containing more than 800 species of herbs and vines, including many familiar plants such as squash, melons, and pumpkins. In most species, flowers are unisexual and radially symmetrical, with parts in multiples of five. Unisexual species typically produce male and female flowers at different times during the growing season, necessitating cross-pollination, which is brought about by pollinators (Figure 23.14d). The fruit is a berry containing many seeds. Stems are often five-angled, with coiling tendrils.

The Salicaceae, or willow family, is a group of eudicots consisting mainly of trees and shrubs, such as willow, pussy willow, poplar, cottonwood, and aspen. Species are pollinated by wind or insects, with flowers usually arranged in elongated, unisexual inflorescences called *catkins,* so named because they resemble a cat's tail (Figure 23.14e). Some species produce scented nectar, while others have scented buds. Flower parts occur in multiples of three to eight. The fruits—capsules, berries, or drupes—release many tiny seeds containing little or no endosperm. In some familiar species, like the willow, the seeds have cottonlike hairs that aid in wind dispersal.

The Lemnaceae, or duckweed family, contain the world's smallest flowering plants (Figure 23.14f). In some species, the entire plant is less than 1 millimeter (0.04 inch) across. Members of the Lemnaceae do not have the typical primary plant body consisting of stems, leaves, and roots. Instead, the tiny plant consists of one or more oval leaflike structures that are a type of reduced stem. Some species are rootless. Duckweeds serve as food for many animals that live in lakes and ponds, but the plants can become a nuisance when their vegetative reproduction gets out of hand. Some systematists categorize the Lemnaceae as a subfamily within a larger herbaceous family of monocots called the Araceae, which include such plants as taro and philodendron. This disagreement is an example of differing approaches of systematists. As discussed in Chapter 16, some are "splitters," favoring a greater number of taxonomic groups, and some are "lumpers," who prefer fewer groups. As with other types of plants, classification of angiosperms is an ongoing effort that involves debate.

Section Review

1. Is a member of a plant family typically identifiable by unique traits? Explain.
2. Pick two of the families described. For each one, explain how someone unfamiliar with the family might tentatively identify whether a plant is a member.

SUMMARY

Flowering plants, or angiosperms, constitute the phylum Anthophyta and are the world's most widespread plants, with over 250,000 species. They are the last major group of plants to evolve, first appearing around 130 to 145 million years ago.

Sexual Reproduction in Flowering Plants

Angiosperms, like gymnosperms, have a dominant sporophyte and dependent gametophyte (pp. 482–484)
Gametophytes are reduced in cell number from those of gymnosperms. Pollen grains develop inside anthers from spores formed by meiotic divisions of microsporocytes. Megagametophytes develop in ovules. Meiosis of the megaspore mother cell produces four megaspores. One survives to produce a megagametophyte. A pollen tube delivers two sperm, resulting in double fertilization, with one sperm uniting with the egg to produce a zygote and the second combining with two polar nuclei to give rise to the endosperm. Each ovule undergoing successful double fertilization potentially develops into a seed.

Self-pollination and cross-pollination are both common in angiosperms (pp. 484–486)
While wind pollination is common in some families, many angiosperms rely on pollinators attracted by flower colors, smell, nectar, and pollen. In moving from plant to plant, they facilitate cross-pollination. Plant–pollinator relationships are examples of coevolution.

The Evolution of Flowers and Fruits

The selective advantages of flowering plants account for their success (pp. 487–488)
The success of flowering plants relates in part to adaptations that avoid desiccation: the presence of vessels and, typically, deciduous leaves. The enclosure of seeds in a fruit protects them from water loss. Flowers often attract pollinators that facilitate cross-pollination. Fruits often attract animals that help disperse the seeds.

Flowers have evolved as collections of highly modified leaves (p. 488)

The flower is a determinate modified shoot that bears sporophylls. Microsporangia are located in the anthers of stamens. Megasporangia develop in the ovaries of carpels. Evolutionary trends include stamens and carpels becoming less leaflike, sepals and petals becoming distinct, floral parts being reduced and fixed, arrangement being whorled rather than spiral, and symmetry becoming bilateral in many species.

The evolution of angiosperms began during the Mesozoic era (pp. 488–494)

The first angiosperms, known as basal angiosperms, appeared around 142 million years ago. Later, other groups of angiosperms arose: magnoliids, monocots, the Ceratophyllaceae, and eudicots. All of these groups are represented by living species, with monocots constituting about 28% of living angiosperm species and eudicots comprising about 69%.

During the Cretaceous period, angiosperms spread rapidly throughout the world (p. 494)

During the late Cretaceous period, angiosperms spread throughout the world in an adaptive radiation. A human-induced migration of angiosperms used as crop plants has occurred since the rise of agriculture around 10,000 years ago.

A Sampling of Angiosperm Diversity

Phylum Anthophyta contains more than 450 families, classified mainly by flower structure (pp. 495–497)

Fruit and flower structure are also commonly used, with stems and leaves being used less frequently. Each family has a large collection of characters that help distinguish its members. In some families, one or more characters can easily be used for classification. In others, less easily observed characters are diagnostic.

Several families illustrate the diversity of floral and fruit structure (pp. 497–500)

The Poaceae (grasses) are wind-pollinated monocots that include almost all grain-producing plants. The Orchidaceae (orchids), the largest family, are monocots with large, showy petals that attract pollinators. The Asteraceae (sunflower family), the largest of the eudicot families, typically have flowers with large composite heads. The Fabaceae (legumes) are the third largest family. The Cucurbitaceae, which include squashes, melons, and pumpkins, typically have unisexual flowers. Flowers of the Salicaceae (willow family) are typically arranged in catkins. The Lemnaceae include duckweeds, the world's smallest flowering plants.

Review Questions

1. Compare and contrast alternation of generations in gymnosperms and angiosperms.

2. Explain why the terms *pollination* and *fertilization* are not synonymous.

3. What is double about "double fertilization," and in what way is the process unique in angiosperms?

4. Explain how coevolution of flowering plants and pollinators involves mutually rewarding relationships.

5. Compare and contrast the selective advantages of gymnosperms and angiosperms.

6. Compare the locations of sporangia in angiosperms and gymnosperms.

7. What does it mean to say that flowers consist of modified leaves?

8. Distinguish between basal angiosperms, magnoliids, monocots, and eudicots.

9. Can the evolution of flowers be described as a trend toward increasing complexity? Explain.

10. What are some factors that facilitated the spread of angiosperms?

11. Many crops are now cultivated throughout the world. Why, then, do researchers continue to investigate centers of origin for various crops?

12. Use examples of varieties in flower structure to illustrate the idea that form fits function.

Questions for Thought and Discussion

1. Why do you think wind-pollinated plants typically do not have colorful flowers?

2. Discuss the advantages and disadvantages of self-pollination and cross-pollination.

3. What is the significance of the discovery of primitive flowering plants such as *Amborella*?

4. Why do you think some "living fossils" such as *Amborella* have changed little from their fossil ancestors? Are surviving basal angiosperms such as *Amborella* the ancestors of other angiosperms? Explain.

5. If the Earth's surface continues to become warmer, what changes would you anticipate in the types of plants that are most successful?

6. Why do you think Darwin described the origin and spread of angiosperms as "an abominable mystery"?

7. Can the spread of crop plants by humans be considered a second adaptive radiation of angiosperms? Explain.

8. Why might convergent evolution complicate the process of classifying flowering plants?

9. Some scientists believe that in the future classification will be done solely by molecular data. Do you agree? Explain.

10. As described in this chapter, the production of a megagametophyte with 8 nuclei in 7 cells occurs in about two-thirds of flowering plant species. Among the remaining one-third, plants in the genus *Oenothera* (evening primroses), for example, produce a megagametophyte with 4 cells, each with a single nucleus. One

of these cells is the egg, two more are synergids, and the fourth is a polar cell. Draw a series of diagrams to show the development of the megagametophyte in *Oenothera* and its subsequent fertilization. How will the endosperm in this example differ from seeds in the "two-thirds majority"?

Evolution Connection

Biologists have speculated that the appearance in the Cretaceous period of smaller herbivorous dinosaurs that traveled in herds probably brought about the conversion of previously forested areas into grasslands. How do you think this might have contributed to the well-documented adaptive radiation of flowering plants at the expense of gymnosperms during this period?

To Learn More

Visit The Botany Place Website at www.thebotanyplace.com for quizzes, exercises, and Web links to new and interesting information related to this chapter.

Baumgardt, John P. *How to Identify Flowering Plant Families: A Practical Guide for Horticulturists and Plant Lovers.* Portland, OR: Timber Press, 1982. The analysis of flower structure, floral diagram, and floral formula will enable readers to place the flowers in the correct family.

Bernhardt, Peter, and John Myers. *The Rose's Kiss: A Natural History of Flowers.* Chicago: University of Chicago Press, 1999. The reader learns how to be a botanist or a gardener.

Moggi, Guido, with Luciano Giugnolini, edited by Stanley Schuler. *Simon and Schuster's Guide to Garden Flowers.* New York: Simon and Schuster, 1983. This book offers the history of flower gardens, gardening techniques, and suggestions on how to create a garden.

Perry, Frances, ed. *Simon and Schuster's Guide to Plants and Flowers.* New York: Simon and Schuster, 1976. Each plant is characterized with regard to place of origin, flowering season, soil type, amount of water and sunlight needed for healthy growth.

Schultes, Richard E., and Siri V. Reis. *Ethnobotany: Evolution of a Discipline.* Portland, OR: Timber Press, 1995. Explains the evolution of ethnobotany and the importance of plant welfare to future generations.

UNIT FIVE

Ecology

24

Ecology and the Biosphere

Saguaro National Park, Arizona.

Abiotic Factors in Ecology

Abiotic factors are physical variables in an organism's environment

The tilt of Earth's axis causes seasons and affects temperatures

The atmosphere circulates in six global cells

The rotation and topography of Earth affect global patterns of wind and precipitation

Ecosystems

The biosphere can be divided into biogeographic realms and biomes

Terrestrial biomes are characterized by rainfall, temperature, and vegetation

Light penetration, temperature, and nutrients are important abiotic factors in aquatic biomes

Earth's deserts are fascinating environments, where the scarcity of water is a predominant physical limitation for organisms. Desert plants have evolved a remarkable variety of adaptations that enable them to deal with this limitation. Many retain water by being barrel shaped or having succulent leaves. Some produce leaves only during rainy seasons, while others, like living stones from South Africa, are mostly buried in the ground. Some desert plants open their stomata only at night, when they take up carbon dioxide and incorporate it into organic acids for use in photosynthesis during the day.

Like plants in all environments, desert plants interact with other organisms. One of the most obvious interactions is with herbivores. Many desert plants, such as *Ocotillo,* protect their stems and leaves from herbivores by producing thorns or spines or by storing poisonous or foul-tasting compounds. Some plants produce tasty fruits that attract animals, which eat the fruits and, in turn, disperse the seeds. Cactus spines can also serve as dispersal agents if they attach parts of the cactus to a passing animal. When the animal cleans itself, the parts may fall to the ground and grow into a new plant.

Many anatomical, reproductive, and chemical traits of organisms have survival value. For some plant traits, such as spines, the survival value is clear. For others, scientists must study the plant and its interactions carefully before proposing a hypothesis about survival value. For example, the fact that the stomata of many desert plants remain closed during the day made no sense until crassulacean acid metabolism (Chapter 8) was

Living stones (*Lithops avcampiae*).

discovered. Likewise, the presence in deserts of seedless vascular plants, which require water for fertilization, was hard to explain until we learned that they become dormant in the absence of water but quickly revive after a rain. One such plant is *Selaginella lepidophylla,* a lycophyte found in deserts in the southwestern United States. Under dry conditions *Selaginella* is brown and looks dead, but within hours after a rain, it turns green, begins photosynthesis, and completes its sexual reproductive cycle. This behavior is the basis for its common name, the Resurrection Plant.

Deserts are just one of a wide variety of environments on Earth in which organisms survive, reproduce, and interact with their surroundings. The scientific study of interactions between organisms and their environment is the topic of this chapter and the next two. Here, we will begin by investigating how the nonliving components of environments affect organisms. Then we will examine major regions of Earth that are characterized by their particular combination of organisms and environmental factors.

Ocotillo, (*Fouquieria splendens*) has large spines that protect the leaves.

The candle plant (*Senecio articulatus*), a desert succulent.

Abiotic Factors in Ecology

A cross section through Earth shows that the planet is like a giant spherical rock, with a solid center surrounded by a liquid core, a thick, rocky mantle, and a thin crust. Life exists only near the outer edge of the crust—in the oceans that cover three-quarters of the crust, in the sediments beneath the oceans, and, on the continents, in and slightly above the very thin surface layer of ground-up crust called *soil*. **Ecology** is the study of this life-sustaining environment and of the organisms within it. The word *ecology* is derived from two Greek words: *oikos,* meaning "the family household," and *logos,* meaning "the study of." Thus, ecology is literally the study of the households of organisms. It includes investigations of the living and nonliving components of the environment and of the interactions between them.

Ecology is a large and complex field of biology with a wide focus and a broad scope. For example, an ecologist interested in oak trees can look at individual trees or populations and must consider how oak trees interact with the different parts of their environment. Other plants may compete with trees for light, soil nutrients, and water. Animals may feed on the trees while providing soil nutrients that help them grow. Viruses, bacteria, or fungi may infect the trees, causing disease. Temperature, moisture, seasonal changes in light intensity, and disturbances such as fires may affect the growth and reproductive success of the trees. Humans may enter the picture and alter the landscape. In terms of the scientific method, ecology is a science with many variables.

Abiotic factors are physical variables in an organism's environment

As far as we know, Earth is the only planet in our solar system that has the physical conditions necessary for life. If Earth were much closer to or farther from the Sun, temperatures on Earth would be too high or too low to support most species that now exist on Earth. If Earth did not have an atmosphere containing oxygen, most terrestrial organisms could not survive. The physical components of the environment, known as *abiotic factors,* provide both challenges and opportunities for organisms.

Temperature

Some animals, particularly birds and mammals, can keep their bodies warmer or cooler than their surroundings through metabolism or behavior. However, most organisms, including plants, have a temperature that is very close to the temperature of their surroundings, called the

Figure 24.1 Timberline. At elevations above the timberline, such as this one at Berthoud Pass, Colorado, the average temperature is too low for trees to grow.

ambient temperature. As a result, there is a narrow range of ambient temperatures in which most organisms can survive. Although some prokaryotes thrive in hot springs at 60–80°C, these organisms are the exception rather than the rule. Very few plants can continue normal metabolism at temperatures below 0°C, the freezing point of water, or above 45°C, the temperature at which most proteins denature. Plants can control their temperature somewhat with anatomical structures, such as leaf hairs, and by mechanical adjustments, such as changing leaf angles, but their temperature is largely determined by the ambient temperature.

Earth's atmosphere traps heat. As you go up in elevation, the atmosphere becomes thinner and the temperature drops. Above a certain elevation, the average temperature is too low for the growth of trees. The upper limit of tree growth is called the *timberline* or tree line (Figure 24.1). In northern Colorado, for instance, the timberline occurs at about 3,500 meters (11,500 feet), but in other locations it may be higher or lower than that. As you will see below, the elevation of timberline depends on several factors, including latitude.

Water

Most organisms are considerably more than 60% water, and the chemical reactions that maintain life occur in aqueous solution. One of the principal obstacles early life-forms encountered as they moved onto dry land was the problem of desiccation. The first terrestrial plants required water for their flagellated sperm to fertilize their eggs and for their embryos to develop without drying out. Pollen tubes enabled seed plants to do without water as a medium for the movement of flagellated sperm, and seeds allowed the embryos of these plants to develop on dry land. Plants have many other anatomical and physiological modifications that help them avoid desiccation. In vascular plants, for example, the xylem transports water throughout the plant, from the roots to the tips of the leaves.

Sunlight

Light from the Sun is necessary for photosynthesis. In a dense forest, less than 5% of the light that can be used in photosynthesis reaches the forest floor. Consequently, only shade-tolerant plants grow on the forest floor. In clear water, every meter (3.3 feet) of depth absorbs 45% of red light and 2% of blue light. Therefore, photosynthesis in water is restricted to the zone near the surface, and photosynthetic pigments that absorb blue light become relatively more effective as depth increases. On land, organisms must protect themselves from the mutagenic effects of ultraviolet radiation and from the excessive heat produced by infrared radiation. These two forms of solar radiation lie at opposite ends of the visible spectrum.

Wind

Wind can be a severe stress on plants and animals. It causes desiccation by increasing the rate of evaporation, and it accelerates heat loss by organisms that are warmer than their environment (the wind-chill effect). Wind also places structural limits on the size and shape of organisms. The force of wind on terrestrial organisms is much like the force of tides and currents on creatures that live in the ocean or in large lakes.

Soil

Soil contains inorganic ions, which are taken up by plant roots and transported in the xylem. For example, the nitrogen-containing ions ammonium (NH_4^+) and nitrate (NO_3^-) are used in large quantities by plants for the synthesis of amino acids, nucleotides, photosynthetic pigments, and other molecules. Mineral nutrition was discussed in detail in Chapter 10. Some soils contain insufficient quantities of necessary ions. This can become a problem particularly in agricultural soils from which crops remove ions. Often, fertilizers are added to the soil to replace the lost ions. Excessive salinity, acidity, and alkalinity are other major soil problems. At least 25% of the world's agricultural soils are too saline and an additional 25% are too acidic for the growth of most plants. However, some plants are adapted for growth on such "problem" soils. Saltbush (*Atriplex*), for example, excretes excess salt from glands on the surface of its leaves (Figure 24.2).

Figure 24.2 Saltbush (*Atriplex*), a salt-tolerant plant.

Disturbances

Disturbances are forces or events that cause changes in an environment. Many abiotic disturbances are weather-related, such as tornadoes, hurricanes, and floods, which can clear away vegetation. In some environments, fire is a normal, recurring disturbance. Vegetation rapidly reestablishes itself using nutrients returned to the soil by fire. Volcanic activity disturbs some environments. Often, the most severe disturbances are caused by human activities, such as logging, mining, agriculture, and urban development. Weeds frequently establish themselves in disturbed lands (see *The Intriguing World of Plants* box on page 509).

The tilt of Earth's axis causes seasons and affects temperatures

In Chapter 15 we noted that Earth's axis of rotation is tilted 23.5° with respect to the plane of Earth's orbit around the Sun and that this tilt causes the seasons. From about March 21 to September 23, the Northern Hemisphere is tilted toward the Sun and experiences spring and summer, while the Southern Hemisphere is tilted away from the Sun and experiences fall and winter. During this half of the year, days are longer than nights in the Northern Hemisphere and shorter than nights in the Southern Hemisphere. The situation is reversed from September to March.

THE INTRIGUING WORLD OF PLANTS

Weeds

Mullein (*Verbascum thapsus*) is one of the plants described by Linnaeus. Native to Europe, it was brought to the United States in the 1700s by settlers who used it as a medicinal herb and fish poison. During its first year of growth, mullein develops a rosette of soft, fuzzy leaves. During its second or third year, it produces a stalk that grows to a height of 1.5–3 meters (5–10 feet) and bears many yellow flowers. The flowers may generate up to 150,000 seeds per plant. Mullein spreads quickly in sunny locations and is more vigorous than many native North American plants. Many people regard mullein as a noxious weed.

To most of us, weeds are plants that grow in locations where we do not want them to grow. Dandelions in an urban lawn, Russian thistle in a pasture, and crabgrass in a garden are all examples of weeds. Many weeds are well adapted to the conditions in an area and therefore grow well without special care. In contrast, the plants we grow for food or flowers often are not well adapted, and many require irrigation, fertilizer, and pesticides to survive. Weeds frequently invade disturbed lands and rapidly establish themselves as dominant plants, even if they were largely absent before the lands were disturbed. For example, overgrazing, strip min-

Mullein (*Verbascum thapsus*).

ing, and slash-and-burn agriculture usually cause land to be permanently transformed by weeds.

Weeds may be native to an area or, like mullein in North America, they may be nonnative, or exotic. Many exotic plants thrive because they are not attacked by local herbivores or diseases. Ecologists estimate that about 1 in 1,000 exotic plants introduced to a new area is successful enough to become a weed.

Weeds generally grow well in a variety of environments. They typically have high reproductive ability, rapid growth, and short lives. One interesting characteristic of weeds that adds to their success is that many of their seeds do not germinate the year after they are produced. A study in Nebraska found that weed seeds in undisturbed soils could survive up to 40 years before germinating.

Weed science is the study of weeds, including how to control them. Much research in weed science is concerned with the use of herbicides to kill weeds and with strategies to control weeds that have become resistant to herbicides. Weed scientists are also interested in the traits that make weeds so successful. Recently, genetic engineers have begun to identify genes that allow weeds to be invasive and to grow rapidly.

Because of Earth's tilted axis, seasonal changes in day length vary with latitude. The Tropics, the region between the tropic of Cancer (23.5° north) and the tropic of Capricorn (23.5° south), experience the smallest variation in day length. As you move farther from the equator, day length varies more between summer and winter. Parts of Earth north of the Arctic Circle (66.5° north) and south of the Antarctic Circle (66.5° south) have sunlight 24 hours a day during the longest days of summer and no sunlight at all during the shortest days of winter. The duration of continuous sunlight or darkness ranges from 1 day at the Arctic Circle and Antarctic Circle to 6 months at each pole. Organisms that live in these regions have anatomical and

physiological adaptations that enable them to tolerate such wide variations in day length throughout the year.

The Tropics receive the most direct sunlight throughout the year and therefore have the highest average annual temperature. As you move north or south of the Tropics, the average temperature decreases because the Sun's radiation strikes Earth at a more oblique angle and is spread out farther (Figure 24.3). This relationship between latitude and average temperature is reflected in the distributions of plants, both wild and cultivated. For example, citrus trees grow reliably only in the southernmost regions of the continental United States, such as southern Florida, southern California, Arizona, and Texas. Farther

Average annual minimum temperature

°C	°F	
Below −46°	Below −50°	
−46° to −43°	−50° to −45°	
−43° to −40°	−45° to −40°	
−40° to −37°	−40° to −35°	
−37° to −34°	−35° to −30°	
−34° to −32°	−30° to −25°	
−32° to −29°	−25° to −20°	
−29° to −26°	−20° to −15°	
−26° to −23°	−15° to −10°	
−23° to −21°	−10° to −5°	
−21° to −18°	−5° to 0°	
−18° to −15°	0° to 5°	
−15° to −12°	5° to 10°	
−12° to −9°	10° to 15°	
−9° to −7°	15° to 20°	
−7° to −4°	20° to 25°	
−4° to −1°	25° to 30°	
−1° to 2°	30° to 35°	
2° to 4°	35° to 40°	
Above 4°	Above 40°	

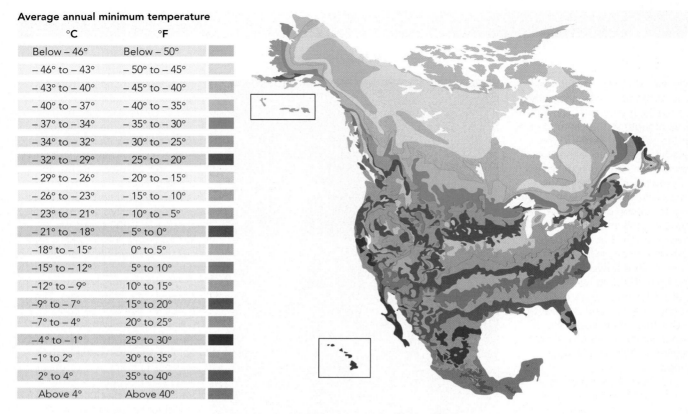

Figure 24.3 Relationship between latitude and average temperature. This map from the U.S. Department of Agriculture shows that average winter low temperatures in North America generally decrease from south to north.

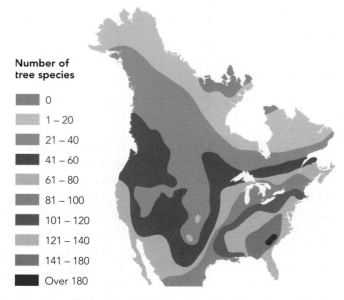

Number of tree species

- 0
- 1 – 20
- 21 – 40
- 41 – 60
- 61 – 80
- 81 – 100
- 101 – 120
- 121 – 140
- 141 – 180
- Over 180

Figure 24.4 Relationship between latitude and number of tree species. In general, more tree species exist at lower latitudes. This relationship also applies to other types of organisms and other continents.

from the Tropics, lower temperatures are correlated with shorter growing seasons and less yearly total photosynthesis. This may help to explain why the number of species generally declines with distance from the equator (Figure 24.4).

Paradoxically, although the number of species generally decreases with increasing latitude, the number of offspring of individual plants and animals seems to increase at higher latitudes. This effect was first noted in birds, but it also occurs in many plants, such as North American cattails. The common cattail (*Typha latifolia*) is found from the Arctic Circle to the equator. The narrow-leaved cattail (*T. angustifolia*) grows only in the north. The southern cattail (*T. domingensis*) is found only in the south. Both the narrow-leaved cattail and northern populations of the common cattail produce more rhizomes than the southern cattail and southern populations of the common cattail. Several explanations have been proposed for this variation in the number of offspring with latitude. For cattails, one explanation is that the colder, harsher winters in the north kill more rhizomes, so plants that produce many rhizomes are more likely to survive.

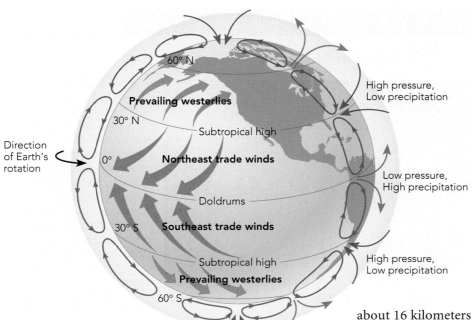

Figure 24.5 Global patterns of air circulation and precipitation. Six air-circulation cells wrap around Earth. Near the equator and at about 60° north and 60° south latitude, air rises, producing low-pressure regions with high precipitation. At the poles and at about 30° north and 30° south latitude, air falls, producing high-pressure regions with low precipitation. As a result of Earth's rotation, surface winds that blow toward the equator are deflected to the west, and surface winds that blow toward the poles are deflected to the east.

Because average annual temperature decreases with latitude, timberline generally occurs at lower elevations as you move from the equator toward either pole. For example, timberline occurs at about 4,700 meters (15,400 feet) in the Sierra Madre in central Mexico (19° north), at about 3,000 meters (9,800 feet) in the Sierra Nevada in California (38° north), and at about 1,200 meters (3,900 feet) in the Coast Mountains in southern Alaska (60° north). In the Arctic, there are no trees, so timberline is effectively at sea level.

The atmosphere circulates in six global cells

The heating of Earth's atmosphere by the Sun causes the air to circulate in six large belts, or cells, that run parallel to the equator. There are three cells in the Northern Hemisphere and three in the Southern Hemisphere (Figure 24.5). The positions of the cells shift from north to south somewhat with the seasons. Notice that in each cell both low-altitude and high-altitude airflow occurs.

To explain how these cells arise, let us begin with the two that are closest to the equator. Near the equator, the intense solar radiation heats the air, making it rise. The rising air leaves a low-pressure region at the surface with light winds. This region is called the *doldrums* by sailors. As the air near the equator rises, it cools, and water vapor in the air condenses and falls as rain, supporting the growth of lush, equatorial rain forests. At a height of

about 16 kilometers (10 miles), the rising air mass splits and moves toward each pole, cooling and becoming denser as it gets farther from the equator. At about 30° north and south latitude, the high-altitude air is dense enough to begin falling. As it falls, it warms and creates two high-pressure regions at the surface, called *subtropical highs*. Not much rainfall occurs in these regions because the falling air is not saturated with moisture. As a consequence, many of Earth's large deserts are located near 30° north and 30° south latitude. At the surface, some of the falling air moves back toward the low-pressure region near the equator, completing the circulation of air in these two cells. This surface air movement toward the equator is known as the *trade winds*. Periodic disruptions of the trade winds are responsible for the climatic phenomena known as El Niño and La Niña (see the *Conservation Biology* box on page 512).

Some of the air that falls in the subtropical highs moves toward the poles instead of the equator. At around 60° north and 60° south latitude, this poleward-flowing air encounters cold air flowing from the poles. Here, the two air masses converge and rise, creating another region of low pressure. The rising air cools and releases water as precipitation, creating conditions favorable for temperate forests, such as those that were once extensive in northern North America and Europe. At high altitudes, the rising air splits. Some of it moves toward the equator, completing the two cells that lie between 30° and 60°. The rest of the rising air moves toward the poles, completing the two cells that lie between 60° and the poles. Like the areas near 30° north and south latitude, the poles are regions of high pressure at the surface and scant precipitation.

CONSERVATION BIOLOGY

El Niño and La Niña

Approximately once every 5 years, an unusually warm ocean current arrives on the coast of Ecuador and Peru during the Christmas season. Because of the timing of this event, the current came to be called El Niño (Spanish for "the child"), in reference to the infant Jesus Christ. El Niño is part of a set of oceanic and atmospheric phenomena now known as the El Niño southern oscillation, which is initiated by a change in the trade winds.

Normally, the trade winds in the southern Pacific Ocean blow from east to west, pushing warm surface water into the western Pacific, where the water may be a half meter (1.5 feet) higher than in the eastern Pacific. The warm water lowers the atmospheric pressure in the western Pacific, causing the air to rise and triggering heavy precipitation in southeastern Asia and northern Australia. In the eastern Pacific, the warm surface water that moves west is replaced by cool, nutrient-rich water that rises from the ocean depths. High atmospheric pressure and sparse precipitation are typical of this area.

During an El Niño, however, the trade winds weaken or may even change direction. Warm water moves back into the eastern Pacific, and cold water wells up in the western Pacific. As a result, the climatic conditions typical of the eastern and western Pacific reverse: A low-pressure area forms in the eastern Pacific, bringing heavy rains to South America, while high pressure and droughts occur in southeastern Asia and Australia. Other climatic effects are felt throughout the rest of the world. For example, southern Alaska, western Canada, and the extreme northern portion of the contiguous United States experience abnormally warm winter weather, while the Gulf Coast

El Niño.

La Niña.

and the southeastern United States are cooler and wetter than normal.

When an El Niño ends, the trade winds return in force, producing a larger-than-normal region of cold water in the eastern Pacific. High atmospheric pressure again predominates in the area, causing decreased rainfall in South America. The resulting rebound effect on climate is called La Niña.

Exceptionally strong El Niños occur about once or twice a century. The last, in 1982–83, caused more than $8 billion in damage worldwide. Some of the damage was due to floods, fires, and windstorms that destroyed buildings and roads, killed livestock, and devastated crops.

Severe El Niños also affect wild populations of organisms. As the water in the eastern Pacific warms, many fish move out of the area in search of colder water. Local animals that depend on fish for food, such as fur seals, must swim farther and dive deeper to feed. During the 1982–83 El Niño, the scarcity of fish in the eastern Pacific led to a 50% mortality in seal colonies. In Utah, the Great Salt Lake became half as salty as normal due to increased rainfall. The lake was then invaded by a predatory insect (*Trichocorixa verticalis*), which ate about 90% of the brine shrimp population. Because brine shrimp eat algae, the decline in the brine shrimp population caused an algal bloom that clouded the lake.

People tend to make agricultural and horticultural decisions based on average or best climate years. Conservation biologists, who are called on to advise decision makers regarding land use and restoration projects, must be aware of the climatic effects of El Niño and La Niña when they make their recommendations.

The rotation and topography of Earth affect global patterns of wind and precipitation

Although the Sun's energy is responsible for producing winds, the rotation of Earth on its axis affects their direction. Earth rotates from west to east. That is, if you were to look directly down on the North Pole from space, Earth would appear to turn counterclockwise. Points on Earth's surface rotate at different speeds, depending on their latitude: The speed of rotation is 465 meters per second (1,049 miles per hour) at the equator, 403 meters per second at 30° latitude, 233 meters per second at 60° latitude, and 0 meters per second at the poles. Because the equator rotates faster than the poles, winds blowing toward the equator are deflected to the west (Figure 24.5). Thus, the trade winds blow from the northeast to the southwest in the Northern Hemisphere and from the southeast to the northwest in the Southern Hemisphere. Conversely, winds blowing toward the poles, such as the surface winds in the cells between 30° and 60° latitude, are deflected to the east. Since winds are named according to the direction they come from, these winds are commonly called the *prevailing westerlies.*

Earth's topographic features, particularly mountain ranges, alter the basic patterns of air circulation and precipitation. In Chapter 15, you learned that the eastern slopes of mountain ranges in western North America are often in a rain shadow because air masses that move across the continent from west to east—the prevailing westerlies—lose moisture as precipitation on the western slopes. For example, Eugene, Oregon, is located west of the Cascade Range and receives 118 centimeters (46.6 inches) of precipitation per year, while Bend, Oregon, which is east of the Cascades and at the same latitude, receives only 30.5 centimeters (12.0 inches). Bend is in a large rain shadow that covers eastern Washington and Oregon as well as much of Nevada, Utah, and Arizona.

The prevailing westerlies pick up some additional moisture as they continue east from the Cascades and Sierra Nevada. As they rise over the Rocky Mountains, they again cool and produce precipitation on the western slopes. The amount of precipitation is less than on the western slopes of the Cascades and Sierra Nevada, however, since the air does not contain as much moisture. As an example, Driggs, Idaho, which is on the western slope of the Rockies and at nearly the same latitude as Eugene and Bend, receives 39.4 centimeters (15.5 inches) of precipitation per year. Worland, Wyoming, east of the Rockies and in another rain shadow, receives just 18.4 centimeters (7.25 inches).

Section Review

1. List the abiotic factors that place physical limitations on living organisms.
2. Why does most photosynthesis in water take place near the surface?
3. Which areas on Earth receive the most direct sunlight throughout the year?
4. Why are many of Earth's large deserts located near 30° north and 30° south latitude?
5. Describe the effect of Earth's rotation on winds that blow toward the equator and on winds that blow toward the poles.

Ecosystems

In the first part of this chapter, we defined ecology as the study of the interactions that organisms (biotic factors) have with each other and with the nonliving (abiotic) components of their environment. An **ecosystem** consists of all the organisms and all the abiotic factors in a given environment. The biotic and abiotic components of an ecosystem have characteristic features. In many desert ecosystems, for example, water is scarce and temperatures are high. Plants and animals in the desert have structural adaptations that minimize water loss.

Ecosystems can be small or large. A rotting log can be considered an ecosystem, as can the forest that contains it. However, most ecologists reserve the term *ecosystem* for larger units. The largest ecosystem is the biosphere, the sum of all ecosystems on Earth.

The biosphere can be divided into biogeographic realms and biomes

Broad geographic areas that are characterized by distinctive collections of organisms are known as **biogeographic realms.** As Figure 24.6 shows, Earth's biogeographic realms correspond roughly to its continents. Because the continents have been separated for millions of years, unique flora and fauna have evolved on different continents, even in ecosystems with similar abiotic factors. For example, the deserts on two continents will have different species of plants and animals, even though the average temperature and annual rainfall may be the same in both deserts.

Major types of terrestrial and aquatic ecosystems that span large areas are called **biomes.** In contrast to biogeographic realms, which are mostly confined to a single

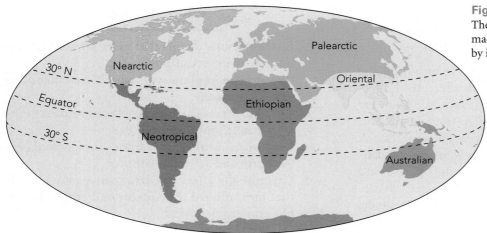

Figure 24.6 Biogeographic realms. The terrestrial part of the biosphere is made of six broad areas, each distinguished by its unique assortment of organisms.

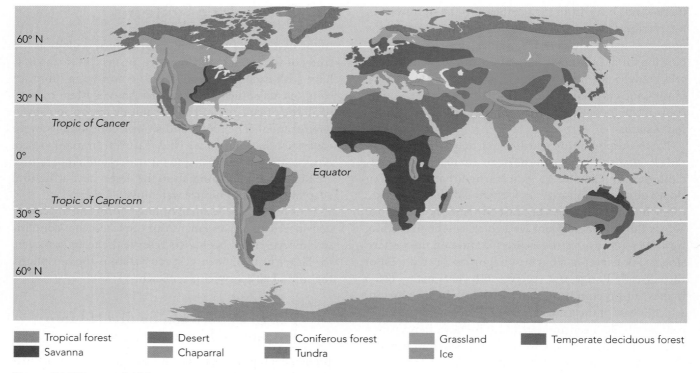

| Tropical forest | Desert | Coniferous forest | Grassland | Temperate deciduous forest |
| Savanna | Chaparral | Tundra | Ice | |

Figure 24.7 Terrestrial biomes. The actual boundaries between biomes are generally less distinct than they appear on this map. In many biomes, human activity has altered the original vegetation that is characteristic of the biome.

continent, terrestrial biomes are widespread across Earth's continents, and some are found on every continent (Figure 24.7). Terrestrial biomes are defined largely on the basis of their forms of vegetation. For example, savannas have extensive grasslands punctuated by occasional tall trees, while tropical forests have many species of plants growing in different horizontal layers. (We will investigate these and other terrestrial biomes in more detail in the next section.) Each terrestrial biome has a char-

acteristic pattern of abiotic factors, including temperature and precipitation, which can be represented in a diagram called a *climograph* (Figure 24.8). These factors, along with the evolutionary history of specific regions, determine the types of plants and other organisms that can live in the biome.

A knowledge of biomes is useful as a general classification of ecosystems in the same way that a taxonomic classification provides a basis for studying different groups of

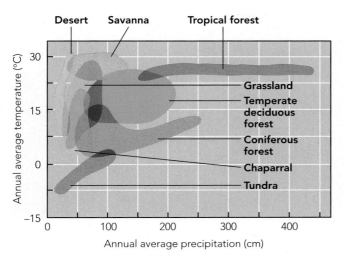

Figure 24.8 A climograph for terrestrial biomes. The average temperature and precipitation of a biome are two factors that distinguish it from other biomes.

Figure 24.9 Tropical forest. Little light penetrates the dense canopy in this tropical rain forest in Borneo.

organisms. In Chapter 25, we will consider some of the interactions that organisms have with each other and with the abiotic factors in their environment. Similar interactions take place in different forms in all biomes. For this reason, many ecologists today place less emphasis on describing biomes than on explaining processes that are common to all biomes.

Terrestrial biomes are characterized by rainfall, temperature, and vegetation

Ecologists have come up with several different systems for classifying terrestrial biomes. The one we will use in this book contains relatively few biomes. Other systems split some of the biomes described below into two or more on the basis of their geographic locations or the specific types of plants they contain.

Tropical Forests
Tropical forests include dry forests, which are found in areas with low annual rainfall; seasonal forests, which undergo an annual drought of several months; and rain forests, which typically have an annual rainfall of 200–400 centimeters (80–160 inches; Figure 24.9). All tropical rain forests are found within 30° latitude of the equator. The three largest are in the Amazon River basin of South America, in South Asia from India to New Guinea, and in central and western Africa. Like most biomes, tropical rain forests are quite diverse, and many different types can be recognized.

Typically, the vegetation in a tropical rain forest is divided into several vertical layers. The highest, called the *emergent layer,* is composed of a few tall trees, some of

which may reach heights of 40–60 meters (130–200 feet). Below that is the dense *canopy,* formed by the branches and leaves of shorter trees. Shrubs and small trees make up the *understory,* which is shaded by the canopy. Seedlings occupy the lowest layer, the *forest floor.*

The most interesting feature of tropical rain forests is the biological diversity they support. One-half of all known plant and animal species are found there. A single 10-square-kilometer (4-square-mile) area may contain 1,500 species of flowering plants and 750 species of trees. As is mentioned more than once in this book, most unprotected tropical rain forests will probably disappear in the first half of this century due to human activities.

Savannas
Savannas are found worldwide, although the most familiar ones are in Africa (Figure 24.10). They have in common a flat terrain, annual rainfall of 50–200 centimeters

Figure 24.10 Savanna. In this savanna in Kenya, the dominant plants are grasses and scattered trees.

Figure 24.11 Grassland. This tallgrass prairie in the Maxwell Game Preserve, Kansas, is a remnant of the grasslands that once blanketed vast areas in the interior of North America.

Figure 24.12 Desert. Arizona's Saguaro National Park, named for the saguaro cactus that is common throughout the park, is part of the Sonoran Desert.

(20–79 inches), and grass as the major plant type. Occasional large trees and shrubs also grow in savannas, particularly in regions with available groundwater. Seasonal fires return nutrients to the soil, a process that is especially important in savannas because of their generally low level of soil nutrients.

Grasslands

Like savannas, grasslands are typically found in regions with flat or rolling terrain and have grass as their major form of vegetation (Figure 24.11). The main difference between grasslands and savannas is in the amount of precipitation they receive: Grasslands get an average of 25–80 centimeters (10–31 inches) of rain a year. The amount of rainfall is the chief factor that determines the mass of plant material a grassland produces. As you learned in Chapter 15, the meristem in grasses is located at or below the soil's surface. Thus, the dividing cells that produce the leaves of grasses are protected somewhat from being eaten by grazing animals.

Grasslands around the world vary somewhat in annual rainfall and dominant plant species and are given regional names: tallgrass prairies and shortgrass plains in North America, pampas in South America, steppes in Europe and Asia, and veldts in Africa. At one time, grasslands of various types covered 42% of Earth's land surface. Today, they cover only about 12%, as large areas that were originally grassland have been converted to farmland.

Deserts

Deserts receive between 0 and 25 centimeters (10 inches) of rain per year (Figure 24.12). They occupy about a quarter of Earth's land surface and, as you read earlier, are found mostly near 30° north and south latitude.

Deserts can be either hot or cool, and most experience wide temperature variations, even within a single day. Since the rate of evaporation increases with temperature, hot deserts are usually a more extreme environment for both plants and animals.

North America has four desert regions: the Chihuahuan Desert, in north-central Mexico, southern New Mexico, and western Texas; the Sonoran Desert, in northwestern Mexico, southeastern California, and southwestern Arizona; the Mojave Desert, in southeastern California, western Arizona, and southern Nevada; and the Great Basin Desert, which covers much of the area between the Sierra Nevada and Cascades to the west and the Rockies to the east. In hot North American deserts—the Chihuahuan, Sonoran, and Mojave—plants flower after either winter rains from the Pacific Ocean or summer rains from the Gulf of Mexico. Because rainfall is so uncertain, in some years the plants do not flower at all.

Chaparral

Also called shrub or scrub biomes, chaparral generally has rainfall comparable to that of grasslands (Figure 24.13). The chaparral of California is typical of scrub biomes found in so-called Mediterranean climates, which are characterized by hot, dry summers and cool, moist winters. Such climates are usually found around 32° to 40° north and south latitude. Chaparral is inhabited by plants called *sclerophylls* (from the Greek words *scleros*, meaning "hard," and *phyll*, meaning "leaf"), which have small leaves with a thick, waxy cuticle.

Temperate Deciduous Forests

Deciduous forests in Earth's temperate regions—between the tropic of Cancer and the Arctic Circle and between the

Figure 24.13 Chaparral. Small-leafed shrubs are the predominant vegetation of chaparral, such as the Los Padres National Forest, California.

Figure 24.14 Temperate deciduous forest. Plants in these forests experience four distinct seasons, and most, such as these trees in Great Smoky Mountain National Park, North Carolina, begin to change color in the fall.

tropic of Capricorn and the Antarctic Circle—typically have four distinct seasons (Figure 24.14). These forests, which are generally dominated by dicotyledonous flowering plants, are found in North America, Europe, and Asia. Those of Europe are mostly gone. In the United States, the largest surviving temperate deciduous forest is in the Appalachian Mountains from Pennsylvania to Alabama, where there are more species of woody and herbaceous plants than anywhere except for tropical rain forests. Trees common in North American temperate deciduous forests include oak (*Quercus*), maple (*Acer*), basswood (*Tilia*), and hickory (*Carya*).

Temperate deciduous forests also exist along rivers and lakes inside other biomes, such as deserts. **Riparian** (riverbank) woodlands in the desert play an important role in maintaining diverse species of animals and plants.

Figure 24.15 Coniferous forests. Tall conifers, such as western hemlock and Douglas fir, are common in the temperate rain forests of the Pacific Northwest.

For example, the Rio Grande, which forms the border between Texas and Mexico, flows through the Chihuahuan Desert. The woodlands along this river serve as a haven for resident and migratory birds. More than 450 species of birds are found in the area, the most found in any one location in the United States.

Coniferous Forests

Several types of coniferous forests are found throughout the world. The temperate rain forests of the Pacific Northwest (Figure 24.15) are dominated by conifers such as western hemlock (*Tsuga heterophylla*), Pacific silver fir (*Abies amabilis*), Douglas fir (*Pseudotsuga menziesii*), and coast redwood (*Sequoia sempervirens*). Old-growth stands of temperate rain forests are extremely complex ecosystems that have developed over hundreds of years. Because the annual rainfall of 250 centimeters (100 inches) or more favors the growth of extremely large trees, these stands have been extensively logged and exist now in only a few places outside national parks and preserves. Land from which the trees have been removed can be reseeded to produce a new crop of trees, but unless the leaves and bark of the old trees are returned to the soil to decay and release nutrients, the second crop of trees will grow much more slowly than the first. In addition, when forest fires burn the waste timber left after logging, reseeding is often unsuccessful because soil microflora and microfauna have been destroyed. Most logging companies are unwilling to wait the 200 or more years required to regenerate anything like the original forest.

Northern coniferous forest, also known as boreal forest or **taiga,** is the largest uniform biome on Earth, occupying about 11% of Earth's surface. In North America, the taiga covers much of Alaska and Canada as well as parts of New England. Dominant plants in the taiga are flowering shrubs and herbs and a few species of conifers. In the northern taiga, the trees are short due to poor soil nutrition and subterranean permafrost, a layer of frozen soil that never thaws. Summers in the taiga are short, cool, and moist, and winters are long, very cold, dry, and windy. Fires are important, periodic events, with regrowth of vegetation requiring decades or even centuries. In recent years, the taiga has been extensively exploited by humans for its trees and minerals. Because of the persistent, harsh environmental conditions, the taiga recovers from timber harvesting and other human activities very slowly. More often, it is simply replaced by tundra (discussed below).

Mountainous or montane coniferous forests are found at high altitudes, where the cold, frequently snowy winters are much better tolerated by conifers than by flowering trees. These forests appear uniform at first glance but are actually divided into different zones based on elevation, precipitation, and dominant trees. In Washington's Cascade Mountains, for example, the lowest zone is inhabited by firs, pines, and red cedars as well as flowering trees such as maples and alders. At higher elevations, noble fir (*Abies procera*) and grand fir (*Abies grandis*) are common. The trees that grow right at timberline are small and wind-distorted and are known as the *krummholz*, or "crooked wood." These zones occur at lower elevations on the cooler, wetter western slopes of the mountains than on the warmer, drier eastern slopes.

In the southeastern United States, coniferous forests composed of loblolly pine (*Pinus taeda*), longleaf pine (*P. australis*), and slash pine (*P. elliottii*) are common. These forests become established on nutrient-poor, sandy soil and eventually are replaced by deciduous forests. Under natural conditions, fires ensure that the coniferous forests are a regularly occurring feature of the landscape.

A few coniferous forests are made up of deciduous conifers such as larch (*Larix occidentalis*), which is widespread in the northern Rocky Mountains. Bald cypress (*Taxodium distichum*) is a deciduous conifer common in the southern United States.

Tundra

Arctic tundra is a treeless, frigid plain found in the very northern regions of North America, Europe, and Asia (Figure 24.16). It constitutes around 20% of Earth's land area. Temperature is the most important abiotic factor regulating plant growth in arctic tundra. Permafrost lies beneath the surface of the soil, and temperatures at the surface may be above freezing for only a few hours a day, even during the summer growing season, which lasts less than 2 months. Mosses and a few species of flowering plants and shrubs are

Figure 24.16 Tundra. Mosses and shrubs are the major types of plants in the tundra. The arctic tundra shown here is in Denali National Park, Alaska.

the dominant vegetation. Tundra plants store as much as 94% of their biomass underground in roots or rhizomes. North of 75° latitude, annual precipitation averages less than 25 centimeters (10 inches). What little precipitation does fall remains in the surface layers atop the permafrost, and because of the low temperatures, evaporation is slow. Consequently, arctic tundra has numerous lakes, bogs, and areas of soggy soil punctuated by higher spots that are quite dry and desertlike.

Alpine tundra is found above timberline on mountains. It is similar to arctic tundra except that permafrost is lacking, the growing season is somewhat longer, and winter temperatures are slightly higher.

Light penetration, temperature, and nutrients are important abiotic factors in aquatic biomes

Aquatic biomes cover about three-quarters of Earth's surface. They include various freshwater ecosystems, oceans, and biomes at the interface between fresh water and salt water.

Lakes and Ponds

Lakes are depressions that contain water. Most are natural; some are made by humans. As a biome they have well-defined boundaries. Lakes are commonly divided into zones based on light penetration and distance from shore (Figure 24.17). The *photic zone* is the upper part of a lake, where there is sufficient light for photosynthesis. Below that is the *aphotic* or *profundal zone,* where very little or no light penetrates. The photic zone can be further divided into the *littoral zone,* which consists of shallow water close to shore where rooted and floating plants

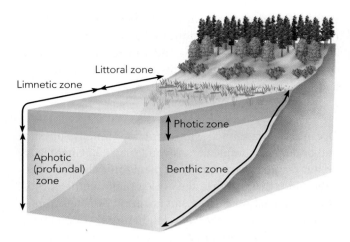

Figure 24.17 The zones in a lake. The photic and aphotic zones are based on light penetration. The littoral and limnetic zones are based on distance from shore. The benthic zone is the bottom of the lake at all depths.

grow; and the *limnetic zone,* which is the open water farther from shore. The plants and planktonic algae and bacteria in a lake support a complex ecosystem that extends into all zones of the lake. The remains of organisms that die in the upper levels eventually settle on the substrate at the bottom of the lake, an area known as the *benthic zone.*

Lakes that are shallow and nutrient rich are said to be **eutrophic** (from the Greek word *eutrophos,* meaning "well-nourished;" Figure 24.18a). These are the lakes that turn murky green in summer due to algal blooms, which you read about in Chapter 18. When the algae die, they are consumed by bacteria on the bottom of the lake. The massive increase in bacterial activity may deplete the oxygen supply to the point where aerobic organisms cannot live beyond a certain depth. Consequently, summer fish kills sometimes occur in eutrophic lakes. In contrast, deep, nutrient-poor lakes are called **oligotrophic** lakes (from Greek words meaning "few nutrients;" Figure 24.18b). The water is clear and there are fewer organisms than in eutrophic lakes, although the number of species may be similar in both types of lakes.

The density of water varies with temperature. As water cools, the hydrogen bonds between water molecules become more stable, and density increases. At 4°C, water reaches its greatest density. The density then decreases as the temperature drops below 4°C. At 0°C, liquid water solidifies, forming ice. Because its density is less than that of liquid water, ice floats.

The changing density of water with temperature causes a seasonal mixing of waters in temperate-zone freshwater lakes. In summer, the water near the surface of a lake absorbs heat. This results in a stable summertime pattern in which warmer, less-dense water is on top of cooler, denser water. With the arrival of cold weather in fall, the lake's upper layers gradually cool until they become colder and denser than the bottom layers. The upper layers then sink to the bottom of the lake and lower

(a)

(b)

Figure 24.18 Eutrophic and oligotrophic lakes. (a) A shallow eutrophic lake in Oxford, England. **(b)** An oligotrophic lake in Glacier National Park, Montana.

layers rise to the surface, a phenomenon known as *turnover*. In winter, when the surface of the lake is covered with ice, another stable pattern is formed, with the water temperature ranging from near 0°C just below the surface to 4°C at the bottom. Because ice forms at the top of the lake rather than throughout, animals, plants, and microorganisms can survive winter in the lake. In spring, the ice melts, and the surface water warms to 4°C and sinks, causing a second turnover. The turnovers that occur during fall and spring mix nutrients in the lake.

Freshwater Wetlands

Wetlands include flooded meadows, dominated by grasses and herbaceous plants; marshes, dominated by herbs and reeds (hollow-stemmed grasses); swamps, dominated by woody plants; and bogs, dominated by acid-forming sphagnum mosses and small, acid-tolerant shrubs. Soil conditions in wetlands range from permanently or periodically flooded to permanently or periodically saturated with water. Wetlands are highly varied and complex ecosystems and, because of the many variables that affect them, are not well understood. Historically, people have tended to view wetlands as areas to be drained. Over the past two centuries, more than half of the wetland area in the lower 48 states has been lost. Because mosquitoes develop in water, the disappearance of wetlands has made the mosquito-borne disease malaria uncommon in the United States. In recent years, however, the value of wetlands has become clearer. Wetlands hold large amounts of water and can provide significant buffers against flooding. They also function as water filtration and purification systems and provide important wildlife habitat.

Streams and Rivers

Streams and rivers are ecosystems with flowing water. The size of the waterway and the velocity of the water are two important variables that determine the kinds of organisms that inhabit these bodies of water. The climate of the region and periodic disturbances such as drying and flooding are also important factors.

Human influences on streams and rivers include pollution, silting, and flow regulation. The causes of pollution are many and varied. Silt consists of fine particles of soil that remain in suspension or gradually settle out. Silting is caused by logging, mining, and other activities that clear vegetation or disturb soil. Flow regulation can involve the removal of water for irrigation or the building of dams. While dams provide economic benefits in terms of electric power, irrigation water, and flood control, they have a significant environmental cost. Reservoirs that form upstream of dams accumulate silt and nutrients and

frequently become eutrophic. In tropical areas, these reservoirs may provide additional places in which parasitic diseases such as malaria and schistosomiasis can become established. Below dams, nutrient levels are low, and flow is typically reduced.

Oceans

Oceans are vast saltwater biomes with immense physical and biological complexity. As you learned in Chapter 18, oceans contain many species of planktonic algae and bacteria, which carry out half the world's photosynthesis. Most of these microscopic organisms are eaten by heterotrophic plankton, called *zooplankton*. Higher up the food chain, both phytoplankton and zooplankton are consumed by *nekton*, the collective term for animals such as fish and whales that can move independently of ocean currents.

Oceans are divided into zones similar to those in lakes. Closest to shore is the *intertidal zone*, which is the region between high tide and low tide (Figure 24.19). Thus, the intertidal zone is under water part of the time and exposed to air the rest of the time. Intertidal zones are characterized by vigorous wave action during the changing of

Figure 24.19 Intertidal zone. The tidepools in this rocky intertidal zone in Olympic National Park, Washington, are inhabited by many species of algae and invertebrates.

tides, which occurs four times a day in most locations. The substrate consists of sand in some intertidal zones and rocky tidepools in others. The latter provide attachment sites for a variety of brown, green, and red algae and marine invertebrates, as well as a few species of grasses and other marine aquatic angiosperms.

Beyond the intertidal zone are the *neritic zone,* which lies over the continental shelves, and the *oceanic zone,* which begins at the edge of the continental shelves and reaches very great depths. In temperate regions, more phytoplankton are found in the shallow, warmer waters of the neritic zone.

Estuaries and Salt Marshes

An estuary is a partially enclosed coastal area where fresh water from a river mixes with salt water from the ocean. If the tide is in, salt water may extend up the river for a considerable distance. If the tide is out, fresh water may push far out into the ocean. Rivers deposit large amounts of nutrients in estuaries. These nutrients are responsible for rich collections of plant and animal life. A few flowering plants, in particular eel grass (*Zostera marina*), are common and important in estuaries.

Salt marshes form on the alluvial plains surrounding estuaries and around sand bars and islands (Figure 24.20). The water in a salt marsh is usually brackish, or partly salty, and changes in salinity as the tide goes in and out. Meandering waterways and salt pans, where evaporation increases salt concentration, are common in salt marshes. A number of flowering plants, including mangrove trees and salt-tolerant grasses of the genera *Spartina* and *Distichlis,* thrive in salt marshes. These plants are of great interest to genetic engineers who are attempting to increase the salt tolerance of agricultural crops.

Figure 24.20 Salt marsh. Salt marshes, such as this one in Cape Elizabeth, Maine, are dominated by salt-tolerant grasses.

Salt marshes are also inhabited by algae and microscopic saprobes, chiefly bacteria. These organisms serve as the basis of food chains that involve aquatic invertebrates as well as many kinds of vertebrates, both aquatic and terrestrial. Food chains in salt marshes are complex and not well understood.

Section Review

1. **What is the biosphere?**
2. **Describe a chaparral biome.**
3. **What is a riparian woodland?**
4. **What is taiga?**
5. **Contrast eutrophic and oligotrophic lakes.**

SUMMARY

Abiotic Factors in Ecology

Abiotic factors are physical variables in an organism's environment (pp. 507–508)

Most organisms have a temperature close to that of their surroundings and are more than 60% water. Photosynthetic organisms face the challenge of absorbing sufficient sunlight for photosynthesis while avoiding the damaging effects of solar radiation. Wind increases the rate of evaporation and the rate of heat loss from terrestrial organisms. Soil contains inorganic ions taken up by plant roots. Disturbances are forces or events, such as tornadoes, hurricanes, floods, fires, and various human activities, that cause changes in an environment.

The tilt of Earth's axis causes seasons and affects temperatures (pp. 508–511)

From mid-March to mid-September, the Northern Hemisphere is tilted toward the Sun and experiences spring and summer. During the rest of the year, the Northern Hemisphere is tilted away from the Sun and experiences fall and winter. Seasonal changes in day length are greatest at the poles and smallest in the Tropics. The average annual temperature is lowest at the poles and highest in the Tropics.

The atmosphere circulates in six global cells (pp. 511–512)

Near the equator, air rises and atmospheric pressure is low. Around 30° north and south latitude, high-altitude air falls and atmospheric pressure is high. Two more low-pressure regions exist near 60° north and south latitude, and two more high-pressure regions exist at the poles. Air circulates in six cells between low-pressure regions, where precipitation is heavy, and high-pressure regions, where precipitation is light.

The rotation and topography of Earth affect global patterns of wind and precipitation (p. 513)

Because the equator rotates faster than the poles, winds blowing toward the equator are deflected to the west, and winds blowing toward the poles are deflected to the east. Mountain ranges affect air circulation and precipitation. In western North America, the western slopes of mountain ranges usually receive more precipitation than the eastern slopes.

Ecosystems

An ecosystem consists of all the organisms and abiotic factors in a given environment. Ecosystems can be small or large. The largest ecosystem is the biosphere.

The biosphere can be divided into biogeographic realms and biomes (pp. 513–515)

Biogeographic realms are broad geographic areas that are characterized by distinctive collections of organisms. Biomes are major types of terrestrial and aquatic ecosystems that span large areas. Terrestrial biomes are defined largely on the basis of their vegetation.

Terrestrial biomes are characterized by rainfall, temperature, and vegetation (pp. 515–518)

Terrestrial biomes include tropical forests, savannas, grasslands, deserts, chaparral, temperate deciduous forests, coniferous forests, and tundra. Tropical rain forests are usually divided into vertical layers of vegetation and are home to half of all known plant and animal species. Both savannas and grasslands have grass as their major form of vegetation, but savannas also have occasional large trees and shrubs. On average, grasslands receive less precipitation than savannas. Deserts receive very little annual precipitation, can be either hot or cool, and often experience wide temperature variations. Chaparral, or scrub biomes, are characterized by hot, dry summers and cool, moist winters. Temperate deciduous forests typically have four distinct seasons and are generally dominated by dicotyledonous flowering plants, such as oak, and maple. Coniferous rain forests have high annual rainfall and extremely large trees, such as western hemlock and Douglas fir. Northern coniferous forest, or taiga, features a few species of conifers as well as flowering shrubs and herbs, while coniferous forests of the southeastern United States are commonly composed of pines. Tundra is a treeless, frigid plain found in the very northern regions of the Northern Hemisphere and above timberline on mountains.

Light penetration, temperature, and nutrients are important abiotic factors in aquatic biomes (pp. 518–521)

Aquatic biomes include lakes and ponds, freshwater wetlands, streams and rivers, oceans, estuaries, and salt marshes. Lakes are divided into zones based on light penetration and distance from shore. Eutrophic lakes are shallow and nutrient rich; oligotrophic lakes are deep and nutrient poor. In temperate-zone freshwater lakes, turnover mixes surface and bottom waters during fall and spring. Freshwater wetlands, including flooded meadows, marshes, swamps, and bogs, have soils that are periodically or permanently flooded or saturated with water. Streams and rivers have flowing water; their size and the velocity of the water flowing through them determine the kinds of organisms they support. Oceans are vast saltwater biomes and are divided into zones similar to those in lakes, including the intertidal, neritic, and oceanic zones. Estuaries are partially enclosed coastal areas where fresh water and salt water mix. High in nutrients, estuaries support rich collections of plant and animal life. Salt marshes form around estuaries, sand bars, and islands and are inhabited by mangrove trees, salt-tolerant grasses, and various microorganisms that serve as the basis of complex food chains.

Review Questions

1. How does saltbush deal with excess salt in the soil?
2. When during the year is the Northern Hemisphere tilted away from the Sun? When are days shorter than nights in the Southern Hemisphere?
3. At what latitudes are most regions of heavy precipitation found? Why?
4. Explain the effect of Earth's rotation on the trade winds.
5. What is a rain shadow? Where are most rain shadows found?
6. What is the difference between an ecosystem and a biome?
7. Describe the vertical layers in a tropical rain forest.
8. Why has there been a decrease in the area covered by grasslands?
9. Where in the United States would you find temperate deciduous forests?
10. What is the *krummholz,* and where is it found?
11. Describe the characteristics of arctic and alpine tundra.
12. At what temperature is the density of water greatest?
13. What is turnover in a lake, and when does it occur?
14. Why has malaria become uncommon in the United States over the past two centuries?
15. What is an estuary?

Questions for Thought and Discussion

1. If you lived in an area of low rainfall, how would you determine if this climate pattern was caused by latitude or by a rain shadow?

2. Imagine you find yourself in a wild, natural location somewhere on Earth in late June. How could you use the abiotic factors and vegetation in the area to determine your approximate latitude and whether you were north or south of the equator?

3. Do you think that knowledgeable ecologists could establish successful new, artificial biomes using plants from around the world in a particular geographical area? Explain why or why not.

4. What is the smallest ecosystem you know about?

5. Consider a grassland that is overgrazed by cattle for many years. Excessive pumping of groundwater lowers the water table beneath the grassland. Is the grassland likely to change into a different biome? Explain why or why not.

6. Draining wetlands eliminates habitat for many plants and animals, including mosquitoes. How can mosquito-borne diseases such as malaria be controlled without habitat destruction?

7. Draw a series of cross-sectional views of a deep lake to show the sequence of changes in water density layers throughout one year to show stratification changes and turnover.

Evolution Connection

Which unique selective pressures face organisms that live in estuarine and salt marsh environments and how might plants adapt to these pressures?

To Learn More

Visit The Botany Place Website at www.thebotanyplace.com for quizzes, exercises, and Web links to new and interesting information related to this chapter.

Abbey, Edward. *Desert Solitaire.* New York: Ballantine Books, 1991. This fascinating and classic book tells about Abbey's three seasons as a park ranger in Arches National Park in Utah. Like all of Abbey's books, it is well worth reading.

Barbour, Michael G., Jack H. Burk, Wanna D. Pitts, Frank S. Gilliam, and Mark W. Schwartz. *Terrestrial Plant Ecology,* 3rd ed. San Francisco: Benjamin Cummings, 1999. This book covers the entire extent of modern plant ecology.

Brower, Kenneth. *The Winemaker's Marsh: Four Seasons in a Restored Wetland.* San Francisco: Sierra Club Books, 2001. Brower tells the story of winemaker Sam Sebastiani, who restored 90 acres of a hayfield to a wetland that is now home to 156 species of birds.

Burroughs, John, and Richard Fleck, ed. *Deep Woods.* Syracuse: Syracuse University Press,1998. John Burroughs popularized the nature essay in American literature. He wrote between 1871 and 1912, visited Yellowstone with President Theodore Roosevelt, and hiked the Grand Canyon.

Leopold, Aldo. *A Sand County Almanac.* New York: Ballantine Books, 1990. Originally published in 1949, this book is a must read for anyone interested in nature writing and has influenced many writers and ecological activists. Leopold wrote the book while living in a summer shack along the Wisconsin River.

Muir, John. *My First Summer in the Sierra.* East Rutherford, NJ: Penguin, 1997. First published in 1911, the book details John Muir's observations as a sheepherder in the Sierras in 1869, when he was a young Scottish emigrant and before he became a famous naturalist. This book is part of the Penguin Nature Classic series, which was republished in the 1990s. The series includes eight interesting books about naturalists who lived and worked in specific biomes.

Murray, Peter. *Deserts (Biomes of Nature).* Chanhassen, MN: Child's World, 1996. Although written for children ages 9–12, the nine books in this series are interesting to people of any age. Each book describes a specific biome and presents several activities related to it.

Zwinger, Ann. *The Mysterious Lands: A Naturalist Explores the Four Great Deserts of the Southwest.* Tucson: University of Arizona Press, 1996. This is one of many great books of nature writing from a prominent writer.

25

Ecosystem Dynamics: How Ecosystems Work

Yarrow and fireweed, Mt. St. Helens National Volcanic Monument.

Populations

The reproductive characteristics of plants create challenges in studies of plant populations

The distribution of plants in a population may be random, uniform, or clumped

Age distributions and survivorship curves describe the age structure of populations

The growth of populations over time is limited by environmental resources

The growth of plant populations depends on reproductive patterns

Interactions Between Organisms in Ecosystems

Commensalism and mutualism are interactions in which at least one species benefits

Predation, herbivory, and parasitism are interactions in which at least one species is harmed

Plants compete for resources with members of their own and other species

Communities and Ecosystems

Communities can be characterized by species composition and by vertical and horizontal species distribution

Apparently uniform environments are often composed of different microenvironments

A moderate level of disturbance can increase the number of species in an ecosystem

Ecological succession describes variation in communities over time

The energy stored in photosynthetic organisms passes inefficiently to other organisms in the same ecosystem

Biological magnification increases the concentration of some toxic substances at higher trophic levels

Water and nutrients cycle between biotic and abiotic components of ecosystems

Human activity has fragmented stable ecosystems into distinctive patches

The ponderosa pine (*Pinus ponderosa*) has a yellowish-brown to reddish-brown trunk with thick, fire-resistant bark. In older trees, the bark smells like vanilla. Five subspecies of ponderosa pine occur in the tree's natural range, the foothills of the Rocky Mountains from northern Mexico to southern Canada. Although ponderosa pines sometimes grow in dense, old-growth stands, they exist most often as widely separated individuals in the transition zone between grasslands and forested mountains. Other plants found occasionally with ponderosa pines include various types of conifers, such as junipers, blue spruce, lodgepole pine, and Douglas fir, as well as aspen and cottonwood in areas with more moisture.

Under natural conditions, low-intensity ground fires burn through ponderosa pine forests about every 3 years. Typically, the grasses and some shrubs and mature trees survive, while smaller trees and shrubs are destroyed. Two or three times a century, more severe fires occur, which may destroy most trees. Between 1850 and the early 1900s, harvesting of ponderosa pines for construction, fencing, railroad ties, and mine timbers devastated natural stands of the tree in many regions. Since then, the establishment of National Forests has markedly increased the extent of ponderosa pine forests. However, the practice of suppressing all fires

Ponderosa pines.

has allowed shrubs and smaller trees to grow unchecked, dramatically elevating the danger of large-scale fires.

The seeds of ponderosa pines are favored by several species of squirrels, including Albert's squirrel (*Sciurus alberti*), which hides seeds in the ground, and by birds such as Clark's nutcracker (*Nucifraga columbiana*), which hides seeds in nooks and crannies of rocks and trees. The hidden seeds serve as food caches during winter. Deer and elk forage in ponderosa pine forests, eating grasses and shrubs and gnawing on the bark of younger trees during periods of heavy snow. The males use older trees as rubbing posts to remove velvet from their antlers. Coyotes, mountain lions, and bears pass through the forest on a regular basis to obtain food.

In this chapter, we will explore how organisms such as ponderosa pines interact with members of their own species, with other species, and with the abiotic features of their environment. We will examine the dynamics that occur in groups of interacting organisms and how these groups fit into ecosystems, such as the coniferous forests of the Rocky Mountains.

Clark's nutcracker.

Albert's squirrel.

Populations

Rather than studying the biosphere as a whole, ecologists often focus on smaller ecosystems. Each terrestrial ecosystem has a predominant group of plants and associated species of animals and other organisms, as well as distinctive abiotic features. As you learned in Chapter 24, the location of an ecosystem on the globe determines many of its abiotic features, such as its temperature range, the direction and intensity of prevailing winds, and the length of its seasons. The biotic components of an ecosystem interact with each other and with the abiotic components in various ways.

An ecosystem's biotic components consist of populations of organisms. A **population** is a group of interbreeding organisms of the same species in the same location. An ecosystem might have many populations of a particular organism or just one. If a population is reproductively isolated from other populations, it may be considered a species or at least on the pathway leading to speciation (Chapter 15). Populations of the same species may also be found in more than one ecosystem.

The reproductive characteristics of plants create challenges in studies of plant populations

Most animals, bacteria, and unicellular algae live as individuals and are treated as single units in studies of populations. Parameters of population study, such as age distribution, population density, distribution in time, and rates of birth, death, and population growth, are easily defined for such organisms, and the results can be analyzed readily by standard statistical methods.

However, plant populations are more complex, and plant population ecology is still a young and developing field. While some plants are clearly individuals, others are part of a collective organism. Many plants reproduce vegetatively by sending up shoots that form on underground roots. The largest organism in the world may well be a clone of connected and genetically identical aspen trees in western North America. One such clone discovered near the Wasatch Mountains of Utah is spread across 80 hectares (200 acres). Should it be considered one plant or many? If we treat it as one plant, how should we define its age? In most physiological respects, each tree behaves as an individual. Even in a plant that is not part of a clone, some parts may die while others continue to live. Since plants grow at apical meristems, in a sense each meristem is potentially an individual plant.

In seed plants, the formation of a seed can lead to a new individual, but the seed must germinate. Many factors influence how many seeds a plant produces each year,

and seed production by plants in the wild is often difficult to estimate. Furthermore, variations in wind or the presence of animals can have a marked influence on the number of seeds that germinate. Some seeds lie dormant and may germinate over a period of several years or not at all.

Even defining a population and a species is difficult in plants, as you learned in Chapter 15. Plants interbreed across species and even across genera more readily than animals do. For example, wheat ($2n = 42$) arose in the wild from a combination of the genomes of three species (each $2n = 14$) in two separate wide crosses (Chapter 14). This sort of transgeneric cross is not unusual in the plant kingdom. Some evolutionary biologists argue that plants' ability to interbreed more readily means that speciation is more complex in plants than it is in animals.

The distribution of plants in a population may be random, uniform, or clumped

Like other organisms, plants are distributed in one of three basic patterns: random, uniform, or clumped (Figure 25.1). Random distributions frequently characterize plants with seeds that are light and spread by the wind, such as dandelions. This pattern also predominates where conditions favoring growth are themselves distributed randomly. A well-groomed lawn is an example of a uniform, or evenly spaced, distribution. Pine trees sometimes form uniform distributions in forests by shading nearby seedlings and thus preventing their growth. Other plants attain a uniform distribution by producing compounds that inhibit seed germination in a circular pattern around them. Known as **allelopathy,** this inhibition reduces competition for water and soil nutrients. A clumped distribution may result from vegetative reproduction by plants or from the short-range dispersal of heavier, less mobile seeds. Thus, the distribution pattern of a particular plant population can provide important information about the plant's lifestyle and life history.

Distribution patterns are highly dependent on scale. In fact, it is possible for plants to be distributed uniformly on a small scale, randomly on an intermediate scale, and in a clumped pattern on a large scale. For example, a single wild strawberry plant sends out runners, producing a uniform cluster of plants on a small scale. On an intermediate scale, clusters may be randomly distributed. On a large scale, groupings of clusters may be clumped in areas where soil conditions and light intensity are best for the species.

Populations of the creosote bush (*Larrea tridentata*) change their distribution pattern over time. Because only certain places are suitable for seed germination, creosote bushes begin life in a clumped distribution. Competition

(a)

(b)

(c)

Figure 25.1 Plant distribution patterns. (a) The deciduous trees in this forest grow in a random distribution. (b) In this pine forest the trees are uniformly distributed. (c) Bear grass is growing in a clumped distribution.

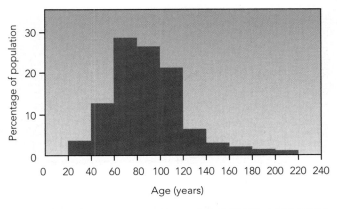

Figure 25.2 An age-distribution graph. The population of oak (*Quercus*) trees represented in this graph consists mostly of middle-aged individuals. No new trees have entered the population in the past 20 years.

occurs within a clump, giving rise to a random distribution. Finally, the root systems of larger plants compete, giving rise to a uniform distribution. A detailed study of creosote bushes by Donald Phillips and James MacMahon at Utah State University demonstrated that the plants' root systems rarely overlap and are not circular, suggesting that there is competition between neighboring plants.

Age distributions and survivorship curves describe the age structure of populations

Different species of plants vary widely in longevity. Many plants live just one growing season, while others live thousands of years. Between these extremes are plants such as biennials, which produce only vegetative growth during their first year and flower in their second year. Age-distribution graphs indicate the relative number of individuals of different ages in a population (Figure 25.2). These graphs can reveal not only the most common age in a population but also which age has the highest death rate.

Survivorship curves show how the death rate in a population correlates with age (Figure 25.3). The curves are

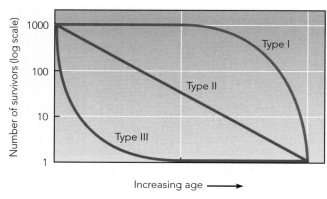

Figure 25.3 Survivorship curves. In each curve, a steep downward deflection represents a sharp decrease in the number of survivors (the death rate is high). A relatively flat region represents a period when the number of survivors remains fairly steady (the death rate is low).

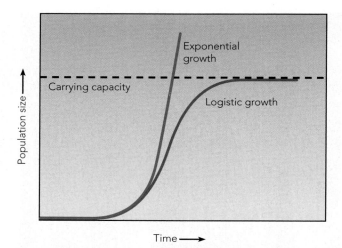

Figure 25.4 Exponential and logistic population growth. Populations exhibit exponential growth when resources are unlimited. In environments with limited resources, growth is logistic: it slows as the population approaches the carrying capacity of the environment.

divided into three basic types. In populations with a Type I curve, the death rate is very low for young and middle-aged individuals and but increases sharply in old age. Populations with a Type II curve have a death rate that is the same for all ages. A Type III curve describes populations with a death rate that is very high in the young and very low in middle-aged and old individuals.

In plants, an increased death rate among older individuals (a Type I curve) is easy to understand, since old plants have accumulated wear and tear. Moreover, natural selection has no mechanism for favoring the survival of plants that have lived past their reproductive age. In contrast, an increased death rate in seedlings and young plants (a Type III curve) may seem curious at first. Natural selection should favor adaptations that increase the survival of young plants, which will soon be reproductive. However, young plants have a wide variety of genotypes, which are exposed to natural selection. Also, plants are more vulnerable to herbivores and harsh growing conditions, such as thin soil and shading by other plants, during germination, seedling growth, and establishment of a mature plant. In addition, a young plant with its shallow root system may be unable to obtain sufficient water and nutrition.

The growth of populations over time is limited by environmental resources

Demography is the study of changes in population size over time. Bacteria and other unicellular organisms are often used to develop population-growth models because they can be studied easily in the laboratory, where variables can be controlled. Also, since division of a unicellular organism immediately produces new organisms, there are no complex embryological or developmental periods

to consider. For plants, particularly in the field, demographic studies must include many variables. For example, a population of lodgepole pines (*Pinus contorta*) will increase with the number of seeds produced, the number of seeds released by the heat of forest fires, and the amount of sunlight, soil nutrients, and rainfall. Disease organisms, herbivores, and seasonable variables also play a role.

Any population will increase in size if its reproductive rate—the rate at which new individuals are added to the population by reproduction—exceeds its death rate. In a hypothetical ideal environment with unlimited resources, populations expand rapidly, exhibiting what is known as *exponential growth* (Figure 25.4). The growth rate of a population under these conditions, denoted as r_{max}, is the maximum growth rate of which the species is physiologically capable.

In any real environment, however, resources are limited. As populations continue to grow, each individual's share of the available resources gets smaller and smaller. As a result, the growth of the population slows. Growth under these conditions is called *logistic* or *density-dependent growth* (see Figure 25.4). To return to the example of lodgepole pines, the shade produced by mature trees slows population growth because seedlings of this species grow poorly in shade. Light becomes a limiting resource, and increasing population density reduces the rate of population growth. At some point, populations reach a maximum size that the environment's resources can support, and further population growth stops. This size is called the **carrying capacity** of the environment and is represented by K.

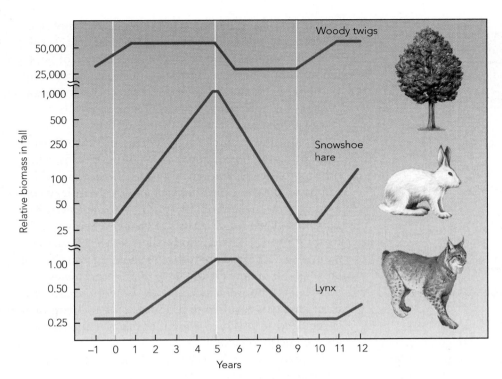

Figure 25.5 Linked oscillations in populations. Increases and decreases in the snowshoe hare population (measured by relative biomass) follow changes in the hares' main winter food supply, woody twigs. Similarly, changes in the lynx population parallel changes in the hare population.

On reaching the carrying capacity, some populations oscillate in size around that value. Frequently, several species are involved in these oscillations. For example, Figure 25.5 shows that oscillations in the population of Canadian snowshoe hares (*Lepus americanus*) are linked to changes in the abundance of their chief winter food, woody twigs (browse), and their main predator, the lynx (*Lynx canadensis*). When woody twigs are abundant, the population of hares increases after a delay of 1 year. After another year, the lynx population expands. An increase in the hare population reduces their food supply, and as vegetation becomes scarcer, the hares produce fewer offspring and are more likely to starve or to be eaten by predators, causing the hare population to crash. This allows the vegetation to rebound, which, in turn, provides food for more hares.

The growth of plant populations depends on reproductive patterns

Natural selection favors different patterns of reproduction under different environmental conditions and population sizes. In environments where individuals face little competition and population size is far below the carrying capacity, selection favors traits that lead to rapid reproduction (a large r_{max}). Such traits include quick maturation and the production of many offspring. Fern sporophytes, for example, produce millions of spores, of which only a few will survive to produce gameto-

phytes. Selection for traits that maximize the reproductive rate of populations in uncrowded environments is called *r*-selection. Other traits of *r*-selected populations include a short life span and often a high death rate.

In populations that are close to the carrying capacity, natural selection favors traits that enable individuals to compete successfully for resources and to use resources efficiently. Selection under these conditions is called **K-selection,** or density-dependent selection. *K*-selected populations produce only a few offspring and have adaptations that increase the likelihood that each offspring will survive. For example, coconut palms produce only a few seeds each year, and the fruit that bears each seed contains a large amount of endosperm, which nourishes both the embryo and the seedling. Table 25.1 compares some of the characteristics of *r*-selected and *K*-selected populations.

Plants also differ in how frequently they reproduce and at what age reproduction begins. Many plants reproduce every year of their lives beginning with the first year. Others, including many trees, reproduce every year but only after they are at least a few years old. Still others, such as *Yucca*, practice "big-bang" reproduction, flowering just once during a life of many years and dying after one reproductive cycle.

The number and size of seeds produced by plants may vary with environmental conditions. Plants that colonize widely separated environments often produce large numbers of small, windblown seeds, most of which do not

Table 25.1 Characteristics of *r*-selected and *K*-selected Populations

Characteristic	*r*-selected populations	*K*-selected populations
Maturation time	Short	Long
First reproduction	Early	Late
Number of offspring per reproduction	Many	Few
Number of reproductions per lifetime	Sometimes only one	Often several
Size of offspring or seeds	Small	Large
Death rate	Often high	Usually low
Life span	Short	Long

Source: Adapted from E. R. Pianka, *Evolutionary Ecology,* 4th ed. New York: Harper & Row, 1987.

survive. Plants in stable environments with good growing conditions generally produce larger seeds with a higher survival rate. Some plants produce two sizes of seeds on each individual. In desert-dwelling asters, for example, the outer seeds are dispersed nearby while the inner seeds, which often have a pappus or plume that can catch the wind, are dispersed far away. Seeds that land near the parent plant usually remain dormant, while those that land far away typically germinate quickly. Purslane speedwell (*Veronica peregrina*) produces a small number of heavy seeds in moist environments (Figure 25.6). When conditions are drier and speedwell must compete with grasses for growing space, it grows taller and produces lighter seeds that are carried farther from the parent plant.

It is important to understand that plants do not sense that they must produce lighter seeds that can reach a suitable environment for germination. Existing plant populations simply have seed-production and dispersal traits that have been favored by natural selection over many generations. Alternative traits that have proven less successful are not represented in the population.

As you know, some angiosperms produce flowers containing the reproductive parts of both sexes, while others produce separate male and female flowers, which can be on the same or different individuals. At least one angiosperm, the jack-in-the-pulpit (*Arisaema triphyllum*), can change sexes as its size changes. Because jack-in-the-pulpits are nonwoody, their size may fluctuate from year to year, depending on growing conditions. Small plants are males, and large plants are females. The relationship between size and sex is adaptive because more energy is required to produce seeds and fruit than is required to produce pollen, and larger plants have greater energy reserves.

Animals have evolved elaborate behavioral and physiological mechanisms that help individuals select mates. Plants, of course, are usually rooted to the ground, and male gametes, which develop from pollen grains, are transferred indirectly from one plant to another. The stigma of a flower may receive pollen from numerous individuals, some of which may be of different species. As you learned in Chapter 15, the chemical environment of the stigma determines which pollen grains germinate and which do not. Plants frequently have pollen-incompatibility genes that prevent some pollen grains from germinating.

Many plants, including most conifers, are wind pollinated. Some conifers release pollen only when female cones of the same species are receptive. Many flowering plants depend on insects, bats, or birds to transfer pollen from one individual to another. The relationship between particular species of plants and pollinating animals is often quite elaborate, and some plants, such as the traveler's palm discussed in Chapter 5, are pollinated by a single species of animal. The danger in this relationship is that the survival of the plant species depends on that of the pollinator. The advantage is that the flowers receive considerable attention from one pollinating species and much less pollen from other species, so the stigmas are exposed to a higher proportion of pollen that can germinate.

Figure 25.6 Purslane speedwell (*Veronica peregrina*). This species produces short plants with heavy seeds and tall plants with light seeds. The two types are produced randomly, but a particular environment may favor one or the other.

Section Review

1. Why is it more difficult to estimate the population size of aspen trees than of deer?
2. Distinguish between Type I and Type III survivorship curves.
3. Define *carrying capacity*.
4. What is the principal difference between *r*-selection and *K*-selection?

Interactions Between Organisms in Ecosystems

Plants are not hermits. They live with and interact with other organisms in various ways. A plant's chances of survival are affected by these interactions, which have been shaped by evolution. For example, many plants produce alkaloids and other compounds that are bitter tasting or poisonous to herbivores. Often, these compounds are contained in trichomes, or leaf hairs, which are the first part of a plant that a herbivore consumes. Some herbivorous insects have evolved a resistance to these compounds, while others have evolved a behavioral avoidance of the plants that produce them. At each level, random mutations have resulted in increased fitness in the organisms that have them.

Commensalism and mutualism are interactions in which at least one species benefits

Plants and other organisms sometimes interact in ways that are beneficial to one or both species. **Commensalism** is an interaction between two species in which one benefits while the other is unaffected. An epiphytic plant living in the top of a tree in a rain forest is an example of commensalism. The epiphyte benefits greatly, but the tree is neither helped nor harmed (unless the epiphyte grows so large that its weight causes limbs to break). In another example of commensalism, saguaro cactus (*Cereus gigantea*) seedlings are usually found growing in the shade of "nurse plants," where the temperature is lower and the soil is moister.

Mutualism is an interaction between two species in which both species benefit (Chapter 4). For flowering plants, pollination by animals is often mutualistic. The pollinators gain nectar and pollen as a food source; the plant gains transportation for its pollen, which enables

cross-pollination. Two other important mutualisms involving plants occur in the soil. Nitrogen-fixing bacteria infect the roots of some plants, providing the plants with nitrate, a key nutrient (Chapter 10). Mycorrhizal associations between fungi and plant roots increase the plants' ability to absorb water and minerals (Chapter 19). In both of these mutualisms, the nonphotosynthetic partner benefits by receiving some of the organic compounds produced by the plant (see *The Intriguing World of Plants* box on page 532).

Predation, herbivory, and parasitism are interactions in which at least one species is harmed

Exploitation occurs when two species interact and one is harmed while the other benefits or is harmed less. Exploitation includes predation, herbivory, and parasitism. In predation, one organism (the predator) feeds on another (the prey), often killing it in the process. Like any other organism, plants are subject to attack by disease organisms in the form of fungi, bacteria, and protists, which are usually predatory in nature. Specific plant diseases were discussed in Chapters 17 and 19.

In herbivory, an animal feeds on a plant but usually does not kill it. Herbivores may be generalists, which eat a variety of plants, or specialists, which eat a specific type of plant. In Chapter 15, you learned that plants such as grasses have evolved a response to herbivory in which the shoot apical meristem, which produces new leaves, is at the very bottom of the plant. In this location, the meristem generally escapes being eaten and can regenerate a photosynthetically active plant after the upper parts of the plant are consumed. All plants respond to the loss of apical meristems by producing axillary shoots that make the plant bushier. Some plants also have prickles, spines, and thorns, which deter many herbivores.

Normally, herbivores and the plants they eat are able to coexist despite the negative effect of herbivory on plants. Typically, the numbers of herbivores and the numbers of plants are related. One example we noted earlier in this chapter: When the supply of woody twigs decreases, the number of snowshoe hares falls. The question for ecologists is whether the drop in the supply of woody twigs is due to abiotic factors, such as moisture and temperature, or to an overabundance of hares. Clearly, both factors might be involved. Similar interactions connect the algal biomass of a stream with the number of grazing caddisflies (*Helicopsyche borealis*). Introduction of caddisflies to a stream soon initiates a drastic decline in algal

THE INTRIGUING WORLD OF PLANTS

Ant Plants

Mutualisms occur between many types of ants and plants. The plants are known as *myrmecophytes,* from the Greek words *myrmeko,* meaning "ant," and *phyton,* meaning "plant." Myrmecophytes provide food or shelter for the ants, while the ants provide protection of nutrients for the plant.

Myrmecophytes include certain species of Central and South American *Acacia* trees. Large, hollow thorns at the base of each of the tree's leaves are inhabited by stinging ants. The trees produce nectar and protein-rich Beltian bodies, which are consumed by the ants. The ants keep herbivores away from the trees and remove debris, fungi, and other plants that grow nearby and might shade the *Acacia.* The services performed by the ants are essential for the survival of the *Acacia.* When the ants inhabiting a tree are poisoned, the tree dies.

Acacia thorn and ants.

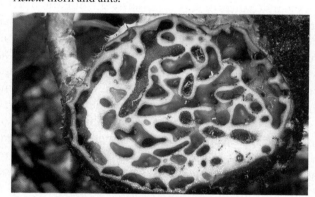

Section through a *Myrmecodia* hypocotyl.

Plants in the genus *Myrmecodia* house ants in the hypocotyl, or embryonic stem, which swells as the ants carve out tunnels and chambers inside. In some species, the chambers are large enough to house lizards or frogs. The ants deposit their waste in specialized chambers, the walls of which are lined by tiny bumps that absorb nutrients in the waste. Many myrmecophytes grow as epiphytes rather than in soil. Therefore, they typically lack sufficient nitrogen without the presence of the ants. The ants protect the plants from herbivorous animals and sometimes play quite an active role in killing or driving away insect larvae.

---

biomass, which then triggers a drop in the number of caddisflies.

Sometimes, herbivory can severely reduce plant populations. For example, if too many cows are living in an area, overgrazing can occur. This happened in southern Texas in what is now Big Bend National Park. Early ranchers lowered the water table by irrigating and destroyed the fragile grasslands by allowing overgrazing. The ranchers based their estimate of the number of cattle per hectare that the grasslands could support—the carrying capacity of the environment—on what was possible during years of above-average rainfall.

Parasitism is a relationship in which one organism feeds on another organism that is still alive. Parasitism by plants is relatively uncommon. Of the approximately 250,000 species of flowering plants, about 3,000 are partial or complete parasites on other plants. Parasitic plants usually have little or no chlorophyll and therefore cannot carry out photosynthesis; they obtain carbohydrates from

their hosts. Some parasitic plants, such as dodder (*Cuscuta salina*) and mistletoes (*Arceuthobium, Phoradendron,* and other genera), form specialized structures called *haustoria* that grow into their hosts' tissues (Figure 25.7). Others, including Indian pipe (*Monotropa uniflora*), absorb carbohydrates from the roots of other plants by way of mycorrhizal fungi.

Plants compete for resources with members of their own and other species

Plants growing in the same area compete for light, water, and mineral nutrients. Intraspecific competition—competition between individuals in the same species—is probably most common at the seedling stage. Such competition results in what is called "self-thinning." Hundreds or thousands of seedlings may germinate in an area that will eventually be occupied by just one plant. As the seedlings grow, the more vigorous ones survive,

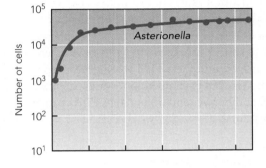

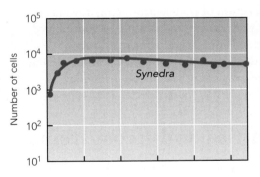

Figure 25.7 Parasitic plants. (a) Threadlike dodder (*Cuscuta gronovii*) growing on a host plant. **(b)** Mistletoe growing on a Jeffery pine.

(b)

(a)

and the others die. In forests, competition can continue for many years, as even large trees compete for resources. Considering that large trees can live hundreds of years, they are probably better at defending their territory than animals are.

Interspecific competition—competition between individuals of different species—can result in the elimination of one species or in the coexistence of both species. The **competitive exclusion principle** states that if two species live in the same area and compete for exactly the same resources, one species will eventually be eliminated from the area. Figure 25.8 illustrates this principle for two species of diatoms. As you read in Chapter 18, diatoms are aquatic, unicellular algae that use silica in the surrounding water to build their cell walls, or frustules. When both species are grown together in a vessel containing a limited amount of silica, only one of the species survives.

There are many cases in which two or more species seem to be using the same resources in the same area, in apparent violation of the competitive exclusion principle. However, close observation will reveal that they actually differ slightly in their use of resources. For example, smartweed (*Polygonum pensylvanicum*), Indian mallow (*Abutilon theophrasti*), and bristly foxtail (*Setaria faberii*) all colonize prairie soil that is no longer being farmed. An examination of their root structures shows that each species draws water and mineral nutrients from a different depth of soil.

Competition for nutrients may explain the finding in numerous experiments that the number of species in an ecosystem decreases as nutrient supplies increase. One

When *Asterionella formosa* and *Synedra ulna* are grown separately in flasks containing a limited amount of silica, each species reaches a stable population level.

When the two species are put in competition in the same flask, *Synedra* thrives while *Asterionella* dies out.

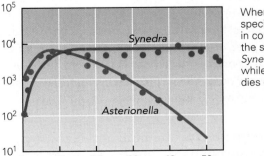

Figure 25.8 Competitive exclusion in two species of diatoms.

study in a rain forest in Ghana, Africa, found that the number of plant species per hectare varied from 2,000 to 100, depending on whether soil fertility was low or high. In another study, conducted at Rothamsted Experimental Station in England, a grassland was fertilized from 1856 to 1949. During that time, the number of plant species declined from 49 to 3. Research has also shown that, even though high nutrient levels lead to a small number of species, the productivity of those species is high. The increase in productivity is easy to understand, but the decrease in diversity is harder to explain. One popular hypothesis is that, with competition for nutrients reduced or removed as a factor, plants compete chiefly on the basis of their ability to use available light. The most efficient species become dominant.

Sometimes competition causes one or both of two competing species to alter their resource use or their tolerance to certain abiotic factors. For instance, wild radish (*Raphanus raphanistrum*) and spurry (*Spergula arvensis*) have essentially the same range of optimal soil pH when grown separately. When the two are put in competition, however, spurry grows best in soil at the lower end of its range. This change minimizes competition between the two species.

Section Review

1. Is the interaction of pollen-producing flowers and pollinating organisms an example of commensalism or mutualism? Explain your reasoning.
2. How do dodder and Indian pipe obtain carbohydrates?
3. What is the competitive exclusion principle?

Communities and Ecosystems

A **community** is a group of species that inhabit a given area. Thus, communities are the biotic components of ecosystems. As you learned in Chapter 24, ecosystems can be small—a pond or even a single rock, for example—or as large as the entire biosphere. Larger ecosystems typically include a number of different communities. Community ecology focuses on the interactions between members of a community and on how these interactions control the types of species found as well as their abundance and diversity.

Communities can be characterized by species composition and by vertical and horizontal species distribution

A community is frequently characterized by one or more **dominant species,** which are those species that have the most individuals, greatest biomass, or some other indicator of importance in the community. In the mountain forests of Colorado, for example, the ponderosa pine is a dominant species at certain elevations.

Most communities also often have a **keystone species,** a species that has a strong effect on the structure of the community, even though the species itself may not be particularly abundant (see the *Conservation Biology* box on page 535). If a keystone species is removed from a community, substantial changes in the community may occur. In the ponderosa pine community, the predominant grasses could be considered keystone species. Removal of the grasses would reduce the populations of large and small herbivores, which would in turn shrink the food supply for carnivores.

Plant communities also have a defining physical structure that is often based primarily on the type and height of the plants. Vertical layering is characteristic of many forest communities (Chapter 24). In a forest, plant layering starts on the ground with grasses and short-lived herbs. Persistent shrubs provide a second layer, particularly in well-lit areas. Saplings, subcanopy trees, and canopy trees are additional, higher layers. In some forests, trees support several layers of epiphyte plants, the number of which depends on the amount of light and precipitation. Frequently, trees also define the vertical layers occupied by animals. For example, in a famous study on Mount Desert Island, Maine, in the 1950s, Robert MacArthur found that each of several warbler species hunted for insects at a different vertical location in spruce trees.

Horizontal patterns are also common in communities. If you walk through a field, you encounter patches composed of different types of plants. In a forest, gaps in the forest canopy allow different species of short plants to become established. Periodic fires and other disturbances often profoundly affect horizontal patterns, at least in the short term.

The requirements and habits of individual plants can affect community structure. Earlier in this chapter, you learned that creosote bushes are distributed in a clumped pattern as seedlings and in a uniform pattern as mature plants. Individual creosote bushes sometimes reproduce vegetatively around the outside of the plant. Over time, this method of reproduction may produce a ring of genetically identical bushes. If you encountered such a ring in the desert communities where creosote bushes live, it would appear as an anomaly in the uniform distribution of the plant. While vegetative reproduction may explain the pattern, the question for the ecologist remains: Why do some plants form rings while most do not? Perhaps the mechanism of seed dispersal leads to a ring of daughter

CONSERVATION BIOLOGY

Figs in the Forest

The 2,000 species of trees, vines, and shrubs in the genus *Ficus* are commonly known as *figs*. Figs are especially numerous in tropical rain forests, where several species may be found in the same few square meters. In many tropical-forest communities, figs are keystone species. Their disappearance would remove a prominent food supply and allow other plant species to assume increased dominance in the forest.

Figs are involved in a number of interesting mutalisms that promote pollination and seed dispersal. A fig fruit begins as an inside-out inflorescence, known as a *syconium,* which contains tiny male and female flowers. The flowers are pollinated by female wasps that are attracted to the fig's scent. The wasps squeeze through an opening in the syconium, often losing their wings and antennae in the process. Inside, the wasps lay eggs in some of the flowers, which develop into a swollen gall that provides food for the young wasps that hatch from the eggs. The wasp offspring complete their development and mate inside the syconium. The males die shortly afterward, but the females leave the syconium, taking pollen with them to another fig.

A strangler fig, *Ficus leprieuri.*

The pea-sized to apple-sized fruits of the fig are eaten by many rainforest animals, including fish, birds, monkeys, pigs, deer, rodents, and bats. Because bats frequently cover large areas of the forest in search of food, they spread fig seeds widely across their range, thereby helping expand tropical forests.

The strangler fig (*Ficus leprieuri*) has evolved a life history that enables it to compete effectively for light, a limiting factor for plant growth in the forest. The fig's seeds are deposited in animal droppings on the branches of trees high in the canopy. After germination, the fig grows as an epiphyte, obtaining nutrients from leaf litter and other debris that accumulate on the branches. It sends out many thin roots that wrap around the host tree's trunk. On reaching the ground, the roots obtain additional nutrients, causing the roots to thicken and the shoot apical meristem to begin a more rapid growth. The fig then begins to compete with the host tree for light and soil nutrients, while the fig's thick roots prevent the host tree from expanding in girth, essentially strangling the host. Eventually, the host tree dies, and the fig is left in its place. Very old strangler figs stand on hollowed-out webs of roots.

---

plants surrounding the parent, or perhaps an abiotic factor, such as soil moisture or fertility, is altered in the region of a ring. For instance, the advancing mycelia of basidiomycete fungi may alter soil fertility at the margin of mycelial growth, where a circle of mushrooms called a *fairy ring* forms (Chapter 19).

The characteristics of populations impact community structure. Consider the lodgepole pine again. Mature cones release their seeds when heated by fire. The fire also clears an area in the forest where the seeds can germinate and the seedlings, which require direct sunlight, can grow successfully. A mature forest of lodgepole pine produces dense shade that prevents new seedlings of that species from successfully establishing themselves. The mature forest may persist for many years, or seedlings of shade-tolerant species may grow and eventually shade and kill the lodgepole pines. In addition, fire, wind, or disease may produce gaps in the forest. Some gaps may be dominated by lodgepole pine seedlings, while others may be occupied by quaking aspen (*Populus tremuloides*), which also requires direct sunlight. An ecologist might want to determine why some gaps are filled by pine and others by aspen. One hypothesis might be that only gaps created by fire are repopulated by pines, but other abiotic factors, such as the type and depth of soil and the availability of water, could be involved as well.

Abiotic factors are principal determinants of community structure. Compared to temperate deciduous forests, for example, tropical rain forests receive more precipitation and more intense sunlight and grow in poorer soil. All of these factors play a role in determining the canopy structure and species diversity that are characteristic of each forest. The canopy in a deciduous forest in eastern North America changes with the seasons and has two basic layers: tall trees (such as tulip poplar, *Liriodendron tulipifera*) and understory (including trees such as dogwood,

Cornus spp., and tall shrubs). In contrast, the canopy in a tropical rain forest is relatively constant throughout the year and is more complex and multilayered. Also, many more species of trees, other plants, and animals are found in a tropical rain forest.

Apparently uniform environments are often composed of different microenvironments

As one moves across an ecosystem, it is not surprising to find a number of different communities. For example, a desert has dry regions and occasional oases, or moist areas where the water table is close to the surface. If a river flows through the desert, different plants and animals will be associated with its course even though it may flow only occasionally. As another example, grasslands may have rocky outcroppings, which support more drought-tolerant forms of vegetation, as well as clumps of trees or bushes in ravines, where there is more water.

Ecosystems also have large regions in which the physical environment appears quite uniform. For many years, it was difficult for ecologists to explain how these environments could support as many species as they do. Some areas of what appear to be uniform rain forest, for instance, contain more than 250 tree species per hectare. How can so many species coexist in an environment that seems to be uniform, given the competition for limited resources that commonly occurs within and between species? The answer is that apparently uniform environments are actually more complex than they appear. As a result, species that were at first thought to be in direct competition really are not. Several lines of research support this idea.

For example, oceans and lakes were originally considered homogeneous environments in which nutrients are uniformly distributed. However, wind, currents, and temperature differences cause nutrient levels to vary considerably in different parts of oceans and lakes. Figure 25.9 shows the range in silica concentration in the surface waters of Pyramid Lake, Nevada. Differences in the concentration of silica affect the distribution of freshwater diatoms (*Asterionella* and *Cyclotella*) in the lake. *Asterionella* dominates where the silica concentration is high, and *Cyclotella* dominates where it is low. Therefore, a lake such as Pyramid Lake can present a number of different environments based on the level of a single nutrient.

Terrestrial environments are even more complex, especially with respect to the distribution of soil nutrients and moisture. This complexity creates a variety of microenvironments that give the competitive edge to particular species. For instance, two species of *Galium*, a herbaceous

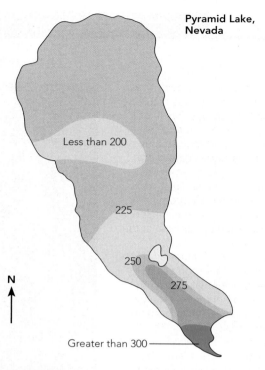

Figure 25.9 Silica concentration in surface waters of Pyramid Lake, Nevada. The concentration (in µg/L) is highest at the south end of the lake, where Truckee River enters the lake.

meadow plant known as *bedstraw,* grow best on quite different types of soil: *G. sylvestre* on alkaline soil and *G. saxatile* on acidic soil. In some locations, the two soil types can be found as patches in the same general area.

When we discuss variations in temperature, nutrient levels, pH, or other environmental factors that may favor one species over another, we are talking about different niches. A **niche,** as formally defined in 1959 by the Yale ecologist G. E. Hutchinson, is the combination of all the physical and biological variables that affect an organism's success. A plant's niche typically includes variables such as temperature range, moisture level, soil type, and seasonal variation. The niche of a lodgepole pine, for example, includes full sunlight, relatively low temperatures, semidry woodland conditions, and well-drained, rocky soil. We could add other conditions, such as the optimum levels of specific nutrients, to more thoroughly define the niche of this tree. A niche also includes an organism's **habitat,** the location where it lives. Mosses live in moist, shaded habitats, whereas sunflowers prefer sunny, dry habitats.

Niches are differentiated on the basis of particular biotic and abiotic characteristics. Each characteristic can be thought of as a point on an axis. For example, one axis might represent low to high rainfall, while a second might represent low to high nitrogen level. Each species' niche consists of a unique collection of axis points. Consider-

Figure 25.10 Effect of moderate disturbance on species diversity. Prairie dog colonies disturb grasslands, creating new microenvironments where forbs and shrubs, as well as grasses, can grow.

ing the number of biotic and abiotic factors in an ecosystem, it is not surprising that there are many different niches even in seemingly uniform environments. This variation appears to account for the surprisingly large number of species found in many ecosystems. Often, a difference in just one key factor, such as silica concentration for diatoms or soil pH for bedstraw, is enough to put two otherwise similar species into different niches.

A moderate level of disturbance can increase the number of species in an ecosystem

The number of species generally decreases with increases in elevation and with distance from the equator (Chapter 24). This seems to imply that warm climates with less seasonal diversity promote species diversity. On the other hand, within a given region, moderate disturbance increases the number of species. This probably occurs because the disturbance creates new microenvironments that can sustain additional species. A good demonstration is provided by prairie dog colonies (Figure 25.10). Regions "disturbed" by these colonies have bare patches, mounds of dirt, and areas where the animals have eaten some species of plants. Each of these areas supports a different community of plants, including the grasses found in nearby undisturbed areas, forbs (herbs that are not grasses), and shrubs.

Ecological succession describes variation in communities over time

Many ecosystems undergo a gradual change in the communities they support. This change is called **ecological succession.** Often, ecological succession follows some

kind of disturbance that eliminates species from an ecosystem or makes a new environment available for colonization by organisms. Ecologists distinguish two types of succession: primary and secondary.

Primary succession describes changes in communities over time in areas that are initially devoid of almost all life and where soil has not yet formed. For example, when a glacier retreats, it leaves behind moraines (long ridges of rocks deposited by the glacier) that contain no organisms except some bacteria. Primary succession can also occur after volcanic eruptions in which the lava or other ejected material creates a new island in the ocean or buries the surrounding area on land (see the *Evolution* box on page 538).

Primary succession often begins with lichens (Chapter 19) and mosses (Chapter 20) that can establish a successful existence on bare rock. Lichens secrete acidic substances, which break down rock. Water seeps into small cracks in the rock and expands when it freezes, further breaking down the rock and providing growing spots for mosses, which themselves expand and contract with the availability of water. These processes gradually form small pockets of soil, where the seeds of small herbs and shrubs can germinate. Eventually, the buildup of organic matter in larger pockets allows trees to become established. Tree roots often serve as major engines to further split rocks. In many cases, succession does not strictly follow this pattern. Because primary succession can begin on a variety of substrates—exposed rock, mudflats, sandbars, glacial moraines, and lava—the progression of organisms may be different, even in the same ecosystem. Eventually, however, primary succession results in the formation of a **climax community,** a community that will remain relatively stable unless it is upset by another disturbance. The

EVOLUTION

Primary Succession after a Volcanic Eruption

May 18, 1980, started out quietly in southwestern Washington State. Suddenly, at 8:32 A.M., the top and one side of Mount St. Helens were blown away by a huge volcanic eruption. The ground shook with the force of a magnitude 5.1 earthquake, and the sky was filled with clouds of smoke and ash. The eruption devastated more than 500 km² (200 mi²) of healthy coniferous forest and left a barren expanse of ash and debris in its place.

The reinvasion of life at Mount St. Helens has been rapid because surrounding areas supplied seeds and because patches of vegetation survived within the devastated area. The first plants to enter the site were pioneer

Mount St. Helens.

species, plants that grow, reproduce, and disperse quickly. One of the early inhabitants, fireweed (*Epilobium*), is well adapted for growth in disturbed, sunny environments. The plant is named after its habit of being among the first plants to grow after a forest fire. The next stages of succession at Mount St. Helens include the establishment of other annual plants, perennial herbs and grasses, shrubs, pines and other softwood trees, and eventually hardwood trees. Some of these plants have already appeared, but the entire process will likely take hundreds of years to unfold.

---

building of a climax community through primary succession can take hundreds or even thousands of years (Table 25.2).

Glacier Bay, Alaska, provides a specific example of primary succession (Figure 25.11). When Captain George Vancouver visited the area in 1794, there was no bay, but rather a thick sheet of ice that ended at the ocean. By 1879, John Muir found open water in Glacier Bay and estimated that the glaciers had retreated 30 to 40 kilometers (18.64–24.85 miles) since Vancouver's visit. The exposed land between the bay and the glacier was covered with plants but had no trees. Since Muir's visit, scientists have documented the con-

tinued retreat of the glaciers and the progress of primary succession in the region. Their studies have revealed that succession at the edge of the bay occurs in several stages:

◆ As it retreats, the glacier leaves behind a variety of microenvironments, which support many small pioneer communities during the first 20 years. Prominent plants include horsetail (*Equisetum varietaum*), willow herb (*Epilobium latifolium*), willow (*Salix*), cottonwood (*Populus balsamifera*), mountain avens (*Dryas drummondi*), and Sitka spruce (*Picea sitchensis*).

◆ By 30 years, a secondary community forms in which the principal plants are shrubs in the genus *Dryas*, including mountain avens. Other plants from the pioneer community are still found among the mats produced by the shrubs.

◆ By 40 years, larger shrubs dominate, in particular alder (*Alnus*). *Populus* and *Picea* are also important community members.

◆ By 75 years, a forest community composed principally of *Picea* and two species of hemlock (*Tsuga*) has come to dominate. The understory consists of mosses, herbs, and seedlings of other trees. These species will form a climax community sometime between 100 and 200 years after glacial retreat. Low-lying areas nearby follow a different successional path and end as a climax community called *muskeg*, which is composed of alternating peat bogs and meadows.

Table 25.2 Length of Time for Various Climax Communities to Form

Climax community	Starting condition	Years to climax
Rain forest	Fresh lava in Hawaii	400
Pine scrub forest	Bare granite in Georgia	700
Spruce–hemlock forest	River terrace sediment in Washington	750
Deciduous forest	Sand dunes in Michigan	1,000
Sagebrush–bitterbrush desert scrub	Inland dunes in Idaho	1,000–4,000
Moss–birch–tussock grass tundra	Glacial debris in Alaska	5,000

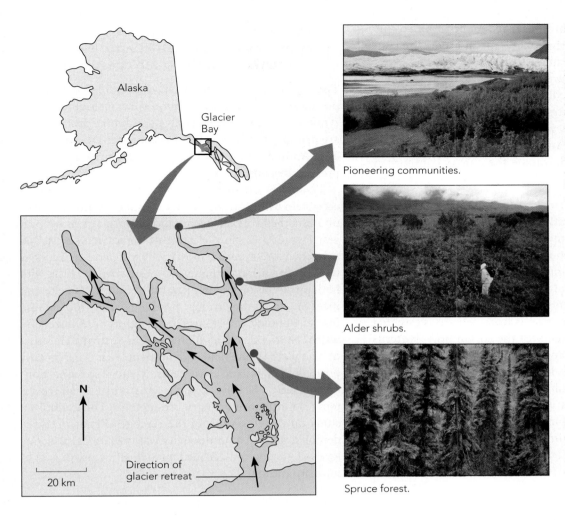

Figure 25.11 Primary succession in Glacier Bay, Alaska. Retreating glaciers leave behind barren moraines, which support a succession of plant communities. By about 40 years after a moraine is exposed, alders and other shrubs are dominant. After another 35 years, a spruce forest is well established.

Pioneering communities.

Alder shrubs.

Spruce forest.

- ◆ Between 250 and 1,500 years after the retreat of the glacier, the number of species gradually increases. This increase, which is characteristic of primary succession, can be seen in Glacier Bay by studying a number of sites around the bay.

Models to explain primary succession differ over the question of whether early species in succession prepare the way for or inhibit the establishment of later species. Researchers have accumulated evidence that supports each possibility. To an extent, the plants that participate in succession represent the species that were available to reseed an area. At each stage of succession, competition occurs between the species that are present and therefore eligible to dominate the region.

Secondary succession takes place where a community has been removed by a disturbance but the soil remains intact. For example, if a farm field is harvested and then left uncultivated or a forest is logged and not replanted, a series of plant and animal communities will occupy the land over time. Although secondary succession often follows human activity, it can also occur after a disease outbreak,

windstorm, fire, minor flood, or climate disturbance such as El Niño alters the species composition of a community.

Secondary succession is generally more rapid than primary succession. As an example, consider the change in species composition of eastern deciduous forests caused by chestnut blight (Chapter 19). In the eighteenth century, American chestnuts (*Castanea dentata*) made up at least 25% of the forests from Maine to Mississippi. Around 1900, the fungus responsible for chestnut blight (*Cryphonectria parasitica*) was introduced to the United States, and within 30 years all mature chestnut trees were gone from North American forests. In the same forests today, the dominant species are hickory, oak, maple, and cherry, depending on the region.

Studies of secondary succession in abandoned fields in the Piedmont Plateau of North Carolina demonstrated the following stages:

- ◆ Crabgrass (*Digitaria sanguinalis*) and horseweed (*Erigeron canadense*) colonize the fields during the first year.
- ◆ During the second year, either aster (*Aster pilosis*) or ragweed (*Ambrosia artemisiifolia*) dominates.

◆ Around 4–5 years, broomsedge (*Andropogon virginicus*) becomes the principal plant, and there are scattered shrubs and small trees.

◆ By 15 years, a pine forest is the most notable botanical component. Pine seedlings require full sunlight, so the understory of the forest is composed of oak (*Quercus*) and hickory (*Carya*), which grow well in shade.

◆ By 150 years, oak and hickory are the major tree species. They will dominate the climax community that eventually forms.

All succession moves toward an end point that is determined by the global location of the community. As succession progresses, common stages are observed in all ecosystems. The total biomass of the community increases. In some climax communities, such as tropical rain forests, a considerable fraction of available mineral resources is incorporated in living plants and in decaying plant material, or detritus. Release of nutrients from detritus is essential for the continued existence of the community. Patterns found in vegetation change as well. For example, trees in early successional stages have many, small, randomly oriented leaves and are multilayered; that is, leaves occur up and down new branches, and some leaves shade others. Aspens, alders, and some pines are examples. Trees in climax forests have smaller numbers of larger leaves and are monolayered; that is, leaves occur at the tips of new branches, where they are not shaded by other leaves.

The energy stored in photosynthetic organisms passes inefficiently to other organisms in the same ecosystem

The organisms in an ecosystem are classified as primary producers and consumers. Autotrophs, including photosynthetic organisms (plants, algae, and some prokaryotes) are primary producers of organic matter and stored energy. Animals, fungi, and heterotrophic prokaryotes and protists are consumers.

Ecologists calculate the primary productivity of an ecosystem by measuring the dry weight of plants and other photosynthetic organisms produced per square meter each year. Figure 25.12 shows clearly that tropical rain forests and temperate forests are the most productive terrestrial ecosystems on Earth, largely because productivity increases with both precipitation and temperature. (Cultivated fields can meet or exceed the primary productivity of tropical rain forests, but only with considerable inputs of fertilizer and water.) Nutrient levels also influence productivity. The effect of high temperatures and high nutrient levels is particularly noticeable in aquatic ecosystems, where these conditions stimulate algal blooms (Chapter 18). Animals in an ecosystem also influence primary productivity. For example, a study on the grasslands of the Serengeti Plains in Africa demonstrated that productivity was greatest at moderate levels of grazing and declined under low-intensity or high-intensity grazing.

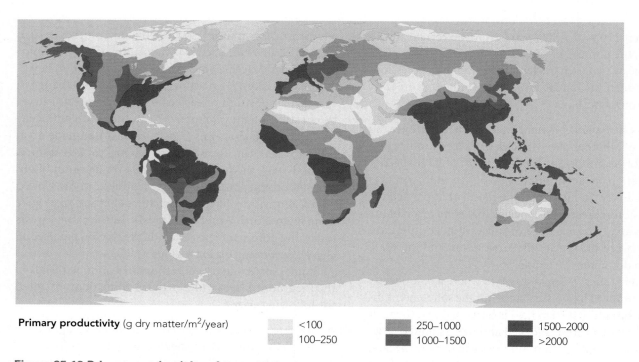

Primary productivity (g dry matter/m²/year)

<100	250–1000	1500–2000
100–250	1000–1500	>2000

Figure 25.12 Primary productivity of terrestrial ecosystems. Primary productivity is greatest in tropical and temperate regions.

As we discussed in Chapter 9, the chemical energy contained in an organism is determined by burning the organism in a calorimeter and measuring how much energy is released. On average, plants and other primary producers convert only about 1% of the visible light that reaches them into chemical energy. In other words, primary producers store about 10,000 joules (the metric energy unit) of chemical energy for every million joules of solar energy available to them. Consumers convert about 10% of the chemical energy they consume into biomass. Thus, for every 10,000 joules of plant material that enters a food chain, a primary consumer (a herbivore) stores about 1,000 joules of biomass, a secondary consumer (a carnivore that eats herbivores) stores about 100 joules, and a tertiary consumer (a carnivore that eats other carnivores) stores about 10 joules. These relationships can be represented by a productivity pyramid (Figure 25.13). Each stage in the pyramid—primary producer, primary consumer, and so on—is called a *trophic level.*

Productivity pyramids are the basis of the observation that agriculture is more efficient and can feed more people if people eat plant products, like rice, instead of animal products, like beef. Of course, such efficiency considerations depend on what is eaten. Rice grains are almost completely digestible by humans, so approximately ten times as many people can be fed per agricultural hectare if rice is produced instead of beef. However, much plant material contains a large amount of cellulose, which humans cannot digest. If broccoli or spinach is on the menu instead of rice, the efficiency of a vegetarian diet is considerably lower.

Biological magnification increases the concentration of some toxic substances at higher trophic levels

Certain toxic substances accumulate in ecosystems and become more concentrated at successively higher trophic levels,

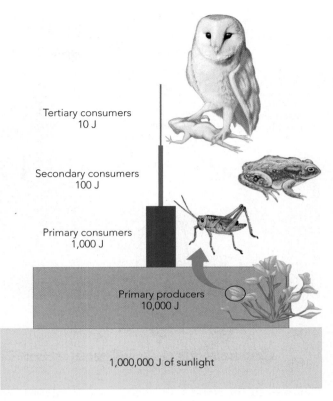

Tertiary consumers
10 J

Secondary consumers
100 J

Primary consumers
1,000 J

Primary producers
10,000 J

1,000,000 J of sunlight

Figure 25.13 A productivity pyramid. Primary producers convert about 1% of the solar energy they receive into biomass. Each higher trophic level transfers about 10% of the energy it consumes to the next level.

a process called *biological magnification.* One such substance is mercury. Mercury is used to extract gold from ore and to produce plastics, and at one time it was simply discharged into rivers and the ocean after use. Mercury-laden bentonite clays used in oceanic oil drilling also release mercury into the water. Bacteria incorporate mercury into an extremely poisonous organic compound, methylmercury, which accumulates in other organisms, particularly certain species of fish.

Nondegradable, lipid-soluble pesticides, such as DDT, accumulate in the fatty tissues of animals and are subject to biological magnification. In birds that feed at the top of food chains, DDT interferes with the deposition of calcium in eggshells,

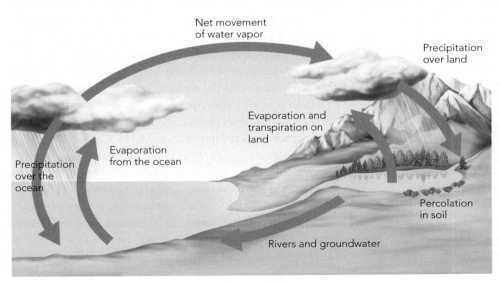

Net movement of water vapor

Precipitation over land

Evaporation and transpiration on land

Evaporation from the ocean

Precipitation over the ocean

Percolation in soil

Rivers and groundwater

Figure 25.14 The water cycle. Evaporation exceeds precipitation over the oceans, but the reverse is true over land.

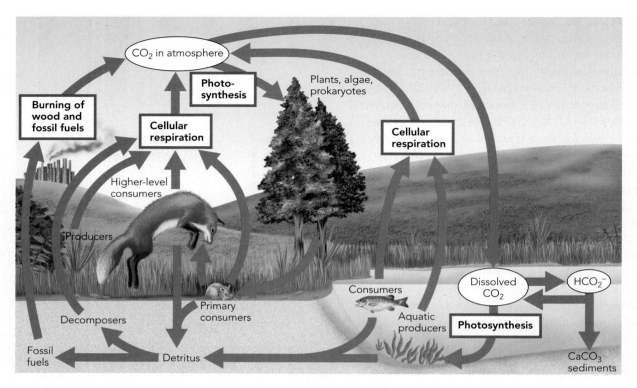

Figure 25.15 The carbon cycle. Photosynthesis and cellular respiration are primarily responsible for carbon cycling through ecosystems. They balance each other, but the burning of wood and fossil fuels is causing a steady increase in the concentration of CO_2 in the atmosphere.

resulting in fragile eggs that break before the young are ready to hatch. Before DDT was banned in the United States in 1973, it caused serious declines in the populations of bald eagles, ospreys, and brown pelicans. Since then, all three populations have recovered.

Because plants are at the base of food chains, they usually contain relatively low concentrations of toxic substances that accumulate through biological magnification. Consequently, you might expect a vegetarian diet to pose a lower risk of ingesting many types of toxic substances. This is not a general rule, however, since vegetables and fruits produced for human consumption are typically treated with more herbicides and pesticides than grasses, which are fed to cattle produced for meat.

Water and nutrients cycle between biotic and abiotic components of ecosystems

As plants grow, they take up water and minerals from the soil and carbon dioxide (CO_2) from the atmosphere. The elements in these substances are incorporated into the plants' structure and pass from there into the structure of primary, secondary, and tertiary consumers. When any of these organisms dies, bacteria and fungi break down its structure,

ultimately releasing water and minerals back into the soil and CO_2 back into the atmosphere. Thus, water, carbon, and minerals cycle continuously between organisms and the nonliving components of ecosystems.

Water enters the atmosphere through evaporation from the oceans and other bodies of water and through transpiration, and it returns to the oceans and the land in the form of precipitation (Figure 25.14). Some of the precipitation that falls over land is carried back to the oceans by rivers, and some percolates into the soil, where it is bound to soil particles. Water that enters the soil, called *groundwater,* flows back to the sea, a journey that may take thousands of years. Large quantities of groundwater are also pumped to the surface for human use.

In the carbon cycle, carbon in CO_2 is incorporated into organic compounds during photosynthesis in plants, algae, and certain prokaryotes (Figure 25.15). Carbon returns to the atmosphere as CO_2 through cellular respiration in producers, consumers, and decomposers and through the burning of wood and fossil fuels. Terrestrial producers obtain CO_2 directly from the atmosphere, where it is a minor component (0.04%). Aquatic producers use dissolved CO_2, which is in equilibrium with dissolved bicarbonate (HCO_3^-) ions and atmospheric CO_2. More than 90% of the carbon on Earth is

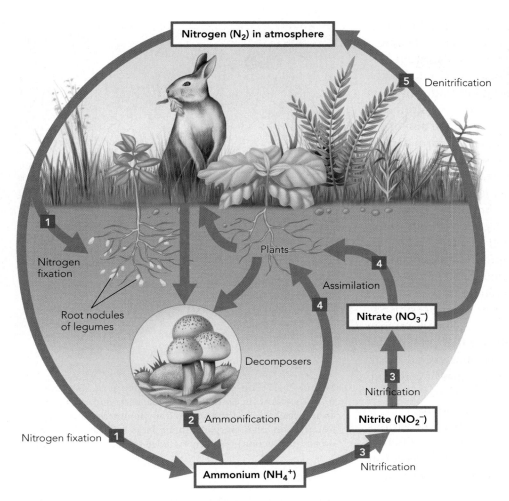

Figure 25.16 The nitrogen cycle. Different groups of soil bacteria are essential components of pathways 1, 2, 3, and 5, which are explained in the text. Fungi also play an important role in pathway 2.

at the bottom of the oceans, in calcium carbonate ($CaCO_3$) sediments formed from the shells of marine organisms.

Minerals such as nitrogen and phosphorous exist as dissolved ions in oceans, lakes, and rivers and are bound to soil particles. Nitrogen gas (N_2) is also the largest component of the atmosphere (78%). In fact, more than 99.9% of Earth's nitrogen is in the atmosphere. Figure 25.16 illustrates the five pathways of the nitrogen cycle. **1** In *nitrogen fixation,* N_2 gas from the atmosphere is converted to ammonia (NH_3) by bacteria in soil and in the root nodules of legumes. This is the only pathway by which nitrogen is added to organic compounds. Ammonia reacts with water in the soil, producing ammonium (NH_4^+). **2** In *ammonification,* decomposers also release NH_4^+ into the soil as they break down the remains of dead organisms. **3** In *nitrification,* soil bacteria convert NH_4^+ to nitrite (NO_2^-) and nitrate (NO_3^-). **4** In *assimilation,* plants take up both NH_4^+ and NO_3^- from the soil and incorporate the nitrogen into amino acids, nucleotides, and other organic compounds, which are passed up the food chain to consumers. **5** In *denitrification,* soil

bacteria convert NO_3^- to N_2, which is released back into the atmosphere.

Human activity has fragmented stable ecosystems into distinctive patches

Landscape ecology is a relatively new and developing area. One of the chief interests of landscape ecologists is the fragmentation of existing ecosystems caused by humans. Three principal developments have contributed to the formation of ecosystem fragments. The first is deforestation, which has proceeded at an especially rapid rate during the last 100 years. With the exception of Canada and Russia, most industrialized nations have almost none of their original, old-growth forests remaining. Forests in many developing nations are almost as devastated. Where less deforestation has occurred, forested corridors may connect forest fragments. The second development contributing to ecosystem fragmentation has been agriculture. Approximately one-third of global land has been converted to farmland. The third factor fragmenting ecosystems is the growth of cities, towns, and subdivisions, which sometimes encroach onto relatively undisturbed ecosystems and other times replace farmland.

As we travel through most regions of the United States, we encounter a mosaic of different kinds of ecosystem fragments, or landscape elements, which flow together to create the overall landscape. Some are a portion of the original ecosystem, usually in a modified form. Others are made by humans or represent secondary succession after human modification. For example, we may pass through an inner city that has occasional planned areas of vegetation. Perhaps a city park breaks the monotony. Some human dwellings have no vegetation, while others may have a balcony for container plants or perhaps a grassy yard and a garden. Here and there we encounter vacant lots with thriving weeds. Outside the city, we enter farmland, where cultivated fields alternate with houses. In some areas, grazing, timber harvesting, or fire suppression has altered the natural vegetation. Finally, we may come to a national park, established to preserve an ecosystem as it existed before Columbus visited North America. Even here, exotic plant species may be common, and major carnivores may no longer be found, allowing herbivores to overgraze and alter the vegetation.

Landscape ecologists seek to understand the various communities that develop in the multitude of different landscape elements. Some landscape elements represent a combination of biotic and abiotic factors not encountered before in nature. Often, an original population is fragmented into smaller populations that live in isolated patches. Landscape ecologists also ask questions about the effect of patch size on population size and density and on species diversity. For example, the number of mammal species declines markedly as a habitat is fragmented and as patch size decreases. When parks and other preserves are considered, it is important to determine if the size of the preserve and its types of use will allow native species to sustain large enough populations to survive. If timber harvesting is permitted in a national forest, will a particular plant or animal find enough suitable habitat to prosper? Will a long, narrow national park provide enough space to sustain the communities that were present before humans arrived? What is the nature of boundary effects where the national park borders on farm and ranch lands? The zoned reserve concept of landscape management seeks to surround conservation areas with compatible human activities (Figure 25.17).

Landscape ecologists can also analyze ecosystem patches to determine if the range of an endangered species lies

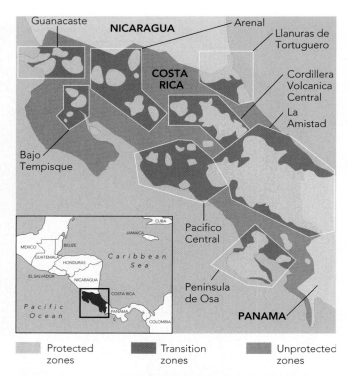

Figure 25.17 The zoned reserve concept of landscape management. In Costa Rica, national parks and other protected areas (beige) are mostly separated from unprotected areas (orange) by transition zones (purple). Mining, large-scale farming, and new urban construction are largely excluded from the transition areas.

within a protected area. Based on such analyses, ecologists can recommend that the boundaries of the area be altered or that safe corridors be established to connect it with other protected areas. For animals, a safe corridor might have fences along highways or underpasses specifically for wildlife. For plants, a safe corridor might be a region where natural vegetation is allowed to grow without mowing or grazing in an area large enough to ensure that pollination and seed dispersal can connect what would otherwise be isolated populations.

Section Review

1. **Distinguish between a dominant species and a keystone species.**
2. **Describe the typical events of primary succession.**
3. **Give examples of primary producers and primary consumers.**
4. **Typically, what percentage of energy is transferred from one trophic level to the next?**

SUMMARY

Populations

A population is a group of interbreeding organisms of the same species in the same location. An ecosystem may have one or many populations of a particular organism.

The reproductive characteristics of plants create challenges in studies of plant populations (p. 526)
Studying plant populations is complicated because many plants are part of a collective organism produced through vegetative reproduction, because seed production and germination in the wild are highly variable, and because interbreeding makes it difficult to define populations and species of plants.

The distribution of plants in a population may be random, uniform, or clumped (pp. 526–527)
Plants that produce light, wind-dispersed seeds are often randomly distributed. Plants that shade nearby seedlings or that inhibit seed germination around them are often uniformly distributed. A clumped distribution may result from vegetative reproduction or the short-range dispersal of heavier seeds. The same plant species may be distributed differently on a small, intermediate, or large scale or at different times during its life.

Age distributions and survivorship curves describe the age structure of populations (pp. 527–528)
Age-distribution graphs indicate the relative number of individuals of different ages in a population. Survivorship curves show how the death rate in a population is correlated with age.

The growth of populations over time is limited by environmental resources (pp. 528–529)
In an ideal environment with unlimited resources, populations grow exponentially. When resources are limited, however, populations exhibit logistic, or density-dependent, growth. At some point, populations reach a maximum size, called the *carrying capacity*, or *K*.

The growth of plant populations depends on reproductive patterns (pp. 529–530)
Selection for traits that maximize the reproductive rate of populations in uncrowded environments is called *r*-selection. Selection for traits that enable individuals to compete successfully for resources and to use resources efficiently is called *K*-selection. Plants differ in how frequently they reproduce, at what age they begin reproducing, the number and size of seeds they produce (in seed plants), and whether they have separate male and female flowers (in angiosperms).

Interactions Between Organisms in Ecosystems

Commensalism and mutualism are interactions in which at least one species benefits (p. 531)
Commensalism is an interaction between two species in which one benefits while the other is unaffected. Mutualism is an interaction in which both species benefit.

Predation, herbivory, and parasitism are interactions in which at least one species is harmed (pp. 531–532)
In predation, one organism feeds on and kills another organism. In herbivory, an animal feeds on a plant but usually does not kill it. In parasitism, one organism feeds on another that is still alive.

Plants compete for resources with members of their own and other species (pp. 532–534)
Intraspecific competition in plants results in self-thinning, while interspecific competition can result in the elimination of one species or in the coexistence of both species. According to the competitive exclusion principle, if two species live in the same area and compete for exactly the same resources, one species will eventually be eliminated from the area.

Communities and Ecosystems

A community is a group of species that inhabit a given area.

Communities can be characterized by species composition and by vertical and horizontal species distribution (pp. 534–536)
Dominant species in a community are the most important based on some indicator, such as biomass or number of individuals. Keystone species have a strong effect on the structure of a community. In many forest communities, plant types are vertically layered. Fires and other disturbances can affect horizontal distributions of plants.

Apparently uniform environments are often composed of different microenvironments (pp. 536–537)
Many environments that appear to be uniform are actually quite complex. As a result, species that may seem to be in direct competition are not. Species can avoid direct competition by having different niches, which are combinations of physical and biological variables that affect their success.

A moderate level of disturbance can increase the number of species in an ecosystem (p. 537)
Within a given region, moderate disturbance increases the number of species, probably by creating new microenvironments that can sustain additional species.

Ecological succession describes variation in communities over time (pp. 537–540)

Primary succession, which occurs in areas that are initially devoid of almost all life and where soil has not yet formed, often begins with lichens and mosses that can establish themselves on bare rock. Secondary succession takes place where a community has been removed by a disturbance, such as harvesting, a disease outbreak, or a windstorm, but the soil remains intact. Both types of succession eventually result in the formation of a climax community.

The energy stored in photosynthetic organisms passes inefficiently to other organisms in the same ecosystem (pp. 540–541)

Plants and other primary producers convert about 1% of the visible light that reaches them into chemical energy. Consumers convert about 10% of the chemical energy they consume into biomass.

Biological magnification increases the concentration of some toxic substances at higher trophic levels (pp. 541–542)

Certain toxic substances accumulate in ecosystems and become more concentrated at successively higher trophic levels. Such substances include mercury, which is incorporated into the extremely poisonous compound methylmercury by bacteria, and the pesticide DDT, which accumulates in the fatty tissues of animals.

Water and nutrients cycle between biotic and abiotic components of ecosystems (pp. 542–543)

Water enters the atmosphere through evaporation from the oceans and other bodies of water and through transpiration, and it returns to the oceans and the land as precipitation. Carbon in CO_2 is incorporated into organic compounds during photosynthesis in producers, and it returns to the atmosphere as CO_2 through cellular respiration and the burning of wood and fossil fuels. The nitrogen cycle involves five pathways: (1) nitrogen fixation, the conversion of atmospheric nitrogen to ammonium; (2) ammonification, the release of ammonium by decomposers; (3) nitrification, the conversion of ammonia to nitrite and nitrate; (4) assimilation, the uptake of ammonia and nitrate by plants; and (5) denitrification, the conversion of nitrate to nitrogen gas.

Human activity has fragmented stable ecosystems into distinctive patches (pp. 543–544)

Deforestation and agriculture are the two principal developments that have contributed to ecosystem fragmentation. Landscape ecologists study the communities that develop in ecosystem fragments and try to understand how the size of the fragments affects population size and density and species diversity.

Review Questions

1. What is a population?
2. Give examples of random, uniform, and clumped plant distribution patterns.
3. What is allelopathy?

4. What does a survivorship curve show?
5. How does exponential growth of a population differ from logistic growth?
6. What is "big-bang" reproduction?
7. How could mate selection occur after pollination in plants?
8. What is the difference between commensalisms and mutualism? Use examples.
9. Distinguish the types of exploitation that can occur between organisms.
10. How does soil fertility relate to species diversity?
11. What is a community?
12. In a ponderosa pine community, what is the dominant species, and what is the keystone species?
13. Why are apparently uniform environments, such as some rain forests, able to support large numbers of species?
14. What is a niche?
15. How does moderate disturbance affect species diversity?
16. Contrast primary and secondary succession.
17. Distinguish between primary, secondary, and tertiary consumers.
18. What is meant by the statement that energy passes between trophic levels with an average efficiency of 10%?
19. Describe the nitrogen cycle in a typical terrestrial ecosystem.
20. What are the two main contributors to ecosystem fragmentation?

Questions for Thought and Discussion

1. Some researchers believe that ecology is the most complex biological science because it has the greatest number of variables. Do you agree or disagree? Defend your answer.
2. If a mountainside has 10,000 aspen trees divided into 100 clones, are there 10,000 aspen plants or 100?
3. Suppose someone tells you that humans started out as a *K*-selected species and have become an *r*-selected species. Is there any truth to this statement? Explain why or why not.
4. A bear eats wild raspberries in a forest. What type of interaction is this? What type of interaction is represented by a family that harvests raspberries and takes them home to make jelly?
5. Is an oasis in the desert an ecosystem or a community?
6. Imagine walking from a well-tended vegetable garden to a somewhat weedy lawn and then into a vacant lot. Discuss the communities you might encounter as you walk.
7. Try to describe all of the conditions and features of your personal niche. Include descriptors besides those that specify food, water, and shelter.
8. In a vacant lot, you discover areas of grass, several patches of wildflowers, small shrubs taking hold, puddles, and several patches of cactus. What is happening in the lot?

9. Olympic National Park in Washington State maintains a thin strip of forest—in some places only a few hundred meters wide—to protect the wild, natural shoreline. How could a forest protect a shoreline?

10. Examine Figure 25.8, which shows data from Tilman's research on competitive exclusion in two species of diatoms. Two aspects of Tilman's work that are not shown on this figure are (1) both diatom species lower the concentration of silica in the water through time, and, in consequence (2) the eventual carrying capacity of *Synedra* cultured in mixture with *Asterionella* is slightly lower than its carrying capacity when cultured alone. Redraw these graphs to show these two phenomena; it is suggested that you indicate silica concentration on the right-hand vertical axis of each figure and use unfilled circles as the graph symbol.

Evolution Connection

What are likely advantages and disadvantages of each of the generalist and specialist modes of herbivory? Which characteristics of plants will most likely select for the evolution of a progressively more specialist mode in a population of herbivores through time?

To Learn More

Visit The Botany Place Website at www.thebotanyplace.com for quizzes, exercises, and Web links to new and interesting information related to this chapter.

Davis, Wade. *One River: Explorations and Discoveries in the Amazon Rain Forest.* Riverside: Simon and Schuster, 1996. Davis is an ethnobotanist interested in the drugs and other medicines provided by plants. He is also a good storyteller, and his book is full of the history, adventure, and magnificence of the rain forest.

Durning, Alan T. *This Place on Earth 2001: Guide to a Sustainable Northwest.* Seattle: Northwest Environment Watch, 2001. From Northern California to British Columbia, the Pacific Northwest has vast resources but many problems—its forests have been mostly cut, and population and pollution are on the increase. The author surveys the problems and presents solutions applicable throughout the world.

Hertsgaard, Mark. *Earth Odyssey: Around the World in Search of Our Environmental Future.* New York: Broadway Books, 1999. Hertsgaard traveled around the world, asking people about the environment and studying and analyzing ecosystems. He presents a fascinating combination of personal accounts, horror stories, and hope for the future.

Matthiessen, Peter. *The Cloud Forest: A Chronicle of the South American Wilderness.* East Rutherford, NJ: Penguin, 1996. In this well-written story, Matthiessen describes his explorations, his encounters with bandits, and the surprising discoveries he made while traveling through 10,000 miles of wilderness.

National Park Service. *Glacier Bay: A Guide to Glacier Bay National Park and Preserve, Alaska.* Washington, DC: U.S. Government Printing Office, 1983. This guidebook contains interesting material on the park's history, both human and natural.

Quammen, David. *The Song of the Dodo: Island Biogeography in an Age of Extinction.* New York: Touchstone Books, 1996. A well-written and interesting account of the role of evolution in the flora and fauna of island chains.

26
Conservation Biology

Urban sprawl near Las Vegas, Nevada.

Human Population Growth

Human population is increasing exponentially

Increased food production will involve genetically altered plants, improved growing practices, and more efficient food distribution systems

Human Impacts on Ecosystems

The presence and activities of large human populations disturb ecosystems

The geographic information system provides a new tool to record changes in ecosystems

The Future

The future of human interactions with ecosystems can be modeled on worst-case or best-case scenarios

Achieving a best-case scenario for the biosphere would involve a marked reversal of current trends

A number of problems would have to be overcome to reverse current trends of ecosystem destruction and modification

It is important to establish models of success in promoting ecosystem restoration

The Earth before humans and the Earth of today are quite different. In part, the difference is due to the several million years of evolution that have occurred since the origin of the genus *Homo* and its predecessor, *Australopithecus,* in Africa. In recent centuries, though, humans have begun to change the world at an accelerating pace, and the change is evident almost everywhere on the planet. Much of the natural landscape of Earth has been altered radically by forestry, agriculture, and the growth of urban areas, which contain the bulk of a burgeoning human population. Even when no visible evidence of our species is seen, one has but to sample the air and water to find chemical signs of our presence. One example is caffeine, a chemical compound found naturally in coffee and tea and added to some sodas and other beverages. Caffeine is an excellent indicator of organic contamination associated with human wastewater. Even though sewage treatment removes up to 99% of caffeine from wastewater, the chemical now occurs in trace amounts in even the most remote parts of the ocean.

Of course, as a product of evolution by natural selection, humans are part of the natural order. We are trophically linked to other organisms, and we influence them and are influenced by them. However, because of our large brains and our ability to alter our surroundings, we are different from any other species in the extent of our impact on ecosystems. This is especially true now that the human population exceeds 6 billion. Thus, it is easy to understand the view of some people that humans act more as destroyers of the natural order than as a part of it. Civilization causes rapid change, which can be viewed as massive disturbance. As you learned in Chapter 25, it takes hundreds or thousands of years after a large disturbance for an ecosystem to go through the stages of primary or sec-

ondary succession to reach a climax community. One of the unique things about the changes caused by human activity is that they alter the biotic and abiotic features of the landscape so extensively that a return to the original climax communities often will never occur, even with human intervention.

The biotic and abiotic factors of an ecosystem are linked in many ways. Changes introduced by humans have the effect of breaking many links in existing ecosystems, resulting in ecosystem destruction and species extinction. Removing just one species often has multiple effects that are difficult to understand without considerable study. The immediate effects of our many activities are possible to record, but the long-term consequences are beyond our ability to accurately predict or to alter.

National parks, forests, and wilderness areas, as well as private preserves, are ways of maintaining selected natural ecosystems in as undisturbed a condition as possible. These preserves protect remnants of original biomes and allow disturbed areas to return slowly to their original, natural state. For example, Big Bend National Park in western Texas preserves part of the Chihuahuan Desert along the Rio Grande, below the lower elevation of the Chisos Mountains. Desert, riparian areas, and high mountains give the park considerable diversity in terms of ecosystems and communities.

Big Bend National Park.

In this chapter, we will examine human population growth and its effects on air, water, natural habitats, biodiversity, and extinction. We will also examine worst-case and best-case scenarios for the future. Worst-case scenarios involve continued human population growth at or above its current rate. Best-case scenarios depend on stabilizing human population and make use of the new science of conservation biology, which is based on an increased understanding of the principles of ecology.

Human Population Growth

Both human numbers and human activities contribute to environmental problems and ecosystem damage. A lack of resources to sustain human populations may ultimately become the predominant factor stabilizing population growth.

Human population is increasing exponentially

For thousands of years, since the beginning of agriculture and animal husbandry around 10,000 years ago, human population increased very slowly (Figure 26.1). In the past few centuries, however, the population increase has become exponential. The United Nations estimates that world population will grow to somewhere between 7.3 billion and 10.7 billion by 2050. Estimates of the human population Earth can support, known as Earth's carrying capacity, vary considerably and depend on assumptions about the future of agriculture and the quality of life of Earth's human inhabitants.

In Chapter 15, you read that Thomas Malthus observed in 1798 that all species have the ability to overproduce offspring. Excessive human reproduction, he contended, would lead to much human suffering due to famine, disease, and war, as people struggled to use inadequate resources to feed and care for their ever-increasing numbers. These effects of overpopulation are obvious in the developing world today, where human population is increasing most rapidly. Currently, world agriculture produces more than enough food for everyone, but because of political and distribution issues, hunger is a constant

companion for close to 1 billion people. As the world population continues to increase, modern agriculture will eventually be unable to supply enough food. Water shortages may provide an even earlier fulfillment of Malthus's predictions. In many areas, there is too little water for agricultural production sufficient to meet local food needs, and drinking water is contaminated with pollutants and disease-causing organisms.

Human population will decrease only after birthrates are reduced below the death rates. Although many countries have successfully lowered their birthrate, modern medicine has also lowered death rates. In some countries, such as Japan and Sweden, birthrates are now close to or lower than death rates. In many other countries, however, birthrates are still considerably above death rates.

The age structure of populations in individual countries is also an important consideration in programs of population control. Countries in which a large fraction of the population consists of individuals at or younger than the reproductive age have an increasing population already built into their future (Figure 26.2). Continued population growth in the developing world is influenced by a complex matrix of interacting factors, including cultural views of birth control, the importance of children in the workplace and in caring for aging parents, and societal pressures to import the technological, consumer-oriented way of life found in developed countries.

Increased food production will involve genetically altered plants, improved growing practices, and more efficient food distribution systems

The ability of plant biologists, from geneticists to agronomists, to provide enough food for the human population is one important aspect of the interaction of humans with the biosphere. In the 1940s through the 1960s, traditional plant breeding led to the "green revolution" and its varieties of the world's major cereals: wheat, corn, and rice. These varieties were designed to channel more of the plants' photosynthetic production to grain and less to leaves. They require high inputs of fertilizers, pesticides, and water to achieve maximum yields, so production has an economic price. Research to produce high-yielding green revolution varieties occurred principally at international agricultural centers under the Consultative Group for International Agricultural Research (CGIAR). Green revolution wheat and corn were pioneered at the Centro International de Mejoramento de Maize y Trigo (CIMMYT) in Mexico, while green revolution rice was developed at the International Rice Research Institute (IRRI)

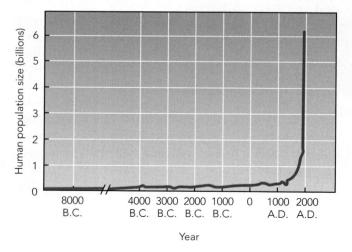

Figure 26.1 Human population growth. The exponential phase of human population growth began only a few hundred years ago.

Zero growth/decrease
Italy

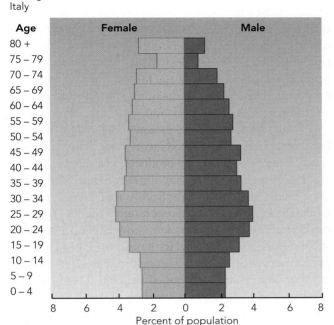

Rapid growth
Kenya

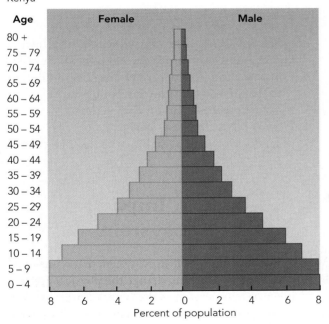

Figure 26.2 Age structure of human populations in two countries. Italy's population is stable and is fairly evenly distributed in all age groups. In Kenya, a large proportion of the population is young and will likely cause the population to grow for another 40 years.

in the Philippines. Use of green revolution varieties enabled many countries to substantially increase food production to keep pace with rapidly growing populations. Mexico, for example, went from a wheat-importing country in 1944 to a wheat-exporting country in 1964, despite a considerable increase in its population.

Will biotechnology be able to effect a similar increase in yields to help solve existing and future food shortages? Most experts suspect that the yield per plant may not increase dramatically but that overall production will expand over the next two decades, due to the introduction of varieties that are resistant to diseases and pests or tolerant of drought, salinity, acidity, and other soil stresses (Chapter 14). Crop plants with these traits would effectively increase the amount of land that is suitable for agriculture. However, some countries, including most of Europe, have rejected genetically modified (GM) plants despite their increased yields and improved nutrition. GM crops must be adequately tested, and the public must be educated about both the advantages and the potential drawbacks of these crops. The situation is similar to that encountered when vaccines were first introduced to curtail the spread of dangerous diseases. Vaccines have both good and bad features, but overall they have saved many lives and prevented much disease.

In addition to developing genetically improved crops, several other practices can support future increases in food production. One such practice is improving the genetic diversity of crops. Modern agriculture frequently uses just one or a few varieties of each crop plant that have particularly useful traits and respond well to fertilization, irrigation, and pesticides. However, this reduction in genetic variability makes much of the crop vulnerable to damage from a single, well-adapted pathogen or herbivore. In 1846–47, for example, the fungus *Phytophthora infestans* wiped out the potato crop in much of northern Europe in just a few weeks. The results were particularly devastating in Ireland because the population relied heavily on potatoes for food. A million Irish died of starvation and an equal number emigrated, many of them to the United States. In 1970, more than 15% of the corn crop in the United States was destroyed by the fungus *Cochliobolus heterostrophus*, which causes corn blight. Growers used only a few varieties of corn, many of which were genetically related. As a result, many plants were susceptible to fungal infection when a new race of the fungus arose by mutation.

Farmers have discovered that building complexity into their cultivated ecosystems can have beneficial results. In the practice known as *multicropping*, several crops are planted either simultaneously or sequentially throughout the year or in alternating years. One example of multicropping is *crop rotation*, in which different crops are planted in alternate years. Sometimes wheat is alternated with a nitrogen-fixing legume, such as clover. In another example of multicropping, called *intercropping*, rows of

Figure 26.3 Intercropping. In this farm field in Cuba, bananas are intercropped with cabbages and other crops.

different crop plants alternate with each other in the same field at the same time. For instance, fruit trees may alternate with several rows of beans and several rows of potatoes, or rows of soybeans may alternate with rows of barley and rows of corn (Figure 26.3). Intercropping can make a variety of food crops available throughout the year if the crops have different harvest times. A well-designed multicropping system can still take advantage of mechanized planting, cultivation, and harvesting.

By presenting a smaller target, multicropping slows the spread of pathogens and herbivores that are attracted to large, single-crop fields. In some multicropping situations, plants that produce substances that repel or attract herbivores are grown near food crop plants. For example, marigolds (*Tagetes* spp.) produce volatile compounds that repel many insects, whereas parsley (*Petroselinum* spp.) attracts butterflies and moths, whose larvae eat the parsley and ignore other plants. Most family gardens provide simple examples of multicropping in action.

The use of herbivore-resistant crop varieties and multicropping are two aspects of *integrated pest management* (IPM), a system that employs a range of strategies to protect plants from herbivores and disease. IPM specialists try to find the crops that are best suited for particular agricultural regions, and they work to modify existing practices that attract crop pests. For example, an IPM approach for dealing with corn pests might be to plant corn at different times in adjacent fields, so that the entire crop is not attacked by a particular insect that hatches at a certain time. IPM also includes the use of biological control agents, such as wasps that lay eggs in caterpillars or ladybird beetles that eat aphids.

In Chapter 25, we noted that approximately 10% of the energy in a trophic level is passed to the next trophic level in an ecosystem. This means that humans capture about 10% of the energy in plants when they eat plants but only about 1% of plant energy when they eat meat. Consequently, a population of humans who eat primarily plants can be supported by less agricultural land than a population of humans who eat meat. Thus, eating plants instead of meat—eating "lower on the food chain"—is an effective method for increasing the number of people who can be fed on a given amount of land.

The organization of food production at a local level can increase the availability of food just as much as an increase in food production itself. Local production eliminates the cost and energy expenditure of long-distance transportation and other intermediate steps in the field-to-market movement of food. Home gardens and regional farmer's markets are obvious examples of local production. Growers' cooperatives enable producers to meet local food needs and to get the best price for excess production. Throughout the history of agriculture, farmers have developed varieties of crop plants, called *land races,* that grow particularly well under local conditions (see the *Conservation Biology* box on page 553).

Section Review

1. **What are some likely consequences of allowing human population growth to continue unchecked?**
2. **Contrast the age structure of a stable population with that of a rapidly growing population.**
3. **What are green revolution crop varieties?**
4. **What is integrated pest management?**
5. **What does it mean to "eat lower on the food chain"?**

Human Impacts on Ecosystems

Most people have never seen an ecosystem that is unaltered by human activity. Humans change the physical appearance of ecosystems as well as the types of organisms in them. We alter the natural flow of energy in ecosystems and change the distribution and quantity of abiotic components, such as water and minerals. We also introduce new and often damaging biotic and abiotic agents.

The presence and activities of large human populations disturb ecosystems

Visitors to a national park unavoidably have an adverse effect on the ecosystems of the park. Hiking along a trail

CONSERVATION BIOLOGY

Land Races and Seed Banks

Land races are local populations of plants that have been carefully selected by farmers for hundreds or even thousands of years. These populations may be found throughout a geographic region or just in a particular valley or on a certain mountain. Through selection, each land race has come to have certain alleles that support successful growth and reproduction in a specific area. Often, the alleles confer resistance to particular diseases or herbivores. The genetic constitution of a land race may also provide an adaptation to local climatic conditions or result in a plant that appeals to local tastes. As modern agriculture tends toward monoculture (growing one or a few varieties of a plant over wide areas), land races have been ignored, and many have become extinct.

The preservation of land races is one important function of seed banks. Seed banks store seeds from different natural and agricultural varieties of plants. Seeds obtained from the same species in different geographic locations may turn out to be genetically different, and seed banks help to prevent the loss of these differences. Preserving the genetic diversity of local plant populations is especially important in the case of crop plants. For endangered plant species, the seeds kept in a seed bank are a form of insurance against extinction.

The National Seed Storage Laboratory for the United States is located in Fort Collins, Colorado. In the facility, 1.5 million seed samples from all over the world are stored dry at −18°C (0°F) or preserved in liquid nitrogen at −196°C (−321°F). Each sample remains viable for 20–50 years and

The National Seed Storage Laboratory.

is regularly tested and replaced. In 2000, the Millennium Seed Bank opened in Great Britain. Devoted to saving endangered plants, the facility will store the seeds of more than 24,000 species by 2010. Many other seed banks throughout the world maintain collections of locally important germplasm.

It is especially important to protect seed banks during periods of political instability. During the war in Afghanistan in 2002, the Afghan seed bank was ransacked by looters, who dumped the seeds on the floor and kept the plastic and glass containers in which the seeds were kept.

---

rapidly wears through surface vegetation and soil and turns the trail into a conduit for water, thus increasing water loss and erosion. Air pollution from automobiles, recreational vehicles, and buses alters the growth of lichens and plants. Animals are killed on roads. Well-meaning tourists feed chipmunks, birds, and deer, making them dependent on handouts. Large carnivores, formerly a part of the ecosystem, are kept out or reduced in number for visitor safety. As a result, the herbivore population increases, and some species of plants become endangered or are eliminated.

The effects of human activities on a natural area such as a national park are mostly controlled and often correctable. Unfortunately, they represent only a minor example of the harmful effects of human populations on natural ecosystems. In the rest of this section, we will examine ten areas in which human activities have had a major negative impact on ecosystems.

Fire Suppression

Until the twentieth century, forest fires were a normal abiotic factor in many ecosystems. Humans began suppressing forest fires not only to save economically valuable timber but also to protect houses and other property. This policy has been the norm in most countries for decades. When fires are allowed to burn, they release bound-up nutrients and create a mosaic of diverse habitats for herbaceous plants, trees, and animals. In such a mosaic, future fires do not easily cause widespread destruction. In contrast, forest fires that occur after years of fire suppression are often very destructive. For example, in the Tolan Creek watershed of the Bitterroot National Forest in Montana, 23 forest fires of varying sizes occurred between 1734 and 1900. None occurred in the 100 years after 1900 because of fire suppression. As a result, the forest there became uniform and dense with considerable fuel for fire. In the summer of 2000, a fire started by lightning burned

40% of the watershed. Huge fires also burned through Yellowstone National Park in 1988 because naturally occurring fires had been suppressed in the park for many years (Figure 26.4).

Logging

Eighty percent of Earth's original forests have already been cleared or broken into fragments. To obtain sustainable yields in forested areas, tree removal must be scheduled to allow for regrowth. In other words, if trees require 100 years to reach maturity, then 1% or less of a mature forest should be cut each year. Unfortunately, current world demands for timber exceed what can be supplied by sustainable production, so forest resources in many areas are harvested much more rapidly than they regrow. In the temperate rain forests of the Pacific Northwest, for example, timber has been extensively overharvested, resulting in bald mountains and a loss of employment for people in forest communities, which is only partially replaced by tourism in the area's few national parks.

Even if practices leading to sustained yields are followed, secondary-growth forests are different from primary-growth, or virgin, forests. If the bark and leaves are not returned to the soil, the nutrient content of the soil is reduced, causing secondary-growth forests to grow more slowly and have a lower yield. Moreover, the complex ecosystem of a virgin forest is difficult to restore. Scientists who studied the upper levels of Sitka spruce forests on Vancouver Island discovered numerous microenvironments that supported the growth of lichens, mosses, liverworts, and accompanying animal species (Figure 26.5). In one study, they found 300 new species of arthropods in just one of the microenvironments. These microenvironments apparently do not reform in the treetops of secondary-growth forests, no matter how long the forests are allowed to grow.

Deforestation due to forest fires and logging causes a 30–40% increase in water runoff on land. The effects include mudslides, soil erosion, flooding, and silting of rivers and lakes, resulting in fish kills and water that is unsuitable for drinking. The loss of particular nutrients from the soil increases from 4- to 60-fold.

Depletion of Water Supplies

About half of the U.S. population obtains water from underground sources called *aquifers*. These enormous reservoirs can store water for thousands of years. Aquifers lose water through springs and wells dug by humans, and they gain water when precipitation trickles through soil and rocks and enters the underground supply. However, many aquifers are experiencing a net loss of water as people pump water out more rapidly than it can be replaced. As an example, consider the High Plains Aquifer (also known as the Ogallala Aquifer), which underlies parts of eight midwestern states (Figure 26.6). Currently, the rate at which water is being extracted from the aquifer is about 10,000 times the rate at which water is entering the aquifer. In some areas, the water level in the aquifer fell more than 30 meters (100 feet) between 1950 and 1980

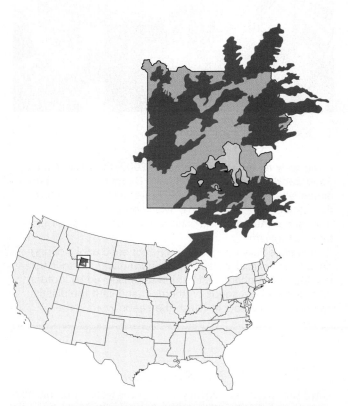

Figure 26.4 Fires in Yellowstone National Park. Numerous fires burned 45% of the park's 900,000 hectares (2.2 million acres) in 1988. A number of research projects are monitoring the secondary succession that will lead to recovery of these areas.

tion expected to reach 264 million by 2025, these countries must increase agricultural production using water from the Nile River, which flows through all three nations. Countries that are farther upstream can exercise greater control of a river, but geographical position is only one aspect of a complex political reality. Water shortages for agriculture and cities are bound to be the cause of many future political disputes.

About 40% of the world's food comes from the 17% or so of agricultural land that is irrigated. Seventy percent of available global fresh water is already used in agriculture. We extract so much water from rivers that many large rivers, such as the Nile, the Colorado River in the southwestern United States, and the Yellow River in China, run dry before they reach the sea.

Figure 26.5 Working in the treetops. Researchers use ladders, long-handled nets, and sure footing to study the biological diversity in the upper branches of this Sitka spruce.

and another 12 meters (40 feet) between 1980 and 1994. In Texas, where the population may double by 2050, water from the aquifer is piped to Dallas, Houston, and other cities. Many farmers in the Midwest now find irrigation uneconomical due to the increased costs of pumping water out of the aquifer.

Depletion of aquifers is not restricted to the United States. In the Middle Eastern country of Yemen, for example, the aquifer that provides water for the capital city of Sana'a and nearby farms is dropping by 6 meters (20 feet) per year and will be depleted by 2010. At that time, the government will have to pipe in fresh water produced by desalination of seawater—an expensive process—or relocate the city.

Because of water shortages, most Middle Eastern countries currently import 40–90% of their grain. In northeastern Africa, serious grain shortages plague Egypt, Ethiopia, and Sudan. To support a burgeoning popula-

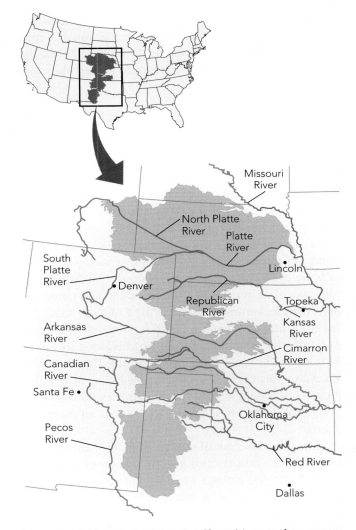

Figure 26.6 The High Plains Aquifer. This vast subterranean reservoir supplies about 30% of the groundwater used for irrigation in the United States.

World supplies of fresh water could be used much more efficiently if plants were genetically engineered for decreased transpiration. Also, much less water would be wasted if five simple irrigation practices were followed:

◆ Using drip irrigation. By delivering water to plants but not to the surrounding soil, drip irrigation can reduce water use by 30–70% while increasing crop yields by 20–90%.

◆ Using water-efficient sprinklers. On the Texas High Plains, such sprinklers have increased irrigation efficiency by 90% and improved yields by 10–15%.

◆ Irrigating at night. Because of lower temperatures, less water is lost to evaporation at night.

◆ Improving irrigation canals. By preventing leaks and excess irrigation, such improvements have markedly decreased water use in several studies.

◆ Reusing wastewater. In Israel, 65% of domestic wastewater is reused for irrigation. In Texas, 115 counties reused 610 million liters (160 million gallons) of municipal wastewater per day in 1998. In Lubbock and Amarillo, enough water was recycled to meet the needs of 100,000 people.

Water Pollution

More than 1 billion people currently lack access to safe drinking water. To be safe for human consumption, water must be free of disease-causing organisms and toxic substances and should have a low concentration of dissolved salts—less than 500 parts per million, according to many health experts. Some of the salts in freshwater supplies come from natural processes, such as the uptake of salts from soil and rocks by rivers. Additional salts are contributed by human activities, including fertilization of farm fields and the collection and treatment of human and animal wastes. In the Colorado River, for example, half of the salt loading is natural while half is due to human activity. At the head waters, in Rocky Mountain National Park in Colorado, the Colorado River contains nearly 50 ppm dissolved salts. By the time it enters Mexico, the salinity level has increased to more than 1,200 ppm, which is toxic to most animals and plants.

Human production of synthetic fertilizer and the cultivation of nitrogen-fixing plants have doubled the amount of fixed nitrogen (nitrate, nitrite, and ammonium) available in the biosphere. Humans produce artificially fixed nitrogen both for fertilizer and as the basis for most conventional explosives. Ultimately, a considerable amount of this nitrogen ends up in bodies of water, where it fuels the growth of algae and bacteria.

The discharge of untreated human and domestic animal wastes into aquatic ecosystems also stimulates algal and bacterial growth and introduces disease-causing organisms. Although sewage treatment plants can eliminate the health risks posed by sewage, many regions of the world lack adequate treatment. Ninety percent of Asia's wastewater is untreated.

Historically, people have regarded rivers as free dumping sites for wastes and unwanted chemical by-products of manufacturing. Almost everyone lives downstream from someone else, so the effects of such dumping are magnified as rivers flow to the ocean. Downstream, many rivers are now too polluted for safe human consumption or even for irrigation. The "dead zone" in the Gulf of Mexico is a region the size of New Jersey where most of the dissolved oxygen has been removed by bacteria that decompose dead algae. The algae result from blooms caused by excessive fertilizer runoff from midwestern farmland that drains into the Mississippi River.

Air Pollution

Air pollution was a factor in the biosphere long before humans evolved. Many natural cycles introduce molecules besides oxygen, nitrogen, and carbon dioxide into the atmosphere. However, human activity has changed the natural balance in two ways. First, the burning of fossil fuels—coal, oil, and natural gas—has markedly increased the concentration of toxic pollutants. Second, humans have introduced a wide array of molecules that were either absent or very rare on Earth before human civilization. Chlorine monoxide (ClO), which contributes to ozone destruction, is an example.

Sulfur dioxide (SO_2) is the largest contributor to air pollution. Natural sources, such as volcanoes, weathering of rocks, and chemical activity of bacteria, are responsible for about 60% of SO_2 production. Humans contribute the other 40% by burning fossil fuels. The world's industrialized countries release well over 90,000 kilograms (100 tons) of sulfur each year into the atmosphere. Around sources of sulfur emission, plants are killed outright. Severe damage can be monitored for hundreds of kilometers downwind from such sources, and global damage occurs as the polluting gases mix with the rest of the atmosphere. As you learned in Chapter 19, lichens are particularly good indicators of air pollution because they absorb a substantial portion of their minerals and water from the air.

Forest fires release toxic pollutants, such as SO_2, nitrogen dioxide, and carbon monoxide into the atmosphere. Extensive forest fires in Indonesia in 1997 created a toxic haze that covered a large region of Southeast Asia. The fires, which were started by people clearing land for plantations or other forms of agriculture, spread rapidly to the surrounding forest in the midst of an El Niño–created drought. The haze and ash affected the health of people and wildlife, reduced primary productivity by photosynthetic organisms throughout the region, and increased the contamination of surface waters when the rains finally came.

Ozone is a minor atmospheric gas, but it is important because it shields Earth's surface from mutation-inducing ultraviolet (UV) radiation. In the upper atmosphere, solar radiation produces ozone (O_3) from molecular oxygen (O_2). The normal concentration of ozone in the atmosphere is a balance between the rates of production and destruction. However, pollutants such as chlorofluorocarbons (CFCs) decrease the ozone concentration in the upper atmosphere. The light-activated breakdown of CFCs produces ClO, which catalyzes the destruction of ozone. Consequently, the same ClO molecule can promote the destruction of many molecules of ozone. CFCs were once commonly used as propellants in aerosol cans and as refrigerants in air conditioners and refrigerators. Although the production of CFCs is now banned in the United States, these compounds are still used by some other countries.

So far, ozone depletion in the atmosphere has been noted chiefly over Antarctica, but a global decrease in the concentration of ozone may not be far behind (Figure 26.7). As the ozone concentration falls, more UV radiation reaches Earth's surface. Plants exhibit a wide range of sensitivity to UV radiation. The epidermal cells of some plants, particularly tropical and alpine plants, screen it out rather effectively. In other plants, UV radiation can induce mutations, inhibit photosynthesis and growth, and reduce yields. Around half of our crop plants appear to be adversely affected by increased exposure to UV radiation, while the other half are relatively unaffected.

Air pollution often suppresses precipitation. All particles in the atmosphere serve as centers of raindrop formation. Air pollution adds many small particles to the atmosphere, causing the formation of very small raindrops, which fall but evaporate before reaching the ground. Because the amount of precipitation is an important abiotic determinant of the nature of ecosystems, this effect of air pollution may have a significant impact on the biosphere.

Global Warming

The extensive burning of fossil fuels has contributed markedly to global warming due to the greenhouse effect (Chapter 9). Greenhouse gases, which retain thermal energy in the atmosphere, include water vapor, carbon dioxide, nitrous oxide, methane, and ozone. Nitrous oxide absorbs 320 times as much heat as carbon dioxide, while methane absorbs 25 times as much. Therefore, small concentrations of nitrous oxide and methane have large greenhouse effects. Most nitrous oxide is produced by the action of soil bacteria. The four largest sources of methane are bacterial action in wetlands, production and use of fossil fuels, digestion in herbivorous animals such as cows, and rice cultivation in paddies with significant amounts of bacterial decay.

Scientists have begun to report significant alterations in ecosystems because of global warming. In Great Britain, for example, the first flowering of 385 plant species occurred an average of 4.5 days earlier during the past decade than during previous decades, indicating earlier springs and shorter winters. In Washington, DC, a study of 100 plant species over 30 years showed that 89% flowered earlier, by an average of 2.4 days. Surveys of butterflies and birds have revealed that the ranges of many species are moving toward the poles, an indication that temperate regions are becoming warmer.

In 1997, the United Nations Framework Convention on Climate Change adopted the Kyoto Protocol, which limits industrialized countries' net production of carbon dioxide and other greenhouse gases. As of 1997, the United States produced 5,500 kilograms (6.1 tons) of carbon dioxide per inhabitant, compared with 290 kilograms (0.32 ton) for India. Because plants remove carbon dioxide from the atmosphere during photosynthesis, reforestation is one method that can be used to reduce net

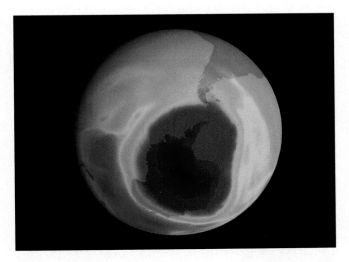

Figure 26.7 The Antarctic ozone hole. The amount of ozone in the upper atmosphere is expressed in Dobson units (DU). Normal values range from about 250 to 500 DU. The ozone hole is defined as the area having less than 220 DU, shown as blue and purple in this satellite image made in October 1998.

greenhouse gas production. By 2003, 113 countries had ratified the Kyoto Protocol, and 17 countries that participated in the convention, including the United States, had not.

Acid Precipitation

Normal precipitation is slightly acidic, having a pH of about 5.5 because of the dissolved carbon dioxide it contains. Various sulfur and nitrogen compounds, produced almost exclusively by the burning of fossil fuels, readily dissolve in water droplets in the atmosphere, making precipitation more acidic. In the United States, most acid precipitation falls in the Northeast (Figure 26.8), due to air pollution generated by industries in the region as well as pollution carried in by prevailing winds from the west. Much of the Northeast receives precipitation with an average pH of 4.5 or lower—that is, at least ten times as acidic as normal precipitation. Acid precipitation lowers the resistance of plants to insect pests and disease and alters the species composition of lakes and streams by reducing the pH of these bodies of water.

Exotic Species

Global travel and trade by humans are responsible for the planned and accidental introduction of thousands of species to new locations. In their new homes, such species are known as *exotic species*. Those that are well adapted to their new environment and cause significant environmental or economic harm are regarded as invasive. Invasive plants lower crop production by competing for water, light, and minerals and by interfering with harvesting. None of Earth's biomes is free from exotic species. North America now has 20% more plant species than it had in 1492. In Florida, nearly 30% of plants are exotics. In Great Smoky Mountains National Park, 400 of the park's 1,500 species of vascular plants are exotics, and 10 are classified as dangerously invasive. According to U.S. government estimates, the damage done by exotic species amounted to $125 billion in 2001 (see *The Intriguing World of Plants* box on page 559).

One example of an invasive exotic plant is Russian thistle (*Salsola kali*), also known as *tumbleweed* (Figure 26.9). It was probably introduced to the western United States accidentally in the 1800s as contaminating seeds in packages of flax or wheat seeds brought by European immigrants. Russian thistle is an annual that forms a large, spherically shaped bush. The bush dies in late summer, breaks off at the roots, and is carried by the wind as a "tumblin' tumbleweed," scattering thousands of seeds as it goes. This adaptation allowed the plant to spread rapidly through the western and midwestern states. By

Average annual pH of precipitation in the United States

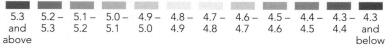

| 5.3 and above | 5.2 – 5.3 | 5.1 – 5.2 | 5.0 – 5.1 | 4.9 – 5.0 | 4.8 – 4.9 | 4.7 – 4.8 | 4.6 – 4.7 | 4.5 – 4.6 | 4.4 – 4.5 | 4.3 – 4.4 | 4.3 and below |

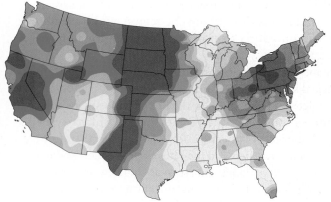

Figure 26.8 Acid precipitation in the United States. As this map from 2001 shows, the most acidic precipitation in the contiguous 48 states falls in the Northeast.

the early 1900s, the tumbleweed had become as much a part of western American lore as the cowboy. Russian thistle is drought tolerant and effective at extracting nitrogen from soil. It competes particularly well in fields with crop plants, which require more water and fertilizer in its presence. The exotic plant grows large enough to shade many crops, reducing photosynthesis, and it hosts insect pests such as the beet leafhopper, which carries curly top virus to sugar beets, tomatoes, and many other crops. Although Russian thistle is not generally important as an agricultural plant, it was used as an inferior cattle food called "Hoover Hay" during the drought that preceded and accompanied the Great Depression.

Figure 26.9 Russian thistle. The "Tumblin' Tumbleweed" is an exotic plant with a unique method of seed distribution involving the entire plant.

THE INTRIGUING WORLD OF PLANTS

Kudzu

Pueraria montana, or kudzu, is an invasive exotic plant that covers 3 million hectares (7 million acres) in the United States as far north as Pennsylvania. A member of the pea family, kudzu is a climbing, semiwoody, perennial vine native to Asia. The plant was introduced to the United States in 1876, when it was displayed in the Japanese Pavilion at the Philadelphia Centennial Exposition. During the depression years of the 1930s, kudzu was planted widely in the southern United States to reduce persistent soil erosion. It was also promoted as a forage crop for cattle that restored soil fertility because of its association with nitrogen-fixing bacteria.

In the United States, kudzu encounters no pests or diseases and grows rapidly—30 centimeters (1 foot) per day when conditions are optimal. Fleshy storage roots can weigh as much as 180 kilograms (400 pounds). The plant flowers in late summer and produces brown, flattened, hairy seedpods, each of which contains between three and ten hard seeds. In the southern United States Kudzu grows well in disturbed soils, around the outsides of forests and woodlots, and along the sides of roads. It kills trees and other plants by crushing them with its weight or by blocking light for photosynthesis. In many areas it has completely covered fields and buildings. In 1972, the U.S. Department of Agriculture declared kudzu a weed.

The cost of even attempting to control kudzu is high. Power companies alone spend more than $2 million a year to control this exotic pest. Herbicides can kill the plant, but applications for 4–10 years are necessary. One herbicide actually causes the plant to grow more rapidly! Various avenues of biological control have been investigated as alternatives to herbicides. In Asia, a leaf-feeding beetle and a species of sawfly that eats only kudzu were discovered, as well as two kinds of weevils that feed exclusively on the vine. In addition, at least one fungal pathogen that attacks kudzu has been identified.

The story of kudzu in the United States is an excellent example of the problems that can arise when a species is introduced to a new area and allowed to become established. It also illustrates the many difficulties that may be encountered in attempts to control exotic species once they gain a foothold.

Kudzu overtaking an abandoned cabin in Georgia; winter, spring, mid-summer, and early fall.

Habitat Fragmentation

Natural habitats are broken into small fragments by agriculture, forestry, mining, urbanization, and transportation links such as highways. Recall from Chapter 25 that habitat fragments, called *patches,* often cannot support communities that are truly representative of the original habitat. For example, in tropical rain forest fragments, large trees die at a higher rate than small trees, although large and small trees have the same mortality rate in the interior of unfragmented rain forests. The increased mortality rate of large trees in small forest fragments is due to increased exposure to turbulent winds and parasitic woody vines and higher water stress, all of which are associated with growing near the edges of fragments.

Habitat fragments are frequently too small to support large predators, such as mountain lions, that range widely in search of food and mates. When large predators are absent, the population of herbivores increases, putting greater pressure on plant populations. This type of situation has been studied with regard to wolf and moose populations in Isle Royale National Park in Michigan. Large habitats allow for populations large enough to avoid genetic drift and for the species complexity needed to maintain the habitats. In some cases, habitat fragments connected by corridors can function effectively as larger habitats.

Extinction

During the seventh century B.C., the Greeks discovered that silphion, a plant in the giant fennel family, could be used as a birth control agent. Within 40 years of this discovery, silphion was difficult to find in the wild due to overharvesting. A related species (*Ferula asafoetida*) became a less effective substitute. Silphion was driven to extinction, while *F. asafoetida* survives as the source of a flavoring in Worcestershire sauce.

An estimated 15–20% of all plant species will become extinct in the next quarter century. Almost 1,000 species of trees are critically endangered. The highest rates of extinction are in tropical rain forests, which have the most species per unit area of land. Tropical rain forests are being rapidly destroyed in the quest for forest products and to create agricultural land.

Biologists have identified 25 "hotspots" around the world where biodiversity is high and extinction is rampant (Figure 26.10). All of these areas are either tropical forests or dry shrubland. Although biodiversity hotspots have just 6% of Earth's land area, they contain up to 33% of plant and vertebrate species. They are also frequently areas of dense human habitation or high levels of human activity. Biodiversity hotspots are regions where conservation efforts would have the most effect in preserving species.

One biodiversity hotspot is Madagascar, an island country off the east coast of Africa (hotspot #9 in Figure 26.10). Most of the animal species and 81% of the plant species on Madagascar are endemic; that is, they are found nowhere else on Earth. Currently, more than 90% of the island has been deforested, with drastic results for the environment and for the people who live there. The ecosystem of this island can never be reformed, partly because of the extent and intensity of habitat destruction and partly because of the island's large human population.

In 2000, an eight-nation team of scientists identified the five most important forces that decrease global biodiversity:

- Changes in land use, especially deforestation and conversion of natural ecosystems to agricultural land
- Changes in climatic factors, including precipitation and temperature
- Addition of nitrogen to water, chiefly from artificial fertilizers, human and animal wastes, and auto emissions
- Introduction of exotic species
- Increases in the concentration of atmospheric carbon dioxide

It is clear that many species will become extinct in the next few decades unless drastic measures are taken. Many more species are in imminent danger of reaching their critical minimum population sizes, at which the danger of extinction is considerably increased. Two such plant species are American ginseng (*Panax quinquefolius*) and wild leek (*Allium tricoccum*), both native to the deciduous forests of eastern North America. American ginseng is prized as an herbal remedy, and wild leek is collected for food. Overharvesting of these plants has led to declines in many native populations and the disappearance of some. Computer models indicate that a population of American ginseng must have at least 170 plants to survive, and a population of wild leek must have between 300 and 1,030 plants. Relatively few native populations have the minimum number, and these populations are often widely separated from each other. Therefore, the future of both species in the wild appears to be completely dependent on human action.

The geographic information system provides a new tool to record changes in ecosystems

The *geographic information system* (GIS) is a computer system that combines and analyzes geographic data, such

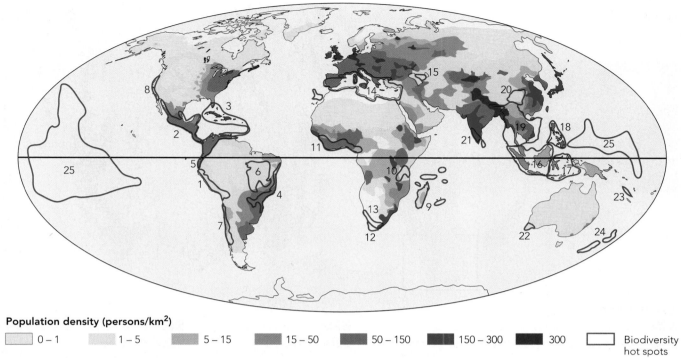

Population density (persons/km²)

0 – 1 | 1 – 5 | 5 – 15 | 15 – 50 | 50 – 150 | 150 – 300 | 300 | □ Biodiversity hot spots

Figure 26.10 Global biodiversity hotspots. Some of these areas of high species diversity are in regions densely populated by humans.

as topography, soil type, temperature, precipitation, vegetation type, roads, and population density. Each data set is portrayed on a map, and the maps are then graphically layered upon each other to create a multidimensional map. For example, the GIS map in Figure 26.11 combines data on native vegetation, type of rock substrate, and the locations of protected areas in southern Madagascar.

Conservation biologists use GIS to monitor ecosystems, to help plan the management of natural resources, and to study species diversity and environmental changes over time. In addition, GIS has become a major resource that helps biologists who study threatened and endangered species evaluate critical areas for land protection.

As an example of the application of GIS in conservation biology, consider the recent story of the rarest orchid in North America, *Isotria medeoloides* (Figure 26.12). Known as the *small whorled pogonia* or *green fiveleaf orchid*, *I. medeoloides* may grow up to 25 centimeters (10 inches) high in acidic soils of dry, deciduous forests where trees are 45–80 years old. Habitat fragmentation, overharvesting by plant collectors, and other factors contributed greatly to a decline in the populations of this species. Researchers evaluated sites in Maine and New Hampshire where *I. medeoloides* occurred and recorded the soil type, slope of land, and other vegetation at each site. They then used GIS to cre-

ate a map showing these sites as well as others where the plant would likely be found. Preservation of these sites has led to a resurgence in *I. medeoloides* populations. In 1996, the plant's federal status was changed from endangered to threatened, a step in the right direction for plant species conservation.

GIS is changing the way people view the world. It is transforming maps from flat images into highly interactive digital models equipped to answer important research questions. Aided by GIS technology, plant ecology and conservation have begun to take an aggressive approach to preserving habitats and species.

Section Review

1. **What problems are introduced into forest environments by suppression of naturally occurring forest fires?**

2. **What are aquifers, and how does human use affect them?**

3. **How does air pollution affect the amount and pH of precipitation?**

4. **What is a biodiversity hot spot?**

5. **What is GIS, and how can it help conservation biologists protect endangered species?**

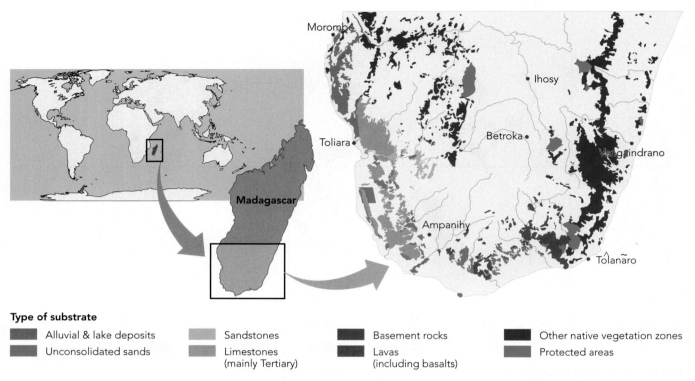

Type of substrate

- Alluvial & lake deposits
- Unconsolidated sands
- Sandstones
- Limestones (mainly Tertiary)
- Basement rocks
- Lavas (including basalts)
- Other native vegetation zones
- Protected areas

Figure 26.11 GIS mapping. This map of southern Madagascar shows the locations of remaining native plants on different types of rock substrates. Very few of these locations lie within protected areas, represented in blue.

Figure 26.12 The green fiveleaf orchid, *Isotria medeoloides*. GIS technology helped remove this extremely rare plant from the endangered species list.

The Future

One goal of studies in conservation biology is to lower the overall impact of human activity on the natural world. Conservation biologists also work to develop future policies that will preserve diversity and retain substantial examples of the world's naturally occurring biomes. The success of such policies depends on the extent of our understanding of ecology and of how humans impact ecosystems.

The future of human interactions with ecosystems can be modeled on worst-case or best-case scenarios

If human population continues to increase at the current rate or at an increased rate, a worst-case scenario will very likely be realized. Under these conditions, the population will soon reach its carrying capacity in terms of important resources, such as fresh water, food, and suitable living space. Millions of people in developing countries already live under carrying-capacity conditions, with inadequate diets, impure water, insufficient medical care, and substandard housing. Political considerations will probably continue to divide the world into have and have-not nations that compete for an ever-decreasing resource base. Increasing pollution of the atmosphere and the oceans will lower the quality of life worldwide. Natural ecosystems, which existed on Earth for millions of years, will be replaced by totally human-produced landscapes. Pandemics of uncontrolled human disease will become increasingly common.

Figure 26.13 Restoring a polluted river. (a) In 1960, the Nashua River in Massachusetts was polluted with sewage and red dye from paper mills. **(b)** By 1993, the river was clean and had been restocked with fish.

Best-case scenarios depend on population stabilization, increased food production, decreased pollution of air and water, and increased preservation of renewable resources. World political stability and cooperation between nations would also be important factors in stabilizing or improving habitats. Ecosystem preservation and restoration would be given priority status worldwide. The studies of ecosystem dynamics (Chapter 25), landscape ecology, and conservation biology would assume greater importance.

Achieving a best-case scenario for the biosphere would involve a marked reversal of current trends

Stabilizing and then reducing the effects of humans on the planet would require a number of changes. One of the most important changes is that fossil fuels and wood would have to be replaced by alternative sources of energy. As of 2003, we have used half the world's known petroleum reserves. According to current estimates, the remaining reserves will be 90% depleted by 2050. As oil becomes scarce, the use of other fossil fuels, including oil shale, coal,

and peat, will likely increase, but at great environmental cost. Obtaining these fuels would involve considerable destruction to ecosystems, and increasing the use of poorer-quality coal reserves would add greatly to air pollution.

Alternative energy sources will need to be rapidly and more fully developed to replace dwindling supplies of fossil fuels. Wind, harvested by turbines, and water, trapped behind dams or moving as the result of ocean tides, can be used to generate electricity. Solar energy is abundant and free. Because plants trap solar energy in photosynthesis, they can serve as a source of energy. Virtually any plant material can be burned to produce heat or fermented by yeast to make ethanol (Chapter 9), which is already used as a pollution-reducing additive in gasoline in winter. Other photosynthetic organisms are being investigated as sources of hydrogen, a nonpolluting fuel that yields only water when burned (Chapter 18).

Air and water pollution would have to be stabilized and then reduced. With respect to water pollution, this change could be effected partly through the reduced use of fossil fuels and partly through direct efforts to clean rivers and lakes (Figure 26.13).

Conservation areas much larger than those that currently exist would have to be set aside for the purpose of preserving Earth's primary ecosystems. There is a critical need for such areas in tropical rain forests, where the highest plant diversity is found. Multiuse areas, in which some amount of human activity (such as forestry, mining, and grazing) were allowed, probably would not be compatible with this goal. Human access to preserved natural areas would be permitted but strictly regulated.

The harvesting of trees would have to be placed on a strict plan of sustainable yields in areas devoted to the growing of trees for harvest. Plant biotechnology might be helpful in increasing the yields of these trees. Such a plan would considerably increase the price and reduce the availability of wood products. As a result, paper packaging and newspapers would probably be items of historical interest, and recycling of wood and wood products would be an important activity. The world's remaining intact natural forest ecosystems would have to be maintained and, in some cases, considerably expanded.

Agricultural production would have to be increased on existing farmland through practices such as multi-cropping, integrated pest management, and selective breeding but without major increases in fertilizer, insecticide, or herbicide use. Genetic engineering of crop plants could increase their nutritional value as well as their ability to grow on stressed soils and in the presence of what were formerly disease-causing organisms.

International treaties regulating the burning of fossil fuels, harvesting of trees, emission of industrial pollutants, and preservation of existing ecosystems would have to be written, signed, and enforced. People in all countries would have to be educated about the importance of ending ecosystem destruction and about the mechanisms by which this goal could be achieved. The participation of citizens in efforts to decrease the use of fossil fuels, reduce and reverse pollution, and restore habitats would have to be encouraged and facilitated.

A number of problems would have to be overcome to reverse current trends of ecosystem destruction and modification

A lack of knowledge about critical processes in ecosystems and communities and about the effects of human activities on ecosystems and habitats hinders the development of action plans for conservation. Biotechnology provides one starting point for understanding ecosystems because it allows us to determine the effect of particular point mutations on the biochemistry, physiology, and anatomy of organisms (see the *Biotechnology* box on page 565). Genomics and proteomics can provide information about specific genetic changes and the selective advantage or disadvantage they give to an individual and ultimately to a population.

The cost of proposed efforts to limit ecosystem destruction would be astronomical, and the benefits of such efforts are largely in the distant future. Political leaders are elected mainly on the basis of their promises about improvements in the present and near future. Therefore, the political will to propose and implement needed changes may be lacking. Most countries deal with problems when they become immediate and serious rather than planning to avoid future catastrophes. A history of crisis management needs to be replaced by a plan to avoid crises.

Existing biological diversity needs to be preserved both in nature and in agriculture. Reducing current rates of extinction and of ecosystem fragmentation and destruction is a goal that has no clear end point. In part, this is due to our inability to accurately predict changes caused by specific abiotic factors, such as a certain concentration of greenhouse gases. It also reflects the lack of consensus about what level of pollution and ecosystem destruction is acceptable given available resources and predicted worldwide effects.

It is important to establish models of success in promoting ecosystem restoration

Many citizens and political leaders are interested in ecosystem conservation but do not know how to proceed. Public awareness of the importance of photosynthetic organisms in the biosphere and of how biotic and abiotic factors interact in ecosystems is often limited. However, successful and well-publicized models of conservation and ecosystem stabilization can be very useful in raising awareness.

The U.S. National Park Service (NPS) and its system of national parks, national monuments, and other sites have served as a model of conservation for other countries. The NPS manages 33.8 million hectares (83.6 million acres), of which approximately 60% is in Alaska. Most of NPS land is dedicated to the preservation of ecosystems as they existed prior to the arrival of Europeans in North America. Yet, many national parks are probably too small to adequately protect the communities and ecosystems they contain. Despite its popularity with the public, the NPS is drastically underfunded, and most national parks are in decline even as the annual number of visitors steadily increases. Both the extent and the funding of the national park system need to be increased.

BIOTECHNOLOGY

Recreating Lost Worlds Through Genomics and Proteomics

The premise behind the books and movies, *Jurassic Park* and *The Lost World*, was that Jurassic mosquitoes fed on the blood of dinosaurs and therefore contained dinosaur blood in their stomachs. Some of the mosquitoes were trapped in tree sap, which hardened, fossilized, and became amber. Millions of years later, scientists removed dinosaur blood from the mosquitoes in amber, isolated white blood cells from the blood, and extracted DNA from the cells. They then used the DNA to create zygotes that developed into dinosaurs. Missing segments of dinosaur DNA were filled in with genes from present-day organisms. The problem with this premise is that fossilized DNA is usually so damaged that determining its original nucleotide sequence would likely be impossible.

However, modern biotechnology suggests another approach that might eventually succeed. As data from genomics and proteomics continue to accumulate, we will soon have libraries detailing the biochemical, physiological, morphological, and behavioral changes caused by a large number of point mutations. Scientists already know the phenotypic effects of thousands of point mutations (SNPs) in *Arabidopsis,* and they are beginning to identify the specific proteins coded for by the mutated genes and the biochemical effects of some of those mutations. Many SNPs have dramatic phenotypic effects. For example, point mutations in the *cin* gene cause *Arabidopsis* plants to develop round, crinkled leaves instead of flat, elongated leaves. The *cin* gene codes for a protein that interrupts the cell cycle, leading to division and elongation in certain leaf cells. The protein belongs to a family of plant proteins that function as transcription factors, proteins that bind to DNA and influence the activity of other genes (Chapter 13).

A late Jurassic landscape.

As the genomic sequences of more organisms are determined, more sophisticated computer programs will be needed to deal with the immense amount of data that are generated. Bioinformatics, the science of analyzing such data, will be able to make accurate predictions about the biological effects of specific changes in nucleotide sequence as data accumulate.

Using plant tissue culture, we already have the technology to recreate extinct plant species, in cases where the extinction was recent enough that sufficient preserved plant material containing DNA can be found. A machine that can use the complete nucleotide sequence of an organism to generate a complete phenotypic "picture" of the organism is easily in the realm of future possibility. Eventually, computer programs will be able to determine the genomic sequences necessary to replicate any number of long-extinct or hypothetical organisms with particular characteristics. Imagine, for instance, a program that could rapidly analyze all the possibilities for how an accumulation of point mutations led to the evolution of liverworts from green algae! Scientists can test the accuracy of these programs by using examples of evolution and speciation that have occurred relatively recently, such as the evolution of C_4 plants from C_3 plants.

Since photosynthetic organisms are at the base of nearly all food chains, simply producing dinosaurs, woolly mammoths, or saber-toothed tigers will not be enough to ensure the survival of recreated, extinct ecosystems. Reproducing the extinct plants will be just as important, if not more so. Scientists might be able to recreate the first land plants as well as the giant lycophytic trees of the Carboniferous period.

Most other countries have a portion of their land area in some sort of protected category, but the degree of protection is highly variable. National parks in many countries are more like national forests in the United States, in that many activities counterproductive to ecosystem preservation, such as logging, mining, and hunting, are allowed and encouraged within the park boundaries.

In a few cases, neighboring countries have formed or are investigating the formation of national parks on both sides of a shared border to protect an ecosystem that spans the border. Glacier National Park in Montana and Waterton Lakes National Park in Alberta provide one example (Figure 26.14). Farther south, Big Bend National Park in Texas and two protected areas in Mexico— Maderas del Carmen in Coahuila and Cañon de Santa Elena in Chihuahua—are "partners in protection" of the Chihuahuan Desert. These two areas in Mexico may some day be more fully defined as national parks.

Because of their numerous visitors and limited funding, national parks may not represent the most effective

(a)

(b)

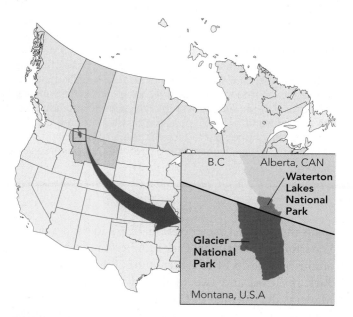

Figure 26.14 An international park. (a) Glacier National Park in the United States and **(b)** Waterton Lakes National Park in Canada were combined in 1932 to form the world's first international park.

way to preserve natural areas. In the United States, the National Wilderness Preservation System, created by Congress in 1964, preserves 42 million hectares (104 million acres) in 628 wilderness areas. Commercial activities, motorized access, and structures are highly restricted within these areas, and there are often limits on the number of people who are allowed to enter. Some wilderness areas lie within national parks or national forests.

Federal agencies that protect natural areas cooperate with state and local park systems and private conservation agencies. One such agency is The Nature Conservancy, a nonprofit corporation founded in 1951 by a small group of scientists who resolved to take direct action to save threatened natural areas. The main goal of The Nature Conservancy is to purchase land that, if protected, will preserve key and endangered habitats required to maintain the biodiversity of a region. Currently, The Nature Conservancy protects 6 million hectares (15 million acres) in the United States and 40 million hectares (101 million acres) in other countries. It frequently transfers acquired land to other protective agencies, including the NPS. In some areas, The Nature Conservancy is involved in habitat restoration. In 1998, for example, it purchased a 13,000-hectare (32,000-acre) ranch in Oklahoma's Osage Hills and established the Tallgrass Prairie Preserve, with the aim of recreating a tallgrass prairie where both bison and fires will again be normal parts of the ecosystem.

Many states now have laws that enable individuals to put their land in trust so they can have a financial return on the land without giving it over to development or paying excessive ownership taxes. This is an example of a program that provides an economic incentive for actions that preserve ecosystems.

Section Review

1. Describe a worst-case scenario of future human interactions with ecosystems.
2. What changes in energy sources, conservation areas, and tree harvesting would have to be made to achieve a best-case scenario?
3. List the problems that would have to be solved to reverse the current destruction of ecosystems.
4. To what goal is most National Park Service land dedicated?

SUMMARY

Human Population Growth

Human population is increasing exponentially (p. 550)
Human population has been growing exponentially for the past few centuries and may exceed 10 billion by 2050. In much of the developing world, the population has already reached the carrying capacity. Although some countries have birthrates that are close to or lower than death rates, many others have high birthrates and a large fraction of the population at or younger than the reproductive age.

Increased food production will involve genetically altered plants, improved growing practices, and more efficient food distribution systems (pp. 550–552)
The green revolution in the 1940s through the 1960s led to the development of high-yielding varieties of wheat, corn, and rice, which enabled countries to increase food production to keep pace with population growth. Future increases in yields will likely come from genetically engineered plants that have greater resistance to diseases and pests and tolerance for soil stresses; greater genetic diversity of crops; multicropping practices, such as crop rotation and intercropping; integrated pest management; and the organization of food production at a local level.

Human Impacts on Ecosystems

The presence and activities of large human populations disturb ecosystems (pp. 552–560)
Suppression of forest fires over many years leads to the occurrence of more destructive fires. Logging has destroyed or fragmented 80% of Earth's original forests and contributes to an increased frequency of mudslides, soil erosion, flooding, and silting of rivers and lakes. Overuse of water has lowered the water level in many aquifers, but implementation of simple irrigation practices, such as drip irrigation and reusing wastewater, could make water use more efficient. The discharge of untreated human and domestic animal wastes into rivers and lakes is a major source of water pollution; many regions of the world lack adequate water treatment. The burning of fossil fuels and the release into the atmosphere of molecules that were absent or rare on Earth before human civilization have worsened air pollution. Pollutants such as chlorofluorocarbons decrease the ozone concentration in the upper atmosphere. The burning of fossil fuels has also contributed to global warming due to the greenhouse effect and to acid precipitation. The introduction of exotic species causes significant environmental and economic harm, due in part to reduced crop production. The fragmentation of natural habitats by agriculture, forestry, and other human activities leads to changes in community structure. An increase in the rate of extinction, especially in tropical rain forests, is causing a decrease in global biodiversity.

The geographic information system provides a new tool to record changes in ecosystems (pp. 560–561)
The geographic information system (GIS) combines and analyzes geographic data, such as topography, soil type, and population density, on multidimensional maps. Conservation biologists can use these maps to monitor ecosystems and plan the management of natural resources.

The Future

The future of human interactions with ecosystems can be modeled on worst-case or best-case scenarios (pp. 562–563)
Under a worst-case scenario, human population will soon reach its carrying capacity, increased air and water pollution will lower the quality of life worldwide, natural ecosystems will be replaced by artificial landscapes, and pandemics of uncontrolled disease will become more common. Under a best-case scenario, human population will stabilize, air and water pollution will decrease, food production and the use of renewable

resources will increase, and ecosystem preservation and restoration will be given priority status.

Achieving a best-case scenario for the biosphere would involve a marked reversal of current trends (pp. 563–564)
Achieving a best-case scenario will require replacing fossil fuels and wood with alternative sources of energy; reducing air and water pollution; setting aside larger conservation areas; harvesting trees on a strict plan of sustainable yields; increasing agricultural production through practices such as multicropping, integrated pest management, and genetic engineering; and educating people about the importance of ending ecosystem destruction.

A number of problems would have to be overcome to reverse current trends of ecosystem destruction and modification (p. 564)
A lack of knowledge about how human activities affect ecosystems processes hinders the development of conservation plans. Efforts to limit ecosystem destruction would be very expensive, and most of the benefits of such efforts would be in the distant future. The goal of reversing current rates of extinction and of ecosystem fragmentation and destruction has no clear end point.

It is important to establish models of success in promoting ecosystem restoration (pp. 564–567)
The system of national parks and other sites administered by the U.S. National Park Service has served as a model of conservation for other countries. Private conservation agencies, such as The Nature Conservancy, also work to maintain biodiversity by preserving key habitats. State laws that enable individuals to put their land in trust provide an economic incentive for actions that preserve ecosystems.

Review Questions

1. How does the age structure of a population affect efforts to achieve population control?
2. How can multicropping improve food production?
3. What are some examples of biological control agents?
4. Why did the Bitterroot National Forest have disastrous forest fires in the summer of 2000?
5. Explain how sustainable forestry is achieved.
6. What change is happening to the High Plains Aquifer, and how does that change affect agriculture in the United States?
7. The Nile River flows through what three countries? What are the political consequences of this geographical situation?
8. What percentage of the world's food comes from irrigated agricultural land? List several ways in which the efficiency of agricultural irrigation could be improved.
9. Why is ozone an important component of the upper atmosphere?

10. What is another name for Russian thistle? Why has it spread rapidly through the western United States?
11. How does habitat fragmentation affect the composition of communities?
12. Why should areas that preserve natural habitats be large?
13. What is silphion? Why did it become extinct?
14. Why is Madagascar considered a biodiversity hotspot?
15. List some components of a best-case scenario for ecosystem stabilization.

Questions for Thought and Discussion

1. Many products sold in the United States are made in developing countries, where wages are lower. At the same time, food crops for export are often grown in place of plants for local consumption. What would the world be like if everyone in the world received at least the minimum wage that is paid in the United States?
2. Some people have suggested that all existing water shortages could be remedied by massive desalination of ocean water or by towing icebergs from polar regions. What are the strengths and weaknesses of each plan?
3. Given the rapid depletion of Earth's petroleum reserves, what strategy should the United States have for developing alternative energy sources? How would you pay for the development of these sources?
4. Should national parks have different purposes and uses than wilderness areas?
5. When Shenandoah National Park was created, many families lost their land and were relocated outside the park. Should the government continue to relocate people to preserve important ecosystems?
6. Would you favor a program in which existing national parks would be enlarged through the purchase of neighboring land? Would you support such a program if it were paid for by a 1% sales tax?
7. Federal and state governments have a limited amount of tax money to devote to environmental concerns. Make a priority list of how you think the money should be spent.
8. Using Figure 26–1 as a starting point and extending the horizontal axis to the year 2250, diagram how you think the human population will change in the next 250 years under the worst- and best-case scenarios discussed in this chapter.

Evolution Connection

Which do you think is the more powerful force for evolutionary change among plants on earth today, natural selection or artificial selection via human activity? Explain your reasons.

To Learn More

Visit The Botany Place Website at www.thebotanyplace.com for quizzes, exercises, and Web links to new and interesting information related to this chapter.

Angier, Matalie, ed. *The Best American Science and Nature Writing 2002.* New York: Mariner Books, 2002. This very interesting and informative anthology has a variety of styles, from conventional to controversial. The nature writing in particular celebrates wild areas and reminds us of the importance of conserving them.

Crichton, Michael. *Jurassic Park.* New York: Ballantine Books, 1999 (reissue). See the movie and read the book. They are different, and both are interesting.

Crichton, Michael. *The Lost World.* New York: Ballantine Books, 1996. This sequel to *Jurassic Park* is just as good.

Ewing, Rex. *Power with Nature: Solar and Wind Energy Demystified.* Masonville, CO: PixyJack Press, 2003. A very readable, interesting introduction to renewable sources of energy.

Primack, Richard B. *Essentials of Conservation Biology, Third Edition.* New York: Sinauer Associates, 2002. This excellent introductory text combines theory and basic research with numerous examples.

Stone, Richard. *Mammoth: The Resurrection of an Ice Age Giant.* Cambridge, MA: Perseus Publishing, 2001. The proposal presented in this book is to recreate the Pleistocene world of the mammoth using frozen mammoth tissue, found occasionally in the Arctic. This proposal could work.

Weddell, Bertie J. *Conserving Living Natural Resources: In the Context of a Changing World.* London: Cambridge University Press, 2002. This book is an introduction to the management of biological resources from three possible perspectives.

Weidensaul, Scott. *The Ghost with Trembling Wings: Science, Wishful Thinking, and the Search for Lost Species.* New York: North Point Press, 2002. Stories about what happens when a supposedly extinct species makes an appearance.

Wilson, E. O. *The Future of Life.* New York: Knopf, 2002. The Pulitzer Prize–winning naturalist combines dire warnings of extinctions with interesting stories and explains how the survival of particular species relates to economics.

Basic Chemistry*

Matter

The concept of **matter** includes anything that occupies space and has mass—this book, the food we eat, the water in which we bathe, the oxygen we breathe. The study of chemistry is basically the study of those pure substances called **chemical elements,** which occur alone or in various combinations in the universe, and how those elements interact and change in composition.

At the present time, 112 different elements have been discovered, of which approximately 88 occur naturally on Earth, with the remaining created in the laboratory. Of these natural elements, only a few occur in pure form, for example, hydrogen, oxygen, carbon, nitrogen, gold, silver, and copper, while the others are found combined chemically.

Atoms

The smallest particle of a chemical element that can exist and still retain its unique composition is called an **atom** (from the Greek *atomos*, which means "indivisible"). An atom consists of a core, called the **nucleus,** plus three kinds of subatomic particles, each of which differs in mass, electrical charge, and location within the atom. **Protons,** which have one or more positive electrical charges, and **neutrons,** which have no electrical charge and are neutral, are both found within the nucleus. **Electrons,** which have one or more negative electrical charges, form a cloud around the nucleus (Figure A.1).

For each atom, the number of protons and electrons is always equal. The number of protons in an element is

Table A.1 Atomic structures of some chemical elements commonly found in plants

Element	Chemical symbol	Number of protons*	Number of neutrons	Number of electrons	Mass number
Hydrogen	H	1	0	1	1
Carbon	C	6	6	6	12
Nitrogen	N	7	7	7	14
Oxygen	O	8	8	8	16
Sodium	Na	11	12	11	23
Magnesium	Mg	12	12	12	24
Phosphorus	P	15	16	15	31
Sulfur	S	16	16	16	32
Chlorine	Cl	17	18	17	35
Potassium	K	19	20	19	39
Calcium	Ca	20	20	20	40
Iron	Fe	26	30	26	56

*Atomic number

called its **atomic number.** The *combined* number of protons and neutrons in an atom is called its **mass number.** The atomic structure of some elements most common to plants is displayed in Table A.1.

Isotopes

An atom with the same number of protons but with a different number of neutrons is called an **isotope.** All isotopes of an element have the same atomic number, meaning they have the same number of electrons surrounding their nuclei. Further, because isotopes of the same element have the same number of protons and the same number of electrons, they have the same chemical and physical properties, though with some variations. These variations are attributable to the differing number of neutrons present in some atoms. For instance, while all atoms of the same element hydrogen have 1 proton—that is what makes the element hydrogen and not something else—the number of neutrons can vary from one hydro-

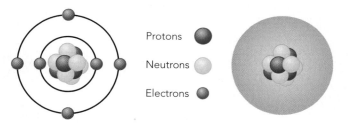

Protons
Neutrons
Electrons

Figure A.1 Two commonly used models of a carbon (C) atom.

*Sources for this Appendix include the following texts published as imprints of Pearson Education, Inc.: Timberlake, *General, Organic, & Biological Chemistry,* Addison Wesley, 2002; Marieb, *Essentials of Human Anatomy & Physiology, Seventh Edition,* Benjamin Cummings, 2003; and Campbell, *Biology: Concepts & Connections, Fourth Edition,* Benjamin Cummings, 2003.

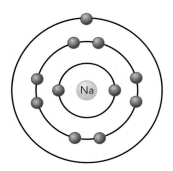

Figure A.2 Electron shell arrangement of a sodium (Na) atom.

gen atom to another, thus changing the atom's mass number. The isotope (^{2}H), called Deuterium (D), for instance, has 1 neutron, which makes its mass about twice that of an ordinary hydrogen atom (or, in other words, D is 1.998 times heavier). The different numbers of neutrons cause the mass of these atoms to be relatively large.

Electron Arrangements and Energy Levels

Most of the mass of an atom is in its nucleus, where the protons and neutrons are located. However, the nucleus is only a small region of the atom. The remaining space is virtually empty except for the continually moving, negatively charged electrons (see Figure A.1). This constant movement indicates that electrons possess energy, although they do not all have the same level of energy. Electrons of similar energies occupy energy levels called **shells** (Figure A.2). The capacity of each shell is different, with the shell closest to the nucleus containing both the lowest number of electrons and those with the lowest amount of energy. Electrons in the outermost shell have the highest energy. Energy levels are filled in order, the first before the second, the second before the third, and so forth. The first (lowest) energy level can hold two electrons at most. The second energy level can hold up to eight electrons, the third can hold up to 18, the fourth up to 32 (Table A.2). As each level fills, additional electrons are forced to occupy the next higher level. The electrons closest to the nucleus, with their negative charge, are most strongly attracted to the positive charge of the protons within it. Those electrons farther away are less attracted to the positive charge within the nucleus and so are more likely to react chemically with other atoms.

Molecules and Compounds

A molecule consists of a collection of atoms bound together. Some elements exist in molecular form; for instance, each molecule of oxygen consists of two atoms. The

same is true of hydrogen. When a hydrogen atom combines with another hydrogen atom, we have two molecules of hydrogen, or H_2. This chemical reaction—that is, the process of chemical change—can be expressed as

$$H + H \rightarrow H_2$$

The **product,** hydrogen, is expressed using its atomic symbol on the left, followed by a reaction arrow and the **reactant,** H_2 on the right. Chemical reactions are discussed in further detail on page A-5.

If two or more atoms of *different* elements combine, this combination is called a **compound.** For example, a molecule of water, or H_2O, consists of two hydrogen atoms and one oxygen atom. This chemical reaction can be expressed as

$$2H + O \rightarrow H_2O$$

The notation 2H on the left indicates two unjoined atoms, while H_2 on the right indicates that two hydrogen atoms are joined to one oxygen atom to form one water molecule.

Organic vs. Inorganic Compounds

All chemicals fall into one of two classes of compounds. **Organic compounds** contain carbon. Included in this

Table A.2 Electron shell arrangements for some chemical elements commonly found in plants

Element	Symbol	Atomic number	Number of electrons in shell* 1	2	3	4
Hydrogen	H	1	1			
Carbon	C	6	2	4		
Nitrogen	N	7	2	5		
Oxygen	O	8	2	6		
Sodium	Na	11	2	8	1	
Magnesium	Mg	12	2	8	2	
Phosphorus	P	15	2	8	5	
Sulfur	S	16	2	8	6	
Chlorine	Cl	17	2	8	7	
Potassium	K	19	2	8	8	1
Calcium	Ca	20	2	8	8	2
Iron**	Fe	26	2	8	16	

*Remember that the atomic number (number of protons) and the total number electrons for each atom are always equal.

**Notice that iron is the only element in this list that completely fills its outermost shell.

class are carbohydrates, lipids, proteins, and nucleic acids. These compounds originally were called *organic* because it was once assumed that they could be formed only by living organisms; that assumption has since been proven incorrect. With the exception of the gas carbon dioxide (CO_2), and carbonates, which include chalk and are treated as carbon compounds, **inorganic compounds** lack carbon and are usually smaller molecules, such as water, salts, and many acids and bases.

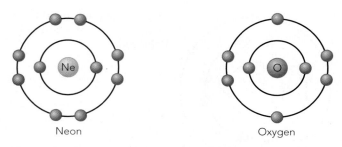

Figure A.3 Two models that illustrate electron shell arrangements for chemically inert (left) and chemically reactive (right) atoms. To fill its valence shell, oxygen must bond with another atom.

Chemical Bonds

When atoms combine with other atoms, an energy relationship, or chemical bond, forms. There are three types of chemical bonds discussed here that are of particular significance in living organisms, including plants. The only electrons involved in bonding are those in an atom's outermost shell, which is called a *valence shell*. It is the electrons in this shell that determine how an atom reacts chemically. To form a chemical bond, an atom must contain an octet of 8 electrons in its valence shell or share the electrons of another atom to fill its valence shell. This is referred to as the "Rule of 8" or the "Octet rule." If the valence shell is filled, the element is said to be inert or chemically unreactive; if the valence shell is not filled, the element is chemically reactive, interacting with other atoms, thus gaining, losing, or sharing electrons to fill its valence shells (Figure A.3). It can be said that atoms interact in such a way that allows their valence shells to be filled.

Types of Chemical Bonds

Ions

One way to fill a valence shell is to transfer one or two electrons from one atom to another. The gain or loss of electrons produces charged atoms called **ions.** When an atom loses an electron, it becomes positively charged. When an atom gains an electron, it becomes negatively charged. Ions with a positive charge are called **cations,** while negatively charged ions are called **anions.** Sodium (Na^+) and Chloride (Cl^-) are two common ions that have opposite charges, which means they attract one another (Figure A.4). The result when these two ions combine is the compound sodium chloride (NaCl), which is known to us in our daily lives as table salt. The mutual attraction between two such ions is called an **ionic bond.**

Covalent Bonds

When one or more pairs of electrons are shared by atoms, **covalent bonds** are formed. In this way, the outer

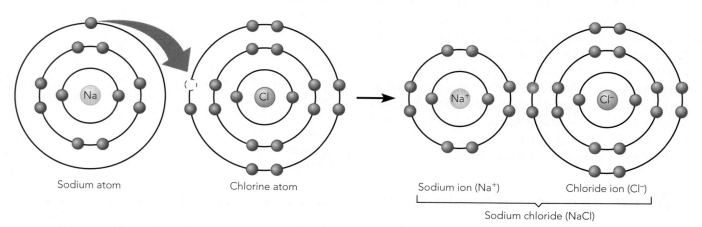

Figure A.4 Formation of an ionic bond. Both the sodium atom and the chlorine atom have incomplete valence shells. When the sodium atom shares an electron with the chlorine atom, both gain stability as ions.

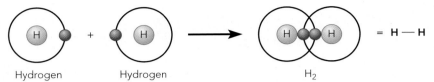

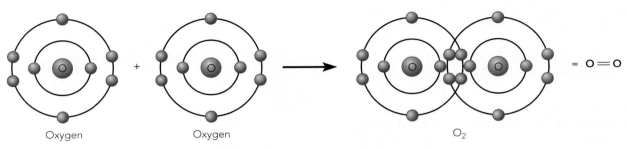

(a) Formation of a single covalent bond; two hydrogen atoms share one electron pair

(b) Formation of a double covalent bond; two oxygen atoms share two electron pairs

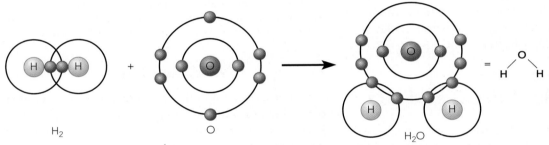

(c) Formation of a single covalent bond between two different elements, resulting in a compound

Figure A.5 Formation of covalent bonds.

valence shells of atoms are filled and thereby gain stability. A **single covalent bond** is formed when two or more atoms share one electron pair. For instance, a single covalent bond connecting two hydrogen atoms, each with its single electron, forms two molecules of hydrogen (Figure A.5a). **Double covalent bonds** are formed when two or more atoms share two electron pairs, as shown in Figure A.5b. As we saw in the previous section on compounds, when two or more atoms of one element share one one or more electron pairs of a *different* element, a compound is formed. The number of bonds is determined by the number of electron pairs shared. For example, in a water molecule (H_2O), which is a compound, each single covalent bond consists of 2 electrons each from the oxygen and the hydrogen atoms (Figure A.5c).

Electronegativity and Bond Polarity

In Figure A.5a and b, the electrons are shared *equally,* because both atoms are equally attracted to one an-

other. In other words, each atom has the same **electronegativity,** which is defined as the relative ability of a bonded atom in a molecule to attract shared electrons. The greater the electronegativity of an atom, the greater its ability to attract electrons to itself. Covalent bonds, shared by two or more atoms of the same element are considered *nonpolar.*

In contrast, electron pairs in a compound may be shared unequally, because one of the atoms has a stronger attraction for the pair of electrons than the other. Water (H_2O) is a good example of such a compound (Figure A.5c).

The water molecule is formed by two hydrogen atoms binding covalently to a single oxygen atom. Each hydrogen atom shares an electron pair with the oxygen atom, but in this case, oxygen is more electronegative than hydrogen, and thus attracts and gains a greater share in the bonding electron pair (Figure A.6). In this situation, the bond is called a **polar covalent bond,** which means that because the shared electrons are pulled closer to the atom

Figure A.6 Water molecule.

that is more electronegative, a water molecule has one partial negative pole and two partially positive poles.

Hydrogen Bonds

A **hydrogen bond** is formed by a partially positive hydrogen atom lying between two strongly electronegative atoms with single pairs of electrons, such as oxygen, nitrogen, or fluoride. The hydrogen bond, which is denoted by dots to distinguish it from a true covalent bond, is common between water molecules (Figure A.7).

Hydrogen bonds are also important in maintaining the structure of macromolecules such as proteins, nucleic acids (one of which is DNA), and carbohydrates. Macromolecules are discussed in Chapters 2 and 7.

Water, Acids, Bases, and pH

Water makes up more than 70–80% of the mass of most plants. Its cohesiveness, its high heat vaporization, and its versatility as a solvent are a result of its chemical structure, which is based upon hydrogen bonding. We can see in Figure A.7 that the oxygen atoms attract the hydrogen atoms of neighboring water molecules. Individual hydrogen bonds are weak, and are thus formed, broken, and reformed again in split seconds; however, the cohesiveness of the bonds, which result from the strong attraction of the oxygen atoms for the hydrogen atoms, holds many hundreds of water molecules together at the root of a plant. What this means for plants is that water can be transported upward against the pull of gravity from their roots to their leaves, where the water is emitted as vapor into the environment.

The polarity of water's molecules is responsible for its versatility as a solvent. As the solvent in all cells, including plant sap, water dissolves a variety of solutes that are necessary for life. Water molecules remain intact in the aqueous solutions of most organisms; however, some water molecules break apart into hydrogen ions (H^+) and hydroxide ions (OH^-). Hydrogen ions, which consist of the hydrogen nucleus only, are called *protons*. It is crucial in maintaining proper chemical processing functions within organisms that the right balance of H^+ ions and OH^- ions exists.

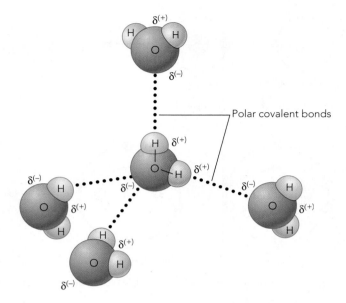

Figure A.7 Polar covalent bonds. Hydrogen bonds between water molecules are polar because the positively charged hydrogen electrons bind with the strong negatively charged oxygen atom.

Acids and Bases

A chemical compound that *increases* the relative number of H^+ ions in solutions is called an **acid,** which is also known as a **proton donor.** When an ionic substance is dissolved in water, it may change the relative numbers of H^+ and OH^- ions such that the concentration of H^+ no longer equals the concentration of OH^-. For example, if hydrogen chloride (HCl) is dissolved in water, it breaks up into H^+ and Cl ions. As a consequence, H^+ will exceed OH^-. A solution is acidic when H^+ exceeds OH.

A chemical compound that *decreases* the number of H^+ ions in solutions is called a *base,* which is also known as a **proton acceptor.** As with acids, when an ionic substance is dissolved in water, it may change the relative numbers of H^+ and OH^- ions. If sodium hydroxide is dissolved in water, it breaks up into Na^+ and OH^- ions, so that the concentration of OH exceeds H^+. A solution is basic (or alkaline) when OH exceeds H^+.

pH

The degrees of acidity are expressed by means of the **pH** (from the German *potenz Hydrogen,* which means the "power of hydrogen") scale, which is based upon the number of hydrogen ions in solution (Figure A.8). The concentration (expressed in moles per liter) of hydrogen ions and the corresponding hydroxide ion concentration are indicated for each pH value noted. At a pH of 7, the concentrations of H^+ ions and OH^- ions are equal and the

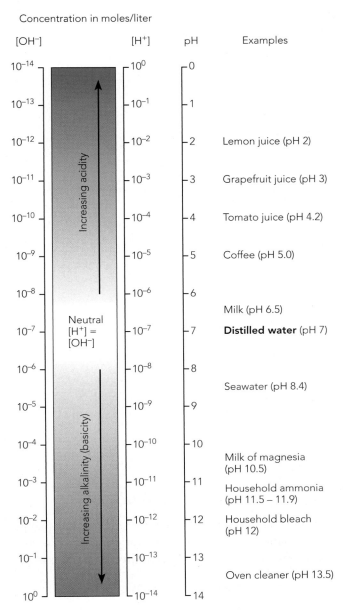

Concentration in moles/liter

[OH⁻]	[H⁺]	pH	Examples

Increasing acidity

Neutral [H⁺] = [OH⁻]

Increasing alkalinity (basicity)

$[OH^-]$: 10^{-14}, 10^{-13}, 10^{-12}, 10^{-11}, 10^{-10}, 10^{-9}, 10^{-8}, 10^{-7}, 10^{-6}, 10^{-5}, 10^{-4}, 10^{-3}, 10^{-2}, 10^{-1}, 10^{0}

$[H^+]$: 10^{0}, 10^{-1}, 10^{-2}, 10^{-3}, 10^{-4}, 10^{-5}, 10^{-6}, 10^{-7}, 10^{-8}, 10^{-9}, 10^{-10}, 10^{-11}, 10^{-12}, 10^{-13}, 10^{-14}

pH: 0, 1, 2, 3, 4, 5, 6, 7, 8, 9, 10, 11, 12, 13, 14

Examples:
- Lemon juice (pH 2)
- Grapefruit juice (pH 3)
- Tomato juice (pH 4.2)
- Coffee (pH 5.0)
- Milk (pH 6.5)
- **Distilled water** (pH 7)
- Seawater (pH 8.4)
- Milk of magnesia (pH 10.5)
- Household ammonia (pH 11.5 – 11.9)
- Household bleach (pH 12)
- Oven cleaner (pH 13.5)

Figure A.8 pH scale.

solution is neutral. A solution below pH7 is acidic, while a solution above pH7 is basic. Note that each pH unit represents a tenfold change in the concentration of H⁺ in the solution.

Chemical Reactions

A chemical reaction is the process of chemical change in matter when atoms combine to form molecules. Reactions occur in nature, in the laboratory, and in biological systems. The number and the arrangement of an atom's subatomic particles, particularly electrons, determine its chemical properties and its behavior. Atoms are neither created nor destroyed in a chemical reaction; rather, by virtue of making and breaking bonds, they simply rearrange matter (see page 572). We can see this if we observe a reaction in a container, where we will find that there is no change in mass. This preservation of mass is called **the law of conservation of mass.** Chemical equations are used to express both the qualitative change that occurs in a reaction, as well as the quantitative expression of this law. For example, the equation

$$2Na + 2H_2O \rightarrow 2\,NaOH + H_2$$

which gives us four H atoms, two Na atoms, and two O atoms on each side. There are recognizable patterns to most chemical reactions, and this is one of the most common (only three will be discussed here). This is an example of an **exchange reaction,** which involves both breaking and making bonds. Here sodium and water react to form sodium hydroxide and hydrogen.

A second reaction pattern is one that occurs when two or more atoms or molecules combine to form a larger, more complex molecule. This type of reaction always involves bond formation, such as the joining of hydrogen and oxygen to form the product water, which is expressed as

$$2H + O \rightarrow H_2O$$

A third common reaction occurs when bonds are broken, that is, when a molecule is broken down into smaller molecules, atoms, or ions. For example, during respiration, a carbohydrate (glucose) is oxidized in body cells to produce carbon dioxide, water, and energy, which can be expressed as

$$C_6H_{12}O_6 + 6O_2 \rightarrow 6CO_2 + 6H_2O + energy$$

Metabolic Pathways

The sum of the chemical reactions occurring within the cells of living organisms, such as plants, is known as **metabolism.** Enzymes, which are protein molecules, speed up chemical reactions by forming a temporary association with the molecules that are reacting, which then become the substrate (the molecule on which the enzyme acts) for the next reaction. Thus, within a cell, chemical reactions are linked together in a series, which constitutes a **metabolic pathway.** There are various metabolic pathways, each of which serve different functions in the cell. There are three common types of

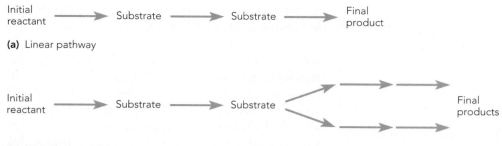

(a) Linear pathway

(b) Branching pathway

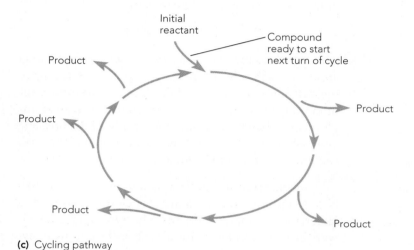

(c) Cycling pathway

Figure A.9 Models of three metabolic pathways.

metabolic pathways, although frequently identical reactions occur in different pathways that are catalyzed by different enzymes. The three types of metabolic pathways discussed here are **linear, branching** (in an otherwise linear pathway), and **cyclic,** which always regenerates the starting compound (Figure A.9). In a branching pathway, the product of a reactant may proceed along either of two pathways. In a cyclic pathway, the cycle may begin with simple molecules and produce larger ones, or it may begin with large molecules and break into simpler ones. Oxidation-Reduction reactions, which include both photosynthesis and respiration and are discussed in Chapter 7, involve cyclic pathways.

Chemical Equilibrium

In many reactions, the products can interact and change back into the reactants. In other words, the reaction occurs in both the forward and the reverse directions. For example

$$2SO_2 + O_2 \rightarrow 2SO_3$$
$$\leftarrow$$

Eventually the rate of the reverse reaction becomes equal to the rate of the forward reaction, at which time the reaction has reached **chemical equilibrium.** At equilibrium, both have the same *proportions* of reactants and product.

Metric Conversions

Measurement	Unit and Abbreviation	Metric Equivalent	Metric to English Conversion Factor	English to Metric Conversion Factor
Length	1 kilometer (km)	= 1000 (10^3) meters	1 km = 0.62 mile	1 mile = 1.61 km
	1 meter (m)	= 100 (10^2) centimeters = 1000 millimeters	1 m = 1.09 yards 1 m = 3.28 feet 1 m = 39.37 inches	1 yard = 0.914 m 1 foot = 0.305 m
	1 centimexter (cm)	= 0.01 (10^{-2}) meter	1 cm = 0.394 inch	1 foot = 30.5 cm 1 inch = 2.54 cm
	1 millimeter (mm)	= 0.001 (10^{-3}) meter	1 mm = 0.039 inch	
	1 micrometer (μm) [formerly micron (μ)]	= 0.000001 (10^{-6}) meter		
	1 nanometer (nm) [formerly millimicron (mμ)]	= 0.000000001 (10^{-9}) meter		
	1 angstrom (Å)	= 0.0000000001 (10^{-10}) meter		
Area	1 square meter (m^2)	= 10,000 square centimeters	1 m^2 = 1.1960 square yards 1 m^2 = 10.764 square feet	1 square yard = 0.8361 m^2 1 square foot = 0.0929 m^2
	1 square centimeter (cm^2)	= 100 square millimeters	1 cm^2 = 0.155 square inch	1 square inch = 6.4516 cm^2
Mass	1 metric ton (t)	= 1000 kilograms	1 t = 1.103 ton	1 ton = 0.907 t
	1 kilogram (kg)	= 1000 grams	1 kg = 2.205 pounds	1 pound = 0.4536 kg
	1 gram (g)	= 1000 milligrams	1 g = 0.0353 ounce 1 g = 15.432 grains	1 ounce = 28.35 g
	1 milligram (mg)	= 0.001 gram	1 mg = approx. 0.015 grain	
	1 microgram (μg)	= 0.000001 gram		
Volume (solids)	1 cubic meter (m^3)	= 1,000,000 cubic centimeters	1 m^3 = 1.3080 cubic yards 1 m^3 = 35.315 cubic feet	1 cubic yard = 0.7646 m^3 1 cubic foot = 0.0283 m^3
	1 cubic centimeter (cm^3 or cc)	= 0.000001 cubic meter = 1 milliliter	1 cm^3 = 0.0610 cubic inch	1 cubic inch = 16.387 cm^3
	1 cubic millimeter (mm^3)	= 0.000000001 cubic meter		
Volume (liquids and gases)	1 kiloliter (kl or kL)	= 1000 liters	1 kL = 264.17 gallons	1 gallon = 3.785 L
	1 liter (l or L)	= 1000 milliliters	1 L = 0.264 gallon 1 L = 1.057 quarts	1 quart = 0.946 L
	1 milliliter (ml or mL)	= 0.001 liter = 1 cubic centimeter	1 ml = 0.034 fluid ounce 1 ml = approx. $\frac{1}{4}$ teaspoon 1 ml = approx. 15–16 drops (gtt.)	1 quart = 946 ml 1 pint = 473 ml 1 fluid ounce = 29.57 ml 1 teaspoon = approx. 5 ml
	1 microliter (μl or μL)	= 0.000001 liter		
Time	1 second (s)	= $\frac{1}{60}$ minute		
	1 millisecond (ms)	= 0.001 second		
Temperature	Degrees Celsius (°C)		°F = $\frac{9}{5}$ °C + 32	°C = $\frac{5}{9}$(°F − 32)
Land Area Conversions	1 hectare	= 0.01 square kilometers		= 2.4 acres
	1 acre			= 43,560 square feet
	1 square mile	= 258.9 hectares		= 2.59 square kilometers
Pressure Conversions	1 atmosphere			= 101.32 kilo pascals
				= 760 mm mercury
				= 22.4 molal

Classification of Life

This appendix presents the taxonomic classification used for living organisms in this book. The classification system utilizes three domains, two of which, Archaea and Bacteria, contain prokaryotes and one, Eukarya, contains eukaryotes. I have chosen not to break the prokaryotic domains into kingdoms, primarily because these are not discussed in this text, and also because a recent influx of molecular information has stimulated considerable discussion of classification within each prokaryotic kingdom.

The Kingdom Protista is also in a state of transition. It is not a phylogenetic group and should be split into several different kingdoms. Several candidate kingdoms have been proposed and research is underway by systematists.

Also, although I make note in the text that some or all of the green algae are now frequently classified with plants, agreement on exact names and taxonomic ranks has not yet occurred. Some systematists have proposed the kingdom Viridiplantae to include all green algae and plants or the kingdom Streptophyta to include charophyceans and plants. I continue to use Kingdom Plantae until the systematic dust settles.

Within the plant kingdom, I note the four traditional phyla of seedless vascular plants. While systematists generally agree that psilotophytes and sphenophytes should be in the same phylogenetic group as ferns, naming and ranking of the resulting taxonomic categories has not been formally agreed upon.

DOMAIN ARCHAEA

DOMAIN BACTERIA
Includes cyanobacteria and other groups.

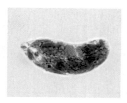

DOMAIN EUKARYA
In Kingdom Protista, this text includes only algae. Nonphotosynthetic protists are not included in this classification.

Kingdom Protista
  Phylum Euglenophyta (euglenoids)
  Phylum Dinoflagellata (dinoflagellates)
  Phylum Bacillariophyta (diatoms)
  Phylum Xanthophyta (yellow-green algae)
  Phylum Chrysophyta (golden-brown algae)
  Phylum Cryptophyta (cryptomonads)
  Phylum Prymnesiophyta (haptophytes)
  Phylum Rhodophyta (red algae)
  Phylum Phaeophyta (brown algae)
  Phylum Chlorophyta (green algae)

Kingdom Plantae
"Bryophytes"
  Phylum Hepatophyta (liverworts)
  Phylum Anthocerophyta (hornworts)
  Phylum Bryophyta (mosses)

"Pteridophytes" (seedless vascular plants)
  Phylum Lycophyta (lycophytes)
  Phylum Psilotophyta (whisk ferns)
  Phylum Sphenophyta (horsetails)
  Phylum Pterophyta (ferns)
"Seed plants"
  "Gymnosperms"
    Phylum Ginkgophyta (ginkgo)
    Phylum Cycadophyta (cycads)
    Phylum Gnetophyta (gnetae)
    Phylum Coniferophyta (conifers)
  "Angiosperms"
    Phylum Anthophyta (flowering plants)

Kingdom Fungi
  Phylum Chytridiomycota (chytrids)
  Phylum Zygomycota (zygomycetes)
  Phylum Ascomycota (sac fungi)
  Phylum Basidiomycota (club fungi)
  Deuteromycetes (imperfect fungi)
  Lichens (symbiotic associations of fungi and algae)

Kingdom Animalia

Phylum Porifera (sponges)

Phylum Cnidaria (cnidarians)

    Class Hydrozoa (hydrozoans)

    Class Scyphozoa (jellies)

    Class Anthozoa (sea anemones and coral animals)

Phylum Ctenophora (comb jellies)

Phylum Platyhelminthes (flatworms)

    Class Turbellaria (free-living flatworms)

    Class Trematoda (flukes)

    Class Monogenea (monogeneans)

    Class Cestoidea (tapeworms)

Phylum Bryozoa (bryozoans)

Phylum Phoronida (phoronids)   } "Lophophorate phyla"

Phylum Brachiopoda (brachiopods)

Phylum Rotifera (rotifers)

Phylum Nemertea (proboscis worms)

Phylum Mollusca (mollusks)

    Class Polyplacophora (chitons)

    Class Gastropoda (gastropods)

    Class Bivalvia (bivalves)

    Class Cephalopoda (cephalopods)

Phylum Annelida (segmented worms)

    Class Oligochaeta (oligochaetes)

    Class Polychaeta (polychaetes)

    Class Hirudinea (leeches)

Phylum Nematoda (roundworms)

Phylum Arthropoda (This review groups arthropods into a single phylum, but some zoologists now split the arthropods into multiple phyla.)

Class Arachnida (spiders, ticks, scorpions)

Class Diplopoda (millipedes)

Class Chilopoda (centipedes)

Class Crustacea (crustaceans)

Class Insecta (insects)

Phylum Echinodermata (echinoderms)

    Class Asteroidea (sea stars)

    Class Ophiuroidea (brittle stars)

    Class Echinoidea (sea urchins and sand dollars)

    Class Crinoidea (sea lilies)

    Class Concentricycloidea (sea daisies)

    Class Holothuroidea (sea cucumbers)

Phylum Hemichordata

Phylum Chordata (chordates)

    Subphylum Urochordata (urochordates: tunicates)

    Subphylum Cephalochordata (cephalochordates: lancelets)

    Subphylum Vertebrata (vertebrates)

        Class Myxini (hagfishes)

        Class Cephalaspidomorphi (lampreys)

        Class Chondrichthyes (cartilaginous fishes)

        Class Actinopterygii (ray-finned fishes)

        Class Actinistia (lobe-finned fishes)   } "Osteichthyes" (bony fishes)

        Class Dipnoi (lungfishes)

        Class Amphibia (amphibians)

        Class Testudines (turtles)

        Class Lepidosauria (lizards, snakes, tuataras)   } "Reptiles"

        Class Crocodilia (crocodiles, alligators)

        Class Aves (birds)

        Class Mammalia (mammals)

CREDITS

Illustration Credits

The following figures are adapted from Neil A. Campbell and Jane B. Reece, *Biology,* 6th ed., San Francisco, CA: Benjamin Cummings © 2002 Pearson Education, Inc.: **3.4, Chapter 9 Conservation Biology box inset, 15.16.**

The following figures are adapted from Neil A. Campbell, Jane B. Reece, and Lawrence G. Mitchell, *Biology,* 5th ed., Menlo Park, CA: Benjamin/Cummings © 1999 Addison Wesley Longman, Inc.: **2.8, 12.1.**

Chapter 11 The Intriguing World of Plants box inset Adapted from M.W. Nabors and A. Lang, 1971. "The Growth Physics and Water Relations of Red-Light-Induced Germination in Lettuce Seeds." *Planta* 101: 1–25.

12.18 Adapted from photo by W. Ellis Davies.

15.10 Adapted from J.N. Clatworthy and J.L. Harper, 1962. "The Comparative Biology of Closely Related Species Living in the same Area. V. Inter-and Intraspecific Interference within Cultures of *Lemna spp* and *Salvinia natans.*" *Journal of Experimental Botany* 13: 30.

Chapter 18 Biotechnology box inset Adapted from A. Melis, et al., 2000. "Sustained Photobiological Hydrogen Gas Production upon Reversible Inaction of Oxygen Evolution in the Green Alga *Chlamydomonas reinhardtii.*" *Plant Physiology* 122: 131.

25.5 Adapted from L.B. Keith, 1974. "Some Features of Population Dynamics in Mammals." *Proceedings of the International Congress of Game Biologists (Stockholm)* 11: 17–58.

25.8 Adapted from Tilman et al., 1981. "Competition and Nutrient Kinetics Along a Temperature Gradient: An Experimental Test of a Mechanistic Approach to Niche Theory." *Limnology and Oceanography* 27: 1025, 1027.

26.2 Adapted from J.A.J. McFalls, 1998. "Population: A Lively Introduction." *Population Bulletin* 53: 52.23.

26.10 Adapted from Cincotta, et al., 2000. "Human Population in the Biodiversity Hotspots." *Nature* 404: 990.

Photo Credits:

Unit Opening Photos: Unit I Michael Clayton **Unit II** Johnathan Smith, Cordaiy Photo Library Ltd./CORBIS **Unit III** John S. Heywood **Unit IV** Tom Bean/CORBIS **Unit V** Keren Su/Stone

Chapter 1 opening photo Michael S. Yamashita/CORBIS **page 2 top left** Jonathan Nourok/PhotoEdit, **right** Benjamin Cummings, **bottom** Gerard Lacz/Peter Arnold, Inc. **1.2** Erich Lessing/Art Resource, NY **1.3** Walter Bibikow/Taxi **Plants & People, page 5 left** Eisenhut & Mayer/FoodPix, **right** Denis Waugh/Stone **1.4 top** Gianni Dagli Orti/CORBIS **1.4 bottom** Dorling Kindersley **The Ingriguing World of Plants, page 7** Dorling Kindersley **1.5a** Andrea Pistolesi/The Image Bank **1.5b** Gregory G. Dimijian/Photo Researchers, Inc. **Conservation Biology, page 9** Roger Peterson, USDA Forest Sevice **1.6** Peter Berger, Institut für Biologie, Freiburg **Biotechnology, page 10** Dorling Kindersley **1.7** Caterina, M. J., Leffler, A., Malmberg, A. B., Martin, W. J., Trafton, J., Petersen-Zeitz, K., Koltzenburg, M., Basbaum, A. I., and Julius, D. (2000) Im-

paired Nociception and Pain Sensation in Mice lacking the Capsaicin receptor. Science 288: 306-313. **1.10a** John Shaw/Tom Stack & Associates **1.10b** Digital Vision **1.10c** Steve Terill/CORBIS **1.10c inset** Robert & Lorri Franz/CORBIS **1.10d** Lindsay Hebberd/CORBIS **1.11** National Portrait Gallery, London **1.12a,b** Malcolm Wilkins **1.12 inset** Library of Congress **1.14** Michael Rosenfeld/Stone

Chapter 2 opening photo Ann Clemens/UTEX Culture Collection of Algae, University of Texas at Austin **page 26 left** New York Academy of Medicine Library, **right** Carolina Biological/Visuals Unlimited **2.1a** Barry Runk/Stan/Grant Heilman Photography **2.1b** Eldon Newcomb **2.1c** David Scharf/Peter Arnold, Inc. **Plants & People, page 28 left** Alan Shinn, **right** Hooke, Robert, Micrographica, London, 1665 **2.3 left inset** Dennis Strete, Benjamin Cummings **2.3 left** Paul Kugrens **2.3 right inset** Paul Kugrens **2.7 top** Dorling Kindersley **2.7 middle** Graham Kent **2.7 bottom** E. H. Newcomb & W.P.Wergin/Biological Photo Service **2.8** Daniel S. Friend, Harvard Medical School **2.9** E. H. Newcomb **Biotechnology, page 42** Dorling Kindersley **2.16 all** Ed Reschke/Peter Arnold

Chapter 3 opening photo Gary Randall/Taxi **page 49** CORBIS **3.1a** Michael Clayton **3.1b** Ken Wagner/Visuals Unlimited **3.1c** Brian Capon **3.2** Graham Kent **3.3a,b** Graham Kent **3.3 inset** Dorling Kindersley **The Intriguing World of Plants, page 53 left** Dorling Kindersley, **right** Jacqui Hurst/CORBIS **3.5** Michael Clayton **Plants & People, page 55 left** Dorling Kindersley, **right** Library of Congress **3.6a** Graham Kent **3.6c** Scanned from book: H. A. Core, W. A. Cote, and A. C. Day, Wood: structure and Identification, 2nd edition, Syracuse U. Press, 1979 **3.7b,c** Scanned from book: H. A. Core, W. A. Cote, and A. C. Day, Wood: structure and Identification, 2nd edition, Syracuse U. Press, 1979 **3.8a,c** Graham Kent **3.9** Buddy Mays/CORBIS **3.10a** Brian Capon **3.10b** Graham Kent **3.13b** Michael Clayton **3.13c** Graham Kent **3.14 both** Malcolm Wilkins, University of Glasgow **3.16a,b** Dorling Kindersley **3.16c** Photographer's Choice

Chapter 4 opening photo Sharon Guynup/The Image Bank **page 71** Dorling Kindersley **Plants & People, page 87 left** Dorling Kindersley, **right** Richard A. Cooke/CORBIS **4.1a** Barry Runk/Stan/Grant Heilman Photography **4.1b** Dorling Kindersley **4.2** Michael Clayton **4.3a,b** Ed Reschke **4.4 all** Michael Clayton **4.6a** Dorling Kindersley **4.6b** James Strawser/Grant Heilman Photography **4.6c** Dorling Kindersley **4.6d** Australian Picture Library/CORBIS **4.6e** Ann Hirsch, Botanical Society of America **4.6f** Dorling Kindersley **The Intriguing World of Plants, page 79 top** Daniel Nickrent, **bottom** Brian Capon **4.7b** Stanley L. Flegler/Visuals Unlimited **4.7c** George Barron **4.8 top** Ed Reschke **4.8 bottom** Adrian Bell **4.10c** Michael Clayton **4.10d** Carolina Biological/Visuals Unlimited **4.10e** Graham Kent **4.14 all** Dorling Kindersley **4.15 both** Graham Kent **4.17 both** Graham Kent **4.19a** J. C. Revy/Phototake **4.19b** CORBIS **4.19c** Rex Butcher/Stone **4.22a,b** Dorling Kindersley **4.22c** John N. Trager **4.22d** Brian Capon **4.22e** Bryan Bowes **Evolution, page 95 top** Fritz Polking; Frank Lane Picture Agency/CORBIS, **bottom** Dorling Kindersley

Chapter 5 opening photo Sally A. Morgan, Ecoscene/CORBIS **page 100 top** AEF/Imagebank, **bottom** David R. Parks/Missouri Botanical Garden **Plants & People, page 102 both** Dorling Kindersley **5.6 all** Dorling Kindersley **5.13a** Dorling Kindersley **5.13b** Michael Clayton **05.13c** Brian Capon **5.14** David Muench/CORBIS **5.15** Dorling Kindersley **The Intriguing World of Plants, page 111 left** David W. Stahle, **right**

National Park Service, Colonial National Historical Park, Jamestown collection **5.16 left** Dorling Kindersley **5.16 right** Graham Kent **5.17** Paul Chesley/Stone **5.18** Grant Heilman/Grant Heilman Photography **5.19** Berkeley Mills **5.20 both** Gibson Guitars **5.21** Graham Kent **Plants & People, page 116** Michael S. Yamashita/CORBIS **5.22** Landmann-Benali/Liaison **5.23** Charles O'Rear/CORBIS **Conservation Biology, page 118 top** Natalie Fobes/CORBIS, **bottom** Morbark, Inc.

Chapter 6 opening photo Dorling Kindersley **page 123 top left, bottom left, top right** Dorling Kindersley, **bottom right** Travis Amos **6.1a** Phil Schermeister/National Geographic Image Collection **6.1b** Jerome Wexler/Photo Researchers, Inc. **6.1c** Dorling Kindersley **6.1d** Richard Cummins/CORBIS **6.1e** Dorling Kindersley **Biotechnology, page 137 top** Dorling Kindersley, **bottom** Twain Butler **The Intriguing World of Plants, page 138 all** Dorling Kindersley **6.14a** Photo Researchers, Inc. **6.14b** Graham Kent **6.14c** Ed Young/CORBIS **6.14d** Dorling Kindersley **6.15e** Martin Harvey; Gallo Images/CORBIS **6.14f** Microscopy by Brigitte Krückl. Courtesy of Veronika Mayer, Institute of Botany, University of Vienna

Chapter 7 opening photo Dorling Kindersley **The Intriguing World of Plants, page 150** Digital Vision/CORBIS **7.7c** SETOR Image Gallery, University of Ottawa **Biotechnology, page 154 both** Dorling Kindersley **7.8** Digital Vision/CORBIS **7.14b** Dorling Kindersley

Chapter 8 opening photo Gentl & Hyers/FoodPix **page 176 top** Richard Hamilton Smith/CORBIS **bottom** Catherine Karnow/CORBIS **8.1 top** Natural World/CORBIS **8.1 middle, bottom** Ann Clemens/UTEX Culture Collection of Algae, University of Texas at Austin **The Intriguing World of Plants, page 178 left** Gary W. Carter/CORBIS, **right** Joe McDonald/CORBIS **8.9** Lawrence Berkeley National Laboratory **Evolution, page 189** Dorling Kindersley **8.14a** David Muench/CORBIS **8.14b** Dave Bartruff/CORBIS

Chapter 9 opening photo David Muench/CORBIS **page 196** Reuters NewMedia Inc./CORBIS **The Intriguing World of Plants, page 207** William Banner **9.10a** Mason Morfit/Taxi **9.10b** Stephen J. Kron, U of Chicago

Chapter 10 opening photo George Grall/National Geographic Image Collection **page 214** ArsNatura **The Intriguing World of Plants, page 219** Barry L. Runk/Grant Heilman Photography, Inc. **Biotechnology, page 221** The Image Bank **10.6a** Graham Kent **10.9 all** M. H. Zimmerman courtesy of Professor P. B. Tomlinson, Harvard University **Plants & People, page 228** CORBIS

Chapter 11 opening photo Kim Taylor **page 236 top** Raymond Gehman/CORBIS, **middle and bottom** Dorling Kindersley **Biotechnology, page 241 all** Murray Nabors **11.4** Fred Jensen, Kearney Agricultural Center **11.5** Janet Braam, from *Cell* 60 (9 February 1990):cover. ©1990 by Cell Press, **11.9** Malcolm Wilkins, University of Glasgow **11.12** David Muench/CORBIS **11.13a** Malcolm Wilkins **11.14** Michael Evans, Ohio State University **11.15** Dorling Kindersley **11.16a,b** David Sieren/Visuals Unlimited **11.16c** From K. Esau, Anatomy of Seed Plants, 2nd ed. (New York: John Wiley and Sons, 1977), fig. 19.4, p. 358 **11.17** Richard Kirby, David Spears Ltd/Photo Researchers, Inc. **Evolution, page 253** Jane Grushow/Grant Heilman Photography, Inc.

Chapter 12 opening photo Jane Grushow/Grant Heilman Photography **page 262** Antonio Montanier/Smithsonian Institution **Plants & People, page 263** Bettmann/CORBIS **The Intriguing World of Plants, page 270** Wally Eberhart/Visuals Unlimited **12.11** Dorling Kindersley **12.13** Photodisc

Chapter 13 opening photo Juergen Berger, **The Intriguing World of Plants, page 291** Julian Schroeder, UCSD **Biotechnology, page 293** Incyte Pharmaceuticals, Inc., Palo Alto, CA, from R. F. Service, Science (1998) 282:396-399, with permission from Science **13.14a** Juergen Berger **13.14b** Leslie Sieburth **13.15a** Bettmann/CORBIS **13.15b** Virig-

ina Walbot, Stanford University **13.15c** Evelyne Cudel-Epperson, MSU **13.16a** P. S. Springer, etal. Gene Trap Tagging of PROLIFERA, an Esential MCM2-3-5-Like Gene in Arabidopsis. Science 268: 877, 12 May 1995, Fig. 1 Part B **13.16b** Heiko Schoof, Technical University of Munich, Munich, Germany (computer image manipulation), Dimitris Beis and Ben Scheres, Utrecht University, Netherlands (microscopy)

Chapter 14 opening photo Mauro Fermariello/Science Photo Library **page 300 top right** Tim McCabe, ARS, USDA, **bottom left** Ray Kriner/Grant Heilman Photography **14.1** Keith V. Wood/Visuals Unlimited **14.6** Bayer AG **14.8** R. Manshardt, University of Hawaii **14.9** Monsanto **Biotechnology, page 310** Salt Tolerance Conferred by Overexpression of a vacuolar Na+/H+ Antiport in Arabidposis by Maris P. Apse, Gilad S. Aharon, Wayne A. Snedden, and Eduardo Blumwald. Science 1999. August 20; 285: 1256-1258. The figure is figure 3, 14. **14.10** Bill Santos Photography **The Intriguing World of Plants, page 312** Use of plant roots for phytoremediation and molecular farming by Doloressa Gleba, Nikolai V. Borisjuk, Ludmyla G. Borisjuk, Ralf Kneer, Alexander Poulev, Marina Skarzhinskaya, Slavik Dushenkov, Sithes Logendra, Yuri Y. Gelba, and Ilya Raskin. PNAS 96: issue no 11 pp. 5973-5977, May 25, 1999. Figure 1

Chapter 15 opening photo Michael Fogden/DRK Photo **page 324** Dorling Kindersley **15.1a** Colin Keates/Natural History Museum/Dorling Kindersley **15.1b** George H. H. Huey/CORBIS **Plants & People, page 326 left** Bettmann/CORBIS, **right** Hulton Archive Photos/Getty **15.3** Roger Garwood & Trish Ainslie/CORBIS **15.7** Wayne Lawler, Ecoscene/CORBIS **15.9 left** George Loun/visuals Unlimited **15.9 right** Dorling Kindersley **15.12** J. Antonovics/Visuals Unlimited **15.13** Laura Sivell, Papilio/CORBIS **Evolution, page 339 all** Conley McCullin **15.14 left** D. Robert & Lorri Franz/CORBIS **15.14 right** James Randklev/CORBIS **15.17** Graham Day, www.habitas.org.uk

Chapter 16 opening photo Dorling Kindersley **page 348 top left** Hal Horwitz/CORBIS, **top right** Eric Crichton/CORBIS, **bottom, both** Dorling Kindersley **16.1 left** Archivo Iconografico, S.A./CORBIS **16.1 right** Dorling Kinderley **The Intriguing World of Plants, page 351 left, both** Dorling Kindersley, **right** Pat O'Hara/CORBIS **Plants & People, age 352** Hunt Institute for Botanical Documentation, Carnegie Mellon University **16.4** Dept. of Phanerogamic Botany, Swedish Museum of Natural History Thomas Karlsson, Curator **16.5 left** Graham Kent **16.5 middle** Tom McHugh/Photo Researchers, Inc. **16.5 right** Anthony Bannister, Gallo Images/CORBIS **16.7 top right** Dorling Kindersley **16.7 others** PhotoDisc **16.8** M. I. Walker/Photo Researchers **16.10** Darrell Gulin/CORBIS **16.11 all** Gerald Carr

Chapter 17 opening photo Brad Mogen/Visuals Unlimited **page 371** Matt Lee **Plants & People, page 372** N. Thomas/Photo Researchers, Inc., **inset** K. Murti **17.3a** Dr. Dennis E. Mayhew, Plant Health and Pest Prevention Services, Sacramento, CA **17.3b** Mike Davis, UC Statewide IPM Project. © Regents, University of California **17.4a** Arden Sherf, Department of Plant Pathology, Cornell University **17.4b** Dr. Dennis E. Mayhew, Plant Health and Pest Prevention Services, Sacramento, CA **17.5** Karl Maramorosch **17.6 all** David M. Phillips/Visuals Unlimited **Conservation Biology, page 378 left** Inga Spence/Visuals Unlimited, **right** Doug Sokell/Visuals Unlimited **17.7** Sue Barns **The Intriguing World of Plants, page 379** Claire S. Ting, Department of Biology, Williams College; J. King, S. W. Chisholm; 1999 **17.8** Lowell L. Black, Asian Vegetable Research and Development Center **17.9** Hubert de Franqueville, CIRAD **17.10** Exxon Corporation

Chapter 18 opening photo Ralph A. Clevenger/CORBIS **page 385 left** Georgette Douwma/Taxi, **right** Gerry Ellis/Minden Pictures **18.2** Ann Clemens/UTEX Culture Collection of Algae, University of Texas at Austin **18.3** J. Woodland Hastings, Hastings Lab, Harvard University **18.4** Stephen Frink/CORBIS **18.5a** Carleton Ray/Photo Researchers, Inc. **18.5b** Frank Borges Llosa **18.6** Dr. Stanley Flegler/Visuals Unlimited **18.7** Ann Clemens/UTEX Culture Collection of Algae, University of Texas at Austin

18.8 Carolina Biological/Visuals Unlimited 18.9 CSIRO Marine Research/Visuals Unlimited 18.10 Vita Pariente, College Station, Texas. Scanning electron microscopy performed at the Texas A&M University Electron Microscopy Center. 18.13a Ann Clemens/UTEX Culture Collection of Algae, University of Texas at Austin 18.13b Brandon D. Cole/CORBIS 18.13c Douglas P. Wilson, Frank Lane Picture Agency/CORBIS **The Intriguing World of Plants, page 397** Gerald and Buff Corsi/Visuals Unlimited 18.14 Ann Clemens/UTEX Culture Collection of Algae, University of Texas at Austin 18.15 Ann Clemens/UTEX Culture Collection of Algae, University of Texas at Austin 18.17a T. Mellichamp/Visuals Unlimited 18.17b John D. Cummingham/Visuals Unlimited.

Chapter 19 opening photo Felix Labhardt/Taxi **page 405 top** PhotoDisc, **bottom** Jim Brandenburg/Minden Pictures 19.1c Mark Moffett/Minden Pictures 19.2 George Barron 19.4 Martha J. Powell and Peter Letcher 19.5 **top** Barry Runk/Stan/Grant Heilman Photography 19.5 **bottom** Silver Burdett Ginn,Pearson Education **The Intriguing World of Plants, page 410** Darlyne A. Murawski/National Geographic Image Collection 19.7a,b Dorling Kindersley 19.9a Viard/Jacana/Photo Researchers, Inc. 19.9b Michael P. Gadomski/Photo Researchers, Inc. 19.10 **left** David Scharf/Peter Arnold, Inc. 19.10 **right** Jack M. Bostrack/Visuals Unlimited **Conservation Biology, page 414 top** Robert L. Anderson/USDA Forest Service, www.forestryimages.org, **bottom** M. F. Brown and H. G. Brotzman 19.11a Frank Young, Papilio/CORBIS 19.11b Michael Fogden/DRK Photo 19.11c Michael P. Gadomski/Photo Researchers, Inc. 19.11d Ed Reschke/Peter Arnold, Inc. 19.11e Michael & Patricia Fogden/CORBIS 19.11f Dorling Kindersley **Plants & People, page 417** Ed Young/CORBIS 19.13 Rob Simpson/Visuals Unlimited 19.14a Dorling Kindersley 19.14b Matt Meadows/SPL/Photo Researchers, Inc. 19.15a Natural Resources Canada, Canadian Forest Service 19.15b Holt Studios/Photo Researchers, Inc. 19.16 Brad Mogen/Visuals Unlimited 19.17 **all** Stephen Sharnoff 19.18a,b Mark Moffett/Minden Pictures

Chapter 20 opening photo M. L. Sinibaldi/CORBIS **page 425 top** Stephanie Maze/CORBIS, **bottom** Chris Lisle/CORBIS 20.1a Ken Wagner/Phototake 20.1b David T. Hanson, photographed by Claudia Lipke 20.1c Fritz Polking, Frank Lane Picture Agency/CORBIS **Evolution, page 427** Visuals Unlimited 20.2 MCC-NIES 20.4 Paul G. Davison, University of North Alabama 20.5 **left** A. J. Silverside, Biological Sciences, University of Paisley 20.5 **right** Alan Hale 20.6 **both** The Hidden Forest, hiddenforest.co.nz 20.7 **left** http://emmakael.free.fr 20.7 **right** Wolfgang Amri, naturaquarium.at 20.9 Specimen from Lyon Collection, Department of Geology, University of Aberdeen, Scotland 20.10a Dorling Kindersley 20.10b Andrew Syred/SPL/Photo Researchers, Inc. 20.11 J. D. Sleath **The Intriguing World of Plants, page 436** Steven Jessup, Southern Oregon University 20.12 **all** The Hidden Forest, hiddenforest.co.nz

Chapter 21 opening photo Konrad Wothe/Minden Pictures **page 441 top right** Peggy Greb/ARS/USDA, **left** Dorling Kindersley, **bottom right** Frank Siteman/Stone 21.1 The Field Museum of Natural History 21.5 From "A new species of Psilophyton from the lower Devonian of northern New Brunswick," by Jeffrey B. Doran. Published in Canadian Journal of Botany, 1980. 58: 2241-2262. Page 245, Figure 12 21.6a Murray Fagg, Australian National Botanic Gardens 21.6b Sally A. Morgan, Ecoscene/CORBIS 21.6c Milton Rand/Tom Stack & Associates, Inc. 21.6d Dorling Kindersley 21.7 Frank Landis, University of Wisconsin-Madison 21.9a Barry Runk/Stan/Grant Heilman Photography 21.9b Jane Grushow/Grant Heilman Photography 21.9c Murray Fagg, Australian National Botanic Gardens 21.12 **both** Barry Runk/Stan/Grant Heilman Photography 21.16 Inga Spence/Visuals Unlimited 21.18 **all** Dorling Kindersley 21.20 ?, 21.20 **both** Murray Fagg, Australian National Botanic Gardens 21.21 Robert Calentine/Visuals Unlimited

Chapter 22 opening photo Darrell Gulin/CORBIS **page 462 top** Doug Sokell/Visuals Unlimited, **left** Travis Amos, **right** Michael and Patricia Fogden/Minden Pictures 22.7 **left** Eric Crichton/CORBIS, **right** Dorling Kindersley **Biotechnology, page 472 both** USDA 22.8 **left** Gunter Marx Photography/CORBIS, **right** Dorling Kindersley 22.9a Dorling

Kindersley 22.9b W. John Hayden 22.10 PhotoDisc **The Intriguing World of Plants, page 474** AFP/CORBIS 22.11 **left** Dorling Kindersley, **right** Gerry Ellis/Minden Pictures 22.12a Nicole Duplaix/Omni-Photo Communications, Inc. 22.12b Fred Spiegel 22.13a Dorling Kindersley 22.13b Barry Runk/Stan/Grant Heilman Photograpy 22.13c Grant Heilman Photography 22.14a,b Dennis Woodward 22.15 Michael Clayton 22.16a,b Thomas Schoepke.

Chapter 23 opening photo Frans Lanting/Minden Pictures **page 281 top** Dorling Kindersley, **middle left** David Scott, http://theferocactus.free.fr **bottom left** David Sieren/Visuals Unlimited, **bottom right both** Peggy Grebb, Agricultural Research Service, USDA **Biotechnology, page 485** Albert Normandin/Masterfile **Evolution, page 496 all** Dorling Kindersley **The Intriguing World of Plants, page 498** John T. Atwood, Stig Dalström, and Ricardo Fernandez. Phragmipedium kovachii, A New species from Peru. Selbyana, The Journal of the Marie Selby Botanical Gardens, pp1-4. 23 Supplement, 2002. Photo by Lee Moore. 23.2 **top** Merlin D. Tuttle, Bat Conservation International 23.2 **middle** Louis Quitt/Photo Researchers, Inc. 23.2 **bottom** Mark Moffett/Minden Pictures 23.5a Else Marie Friis, Kaj Raunsgaard Pedersen, and Peter R. Crane.Fossil evidence of water lilies (Nymphaeales) in the Early Cretaceous. *Nature* 410, 357–360 (2001); doi:10.1038/35066557. 23.5b David Dilcher and Ge Sun 23.6a Andrew Syred/SPL/Photo Researchers, Inc. 23.6b CNRI/SPL/Photo Researchers, Inc. 23.8a Stephen McCabe 23.8b Dorling Kindersley, 23.9a Dorling Kindersley, 23.9b Gerald D. Carr 23.11 Gerald D. Carr 23.12 Dorling Kindersley 23.14a Dorling Kindersley 23.14b Tim Fitzharris/Minden Pictures 23.14c Alan and Linda Detrick/Grant Heilman Photography 23.14d,e Gerald D. Carr 23.14f Wayne P. Armstrong

Chapter 24 opening photo Art Wolfe/Stone **page 506 top** Dorling Kindersley, **bottom left** Frans Lanting/Minden Pictures, **bottom right** Dorling 24.1 Dave Costner/Geosurv Inc. 24.2 George H. H. Huey/CORBIS Kindersley **Intriguing World of Plants, page 509** Dorling Kindersley **Conservation Biology, page 512** NOAA 24.9 Frans Lanting/Minden Pictures 24.10 Wolfgang Kaehler/CORBIS 24.11 Philip Gould/CORBIS 24.12 Joe McDonald/CORBIS 24.13 Charles Mauzy/CORBIS 24.14 Jim Brandenburg/Minden Pictures 24.15 Steve Terrill/CORBIS 24.16 Darrell Gulin/CORBIS 24.18a Nick Hawkes; Ecoscene/CORBIS 24.18b Michael T. Sedam/CORBIS 24.19 Scott T. Smith/CORBIS 24.20 Ron Thomas/Taxi

Chapter 25 opening photo Charles Mauzy/CORBIS **page 525 top** Michael Orton/Stone, **bottom left** Andrew Wilson, **bottom right** Tom and Pat Leeson/Photo Researchers, Inc. 25.1a James Randklev/CORBIS 25.1b Charles Mauzy/CORBIS 25.1c Carr Clifton/Minden Pictures 25.2 Kent Foster/Photo Researchers, Inc. 25.6 **left** Dan Tenaglia, www.missouriplants.com 25.6 **right** Shawn Askew **The Intriguing World of Plants, page 532 top** Michael and Patricia Fogden/CORBIS, **bottom** Christian Puff 25.7a Raymond Coleman/Visuals Unlimited 25.7b Inga Spence/Visuals Unlimited **Conservation Biology, page 535** Fritz Polking/Visuals Unlimited 25.10 Steve Harper/Grant Heilman Photography **Evolution, page 538** Layne Kennedy/CORBIS 25.11 **top** Charles Mauzy/CORBIS 25.11 **middle** Tom Bean/DRK Photo 25.11 **bottom** Tom Bean/CORBIS

Chapter 26 opening photo James Marshall/CORBIS **page 549** Tim Fitzharris/Minden Pictures 26.3 Mark Edwards/Peter Arnold **Conservation Biology, page 553** Scott Bauer, ARS/USDA 26.4 Jonathan Blair/CORBIS 26.5 Mark Moffett/Minden Pictures 26.7 NASA **The Intriguing World of Plants, page 560** Jack Anthony 26.9 Phil Schermeister/CORBIS 26.12 James Henderson 26.13 Nashua River Watershed Association **Biotechnology, page 566** Chip Clark 26.14 **top** Josh Mitchell/Stone 26.14 **bottom** Michael Melford/The Image Bank

Appendix C: Archaea K.O. Stetter, R. Huber, and R. Rachel, U. of Regensburg **Bacteria** Oliver Merkes/Nicole Ottawa/Photo Researchers, Inc. **Eukaryote** Ann Clemens/UTEX Culture Collection of Algae, University of Texas at Austin **Plant** Dorling Kindersley **Fungi** Dorling Kindersley **Animal** PhotoDisc

Glossary

abscisic acid (ABA): A *hormone* that controls the opening and closing of *stomata* by causing water to enter or leave *guard cells*. Also establishes dormancy in seeds and other plant organs.

abscission zone (Latin, "to cut off"): The region of petiole where a leaf separates from a *deciduous* plant.

absorption spectrum: The measure of a pigment's ability to absorb various wavelengths of light.

accessory fruit: A fruit in which other flower parts besides the ovary become part of the fruit.

accessory pigment: A *pigment* molecule that aids another molecule, often by transmission of light energy; for example, *chlorophyll* b and carotenoids transfer light to *chlorophyll* a in photosynthesis.

acclimation: Preparation of a plant for inclement environmental conditions.

achene: An *indehiscent,* dry fruit similar to a small nut with a hard, thin *pericarp* and single seed; forms from a single *carpel.* Sunflowers produce achenes.

actin: A globular protein that associates in polymers to cause movement or changes in cell shape.

actinomorphic flower (Greek, *aktis,* "ray"): A ray-shaped or *regular flower.*

action spectrum: Profile of how effectively different wavelengths of light promote a process such as photosynthesis. Measured by shining light on intact chloroplasts and measuring oxygen output.

activation energy: The initial input of energy needed to start a chemical reaction.

active site: Specifically shaped region of an *enzyme* (E) where a *substrate* (S) binds to produce an *enzyme-substrate complex* (ES).

active transport: An energy-requiring process, generally using ATP, in which small molecules are transported across a membrane up a concentration gradient; uses one or more transport proteins. Compare with *facilitated diffusion.*

adaptive radiation: A type of rapid evolution that occurs when a species moves into a previously unoccupied environment, such as an island, or an occupied environment that still has many opportunities for the species to succeed. See also *punctuated equilibrium.*

adhesion: The attraction between different kinds of molecules. See also *cohesion.*

adventitious root: Roots that arise from unusual places, such as a stem.

aerial root: A modified adventitious root that arises from stem tissue and provides additional support for a plant. Commonly found on *epiphytes* such as orchids and as prop roots of corn plants.

aerobic: Describing a reaction or series of reactions that uses oxygen. See also *anaerobic.*

aggregate fruit: A fruit that originates from one flower with many separate *carpels;* for example, blackberries, strawberries, and magnolias.

algal bloom: The rapid expansion of a population of algae when conditions are optimal for their asexual reproduction. With dinoflagellates, commonly known as "red tide."

alkaloid: A ringed hydrocarbon compound containing nitrogen as a part of at least one of the rings.

allele: One of the variant forms of a *gene* that codes for a *trait.* A *diploid* cell has one allele from each parent.

allelopathy: The chemical inhibition that an individual or group of plants exerts on other plants.

allopatric: Descriptive of populations that are distinctly separate on a map. See also *sympatric.*

allopolyploidy: A type of *polyploidy* that results from breeding between two different species followed by chromosome doubling.

alternate: A leaf arrangement with one leaf per node.

alternation of generations: The alternation of *sporophyte* and *gametophyte* life-forms during sexual reproduction in all plants.

anaerobic: Describing a reaction or series of reactions that does not use oxygen. See also *aerobic.*

anagenesis: The transformation of one species into another; also called phyletic evolution. See also *cladogenesis.*

analogy: A similarity in structure or function between two species that are not closely related by evolution. See *homology.*

anaphase: The third phase of *mitosis,* during which the sister *chromatids* of each *chromosome* move apart, so that each chromatid is now a separate chromosome.

anaphase I: A phase of *meiosis* in which *homologous chromosomes* separate and move to opposite poles of the dividing cell; reduces the original number of chromosomes by half. Mendelian segregation occurs during anaphase I.

androecium (Greek, "the house of man"): Collection of *stamens* on a flower. See also *gynoecium.*

aneuploidy: The condition in which a cell has too few or too many copies of particular *chromosomes.* See *nondisjunction.*

angiosperm (Greek, *angion,* "container," *sperma,* "seed"): A flowering plant, whose seeds are enclosed in ovaries, which are called fruits when they mature. Compare with *gymnosperm.*

annual: A plant that completes its life cycle during a single growing season and then dies. Compare with *biennial* and *perennial.*

annulus: A backbone-like line of cells with thickened walls surrounding fern *sporangia;* aids in spore dispersal.

anther: A structure on a *stamen* composed of two lobes and four pollen sacs.

anther culture: A form of *tissue culture* in which anthers of flowers are placed on a medium that causes *pollen* to develop directly into plants without fertilization.

antheridiophore: A structure that holds male *gametophytes* of the liverwort *Marchantia;* has a flattened, disk-shaped top embedded with *antheridia.* See also *archegoniophore.*

antheridium (plural, *antheridia*): A male *gametangium* of a *bryophyte,* fern, or other non-seed-producing plant; contains sperm produced by mitosis.

anticlinal division: The division of cells perpendicular to a surface. Compare with *periclinal division.*

anticodon: A triplet of *nucleotides* on the middle loop of a *tRNA* molecule; complementary to one of the codons in *mRNA.* See *translation.*

apex (plural, *apices*): The tip of a root or shoot.

apical dominance: The suppression of *axillary bud* growth by *auxin* produced by a terminal bud.

apical meristem: A *meristem* at the tip of a shoot or root, the site of *primary growth.* See also *primary plant body.*

apomictic (Greek, "away from the act of mixing"): Describing a seed formed asexually.

apoplastic transport (Greek, *apo,* "away from"): The movement of molecules within the cell walls throughout a plant. Compare *symplastic transport.*

aquifer: An underground deposit of groundwater located in or below the deepest *soil horizon.*

archegoniophore: An umbrella-shaped stalk with a top bearing female gametophores of the liverwort *Marchantia.* See also *antheridiophore* and *calyptra.*

archegonium (plural, *archegonia*): The flask-shaped female *gametangium* of a *bryophyte* or other non-seed-producing plant; contains one egg produced by mitosis. Archegonia are also found in some gymnosperms.

ascocarp or **ascoma:** The fruiting body of ascomycetes.

ascogonium: The oogonium or female gametangium in ascomycetes.

ascus (plural, *asci*): A saclike structure containing ascospores formed within an *ascocarp.*

asexual reproduction: The process by which a single parent produces offspring that are identical to itself. Compare with *sexual reproduction.*

ATP (adenosine triphosphate): An organic molecule that is the main energy source for cells; *mitochondria* break down sugar to store its chemical energy in ATP. Photosynthetic *light reactions* produce ATP, which is used in the *Calvin cycle.*

ATP synthase: An enzyme that uses energy from *chemiosmosis* to add an inorganic phosphate (P_i) to ADP to form *ATP,* in a process called *photophosphorylation.*

autotroph ("self-feeder"): A plant that can make its own food through photosynthesis. See also *heterotroph.*

auxin: The first plant *hormone* discovered; produced in or near *apical meristems* suppresses the growth of *axillary buds;* stimulates growth of plant cells. Chemically, auxin is indoleacetic acid. See *apical dominance.*

axillary bud: A bud that forms in the upper angle or axil where the *petiole* joins the *stem;* grows to become a new *shoot.*

bark: All tissues outside the *vascular cambium;* the part of a *stem* or *root* surrounding the wood. See *inner bark* and *outer bark.*

basidia (singular, *basidium*): Large, club-shaped cells that form at the ends of the dikaryotic hyphae inside a *basidiocarp.* Nuclear fusion and meiosis occur inside basidia to produce *basidiospores.*

basidiocarp or **basidioma:** The aboveground portion of a mushroom composed of dikaryotic hyphae and producing basidia.

basidiospore: A haploid spore that produces new haploid *mycelia* when it germinates.

berry: A simple, fleshy fruit that may originate from one or several *carpels;* for example, tomatoes, grapes, and bananas.

biennial: A typically *herbaceous* plant that usually requires two growing seasons to complete its life cycle. It produces flowers and seeds in the second growing season. Compare *annual* and *perennial.*

binomial: The two-part name of a species; consists of a genus name and descriptive *specific epithet.*

biogeographic realm: A broad geographic area characterized by distinctive collections of organisms, typically confined to a single continent. See also *biome.*

biogeography: The study of where particular species of organisms occur and when they colonized a particular region.

biome: One of several major types of terrestrial and aquatic ecosystems that span large areas—for example, forest, savannah, grassland, desert—and are characterized by particular types of vegetation.

bioremediation: The use of *prokaryotes* to clean up oil spills and to decontaminate soils containing pesticides or other toxic substances.

biosphere: The thin layer of air, land, and water at Earth's surface that is occupied by living organisms.

bisexual flower: Has both *stamens* and *carpels.* Also called *perfect flower.*

blade: One of several flattened structures on a brown alga's *thallus* that provides most of the surface area for photosynthesis; the flattened portion of a *leaf.*

bolting: The rapid production of a long flowering stalk, caused by *gibberellin.*

botany: The scientific study of plants.

bract: A modified leaf at the base of a flower.

brassinosteroid: A plant steroid *hormone* that has effects similar to those of *auxin.*

bryologist: A scientist who studies *bryophytes:* liverworts, hornworts, and mosses.

bryophyte (Greek, *bryon,* "moss," *phyton,* "plant"): A group of small, nonflowering plants that evolved from algae-like ancestors around 450 to 700 million years ago; mosses, hornworts, and liverworts.

bulb: A modified stem structure in which starch accumulates in thickened, fleshy leaves attached to the stem. Compare to *corm* and *tuber.*

bundle sheath cell: These cells surround the vascular bundle in flowering plants. In C_4 *plants,* they are large and photosynthetic and are the site of *Calvin cycle* reactions.

buttress root: Flared root that extends from a tree trunk to provide stability in thin soils.

C$_3$ plants: Plants that carry out only the *Calvin cycle* for *carbon fixation.* Such plants make a three-carbon product as the first organic product of carbon fixation and revert to *photorespiration* on hot, dry days. Rice, wheat, and soybeans are common C_3 plants. Compare C_4 *plants* and *crassulacean acid metabolism*.

C_4 pathway: Addition to the *Calvin cycle* used by C_4 *plants* common in hot, dry regions; binds CO_2 into four-carbon (C_4) compounds, which are then used to supply an increased concentration of CO_2 to the Calvin cycle. See also C_3 *plants* and *crassulacean acid metabolism*.

C_4 plants: Plants adapted to hot, dry climates, when CO_2 availability to chloroplasts is limited by partially closed stomata. In specialized mesophyll cells, an enzyme adds a molecule of CO_2 to a three-carbon compound to form a four-carbon product. This product is passed into *bundle sheath cells*, where the *Calvin cycle* reactions of photosynthesis occur. Common C_4 plants are sugarcane and corn.

callose: A carbohydrate molecule formed around the sieve plate by a damaged *sieve-tube member*.

callus: A mass of undifferentiated cells used in *tissue culture*, which is stimulated to differentiate into the tissues and organs of the complete plant under the influence of hormones in a culture medium.

Calvin cycle: Photosynthetic reactions that assemble simple three-carbon sugars, using ATP and NADPH from the *light reactions* and CO_2 from the air; takes place in the *stroma* of chloroplasts. To make one three-carbon sugar molecule, 3 CO_2, 9 ATP, and 6 NADPH are required. See *rubisco*.

calyptra: A thin, veil-like structure that develops from the *archegonium* of some liverworts and mosses and partially covers the capsule or sporangium.

calyx: A group of *sepals* around a flower bud. See also *corolla* and *perianth*.

CAM plants: Plants (usually tropical) that use *crassulacean acid metabolism* (*CAM*) for *carbon fixation* at night. Carbon fixation and *Calvin cycle* reactions occur in the same cells at different times. Typical CAM plants are succulents in the family Crassulaceae, many cacti, and pineapples. These plants close their stomata during the day and open them at night. Compare C_3 *plants* and C_4 *plants*.

capsule: A *dehiscent* fruit that may split in several different ways, depending on the species. All capsules develop from at least two *carpels*. Examples include poppies, irises, and orchids.

carbohydrate: A *macromolecule* composed of carbon, hydrogen, and oxygen in CH_2O subunits. Sugars are carbohydrates that supply and store energy and can serve as building blocks for larger carbohydrates, such as cellulose in a plant cell wall or starch, which stores energy in seeds.

carbon fixation: A process which fixes or binds carbon from CO_2 into a three- or four-carbon organic molecule.

carpel (Greek, *karpos*, "fruit"): The ovule-bearing, female part of a flower. Carpels collectively are called the *gynoecium*. See also *ovary, pistil, stigma,* and *style*.

carrying capacity: The maximum size *population* that an environment's resources can support.

caryopsis (plural, *caryopses*): Also called a grain; a single, dry, *indehiscent*, *achene*-like fruit with a hard *pericarp* joined firmly to the seed coat.

catalyst: A substance that increases the rate of a chemical reaction without being changed itself by the reaction. Enzymes act as catalysts in living systems. See also *active site, cofactor, enzyme-substrate complex, induced fit,* and *substrate*.

cation exchange: A process in which hydrogen ions secreted by roots change places with mineral cations bound to soil particles.

cell cycle: The sequence of events from the time a cell first arises as a result of cell division until the time when that cell itself divides. See also *meiosis* and *mitosis*.

cell layer model: An explanation of shoot growth that describes the initials of the shoot apical meristem as forming several cell layers. See also *corpus* and *tunica*. Compare with *zone model*.

cell membrane: See *plasma membrane*.

cell plate: During *mitosis*, two new *plasma membranes* and *cell walls* that form between the nuclei and in the center of the *phragmoplast*. The cell plate gradually extends to divide the cell into two daughter cells.

cell theory: Three general conclusions about structure and function of living organisms, developed in the mid-nineteenth century: All organisms are made up of one or more cells, the cell is the basic unit of structure of all organisms, and all cells arise only from existing cells.

cellulose: The principal building block of plant cell walls, consisting of chains of glucose molecules.

central mother cell zone: A region of the shoot *apical meristem* containing cells that divide infrequently. See also *peripheral zone* and *pith zone*.

centromere: A region of constricted DNA in a *chromosome* that joins *chromatids*. See *prophase*.

centrosome: A microtubule-organizing center, important during *prophase* of the *cell cycle*.

character: An inherited feature that is observable or measurable, such as height, flower color, or seed shape and has two or more distinguishable *traits* such as tall or short, red or white, and wrinkled or smooth. Characters are related to genes and traits to *alleles* of specific genes.

chemiosmosis (Greek, *chemi-,* "chemical," *osmos,* "a push"): A process of moving H^+ ions across the *thylakoid* membrane during photosynthesis or respiration, resulting in a release of energy. See also *ATP synthase* and *photophosophorylation*.

chemoautotroph: An organism, primarily bacteria, that does not rely on photosynthesis but instead gets carbon from CO_2 and energy from inorganic chemicals.

chemoheterotroph: An organism that obtains both energy and carbon from organic compounds from other organisms.

chitin: A nitrogen-containing carbohydrate similar in structure to cellulose that makes up the cell walls of fungi and exoskeletons of arthropods such as insects.

chlorenchyma cell: A specialized *parenchyma cell* where photosynthesis takes place.

chlorophyll *a*: Blue-green photosynthetic *pigment* directly involved in the *light reactions;* absorbs light from the blue-violet and red ranges of the spectrum. See also *chlorophyll* b.

chlorophyll *b*: Yellow-green photosynthetic *pigment* that acts as an *accessory pigment* by transmitting light energy to *chlorophyll* a molecules. Differs in structure from chlorophyll *a* by only a few atoms.

chloroplast (Greek, *chloros*, "greenish yellow"): Organelle that contains green chlorophyll pigments; the site of photosynthesis in plant cells. See *grana*, *stroma*, and *thylakoid*.

chromatids: Sister strands of DNA produced during *S phase* of the *cell cycle*. Joined together by a narrow region called a *centromere*. See also *centrosome*.

chromoplast (Greek, *chroma*, "color"): A *plastid* containing pigments, responsible for the yellow, orange, or red colors of many leaves, flowers, and fruits. Compare with *leucoplast*.

chromosomal mutation: A mutation that involves more than one *nucleotide;* may be a deletion, duplication, inversion, or translocation.

chromosome (Greek, *chroma*, "color," *soma*, "body"): A complex, threadlike structure composed of DNA and associated proteins and located in the nucleus of a cell. Each chromosome consists of many *genes*, sections of DNA, containing sequences of nucleotides that contain the code for making a particular protein.

cilium (plural, *cilia*): Short external propelling appendage of a cell, composed of *microtubules*. Compare with *flagellum*.

circadian rhythm (Latin, *circa*, "about," *dies*, "day"): A biological cycle of approximately 24 hours.

cisternae (singular, *cisterna*): Flattened, interconnected sacs that form the exterior surface of the *endoplasmic reticulum*.

clade: A branch of a *cladogram* that consists of an ancestor and all its descendents, all of whom share one or more *characters* that make them unique as an evolutionary branch. See *monophyletic*.

cladistics (Greek, *klados*, "branch"): A method of classifying organisms according to the order in time when *homologies* were inherited.

cladogenesis: The evolution of one species into two species, also called branching evolution. See also *anagenesis*.

cladogram: A tree-like branching diagram that shows evolutionary relationships. See also *phylogenetic tree*.

class: A taxonomic group higher than *order* and lower than *phylum*. For example, the Class Charophyceae contains green algae that are related to higher plants.

clay: The smallest soil particles, smaller than 0.002 millimeter in diameter. See also *sand, silt*, and *soil horizon*.

climax community: A community that will remain relatively stable unless upset by a major disturbance.

cline: A variation in *phenotype* that occurs along with a gradation in some measurable feature of the environment.

clone: The genetically identical offspring of a single parent, produced by asexual (vegetative) reproduction.

codon: A triplet of *nucleotides* in a *DNA* sequence that codes for an amino acid or gives a "start" or "stop" signal. See *exon* and *intron*.

coenocytic: The life habit of some yellow-green algae and other *protists,* consisting of a single cytoplasmic mass that contains many nuclei, with no internal partitions separating the nuclei.

coenzyme: A *cofactor* that is a nonprotein organic compound, such as a vitamin.

coevolution: Descriptive of the related paths of development in different species, such as pollinators and plants, in which adaptations in one species have a selective effect on adaptations in another species.

cofactor: A small nonprotein molecule that binds to an *enzyme* or *substrate* and assists a chemical reaction. See also *coenzyme*.

cohesion: Attraction between identical molecules; cohesion between water molecules causes water to have a high boiling point. See also *adhesion*.

collenchyma cell (Greek, *kolla*, "glue"): A living, elongated cell that provides flexible support to a plant. Compare *parenchyma cell* and *sclerenchyma cell*.

commensalism: An interaction between two species in which one benefits while the other is unaffected; for example, an epiphytic plant living in the top of a tree in a rain forest.

community: A group of species that inhabit a given area; the biotic components of ecosystems.

companion cell: A nucleated cell adjacent to a *sieve-tube member* that can supply it with proteins.

competitive exclusion principle: An observation of ecosystem dynamics, which states that if two species live in the same area and compete for exactly the same resources, one species will eventually be eliminated from the area.

complete flower: A flower that contains all four types of modified leaves: *carpels, petals, sepals*, and *stamens*.

complex tissue: Group of cells of several cell types. See also *simple tissue*.

compound leaf: A leaf in which the blade is divided into leaflets. Compare *simple leaf*.

concentration gradient: A transition between regions of higher and lower *solute* concentration. See *diffusion*.

conidium (plural, *conidia*): An asexual spore of ascomycetes and some basidiomycetes. See also *ascogonium*.

conservation biology: A multidisciplinary field of science that studies the impact of human activities on all facets of the environment.

continental drift: According to *plate tectonics*, the slow movement of both seafloor plates and continental plates over the surface of the Earth.

contractile root: A root that can shorten to pull a plant deeper into the soil.

control element: A segment of noncoding *DNA* where *transcription factors* can bind and exert control over the expression of one or more *genes*.

convergent evolution: The convergence of different evolutionary paths, resulting in a similarity in a particular *character* among plants that are not closely related. For example, desert plants known as cacti come from several different, unrelated families.

cork: Tissue that forms to the outside of the cork cambium and consists of dead cells when mature; also known as *phellem* (Greek, *phellos*, "cork").

cork cambium: Secondary growth, or tissue that produces new dermal tissue; also known as *phellogen* (Greek *phellos*, "cork", *genos*, "birth"). Compare *vascular cambium*.

corm: An underground food storage stem shaped like a *bulb* but consisting primarily of stem tissue rather than thick leaves.

corolla: A group of petals on a flower. See also *perianth*.

corpus: In a shoot apex, shoot initial layers underlying the *tunica,* roughly equivalent to the *central mother zone,* the inner parts of the *peripheral zone,* and the *pith zone.*

cortex: Ground tissue formed between dermal and vascular tissue.

cotyledon: The first *leaf* or leaves of a developing plant embryo; stores food for the germinating seed and may be thickened or fleshy. See also *epicotyl* and *hypocotyl.*

crassulacean acid metabolism (CAM): A variation of the C_4 pathway in which *CAM plants* take up CO_2 at night using the C_4 *pathway* and then carry out *Calvin cycle* reactions to make sugars during the day. The two processes occur in the same cells but at different times. CAM is common in succulent desert plants, helping them to conserve water during the day and to avoid using *photorespiration.*

crista (plural, *cristae*): The infoldings of the *mitochondrion*'s inner membrane. See also *matrix.*

cross section: A horizontal cut at a right angle to the long axis of a structure; also known as a *transverse section.*

crossing over: The exchange of chromosome segments due to overlap of chromatids during the interphase before *prophase I* of *meiosis.*

cross-pollination: In angiosperms, the transfer of pollen from the anther of one plant to the stigma of another plant of the same species.

cuticle: A layer outside the cell wall composed of wax and a fatty substance called cutin, which helps limit water loss.

cyclosis: See *cytoplasmic streaming.*

cytokinesis ("cell movement"): During the *cell cycle,* separation of *cytoplasm* and the new nuclei into *daughter cells.*

cytokinin: A *hormone* synthesized in roots that controls cell division and differentiation; counteracts *apical dominance* and delays aging of leaves.

cytoplasm (*cyto,* "cell," *plasm,* "formed material"): All the parts of the cell within the *plasma membrane* except the *nucleus.*

cytoplasmic inheritance: Inheritance caused by genes on small chromosomes in mitochondria and chloroplasts; also known as maternal inheritance because the *egg* carries *cytoplasm* with organelles to the next generation, but the *sperm* does not.

cytoplasmic streaming: Circular motion of a cell's contents around its central vacuole, caused by *microfilaments.* Also called *cyclosis.*

cytoskeleton ("cell skeleton"): Made up of threadlike proteins: *microtubules, microfilaments,* and *intermediate filaments.* See also *cytosol.*

cytosol: The fluid part of *cytoplasm.*

daughter cells: New cells formed by division of a single cell. See *cell cycle, cell plate, cytokinesis, interphase, mitosis,* and *phragmoplast.*

day-neutral plant: A plant that flowers regardless of the day length.

deciduous (Latin, "to fall off"): An adjective describing plants that lose all their leaves at certain seasons of the year. See also *abscission zone.*

deductive reasoning: Reasoning from the general to the specific. Compare with *inductive reasoning.*

dehiscent fruit (Latin, "to split open"): A dry fruit that splits open at maturity to shed seeds. See also *capsule, follicle, legume,* and *silique.*

dehydration synthesis: A chemical reaction that links *monomers* into a *polymer* by removing a molecule of water; also called a dehydration or condensation reaction.

denaturation: Disruption of the *tertiary structure* of a *protein.*

dendrochronology (Greek, *dendron,* "tree," *chronos,* "time"): The science of tree-ring dating and climate interpretation.

density: The amount of matter per a unit of volume, related to a wood's hardness. See also *specific gravity.*

derivative: A daughter cell that is pushed out of the *meristem* and either divides again or begins elongation and *differentiation.* Its sister cell remains as an *initial.*

dermal tissue system (Greek, *derma,* "skin"): The outer protective covering of a plant, derived from *parenchyma cells.* See *epidermis* and *periderm.*

desmotubule: The connecting tubule of endoplasmic reticulum between cells; located in the *plasmodesma.*

determinate growth: A growth pattern in which an organism or tissue grows for a limited time, typical of animals and floral meristem. Compare with *indeterminate growth.*

dicot: Flowering plants with two cotyledons; for example, beans, peas, sunflowers, roses, and oak trees. Compare with *eudicot* and *monocot.*

dictyosome (Greek, *diktyon,* "to throw"): A collection of flat, membrane-bound sacs that serve to modify molecular components secreted from the cell; the *Golgi stacks* in a plant cell.

differentiation: The processes by which an unspecialized cell develops into a specialized cell.

diffusion: Tendency of molecules to spread out spontaneously within the available space, from a region of higher solute concentration to a region of lower concentration. See *concentration gradient.*

dihybrid cross: A cross of pure-breeding plants that differ in traits of two *characters.* For example, if the characters are height and seed shape, a tall plant with smooth seeds might be crossed to a short plant with wrinkled seeds. See also *monohybrid cross.*

dikaryotic ("two nuclei"): Descriptive of a *mycelium* formed by *plasmogamy;* contains two different *haploid* nuclei per cell. Also called *heterokaryotic.*

dioecious (Greek, "two houses"): Describing a plant with male and female flowers on different plants; for example, willows. See also *monoecious.*

diploid (Greek, *diplous,* "double"): A cell with two sets of chromosomes; symbolized $2n$. See also *haploid* and *polyploidy.*

directional selection: A method of changing the frequency of *phenotypes* in a population by favoring individuals that have one extreme phenotype. See also *diversifying selection* and *stabilizing selection.*

disaccharide: A molecule formed by linking two monosaccharide, or sugar, molecules.

diversifying selection: Splits a population into two parts by favoring individuals at both extremes of the phenotypic range; decreases the frequency of individuals with intermediate phenotypes. See also *directional selection* and *stabilizing selection.*

DNA (deoxyribonucleic acid): A double-helical molecule that contains coded genetic information for an organism. Composed of nucleotides, each of which contains a phosphate group, a sugar molecule (deoxyribose), and one of four different bases. See also *RNA*.

DNA ligase: An enzyme that links DNA fragments to form the lagging strand during replication; also used in preparing *recombinant DNA*.

domain: The most general, highest taxonomic category of organisms. For example, Domain Eukarya contains all eukaryotic organisms.

dominant species: Those species that have the most individuals, greatest biomass, or some other indicator of importance in the community. See also *keystone species*.

dominant trait: The visible *trait* in the F_1 generation of a Mendelian cross.

double fertilization: A defining feature of flowering plants, in which one sperm combines with the egg and a second with the polar nuclei.

double helix: Characteristic of the structure of *DNA* molecules, in which two chains of *nucleotides* wrap around each other and attach with hydrogen bonds between the bases.

drupe: A simple fleshy fruit that develops from flowers with *superior ovaries* and one *ovule*. Examples of drupes are olives, peaches, and almonds.

durability: The extent to which wood is resistant to breakdown and decay by fungi, bacteria, and insects.

ecological succession: A gradual change in the communities that ecosystems support.

ecology (Greek, *oikos*, "the family household," *logos*, "the study of"): The study of Earth's environment and its organisms.

ecosystem: The interacting collection of all the organisms and all the nonliving components in an environment.

ectomycorrhizae: A *mutualistic* association between roots and fungi in which the fungi do not penetrate plant roots. Compare with *endomycorrhizae*.

egg: A female reproductive cell.

elater: An elongated cell in a liverwort's sporangium. Elaters absorb water, which causes them to twist and turn, assisting in the dispersal of spores. Also, in horsetails, elaters are bands of tissue attached to spores that takes up water and moves spores around, again to assist dispersal.

electron microscope: Developed in 1939, it focuses electrons (instead of visible light) with magnetic (instead of glass) lenses. See also *light microscope, scanning electron microscope,* and *transmission electron microscope*.

electron transport chain: A series of electron carriers (such as *NADH, NADPH, FADH₂,* and cytochromes) that pass an electron through a series of *oxidation-reduction reactions* during *photosynthesis* and also during *respiration*.

electroporation: A method of inserting cloned *DNA* into a plant cell using a brief pulse of electric current.

embryo: The product of a sperm-fertilized egg that develops into an adult organism. See also *seed*.

enation (Latin, *enatus*, "to rise out of"): A small, scale-like, nonvascularized flap of green tissue that serves in place of a leaf in some whisk ferns (psilotophytes).

endergonic ("energy inward"): A description of a chemical reaction that requires a net input of free energy. See also *exergonic*.

endocarp: Inner parts of a *pericarp*.

endocytosis: A process by which plant cells take up large molecules. Compare *exocytosis*.

endodermis: The layer of cells around the *stele* that regulates the flow of substances between cortex and vascular tissue. See also *pericycle*.

endomycorrhizae: A type of *mutualistic* association in which the fungi penetrate plant roots and produce branching structures that press up against plant cell membranes to obtain nutrients. See also *ectomycorrhizae*.

endoplasmic reticulum (ER) (Latin, "within the plasm," and "little network"): A network of connected membranes throughout the *cytoplasm*. The ER, which is formed from and is continuous with the outer *nuclear envelope*, serves as a synthesis and assembly site for making *proteins, lipids,* and other molecules. See *rough endoplasmic reticulum* and *smooth endoplasmic reticulum*.

endosperm: Storage tissue surrounding the embryo in flowering plants; provides nourishment to the developing embryo.

endosporic: Descriptive of gametophyte development in spike mosses, taking place mostly inside the spore wall.

endosymbiotic theory ("living together within"): A theory suggesting that the ancestors of some organelles evolved as a result of prokaryotic cells ingesting other prokaryotic cells.

energy: The capacity to perform work. See also *first law of thermodynamics, kinetic energy, potential energy,* and *second law of thermodynamics*.

energy coupling: The pairing of an *exergonic* reaction with an *endergonic* reaction so that the overall reaction will occur spontaneously.

entropy: The degree of disorder in a sample of matter.

enzyme: A *protein* that helps regulate chemical reactions in a cell.

enzyme-substrate complex (ES): See *active site*.

epicotyl: A portion of the embryonic plant stem over the *cotyledon* that develops from the *plumule*. See also *hypocotyl*.

epidermis: The single, outer layer of protective *dermal tissue* formed in a plant's first year of growth and in all subsequent new tissue. See also *periderm*.

epiphyte (Greek, *epi-*, "upon," *phyton*, "plant"): A plant that grows on another plant for support but nourishes itself.

epistasis: A situation in which one *gene* interacts with and alters the effect of another.

equilibrium: A random, equal distribution of substances or organisms.

ergot: A disease of grains such as wheat and barley caused by the ascomycete *Claviceps purpurea*. Ergotized grain is responsible for episodes of hallucinations and serious illness in people who consume products made from it.

essential amino acids: Amino acids that the human body cannot make; must be obtained from the diet.

ethylene: A gas that acts like a *hormone,* causing responses to mechanical stress and stimulating aging responses such as fruit ripening and leaf abscission.

eudicot ("true" dicot): The largest clade of living angiosperms.

eukaryote (Latin, "true nucleus"): An organism whose cells have nuclei. Eukaryotes include plants, animals, fungi, and algae. Compare with *prokaryote.*

eustele: The arrangement of *vascular bundles* in a circle around the pith; common in most gymnosperm and dicot stems. Compare *siphonostele.*

eutrophic (Greek, *eutrophos,* "well-nourished"): Descriptive of a shallow, nutrient-rich lake. See also *oligotrophic.*

evolution: A change in the frequency of an allele in a population over time.

exergonic ("energy outward"): Description of a chemical reaction that has a net release of free energy into the surroundings. See also *endergonic.*

exocarp: The outer part, often the skin, of a *pericarp.*

exocytosis: A process by which large molecules and multimolecular components leave plant cells by means of the fusion of membrane-bound vesicles with the plasma membrane. See also *endocytosis.*

exon: The section of a *gene* that codes for a *protein.* See also *intron.*

exosporic: Descriptive of gametophyte development in most seedless vascular plants and bryophytes, taking place outside the spore wall.

facilitated diffusion: A passive process in which water-soluble molecules are assisted by transport proteins to diffuse through a *plasma membrane.* See also *active transport.*

family: A taxonomic group higher than genus and lower than order. In plants most family names end in –*aceae,* as in Solanaceae.

fermentation: An *anaerobic* pathway to break down pyruvate that takes place completely within the *cytosol.* Fermentation follows *glycolysis* and yields ethanol or lactic acid.

fertilization: The joining of two *gametes* to form a *zygote.*

fiber: An elongated *sclerenchyma cell* with thick secondary walls reinforced with *lignin;* common in tree trunks. See also *sclereid.*

fibrous root system: A type of root system common in seedless vascular plants and grasses; characterized by many similar-sized, small and short roots. Compare with *taproot system;* see also *adventitious root.*

fiddlehead: The coiled immature frond of a fern.

first filial (F₁) generation (Latin, *filius,* "son"): The offspring of a *monohybrid cross.* See also *second filial (F₂) generation.*

first law of thermodynamics: States that *energy* can be transformed to other forms of energy, but it cannot be created or destroyed.

flagellum (plural, *flagella*): A long external propelling appendage of a cell; eukaryotic flagella are composed of microtubules. Compare with *cilium.*

florigen: A hypothetical substance that promotes flowering in plants; possibly a mixture of hormones, which are made in leaves in response to inductive day lengths and transported to vegetative shoot apices, which are transformed into floral meristems.

fluid mosaic model: The structure of the *plasma membrane* consists of a double layer of phospholipid molecules. These long molecules have a water-soluble and a water-insoluble end. The water-soluble end faces the external and internal surfaces of the membrane. Proteins are associated with either side of the membrane or may pass entirely through it.

follicle: A simple, *dehiscent,* dry fruit that has single *carpels* that split along one seam to release seeds. Examples are milkweed, columbines, and magnolias.

food chain: A sequence of food transfer from one organism to the next, beginning with a food producer.

frameshift mutation: An insertion or deletion *point mutation* in *DNA* that causes *codons* to shift.

frond: A sporophyte *megaphyll* (leaf) of a fern. Site of spore production.

fusiform initial (Latin, "tapered toward each end"): An *initial* that arises within *vascular bundles* and produces new *xylem* and *phloem* cells. See also *ray initial.*

G₁ phase ("first gap"): A relatively long first part of *interphase* in the *cell cycle,* when the cell grows, develops, and functions as a particular cell type.

G₂ phase ("second gap"): A section of *interphase* following *S phase,* during which the cell continues normal functioning and prepares for cell division.

gall: A swollen region of plant tissue infected with a rust such as corn smut or by *Agrobacterium tumefaciens,* which causes crown gall disease. Galls are also caused by insects, which lay eggs inside them.

gametangium (plural, *gametangia*): A single-celled or multicellular structure that produces gametes. See *antheridium* and *oogonium.*

gamete (Greek, *gamein,* "to marry"): A haploid *sperm* or *egg* cell. See also *embryo* and *zygote.*

gametophore: A structure formed by a bryophyte's protonema; bears either leafy or flattened *gametangia.*

gametophyte ("gamete-producing plant"): One of two multicellular forms of a plant; consists of *haploid* cells. See also *alternation of generations* and *sporophyte.*

gel electrophoresis: A procedure that sorts DNA fragments by size as they move through a polymeric gel in response to an electric current. The negatively charged phosphate groups of DNA result in DNA fragments being attracted to the positively charged pole (cathode).

gemma (plural, *gemmae;* Latin, *gemma,* "bud"): A brood body common in liverworts and mosses; a small multicellular body that grows into a new *gametophyte* when detached from the parent plant.

gene: A specific sequence of DNA nucleotides coding for one protein.

gene cloning: The process of making multiple copies of *recombinant DNA.*

gene flow: The movement of alleles from one population to another as the result of cross pollination or some other form of crossing.

gene library: A collection of DNA clones containing *plasmids* with different segments of foreign *recombinant DNA.*

gene trapping: Use of specialized *transposons* containing reporter genes to inactivate *genes* affecting development. Also called transposon tagging.

genetic drift: A phenomenon in small populations of organisms showing that the frequency of *alleles* can change over generations due to chance.

genetic engineering: The process of moving and modifying genes to produce plants with desired traits.

genome: The complete description of an organism's *DNA;* all the genes and chromosomes necessary to produce an organism. For example, the genome of the garden pea has 14 chromosomes of 7 types.

genomics: The science of determining the *nucleotide* sequence of whole *genomes.* See also *proteomics.*

genotype: The combination of *alleles* that an organism has (such as *PP, pp,* or *Pp*). Compare with *phenotype.*

genus (plural, *genera*): A taxonomic group higher than species and lower than family. For example the garden pea, *Pisum sativum,* is in the genus *Pisum.* Genera are capitalized and italicized.

germination: The process of a seed's sprouting in which the first event is protrusion of the radicle or embryonic root through the seed coat. In a more general sense, germination is the beginning of active growth of a spore or seed.

gibberellin: Member of a group of plant *hormones* affecting cell elongation and seed germination.

girdling: Removal of the entire bark in a complete ring around a tree. Girdling disrupts phloem transport and kills the tree.

glycolysis (Greek, *glyco,* "sweet or sugar," *lysis,* "splitting"): A series of ten *anaerobic* enzymatic reactions taking place in the *cytosol* to split a six-carbon sugar (glucose) into two molecules of pyruvate and produce two molecules of *ATP;* the beginning steps of *respiration.* Glycolysis produces ATP by *substrate-level phosphorylation.* Glycolysis is followed by reactions of the *Krebs cycle* or by *fermentation,* depending on whether oxygen is present or not.

glyoxysome: A type of *microbody* whose enzymes assist in converting stored fats into sugars, particularly important in germinating seeds.

Golgi apparatus or **Golgi complex:** All the unconnected *dictyosomes* in a cell. The Golgi apparatus of a cell aids in modification and transfer of materials to be secreted from the cell through the cell membrane. Finishes synthesis of some products of the *rough endoplasmic reticulum* and produces some noncellulose polysaccharides.

grain: Of lumber, the overall alignment of the conducting cells of *xylem;* can be straight-grained, cross-grained, spiral-grained, or interlocking-grained.

grana (singular, *granum*): Stacks of *thylakoids* in *chloroplasts.* See also *stroma.*

gravitropism: Growth toward or away from gravity.

ground meristem: A part of the root and shoot *apical meristem* that produces the *ground tissue system.*

ground tissue system: A fundamental tissue system, consists of all tissues other than the *vascular tissue system* and the *dermal tissue system.* Its cells carry out *photosynthesis* and store nutrients. See also *cortex* and *pith.*

guard cell: One of two epidermal cells on either side of a leaf pore; the combination of pore plus guard cells constitutes a *stoma.*

guttation: The process by which water pushed into a stem by root pressure may end up leaving the leaves as droplets through specialized epidermal regions.

gymnosperm (Greek, *gymnos,* "naked," *sperma,* "seed"): A nonflowering seed plant that first evolved around 365 million years ago. Their most familiar modern descendents are conifers. Compare with *angiosperm.*

gynoecium (Greek, "the house of woman"): A collection of *carpels* on a flower.

habitat: The location where a plant lives; for example, mosses live in moist, shaded habitats, whereas sunflowers prefer sunny, dry habitats.

half-life: The time required for half of a sample of a radioactive isotope to decay.

haploid (Greek, *haplous,* "single"): A cell with a single set of chromosomes; symbolized *n.* See also *diploid* and *polyploidy.*

hardwood: Fibrous, durable wood, often from dicot trees such as hickory, maple, and oak.

haustorium (plural, *haustoria*): Parasitic root that penetrates stems and roots of other plants to obtain water, minerals, and organic molecules.

heartwood: Older, nonconducting rings of *xylem* at the center of a tree's trunk or roots. See also *sapwood.*

heliotropism: Sun tracking; heliotropism describes flowers or leaves that either follow the sun or avoid the sun throughout the day.

hemicellulose: A cell wall component that resembles cellulose but is less structurally ordered.

herbaceous: Descriptive of a nonwoody plant, with little or no *secondary growth.*

hesperidium (plural, *hesperidia*): A fruit type similar to a berry, but with a leathery skin that produces fragrant oils; for example, citrus fruits.

heterokaryotic ("different nuclei"): See *dikaryotic.*

heteromorphic: Descriptive of alternating generations in which the sporophyte and gametophyte are very different in appearance. Compare with *isomorphic generation.*

heterosporous: Producing two types of spores: megaspores and microspores. See also *homosporous.*

heterotroph ("other feeder"): An organism, such as an animal, that obtains food from other organisms. See also *autotroph.*

heterozygous: Description of a plant that has two different *alleles* for a *gene.* Compare with *homozygous.*

holdfast: The rootlike part of a brown alga's *thallus* that anchors it to a substrate.

homeotic gene: A *gene* that controls the body plan of an organism by directing certain organs to form in the right places during development.

homologous chromosome: One of a pair of chromosomes that come from the fertilization of an *egg* by a *sperm.* Homologous chromosomes have genes for the same *characters.*

homology: A similarity between two plants that may have inherited the trait from the same ancestor. Compare *analogy.*

homosporous: Producing one type of spore, descriptive of many seedless vascular plants such as bryophytes. Compare with *heterosporous.*

homospory: The production of one size of spore, which can result in separate male and female gametophytes or in bisexual gametophytes, depending on the species.

homozygous: The description of a plant that has two copies of the same *allele* for a *gene.* Compare with *heterozygous.*

hormone (Greek, *hormon,* "to stir up"): An organic compound in multicellular organisms that causes developmental or growth responses in target cells. Important plant hormones include *auxin, ethylene,* and *giberellin.*

hydroid: A water-conducting cell present in many mosses; collectively called the hadrom. Hydroids resemble *tracheids* but without specialized secondary wall thickenings. See *leptoid.*

hydrolysis: Splitting a larger molecule into two smaller ones in a process that also splits water and adds H$^+$ or OH$^-$ to each product. The reverse of a *dehydration synthesis* reaction.

hydrophilic ("water-loving"): A substance that is soluble in water, such as most simple sugars.

hydrophobic ("water-fearing"): A substance that is insoluble in water, such as a lipid.

hydroponics (Greek, *hydo,* "water," *ponos,* "labor"): Soil-less gardening in which mineral nutrients normally supplied by the soil are mixed into a liquid solution used to water the plant roots.

hydrotropism: Growth toward or away from water.

hypertonic (Greek, *hyper,* "above"): Describing a solution with higher *solute* concentration than another. See also *hypotonic, isotonic,* and *osmosis.*

hypha (plural, *hyphae*): Long filament of cells that makes up the body of a fungus.

hypocotyl: A portion of the embryonic stem located under the *cotyledon* and above the *radicle.* See also *epicotyl.*

hypothesis: A tentative answer to a question, attempting to link together data in a cause-and-effect relationship; an educated guess that can be tested. See also *theory.*

hypotonic (Greek, *hypo,* "below"): Describing a solution with lower *solute* concentration than another. See also *hypertonic, isotonic,* and *osmosis.*

imbibition: A passive process in which a dry seed takes up water to begin germination.

imperfect flower: See *unisexual flower.*

incomplete dominance: A type of inheritance in which *characters* are not controlled by one *dominant* and one *recessive allele.*

incomplete flower: A flower that lacks one or more of the four types of modified leaves: *carpels, petals, sepals,* and *stamens.*

indehiscent: Describes a dry fruit that remains closed at maturity. Examples are *achenes, caryopses, nuts, samaras,* and *schizocarps.*

indeterminate growth: Unlimited growth throughout the life of a plant. Many vegetative meristems have indeterminate growth. Compare with *determinate growth.*

induced fit: Binding of an *enzyme* to a *substrate* that alters the shape of the *active site.*

inductive reasoning: A thought process starting with specific observations proceeding to general conclusions based on those observations. Compare with *inductive reasoning.*

indusium (plural, *indusia;* Latin, "tunic"): An umbrella-like structure covering a fern *sorus* of a fern leaf.

inferior ovary: Flower parts are attached above the ovary. See also *semi-inferior ovary* and *superior ovary.*

inflorescence: A group of flowers, with a definite arrangement, borne on a *peduncle.*

ingroup: A group of organisms to study in creating a *cladogram.* Compare *outgroup.*

initial: A *meristematic cell* that remains within a meristem as a source of new growth.

inner bark: Tissue consisting of living secondary *phloem,* dead phloem between the *vascular cambium* and the currently active, innermost *cork cambium,* and any remaining *cortex.*

intercalary meristem: A region of dividing cells at each *internode* that allows the stem to grow rapidly all along its length; common in grasses.

intermediate filaments: A component of the *cytoskeleton,* thicker than *microfilaments* but thinner than *microtubules;* formed from linear proteins. Intermediate filaments are involved in holding the nucleus in its permanent position in the cell and controlling the shape of the nucleus.

internode: A section of a stem between attachment site of leaves. See *node.*

interphase: Long portion of the *cell cycle* in which cells are preparing to divide. See G_1 *phase,* G_2 *phase,* and *S phase.*

intron: The section of a *gene* that interrupts or separates coding regions. Introns are unexpressed segments of genes. See also *exon.*

irregular flower: A flower with bilateral symmetry; Also called *zygomorphic.*

isogamete: Male and female *gametes* that are identical in appearance in some alga. See *isomorphic generation.*

isomorphic generation: Life form typical of some alga in which gametophyte and sporophyte look nearly identical; see *isogamete.*

isotonic (Greek, *isos,* "equal"): Describing two solutions that have equal *solute* concentrations. See also *hypertonic, hypotonic,* and *osmosis.*

karyogamy: Nuclear fusion; in fungi, karyogamy often occurs considerably after plasmogmany. See also *plasmogamy.*

keystone species: A species that has a strong effect on the structure of the community, even though the species itself may or may not be particularly abundant.

kinetic energy: Energy related to motion.

kinetochore: A complex structure of proteins formed by each *chromatid* at its own *centromere.* Important during cell division.

kingdom: A taxonomic group higher than *phylum* and lower than *domain,* such as Kingdom Plantae.

Krebs cycle: A series of eight enzymatic reactions of *respiration* that generates *ATP* by *substrate-level phosphorylation* as it breaks pyruvate ions into CO_2. Also generates NADH and FADH$_2$. The Krebs cycle follows *glycolysis* and precedes *oxidative phosphorylation.* It takes place in mitochondria and uses oxygen. The Krebs cycle stops if oxidative phosphorylation does not occur and, in that sense, requires oxygen.

K-selection: Common in populations that are close to the *carrying capacity;* favors traits that enable individuals to compete successfully for resources and to use resources efficiently. Also called density-dependent selection. Examples are organisms with long life span, low death rate. See *r-selection.*

lateral meristem: A cylindrical, slightly conical, single layer of meristematic cells that causes thickening of stems and roots in woody plants. See also *secondary growth*.

lateral root: A branch root produced by a *taproot*.

law of independent assortment: Mendel's second law of inheritance; each pair of *alleles* segregates independently during *meiosis*.

law of segregation: Mendel's first law of inheritance; *alleles* segregate during *anaphase I* of *meiosis* and then come together randomly during fertilization.

leaf: The main photosynthetic organ of modern plants.

leaf buttress: A bulge on the flank of a shoot *apical meristem* that appears during leaf development and develops into a leaf primordium.

leaf gap: The region in a *siphonostele* where vascular tissue branches off from the stele to enter a leaf.

leaf primordium (plural, *primordia*): Develops from a small bulge on the side of a shoot *apical meristem*, and develops into a leaf.

leaf trace: A small vascular bundle at each stem node that leaves the main vascular system of the stem and passes through a connecting *petiole* into the *leaf blade*.

leaf vein: A vascular bundle inside a *petiole* or *leaf blade*.

legume: A simple, *dehiscent*, dry fruit similar to a *follicle*. Arising from one carpel with two seams that split the fruit into two halves. Examples are beans, peanuts, and peas.

lenticel: A small opening in the thin cork layer of *outer bark* of stems and roots that allows gas exchange.

leptoid: Food-conducting cell present in many mosses; similar to sieve elements of non-seed-producing vascular plants. Collectively called the leptom. See *hydroid*.

leucoplast (Greek, *leukos*, "white"): A *plastid* that lacks pigments. Compare with *chromoplast*.

lichen: A living association between a fungus and a photosynthetic alga or a cyanobacterium.

life cycle: A sequence of stages leading from the adults of one generation of a species to the adults of the next generation.

light microscope (LM): Uses glass lenses to bend the path of visible light, producing magnified images up to 1,000 times original size. Compare with *scanning electron microscope (SEM)* and *transmission electron microscope (TEM)*.

light reactions: Photosynthetic reactions that occur within *thylakoid* membranes of *chloroplasts*. Inputs are light energy and H_2O; outputs are the chemical energy of ATP and NADPH and O_2 (as a by-product). See also *Calvin cycle* and *chlorophyll* a.

lignin: A rigid molecule that strengthens and stiffens cell walls in vascular plants; the most common polymer in plants after cellulose.

linked genes: *Genes* on a *chromosome* that segregate as a unit during *meiosis*.

lipid (Greek, *lipos*, "fat"): A water-insoluble hydrocarbon and *macromolecule* that stores energy (simple fats) or serves as a building block of membranes (phospholipid). See also *smooth endoplasmic reticulum*.

LM: See *light microscope*.

long-day plant (LDP): Plant that flowers only when the day length is longer than a critical length. Actually flowers when the night length is shorter than a critical length. See also *short-day plant (SDP)*.

lysogenic cycle: A cycle of viral reproduction in which viral genes coding for capsid proteins are not transcribed and new viruses are not produced. Common when host bacterial cells have little food available. In lysogeny viral DNA is closely associated with or incorporated into host DNA.

lytic cycle: A cycle for viral reproduction in which new viral particles are rapidly reproduced. The host cell eventually ruptures and the new viruses are released to infect other cells. Compare *lysogenic cycle*.

macromolecule: A large molecule made up of smaller molecules, such as *carbohydrates, lipids, nucleic acids*, and *proteins*.

macronutrient: An essential chemical element, such as nitrogen (N) or phosphorus (P) used in large amounts for producing the body of the plant and for carrying out essential physiological processes. Compare *micronutrient*.

matrix: The space enclosed by and inside the inner mitochondrial membrane. See also *crista*.

matrix potential: The force with which a soil particle binds water molecules.

megagametophyte: A female gametophyte produced by a megaspore.

megaphyll: A leaf with a highly branched vascular system, most common type of leaf in modern plants, including ferns, gymnosperms and angiosperms.

megasporangia: A sporangium that produces megaspores.

megaspore: A spore that produces a female *gametophyte*. See also *microspore*.

megasporocyte, or megaspore mother cell: a diploid cell that undergoes meiosis to produce haploid megaspores.

megasporophyll: A sporophyll with megasporangia.

meiosis: A type of nuclear division that is involved only in sexual reproduction and results in daughter cells with half the original number of *chromosomes*.

meiosis I: The first of two stages of gamete cell division; results in cells with half the number of chromosomes in the original cell. See also *homologous chromosome, prophase I, metaphase I, anaphase I, telophase I, diploid,* and *haploid*.

meiosis II: The second of two stages of gamete cell division, in which sister *chromatids* of a now *haploid* cell separate.

meristem (Greek, *meristos*, "divided"): Region of meristematic cells that produce new cells by cell division. See also *apical meristem* and *initial*.

meristem culture: A type of *tissue culture* in which the top few millimeters of a shoot *apex* are cultured on a medium that encourages *axillary buds* to develop into a whole plant. Also called shoot tip culture.

meristematic cell: An unspecialized cell that can divide indefinitely to produce new cells.

mesocarp: The middle part of a *pericarp*.

mesophyll (Greek, *mesos*, "middle," *phyllon*, "leaf"): Chlorenchyma ground tissue located between the upper and lower layers of epidermis; site of photosynthesis in leaves. See *palisade mesophyll* and *spongy mesophyll*.

messenger RNA (mRNA): The carrier of genetic messages from *DNA*; mRNA is made during *transcription*.

metabolism: The chemical reactions that occur inside cells.

metaphase: The second phase of *mitosis,* during which chromosomes line up along the metaphase plate in the center of the cell.

metaphase I: The second stage of *meiosis;* similar to metaphase of *mitosis,* except that *tetrads* of *homologous chromosomes,* instead of single chromosomes, move onto the *metaphase plate.*

metaphase plate: An imaginary plane that extends across the diameter of the cell during *metaphase.*

micelle: A crystalline subunit in a *microfibril,* linked by proteins such as *pectins* and *hemicelluloses.*

microbody: A small, membrane-bound spherical organelle about 1 μm in diameter that contains enzymes. See also *glycosome* and *peroxisome.*

microfibril: A cylindrical structure composed of many long cellulose molecules arranged side by side.

microfilament: A long filament in the *cytoskeleton* that moves cells or cell contents and helps determine cell shape. Microfilaments are made from polymers of the globular protein *actin.* Microfilaments are thinner than *microtubules.* See also *cytoplasmic streaming.*

microgametophyte: A male gametophyte produced by a microspore.

micronutrient: An essential chemical element, such as copper (Cu) or zinc (Zn), used in small amounts by a plant. Compare *macronutrient.*

microphyll: A small leaf with a single leaf vein; characteristic of modern lycophytes (club mosses, spike mosses, and quillworts).

micropyle (Greek, *pyle,* "gate"): The opening between integuments in ovules where the pollen tube enters.

microsporangia: Sporangia that produce microspores by meiotic division of microspore mother cells.

microspore: One of two types of spores; produces a male gametophyte. See also *macrospore.*

microsporocyte, or microspore mother cell: Contained in a microsporangium, divides by meiosis to produce microspores.

microsporophyll: A sporophyll with microsporangia; common in spike mosses.

microtubule: A long hollow tube in the *cytoskeleton* that moves cell components such as molecules, organelles, and chromosomes from one place to another. See *cilium, flagellum,* and *tubulin.*

middle lamella (Latin, *lamina,* "thin plate"): A thin layer between the *primary cell walls* of adjacent cells, composed mainly of *pectins.*

mitochondrion (plural, *mitochondria*): An membranous organelle that completes the breakdown of sugar to store its chemical energy in *ATP* (*adenosine triphosphate*). Has DNA that codes for proteins made by the mitochondrion's ribosomes. See also *crista* and *matrix.*

mitosis or **M phase:** A stage of the *cell cycle* when cells divide; consists of *prophase, metaphase, anaphase,* and *telophase.* M phase is typically the shortest phase of the cell cycle, constituting about 10% of total cell division time.

mixotroph: An organism, such as *Euglena,* that can produce organic molecules through photosynthesis (autotrophy) and can absorb or ingest organic molecules (heterotrophy).

molecular clock: A marker, such as cytochrome *c,* or the gene for cytochrome c, used to estimate the extent of evolutionary separation of two species by noting the gradual accumulation of amino acid, or nucleic acid, differences between proteins and genes of different species.

monocot: A flowering plant with one *cotyledon;* for example, orchids, lilies, palms, onions, and grasses. Compare with *dicot.*

monoecious (Greek, "one house"): Describing a plant with male and female *gametophytes* on different flowers of the same plant; for example, pumpkins and corn. See also *dioecious* and *self-pollination.*

monohybrid cross: A breeding in which parents differing in one trait of a particular character are crossed. For instance, if the character is height, a pure-breeding tall parent might be crossed to a pure-breeding short parent. See also *dihybrid cross* and *first filial* (F_1) *generation.*

monomer: A simple building block molecule that makes up a *polymer.*

monophyletic: Characteristic of a *clade,* being a "single tribe" of organisms that evolved from a common ancestor.

monosaccharide: The simplest type of *carbohydrate,* a single sugar molecule with a molecular formula that is usually a multiple of CH_2O.

motor proteins: Use energy in ATP to cause movement of cell structures when associated with *microtubules* and *microfilaments.* Also called "walking molecules."

mRNA: See *messenger RNA.*

mucigel: A slimy polysaccharide that lubricates the passage of roots through soil; produced by outer cells of the *root cap.*

multiple fruit: A fruit that develops from the *carpels* of more than one flower in a single *inflorescence;* for example, pineapples and figs.

mutation: A change in the order or structure of *DNA.* See *chromosomal mutation, frameshift mutation,* and *point mutation.*

mutualistic: Mutually beneficial, as in root associations with other organisms such as soil fungi. See *mycorrhizae.*

mycelium (plural, *mycelia*): Interwoven mass consisting of all the *hyphae* of a particular type in a fungus.

mycology (Greek, *mykes,* "fungus"): The study of fungi.

mycorrhizae (Greek, *mykes,* "fungus," *rhiza,* "root"): *Mutualistic* associations between vascular plant roots and soil fungi. See *ectomycorrhizae* and *endomycorrhizae.*

NADH, NADPH, and FADH₂: Three complex organic molecules that can pick up and release electrons and protons as part of an *electron transport chain.* They move electrons between enzymatic reactions throughout cells.

netted venation: A pattern of leaf veins in most dicots and ferns, also known as reticulate venation, in which leaf veins form branching networks. Compare *parallel venation.*

niche: The combination of all the physical and biological variables that affect an organism's success. A plant's niche typically includes temperature range, moisture level, soil type, *habitat,* and seasonal variation.

nitrogen fixation: The conversion of nitrogen gas into nitrate or ammonium ions by soil bacteria. See also *root nodule*.

node: The point on a stem where a leaf is attached. See also *axillary bud* and *internode*.

nondisjunction: The failure of sister *chromatids* or *homologous chromosomes* to separate during *mitosis* or *meiosis*; common cause of *aneuploidy*.

nucellus (Latin, *nucella*, "small nut"): A *megasporangium* inside the ovule of a seed plant where the megagametophyte or embryo sac develops.

nuclear envelope: Two membranes that surround the nucleus. Pores in the nuclear envelope control movement of substances into and out of the nucleus.

nucleic acid: A large molecule composed of nucleotides, such as DNA and RNA, that holds a cell's genetic information. See *DNA* (*deoxyribonucleic acid*) and *RNA* (*ribonucleic acid*).

nucleic acid probe: A short piece of *RNA* or single-stranded *DNA* that is complementary to a DNA sequence in the *gene* of interest.

nucleolus (plural, *nucleoli*): One of two round structures in a diploid *nucleus* associated with the genes in *chromosomes* that synthesize ribosomal RNAs. Nucleoli synthesize subunits that then come together in the *cytoplasm* to form *ribosomes*.

nucleotide: The basic building block of *nucleic acids,* a nucleotide consists of three parts: base, sugar, and phosphate.

nucleus (plural, *nuclei*): A membrane-bound structure that contains the cell's DNA.

nut: An *indehiscent*, dry, simple fruit with a stony *pericarp* shell; originates from compound *carpels*. For example, acorns and hazelnuts.

nutrition: The processes by which an organism takes in and uses food substances.

oligotrophic (Greek, "few nutrients"): Descriptive of a deep, nutrient-poor lake. See *eutrophic*.

oogonium: Female *gametangium* composed of one cell that contains one or more eggs in some kelp and fungal species.

operculum: The sporangium lid in moss sporophytes; falls off after a layer of cells at its base dries out to release spores.

opposite: A leaf arrangement with two leaves per node.

order: A taxonomic group higher than *family* and lower than *class*. Often ending in *–ales* such as Solanales.

organ: A combination of several types of tissue adapted as a group to perform particular functions. See *leaf, root,* and *stem*.

organelles: Separate cell structures with surrounding membranes of their own. Organelles include chloroplasts, mitochondria, microbodies, dictyosomes. Also called "little organs." See also *chloroplast, endoplasmic reticulum, endosymbiotic theory,* and *ribosome*.

osmosis (Greek, *osmos*, "a push"): Movement of water or other solvent across a selectively permeable membrane.

osmotic potential: Measurement of differing water's tendency to move across a membrane as a result of *solute* concentrations. Also called solute potential.

outer bark: Consists of dead tissue including dead secondary *phloem* and all the layers of *periderm* outside of the most recent *cork cambium*.

outgroup: A species or group of species closely related to an *ingroup*, but not as closely related as the ingroup members are to each other. Used in creating a *cladogram*.

ovary: A structure at the base of a *carpel* in a flower; contains one or more *ovules* and will eventually swell to become part or all of a fruit.

ovule (Latin, *ovulum,* "little egg"): A structure containing an *egg*. After fertilization, the ovule will develop into a seed.

oxidation: The loss or partial loss of one or more electrons.

oxidative phosphorylation: Reactions of cellular respiration that produce *ATP* using energy from NADH rather than light energy. Occurs primarily in the inner membranes of mitochondria using molecules of the electron transport chain. Oxidative phosphorylation produces approximately 34 molecules of ATP per molecule of glucose. Oxygen is the final electron acceptor as O_2 is converted to H_2O. Compare with *photophosphorylation* and *substrate-level phosphorylation*. See also *Krebs cycle*.

paleobotanist: A scientist who studies plant fossils to decipher the evolutionary history of the plant kingdom.

palisade mesophyll (Latin, *palus*, "stake"): Elongated, aligned mesophyll cells beneath the epidermis; contains most of the chloroplasts in a leaf.

parallel venation: A pattern of leaf veins in most monocots and gymnosperms, also called striate venation, in which veins run parallel with each other and the leaf edges. Compare with *netted venation*.

parasite: An organism that feeds on a living host. Compare *saprobe*.

parenchyma cell (Greek, *parenchein*, "to pour in beside"): The most common and least specialized type of cell in most plants. See also *chlorenchyma cell, collenchyma cell,* and *sclerenchyma cell*.

pectin: A jellylike protein, common in the intercellular space between cells; also important in *micelle* formation within cell walls.

peduncle: The stem on which a flower or inflorescence sits. See *receptacle*.

pellicle: A supporting structure beneath a euglenoid's *plasma membrane*; composed of helical bands of protein connected to the *endoplasmic reticulum* by *microtubules*.

pepo: A fruit type similar to a berry but with a thick rind; for example, watermelons, pumpkins, and cantaloupes.

perennial: A plant that grows for many years; may be woody or *herbaceous*.

perfect flower: See *bisexual flower*.

perianth ("around the flower"): All the sterile modified leaves of a flower; *calyx* and *corolla*.

pericarp: The wall of an *ovary*. See also *endocarp, exocarp,* and *mesocarp*.

periclinal division: The division of cells parallel to a surface. Compare with *anticlinal division*.

pericycle: A cell layer immediately encircling the *stele* that gives rise to *lateral roots*. See also *endodermis*.

periderm (Greek, "the skin around"): A protective tissue replacing the *epidermis* of stems and roots of plants that live more than one year. Commonly found in woody plants. Contains products of cork cambium including cork (phellem) and phelloderm.

peripheral zone: A region of the shoot *apical meristem* in the form of a three-dimensional ring around the *central mother zone*. Consists of cells that divide rapidly to become leaf primordia and parts of the stem. See also *pith zone*.

peristome (Greek, *peran*, "to pass through," *stoma*, "mouth"): One or more rings of "teeth" around the exposed opening of a sporangium; aids in spore dispersal in some mosses.

peroxisome: A type of *microbody* that generates and breaks down hydrogen peroxide; involved in photosynthesis and in the conversion of sugars to fats in plants.

petal (Latin *petalum*, "to spread out"): The colorful but sterile modified leaf of a flower. Petals form inside or above the *calyx* on the *receptacle*. Petals are collectively called a *corolla*.

petiole: Thin, stemlike structure that attaches leaf to stem at a *node*.

phellem: See *cork*.

phelloderm (Greek, *phellos*, "cork," *derma*, "skin"): A thin layer of living *parenchyma* cells that forms to the inside of each *cork cambium* layer.

phellogen: See *cork cambium*.

phenolic: One of a group of ringed hydrocarbon compounds lacking nitrogen in their structure. Examples are lignins, flavonoids, and allelopathics.

phenotype: The physical appearance of an organism. Compare with *genotype*.

phloem: A tissue that moves sugars and other organic nutrients from the leaves to the rest of the plant. See also *sap* and *xylem*.

phosphorylation: The transfer of a phosphate group from a molecule of one substance to a different substance. Examples are the formation of glucose-6-phosphate from glucose and ATP in the first step of *glycolosis* and formation of ATP from ADP and inorganic phosphate. See also *oxidative phosphorylation* and *substrate-level phosphorylation*.

photoautotrophs: Organisms that get their energy through photosynthesis; plants, algae, and photosynthetic bacteria.

photodormant: Description of a seed that requires activation by light.

photoheterotroph: An organism that gets its energy from light and its carbon from organic compounds.

photon: A packet of electromagnetic energy; a photon's energy depends on its wavelength.

photoperiodism: A plant's response to the relative lengths of night and day. See *day-neutral plant*, *long-day plant* (*LDP*), and *short-day plant* (*SDP*).

photophosphorylation: The process of forming *ATP* from ADP, using *ATP synthase* and light energy.

photorespiration: A process common in C_3 plants during hot, dry weather when stomata close to prevent dehydration. Photorespiration produces but does not fix CO_2. It uses light and consumes oxygen, but produces no ATP or food. C_4 photosynthesis and *crassulacean acid metabolism* are adaptations in other plant species that minimize photorespiration.

photosynthesis: The process by which plants use solar energy to make their own food, transforming carbon dioxide and water into sugars that store chemical energy. See *Calvin cycle* and *carbon fixation*.

photosystem: One of two light-harvesting units consisting of a *reaction center* and *accessory pigments*; absorbs light energy on the *stroma* side of a *thylakoid* membrane.

phototropism: Growth toward or away from light. See *phytochrome*.

phragmoplast: A cylinder consisting of *microtubules* that are derived from the spindle and aligned between the daughter nuclei. This structure forms a *cell plate*.

phyllotaxy (Greek, "leaf order"): The basic pattern of leaf arrangement. See *alternate*, *opposite*, and *whorled*.

phylogenetic tree: A branching diagram that shows evolutionary relationships over time.

phylogeny: The evolutionary history of related species.

phylum (plural, *phyla*): A taxonomic group higher than *class* and lower than *kingdom*, such as Phylum Coniferophyta.

phytochrome: A photoreceptor that absorbs light and causes developmental effects. See *phototropism*.

phytoplankton: The collection of microscopic, photosynthetic organisms that float freely near the surface of oceans and lakes.

pigment: A light-absorbing molecule; for example, *chlorophyll*.

pinnae: Leaflets of a compound leaf or *frond*.

pistil: An individual *carpel* or a group of fused carpels.

pit: A region lacking the secondary cell wall in a *tracheid* that allows water and minerals to flow from one tracheid to another.

pith: Ground tissue formed inside vascular tissue.

pith zone: A region of the shoot *apical meristem* below the *central mother zone* and *peripheral zone*, that produces cells that become the part of the ground meristem that produces pith.

plain-sawed board: Produced by trimming a *tangential cut* of lumber. Compare *quarter-sawed board*.

plant biotechnology: Efforts to obtain improved plants and plant products using scientific techniques such as genetic engineering and tissue culture.

plasma membrane (Latin, *membrana*, "skin"): The flexible, protective layer surrounding all cells. Also called the *cell membrane* or *plasmalemma* (Greek, *lemma*, "husk"). Controls the movement of water, gases, and other molecules into and out of the cell.

plasmalemma: See *plasma membrane*.

plasmid: A self-replicating, circular *DNA* molecule present in bacteria.

plasmodesma (plural, *plasmodesmata*; Greek, *desma*, "bond"): A channel between adjacent cells to allow material flow between them. Usually contains a connecting *desmotubule* of *endoplasmic reticulum*.

plasmogamy: Cytoplasmic fusion; in fungi plasmogamy is frequently separated temporally from karyogamy. See also *karyogamy*.

plasmolysis: A condition in which a *plasma membrane* shrinks away from its *cell wall*, caused by net flow of water out of the cell.

plastid: A general term for plant organelles involved either in making or in storing food or pigments. See *chloroplast, chromoplast,* and *leucoplast.*

plate tectonics: A unifying theory of modern geology that arose from the work of geologist Alfred Wegener in 1912. See also *continental drift.*

pleiotropy: A type of inheritance in which a single *gene* controls more than one *character.* Compare with *polygenic inheritance.*

plumule (Latin, *plumula,* "soft feather"): An embryonic *shoot.* See also *epicotyl.*

pneumatophore: Provides oxygen for plants in swampy areas. Also known as an air root. Common on mangrove and bald cypress trees.

point mutation: A change in one *nucleotide* in *DNA;* also called *single nucleotide polymorphism* (*SNP*); can be a substitution, an insertion, or a deletion.

polar molecule: A molecule with uneven distribution of positively and negatively charged regions. Water is a typical polar molecule.

pollen: A collection of *pollen grains.*

pollen grain: A male *gametophyte* formed from a spore inside the pollen sacs of an *anther.*

pollination: The process of transferring pollen from the male part of a plant to the female part of a plant; does not immediately result in fertilization.

polyembryony: Production of more than one embryo as a result of the presence of multiple pollen tubes; characteristic of some gymnosperms.

polygenic inheritance: Referring to inheritance in which *characters* are controlled by more than one *gene; phenotypes* often exhibit a continuum of values. See also *pleiotropy.*

polymer: A *macromolecule* composed of repeating structural units called *monomers.*

polymerase chain reaction (PCR): A method of cloning *DNA* fragments enzymatically without using *plasmids* or bacteria.

polypeptide: A polymer of amino acids; a large polypeptide is a *protein.*

polyploidy: A cell having more than the *diploid* number of chromosomes. See also *haploid.*

polysaccharide: A polymer formed by hundreds to thousands of linked *monosaccharide* molecules; usually store energy or provide structural support; for example, starch and cellulose.

pome: A type of *accessory fruit* resembling a berry, with the bulk of the fleshy fruit coming from an enlarged *receptacle* that forms from the swollen end of the *peduncle* or *pedicel.* Some familiar pomes are apples and pears.

population: A group of interbreeding organisms of the same species in the same location.

population genetics: The study of the behavior of *genes* in populations.

potential energy: Stored energy due to an object's position or chemical composition.

pressure-flow hypothesis: A mechanism for *phloem* transport, first proposed by Ernst Munch in 1927.

pressure potential: The pressure of a cell wall around its contents. See also *osmotic potential* and *water potential.*

prickle: A sharp outgrowth of epidermal or cortical cells. See also *spine* and *thorn.*

primary cell wall: A structure composed primarily of cellulose that is formed by growing cells to prevent their exploding due to water uptake. See also *secondary cell wall.*

primary growth: Growth in length of roots and shoots, caused by *meristems* at the tip, or *apex,* of each root or shoot. See *apical meristem.*

primary meristem: A region of cell division that produces the tissues of a primary plant body. See *ground meristem, procambium,* or *protoderm.*

primary metabolite: Essential biochemical components of metabolism in every plant cell; including *carbohydrates, proteins, nucleic acids,* and *lipids.*

primary plant body: The body of a plant produced by shoot and root *apical meristems.*

primary producers: Organisms that make their own food, such as plants and other photosynthetic organisms.

primary structure: The sequence of the amino acids in a *protein.*

primary succession: The set of changes in communities over time in areas that are initially devoid of almost all life and where soil has not yet formed. See also *secondary succession.*

procambium: Root and shoot *apical meristem* that produces *xylem* and *phloem.*

prokaryote (Latin, "before the nucleus"): An organism whose cells do not have an enclosed nucleus, such as bacteria. Compare with *eukaryote.*

promoter: A sequence of several dozen *nucleotide* pairs located at one end of a *gene;* site where RNA polymerase attaches during gene *transcription.*

prophase: First phase of *mitosis,* during which chromosomes shorten and thicken enough to be visible under a light microscope. The nuclear envelope and nucleoli have disappeared.

prophase I: The first, most complex stage of *meiosis.* Similar to *prophase* in *mitosis,* except that *homologous chromosomes* form pairs. See *synapsis* and *tetrad.*

protein (Greek, *proteios,* "holding first place"): A *macromolecule* consisting of one or more chains of amino acids. An organism's proteins define its physical characteristics, serve as structural building blocks, and govern the rates of chemical reactions by serving as *enzymes.* Proteins are coded for by *genes.*

protein kinase: An enzyme that phosphorylates other proteins when activated by a *second messenger* as part of a *signal transduction pathway.*

proteomics: The science of sequencing all of an organism's proteins and understanding their functions.

protobionts: Cell-like structures with various degrees of organization that aggregate spontaneously from mixtures of organic compounds.

protoderm: Root and shoot *apical meristem* that produces a plant's epidermis.

protonema (plural, *protonemata;* Latin, *proto,* "first," and Greek, *nema,* "thread"): A typically threadlike structure formed by a germinated spore; most noticeable in mosses; forms buds that grow into *gametophores.*

protoplast: A plant cell without a cell wall.

protostele (Greek, *proto,* "before"): The simplest, earliest-evolving *stele,* consisting of a solid cylinder comprised of both xylem and phloem.

punctuated equilibrium: Niles Eldredge and Stephen Jay Gould's model of evolution in which long periods of little or no evolutionary change are punctuated by short periods of rapid change. See also *adaptive radiation.*

pyrenoid: A protein-rich structure in the chloroplasts of many algae; contains the enzyme *rubisco.* Also, a region in hornwort and algal chloroplasts containing starch deposits resulting from photosynthesis.

quarter-sawed board: Produced by trimming a *radial cut* of lumber. Compare *plain-sawed board.*

quaternary structure: The spatial arrangement of more than one *polypeptide* chain in a *protein.*

quiescent center (Latin, "to rest"): The spherical center of a root apical meristem that contains the initials.

rachis: An extension of the *petiole* attaching fern *pinnae* to their *frond.*

radial cut: The direction of a lumber cut that passes longitudinally through the center of the stem. Compare *tangential cut.* See also *quarter-sawed board.*

radicle (Latin, *radix,* "root"): The embryonic *root* in a developing plant embryo.

ray initial: An *initial* that arises between *vascular bundles;* often cube-shaped. See also *fusiform initial.*

reactant: A participant in a chemical reaction.

reaction center: A combination of a *chlorophyll* a molecule and a primary electron acceptor, which together absorb light to trigger the *light reaction* of photosynthesis. See also *photosystem.*

reaction wood: Tension or compression wood that develops in trunks or branches that are leaning.

receptacle: A swollen tip at the top of a *peduncle* that bears the parts of a flower. See also *carpel, petal, sepal,* and *stamen.*

recessive trait: A masked *trait* when *heterozygous;* phenotypically observable when *homozygous.*

recombinant DNA: DNA that is a combination of DNA from different sources.

redox reaction: A coupled set of *oxidation* and *reduction* reactions.

reduction: Gain or partial gain of one or more electrons.

regular flower: A flower with radial symmetry; also called *actinomorphic flower.*

respiration: Aerobic process of extracting energy from food. *Glycolysis* reactions take place in the *cytosol. Krebs cycle* reactions and oxidative phosphorylation take place in *mitochondria.* See also *fermentation.*

restriction enzyme: A bacterial enzyme that breaks bonds between specific nucleotides in DNA. Used in genetic engineering to fragment DNA. See *recombinant DNA.*

restriction fragment length polymorphisms (**RFLPs,** pronounced "RIF-lips"): Sections of DNA produced by restriction enzymes.

rhizoid: Slender, branched, tubelike cell or filament of cells that anchors bryophytes to the ground. Also found in fungi where rhizoids grow into the food supply and hold a fungus in place.

rhizome: A horizontal, underground stem. Compare with *stolon.*

ribosome: An organelle formed in the *cytoplasm* that directs the synthesis of *proteins* using genetic instructions in the form of *messenger RNA* (*mRNA*).

riparian: An adjective describing riverbank environments.

RNA (ribonucleic acid): A single helical molecule, similar to DNA, but containing the sugar ribose. RNA plays an important role in directing the synthesis of *proteins.* See *messenger RNA* (*mRNA*) and *transfer RNA* (*tRNA*).

root: An organ that anchors a plant in soil and absorbs water and minerals. See also *root hair.*

root cap: Several layers of cells that protect the root apical meristem as the root pushes between soil particles.

root hair: A specialized *trichome* near the tip of a root, responsible for water and mineral absorption for a plant.

root nodule: A structure on a root where nitrogen-fixing bacteria live.

root system: Consists of all roots, usually belowground.

rough endoplasmic reticulum: A network of membranes derived from the outer nuclear membrane and dotted with protein-synthesizing *ribosomes.* Rough ER makes secretory proteins (hormones) and membrane components.

r-selection: Selection for traits that maximize the reproductive rate of populations in uncrowded environments; examples are organisms with short life span, high death rate. See K-*selection.*

rubisco: The abbreviation for the enzyme ribulose1,5-bisphosphate carboxylase/oxygenase, which adds carbon from CO_2 to another molecule in *carbon fixation.* The most abundant protein in chloroplasts. As an oxygenase, rubisco also catalyzes *photorespiration.*

rule of parsimony: The observation that when more than one *cladogram* can be constructed from a particular set of data, the simplest one is probably correct.

S phase: The part of *interphase* following the G_1 *phase,* when chromosomes replicate to produce two joined strands of DNA called *chromatids.* Also called DNA synthesis.

samara: A simple, dry, *indehiscent* fruit like an *achene* with the addition of a hard, thin, elongated *pericarp* that produces wings around a single seed. Ash and elm seeds are samaras.

sand: The largest of soil particles, 0.02 to 2 millimeters in diameter. See also *clay, silt,* and *soil horizon.*

sap: The contents transported by *xylem* and *phloem.*

saprobe (Greek, *sapros,* "rotten"): An organism that feeds on dead organic matter. Compare *parasite.*

sapwood: Outer rings of xylem that still transport xylem *sap.* See also *heartwood.*

scanning electron microscope (SEM): An SEM bounces electrons off a specimen to reveal the surface structure. An SEM can magnify up to about 20,000 times. See also *light microscope* and *transmission electron microscope.*

schizocarp: A simple, dry, *indehiscent* fruit that occurs in parsley, carrots, dill, and celery, as well as in maples. Schizocarps have a hard, thin *pericarp,* composed of two or more carpels, which opens into two or more parts, each of which contains a seed.

sclereid: A cubical or spherical *sclerenchyma cell.* Commonly found in nutshells and fruit pits.

sclerenchyma cell (Greek, *skleros*, "hard"): A nonliving, structural cell with *secondary cell walls* that are hardened with *lignin*. See also *collenchyma cell*, *fiber*, and *parenchyma cell*.

scutellum: The *cotyledon* of a monocot embryo that is attached to an embryonic axis containing shoot and root *meristems*.

second filial (F$_2$) generation: The offspring of interbred F$_1$ plants.

second law of thermodynamics: States that every transformation of *energy* increases the *entropy* (disorder of matter) in the universe.

second messenger: A cytoplasmic substance produced in a *signal transduction pathway* by binding or absorption of a primary messenger (hormone or light) to a membrane protein. See also *protein kinase*.

secondary cell wall: A thick layer, composed principally of cellulose and lignin, produced by woody plants, between the *primary cell wall* and the *plasma membrane*.

secondary growth: Growth in thickness produced by *lateral meristems*, common in conifers and dicots.

secondary metabolite: A molecule not essential for basic plant growth and development, but plays a role such as providing structural support or protecting a plant from herbivores and disease.

secondary structure: The local twisting and folding, stabilized by hydrogen bonds, of a *polypeptide* chain in a *protein*. Alpha (α) helices and beta (β) pleated sheets are examples of secondary protein structures.

secondary succession: A combination of changes that take place where a community has been removed by a disturbance but the soil remains intact; can follow human activity or natural changes.

seed: A structure that includes a plant embryo and a store of food, packaged together within a protective coat.

seedless vascular plants: The simplest vascular plants, which began to evolve about 450 to 700 million years ago.

selectively permeable: Describes a membrane that transports some molecules but not others across the plasma membrane. Also called partially permeable.

self-pollination: A process that is possible when male and female *gametophytes* are on the same plant or in the same flower. See *monoecious*.

SEM: See *scanning electron microscope*.

semi-inferior ovary: Flower parts are attached halfway up the ovary. See also *inferior ovary* and *superior ovary*.

sepal (Latin, *sepalum*, "covering"): A sterile modified leaf that forms at the bottom of a *receptacle* to protect a flower bud before it opens. Sepals are collectively called a *calyx*.

septa: Internal walls that divide *hyphae* into cells.

sessile: Adjective describing a leaf that lacks a petiole and is attached directly to the stem.

seta (plural, *setae*): The short stalk in bryophytes that bears the sporangium.

sexual reproduction: Fertilization of an egg by a sperm; results in offspring that are different from either parent. Most animals can only reproduce sexually.

shared derived character: A *homology* that is unique to a particular group.

shared primitive character: A *homology* that is not unique to the organisms being studied.

shoot: Any individual *stem* and its *leaves*, as well as any reproductive structures that extend from the stem, such as flowers.

shoot system: All stems, leaves, and reproductive structures of a plant, usually aboveground.

short-day plant (SDP): A plant that flowers when days are shorter than a critical length; actually flowers when night length is longer than a critical length. See also *long-day plant (LDP)*.

sieve cell: A simple water-conducting cell in ferns and conifers that functions much like a *sieve-tube member* in flowering plants.

sieve plate: A feature of a *sieve-tube member*, consisting of cell walls with membrane-lined pores.

sieve tube: A multicellular structure in *phloem* that conducts organic nutrients from the leaves to other parts of the plant.

sieve-tube member: A living cell in the *phloem* of flowering plants, stacked end-to-end to form *sieve tubes*. Nonnucleated at maturity. See also *callose, companion cell*, and *sieve cell*.

signal transduction pathway (STP): A series of steps that links the binding of a receptor to a change in a cell's activity; caused by interaction between a *hormone* or light and a protein on the outside surface of a cell. See *protein kinase* and *second messenger*.

silicate (SiO$_4$$^{-4}$): The most common negative ion in Earth's crust and soil particles.

silique: A dry, *dehiscent* fruit produced by species in the mustard family; consists of two *carpels* that split into two halves, with the seeds found on a central partition between the halves.

silt: Medium-sized soil particles, 0.002 to 0.02 millimeter in diameter. See also *clay, sand*, and *soil horizon*.

simple fruit: A fruit that develops from one *carpel* or from several fused carpels.

simple leaf: A leaf with a single, undivided blade; may be toothed or lobed. Compare *compound leaf*.

simple tissue: A tissue composed of one type of cell. See also *complex tissue*.

single nucleotide polymorphism: See *point mutation*.

siphonostele: A continuous vascular cylinder that surrounds a core of pith in the stems of ferns and horsetails. Compare *eustele*. See also *leaf gap*.

smooth endoplasmic reticulum: A tubular-shaped membrane, derived from the outer nuclear membrane, that makes *lipids* and modifies the structure of some *carbohydrates*.

softwood: Wood with few fibers and no vessels, typically from conifers.

soil horizon: A horizontal profile of soil; lettered from uppermost layer down. See *topsoil*.

soil solution: The combination of water, dissolved mineral ions, and dissolved O_2; source of macro- and micronutrients for plants.

solute: A substance that dissolves in water.

sori (singular, *sorus*): Groups of sporangia, usually on the underside of fertile fern fronds; typically appear as dot-like structures.

sorus (plural, *sori*): A region filled with septate *basidia* on the leaves or stems of plants infected with rusts.

Southern blotting: A method of finding particular nucleotide sequences in a *DNA* sample.

species: Commonly, an organism within a *genus*. Technically, species is designated by a *binomial* consisting of *genus* and *specific epithet*.

specific epithet: The second part of a *bionomial* name of a species.

specific gravity: The ratio of a wood's weight to the weight of an equal volume of water at room temperature. See also *density*.

sperm: A male reproductive cell. See also *sexual reproduction*.

spine: A sharp modified leaf or stipule. See also *prickle* and *thorn*.

spongy mesophyll: Loosely organized photosynthetic cells beneath the epidermis of a leaf.

sporangiophore: In *Rhizopus stolonifer* and other fungi, one of several upright *hyphae*, each of which bears a sporangium at its tip. See *zygospore*.

sporangium (plural, *sporangia*): A hollow structure, derived from one or more cells, that contains spores.

spore: A reproductive cell of plants that can develop into an adult without fusing with another reproductive cell. See also *asexual reproduction* and *sexual reproduction*.

sporophyll: A modified spore-producing leaf; found on flowers and cones and in some non-seed-producing plants. See also *sporangium*.

sporophyte (Greek, "spore-producing plant"): One of two multicellular forms of a plant; consists of *diploid* cells. See also *alternation of generations* and *gametophyte*.

stabilizing selection: A method of reducing variation in a population by eliminating individuals that have extreme *phenotypes*. See also *directional selection* and *diversifying selection*.

stamen: A male, *pollen*-producing part of a flower. Stamens collectively are called the *androecium*. See also *anther* and *carpel*.

statolith: A specialized *plastid* in root cap cells filled with dense starch grains; possible explanation of *gravitropism*.

stele (Greek, "pillar"): The central cylinder of a root or stem, which is surrounded by the *cortex*. See also *pericycle* and *protostele*.

stem: Any part of a plant that supports leaves or reproductive structures.

sticky end: Short, single-stranded sequence at each end of a *DNA* fragment prepared with a *restriction enzyme*. These fragments readily bind to complementary sequences on other DNA fragments produced by the same enzyme.

stigma: A structure at the top of a *carpel* that provides a sticky surface for *pollen*.

stipe: A stemlike, often hollow stalk; part of a brown alga's *thallus*.

stipule: One of two small leaflike flaps on a *petiole* at the basal part of a leaf. Some stipules encircle the petiole.

stolon: A horizontal, aboveground stem, also called a runner. Compare with *rhizome*.

stoma (plural, *stomata*; from the Greek, *stoma*, "mouth"): A pore in a leaf regulated by two *guard cells*; controls movement of water vapor, CO_2, and O_2.

strobilus (plural, *strobili*): A cone consisting of modified sporophylls found in gymnosperms, lycophytes, and sphenophytes.

stroma: The fluid surrounding *thylakoids*; the site of sugar production and storage in *chloroplasts*.

stromatolite: Rock made of layers of fossilized remains of *prokaryotes* dating back as far as 3.5 billion years. Top layer may contain living cells.

style: The middle section of a *carpel* that connects the *stigma* to the *ovary*.

suberin: Waterproof, fatty substance that coats and impregnates *cork* cell walls.

substrate: A *reactant* that is acted upon by an *enzyme*.

substrate-level phosphorylation: Enzymatic production of *ATP* not involving chemiosmotic transport of protons. Occurs during *glycolysis*. Compare with *oxidative phosphorylation* and *photophosphorylation*.

sugar sink: A part of a plant that mainly consumes or stores sugar, such as roots, stems, and fruits.

sugar source: A part of a plant that produces sugar, usually leaves and green stems.

superior ovary: Flower parts are attached to the *receptacle* under the ovary. See also *inferior ovary* and *semi-inferior ovary*.

sympatric: Descriptive of populations that have overlapping ranges but may have different microhabitat preferences. See also *allopatric*.

symplastic transport (Greek, *sym*, "with"): Movement of substances through the interior of cells (within the cytoplasm). Compare *apoplastic transport*.

synapsis: Pairing of *homologous chromosomes* during *prophase I* of *meiosis*. See also *tetrad*.

synteny: Description of plants in different genera that have many regions of *chromosomes* in which *genes* occur in the same order.

systematics: The modern scientific study of the evolutionary relationships among organisms. See *taxonomy*.

taiga: The northern coniferous or boreal forest; the largest uniform *biome* on Earth.

tangential cut: Direction of a lumber cut that is longitudinal but crosses the radius at a right angle instead of passing through the center of the stem. Compare *radial cut*. See also *plain-sawed board* and *veneer*.

taproot system: A root system common in dicots and gymnosperms; features a large main taproot. Compare *fibrous root system*. See also *lateral root*.

taxon (plural, *taxa*): A named group of organisms at any level of hierarchy.

taxonomy (Greek, *taxis*, "arrangement," *onoma*, "name"): The naming and classifying of species, or a category in a formal system of classification.

telome (Greek, *telos*, "end"): One of the twigs of a dichotomous branch. According to telome theory, differential growth of telomes gave rise to many anatomical structures of plants, such as *microphylls*.

telophase: The last phase of *mitosis*, and a reversal of prophase. The nuclear envelope reforms in each cell, chromosomes uncoil, and the spindle disappears.

telophase I: The last stage of *meiosis* in which the cell returns to its pre-meiotic state before entering *meiosis II*.

TEM: See *transmission electron microscope*.

tendril: A slender, coiling structure that attaches a climbing plant to a supporting structure; can be modified leaves or modified stems. See also *thigmotropism*.

tension: A negative pressure on water or solutions; caused in xylem by transpiration through *stomata.*

tension-cohesion theory: An explanation of how transport occurs in *xylem;* relies on *tension, cohesion,* and *adhesion* in a water column and *transpiration* through stomata.

terpenoid: One of numerous plant-produced hydrocarbon compounds formed from two to hundreds of five-carbon isoprene subunits. Terpenoids protect plants by tasting bitter, being poisonous, or being sticky. Also called terpene.

tertiary structure: The overall three-dimensional pattern of folding in a protein, caused by interactions between side groups on amino acids.

testcross: A method of determining the *genotype* of a plant that has a *dominant* phenotype; the plant whose genotype is unknown is crossed with a plant that exhibits the *recessive* phenotype for the *character* in question. Also called a backcross.

tetrad: A structure consisting of four chromatids, formed by the process of *synapsis* during *prophase I* of *meiosis.*

texture: Of lumber, refers to the sizes of the cells in the *xylem* and *phloem* and to the sizes of growth rings; may be coarse, fine, or uneven.

thallus (plural, *thalli;* from Greek, *thallos,* "sprout"): A plant-like body of brown algae. See *blade, holdfast,* and *stipe.*

theory: A hypothesis that is supported on a wide scale or in broader application than a supported hypothesis.

thigmotropism (Greek, *thigma,* "touch"): Growth stimulated by touch; typical of *tendrils.*

thorn: A sharp modified stem that arises from an axillary bud where a leaf joins a stem. See also *prickle* and *spine.*

thylakoid: A membrane-bound sac inside a *chloroplast.* Conversion of solar energy to chemical energy takes place within the membranes of thylakoids.

tissue: A group of identical cells with a common function. See also *complex tissue* and *simple tissue.*

tissue culture: A method of growing a whole plant, plant organs, or plant tissues from cells in an artificial medium containing nutrients and hormones. See *anther culture, callus,* and *meristem culture.*

tissue system: A functional unit of simple and complex tissues. See *dermal tissue system, ground tissue system,* and *vascular tissue system.*

tonoplast: The membrane surrounding a mature cell's central *vacuole.*

topsoil: Uppermost *soil horizon,* contains the smallest soil particles, decaying organic matter, and various organisms.

tracheid: A long, nonliving cell with tapered ends, commonly found in the *xylem* of all vascular plants. See also *pit.*

trait: One of two or more forms of a *character.* In peas, for example, the character of seed color may appear as the trait of green seeds or yellow seeds. See *allele.*

transcription: Of a *gene,* first step of converting genetic code to make protein; consists of making a complementary copy of the *nucleotide* sequence on a section of *DNA* by formation of a section of *messenger RNA.* See also *promoter* and *translation.*

transcription factor: A *protein* molecule that aids attachment of RNA polymerase to a *promoter;* usually stimulates, but may inhibit transcription.

transfer RNA (tRNA): Folded RNA molecule consisting of 70 to 80 *nucleotides* involved in *translation* of genetic information into *protein* molecules. Contains an *anticodon* (*codon* binding site) and an attachment site for amino acids in the growing protein chain. *tRNA* is the translating molecule.

transgenic organism: An organism that contains at least one *gene* from a different kind of organism.

translation: The second step in converting genetic code to make a protein; conversion of a sequence of *nucleotides* in *mRNA* into a sequence of amino acids in a *protein.* See also *transcription* and *transfer RNA.*

transmission electron microscope (TEM): The microscope passes electrons completely through a thin section of tissue. A TEM can magnify objects up to about 100,000 times. See also *light microscope* and *scanning electron microscope.*

transpiration: The loss of water through the pores in leaves, which pulls water and mineral nutrients up from the roots to the leaves.

transport vesicle: A membrane-surrounded structure that holds lipids, proteins, and other substances produced in the *endoplasmic reticulum.* Transport vesicles then separate from the ER and move to the *Golgi apparatus.*

transposon: A piece of *DNA* that can move from one location to another or can produce a copy of itself that moves to another location; first characterized in the 1940s by Barbara McClintock. See also *gene trapping.*

transverse cut: A direction of cutting lumber to give a circular cross section.

transverse section: The same as a *cross section.*

trichome: A hairlike extension of a dermal cell; for example, the long hairs that extend from leaves and from cotton seeds. *Root hairs* are also trichomes.

triple response: The growth response initiated by *ethylene* that includes slowing of stem or root elongation, thickening of the stem or root, and curving to grow horizontally.

triticale: A hybrid of wheat (*Triticum aestivum*) and rye (*Secale cereale*); example of a *wide cross.*

tRNA: See *transfer RNA.*

tropism (Greek, *tropos,* "turn"): A growth response produced by a *hormone* that causes turning toward or away from an external stimulus. See *gravitropism, heliotropism, hydrotropism,* and *phototropism.*

tuber: The underground stem composed primarily of starch-filled *parenchyma* cells that form at the tips of *stolons* or *rhizomes.*

tubulin: A spherical protein that makes up *microtubules.*

tunica: According to the *cell layer model* of shoot growth, the outer layer of cells in an apical meristem; equivalent to the outer part of the *peripheral zone.*

turgid: Swollen or enlarged as a result of being full of water.

turgor pressure: The pressure developed by water uptake in *phloem;* responsible for movement of water and sugar down into root cells.

type specimen: An identified specimen of a plant preserved in a herbarium; can be used to determine whether another specimen is a member of the same species.

unisexual flower: Has either *stamens* or *carpels* but not both. Also called *imperfect flower.*

vacuole (Latin, *vacuus*, "empty"): A large central space in many mature plant cells, filled with water, inorganic ions, proteins, and metabolic waste products. The vacuole helps maintain cell shape by pressing the rest of the cytoplasm contents against the cell wall. See also *tonoplast*.

vascular bundles: Consist of strands of vascular tissue composed of *xylem* and *phloem;* common in stems of all vascular plants.

vascular cambium (Latin, *cambire*, "to exchange."): A *secondary meristem* that produces secondary *xylem* and secondary *phloem*. Compare with *cork cambium*.

vascular plants: Plants with highly organized, efficient cells joined into tubes that transport water and nutrients throughout the plant's body. See also *vascular tissue*.

vascular tissue: Plant cells joined into tubes that transport water and nutrients throughout a vascular plant's body.

vascular tissue system: A continuous system of tissues that conducts water, minerals, and food; consists of *xylem* and *phloem*.

vector: An agent that carries a *gene* from one organism to another.

veneer: A thin lumber section, produced by an angled, continuous *tangential cut*.

vernalization (Latin, *vernus*, "spring"): Use of cold treatment to hasten flowering.

vessel: A continuous water-conducting tube, formed by *vessel elements*.

vessel element: A large, water-conducting cell in the xylem of most flowering plants; transports water and minerals more rapidly than *tracheid* cells.

viroid: A simple plant pathogen consisting of viral form with circular strands of RNA containing 250–370 nucleotides.

walking molecule: See *motor protein*.

water content: The percent of water by weight in wood.

water potential: The sum of a cell's *osmotic potential* and its *pressure potential;* used to predict which way water will tend to flow between a plant cell and its surroundings.

whorled: A leaf arrangement with three or more leaves per node.

wide cross: A crossing of relatively unrelated plants, which occurs occasionally in nature to yield fertile offspring if spontaneous chromosome doubling occurs. For example, wheat arose as the result of two or three natural wide crosses between related plants in different genera.

wood: Secondary *xylem*.

xerophyte (Greek, *xeros*, "dry," and *phyton*, "plant"): A plant that thrives in dry, desert environments.

xylem (Greek, *xylon*, "wood"): The tissue that brings water and mineral nutrients from the roots to the rest of the plant. See also *phloem, sap, tracheid*, and *vessel element*.

zone model: A description of the shoot *apical meristem* as a three-part dome. See *central mother cell zone, peripheral zone*, and *pith zone*. Compare with *cell layer model*.

zone of cell division: The region in plant roots consisting of the root *apical meristem* and the three *primary meristems*.

zone of elongation: The region in plant roots where derivatives stop dividing and begin to grow in length.

zone of maturation: The region in plant roots where cells begin specializing in structure and function into different cell types; region where some epidermal cells form *root hairs*.

zygomorphic (Greek, *zygon*, "yoke" or "pair"): An *irregular flower*, typically bisymmetrical.

zygospore: A spore produced by a fungus such as black bread mold, *Rhizopus stolonifer*.

Index